ELASTOHYDRODYNAMIC LUBRICATION for LINE and POINT CONTACTS

Asymptotic and Numerical Approaches

ELASTOHYDRODYNAMIC LUBRICATION for LINE and POINT CONTACTS

Asymptotic and Numerical Approaches

Ilya I. Kudish

CRC Press
Taylor & Francis Group
Boca Raton London New York

CRC Press is an imprint of the
Taylor & Francis Group, an **informa** business

CRC Press
Taylor & Francis Group
6000 Broken Sound Parkway NW, Suite 300
Boca Raton, FL 33487-2742

First issued in paperback 2017

Version Date: 20130401

ISBN 13: 978-1-138-07396-8 (pbk)
ISBN 13: 978-1-4665-8389-4 (hbk)

Library of Congress Cataloging-in-Publication Data

Kudish, Ilya I.
  Elastohydrodynamic lubrication for line and point contacts : asymptotic and numerical approaches / author, Ilya I. Kudish.
    pages cm
  Includes bibliographical references and index.
  ISBN 978-1-4665-8389-4 (hardback)
  1. Elastohydrodynamic lubrication. I. Title.

TJ1077.5.E43K83 2013
621.8'9--dc23   2013007650

Visit the Taylor & Francis Web site at
http://www.taylorandfrancis.com

and the CRC Press Web site at
http://www.crcpress.com

To Rima

Contents

V Isothermal and Thermal EHL Problems for Line Contacts and Lubricants with Newtonian and Non-Newtonian Rheologies   167

7 Thermal EHL Problems for Line Contacts   171

Preface

Since the ground-breaking numerical study by Petrusevich [1] and approximate analytical studies by Ertel and Grubin [2, 3] of elastohydrodynamic lubrication (EHL) problems were published over sixty years ago, these two approaches, i.e., the direct numerical solution of EHL problems and Ertel-like approximate analysis of EHL problems, completely dominated the field of EHL research. There were a number of different numerical methods developed as well as some analytical variations of the Ertel method. However, most studies of EHL problems were done numerically. Practically all these numerical methods work really well in cases when an EHL contact is lightly to moderately heavily loaded. At the same time, all direct numerical methods applied to heavily loaded isothermal EHL problems suffer from solution instability which results in poor solution convergence and precision in the exit zone of a contact. With the transition from the numerical solution of two-dimensional EHL problems (line contacts) to three-dimensional EHL problems (point contacts) the difficulties just get exacerbated. Therefore, the time has come to understand the roots of most difficulties in direct numerical approaches to solution of EHL problems and provide an effective remedy.

The idea of most direct numerical methods is to take a numerical solver based on a particular numerical procedure (Newton-Raphson method, multilevel multigrid method, fast Fourier transform method, etc.) and apply it more or less uniformly to all points of a lubricated contact region to obtain a solution of an EHL problem without any regard to different physical mechanisms driving the lubrication phenomenon in a particular subregion of the lubricated contact. That creates the situation when a particular numerical method at hand is supposed to be good for all physical mechanisms engaged in a lubricated region. These mechanisms include the effects as different as lubricant fluid flow, elasticity of contact solids, heat generation and transfer, etc. It is reasonable to expect that such numerical algorithms exist, however, for only some moderate conditions which are far from extremes. Nowadays, it is most important for practical applications to be able to properly analyze the extreme cases of heavily loaded lubricated conditions. In these cases in different zones of lubricated contacts different mechanisms show dominance. For example, in the central part of a heavily loaded lubricated contact away from contact boundaries the effects of solid material elasticity dominate the effects of lubricant flow while in narrow zones close to contact boundaries the effects of contact material elasticity and lubricant fluid flow are of the same order of magnitude. Moreover, these narrow zones are extremely important as they give rise for lubricant film thickness in a lubricated contact and, therefore, they cannot be neglected. Thus, it is almost impossible to expect that the same numerical procedure would work equally well in such different in nature contact zones in a heavily loaded EHL contact. It means that it can be expected that such numerical solutions would suffer from at least inadequate precision in some zones of the contact. One of the manifestations of that is the existence of abundance of different formulas for calculation of the lubrication film thickness producing sometimes significantly different values while all of them are obtained as a result of numerical solution of the same exactly isothermal EHL problem for smooth elastic solids and consequent curve fitting.

The EHL problems for line and point contacts are so complex that there is no chance of solving these problems analytically with a given precision. It means that application of numerical methods for solution of EHL problems is a necessity. However, it does not mean that we should disregard the opportunity to extract analytically some valuable information about solution of EHL problems prior to solving them numerically. Extracting analytical information from EHL problems before solving them numerically is very well spent time and effort. This analytical information may help a researcher to fine tune solution methods used and to design better numerical methods or even get some analytical results such as lubrication film thickness formulas. On the other hand, due to the fact that usually numerical solutions of isothermal EHL problems for heavily loaded contacts are unstable, one may try to use some analytical results to make such solutions stable, i.e., to regularize the numerical solution process. Solution instability is a very serious defect of numerical solutions of EHL problems. If a problem numerical solution is unstable it means that this solution does not converge to the solution of the original continuous problem as the grid size approaches zero. Also, it means that the precision of such a solution is not adequate/low, the precision cannot be controlled, and the obtained numerical solution does not really correspond closely to the original continuous problem. On the other hand, a regularized solution process for EHL problems produces stable numerical solutions which converge to their continuous analogs as the grid size approaches zero and the precision of these numerical solutions can be controlled by the degree of regularization. It must be understood that regularized EHL problems are slightly different from the original isothermal EHL problems and the degree of this difference is controlled by the regularization parameter.

The idea of this monograph is to present a coherent approach to line and point EHL contacts based on the application of asymptotic and, subsequently, numerical methods which take into account the specific solution properties obtained analytically using the asymptotic methods. Throughout the monograph EHL problems are first analyzed asymptotically and, after that, as certain knowledge concerning the interrelations of different physical mechanisms taking place in lubricated contacts and the corresponding terms in the problem equations is gained, some specific numerical methods are developed which follow the uncovered asymptotically structure of the problem solution and are free from many drawbacks suffered by most of the existing direct numerical methods applied to heavily loaded EHL contacts. The asymptotic methods in many cases reveal solution analytical structure as well as emphasize the dominant mechanisms involved in the phenomenon in each zone of a lubricated contact. At the same time, the numerical methods based on these asymptotic methods provide the details to such solution structure and validate asymptotic approaches and solutions as well as provide the means to regularize the numerical solutions for isothermal heavily loaded EHL contacts.

In application to EHL problems the asymptotic methods in conjunction with special numerical methods offer certain advantages in comparison with direct numerical methods used for solution of EHL problems. The asymptotic methods allow determination of the structure of the contact region and EHL problem solution, the characteristic sizes of different zones in the contact and the characteristic values of the components of the EHL problem solution in these zones. In particular, the asymptotic approach to EHL problems for heavily loaded contacts provides analytical formulas for such an important design parameter as the lubrication film thickness in which only one constant coefficient has to be determined numerically or experimentally while everything else in these formulas is obtained analytically using easy algebra. By revealing the structure of EHL problem solutions the asymptotic methods allow to reduce the number of input parameters the numerical solution of the asymptotically valid equations depends on, analytically solve the problem for lubricant temperature, reduce solution of thermal EHL problems to solution of generalized isothermal EHL problems, reduce solution of isothermal and thermal point EHL problems for lubricants with Newtonian and non-Newtonian rheologies along "central" lubricant flow streamlines to solution of the corresponding line EHL problems, regularize the numerical solutions for heavily loaded EHL contacts, etc.

This monograph aims at several objectives. First, this monograph not only proposes a robust combination of asymptotic and numerical techniques in application to solution of EHL problems for lightly and heavily loaded line and point contacts but also proposes a reasonably simple and naturally based regularization approach that produces stable solutions in heavily loaded EHL contacts. In addition to that the aim of the monograph is to establish a clear understanding of the processes taking place in heavily loaded line and point EHL contacts, EHL problem solution structure, the concrete ways to determine the important for design parameters such as lubrication film thickness and frictional stresses and forces, and to establish a close link between EHL problems for heavily loaded point and line contacts.

The asymptotic methods applied to EHL problems are based on regular perturbation techniques and methods of matched asymptotic expansions [4, 5]. In [6] the above combination of asymptotic and numerical methods is used for analysis of line EHL problems. The numerical methods are iterative in nature and are based on finite-difference and numerical integration methods as well as certain quasi-linearization approaches.

The entire monograph is based on the following ideas: (a) proposing an alternative to the Ertel and direct numerical methods for solution of problems for heavily loaded EHL contacts such as asymptotic methods followed by specialized numerical techniques, (b) developing a general asymptotic approach to solution of problems for heavily loaded contacts lubricated by fluids with various non-Newtonian rheologies and greases, (c) reducing solution of thermal EHL problems to their certain isothermal analogs, (d) designing a regularization approach to virtually eliminate instability of numerical solutions for isothermal heavily loaded EHL contacts, (e) developing the asymptotic procedure which would allow to reduce solution of heavily loaded point EHL problems to their line EHL problem analogs, (f) formulating and numerically solving EHL problems for line contacts with stress-induced lubricant degradation, (g) properly formulating and solving non-steady lubrication problems, and (h) properly formulating and analyzing mixed friction/lubrication problems.

The book consists of eight parts and twenty three chapters which feature the application of asymptotic and numerical techniques to solution of EHL problems for line and point contacts. In most cases the equations are derived from first principles. Each of the asymptotic and numerical methods is described in detail which makes it easier for a reader to apply them to different problems. Some problems allow for relatively simple approximate (asymptotic) analytical solutions. The problem solutions are presented in the form of simple analytical formulas, graphs, and tables. Every chapter (except for Chapter 2) is provided with exercises of different levels of difficulty. The exercises highlight certain points that are important for understanding the material and mastering the appropriate skills.

Chapters 1 and 2 cover the basic physical properties of lubricating fluids and solids, respectively. Chapter 3 provides an introduction to the basics of asymptotic and perturbation techniques. An introduction to classic plane contact problems of elasticity as well as the approach used for the derivation of the Reynolds and energy equations applicable to lubrication processes are presented in Chapter 4. In Chapter 5 some problems for lightly loaded line and point EHL contacts are considered. The next chapter - Chapter 6 - is devoted to the development of the fundamental framework for the asymptotic approaches to the analysis of the isothermal line EHL problems for heavily loaded contacts lubricated by a Newtonian fluid as well as the validation of the asymptotic approaches. Also, this chapter offers a new numerical approach to solution of asymptotically valid equations in the inlet and exit zones. These numerical methods serve as a basis for regularized numerical solution revisited in Chapter 8. Thermal EHL problems for heavily loaded line contacts lubricated by Newtonian fluids are analyzed in Chapter 7. Chapter 8 presents an effective regularization method which produces stable numerical solutions for isothermal EHL problems for heavily loaded line contacts. Chapters 9 and 10 are devoted to isothermal and thermal EHL problems for heavily loaded line contacts lubricated by non-Newtonian lubricants, respectively. Chapter 11 is devoted to modeling and numerical analysis of line EHL contacts with degrading lubricant.

Chapters 12-17 are concerned with the analysis of heavily loaded point EHL contacts. Different types of lubricant motions are considered such as straight and skewed lubricant entrainment with and without spinning. Isothermal and thermal EHL problems for lubricants with Newtonian and non-Newtonian rheologies are considered. In the inlet and exit zones of heavily loaded point lubricated contacts, along the "central" lubricant flow streamlines the EHL problems are reduced to the asymptotic equations which coincide with or are very similar to the ones in the inlet and exit zones of the corresponding heavily loaded line EHL contacts.

Chapter 18 is devoted to modeling of the Kaneta "dimple" phenomena for soft and hard lubricated solids. In Chapter 19 some scale effects and their application to experimental studies of lubricants with non-Newtonian rheologies are considered. Chapter 20 provides a basic modeling of grease behavior in EHL contacts. Chapter 21 is devoted to proper formulation and solution of some non-steady EHL problems. Chapter 22 deals with the effects of starvation in point EHL problems as well as with mixed friction problems. Finally, Chapter 23 provides some final remarks.

From the standpoint of understanding the application of the asymptotic methodology to EHL problems for heavily loaded contacts as well as understanding the principles the proposed numerical methods are based on the most crucial chapters are Chapters 6, 7, and 12. It is important to mention that in the inlet and exit zones of heavily loaded point EHL contacts along the lubricant "central" flow streamlines the asymptotically valid equations are uniform with those for the corresponding line contacts. It is important to stress that these asymptotic equations obtained for heavily loaded point EHL contacts in their form are independent of the specific solid surfaces motion and depend only on the lubricant rheology, amount of fluid available at the inlet in the contact, and isothermal or thermal lubrication conditions. In other words, solid understanding of lubricant behavior in heavily loaded line EHL contacts will serve the reader well in understanding lubricant behavior in heavily loaded point EHL contacts. The new basic approaches to numerical solution of the EHL problems in asymptotic and original formulations including a regularization technique are presented in Chapters 6 and 8.

The book is offered as an enhancement of the elastohydrodynamic lubrication curriculum for senior undergraduate and graduate engineering and applied mathematics students as well as a reference/guide for researchers and practitioners in the field. For example, experimentalists measuring lubrication film thickness in heavily loaded line and point EHL contacts can use analytically derived formulas for the lubrication film thickness for different lubricants and lubrication regimes which require determination of just one coefficient of proportionality in each case. Engineering students can be offered courses based upon chapters on line EHL problems, point EHL problems, or on some specific problems from line and point EHL problems. On the other hand, applied mathematics students can be offered a course on nontraditional applications of asymptotic and perturbation methods as well as on a regularization approach to solution of inherently numerically unstable problems for heavily loaded lubricated contacts. In addition, the content of the monograph can be used as a basis for designing an introductory/overview course on elastohydrodynamic lubrication.

Chapter 1 was written by M.J. Covitch while Chapters 2-23 were prepared by I.I. Kudish.

The author acknowledges the contributions of Zachary Smith, Applied Mathematics student at Kettering University at the time, for the preparation of many graphs and figures and Bruce Deitz, Kettering University librarian, for helping with literature searches.

References

[1] Petrusevich A.I. 1951. Fundamental Conclusions from the Contact-Hydrodynamic Theory of Lubrication. _Izv. Acad. Nauk, SSSR (OTN)_, 2, 209.

[2] Ertel, M.A. 1945. Hydrodynamic Lubrication Analysis of a Contact of Curvilinear Surfaces. Dissertation. _Proc. of CNIITMASh_, Moscow, 1-64.

[3] Grubin, A.N. and Vinogradova, I.E. 1949. Investigation of the Contact of Machine Components. Ed. Kh. F. Ketova, _Central Sci. Research Inst. for Techn. and Mech. Eng._ (Moscow), Book No. 30 (DSIR translation No. 337).

[4] Van-Dyke, M. 1964. _Perturbation Methods in Fluid Mechanics_. New-York: Academic Press.

[5] Kevorkian, J. and Cole, J.D. 1985. _Perturbation Methods in Applied Mathematics. Applied Mathematics Series, Vol._ 34. New York: Springer-Verlag.

[6] Kudish, I.I. and Covitch, M.J. 2010. _Modeling and Analytical Methods in Tribology._ Chapman & Hall/CRC.

About the author

Ilya I. Kudish is a mathematics professor at Kettering University in Flint, Michigan. Dr. Kudish was born in Russia. He graduated from the Moscow Institute of Physics and Technology (Moscow, Russia) with the Master's Degree in Mathematics, Physics, and Engineering. In 1981 he defended his Ph.D. dissertation in mathematics and physics at St. Petersburg Polytechnic University, St. Petersburg, Russia. In 1991 Dr. Kudish with his immediate family immigrated to the USA.

Dr. Kudish is a Fellow of the American Society of Mechanical Engineers (ASME). For two terms, 2005-2011, he served as an Associate Editor of the ASME *Journal of Tribology*. Dr. Kudish is a recipient of all Kettering University research awards: Outstanding New Researcher, Outstanding Applied Researcher, Outstanding Researcher, and Distinguished Researcher Awards as well as of Rodes and Oswald Professorships. His research was supported by NSF and Kettering University grants. Also, he served as a consultant to Caterpillar, Inc. and as a Visiting Professor at Perdue University, Cardiff University (Cardiff, UK), and INSA (Lyon, France). In 2010, together with M.J. Covitch he published a monograph "Modeling and Analytical Methods in Tribology," CRC Press. Dr. Kudish presented a number of invited talks at conferences and at company meetings. He published over 140 research papers in leading journals and conferences, contributed several chapters to the *Encyclopedia of Tribology*, Springer, 2013, as well as to the *Encyclopedia of Life Support Systems*, EOLSS, published in 2012 under the auspices of UNESCO.

Dr. Kudish's main interests are in the sphere of application mathematical methods to various problems of tribology. Over the years he has made certain theoretical contributions to the fields of elastohydrodynamic lubrication, stress-induced lubricant degradation, contact problems for coated/rough elastic solids, fracture mechanics and fracture mechanics based contact and structural fatigue modeling.

Part I

Basic Properties of Solids and Fluids Involved in Lubricated Contacts

1

Basic Properties and Rheology of Lubricating Oils

1.1 Introduction

The transfer of power from its source of generation to its final destination where useful work occurs is the subject of much study and engineering development. In the case of internal combustion engines, liquid fuel is combusted by compression ignition (diesel) or spark ignition (gasoline). The gas expansion pressure within the combustion chamber drives a piston, which, in turn, translates reciprocal motion into rotary motion via a connecting rod that is attached to a crankshaft. Once the crankshaft is set in motion, pulleys, chains, clutches, bearings, and gears continue the chain of energy transfer to accomplish the ultimate purpose of the engine, be it transportation, pumping, electricity generation, or the like.

At every step of the energy transfer process, solid contacts are set in relative motion, and they often sustain considerable loads acting both normal and tangential to the direction of motion. Unless the contacts are separated by a lubricating film (solid, liquid, or gas), a tremendous amount of frictional heat builds up in the contact zone that can lead to equipment failure in a relatively short period of time. Therefore, the lubricant is one of the most important components of an energy transfer device.

This monograph, in part, is devoted to selected performance attributes of liquid lubricants, particularly mineral oils and synthetic fluids. To fully describe the design features of modern lubricating oils is beyond the scope of this work. However, some elements need to be discussed to serve as the basis for further discussion.

Lubricating oils consist of base oils, which are selected on the basis of viscosity, thermal/oxidative stability, and cost. Chemical additives are also present and provide a number of useful functions. Viscosity modifiers (improvers) are oil-soluble polymers that impart useful rheological characteristics over a wide range of temperatures. Pour point depressants are added to prevent paraffin waxes (present in most refined mineral oils) from impeding oil flow at low temperatures. Dispersants, detergents, anti-wear agents, anti-oxidants, corrosion inhibitors, foam inhibitors, and friction modifiers do what their names imply.

Since viscosity is one of the most important design parameters of a loaded tribological contact, each application requires a lubricating oil with a particular set of rheological properties. To maintain proper lubrication under all operating conditions, the viscosity is required to meet stringent specifications at different temperatures, pressures, and shear rates. It should also be recognized that lubricants are subject to degradation during use. Exposure to high local temperatures, corrosive acids, unburned fuel, contaminants (liquids, solids, and gases), and severe mechanical forces can play havoc on the properties of a carefully designed lubricating oil. The useful life of the lubricant is often limited by its ability to maintain critical performance specifications during use.

A number of fundamental EHL problems for lightly and heavily loaded line and point contacts is considered in the following chapters while modeling of lubricant degradation is presented in detail in [1].

1.2   Rheology Relationships for Lubricating Oils

1.2.1   Definition of Viscosity

The resistance to flow exerted by a fluid under the influence of external force is a phenomenological definition of viscosity. Mathematically, the dynamic viscosity μ is the proportionality constant between shear stress τ and shear rate γ otherwise known as Newton's law:

$$\tau = \mu\gamma. \qquad (1.1)$$

Common units of measure of viscosity are summarized in Table 1.1.

TABLE 1.1

Viscosity units of measure (after Kudish and Covitch [1]).
Reprinted with permission from CRC Press.

	units	comments
Dynamic Viscosity	$mPa \cdot s$	SI unit
Dynamic Viscosity	cP	Centipoise, $1\ cP = 1\ mPa \cdot s$
Kinematic Viscosity	mm^2/s	SI unit
Kinematic Viscosity	cSt	Centistoke, $1\ cSt = 1\ mm^2/s$

Dynamic viscosity is calculated by independently measuring shear stress and shear rate. There are a large number of instruments, known collectively as viscometers, for measuring dynamic viscosity. In most cases, a fixed surface is opposed by a parallel surface moving at a certain velocity. Popular configurations include concentric cylinder and parallel plate devices. The surfaces are separated by a known gap, which is filled with the test lubricant. Shear stress is defined as the force required to maintain velocity divided by the surface area wetted by the lubricant. Shear rate, with units of s^{-1}, equals velocity divided by gap distance. In a constant rate viscometer, the velocity of the moving surface is set by the operator, and the torque on the fixed surface is measured. In a constant stress viscometer, the operator specifies the torque applied to the moving surface, and velocity is monitored.

Tribological contacts consist of two opposing surfaces in relative motion. One can be fixed or it can be moving at a different velocity or in a different direction than the other. Under a given set of conditions, all of the physical quantities relating to dynamic viscosity are correlated (force, contact area, velocity, and gap). Therefore, for all practical purposes, dynamic viscosity is the most appropriate parameter to use to characterize the load-carrying capacity of a lubricant.

Kinematic viscosity η is another common measure of viscosity, which is related to the dynamic viscosity μ as follows: $\mu = \rho\eta$, where ρ is the fluid density. The kinematic viscosity is determined by measuring the time for a fixed volume of fluid to flow through a vertical capillary under the influence of gravity. Consider two fluids of equal dynamic viscosity but of different densities. The force exerted by gravity on a unit volume of fluid is proportional to mass (or density, mass per unit volume). Therefore, the flow time through the capillary will be greater for the higher density fluid, and its kinematic viscosity will be lower than that of the lower density lubricant. To convert kinematic viscosity to dynamic viscosity, the former needs to be multiplied by lubricant density determined at the same temperature. For example, the kinematic viscosity and density of castor oil [2] at $40°C$ is 244 mm^2/s and 0.9464 g/cm^3, respectively. Dynamic viscosity equals 231 $mPa \cdot s$. Because of the low cost and simplicity of measuring kinematic viscosity, it is widely used to classify lubricating fluids into viscosity grades (such as SAE 5W-30 or ISO 46 for example). Its practical applicability in tribology, however, is limited to cases of freely draining contacts under low applied loads.

1.2.2 The Effect of Temperature on Viscosity

We all know from practical experience that a common way to lower fluid viscosity is to increase its temperature. Since viscosity is such a critical design parameter, it is important to be able to quantify its response to temperature. Although a number of theoretical and empirical mathematical models have been proposed for more than 100 years, the most common equation for lubricating oils was proposed by Walther [3, 4].

$$\log[\log(\eta + \theta)] = A - B \log T, \tag{1.2}$$

where η is the kinematic viscosity, T is the absolute temperature in degrees Kelvin and A, B, and θ are empirical constants (θ is usually between 0.6 mm^2/s and 0.8 mm^2/s and can be ignored for high viscosity fluids).

This equation is applicable over a limited temperature range. At low temperatures, paraffinic wax molecules (which are present in most mineral oil base stocks) nucleate and grow to form crystals. The crystals are often needle-like or plate-like having a high aspect ratio. The temperature at which wax begins to precipitate is known as the wax appearance temperature, also called the cloud point because the liquid takes on a turbid appearance [5]. The presence of this second phase causes the viscosity to increase (see Section 1.2.5) at a rate greater than predicted by extrapolation of (1.2) from higher temperatures. The upper temperature limit of the Walther equation is close to the initial boiling point. When the vapor pressure of the lowest molecular weight, most volatile fractions of the base oil approaches atmospheric pressure, microscopic gas bubbles begin forming. Just like wax crystals at low temperature, gas bubbles at high temperature form a second phase which affects viscosity.

Typical cloud point temperatures for API Group I and II base oils are $-20°C$ to $0°C$ as measured by ASTM Method D5771. The lowest boiling temperature of similar base stocks ranges from about $280°C$ to about $400°C$ as determined by ASTM D2887. Modern lubricating oils are rated for operation from about $-40°C$ to about $150°C$, depending upon viscosity grade. Thus, the Walther equation is more practically useful for describing the viscosity temperature relationship of mineral oil lubricants at high temperatures than at low temperatures.

It is common practice in the lubrication industry to measure kinematic viscosity at both $40°C$ and $100°C$. With two data points, constants A and B can be determined, and kinematic viscosity at any other temperature can be calculated. An example is shown in Fig. 1.1 for an SAE 15W-40 engine oil with kinematic viscosities of 14.76 mm^2/s and 108.3 mm^2/s at $100°C$ and $40°C$, respectively. By solving a system of two simultaneous equations (1.2) satisfied at the above two points, we get the values of the Walther constants A and B equal to 7.970 and 3.070, respectively, with θ assigned a value of 0.7 mm^2/s. Clearly, the measured viscosity in the cloud point region is substantially higher than its predicted value, whereas agreement is reasonably good at slightly higher sub-ambient temperatures. The percentage difference between measured and predicted kinematic viscosity data for four lubricating fluids at several temperatures is summarized in Table 1.2. At $0°C$, the measured viscosity agrees within 12% of the extrapolated value. In the cloud point region ($-20°C$), the measured viscosity is more than double the predicted value. Similar trends are observed for the other fluids.

The lubrication industry adopted a term known as Viscosity Index (VI) to describe the response of viscosity to changes in temperature. It is an empirical quantity, which is calculated from kinematic viscosity values measured at $40°C$ and $100°C$ according to ASTM method D2270. The viscosity of high VI fluids, such as paraffinic oils, changes less with temperature than that of low VI fluids,

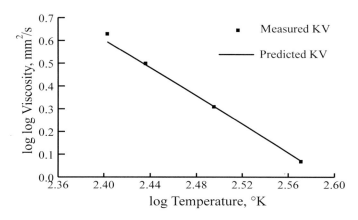

FIGURE 1.1
Measured vs. predicted kinematic viscosity for an SAE 15W-40 engine oil. (after Kudish and Covitch [1]). Reprinted with permission from CRC Press.

such as naphthenic oils. Before the advent of instruments capable of measuring viscosity at low temperatures, VI was viewed as an indicator of flow behavior at sub-ambient temperatures. Today, VI is used primarily as a classification parameter for mineral and synthetic base oils [6].

TABLE 1.2
Percent difference between measured and predicted kinematic viscosity η as a function of temperature for three engine oils and one hydraulic fluid. (after Kudish and Covitch [1]). Reprinted with permission from CRC Press.

Temperature, °C	15W-40 (1)	15W-40 (2)	5W-30	ISO 32
100	0%	0%	0%	0%
40	0%	0%	0%	0%
0	12%	9%	8%	10%
−20	115%	83%	19%	−
−30	−	−	62%	77%

1.2.3   The Effect of Shear Rate on Viscosity

Many formulated fluids, like multi-grade engine oils, do not strictly follow Newton's law and are aptly called non-Newtonian lubricants. In most cases, viscosity decreases as shear rate increases (pseudoplastic behavior), but certain fluids exhibit dilatant flow in which viscosity increases with shear rate. For the purposes of this discussion, only pseudoplastic non-Newtonian fluids will be considered.

The viscosity-shear rate behavior of non-Newtonian fluids consists of three zones (see Fig. 1.2). At both very low and very high shear rates, viscosity appears to be independent of shear rate. These regions are often referred to as the lower and upper Newtonian plateaus, respectively. At intermediate ranges of shear rate, viscosity drops as shear rate increases, often following a simple power law relationship. This region is called the power law zone.

The phenomenological response of viscosity to shear rate can be understood in terms of simple molecular dynamics of a polymer dissolved in a solvent. At equilibrium, a polymer takes on a random coil configuration in solution. The volume occupied by the coil consists of both polymer and associated solvent. Einstein [8] first demonstrated that the relative viscosity of a fluid consisting of spheres suspended in a continuous fluid matrix increases in proportion to the volume fraction of the second phase (see equation (1.5)). When the fluid is set in motion, the polymer coil begins to align in the direction of flow, thus reducing its apparent coil dimension in the plane normal to the flow direction. At very low flow rates, the polymer relaxation time (i.e., the characteristic time it takes to return to its equilibrium configuration when subjected to an external force) is faster than the rate at which the molecule is deformed in the flow field, and the coil size in solution is relatively unperturbed. As flow rate increases, the time scale of fluid deformation approaches that of the polymer relaxation time, defining the onset of the power-law zone. Eventually, at very high shear rates, the polymer chain is extended to its maximum length. This condition describes the onset of the upper Newtonian region.

Consider two fluids containing equal amounts by weight of ethylene-propylene copolymers A and B with weight-average molecular weights of 80,000 and 300,000, respectively (see Section 1.3.1 for a definitions of polymer molecular weight). The time required for a polymer to return to its equilibrium coil size after it is subjected to an external force is dependent upon the cooperative motions of the molecular bonds along the polymer chain. The greater number of bonds per chain, the longer it will take to relax. Thus, polymer B will have a longer relaxation time than polymer A. This means that the onset of power law region will occur at lower shear rates for polymer B than for polymer A. Another consequence of higher molecular weight is higher coil radius. Therefore, the lower Newtonian plateau viscosity of the polymer B fluid will be higher than for the polymer A fluid. The former will also undergo a greater degree of viscosity loss in the power law region because a higher molecular weight polymer in solution is more deformable than its

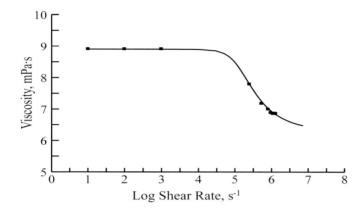

FIGURE 1.2
Shear rate dependence of viscosity (at $373°K$) for an SAE 5W-30 motor oil (data from Sorab et. al [7]).

lower molecular weight analogue (see Fig. 1.1).

1.2.4 The Effect of Pressure on Viscosity

Lubricating oils often experience very high pressures, especially in heavily loaded tribological contacts such as bearings and gears and in hydraulic power transmission systems. Pressures in ball and roller bearings have been reported to be up to $2 - 3\ GPa$ [9], $30 - 2000\ MPa$ in gasoline engine journal bearings [10], and as high as $69\ MPa$ in hydraulic systems [11, 12]. If lubricants were perfectly incompressible, viscosity would be unaffected by pressure. Because of the existence of free volume in the liquid state of most organic substances, lubricating oils are somewhat compressible, and the internal resistance to flow increases with pressure. Since elastohydrodynamic film thickness is governed, in large part, by viscosity, the opposing surfaces in a heavily loaded contact can be better supported by a lubricant under pressure than under ambient conditions.

To a first approximation, viscosity increases exponentially in proportion to pressure according to the Barus equation

$$\mu = \mu_a \exp(\alpha p), \qquad (1.3)$$

where α is the pressure viscosity coefficient and μ_a is the viscosity at atmospheric pressure. The value of α depends upon the chemistry of the fluid as well as temperature. At very high pressures, viscosity increases at a slower rate than predicted, and it can be modeled by a power law relationship

$$\mu = \mu_a(1 + Cp)^n, \qquad (1.4)$$

TABLE 1.3

Pressure viscosity coefficients. (after Kudish and Covitch [1]).
Reprinted with permission from CRC Press.

Lubricating fluid	Temperature ($^\circ C$)	$\alpha \times 10^5$ (KPa^{-1})	Reference
Diethyl-2 hexyl sebacate	0	1.40	[11]
	75	0.64	
Paraffinic mineral oil VI = 99	0	2.00	[11]
	75	1.30	
Paraffinic mineral oil VI = 93	37.8	2.23	[11]
	98.9	1.70	
Naphthenic mineral oil VI = 30	37.8	3.60	[11]
	98.9	2.13	
Polyisobutylene in paraffinic mineral oil	37.8	3.26	[12]
Polymethylmethacrylate in paraffinic mineral oil	37.8	2.45	[12]

where C is a constant at a given temperature and n is equal to 16 [11] for most lubricating oils. Some examples of pressure viscosity coefficients reported for various lubricating oils may be found in Table 1.3.

It is instructive to gain an order-of-magnitude appreciation for the extent to which viscosity increases under pressure. Using the pressure-viscosity coefficient data in Table 1.3 for the 93 VI mineral oil and the 30 VI naphthenic oils, the viscosity at $98.9^\circ C$ increases by a factor of 67% and 90%, respectively, under the influence of 30 MPa pressure.

1.2.5   The Effect of Suspended Contaminants on Viscosity

Lubricating oils are formulated to deal with the reality that they will become contaminated with foreign substances during use. Particulate matter and aqueous acids generated by combustion of fossil fuels, unburned fuel, oxidative degradation of the lubricant, wear particles, and airborne particulates are common examples. Dispersants serve to keep these insoluble particles suspended in the oil and prevent them from forming deposits which can interfere with lubricant flow, heat transfer, and proper operation.

Under extreme conditions, heavy soot contamination of diesel engine oils can thicken the oil to such an extent that the oil cannot be drained out of the vehicle. To prevent this from occurring, a number of strategies have been found to be effective: change oil more frequently or improve the soot dispersing capability of the lubricant via additive design and formulation optimization. The additive design approach inhibits particles from forming large aggregates, which keeps the shape factor low (see equation (1.7)).

High operating temperatures can lead to thermal/oxidative degradation of

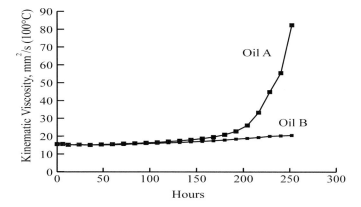

FIGURE 1.3
Kinematic viscosity $(100°C)$ of two SAE 15W-40 oils as a function of time
in the Mack T-11 engine test. (after Kudish and Covitch [1]). Reprinted with
permission from CRC Press.

the lubricating fluid. Introducing heteroatoms like oxygen and nitrogen into
the base oil creates polar species, which are not soluble in most mineral oils or
polyalphaolefin synthetics, often leading to a significant rise in viscosity. To
enable lubricants to withstand high temperatures, certain chemical stabiliz-
ers such as anti-oxidants and detergents are added. Other approaches are to
convert to more inherently-stable base fluids or increase oil change frequency.

Frequent filter inspection and replacement is an effective method for mini-
mizing lubricant contamination with airborne particles such as dust.

Whether the contaminant is a solid or a liquid, it causes the viscosity of the
fluid to increase if it is insoluble in the lubricating oil. The cumulative effect
of insoluble contaminants on engine oil viscosity is illustrated by examining
viscosity data from a Mack T-11 heavy duty diesel engine (see Fig. 1.3).

Although the amount of insoluble soot present in Oils A and B at the end
of the test was nearly equal (7.4% and 8.3%, respectively), the dispersant
additives used in Oil B were far more effective at controlling viscosity rise
than the additive package in Oil A.

Other factors, such as fuel dilution and polymer molecular weight degradation, are responsible for lowering the viscosity of engine oils in field service.
Unburned fuel is soluble in the lubricant and, due to its low viscosity, depresses
viscosity. High molecular weight viscosity index additives often undergo some
level of molecular weight degradation, especially in the initial period of use.
For example, an SAE 5W-30 engine oil formulated with a 35 SSI hydrogenated
styrene-butadiene viscosity modifier (improvers) was run in a New York City
taxi-cab fleet test [13], and kinematic viscosity was measured at $100°C$ on
samples taken during the 8,000 mile oil drain period (see Fig. 1.4). The ini-

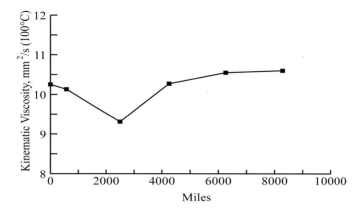

FIGURE 1.4
Kinematic viscosity $(100°C)$ of an SAE 5W-30 oil as a function of time in a Chevrolet Caprice taxi equipped with a 4.3 liter V-6 engine. (after Kudish and Covitch [1]). Reprinted with permission from CRC Press.

tial viscosity loss is attributed to polymer shear; but the steady accumulation of insoluble matter in the oil eventually counteracts the shearing effect, and viscosity begins to rise. An extensive study of polymer molecular weight degradation in heavy duty engine field service [14] confirmed that engine operation mechanically lowers polymer molecular weight as measured by gel permeation chromatography (GPC).

The problem of the effects of non-interacting suspended spherical particles on solution viscosity was first modeled by Einstein [15]

$$\mu_r = 1 + 2.5\varphi, \tag{1.5}$$

where μ_r is the relative viscosity, the ratio of solution dynamic viscosity to that of the solvent, φ is the volume fraction of second phase. This expression was later modified by Brinkman [16] to

$$\mu_r = (1 - \varphi)^{-2.5}. \tag{1.6}$$

The main limitation to these models is the assumption that the dispersed phase consists of non-interacting spheres. In reality, the morphology of diesel engine soot is highly complex; and liquid contaminants, although generally spherical, aggregate into clusters. Therefore, there have been a number of successful attempts to model complex two-phase lubricating fluids by introducing a shape factor s into the Brinkman equation

$$\mu_r = (1 - s\varphi)^{-2.5}. \tag{1.7}$$

Four mathematical expressions that have been proposed for the shape factor are summarized in Table 1.4.

TABLE 1.4
Mathematical forms of the shape factor in equation (1.7), where ρ_s is the density of solid second phase, ρ_0 is the density of oil (continuous) phase, α is the mass fraction of solids in the swollen solid/oil particle, s_j and α are shape factor terms, p is the soot aggregate particle size, p_0 is the primary soot particle size. See references for complete definition of terms. (after Kudish and Covitch [1]). Reprinted with permission from CRC Press.

s	Application	Reference
$(\rho_s/(\alpha\rho_0)) - ((\rho_s/\rho_0) - 1)$	ASTM Sequence IIIC Gasoline Engine	[17]
α	diesel engine soot	[18]
s_j	clusters of uniform spheres	[19]
p^3/p_0^3	diesel engine soot	[20]

Complex particle morphologies trap lubricating fluid within pores and between aggregates. If the particle is partially soluble in the oil phase, it might be highly swollen. If the oil or additive molecules are attracted to the particle surface, a tightly absorbed solvent layer is formed. In all cases, the effective volume fraction of a particle is composed of two contributions: (1) the insoluble substance itself and (2) its associated oil. The latter factor can be ignored in the case of suspended dense non-interacting hard spheres; but most contaminant particles are far from spherical, and they interact with the lubricating oil. The shape factor in equation (1.7) is basically a correction factor, which accounts for non-spherical particle morphology and associated oil. Numerically, it is dependent upon particle shape, state of aggregation, particle/oil compatibility, temperature, and viscosity measurement conditions.

1.2.6 Density

To properly account for the volumetric and elastic response of lubricating oils to changes in temperature and pressure, it is necessary to introduce mathematical expressions for the influence of temperature and pressure on density [21]:

$$\rho_i = \frac{\rho_0}{1 + \beta(T_i - T_0)}, \tag{1.8}$$

where ρ_i is the final density, ρ_0 is the initial density, β is the volumetric temperature expansion coefficient ($63.3 \cdot 10^{-5} \ {}^\circ C^{-1}$ for oil), T_i is the final temperature ($^\circ C$), and T_0 is the initial temperature ($^\circ C$),

$$\rho_i = \frac{\rho_0}{1 - (p_i - p_0)/E}, \tag{1.9}$$

where p_i is the final pressure (Pa), p_0 is the initial pressure (Pa), and E is the bulk modulus fluid elasticity coefficient ($1.5 \ GPa$ for oil).

To illustrate that density, like viscosity, varies more with changes in temperature than pressure, equations (1.8) and (1.9) were used to calculate density at various temperatures and pressures for a mineral oil with $15°C$ density of $938\ kg/m^3$ (see Table 1.5).

TABLE 1.5
The influence of temperature and pressure on density (kg/m^3) of mineral oil. (after Kudish and Covitch [1]). Reprinted with permission from CRC Press.

$Pressure, MPa$	$15°C$	$75°C$	$125°C$	$200°C$
0.02	938.0	903.7	876.9	839.7
1.06	938.7	904.1	877.1	839.8
2.10	939.0	904.7	877.7	840.3
3.01	940.9	906.4	879.3	841.6

1.2.7   Thermal Conductivity

One of the most important functions of a lubricant is to remove heat from the tribological contact zone. High temperatures catalyze wear by lowering lubricant viscosity, which reduces oil film thickness that, in turn, shifts the wear regime from hydrodynamic to mixed to boundary. The ability of a lubricant, or any material for that matter, to remove heat is governed by Fourier's law:

$$q = k\frac{dT}{ds},\tag{1.10}$$

where q is the heat flux per unit area (measured in watts W/m^2), k is the coefficient of thermal conductivity $(W/(m°C))$, dT is the temperature difference across the contact $(°C)$, and ds is the lubricant film thickness (m).

The value of k decreases linearly with temperature. A comparison of the cooling capacity of mineral oil and polyalphaolefin (PAO) synthetic fluids may be found in Table 1.6.

Another equation that can be used to estimate the coefficient of thermal conductivity k of petroleum fluids [23] relates k to density ρ at $15.6°C$:

$$k = \frac{A}{\rho}(1 - BT),\tag{1.11}$$

where k is in units of $W/(m \cdot ° C)$, ρ is in units of kg/m^3, and T is the temperature in $°C$ while constants $A = 132\ W \cdot kg/(m^4 \cdot ° C)$ and $B = 0.00125°C^{-1}$.

To estimate the effect of pressure on k, use equation (1.9) to compute ρ_i at the new pressure and substitute this value for ρ in equation (1.11).

TABLE 1.6

Coefficient of thermal conductivity k as a function of temperature for an engine oil [22] and polyalphaolefin (PAO) synthetic fluids [23]. (after Kudish and Covitch [1]). Reprinted with permission from CRC Press.

Temperature, $°C$	mineral oil	PAO 2	PAO 6	PAO 40
0	0.147	0.138	0.149	0.164
50	0.142	0.135	0.147	0.163
100	0.137	0.132	0.144	0.162
150	0.133	0.129	0.142	0.161

1.3 Polymer Thickening and Shear Stability

1.3.1 Molecular Weight

As stated in Section 1.2.3, polymer molecular weight is a major factor influencing solution rheology. Before proceeding further, a few definitions are in order.

Polymers are chain-like molecules consisting of monomer units of molecular weight w_m linked together through a process known as polymerization consisting of the following steps: initiation, propagation, and termination. The number of monomer units per chain is known as the degree of polymerization (DP, l). A polymer can contain one monomer unit (such as polyethylene or polyisobutylene) or several (such as ethylene-propylene copolymers or polyalkylmethacrylates). One of the differences between a polymer and a pure organic compound is the definition of molecular weight. The molecular weight of the latter is calculated by adding up the atomic weights of the atoms in the molecule. For example, the molecular weight of isobutylene $(C4H8)$ is 56.12 $(4 \times 12.01 + 8 \times 1.01)$. The process of polymerization produces a statistical distribution of chain lengths. For convenience, consider a polymer consisting of a single monomer. The molecular weight w_i of each polymer chain i is $w_m l_i$, where l_i is the number of monomer units in chain i. Let n_i equal the number of chains of molecular weight w_i. Then the number-average and weight-average molecular weights of the polymer M_n and M_w, respectively, are defined as follows:

$$M_n = \sum_i w_i n_i / \sum_i n_i, \tag{1.12}$$

$$M_w = \sum_i w_i^2 n_i / \sum_i w_i n_i, \tag{1.13}$$

where the terms are summed over all values of l_i. The value of M_n correlates with the colligative properties of polymer solutions (vapor pressure lowering, freezing point elevation, osmotic pressure) and is useful for stoichiometric calculations (reactivity of chain ends, etc.). The rheological properties of polymer solutions are better correlated with M_w due in large part to the greater influence of high molecular weight species on bulk flow properties compared to lower molecular weight chains.

1.3.2   Dilute Solution Rheology

For many well-behaved polymer solutions, dilute solution viscosity is adequately described by the Huggins equation

$$\mu_{sp} = [\eta] + k'[\eta]^2 c, \qquad (1.14)$$

where μ_{sp} is the specific viscosity equal to $(\mu - \mu_0)/\mu_0$, c is the polymer concentration (g/dl), k' is Huggins constant, and $[\eta]$ is the intrinsic viscosity, related to molecular size at infinite dilution.

The intrinsic viscosity term $[\eta]$ is related to M_w by the Mark-Houwink relationship

$$[\eta] = k' M_w^a, \qquad (1.15)$$

where k' and a are constants. Experimentally determined values of k' and a have been reported for ethylene-propylene copolymers [24] as well as for a number of other chemistries [25].

1.3.3   Shear Stability

It follows that large macromolecules are more efficient thickeners than low molecular weight polymers. If so, why not formulate multi-viscosity lubricants with the lowest concentration of the highest molecular weight polymers available for sale? The answer is simple. High molecular weight polymers are more susceptible to mechanical degradation than their lower molecular weight analogues. When a chain molecule passes through a tribological contact, it is elongated in the direction of flow. Shear and elongational forces are transmitted to every fluid element of the lubricant, and these forces are additive along the polymer backbone. The longer the backbone, the greater the force and the likelihood of polymer rupture. Polymer degradation can lead to lubricant viscosity dropping below its minimum design limit. In practice, the minimum design limit is understood to be the minimum viscosity of the lubricant's viscosity grade. For example, the kinematic viscosity of an SAE 5W-30 engine oil must be between 9.3 and 12.5 mm^2/s at $100°C$. If the viscosity of this oil stays within these limits over its useful life, it meets "stay-in-grade" performance. If the viscosity drops below 9.3 mm^2/s, several corrective actions can be taken. The oil can be formulated to a higher kinematic viscosity target, or a more shear-stable (lower molecular weight) viscosity modifier can be selected.

To maintain acceptable viscosity retention over the useful life of a lubricating oil, the polymeric viscosity modifier must be designed to strike a balance between thickening efficiency (for economic reasons) and shear stability.

TABLE 1.7

Laboratory tests that simulate viscosity loss due to mechanical shear. (after Kudish and Covitch [1]). Reprinted with permission from CRC Press.

Test	Test method	Application	Relative tendency to affect mechanical degradation
Kurt Orbahn diesel injector rig	ASTM D6278 ASTM D7109 CEC L-14	compression ignition engine oils	moderate
Sequence VIII spark ignition engine	ASTM D6709	spark ignition engine oils	mildest
KRL tapered bearing rig	CEC L-45 DIN 51 350 VW 1437	transmission, gear, and hydraulic system oils	most severe
Sonic shear	ASTM D2603 ATM D5621 JAASO M347	engine, gear, transmission, and hydraulic system oils	mild to severe
FZG gear rig	CEC L-37	gear oils	severe

To quantify the viscosity loss of lubricating oils during use, various industry working groups have developed laboratory methods for simulating shear degradation of polymer-containing fluids. Some lubrication applications impose greater levels of viscosity degradation than others and, therefore, demand the use of lower molecular weight polymer additives. A brief summary of some of the most common shear stability tests may be found in Table 1.7.

When a lubricant is tested in the field or in one of the laboratory tests in Table 1.7, the viscosity change can be expressed in terms of percentage of viscosity loss or SSI (shear stability index). The latter is a measure of the amount of viscosity contributed by the viscosity modified that is lost by mechanical degradation and is computed as follows:

$$SSI = \frac{NOV-SOV}{NOV-BBV} \times 100, \qquad (1.16)$$

where NOV is the new oil viscosity, SOV is the sheared oil viscosity, and BBV is the base blend viscosity (base oil(s) + performance additive(s) without the viscosity modifier present), all measured at the same temperature.

Since the tendency of different field conditions and laboratory tests to degrade polymer molecular weight varies widely, it is important that the correct SSI value is used in a given application. Consider, for example, a single viscosity modifier that is used to formulate a passenger car motor oil, a heavy duty diesel engine oil, and an automotive gear oil. The lubricants are tested in field service; and NOV, SOV, and BBV are measured. SSI values of 15, 25, and 80 are calculated. This illustrates that there is not a unique shear stability index value that can be ascribed to a given viscosity modifier additive. The test conditions used to calculate SSI must be clearly understood to properly match a polymer to a given lubricant application.

Usually, the characteristic SSI associated with a viscosity modifier refers to ASTM method D6278 (30 cycles), but it is always wise to confirm. A subtle but important aspect of SSI calculations is to recognize that the only component that contributes to viscosity loss is the polymer itself. Many viscosity modifiers (improvers) are sold as viscous mixtures of polymers dissolved in mineral oil. The oil component needs to be factored into determination of the base blend viscosity (BBV) in equation (1.16) in order to calculate the true SSI.

1.4   Closure

The concept of viscosity is introduced and shown to be sensitive to temperature and pressure. Classical empirical formulas for modeling these relationships are provided. For fluids containing polymeric additives, viscosity becomes dependent upon shear rate as well. Polymer molecular weight averages are defined and offered as useful measures of thickening efficiency and shear stability. Rheological aspects of lubricant degradation are presented, and mechanisms for viscosity change in automotive and industrial lubricants are discussed.

1.5 Exercises and Problems

1. A Newtonian lubricating fluid having a viscosity of 46.8 $mPa \cdot s$ at $40°C$ fills the gap between two large parallel plates, the upper plate moving at a velocity of 2.16 m/s relative to the fixed lower plate. The shear stress required to maintain this velocity is 123.3 Pa. What is the distance between the plates?

Answer: Shear rate = velocity / gap. Gap = 0.82 mm.

2. A Newtonian transmission oil designed to operate at temperatures less than $85°C$ was used in a new aerodynamic transmission prototype that developed oil temperatures as high as $114°C$. The viscosity of this fluid at $25°C$ and $80°C$ is 66.6 and 11.0 mm^2/s respectively. What is the lowest viscosity this oil is expected to have in the prototype transmission? Assume $\theta = 0.7$.

Answer: Applying equation 2.2.2, and solving simultaneous equations results in $A = 8.1015$ and $B = 3.168$. The viscosity at $114°C$ is equal to 5.6 mm^2/s

3. A rookie tribologist measures the viscosity of an SAE 20W-50 racing oil at two different shear rates. The results were identical. He concludes that the fluid does not contain polymer additives. Why is his conclusion faulty?

Answer: He could have taken two measurements in the lower Newtonian regime (see Fig. 1.2) where a polymer-containing fluid appears to be Newtonian.

4. The polyisobutylene solution in Table 1.3 has a viscosity of 84.3 $mPa \cdot s$ at $100°F$. At the same temperature, what percentage increase in viscosity would be expected at a pressure of 50 MPa?

Answer: Applying equation (1.3) and $\alpha = 3.26 \cdot 10^{-5}$ from Table 1.3, the answer is 410.

5. An engineer examines an oil sample taken from a diesel locomotive. The engine was originally filled with 15.10 mm^2/s ($100°C$) RightTrack II SAE 15W-40 oil. The used oil sample is black in color, and the lab reported that it contains 2.6% soot. There is virtually no dissolved fuel in the sample, yet the kinematic viscosity is unchanged. Provide at least one possible explanation why viscosity did not change.

Answer: (1) Viscosity loss due to polymer degradation exactly balanced viscosity rise due to soot contamination or (2) the oil contained a low molecular weight, highly shear-stable viscosity modifier, but the dispersant additives in the oil prevented significant viscosity increase (as in Oil B in Fig. 1.3).

6. What is the weight-average molecular weight of an A-B block copolymer molecule where the A-block consists of 320 styrene (C_8H_8) monomer units and the B-block consists of 1000 dodecylmethacrylate ($C_{16}O_2H_{30}$) monomer units?

Answer: The molecular weight of styrene is 104.08 and dodecylmethacrylate is 254.16. The molecular weight of the block copolymer is 287,000. In this case, we are dealing with a single molecule where M_n and M_w are, by definition, equal.

7. You are asked to formulate a stay-in-grade SAE 10W-40 engine oil with a 35 SSI viscosity modifier. Given the $100°C$ kinematic viscosity of the oil (base oils and performance additive) without viscosity modifier is $4.64 \ mm^2/s$ and the viscosity range of this viscosity grade is $12.5 \ mm^2/s - 16.3 \ mm^2/s$, what is the minimum new oil viscosity required to stay in grade? Is it feasible?

Answer: $16.7 \ mm^2/s$. Since this target viscosity is greater than the maximum for this viscosity grade, it is not feasible to satisfy design requirements for this lubricant.

References

[1] Kudish, I.I. and Covitch, M.J. 2010. *Modeling and Analytical Methods in Tribology.* Chapman & Hall/CRC.

[2] Forsythe, W.E., ed. 2003. *Smithsonian Physical Tables, Ninth Rev. Ed.*, p. 322, Knovel, Norwich, New York: Smithsonian Institution.

[3] Walther, C. 1931. The Evaluation of Viscosity Data. *Erdol u. Teer* 7:382-384.

[4] Standard Test Method for Viscosity-Temperature Charts for Liquid Petroleum Products. 2004. *ASTM D341 − −03*, ASTM International, West Conshohocken, PA, USA, 1-5.

[5] Webber, R.M., George, H.F., and Covitch, M.J. 2000. Physical Processes Associated with Low Temperature Mineral Oil Rheology: Why the Gelation Index is Not Necessarily a Relative Measure of Gelation. *Soc. Automot. Eng. Tech. Paper Ser.* No. 2000-01-1806.

[6] April 2008. Appendix E - Base Oil Interchangeability Guidelines for Passenger Car Motor Oils and Diesel Engine Oils. *Amer. Petrol. Inst.*, Section E.1.3.

[7] Sorab, J., Holdeman, H.A., and Chui, G.K. 1993. Viscosity Prediction for Multigrade Oils. In *Tribological Insights & Performance Characteristics*, M.J. Covitch and S.C. Tung, eds. *Soc. Automotive Engr.* SP-966:241-252.

[8] A. Einstein, A. 1911. Correction to my paper "Eine neue Bestimmung der Molekldimensionen" or "A New Determination of Molecular Dimensions." *Ann. Phys. (Leipzig)* 34:591-592.

[9] E.V. Zaretsky, ed. 1997. *Tribology for Aerospace Applications.* STLE Publication SP, p. 37.

[10] Bates, T.W. 1990. Oil Rheology and Journal Bearing Performance: A Review. *Lub. Sci.* 2:157-176.

[11] Briant, J., Denis, J., and Parc, G., eds. 1989. *Rheological Properties of Lubricants*, Paris: Editions Technip.:115-120.

[12] So, B.Y.C. and Klaus, E.E. 1980. Viscosity-Pressure Correlation of Liquids, *ASLE Trans.* 23, No. 4:409-421.

[13] Covitch, M.J., Weiss, J., and Kreutzer, I.M. 1999. Low-Temperature Rheology of Engine Lubricants Subjected to Mechanical Shear: Viscosity Modifier Effects. *Lub. Sci.* 11-4:337-364.

[14] Covitch, M.J., Wright, S.L., Schober, B.J., McGeehan, J.A., and Couch, M. 2003. Mechanical Degradation of Viscosity Modifiers in Heavy Duty Diesel Engine Lubricants in Field Service. *Soc. Automot. Eng. Tech. Paper Ser.* No. 2003-01-3223.

[15] Einstein, A. 1906. "Eine neue Bestimmung der Molekldimensionen" or "A New Determination of Molecular Dimensions." *Ann. Phys.* 19:289-306.

[16] Brinkman, H.C. 1952. The Viscosity of Concentrated Suspensions and Solutions. *J. Chem. Phys.* 20:571.

[17] Spearot, J.A. 1974. Viscosity of Severely Oxidized Engine Oil. *Am. Chem. Soc. Div. Petrol. Chem.*:598-619.

[18] Yasutomi, S., Maeda, Y., and Maeda, T. 1981. Kinetic Approach to Engine Oil. 3. Increase in Viscosity of Diesel Engine Oil Caused by Soot Contamination. *Ind. Eng. Chem. Prod. Res. Div.* 20:540-544.

[19] Graham, A.L., Steele, R.D., and Bird, R.B. 1984. Particle Clusters in Concentrated Suspensions. 3. Prediction of Suspension Viscosity. *Ind. Eng. Chem. Fundam.* 23:420-425.

[20] Covitch, M.J., Humphrey, B.K., and Ripple, D.E. 1985. Oil Thickening in the Mack T-7 Engine Test - Fuel Effects and the Influence of Lubricant Additives on Soot Aggregation. *Soc. Automot. Eng. Tech. Paper Ser.* No. 852126.

[21] Density of Fluids - Changing Pressure and Temperature. 2009. *The Engineering Toolbox.* www.EngineeringToolBox.com

[22] Hamrock, B.J. Lubricant Properties. 1994. *Fundamentals of Fluid Friction Lubrication.* New York: McGraw-Hill, 64-65.

[23] Booser, E.R., ed. 1997. *Tribology Data Handbook.* New York: CRC Press, p. 39.

[24] Crespi, G., Valvassori, A., and Flisi, U. 1977. Olefin Copolymers. *Stereo Rubbers*:365-431.

[25] Brandrup, J. and Immergut, E.H. 1975. *Polymer Handbook*, 2nd ed., Section IV, pp. 1-33. New York: John Wiley & Sons.

2

Basic Properties of Elastic Solids

2.1 Introduction

In this chapter we will introduce some basic commonly used properties of elastic solids such as Young's modulus, Poisson's ratio, etc. These material properties are important not only for calculation of material stresses, deformations, and surface displacements but also for studying heat transfer in materials [1, 2].

TABLE 2.1

Density of some solid materials

Material	Density, $[kg/m^3]$
Aluminum and its alloys	$2.7 \cdot 10^3$
Aluminum tin	$3.1 \cdot 10^3$
Brasses	$8.6 \cdot 10^3$
Bronze	$7.5 \cdot 10^3$
Bronze, leaded	$8.9 \cdot 10^3$
Bronze, cast	$8.7 \cdot 10^3$
Bronze, porous	$6.4 \cdot 10^3$
Copper	$8.9 \cdot 10^3$
Copper lead	$9.5 \cdot 10^3$
Iron, cast	$7.4 \cdot 10^3$
Iron, porous	$6.1 \cdot 10^3$
Iron, wrought	$7.8 \cdot 10^3$
Magnesium alloy	$1.8 \cdot 10^3$
Steels	$7.8 \cdot 10^3$
Zinc alloys	$6.7 \cdot 10^3$
Polyformaldehyde	$1.4 \cdot 10^3$
Polyamides	$1.14 \cdot 10^3$
Polyethylene, high-density	$0.95 \cdot 10^3$
Phenol formaldehyde	$1.3 \cdot 10^3$
Alumina (Al_2O_3)	$3.9 \cdot 10^3$
Silicone carbide (SiC)	$2.9 \cdot 10^3$
Silicon nitride (Si_2N_4)	$3.2 \cdot 10^3$

2.2 Material Density

Material density is determined by the material chemical composition and the material molecule packing order. Metals are usually denser than other materials such as polymers. It is due to the fact that usually metal molecules have higher molecular weight and have denser packing while polymer molecules are lighter and branched and it is relatively difficult to get a dense packing of such branched molecules. Some density data on different solid materials are given in Table 2.1

TABLE 2.2

Young's modulus of some materials

Material	Modulus of elasticity, E, [GPa]
Aluminum	62
Aluminum structural alloys	70
Aluminum tin	63
Brasses	100
Bronze	117
Bronze, leaded	97
Bronze, cast	110
Bronze, porous	60
Copper	124
Iron, gray cast	109
Iron, malleable cast	170
Iron, spheroidal graphite	159
Iron, porous	80
Iron, wrought	170
Magnesium alloy	41
Steel, low alloy	196
Steel, medium and high alloys	200
Steel, stainless	193
Steel, high speed	212
Zinc alloys	50
Polyformaldehyde	2.7
Polyamides	1.9
Polyethylene, high-density	0.9
Phenol formaldehyde	7.0
Alumina (Al_2O_3)	390
Cemented carbides	450
Silicon carbide (SiC)	450
Silicon nitride (Si_2N_4)	314

2.3 Material Elastic Properties

Homogeneous elastic materials can be characterized by just two constants: modulus of elasticity (Young's modulus) and Poisson's ratio. Young's modulus is the coefficient of proportionality between stress and strain in uniaxial stress state. Due to tight packing order dense metals usually have higher Young's modulus. For some materials values of Young's modulus are given in Table 2.2.

TABLE 2.3
Poisson's ratio of some solid materials

Material	Poisson's ratio, ν
Aluminum and its structural alloys	0.33
Brasses	0.33
Bronze	0.33
Bronze, porous	0.22
Copper	0.33
Iron, cast	0.26
Iron, porous	0.20
Iron, wrought	0.30
Magnesium alloy	0.33
Steels	0.30
Zinc alloys	0.27
Polyamides	0.40
Polyethylene, high-density	0.35
Alumina (Al_2O_3)	0.28
Cemented carbides	0.19
Silicon carbide (SiC)	0.19
Silicon nitride (Si_2N_4)	0.26

Poisson's ratio ν is a dimensionless characteristic and it represents the degree of dimensional material changes in the direction transverse to the applied uniaxial stress. Poisson's ratio always satisfies the inequality $0 < \nu \leq 0.5$. For some materials values of Poisson's ratio ν are given in Table 2.3.

TABLE 2.4

Thermal diffusivity of some materials at $20°C$

Material	Thermal diffusivity, k, $[W/(m \cdot ° C)]$
Aluminum	209
Aluminum alloys, cast (at $100°C$)	146
Aluminum alloys, silicon (at $100°C$)	170
Aluminum alloys, wrought (at 20 to $100°C$)	146
Aluminum tin	180
Brasses (at $100°C$)	120
Bronze, aluminum (at $100°C$)	50
Bronze, leaded	47
Bronze, cast	50
Bronze, porous	30
Copper (at $100°C$)	170
Copper lead	30
Iron, gray cast	50
Iron, spheroidal graphite	30
Iron, porous	28
Iron, wrought	70
Magnesium alloy	110
Steel, low alloy (at 20 to $200°C$)	35
Steel, medium alloys	30
Steel, stainless	15
Zinc alloys	110
Polyformaldehyde	0.24
Polyamides	0.25
Polyethylene, high-density	0.5
Alumina (Al_2O_3)	25
Silicon carbide (SiC)	15

2.4   Material Thermal Diffusivity and Specific Heat Capacity

If two solids at different temperatures are brought in contact they exchange heat, i.e., the process of heat transfer occurs due to different levels of kinetic energy of molecules of these solids. The measure of a material to conduct/diffuse heat at a constant temperature is the material heat diffusivity/conductivity k. The values of the heat diffusivity for some materials are given in Table 2.4.

The amount of heat transferred to or from a material depends on a number of factors among which are the material nature, the smoothness/roughness of its surface, etc. For solids with smooth surfaces the main factor is the material

TABLE 2.5

Specific heat capacity of some materials at $20°C$

Material	Specific heat capacity, c, $[kJ/(kg \cdot °\,C)]$
Aluminum and its alloys	0.9
Aluminum tin	0.96
Brasses	0.39
Bronzes	0.38
Copper	0.38
Copper lead	0.32
Iron, cast	0.42
Iron, porous	0.46
Iron, wrought	0.46
Magnesium alloy	1.0
Steels	0.45
Zinc alloys	0.4
Cemented carbides	0.7

nature. The amount of heat transferred to of from a solid is proportional to the solid mass and the difference in temperatures. The coefficient of proportionality c in this relationship is the specific heat capacity. In other words, the material heat capacity is numerically equal to the heat transferred to or from a material of unit mass when its temperature is changed by one degree Celsius. For some materials the values of the specific heat capacity are given in Table 2.5.

2.5 Closure

A number of physical characteristics of solid materials such as Young's modulus, Poisson's ratio, coefficient of thermal diffusivity as well as material specific heat capacity are introduced and provided for a number of different materials.

References

[1] Booser, E.R., ed. 1997. *Tribology Data Handbook*. New York: CRC Press, p. 39.

[2] Hamrock, B.J. Lubricant Properties. 1994. *Fundamentals of Fluid Friction Lubrication*. New York: McGraw-Hill, 64-65.

Part II

Asymptotic Methods and Relationships Relevant to Elastohydrodynamic Lubrication Theory

This part of the monograph provides definitions, very basic properties, and permitted operations on asymptotic expansions used in the analysis of lubrication problems. For more detailed information on asymptotic and numerical methods used readers are referred to numerous monographs devoted to these methods.

In addition to that in Part II are introduced all necessary governing equations for physical processes taking place in dry and lubricated line and point contacts. In particular, Navier-Stokes equations get simplified, the expressions for normal elastic displacements of surfaces in line and point contacts are presented, some lubricant rheologies are discussed. Also, for the benefit of the further analysis some classic solutions for plane contact problems of elasticity are considered.

3

Basics of Asymptotic Expansions and Methods

3.1 Introduction

In this chapter we will introduce some basic commonly used notations and operations related to asymptotic expansions. The focus will be on application of these notions and operations to practical problems. A more detailed description of the basics of asymptotic expansions can be found in a variety of books, for example, [1] - [4].

3.2 Ordering, Order Sequences, and Asymptotic Expansions

Asymptotic analysis is a study of the behavior of a certain mathematical object (function, or algebraic, differential, integral equation, integral, etc.) as a parameter or independent variable this object depends on is approaching a certain limit. The asymptotic behavior of a mathematical object as its parameter or variable approaches a certain limit is just the object limiting behavior. We will concentrate on asymptotic behavior as some parameter λ approaches zero or infinity. To discriminate between different asymptotic behaviors, we need to introduce ordering. Suppose we have two functions $f(x, \lambda)$ and $g(x, \lambda)$ determined on an interval $x \in I$ for $\lambda \geq 0$. In many cases it is useful to use two orderings: large O and small o. Let us define these orderings for $\lambda \to 0$. The definitions for $\lambda \to \infty$ are similar.

Large O

For fixed $x \in I$ and $\lambda \to 0$, we say that $f(x, \lambda) = O(g(x, \lambda))$ if there exists such a positive finite value M (generally, dependent on x) that $\mid f(x, \lambda) \mid \leq M \mid g(x, \lambda) \mid$ for $\lambda \to 0$.

Small o

For fixed $x \in I$ and $\lambda \to 0$, we say that $f(x, \lambda) = o(g(x, \lambda))$ if $\lim_{\lambda \to 0} \frac{f(x,\lambda)}{g(x,\lambda)} = 0$. Often, $f \ll g$ as $\lambda \to 0$ (or $\lambda \ll 1$) is used as an equivalent notation.

If $\lim_{\lambda \to 0} \frac{f(x,\lambda)}{g(x,\lambda)} = 1$ for fixed $x \in I$ and $\lambda \to 0$, we say that $f(x,\lambda) \sim g(x,\lambda)$.

If the validity of the relationships $f(x,\lambda) = O(g(x,\lambda))$ or $f(x,\lambda) = o(g(x,\lambda))$ is independent of $x \in I$, then we say that the corresponding relationship is uniformly valid in I.

Here are some examples of the large O and small o definitions.

$$\lambda^2 = O(\lambda), \ \lambda \to 0,$$

$$1 - \cos(\lambda) = O(\lambda^2), \ \lambda \to 0,$$

$$\lambda = O(\lambda^2), \ \lambda \to \infty,$$

$$\ln(1 + \lambda) - \lambda = o(\lambda), \ \lambda \to 0.$$

(3.1)

The following asymptotic estimates are uniformly valid for $x \in I = (0,1]$:

$$\frac{\lambda}{x^2+1} = O(\lambda), \ \lambda \to 0,$$

$$1 - \cos(\lambda x) = O(\lambda^2), \ \lambda \to 0,$$

(3.2)

while on the same interval the asymptotic estimates

$$\frac{\lambda}{x^2} = O(\lambda), \ \lambda \to 0,$$

$$\exp(-\frac{x}{\lambda}) = O(\lambda^2), \ \lambda \to 0$$

(3.3)

are not uniform. The latter is due to the fact that depending on how fast x approaches 0 in relation to λ the above two expressions may be small, large, or of the order of unity as $\lambda \to 0$.

Generally, various operations such as addition, subtraction, multiplication, division, and integration can be performed on order relations. In many cases it is also possible to perform the operation of differentiation on order relations. However, in some cases a straightforward asymptotic estimate of a derivative may lead to an error. Below is an example when it is permissible to use the straightforward approach to the asymptotic estimation of a derivative while $\lambda \to 0$

$$\frac{\lambda}{x^2+1} = O(\lambda), \ x = O(1), \ \Rightarrow \ \frac{d}{dx}\frac{\lambda}{x^2+1} = \frac{O(\lambda)}{O(1)} = O(\lambda), \ x = O(1), \qquad (3.4)$$

and an example when it clearly causes an error for $x \gg \lambda$, $x \in [0,1]$

$$\cos(\frac{x}{\lambda}) = O(1), \ x = O(1), \ \Rightarrow \ \frac{d}{dx}\cos(\frac{x}{\lambda}) = \frac{O(1)}{O(1)} = O(1), \ x = O(1). \qquad (3.5)$$

To produce a correct and accurate asymptotic estimate of a derivative of a function, there is usually need for more detailed knowledge of the function behavior. For more details on asymptotic differentiation of functions see [1].

3.3 Asymptotic Sequences and Expansions

A sequence of functions $\{\varphi_n(\lambda)\}$, $n = 1, 2, \ldots$ is called an asymptotic sequence if $\varphi_{n+1}(\lambda) = o(\varphi_n(\lambda))$ for $\lambda \to 0$ and $n = 1, 2, \ldots$. Asymptotic sequences may have finite or infinite number of members.

If the members of the asymptotic sequence $\{\varphi_n(x, \lambda)\}$ also depend on variable x the asymptotic sequence $\{\varphi_n(x, \lambda)\}$, $n = 1, 2, \ldots$ may or may not be uniformly asymptotic on an interval I. Below, are some examples of uniform on $[0, 1]$

$$\{(\lambda x)^n\}, \ n = 1, 2, \ldots, \ \lambda \to 0,$$

$$\{x^n \ln^{-n} \lambda\}, \ n = 1, 2, \ldots, \ \lambda \to 0,$$

(3.6)

and nonuniform on $(0, 1]$ asymptotic sequences

$$\{(\tfrac{\lambda}{x})^n\}, \ n = 1, 2, \ldots, \ \lambda \to 0,$$

$$\{\exp(-n\tfrac{x}{\lambda})\}, \ n = 1, 2, \ldots, \ \lambda \to 0.$$

(3.7)

Generally, various operations such as addition, subtraction, multiplication, division, and integration can be performed on asymptotic sequences. In many cases it is also possible to perform the operation of differentiation on the asymptotic sequence with respect to the parameter λ or variable x. However, in some cases differentiation of a uniformly valid asymptotic sequence/expansion may not produce a uniformly valid asymptotic sequence/expansion (see examples above and [1]).

Suppose $f(x, \lambda)$ is a function determined for $x \in [0, 1]$ and $\lambda \geq 0$ and $\{\varphi_n(\lambda)\}$, $n = 1, 2, \ldots$ is an asymptotic sequence as $\lambda \to 0$. Then the sum $\sum_{n=1}^{N} f_n(x)\varphi_n(\lambda)$ is called an asymptotic expansion of the function $f(x, \lambda)$ to N terms as $\lambda \to 0$ if

$$f(x, \lambda) - \sum_{n=1}^{N} f_n(x)\varphi_n(\lambda) = o(\varphi_N(\lambda)), \ \lambda \to 0,$$

(3.8)

for $N = 1, 2, \ldots$ If $N = \infty$, we get an infinite asymptotic series

$$f(x, \lambda) \sim \sum_{n=1}^{\infty} f_n(x)\varphi_n(\lambda), \ \lambda \to 0,$$

(3.9)

which may or may not be convergent. Nonetheless, in either case such a series or expansion may provide a very valuable information about the behavior of the function $f(x, \lambda)$ as $\lambda \to 0$. If an asymptotic series is divergent, then we should fix the number of terms N in the asymptotic expansion and consider the expansion behavior for $\lambda \to 0$. One has to keep in mind that the truncation error of the asymptotic expansion in (3.8) is of the order much smaller than function $\varphi_N(\lambda)$, which vanishes as $\lambda \to 0$. An example of a divergent asymptotic series for an integral is given below

$$\int\limits_0^\infty \frac{\lambda e^{-x}}{\lambda + x} dx \sim \sum_{n=0}^\infty \frac{(-1)^n n!}{\lambda^n}, \quad \lambda \to \infty. \tag{3.10}$$

It can be easily estimated that the remainder $R_N(\lambda)$ of the asymptotic series (3.10) satisfies the inequality

$$\mid R_N(\lambda) \mid = \mid \int\limits_0^\infty \frac{\lambda e^{-x}}{\lambda + x} dx - \sum_{n=0}^N \frac{(-1)^n n!}{\lambda^n} \mid < \frac{N!}{\lambda^N}. \tag{3.11}$$

It can be shown that $R_N(\lambda) \to \infty$ for a fixed λ and $N \to \infty$, i.e., the series diverges. On the other hand, $R_N(\lambda) \to 0$ for a fixed N and $\lambda \to \infty$. Therefore, the truncated asymptotic expansion from (3.10) can be successfully used for approximation of this integral for large λ.

A more detailed discussion of the behavior and utilization of convergent and divergent asymptotic expansions can be found in [1] - [4].

An asymptotic expansion is said to be uniformly valid in interval I if the relationship (3.8) holds uniformly in I. Obviously, for that to be true it is sufficient that $f_{n+1}(x) = O(f_n(x))$ as $\lambda \to 0$ and $x \in I$.

By repeated application of definition (3.8), we can uniquely determine the coefficients $\{f_n(x)\}$ of the asymptotic expansion as follows

$$f_k(x) = \lim_{\lambda \to 0} \frac{f(x,\lambda) - \sum\limits_{n=1}^{k-1} f_n(x)\varphi_n(\lambda)}{\varphi_k(\lambda)}, \quad k = 1, 2, \ldots. \tag{3.12}$$

3.4   Asymptotic Methods

Most asymptotic methods can be subjected to a simple classification on regular and singular asymptotic methods. Regular asymptotic methods are applicable to the problems in which the solution can be represented by a uniformly valid asymptotic expansion in the entire solution region. Physically, this situation corresponds to the case when the contribution to the problem solution of the physical mechanism which gave rise to the small parameter of the problem remains small in the entire solution region. Application of regular asymptotic methods is relatively simple.

The situation gets more complex when there is a necessity to apply singular asymptotic methods. In such cases (with few exceptions [2] - [4]) the solution is searched in the form of several non-uniformly valid asymptotic expansions. The reason for that is the change of the leading physical mechanisms contributing to problem solution in different solution regions. Each of these asymptotic expansions is valid in its own region. These regions overlap and allow for matching solutions in these regions. That is the basic concept of the most often used method of matched asymptotic expansions. The particular realization of this method depends on the specifics of the problem. Moreover, after the problem is solved by the method of matched asymptotic expansions, it is easy to construct a uniformly valid approximate solution (see [2]).

We assume that the reader is acquainted with the basics of asymptotic approaches. Therefore, we will not get into details of various asymptotic methods. Many examples of application of different regular and singular asymptotic expansion methods are given in [1] - [4]. The problems considered range from estimating functions and integrals to asymptotic solution of problems for ordinary and partial differential equations. In the following chapters we will use both the regular and singular asymptotic methods. In most cases when we need to use a singular asymptotic method we will use the matched asymptotic expansions method.

3.5 Closure

The ordering definitions of o and O as well as asymptotic sequences and series are introduced. Some operations on asymptotic sequences and series are considered. The application of asymptotic techniques has many different facets. To learn more detail on application of these basic notions a reader is referred to various applications described in [1]-[4].

3.6 Exercises and Problems

1. For $\lambda \to 0$ determine the order of the following expressions: (i) $\ln(1 + \arctan \lambda)$, (ii) $\tan(\frac{\pi}{2} - \lambda^2)$, (iii) $\sqrt[3]{\frac{\ln(1+\lambda)}{1+\lambda}} + \frac{e^\lambda}{1-\lambda}$.

2. For $\lambda \to 0$ determine four-term asymptotic expansions of functions from Problem 1.

3. For $\lambda \to 0$ list the following functions λ^2, $\lambda^{-1/4}$, $\frac{\lambda^2}{\sin \lambda}$, $\lambda \ln \lambda$, $\lambda^{-3/2}$ in the decreasing order of magnitude.

4. For $\lambda \to 0$ determine asymptotic expansions o functions: (i) $\frac{\ln(1+\lambda x^2)}{\arcsin(\lambda x)}$, (ii) $\{1 + \tan(\sqrt{\lambda}x) + e^{\lambda x}\}^{-1}$, (iii) $\frac{\sin(\lambda x) - \lambda x}{\lambda^3 x^3}$. Are these asymptotic expansions uniformly valid for all x? In which regions are these asymptotic expansions non-uniformly valid, i.e., cease to remain asymptotic expansions?

5. For $\lambda \to 0$ find three-term asymptotic expansions of all three solutions of the following algebraic equations: (i) $(1+\lambda)x^3 - (6+\lambda^2)x^2 + (11-\lambda)x - 6 + 2\lambda =$

0, (ii) $\lambda x^3 - x + 1 + \lambda^2 = 0$.

References

[1] Erdeiyi, A. 1956. *Asymptotic Expansions*. New York: Dover Publications.

[2] Van-Dyke, M. 1964. *Perturbation Methods in Fluid Mechanics*. New York-London: Academic Press.

[3] Kevorkian, J. and Cole, J.D. 1985. *Perturbation Methods in Applied Mathematics*. New York: Springer-Verlag.

[4] Nayfeh, A.H. 1984. *Introduction to Perturbation Techniques*. New York: John Wiley & Sons.

4

Basics of the Theory of Elastohydrodynamically Lubricated (EHL) Contacts

---

4.1 Introduction

A lubricated contact is a very complex system. A number of different mechanical, heat generation and transfer, chemical, and failure phenomena are taking place in a lubricated contact. We will focus only on mechanical and heat generation and transfer phenomena. These phenomena include mechanisms of elastic surface displacements and lubricant flows. Further, it is assumed that lubricated contacts always operate within the range of elastic deformations, i.e., conditions of plastic deformations are never realized. The purpose of this chapter is to establish the necessary relationships for surface elastic displacements, equations of fluid motion, and heat generation and transfer. In addition to that, we will establish some classic solutions of contact problems of elasticity for dry contacts which will benefit the development of the theory of elastohydrodynamic lubrication.

4.2 Simplified Navier-Stokes and Energy Equations

Let us consider in a Cartesian coordinate system (x, y, z) and time t a non-steady flow of incompressible viscous fluid with density ρ, viscosity μ, velocity components u, v, w, pressure p and the components of stress tensor p_{xx}, p_{yy}, p_{zz}, $p_{xy} = p_{yx}$, $p_{yz} = p_{zy}$, $p_{zx} = p_{xz}$. Such a flow satisfies the equations [1]-[3]

$$\frac{\partial u}{\partial t} + u\frac{\partial u}{\partial x} + v\frac{\partial u}{\partial y} + w\frac{\partial u}{\partial z} = \frac{1}{\rho}\left(\frac{\partial p_{xx}}{\partial x} + \frac{\partial p_{yx}}{\partial y} + \frac{\partial p_{zx}}{\partial z}\right),$$

$$\frac{\partial v}{\partial t} + u\frac{\partial v}{\partial x} + v\frac{\partial v}{\partial y} + w\frac{\partial v}{\partial z} = \frac{1}{\rho}\left(\frac{\partial p_{xy}}{\partial x} + \frac{\partial p_{yy}}{\partial y} + \frac{\partial p_{zy}}{\partial z}\right), \qquad (4.1)$$

$$\frac{\partial v}{\partial t} + u\frac{\partial w}{\partial x} + v\frac{\partial w}{\partial y} + w\frac{\partial w}{\partial z} = \frac{1}{\rho}\left(\frac{\partial p_{zx}}{\partial x} + \frac{\partial p_{yz}}{\partial y} + \frac{\partial p_{zz}}{\partial z}\right).$$

The mass forces are omitted in equations (4.1) due to the fact that in lubricated contacts they are negligibly smaller than acting stresses. For a fluid with Newtonian rheology for which the components of the stress tensor, pressure, fluid viscosity, and the components of the tensor of shear rates are related as follows [3]:

$$p_{xx} = -p + 2\mu \frac{\partial u}{\partial x}, \ p_{yy} = -p + 2\mu \frac{\partial v}{\partial y}, \ p_{zz} = -p + 2\mu \frac{\partial w}{\partial z},$$

$$p_{xy} = p_{yx} = \mu(\frac{\partial u}{\partial y} + \frac{\partial v}{\partial x}), \ p_{yz} = p_{zy} = \mu(\frac{\partial v}{\partial z} + \frac{\partial w}{\partial y}), \quad (4.2)$$

$$p_{zx} = p_{xz} = \mu(\frac{\partial w}{\partial x} + \frac{\partial u}{\partial z}),$$

equations (4.1) are reduced to classic Navier-Stokes equations if the fluid viscosity μ is constant. If the fluid viscosity varies from one point of the flow to another, the flow equations assume the form (see equations (4.1))

$$\rho(\frac{\partial u}{\partial t} + u\frac{\partial u}{\partial x} + v\frac{\partial u}{\partial y} + w\frac{\partial u}{\partial z})$$

$$= -\frac{\partial p}{\partial x} + 2\frac{\partial}{\partial x}[\mu\frac{\partial u}{\partial x}] + \frac{\partial}{\partial y}[\mu(\frac{\partial u}{\partial y} + \frac{\partial v}{\partial x})] + \frac{\partial}{\partial z}[\mu(\frac{\partial w}{\partial x} + \frac{\partial u}{\partial z})],$$

$$\rho(\frac{\partial v}{\partial t} + u\frac{\partial v}{\partial x} + v\frac{\partial v}{\partial y} + w\frac{\partial v}{\partial z}) =$$

$$-\frac{\partial p}{\partial y} + \frac{\partial}{\partial x}[\mu(\frac{\partial u}{\partial y} + \frac{\partial v}{\partial x})] + 2\frac{\partial}{\partial y}[\mu\frac{\partial v}{\partial y}] + \frac{\partial}{\partial z}[\mu(\frac{\partial v}{\partial z} + \frac{\partial w}{\partial y})], \quad (4.3)$$

$$\rho(\frac{\partial w}{\partial t} + u\frac{\partial w}{\partial x} + v\frac{\partial w}{\partial y} + w\frac{\partial w}{\partial z})$$

$$= -\frac{\partial p}{\partial z} + \frac{\partial}{\partial x}[\mu(\frac{\partial w}{\partial x} + \frac{\partial u}{\partial z})] + \frac{\partial}{\partial y}[\mu(\frac{\partial v}{\partial z} + \frac{\partial w}{\partial y})] + 2\frac{\partial}{\partial z}[\mu\frac{\partial w}{\partial z}].$$

For an incompressible fluid we have to add to equations (4.3) the continuity equation in the form [3]

$$\frac{\partial u}{\partial x} + \frac{\partial v}{\partial y} + \frac{\partial w}{\partial z} = 0. \quad (4.4)$$

To simplify these equations for the case of a lubricated contact of two solids, we need to specify the coordinate system. We will assume that the xy-plane is the plane equidistant from the solid surfaces and the z-axis is directed through the solids upward. For typical lubricated contacts, the characteristic size L_{xy} of the contact in the xy-plane is much larger than the lubrication film thickness L_z, i.e., $L_z/L_{xy} \ll 1$. Moreover, we will assume that the characteristic velocities in the xy-plane and along the z-axis are U_{xy} and U_z, respectively. Let us introduce the following dimensionless variables

$$\bar{t} = t\frac{U_{xy}}{L_{xy}}, \ (\bar{x}, \bar{y}) = \frac{1}{L_{xy}}(x, y), \ \bar{z} = \frac{z}{L_z}, \ \bar{p} = p\frac{L_z^2}{\mu_* U_{xy} L_{xy}},$$

$$(\bar{u}, \bar{v}) = \frac{1}{U_{xy}}(u, v), \ \bar{w} = \frac{w}{U_z}, \ \bar{\mu} = \frac{\mu}{\mu_*}, \quad (4.5)$$

where $\mu_\star$ is the characteristic value of the fluid viscosity. Using the dimensionless variables in the continuity equation, we obtain

$$\frac{\partial \bar{u}}{\partial \bar{x}} + \frac{\partial \bar{v}}{\partial \bar{y}} + \frac{U_z}{U_{xy}}\frac{L_{xy}}{L_z}\frac{\partial \bar{w}}{\partial \bar{z}} = 0. \tag{4.6}$$

To balance the terms of this equation, we assume that

$$\frac{U_z}{U_{xy}}\frac{L_{xy}}{L_z} = 1. \tag{4.7}$$

Therefore, in dimensionless variables equations (4.3) and (4.6) can be reduced to

$$Re(\frac{\partial \bar{u}}{\partial t} + \bar{u}\frac{\partial \bar{u}}{\partial \bar{x}} + \bar{v}\frac{\partial \bar{u}}{\partial \bar{y}} + \bar{w}\frac{\partial \bar{u}}{\partial \bar{z}}) = -\frac{\partial \bar{p}}{\partial \bar{x}} + 2(\frac{L_z}{L_{xy}})^2\{\frac{\partial}{\partial \bar{x}}[\bar{\mu}\frac{\partial \bar{u}}{\partial \bar{x}}]$$

$$+ \frac{\partial}{\partial \bar{y}}[\bar{\mu}(\frac{\partial \bar{u}}{\partial \bar{y}} + \frac{\partial \bar{v}}{\partial \bar{x}})] + \frac{\partial}{\partial \bar{z}}[\bar{\mu}(\frac{\partial \bar{w}}{\partial \bar{x}})]\} + \frac{\partial}{\partial \bar{z}}[\bar{\mu}\frac{\partial \bar{u}}{\partial \bar{z}}],$$

$$Re(\frac{\partial \bar{v}}{\partial t} + \bar{u}\frac{\partial \bar{v}}{\partial \bar{x}} + \bar{v}\frac{\partial \bar{v}}{\partial \bar{y}} + \bar{w}\frac{\partial \bar{v}}{\partial \bar{z}}) = -\frac{\partial \bar{p}}{\partial \bar{y}} + (\frac{L_z}{L_{xy}})^2\{\frac{\partial}{\partial \bar{x}}[\bar{\mu}(\frac{\partial \bar{u}}{\partial \bar{y}}$$

$$+ \frac{\partial \bar{v}}{\partial \bar{x}})] + 2\frac{\partial}{\partial \bar{y}}[\bar{\mu}\frac{\partial \bar{v}}{\partial \bar{y}}]\} + \frac{\partial}{\partial \bar{z}}[\bar{\mu}\frac{\partial \bar{v}}{\partial \bar{z}}] + (\frac{L_z}{L_{xy}})^2\frac{\partial}{\partial \bar{z}}[\bar{\mu}\frac{\partial \bar{w}}{\partial \bar{y}}], \tag{4.8}$$

$$Re(\frac{\partial \bar{w}}{\partial t} + \bar{u}\frac{\partial \bar{w}}{\partial \bar{x}} + v\frac{\partial \bar{w}}{\partial \bar{y}} + \bar{w}\frac{\partial \bar{w}}{\partial \bar{z}}) = -(\frac{L_{xy}}{L_z})^2\frac{\partial \bar{p}}{\partial \bar{z}} + (\frac{L_z}{L_{xy}})^2\{\frac{\partial}{\partial \bar{x}}[\bar{\mu}\frac{\partial \bar{w}}{\partial \bar{x}}]$$

$$+ \frac{\partial}{\partial \bar{y}}[\bar{\mu}\frac{\partial \bar{w}}{\partial \bar{y}}]\} + \frac{\partial}{\partial \bar{x}}[\bar{\mu}\frac{\partial \bar{u}}{\partial \bar{z}}] + \frac{\partial}{\partial \bar{y}}[\bar{\mu}\frac{\partial \bar{v}}{\partial \bar{z}}] + 2\frac{\partial}{\partial \bar{z}}[\bar{\mu}\frac{\partial \bar{w}}{\partial \bar{z}}],$$

$$\frac{\partial \bar{u}}{\partial \bar{x}} + \frac{\partial \bar{v}}{\partial \bar{y}} + \frac{\partial \bar{w}}{\partial \bar{z}} = 0, \tag{4.9}$$

where the effective Reynolds number $Re = \frac{\rho U_{xy} L_{xy}}{\mu_\star}(\frac{L_z}{L_{xy}})^2$ and the ratio $\frac{L_z}{L_{xy}}$ are small. In fact, in a typical journal bearing $L_{xy} = 0.12\ m$, $L_z = 5 \times 10^{-5}\ m$, $U_{xy} = 5\ m/s$, $\rho = 850 N \cdot s^2/m^4$, $\mu_\star = 0.5\ N \cdot s/m^2$ and, therefore, the Reynolds number $Re = 1.77 \cdot 10^{-4}$ and $\frac{L_z}{L_{xy}} = 4.17 \cdot 10^{-4}$. Besides that, we assume that all the dimensionless terms such as $\frac{\partial \bar{u}}{\partial \bar{x}}$, $\frac{\partial \bar{u}}{\partial \bar{y}}$, $\frac{\partial \bar{u}}{\partial \bar{z}}$, etc., are of the order of 1. From the physical point of view, it is clear that at least two mechanisms in the viscous flow are important, i.e., viscous shearing and pressure gradient. Therefore, we need to retain these terms in the simplified equations. Smallness of the Reynolds number Re and ratio L_z/L_{xy} does exactly that. Using these assumptions we obtain that all terms in the right-hand side of the first two equations from (4.3) except for two terms are small. In a similar fashion, evaluating terms of the third equation from (4.8), we obtain that all terms of the equation are negligibly small in comparison with term $\frac{\partial p}{\partial z}$. As a result of this analysis we retain in dimensional form only the largest terms of Navier-Stokes equations as follows

$$-\frac{\partial p}{\partial x} + \frac{\partial}{\partial z}[\mu\frac{\partial u}{\partial z}] = 0, \quad -\frac{\partial p}{\partial y} + \frac{\partial}{\partial z}[\mu(\frac{\partial v}{\partial z}] = 0, \quad \frac{\partial p}{\partial z} = 0. \tag{4.10}$$

Under the same conditions for non-Newtonian fluids we get equations as follows

$$-\frac{\partial p}{\partial x} + \frac{\partial p_{xz}}{\partial z} = 0, \quad -\frac{\partial p}{\partial y} + \frac{\partial p_{yz}}{\partial z} = 0, \quad \frac{\partial p}{\partial z} = 0, \tag{4.11}$$

where the expressions for the tensor components p_{xz} and p_{yz} are determined by the lubricant rheology.

It is important to notice that the terms related to u and v retained in equations (4.10) come from the main terms of the stress tensor components p_{yz}, and p_{zx}. Therefore, the simplified equations of a flow of viscous non-Newtonian fluid in a lubricated contact can be obtained by replacing the second terms in the first two equations from (4.10) by the main terms of $\frac{\partial p_{xz}}{\partial z}$ and $\frac{\partial p_{yz}}{\partial z}$ for that rheology, respectively.

In case of non-isothermal conditions two additional assumptions need to be made. They are (a) the major source of heat generation is due to shear viscous motion in lubricants and (b) the main direction of heat flow is across the lubrication film thickness. These assumptions together with the previously stated ones after estimating terms of the equation allow for significant simplification of the general energy equation. For fluids with Newtonian and non-Newtonian rheologies such a simplified equation can be represented in the form

$$\frac{\partial}{\partial z}(\lambda \frac{\partial T}{\partial z}) = -p_{xz}\frac{\partial u}{\partial z} - p_{yz}\frac{\partial v}{\partial z}, \qquad (4.12)$$

where T is the fluid temperature and λ is the heat conductivity. In equation (4.12), in the stress tensor only the major term components are retained. For example, for Newtonian fluids in equation (4.12) for p_{xz} and p_{yz} we use $p_{xz} = \mu\frac{\partial u}{\partial z}$ and $p_{yz} = \mu\frac{\partial v}{\partial z}$, respectively.

4.3   Some Classic Results for Smooth Elastic Solids

In this section we present some classic contact problem formulations and results for the case of isotropic homogeneous elastic materials. The presentation is limited to only the most basic formulas for spatial and plane problems and to the results for isotropic homogeneous elastic materials occupying half-space and half-plane, which will be used later. The requirements necessary for solution of these problems are usually reduced to minimal smoothness of contact surfaces. There are a couple more complex cases that are considered in Sections 18.2 and 22.3 but not considered here. Besides the solutions presented below, there is a wide variety of solutions for contact problems not related to the analysis presented in this monograph.

4.3.1   Formulas for Elastic Half-Space

Suppose an elastic isotropic homogeneous material with elastic modulus E and Poisson's ratio ν occupies a half-space $z \leq 0$, and it is bounded by the plane $z = 0$. Assuming that in a rectangular coordinate system (x, y, z) the plane $z = 0$ is loaded by pressure $p(x, y)$ distributed over region Ω in the (x, y)-plane

$(p(x, y) = 0$ outside of Ω) the displacements (u, v, w) of the material points along (x, y, z) axes are [4, 5]

$$u(x, y, z) = \tfrac{1+\nu}{2\pi E} \{ z \tfrac{\partial V}{\partial x} - (1 - 2\nu) \int\limits_{-\infty}^{z} \tfrac{\partial V}{\partial x} dz \},$$

$$v(x, y, z) = \tfrac{1+\nu}{2\pi E} \{ z \tfrac{\partial V}{\partial y} - (1 - 2\nu) \int\limits_{-\infty}^{z} \tfrac{\partial V}{\partial y} dz \},$$

$$w(x, y, z) = \tfrac{1+\nu}{2\pi E} \{ z \tfrac{\partial V}{\partial z} - 2(1 - \nu) V \},$$

$$V = \int \int\limits_{\Omega} \tfrac{p(\xi, \eta) d\xi d\eta}{R}, \quad R = \sqrt{(x - \xi)^2 + (y - \eta)^2 + z^2}.$$

(4.13)

Obviously, at the surface $z = 0$ we have

$$w(x, y, 0) = -\tfrac{1-\nu^2}{\pi E} \int \int\limits_{\Omega} \tfrac{p(\xi, \eta) d\xi d\eta}{R}, \quad R = \sqrt{(x - \xi)^2 + (y - \eta)^2}. \qquad (4.14)$$

Let us assume that the frictional stresses applied to the half-space surface and tangential displacements of the half-space surface are negligibly small. Then the formulation of a typical contact problem for a rigid indenter with a smooth bottom of shape $z = f(x, y)$, which is normally pressed into the elastic half-space without friction, is given below [5]

$$-w(x, y, 0) = \delta + \beta_x x + \beta_y y - f(x, y), \quad \int \int\limits_{\Omega} p(\xi, \eta) d\xi d\eta = P, \qquad (4.15)$$

where δ is the normal displacement of the rigid indenter, β_x and β_y are the angles of the indenter rotation about the x- and y-axes caused by the moments M_x and M_y, respectively, P is the normal load applied to the indenter.

Solutions of a number of contact problems with and without axial symmetry and with and without friction can be found in [6]-[9].

4.3.2 Formulas and Results for Elastic Half-Plane

Let us assume that an isotropic homogeneous elastic material with elastic modulus E and Poisson's ratio ν i is bounded by a horizontal plane $z = 0$ and occupies the half-space $z \leq 0$. The half-space boundary $z = 0$ is subjected to pressure $p(x)$ and tangential stress $\tau(x)$, which are uniform along the y-axis parallel to the plane $z = 0$. Then we can consider this as a case of plane deformation when none of the problem parameters depend on the y-coordinate and the whole problem can be considered in just (x, z) plane. The elastic displacements (u, w) of the half-plane boundary $z = 0$ along the (x, z)-axes are determined by the formulas [4]

$$\frac{\pi E}{2(1-\nu^2)} u + C_u = -\frac{1-2\nu}{2(1-\nu)}\pi \int\limits_{-\infty}^{x} p(t)dt + \int\limits_{-\infty}^{\infty} \tau(t)\ln|t-x|\,dt,$$

$$\tag{4.16}$$

$$\frac{\pi E}{2(1-\nu^2)} w + C_w = \int\limits_{-\infty}^{\infty} p(t)\ln|t-x|\,dt + \frac{1-2\nu}{2(1-\nu)}\pi \int\limits_{-\infty}^{x} \tau(t)dt,$$

where C_u and C_w are infinite constants. If the elastic material is represented by an infinitely thick layer of thickness h, then we still can use formulas (4.16) where C_u and C_w are constants proportional to $\ln\frac{h}{a}$, where $\frac{h}{a} \gg 1$ (a is a characteristic size of the contact region) [7].

Assuming that the surface tangential displacements can be neglected the formulation of a typical contact problem for a rigid indenter with a smooth bottom of shape $z = f(x)$, which is pressed into an elastic half-plane with the normal P and tangential T forces per unit length, is as follows [4]

$$-w(x,0) = \delta + \beta_x x - f(x), \quad \int\limits_{\Omega} p(x)dx = P, \tag{4.17}$$

where δ is the normal displacement of the rigid indenter, β_x is the angle of the indenter rotation about the x-axis caused by the moment M_x, Ω is the contact region, and

$$\int\limits_{\Omega} \tau(x)dx = T. \tag{4.18}$$

To complete the contact problem formulation, it is necessary to fix the boundaries of the contact region Ω partially or completely while along the rest of the boundaries to require that pressure vanishes. If the contact region Ω is represented by an interval, then there are only three distinct possibilities: Contact Problem 1 - the end points of the contact interval $(-a, a)$ are fixed; Contact Problem 2 - the boundaries of the contact are free, i.e., $p(\pm a) = 0$; and Contact Problem 3 - one of the boundaries, for example, $x = -a$ is fixed while the other one is free, i.e., $p(a) = 0$. Let us consider contact problems without friction, i.e., $\tau(x) = 0$. Then assuming that angle β_x is known and is included in function $f(x)$ the solution of Contact Problem 1

$$\frac{2(1-\nu^2)}{\pi E} \int\limits_{-a}^{a} p(t)\ln\frac{1}{|t-x|}dt = \delta - f(x), \quad \int\limits_{-a}^{a} p(x)dx = P, \tag{4.19}$$

takes the form

$$p(x) = \frac{1}{\pi\sqrt{a^2-x^2}}\left\{ P + \frac{E}{2(1-\nu^2)} \int\limits_{-a}^{a} \frac{f'(t)\sqrt{a^2-t^2}dt}{t-x} \right\}. \tag{4.20}$$

Calculation of constant δ can be done by substituting $p(x)$ from formula (4.20) into the first equation from (4.19) and setting $x = 0$. The solution of Contact Problem 2, which is reduced to

$$\frac{2(1-\nu^2)}{\pi E} \int\limits_{-a}^{a} p(t)\ln\frac{1}{|t-x|}dt = \delta - f(x), \quad p(\pm a) = 0, \quad \int\limits_{-a}^{a} p(x)dx = P, \tag{4.21}$$

is determined by the formulas

$$p(x) = \frac{E}{2\pi(1-\nu^2)}\sqrt{a^2-x^2}\int_{-a}^{a}\frac{f'(t)dt}{\sqrt{a^2-t^2}(t-x)},$$

$$(4.22)$$

$$P = \frac{E}{2(1-\nu^2)}\int_{-a}^{a}\frac{tf'(t)dt}{\sqrt{a^2-t^2}},\quad \int_{-a}^{a}\frac{f'(t)dt}{\sqrt{a^2-t^2}} = 0.$$

Solution of Contact Problem 3 as well as solutions of the contact problem with linear friction under the conditions of full sliding when $\tau(x) = \lambda p(x)$ (λ is the friction coefficient) can be found in [8] while an approximate solution of the contact problem with stick and slip is presented in [4].

4.4 Closure

The Navier-Stokes equations as well as the energy equation were simplified for the case of small Reynolds number and slow fluid motion. Normal displacements of two- and three-dimensional elastic solids and formulations of classic contact problems of elasticity are considered. The classic solutions of two-dimensional contact problems of elasticity are given.

4.5 Exercises and Problems

1. List and discuss the assumptions used for the derivation of the Reynolds equation for pressure p from Navier-Stokes equations.

2. List and discuss the assumptions used for the derivation of the equation for lubricant temperature T from the energy equation for fluid flow.

3. Obtain formulas (4.22) for the solution of the plane contact problem with both free boundaries from the expression (4.20) for the solution of the plane problem with fixed boundaries.

4. Obtain formulas for the solution of the plane contact problem with one free and another fixed boundaries from the expression (4.20) for the solution of the plane problem with fixed boundaries.

References

[1] Bhushan, B. 1999. *Principles and Applications of Tribology*. Toronto: John Wiley & Sons, Inc., 586-591.

[2] Cheng, H.S. 1980. Fundamentals of Elastohydrodynamic Contact Phenomena. In *Fundamentals of Tribology*, N.P. Suh and N. Saka, eds. Cambridge, MA: The MIT Press, 1009-1048.

[3] Lamb, H. 1995. *Hydrodynamics*. New York: Cambridge University Press, 6th ed.

[4] Galin, L.A. 1980. *Contact Problems in the Theory of Elasticity and Viscoelasticity*. Moscow: Nauka.

[5] Lurye, A.I. 1955. *Spatial Proplems in Elasticity*. Moscow: Gostekhizdat.

[6] Shtaerman, I.Ya. 1949. *Contact Problems in Theory of Elasticity* Moscow-Leningrad: Gostekhizdat.

[7] Vorovich, I.I., Alexandrov, V.M., and Babesko, V.A. 1974. *Non–Classic Mixed Problems in the Theory of Elasticity*. Moscow: Nauka.

[8] Muskhelishvili, N.I. 1966. *Some Fundamental Problems of the Mathematical Theory of Elasticity*. Moscow: Nauka.

[9] Galin, L.A. 1946. Spatial Contact Problems of the Elasticity Theory for Circular Stamps. *J. Appl. Math. Mech.* 11, No. 2:281-284.

Part III

EHL Problems for Lightly Loaded Line and Point Contacts

This part of the monograph is devoted to asymptotic and numerical analysis of lightly loaded line and point EHL contacts when lubricant is a Newtonian or non-Newtonian fluid. The analytical analysis of the problem reveals a clear structure of the solutions in lightly loaded EHL contacts. It produces analytical formulas for the lubrication film thickness and, in some cases, simplified asymptotically valid equations for pressure and gap which can be easily solved. A specialized numerical approach is developed for solution of EHL problems in lightly loaded contacts. The asymptotic solutions are compared to numerical solutions of the EHL problem in its original formulation.

5

Lightly Loaded Lubrication Regimes for Line and Point Contacts

---

5.1 Introduction

This chapter deals with steady isothermal and non-isothermal line and point EHL problems for lightly loaded smooth contacts. A lubricant is considered to be an incompressible non-Newtonian fluid of a rather general rheology. In lightly loaded lubricated contacts, the influence of the elastic properties of the solids in contact on the contact pressure is relatively small. In other words, the problem solution in the entire lubricated contact is close to the solution of the problem for a lubricant flow through a narrow gap of a known shape.

The aim of the chapter is to consider lightly loaded lubricated contact by employing the method of regular perturbations [1, 2], which allows to study the solution behavior in contacts with fluid of rather complex rheology. As a result of application of the above–mentioned method the differential equations for the first two terms of the asymptotic expansion of the problem solution for pressure, gap, and lubrication film thickness are obtained. Moreover, some new two-term analytic formulas for the lubrication film thickness and frictional forces are derived. The obtained relationships are illustrated by examples of fluids with different rheologies. In the special case of Newtonian fluid, the leading terms of the aforementioned asymptotic expansions coincide with the classic Kapitza [3] results. A thermal EHL problem for a lightly loaded contact is briefly considered.

The formulas obtained for the lubrication film thickness and frictional forces can be used in treating experimental results for lightly loaded lubricated contacts as well as for making a judgment on application of a fluid of a particular rheology.

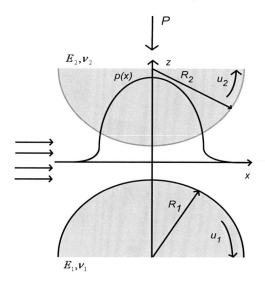

FIGURE 5.1

The general view of a lubricated contact.

5.2 Lightly Loaded Lubrication Regimes for Line Contacts. Problem Formulation

Let us consider two infinite parallel cylinders made of elastic materials with Young's modulus E_1 and Poisson's ratio ν_1 for the lower cylinder of radius R_1 and with Young's modulus E_2 and Poisson's ratio ν_2 for the upper cylinder of radius R_2. In the coordinate system with the x-axis along the contact and the z-axis through the cylinders' centers the cylinders move with surface velocities u_1 and u_2, respectively. The cylinders are separated from one another by a thin layer of incompressible lubricant and pressed one into another by a force P acting along the z-axis. The general view of the lubricated contact is given in Fig. 5.1.

Let us formulate the main problem assumptions. We will assume that the lubrication film thickness is much smaller than the the size of the contact which, in turn, is much smaller than cylinders' radii. The variations of the film thickness $h(x)$ in the contact are small, i.e., $dh/dx \ll 1$. Moreover, we will assume that frictional effects caused by lubricant's viscosity dominate the inertia effects, i.e., the lubricant moves relatively slowly. We will assume that the lubrication process is isothermal and the lubricant viscosity μ depends only on pressure $p(x)$. These are the typical assumptions used in the theory of elastohydrodynamic lubrication (see Hamrock [4]).

Under the above assumptions for a fluid with a rather general rheology, the equations that relate the tangential stress τ to the strain rate $\partial u/\partial z$ in the fluid can be represented in the form [5]

$$\tau = \Phi(\mu\tfrac{\partial u}{\partial z}) \ or \ \mu\tfrac{\partial u}{\partial z} = F(\tau), \tag{5.1}$$

where Φ and F are functions inverses of each other, $u(x, z)$ is the fluid velocity, $\mu(p)$ is the lubricant viscosity being a smooth monotonically increasing function of pressure p. Moreover, function $\Phi(x)$ is an arbitrary smooth monotonically increasing odd function and $\Phi(0) = F(0) = 0$. Therefore, taking into account (5.1) after simplification of Navier-Stokes equations for the fluid flow the EHL problem can be reduced to the following equations [4]

$$\tfrac{\partial}{\partial z}\Phi(\mu\tfrac{\partial u}{\partial z}) = \tfrac{\partial p}{\partial x}, \ \tfrac{\partial p}{\partial z} = 0, \ \tfrac{\partial u}{\partial x} + \tfrac{\partial w}{\partial z} = 0, \tag{5.2}$$

with no-slip and no penetration through the cylinders' surfaces boundary conditions

$$u(x, -\tfrac{h}{2}) = u_1, \ u(x, \tfrac{h}{2}) = u_2,$$

$$w(x, (-1)^j\tfrac{h}{2}) = (-1)^j\tfrac{u_j}{2}\tfrac{dh}{dx}, \ j = 1, 2, \tag{5.3}$$

where $w(x, z)$ is the z-component of the fluid velocity and $z = \pm h(x)/2$ are the upper and lower cylinder surfaces, respectively.

The second equation in (5.2) leads to the conclusion that pressure $p = p(x)$ is independent from z. Integrating the second equation in (5.1) and the first in (5.2) gives

$$u(x, z) = u_1 + \tfrac{1}{\mu} \int\limits_{-h/2}^{z} F(f + t\tfrac{dp}{dx})dt, \tag{5.4}$$

where $f = f(x)$ is the unknown sliding frictional stress which simultaneously represents the tangential stress τ in the midplane $z = 0$.

Using equation (5.4) and the second boundary condition in (5.3), we obtain the equation for $f(x)$ in the form

$$\int\limits_{-h/2}^{h/2} F(f + t\tfrac{dp}{dx})dt = \mu(u_2 - u_1). \tag{5.5}$$

By integrating the third (mass conservation) equation in (5.2) with respect to z from $-h/2$ to $h/2$ and using equation (5.4) and the last pair of boundary conditions in (5.3), we derive the generalized Reynolds equation

$$\tfrac{d}{dx}\Big[\tfrac{1}{\mu} \int\limits_{-h/2}^{h/2} dz \int\limits_{-h/2}^{z} F(f + t\tfrac{dp}{dx})dt + u_1h\Big] = 0. \tag{5.6}$$

By changing the order of integration in equation (5.6) and using equation (5.5), the Reynolds equation can be reduced to the final simplified form

$$\tfrac{d}{dx}\Big[\tfrac{1}{\mu} \int\limits_{-h/2}^{h/2} zF(f + z\tfrac{dp}{dx})dz - \tfrac{u_1+u_2}{2}h\Big] = 0. \tag{5.7}$$

We need to add to equations (5.5) and (5.7) the relationship

$$h(x) = h_e + \frac{x^2 - x_e^2}{2R'} + \frac{2}{\pi E'} \int\limits_{x_i}^{x_e} p(t) \ln \frac{x_e - t}{|x - t|} dt, \qquad (5.8)$$

for gap $h(x)$ [4], which is obtained based on the above assumptions and by replacing the cylinders in the contact by elastic half-planes. In equation (5.8) h_e is the lubrication film thickness at the exit point of the contact x_e, i.e., $h_e = h(x_e)$, x_i is the inlet point of the contact, R' and E' are the effective radius and elasticity modulus, $1/R' = 1/R_1 \pm 1/R_2$ (signs $+$ and $-$ are chosen in accordance with cylinders' curvatures) and $1/E' = 1/E_1' + 1/E_2'$, $E_j' = E_j/(1 - \nu_j^2)$, $j = 1, 2$.

To make the problem complete we need to add to the above equations some conditions. It is reasonable to assume that within the lubricated contact pressure is much higher than the ambient pressure. Moreover, to prevent the lubricant from cavitation at the exit of the contact, we may require the pressure gradient to be zero [4]. Thus, we obtain the following boundary conditions on pressure

$$p(x_i) = p(x_e) = \frac{dp(x_e)}{dx} = 0. \qquad (5.9)$$

The last condition that must be added to the equations is the condition that the integral of pressure over the contact is equal to the applied normal force P, i.e.,

$$\int\limits_{x_i}^{x_e} p(t)dt = P. \qquad (5.10)$$

Therefore, for the given functions μ, Φ, F and constants x_i, u_1, u_2, R', E', and P the EHL problem is reduced to solution of equations (5.5), (5.7)-(5.10) for the functions of the sliding frictional stress $f(x)$, pressure $p(x)$, gap $h(x)$, and two constants - the exit coordinate x_e and exit film thickness h_e.

After the problem is solved the friction force $F_T(z)$ in the lubrication layer can be calculated from the formula

$$F_T(z) = \int\limits_{x_i}^{x_e} \tau(x, z)dx = \int\limits_{x_i}^{x_e} \Phi(\mu \frac{\partial u}{\partial z})dx. \qquad (5.11)$$

Using equation (5.11) we can determine the friction forces $F_T^{\pm} = F_T(\pm h/2)$ on the cylinders' surfaces

$$F_T^{\pm} = F_S \pm F_R, \quad F_S = \int\limits_{x_i}^{x_e} f(x)dx, \quad F_R = \frac{1}{2} \int\limits_{x_i}^{x_e} h(x) \frac{dp}{dx} dx, \qquad (5.12)$$

where F_S and F_R are the sliding and rolling friction forces, respectively.

Let us introduce dimensionless variables related to the flow of a viscous fluid

$$\{x', a, c\} = \{x, x_i, x_e\} \frac{\vartheta}{2R'}, \quad \{z', h'\} = \{z, h\} \frac{1}{h_e}, \quad p' = p \frac{\pi R'}{\vartheta P},$$

$$\mu' = \frac{\mu}{\mu_a}, \quad F' = F \frac{2h_e}{\mu_a(u_1 + u_2)}, \quad \{f', \tau'\} = \{f, \tau\} \frac{2\pi}{h_e P} (\frac{R'}{\vartheta})^2, \qquad (5.13)$$

where $\vartheta = \vartheta(\mu_a, u_1, u_2, R', P, \ldots)$ is a certain dimensionless parameter that depends on the fluid specific rheology (i.e., function F) and is independent from the effective elastic modulus E' and μ_a is the lubricant viscosity at ambient pressure. Some specific forms of parameter ϑ will be considered later.

By integrating equation (5.7) with respect to x from x to x_e and taking into account equation (5.5) and the last of the boundary conditions (5.9), we obtain the problem in dimensionless variables (for simplicity here and further primes are omitted)

$$\frac{1}{\mu} \int_{-h/2}^{h/2} zF(f + z\tfrac{dp}{dx})dz = h - 1, \tag{5.14}$$

$$\int_{-h/2}^{h/2} F(f + z\tfrac{dp}{dx})dz = \mu s_0, \tag{5.15}$$

$$\gamma(h - 1) = x^2 - c^2 + \tfrac{1}{V}\tfrac{2}{\pi} \int_{a}^{c} p(t) \ln \tfrac{c-t}{|x-t|} dt, \tag{5.16}$$

$$p(a) = p(c) = 0, \tag{5.17}$$

$$\int_{a}^{c} p(t)dt = \tfrac{\pi}{2}, \tag{5.18}$$

$$\gamma = \vartheta^2 \tfrac{h_e}{2R'}, \quad V = \tfrac{R'E'}{\vartheta^2 P}, \quad s_0 = 2\tfrac{u_2-u_1}{u_2+u_1}. \tag{5.19}$$

In the above dimensionless variables for the given values of parameters a and V and functions μ and F, the EHL problem is reduced to solution of equations (5.14)-(5.19) for the functions of pressure $p(x)$ and gap $h(x)$ as well as for the values of the dimensionless lubrication film thickness γ and exit coordinate c.

5.2.1 Perturbation Solution of the EHL Problem

Below we will consider application of the method of regular perturbations [1, 2, 6] to solution of the EHL problem (5.14)-(5.19) for a lightly loaded contact. The problem is formulated in the region with unknown boundaries $z = \pm h/2$ and $x = c$ which makes the solution of the problem more complicated. Fortunately, this difficulty can be easily cured by introducing the following substitutions

$$x = \tfrac{1}{2}(c + a) + \tfrac{1}{2}(c - a)\xi, \quad z = \tfrac{1}{2}h\zeta, \quad f(x) = g(\xi), \quad h(x) = H(\xi), \tag{5.20}$$

which allow to reduce the problem equations to the following

$$\frac{H^2}{4\mu} \int_{-1}^{1} tF(g + \tfrac{Ht}{c-a}\tfrac{dp}{d\xi})dt = H - 1, \tag{5.21}$$

$$\int_{-1}^{1} F(g + \tfrac{Ht}{c-a}\tfrac{dp}{d\xi})dt = 2\tfrac{\mu s_0}{H}, \tag{5.22}$$

$$\gamma(H-1) = (\tfrac{c-a}{2})^2\xi^2 + \tfrac{c^2-a^2}{2}\xi + \tfrac{a^2+2ac-3c^2}{4} \tag{5.23}$$

$$+\tfrac{c-a}{\pi V}\int_{-1}^{1} p(t)\ln\tfrac{1-t}{|\xi-t|}dt,$$

$$p(-1) = p(1) = 0, \tag{5.24}$$

$$\int_{-1}^{1} p(t)dt = \tfrac{\pi}{c-a}. \tag{5.25}$$

We will consider the contact being lightly loaded if $V \gg 1$. The proposed definition of a lightly loaded lubricated contact coincides with the notion that the influence of the cylinders elasticity has a small effect on the EHL problem solution.

Equations (5.21) - (5.25) suggest that the solution can be searched in the form of regular power series in $V^{-1} \ll 1$

$$p(\xi) = \sum_{k=0}^{\infty} V^{-k}p_k(\xi), \ H(\xi) = \sum_{k=0}^{\infty} V^{-k}H_k(\xi),$$

$$g(\xi) = \sum_{k=0}^{\infty} V^{-k}g_k(\xi), \ c = \sum_{k=0}^{\infty} V^{-k}c_k, \ \gamma = \sum_{k=0}^{\infty} V^{-k}\gamma_k, \tag{5.26}$$

where functions $p_k(\xi)$, $H_k(\xi)$, $g_k(\xi)$ and constants c_k, γ_k has to be determined.

By using the Taylor series expansion of $F(g + \tfrac{Hv}{c-a}\tfrac{dp}{d\xi})$ centered at $(p_0, H_0, g_0, c_0, \gamma_0)$ and representations (5.26) in equations (5.21) - (5.25), we will be able to obtain a boundary-value problem for nonlinear differential equations for the set of functions and constants $(p_0, H_0, g_0, c_0, \gamma_0)$ and a series of boundary-value problems for linear differential equations for the set of functions and constants $(p_k, H_k, g_k, c_k, \gamma_k)$, $k = 1, \ldots$.

The main advantage of this approach in comparison with the numerical solution is the ability to obtain simple structural asymptotic formulas for the lubrication film thickness and friction forces. Using equations (5.19) and (5.26) we obtain the following two-term formulas for the film thickness

$$h_e = \tfrac{2R'}{\vartheta^2}(\gamma_0 + \tfrac{1}{V}\gamma_1 + \ldots) = \tfrac{2R'}{\vartheta^2}\gamma_0 + \tfrac{2P}{\pi E'}\gamma_1 + \ldots \tag{5.27}$$

and for the components of the friction force

$$F_S = \tfrac{P}{\vartheta}(a_0 + \tfrac{1}{V}a_1 + \ldots) = \tfrac{P}{\vartheta}a_0 + \tfrac{P^2\vartheta}{\pi R'E'}a_1 + \ldots,$$

$$F_R = \tfrac{P}{\vartheta}(b_0 + \tfrac{1}{V}b_1 + \ldots) = \tfrac{P}{\vartheta}b_0 + \tfrac{P^2\vartheta}{\pi R'E'}b_1 + \ldots, \tag{5.28}$$

$$a_0 = \tfrac{\gamma_0(c_0-a)}{\pi} \int\limits_{-1}^{1} g_0(\xi)d\xi, \; b_0 = \tfrac{\gamma_0}{\pi} \int\limits_{-1}^{1} H_0(\xi)\tfrac{dp_0}{d\xi}d\xi,$$

$$a_1 = \tfrac{1}{\pi}[\gamma_0(c_0-a)\int\limits_{-1}^{1} g_1(\xi)d\xi + [\gamma_1(c_1-a)+\gamma_0 c_1]\int\limits_{-1}^{1} g_0(\xi)d\xi],$$

(5.29)

$$b_1 = \tfrac{1}{\pi}[\gamma_0\int\limits_{-1}^{1}[H_1(\xi)\tfrac{dp_0(\xi)}{d\xi}+H_0(\xi)\tfrac{dp_1}{d\xi}]d\xi + \gamma_1\int\limits_{-1}^{1} H_0(\xi)\tfrac{dp_0}{d\xi}d\xi].$$

In the latter formulas the numerical values of coefficients γ_0, γ_1, a_0, b_0, a_1, and b_1 depend on the specifics of functions F, Φ, μ, and constants a and s_0. It follows from equations (5.27) - (5.29) that lubrication film thickness h_e and components F_S and F_R of the friction force depend on E' only starting with the second terms proportional to V^{-1} for $V \gg 1$.

Let us consider some lubricants with different rheologies. First, we will consider a lubricant with power rheology [5, 6], which, under the problem assumptions in dimensional variables, is represented by

$$\tau = \mu^\alpha \mid \tfrac{\partial u}{\partial z}\mid^{\alpha-1}\tfrac{\partial u}{\partial z}, \; \alpha > 0. \tag{5.30}$$

In this case in dimensionless variables $F(x) = 12\gamma^{(1+\alpha)/\alpha}\mid x\mid^{(1-\alpha)/\alpha}x$. By introducing the dimensionless variables (5.13), we obtain

$$\vartheta^2 = \tfrac{R'}{3\mu_a(u_1+u_2)}(\tfrac{P}{\pi R'})^{1/\alpha}, \tag{5.31}$$

while constant V is determined by the expression in (5.19). Therefore, for the power rheology formulas for the lubrication film thickness h_e and components F_S and F_R of the friction force (5.27) - (5.29) can be rewritten in the form

$$h_e = 6\mu_a(u_1+u_2)[\tfrac{\pi R'}{P}]^{1/\alpha}\gamma_0 + \tfrac{2P}{\pi E'}\gamma_1 + \dots,$$

$$F_S = P\sqrt{\tfrac{3\mu_a(u_1+u_2)}{R'}}[\tfrac{\pi R'}{P}]^{1/(2\alpha)}a_0$$

$$+\tfrac{P^2}{\pi R'E'}\sqrt{\tfrac{R'}{3\mu_a(u_1+u_2)}}[\tfrac{P}{\pi R'}]^{1/(2\alpha)}a_1 + \dots, \tag{5.32}$$

$$F_R = P\sqrt{\tfrac{3\mu_a(u_1+u_2)}{R'}}[\tfrac{\pi R'}{P}]^{1/(2\alpha)}b_0$$

$$+\tfrac{P^2}{\pi R'E'}\sqrt{\tfrac{R'}{3\mu_a(u_1+u_2)}}[\tfrac{P}{\pi R'}]^{1/(2\alpha)}b_1 + \dots$$

Obviously, in (5.32) coefficients γ_0, γ_1, a_0, b_0, a_1, and b_1 depend on the specifics of function μ and constants a, s_0, and α.

In case of a Newtonian fluid $\alpha = 1$ and

$$\vartheta^2 = P/[3\pi\mu_a(u_1+u_2)], \tag{5.33}$$

while function $F(x)$ in dimensionless variables has the form

$$F(x) = 12\gamma^2 x. \tag{5.34}$$

By solving equation (5.15), we find

$$f = \frac{1}{12\gamma^2}\frac{\mu s_0}{h}. \tag{5.35}$$

By using (5.35) in (5.12) with proper dimensions, we obtain the analogs of formulas (5.32) for Newtonian lubricants

$$h_e = \frac{\mu_a(u_1+u_2)R'}{P}d_{00} + \frac{P}{E'}d_{01} + \cdots,$$

$$F_S = (u_2 - u_1)\left\{ \sqrt{\frac{\mu_a P}{(u_1+u_2)}}d_{10} + \frac{1}{R'E'}\sqrt{\frac{P^5}{\mu_a(u_1+u_2)^3}}d_{11} + \cdots \right\}, \tag{5.36}$$

$$F_R = \sqrt{\mu_a(u_1+u_2)P}d_{20} + \frac{1}{R'E'}\sqrt{\frac{P^5}{\mu_a(u_1+u_2)}}d_{21} + \cdots,$$

where coefficients d_{00}, d_{01}, d_{10}, d_{11}, d_{20}, and d_{21} depend only on the fluid viscosity μ and inlet coordinate a and are independent of slide-to-roll ratio s_0.

Finally, let us consider a lubricant with Erying rheology [5] for which in dimensional variables

$$\tau = G_1\mu_1 arsh(\frac{\mu}{G}\frac{\partial u}{\partial z}), \tag{5.37}$$

where $\mu_1 = \mu_1(p)$ is a certain dimensionless function of pressure, G and G_1 are constant shear stresses characterizing the fluid rheology. For this rheology constant ϑ can be taken from (5.33) or from

$$\vartheta^2 = \frac{4R'G}{\mu_a(u_1+u_2)}. \tag{5.38}$$

After that it is clear that in formulas for the film thickness (5.27) and for the friction force components (5.29) coefficients γ_0, γ_1, a_0, b_0, a_1, and b_1 depend on constants a, s_0 and the specifics of functions μ and $P/(R'G_1\mu_1)$.

Now, let us consider the realization of the above method and some numerical results for the case of Newtonian rheology. In this case in the dimensionless variables we have

$$F(g + \frac{H\zeta}{c-a}\frac{dp}{d\xi}) = 12\gamma^2(g + \frac{H\zeta}{c-a}\frac{dp}{d\xi}). \tag{5.39}$$

Following the described procedure, we obtain

$$\frac{dp_0}{d\xi} = \frac{\mu_0(c_0-a)(H_0-1)}{2\gamma_0^2 H_0^3}, \quad g_0 = \frac{\mu_0 s_0}{12\gamma_0^2 H_0},$$

$$\gamma_0(H_0 - 1) = (\frac{c_0-a}{2})^2\xi^2 + \frac{c_0^2-a^2}{2}\xi + \frac{a^2+2ac_0-3c_0^2}{4},$$

$$p_0(-1) = p_0(1) = 0, \quad \int_{-1}^{1} p_0(t)dt = \frac{\pi}{c_0-a},$$

$$\frac{dp_1}{d\xi} = \frac{\mu_0(c_0-a)}{2\gamma_0^2 H_0^3}\{(H_0-1)\frac{d\ln\mu_0}{dp_0}p_1 + [3(3-H_0-\frac{2}{H_0})$$

$$+\frac{a(c_0-a)}{\gamma_0}(\xi-1)(2-\frac{3}{H_0})]\frac{c_1}{c_0-a} - 3(1-\frac{1}{H_0})\frac{\gamma_1}{\gamma_0} + (\frac{3}{H_0}-2)Q\}, \tag{5.40}$$

$$g_1 = \frac{\mu_0 s_0}{12\gamma_0^2 H_0}\{\frac{d\ln\mu_0}{dp_0}p_1 - \frac{1}{H_0}\frac{\partial H_0}{\partial c_0}c_1 - (1+\frac{1}{H_0})\frac{\gamma_1}{\gamma_0} - \frac{Q}{H_0},$$

$$H_1 = \frac{\partial H_0}{\partial c_0}c_1 - (H_0-1)\frac{\gamma_1}{\gamma_0} + Q, \quad Q(u) = \frac{c_0-a}{\pi}\int\limits_{-1}^{1}p_0(t)\ln\frac{1-t}{|\xi-t|}dt, \tag{5.41}$$

$$p_1(-1) = p_1(1) = 0, \quad \int\limits_{-1}^{1}p_1(t)dt = -\frac{\pi c_1}{(c_0-a)^2},\dots .$$

Let us consider solution of equations (5.40) and (5.41) for the simplest case of constant viscosity $\mu(p) = \mu_0(p_0) = 1$. It is not difficult to integrate equations (5.40) (see also [3]) and get the solution in the form

$$p_0(x) = \frac{b^3}{8}\{\frac{b^2-3c_0^2}{b^2}[\arctan\frac{x}{b} + \frac{bx}{b^2+x^2}] - \frac{2b\gamma_0 x}{(b^2+x^2)^2} - c^0\},$$

$$c^0 = \frac{b^2-3c_0^2}{b^2}[\arctan\frac{a}{b} + \frac{ab}{b^2+a^2}] - \frac{2ab\gamma_0}{(b^2+a^2)^2}, \tag{5.42}$$

$$b = \sqrt{\gamma_0 - c_0^2}, \quad x = [a + c_0 + (c_0-a)\xi]/2.$$

Constants γ_0 and c_0 involved in (5.42) are determined by a system of nonlinear algebraic equations

$$\frac{b^2-3c_0^2}{b^2}[\arctan\frac{c_0}{b} - \arctan\frac{a}{b} + \frac{bc_0}{\gamma_0} - \frac{ab}{b^2+x^2}] - \frac{2ab\gamma_0}{(b^2+a^2)^2} - \frac{2bc_0}{\gamma_0} = 0,$$

$$\frac{b^2-3c_0^2}{b^2}[\frac{c_0}{b}(\arctan\frac{c_0}{b} - \arctan\frac{a}{b}) - \frac{a(c_0-a)}{b^2+a^2}] + \frac{a-c_0^2}{b^2+a^2} + \frac{2a\gamma_0(c_0-a)}{(b^2+a^2)^2} \tag{5.43}$$

$$= 4\pi b^2.$$

The solution of system (5.43) can be represented in the form

$$p_1(x) = \frac{1}{b^3}\{\frac{3\gamma_1}{b^2}\varepsilon_\gamma(x) + \frac{c_1}{c_0-a}\varepsilon_c(x) + \frac{\gamma_0}{b^2}\varepsilon(x)\}, \quad \varepsilon_\gamma(x) = \frac{\gamma_0}{b^2}A_4(\rho)$$

$$-A_3(\rho), \quad \varepsilon_c(x) = \frac{6\gamma_0(ac_0-\gamma_0)}{b^4}A_4(\rho) + \frac{9\gamma_0-4ac_0}{b^2}A_3(\rho) - 3A_2(\rho)$$

$$+\frac{2a}{b}[2B_2(\rho) - \frac{3\gamma_0}{b^2}B_3(\rho)], \quad \varepsilon(x) = \frac{2b}{\pi}\int\limits_{a/b}^{\rho}\frac{1}{(1+t^2)^3}[\frac{3\gamma_0}{b\gamma^2(1+t^2)} - 2] \tag{5.44}$$

$$\times \int\limits_{a/b}^{c_0/b}q_0(z)\ln\frac{c_0/b-z}{|t-z|}dzdt, \quad A_{n+1}(\rho) = \frac{1}{2n}[\frac{\rho}{(1+\rho^2)^n} - \frac{ab^{2n-1}}{(b^2+a^2)^n}]$$

$$+\frac{2n-1}{2n}A_n(\rho), \quad A_1(\rho) = \arctan\rho - \arctan\frac{a}{b}, \quad B_n(\rho) = \frac{1}{2n}[(\frac{b^2}{b^2+a^2})^n$$

$$-\frac{1}{(1+\rho^2)^n}], \quad \rho = x/b, \quad x = [a + c_0 + (c_0 - a)\xi]/2.$$

From equations (5.44) and the last three equations in (5.41), we find a system of linear algebraic equations for c_1 and γ_1

$$3\varepsilon_\gamma(c_0)\gamma_1 + \frac{b^2}{c_0-a}\varepsilon_c(c_0)c_1 = -\gamma_0\varepsilon(c_0),$$

$$3\int_a^{c_0}\varepsilon_\gamma(x)dx\gamma_1 + \frac{b^2}{c_0-a}\left[\int_a^{c_0}\varepsilon_c(x)dx + \frac{\pi b^3}{2}\right]c_1 = -\gamma_0\int_a^{c_0}\varepsilon(x)dx.$$

(5.45)

By consequently solving systems (5.43) and (5.45), we find constants c_0, γ_0, c_1, and γ_1. After that we can find constants involved in equations (5.36) for the lubrication film thickness and the components of the friction force in the form

$$d_{00} = 6\pi\gamma_0, \quad d_{01} = \frac{2\gamma_1}{\pi}, \quad d_{10} = \frac{\alpha_1}{\sqrt{3\pi}b}, \quad d_{20} = \frac{3}{2\sqrt{3\pi}b\sqrt{2}}\left[\frac{b^2-c_0^2}{b^2}\alpha_1\right.$$

$$\left.+\frac{a\gamma_0}{b^2+a^2} - c_0\right], \quad d_{11} = -\frac{1}{3\pi^2\sqrt{3\pi}b^3}\{\gamma_1\alpha_2 + \frac{2c_1}{c_0-a}[\frac{b^2\alpha_1}{2}$$

$$+(ac_0 - \gamma_0)\alpha_2 - ab\beta_1] + \gamma_0\int_{a/b}^{c_0/b}\frac{Q(t)dt}{(1+t^2)^2}\},$$

(5.46)

$$d_{21} = -\frac{1}{\pi^2\sqrt{3\pi}b^3}\{2\gamma_1(\frac{\gamma_0\alpha_3}{b^2} - \alpha_2) + \frac{c_1}{c_0-a}[\frac{4\gamma_0(ac_0-\gamma_0)\alpha_3}{b^2}$$

$$+(5\gamma_0 - 2ac_0)\alpha_2 - b^2\alpha_1 + 2ab(\beta_1 - \frac{2\gamma_0\beta_2}{b^2})] + \gamma_0\int_{a/b}^{c_0/b}\frac{Q(t)}{(1+t^2)^2}$$

$$\times[\frac{2\gamma_0}{b^2(1+t^2)} - 1]dt\}, \quad \alpha_i = A_i(c_0/b), \quad \beta_i = B_i(c_0/b), \quad i = 1,2,\ldots.$$

The numerical results for c_0, d_{00}, d_{10}, d_{20}, $p_{0max} = \max_{a\leq x\leq c_0} p_0(x)$, $H_{0min} = \min_{a\leq x\leq c_0} H_0(x)$ and c_1, d_{01}, d_{11}, d_{21} versus the inlet coordinate a are given in Tables 5.1 and 5.2, respectively. It follows from Tables 5.1 and 5.2 that as the amount of the entrained lubricant in a lightly loaded lubricated contact increases (which is equivalent to increasing $|a|$, $a < 0$) the film thickness increases while the maximum pressure p_{0max} decreases. The behavior of the rolling and sliding components of the friction force is non-monotonic.

5.2.2   Thermal EHL Problem

Now, let us take a look at a thermal EHL problem for a lightly loaded contact [6]. In addition to the assumptions made earlier, we need to assume that the heat generation in a lubrication layer is primarily due to viscous shear in the lubricant. Moreover, based on the fact that the lubrication film thickness is

TABLE 5.1

Data for c_0, d_{00}, d_{10}, d_{20}, $p_{0max} = \max\limits_{a \leq x \leq c_0} p_0(x)$,

$H_{0min} = \min\limits_{a \leq x \leq c_0} H_0(x)$ versus the inlet coordinate a (after

Kudish [6]). Published with permission from Allerton Press.

a	c_0	d_{00}	d_{10}	d_{20}	p_{0max}	H_{0min}
-0.031	0.014	0.076	3.601	0.037	63.850	0.951
-0.164	0.059	0.572	2.245	0.187	13.181	0.886
-0.554	0.123	1.734	1.792	0.529	4.543	0.835
-0.954	0.148	2.316	1.725	0.757	3.686	0.823
-5	0.170	2.962	1.767	1.285	2.762	0.816
-10	0.171	2.990	1.791	1.375	2.713	0.816

TABLE 5.2

Data for c_1, d_{01}, d_{11}, and d_{21} versus the inlet coordinate a (after Kudish [6]). Published with permission from Allerton Press.

a	c_1	d_{01}	d_{11}	d_{21}
-0.031	0.144	0.006	0.1032	-0.0080
-0.164	0.254	0.022	0.0118	-0.0131
-0.554	0.346	0.052	-0.0019	-0.0155
-0.954	0.366	0.065	-0.0034	-0.0147
-5	0.368	0.073	-0.0036	-0.0121
-10	0.369	0.074	-0.0036	-0.0119

much smaller than the contact size, we can assume that the generated heat is primarily transported across the film in the direction perpendicular to solid surfaces [4]. Using experimental studies we can assume that the lubricant viscosity is a certain function of lubricant pressure and temperature, i.e., $\mu = \mu(p, T)$, where $T = T(x, z)$ is the lubricant temperature.

Based on the above assumption the simplification of the fluid energy equation leads to the problem [4]

$$\frac{\partial}{\partial z}(\lambda \frac{\partial T}{\partial z}) = -\tau \frac{\partial u}{\partial z}, \quad T(x, -\frac{h}{2}) = T_{w1}(x), \quad T(x, \frac{h}{2}) = T_{w2}(x), \qquad (5.47)$$

where $T_{w1}(x)$ and $T_{w2}(x)$ are the temperatures of the lower and upper solids, respectively, and $\lambda = \lambda(p, T)$ is the lubricant coefficient of heat conductivity.

By introducing the following additional dimensionless variables

$$\{T', T'_{w1}, T'_{w2}\} = \{T, T_{w1}, T_{w2}\}/T_0 - 1, \quad \lambda' = \lambda/\lambda_0 \qquad (5.48)$$

(where T_0 and λ_0 are the characteristic lubricant temperature and heat conductivity) and a new dimensionless parameter

$$\kappa = \frac{(u_1 + u_2)P}{\pi \lambda_0 T_0 \vartheta^2} \qquad (5.49)$$

we arrive at the system of equations (for simplicity here and further primes are omitted)

$$\int_{-h/2}^{h/2} \frac{z}{\mu} F(f + z\frac{dp}{dx})dz = h - 1, \ p(a) = p(c) = 0, \tag{5.50}$$

$$\int_{-h/2}^{h/2} \frac{1}{\mu} F(f + z\frac{dp}{dx})dz = s_0, \tag{5.51}$$

$$\frac{\partial}{\partial z}(\lambda \frac{\partial T}{\partial z}) = -\kappa \frac{\gamma^2}{\mu}(f + z\frac{dp}{dx})F(f + z\frac{dp}{dx}),$$

$$T(x, -\frac{h}{2}) = T_{w1}(x), \ T(x, \frac{h}{2}) = T_{w2}(x), \tag{5.52}$$

$$\gamma(h - 1) = x^2 - c^2 + \frac{1}{V}\frac{2}{\pi}\int_{a}^{c} p(t) \ln \frac{c-t}{|x-t|}dt, \tag{5.53}$$

$$\int_{a}^{c} p(t)dt = \frac{\pi}{2}, \tag{5.54}$$

For lightly loaded lubricated contact ($V \gg 1$) the process of solution of system (5.50)-(5.54) is similar to the one described for the isothermal contact. Namely, as in Section 5.2.1 we introduce new variables according to (5.20) and for $V \gg 1$ look for the solution in the form given by (5.26) and

$$T(\xi, \zeta) = \sum_{k=0}^{\infty} V^{-k} T_k(\xi, \zeta), \tag{5.55}$$

where functions $p_k(\xi)$, $H_k(\xi)$, $g_k(\xi)$, $T_k(\xi, \zeta)$ and constants c_k, γ_k has to be determined. As a result we obtain systems of nonlinear and linear differential equations similar to the ones obtained in Section 5.2.1. Therefore, for the considered above rheologies all formulas (5.27), (5.29), (5.27), and (5.36) for film thickness h_e and components F_S and F_R of the friction force are still valid. However, in this case coefficients γ_0, γ_1, a_0, b_0, a_1, b_1, d_{00}, d_{10}, d_{20}, d_{01}, d_{11}, and d_{21} besides earlier mentioned parameters (see the preceding Section) also depend on λ, T_{W1}, and T_{W2}.

5.2.3 Numerical Method for Lightly Loaded EHL Contacts with Non-Newtonian Lubricant

In this section we will consider numerical approach to solution of the EHL problem for lightly loaded contacts with non-Newtonian lubricant. In addition we will compare the numerical and asymptotic solutions and determine the range of parameter V for which the asymptotic solution is sufficiently accurate.

The main advantage of the direct numerical solution is in the fact that the numerical procedure is insensitive to the details of non-Newtonian rheology of the fluid which may limit application of analytical methods. A numerical solution allows to determine the qualitative and quantitative behavior of the solution. However, it does not provide an opportunity to reveal the analytical structure of the solution.

The numerical method for solution of an EHL problem for lightly loaded contacts [7] is designed along the lines of the asymptotic method developed in the preceding section for $V \gg 1$. Because of the problem nonlinearity the solution process is iterative. In particular, it means that the term $\frac{1}{V} \int_{-1}^{1} p(t) \ln \frac{1-t}{|\xi-t|} dt$ describing the surface displacement caused by elastic deformations of the contact solids has a small effect on the problem solution and, therefore, it is always taken from the preceding iteration. The equations of the problem are taken in the form (see equations (5.21)-(5.25))

$$\frac{d}{d\xi}\left[M_0(p,H)\frac{dp}{d\xi} - H\right] = 0, \ p(-1) = p(1) = \frac{dp(1)}{d\xi} = 0,$$

$$M_0(p,H) = \frac{H^2}{4\mu} \int_{-1}^{1} tF(g + \frac{Ht}{c-a}\frac{dp}{d\xi})dt/\frac{dp}{d\xi}, \tag{5.56}$$

$$\int_{-1}^{1} F(g + \frac{Ht}{c-a}\frac{dp}{d\xi})dt = 2\frac{\mu s_0}{H}, \tag{5.57}$$

$$\gamma(H-1) = (\frac{c-a}{2})^2\xi^2 + \frac{c^2-a^2}{2}\xi + \frac{a^2+2ac-3c^2}{4}$$

$$+\frac{c-a}{\pi V} \int_{-1}^{1} p(t) \ln \frac{1-t}{|\xi-t|} dt, \tag{5.58}$$

$$\int_{-1}^{1} p(t)dt = \frac{\pi}{c-a}. \tag{5.59}$$

The main idea of the method is to reduce solution of the problem to solution of the nonlinear system of equations

$$L_1(\gamma, c) = \frac{dp(1)}{d\xi} = 0,$$

$$L_2(\gamma, c) = \int_{-1}^{1} p(t)dt - \frac{\pi}{c-a} = 0, \tag{5.60}$$

by the modified Newton's method [8] for two unknowns γ and c assuming that functions $p(\xi)$, $H(\xi)$, and $g(\xi)$ can be easily calculated. Calculation of functions $L_1(\gamma, c)$ and $L_2(\gamma, c)$ and their approximate derivatives with respect to γ and c for Newton's method is done by calculation expressions $[L_1(\gamma, c) - L_1(\gamma + \triangle\gamma, c)]/\triangle\gamma$, etc.

The general scheme of the iterative process is as follows. For given constants γ, c and function $p(\xi)$ from equation (5.58), we calculate the new approximation for gap $H(\xi)$ and after that from equation (5.57) using Newton's method we determine the new approximation for sliding frictional stress $g(\xi)$. That prepares us for determination of the new approximation of pressure $p(\xi)$ from equations (5.56).

Now, let us consider some details of the method. We will introduce partitions on the intervals $[-1, 1]$ along the ξ-axes $\{\xi_k\}$, $k = 1, \ldots, I$, $\xi_1 = -1$, $\xi_I = 1$ and along the ζ-axes $\{\zeta_n\}$, $n = 1, \ldots, J$, $\zeta_1 = -1$, $\zeta_J = 1$. Nodes $\{\xi_k\}$ are more dense in the zones where $dp/d\xi$ is large while the partition along the ζ-axis is uniform with the step size $\triangle\zeta = 2/J$. Using notations $p_k = q(\xi_k)$, $H_k = H(\xi_k)$, $g_k = g(\xi_k)$, and $\mu_k = \mu(p_k)$ and the approximation for $p(\xi)$ on the interval $[\xi_k, \xi_{k+1}]$

$$p(\xi) = p_k + \frac{p_{k+1} - p_k}{\xi_{k+1} - \xi_k}(\xi - \xi_k)$$

we obtain

$$\int\limits_{-1}^{1} p(t)\ln\frac{1-t}{|\xi-t|}dt \approx \sum_{i=1}^{I-1} N_i(\xi_k), \; k = 1, \ldots, I,$$

$$N_i(\xi) = [p_i - \xi_i\frac{p_{i+1}-p_i}{\xi_{i+1}-\xi_i}][K_1(\xi_{i+1}) - K_1(\xi_i) - K_2(\xi - \xi_{i+1})$$

$$+K_2(\xi - \xi_i)] + \frac{p_{i+1}-p_i}{\xi_{i+1}-\xi_i}[K_3(\xi_{i+1}) - K_3(\xi_i) - K_4(\xi - \xi_{i+1}, \xi) \qquad (5.61)$$

$$+K_4(\xi - \xi_i, \xi)], \; K_1(\xi) = (\xi - 1)[\ln(1 - \xi) - 1],$$

$$K_2(\xi) = \xi[1 - \ln|\xi|], \; K_3(\xi) = \frac{(1-\xi)^2}{2}[\ln(1 - \xi) - \tfrac{1}{2}] + K_1(\xi),$$

$$K_4(\xi, \zeta) = \frac{\xi^2}{2}[\ln|\xi| - \tfrac{1}{2}] + \zeta K_2(\xi).$$

Similarly, using the trapezoidal rule for integration with respect to ζ from (5.57) we find

$$\{F[g_k - \frac{H_k}{c-a}\frac{dp(\xi_k)}{d\xi}] + F[g_k + \frac{H_k}{c-a}\frac{dp(\xi_k)}{d\xi}]\}\frac{\triangle\zeta}{2}$$

$$+ \sum_{i=1}^{J-1} F[g_k + \frac{(-1+i\triangle\zeta)H_k}{c-a}\frac{dp(\xi_k)}{d\xi}]\triangle\zeta = \frac{2\mu_k s_0}{H_k}. \qquad (5.62)$$

Based on the last boundary condition in (5.56), the exact solution of equation (5.57) at $\xi = \xi_I$ is

$$g_I = \Phi(\frac{\mu_I s_0}{H_I}). \qquad (5.63)$$

It is convenient to conduct the solution of equations (5.62) in the sequence from $k = I - 1$ to $k = 1$ by taking the initial approximation for g_k being equal to g_{k+1}.

The discretization of the integral in (5.60) with the help of the trapezoidal rule is obvious.

TABLE 5.3

The dimensionless sliding component $F_S(V, n)$ of the friction force for a fluid with power rheology and constant viscosity (after Kudish and Yatzko [7]). Published with permission from Allerton Press.

V/n	0.75	1	1.25	1.5
128	0.003605	0.005502	0.004649	0.004053
64	0.003520	0.005474	0.004200	0.003021
32	0.003444	0.005436	0.003951	0.002748
16	0.003402	0.005407	0.003817	0.002537

To find a new approximation of the pressure distribution $\{p_k\}$, we integrate the first equation in (5.56) with respect to ξ from $\xi_{k-1/2} = (\xi_{k-1} + \xi_k)/2$ to $\xi_{k+1/2} = (\xi_k + \xi_{k+1})/2$. By approximating the derivatives, we will obtain

$$\frac{M_0(\xi_{k+1/2})}{\xi_{k+1}-\xi_k} p_{k+1} - \left[\frac{M_0(\xi_{k+1/2})}{\xi_{k+1}-\xi_k} + \frac{M_0(\xi_{k-1/2})}{\xi_k-\xi_{k-1}}\right] p_k$$

$$+ \frac{M_0(\xi_{k-1/2})}{\xi_k-\xi_{k-1}} p_{k-1} = H_{k+1/2} - H_{k-1/2}, \ k = 2, \ldots, I-1, \tag{5.64}$$

$$p_1 = p_I = 0.$$

Here values of $H_{k+1/2}$ and $H_{k-1/2}$ are determined using the quadratic interpolation while $M_0(\xi_{k+1/2})$ and $M_0(\xi_{k-1/2})$ are calculated based on the information from the preceding iteration at $\xi_{k+1/2}$ and $\xi_{k-1/2}$, respectively. The latter involves calculation of $p_{k-1/2} = (p_{k-1} + p_k)/2$ and $p_{k+1/2} = (p_k + p_{k+1})/2$ as well as usage of $H_{k+1/2}$ and $H_{k-1/2}$. To avoid the difficulty in calculation of M_0 at a point where $dp/d\xi = 0$ at such a point we take a ratio of M determined at the values of p and H at the above point and at a small value of $dp/d\xi$ and divide it by the latter.

As a result equations (5.64) form a three-diagonal system of linear algebraic equations that can be easily solved for the set of $\{p_k\}$, $k = 1, \ldots, I$, by standard methods. After that the new sets of $dp(\xi_k)/d\xi$ and μ_k are easily found. That completes all necessary calculations for a single iteration. Such iterations are continued until the process converges with the desired precision.

Let us consider some numerical data for lubricants with power and Newtonian rheologies and constant viscosity $\mu = 1$. In dimensionless variables for a fluid with power rheology, we have

$$F(x) = (12\gamma^{n+1})^{1/n} \mid x \mid^{(1-n)/n}, \ n > 0. \tag{5.65}$$

The partition along the ξ-axes is taken as follows: $\xi_k = \xi_{k-1} + d_0\lambda^{k-1}, k = 1, \ldots, I$, $\xi_1 = -1$, $\lambda = d_1^{1/(I-1)}$, $d_0 = 2(1-\lambda)/(1-\lambda^{I-1})$. In particular simulations we used $I = 81$ and $d_1 = 0.05$.

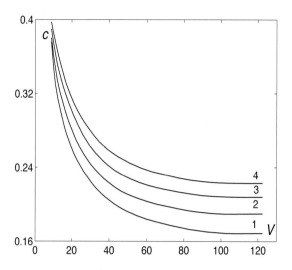

FIGURE 5.2

The dependence of the exit coordinate c on V for $n = 0.75$ (curve 1), $n = 1$ (curve 2), $n = 1.25$ (curve 3), and $n = 1.5$ (curve 4) (after Kudish and Yatzko [7]). Published with permission from Allerton Press.

TABLE 5.4

The asymptotic and numerical values of c, γ, F_S, F_R, and q_{max} for Newtonian fluid with constant viscosity as functions of parameter V (after Kudish and Yatzko [7]). Published with permission from Allerton Press.

V	c	γ	F_S	F_R	q_{max}
		Asymptotic method			
128	0.1890	0.1644	0.005830	0.4466	2.6343
64	0.2071	0.1701	0.005826	0.4451	2.5523
32	0.2432	0.1816	0.005818	0.4423	2.4362
16	0.3154	0.2045	0.005801	0.4366	2.3597
		Numerical method			
128	0.1873	0.1642	0.005502	0.4381	2.6088
64	0.2029	0.1691	0.005474	0.4328	2.5559
32	0.2348	0.1798	0.005436	0.4235	2.4195
16	0.3014	0.2027	0.005407	0.4079	2.2103
8	0.3913	0.2318	0.005371	0.3862	1.9420

The numerical method demonstrated good convergence for $V > 8$. For

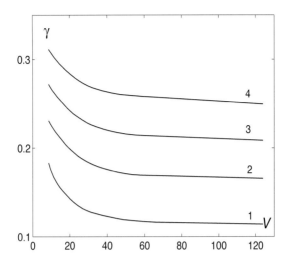

FIGURE 5.3
The dependence of the lubrication film thickness γ on V for $n = 0.75$ (curve
1), $n = 1$ (curve 2), $n = 1.25$ (curve 3), and $n = 1.5$ (curve 4) (after Kudish
and Yatzko [7]). Published with permission from Allerton Press.

decreasing V in the region of $V \leq 8$, the convergence of the method gets
worse and depends on the value of n. It is not surprising as the method was
designed in a manner similar to the regular asymptotic method applicable
only to lightly loaded lubricated contacts represented by values of $V \gg 1$. On
the other hand, the numerical method has a wider range of applicability than
the asymptotic one.

For $a = -10$ and $s_0 = 0.02$ the numerical results for the exit coordinate c,
lubrication film thickness γ, and maximum pressure p_{max} versus parameter V
are presented in Figs. 5.2 - 5.4, respectively. Curves 1, 2, 3, and 4 correspond
to $n = 0.75$, $n = 1$, $n = 1.25$, and $n = 1.5$. It follows from Figs. 5.2 and 5.3
that $c(V)$ and $\gamma(V)$ monotonically decrease as parameter V increases and as
$V \to \infty$ they approach their limiting values, which correspond to the case of
rigid solids. The same figures show that c and γ monotonically increase with
increase of n. The pressure distribution $p(x)$ has one extremum - maximum
located to the right of $x = 0$. The maximum pressure p_{max} monotonically
increases as a function of V and decreases with increase of n (see Fig. 5.4). As
$V \to \infty$ the value of p_{max} approaches its limiting value equal to the one for
rigid solids. Table 5.3 demonstrates monotonic increase of the sliding component of the friction force F_S with parameter V and non-monotonic behavior as
a function of n. To compare the numerical and two-term asymptotic solutions

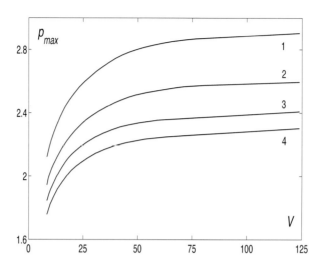

FIGURE 5.4
The dependence of the maximum pressure p_{max} on V for $n = 0.75$ (curve 1), $n = 1$ (curve 2), $n = 1.25$ (curve 3), and $n = 1.5$ (curve 4) (after Kudish and Yatzko [7]). Published with permission from Allerton Press.

(see Section 5.2.1) let us consider the data obtained for a Newtonian fluid ($n = 1$) with $a = -10$ and $s_0 = 0.02$, which is presented in Table 5.4. The data indicate the fact that for $V \geq 16$ the asymptotic solution is in excellent agreement with the numerical one. The above results show that the asymptotic formulas are very accurate and can be used for practical calculations for $V \geq 10$.

5.3   Asymptotic Approach to Lightly Loaded Point EHL Problems

Let us consider a ball rolling and sliding over a grooved raceway. The ball is separated from the raceway by a thin layer of an incompressible viscous fluid described by the Newtonian law. The ball and grooved raceway surfaces move with speeds (u_2, v_2) and (u_1, v_1) while their radii are R_{x2}, R_{y2}, R_{x1}, and R_{y1}, respectively. The ball is loaded by a normal force P. The heat generation can be neglected and the problem can be considered under isothermal conditions.

The equations of the EHL problem for a contact of a ball with a grooved raceway lubricated by an incompressible Newtonian fluid have the form

$$\frac{\partial}{\partial x}\{\frac{h^3}{12\mu}\frac{\partial p}{\partial x}\} + \frac{\partial}{\partial y}\{\frac{h^3}{12\mu}\frac{\partial p}{\partial y}\} = \frac{u_1+u_2}{2}\frac{\partial h}{\partial x} + \frac{v_1+v_2}{2}\frac{\partial h}{\partial y},$$

$$p\mid_\Gamma = 0, \quad \frac{dp}{d\overrightarrow{n}}\mid_{\Gamma_e} = 0,$$

$$h = h_0 + \frac{x^2}{2R_x} + \frac{y^2}{2R_y} + \frac{1}{\pi E'}\int\int_\Omega p(\xi,\eta)\left[\frac{1}{\sqrt{(\xi-x)^2+(\eta-y)^2}}\right.$$

$$\left. - \frac{1}{\sqrt{\xi^2+\eta^2}}\right]d\xi d\eta, \quad \int\int_\Omega p(\xi,\eta)d\xi d\eta = P,$$

(5.66)

where $p(x,y)$ and $h(x,y)$ are the pressure and gap in the contact, respectively, Ω is the contact region while Γ and Γ_e are the boundary of Ω and the exit boundary of Ω where lubricant leaves the contact, respectively, h_0 is the central lubrication film thickness at $(x,y) = (0,0)$, $\overrightarrow{n}$ is the external unit normal vector to contact boundary Γ, R_x and R_y are the effective radii of the contact solids in the $x-$ and $y-$directions, $\frac{1}{R_x} = \frac{1}{R_{x1}} \pm \frac{1}{R_{x2}}$, $\frac{1}{R_y} = \frac{1}{R_{y1}} \pm \frac{1}{R_{y2}}$, E' is the effective elastic modulus of the solid materials, $\frac{1}{E'} = \frac{1-\nu_1^2}{E_1} + \frac{1-\nu_2^2}{E_2}$. The inlet boundary Γ_i of the contact is determined by the inflowing lubricant, i.e., by the relationships

$$\overrightarrow{F}\cdot\overrightarrow{n}\mid_{\Gamma_i} < 0, \quad \overrightarrow{F} = \frac{h}{2}(u_1+u_2, v_1+v_2) - \frac{h^3}{12\mu}(\frac{\partial p}{\partial x}, \frac{\partial p}{\partial y}),$$

(5.67)

where $\overrightarrow{F}$ is the vector of the lubricant flow volume flux.

In the dimensionless variables suitable for a lightly loaded lubricated contact

$$(x',y') = \frac{\vartheta}{2R_x}(x,y), \quad (p',\tau') = \frac{8\pi R_x^2}{3P\vartheta^2}(p,\tau), \quad h' = \frac{h}{h_0}, \quad \mu' = \frac{\mu}{\mu_a},$$

$$\vartheta = \frac{P}{8\pi\mu_a\sqrt{(u_1+u_2)^2+(v_1+v_2)^2}R_x}, \quad \gamma = \frac{\vartheta^2}{2R_x}h_0,$$

(5.68)

$$V = \frac{16\pi R_x^2 E'}{3P\theta_0^3}, \quad \sigma = \frac{u_1+u_2}{\sqrt{(u_1+u_2)^2+(v_1+v_2)^2}}, \quad \delta = \frac{R_x}{R_y},$$

the problem equations can be reduced to (further, primes are omitted at the dimensionless variables)

$$\gamma^2\{\frac{\partial}{\partial x}[\frac{h^3}{\mu}\frac{\partial p}{\partial x}] + \frac{\partial}{\partial y}[\frac{h^3}{\mu}\frac{\partial p}{\partial y}]\} = \sigma\frac{\partial h}{\partial x} + \sqrt{1-\sigma^2}\frac{\partial h}{\partial y},$$

$$p\mid_\Gamma = 0, \quad \frac{dp}{d\overrightarrow{n}}\mid_{\Gamma_e} = 0,$$

$$h = 1 + \frac{x^2+\delta y^2}{\gamma} + \frac{2}{\pi V\gamma}\int\int_\Omega p(\xi,\eta)\left[\frac{1}{\sqrt{(\xi-x)^2+(\eta-y)^2}}\right.$$

$$\left. - \frac{1}{\sqrt{\xi^2+\eta^2}}\right]d\xi d\eta, \quad \int\int_\Omega p(\xi,\eta)d\xi d\eta = \frac{2\pi}{3},$$

(5.69)

γ is the dimensionless central lubrication film thickness, V and ρ are the given dimensionless parameters, θ is a dimensionless parameter. The dimensionless variables are introduced in such a way that for a similar hydrodynamic problem (i.e., for a lubrication problem for rigid solids with $V = \infty$) $p = O(1)$ and $x^2 + y^2 = O(1)$.

Therefore, for given values of parameters V, δ, σ, and the inlet boundary Γ_i we need to find the distributions of pressure $p(x, y)$, gap $h(x, y)$, the central lubrication film thickness γ, and the exit boundary Γ_e.

Lightly loaded lubricated contacts correspond to the cases when parameter $V \gg 1$. Under lightly loaded conditions the solution of problem (5.69) can be searched in the form of power series in V^{-1}. In particular, for the dimensionless film thickness γ the solution has the form (compare with (5.26))

$$\gamma = \gamma_0 + \tfrac{1}{V}\gamma_1 + \dots, \tag{5.70}$$

where γ_0 and γ_1 are dimensionless constants of the order of unity. In dimensional form the expression for the central lubrication film thickness is as follows

$$h_0 = h_{00} + \frac{3P^2}{64\pi^2 \mu_a \sqrt{(u_1+u_2)^2+(v_1+v_2)^2}R_x^2 E'}\gamma_1 + \dots, \tag{5.71}$$

where h_{00} is the lubrication film thickness in a similar contact of rigid solids. Obviously, the second term in the right-hand side of (5.71) is proportional to $1/E'$ which represents the adjustment term that takes into account the elasticity of the solids. Similar expressions can be obtained for the sliding and rolling frictional forces.

Cases of heavily loaded lubricated contacts are considered in the following chapters.

5.4   Closure

Asymptotic and direct numerical solutions for lightly loaded line and point EHL problems with Newtonian and non-Newtonian lubricants are considered. The dominant mechanism and the corresponding to it solution structure are revealed. The asymptotic solutions are obtained in the form of regular asymptotic expansions in negative powers of $V \gg 1$. The asymptotic solutions are compared with the direct numerical solutions.

5.5   Exercises and Problems

1. Explain why the solution of the EHL problem for a line contact searched in the form of (5.26) describes only lightly loaded contacts and cannot possibly describe heavily loaded contacts. Provide arguments based on the physics of the lubrication phenomenon.

2. Determine a two-term asymptotic solution of the isothermal EHL problem for a lightly loaded line contact (i.e., for $V \gg 1$) lubricated by an incompressible fluid with viscosity determined by $\mu = \exp(Qp)$, where $Q = O(1)$. Graph $p(x)$ and $h(x)$ for $V = 5$ and $V = 10$ and compare this solution with the one determined by formulas (5.42)-(5.46) and data from Table 5.1.

3. Derive dimensional two-term asymptotic expressions for the sliding and rolling frictional forces in a lightly loaded point EHL contact.

4. Consider thermal EHL problem for lightly loaded point contact. Show that the structural dimensional formulas for the lubrication film thickness and frictional forces are similar to (5.71).

References

[1] Van-Dyke, M. 1964. *Perturbation Methods in Fluid Mechanics*. New York: Academic Press.

[2] Kevorkian, J. and Cole, J.D. 1985. *Perturbation Methods in Applied Mathematics. Applied Mathematics Series, Vol.* 34. New York: Springer-Verlag.

[3] Kapitza, P.L. 1955. Hydrodynamic Theory of Lubrication during Rolling. *Zhurnal Tekhnicheskoy Fiziki* 25:747-762.

[4] Hamrock, B.J. 1994. *Fundamentals of Fluid Film Lubrication*. New York: McGraw-Hill.

[5] Van Wazen, J.R., Lyons, J.W., Kim, K.Y., and Cowell, R.E. 1963. *Viscosity and Flow Measurment. A Laboratory Handbook of Rheology*. New York-London: Interscience Publishers, John Wiley & Sons.

[6] Kudish, I.I. 1981. Some Problems of the Elastohydrodynamic Theory of Lubrication for a Lightly Loaded Contact. *Mech. of Solids* 16, No. 3:75-88.

[7] Kudish, I.I. and Yatzko, B.G. 1990. Numerical Method for Solution of Problems for Lightly Loaded Contacts with Non-Newtonian Lubricant. *Soviet J. Frict. and Wear* 11, No. 4:594-601.

[8] Fedorenko, R.P. 1978. *Approximate Solution of Optimal Control Problems*. Moscow: Nauka.

Part IV

Isothermal EHL Problems for Heavily Loaded Line Contacts with Newtonian Lubricant

This part of the monograph is devoted to asymptotic and numerical analyses of heavily loaded line EHL contacts when lubricant is a Newtonian fluid. In particular, specific analytical asymptotic approaches are developed in application to pre- and over-critical heavily loaded lubrication regimes. Regimes of starved and fully flooded lubrication are considered. The analytical analysis of the problem reveals a clear structure of heavily loaded EHL contacts. It produces analytical formulas for the lubrication film thickness and simplified asymptotically valid equations in the inlet and exit zones of heavily loaded contacts. A specialized numerical approach is developed for solution of these asymptotic equations. Numerical solution precision and stability/instability properties are analyzed in detail. Finally, the validation of the asymptotic approaches is conducted through the comparison between solutions of asymptotic and original EHL problem equations.

6

Asymptotic Approaches to Heavily Loaded
Lubricated Line Contacts

6.1   Introduction

The science of tribology is dedicated to understanding the fundamental physical and chemical factors that govern friction and wear. As energy conservation and equipment longevity continue to increase in importance, efforts to maximize unit fuel consumption while reducing wear-related failure are becoming more relevant.

Lubricated contacts may be characterized by the ratio of the average lubricant film thickness h and the composite root mean square surface roughness $\sigma = (\sigma_1^2 + \sigma_2^2)^{1/2}$ of both contacting surfaces in the concentrated contact region.

The classical modes of lubrication are depicted schematically in Fig. 6.1. A typical Stribeck curve is shown in Fig. 6.2, which describes wear regimes in terms of the effects of lubricant viscosity, sliding speed, and normal load on friction coefficient. When the lubricant film is sufficiently thick $(h \gg \sigma)$, the opposing surfaces never contact one another. The condition when such contact surfaces are almost non-deformed is known as hydrodynamic lubrication and is the predominant lubrication mode in lightly loaded journal bearings. Since normal loads are relatively small, there is negligible deformation of opposing contact surfaces. Lubrication dynamics are largely dictated by viscous properties of the lubricating film, and the coefficient of friction is directly proportional to viscosity when speed and load are fixed.

As the film thickness h approaches 3-5 times σ the asperities of the opposing surfaces are still not in contact (only rarely asperities brush against each other) [1, 2] normal forces become sufficiently large that elastic deformation of the contacting surfaces occur. This condition is known as elastohydrodynamic lubrication (EHL) because the lubrication regime is controlled by both viscous effects and elastic deformation of the contacting surfaces. The EHL regime is usually encountered in situations in which high loads act over relatively small contact areas such as in ball bearings, roller bearings, and gear teeth. For hard materials, higher loads are required to reach EHL than for softer, deformable materials.

Asperity contact becomes more frequent as h approaches σ. In the mixed

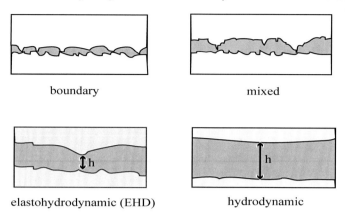

FIGURE 6.1
Pictorial representation of lubrication regimes. (after Kudish and Covitch [3]).
Reprinted with permission from CRC Press.

lubrication regime, the coefficient of friction begins to rise, and a significant
amount of frictional heat is generated in the lubrication zone. Cyclic loading
can lead to wear and pitting failure.

As load further increases or speed decreases, the friction coefficient reaches a
plateau. Unless soft, compliant surface films are formed by reaction of extreme
pressure and/or anti-wear additives with the contact surfaces, asperities can
become welded to one another leading to catastrophic failure. This regime is
known as boundary lubrication.

This chapter describes the analysis of various fluid lubrication problems
associated with EHL for hard contacts, incompressible and compressible lubricants with Newtonian rheology. Analytical and numerical treatments of
selected classic problems of elastohydrodynamic lubrication are covered and
validated through comparison between the asymptotic and numerical solutions of the original EHL equations. Structurally different lubrication regimes
such as pre- and over-critical lubrication regimes are considered using different
asymptotic approaches due to differences in solution structure. Some numerical methods are developed and numerical solutions as well as their precision
are provided. That allows to come up with the proper grid size for numerical solution of asymptotically valid and original EHL problem equations to
provide the desired solution precision. The latter strongly depends of EHL
problem input parameters such as load, speed, contact radii, lubricant ambient viscosity, etc. In addition to that numerical solution stability/instability,
and regularization approaches to solution of isothermal line EHL problems
are considered.

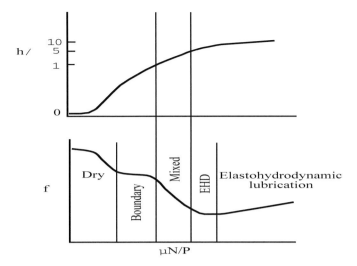

FIGURE 6.2

Stribeck curve relating coefficient of friction (ordinate) to lubricant viscosity μ, speed u, and normal load per unit area P. (after Kudish and Covitch [3]). Reprinted with permission from CRC Press.

6.2 Numerical Solution for Moderately and Heavily Loaded EHL Contacts

To get an idea of the distributions of pressure and gap in moderately and heavily loaded EHL contacts let us first solve the EHL problem numerically. There are a number of serious difficulties in studying EHL problems in general and for heavily loaded regimes, in particular. The difficulties common for all lubrication regimes are related to the problem nonlinearity and a priori unknown exit boundary. In addition to that for heavily loaded lubricated regimes the problem possesses narrow boundary layers near the inlet and exit points of the contact (inlet and exit zones) and in the rest of the contact region, the problem is close to an integral equation of the first kind. It is well known that generally a solution of an integral equation of the first kind is very sensitive to small perturbations and, therefore, unstable. The above indicates that on top of common difficulties typical for EHL problems for heavily loaded regimes we are faced with the necessity to maintain high precision in the inlet and exit zones and to overcome potential solution instability.

Over years a number of different numerical methods were developed [4]-[13]. All these methods work well for lightly and moderately loaded lubricated contacts. However, most of them still have trouble to overcome the potential instability of the problem solution for heavily loaded contacts. The method proposed in this section (see [14]) makes an attempt to overcome this difficulty for moderately to heavily loaded contacts by introducing a naturally occurring regularization along the lines of the methods described in [15]. However, for sufficiently heavily loaded regimes, the method still tends to be unstable and tends to diverge.

6.2.1   Problem Formulation

Let us consider a steady plane isothermal EHL problem for a heavily loaded contact of two moving smooth elastic cylinders (see Fig. 5.1). The axes of the cylinders are parallel, their radii are R_1 and R_2 while the materials of the cylinders have the Young's moduli E_1 and E_2 and Poisson's ratios ν_1 and ν_2, respectively. The cylinders are moving perpendicular to their axes with the surface velocities u_1 and u_2 and are separated by a layer of lubricant. A force P normal to cylinder axes is applied to the cylinders pressing them one into another. The lubricant is represented by an incompressible viscous fluid with Newtonian rheology and viscosity μ. The lubricant viscosity is assumed to satisfy Barus law

$$\mu = \mu_a \exp(\alpha p), \tag{6.1}$$

where μ_a and α are the ambient viscosity and the viscosity pressure coefficient.

Assuming that the thickness of the lubrication layer is much smaller than the size of their contact which, in turn, is much smaller than the cylinder radii we can approximate the surface normal displacements by the displacements of the boundary of an elastic half-plane with the effective elastic parameters. In addition, assuming that cylinders velocities are relatively low we can obtain the following problem equations (for the derivation see the preceding chapter for the rheology function $F(x) = x$)

$$\frac{d}{dx}\left[\frac{h^3}{12\mu}\frac{dp}{dx} - \frac{u_1+u_2}{2}h\right] = 0, \;\; p(x_i) = p(x_e) = \frac{dp(x_e)}{dx} = 0,$$

$$h = h_e + \frac{x^2-x_e^2}{2R'} + \frac{2}{\pi E'}\int\limits_{x_i}^{x_e} p(t)\ln\mid\frac{x_e-t}{x-t}\mid dt, \;\; \int\limits_{x_i}^{x_e} p(t)dt = P, \tag{6.2}$$

were x is the point coordinate in the contact along the x-axis directed along the lubricant flow, $p(x)$ and $h(x)$ are the pressure and gap distributions, respectively, x_i and x_e are the coordinate of the points of contact inlet and exit, respectively, h_e is the the exit film thickness $h(x_e)$, R' and E' are the effective radius of cylinders and the elastic modulus of their materials, respectively, $1/R' = 1/R_1 \pm 1/R_2$ (signs $+$ and $-$ are chosen in accordance with cylinders' curvatures), $1/E' = 1/E_1' + 1/E_2'$, $E_j' = E_j/(1-\nu_j^2)$, $j = 1, 2$.

In (6.2) the boundary conditions follow from the fact that at the contact boundary the pressure is equal to the atmospheric pressure which is negligibly small compared to the pressure developed in the contact. The boundary condition $\frac{dp(x_e)}{dx} = 0$ prevents cavitation at the exit from the contact and is usually called the cavitation boundary condition.

To study the problem in the case of a heavily loaded contact, it is reasonable and convenient to choose the dimensionless variables

$$x' = \frac{x}{a_H}, \; a = \frac{x_i}{a_H}, \; c = \frac{x_e}{a_H}, \; p' = \frac{p}{p_H}, \; h' = \frac{h}{h_e}, \; \mu' = \frac{\mu}{\mu_a} \qquad (6.3)$$

and parameters

$$V = \frac{24\mu_a(u_1+u_2)R'^2}{a_H^3 p_H}, \; Q = \alpha p_H, \; H_0 = \frac{2R'h_e}{a_H^2}, \qquad (6.4)$$

which are scaled based on the parameters of the limiting case of a dry Hertzian contact of two elastic cylinders, namely, based on the maximum Hertzian pressure $p_H = \sqrt{\frac{E'P}{\pi R'}}$ and the Hertzian contact semi-width $a_H = 2\sqrt{\frac{R'P}{\pi E'}}$. In dimensionless variables (6.3) and (6.4) equations (6.2) of the classic EHL problem are reduced to the following system of integro-differential equations (further primes are omitted)

$$\frac{d}{dx}\left[\frac{H_0^2}{V}\frac{h^3}{\mu}\frac{dp}{dx} - h\right] = 0, \qquad (6.5)$$

$$p(a) = p(c) = \frac{dp(c)}{dx} = 0, \qquad (6.6)$$

$$H_0(h-1) = x^2 - c^2 + \frac{2}{\pi}\int_a^c p(t)\ln\left|\frac{c-t}{x-t}\right| dt, \qquad (6.7)$$

$$\int_a^c p(t)dt = \frac{\pi}{2}. \qquad (6.8)$$

In the above equations the dimensionless lubricant viscosity is taken in the form

$$\mu = \mu(p, Q) = \exp(Qp), \qquad (6.9)$$

where Q is the dimensionless pressure viscosity coefficient.

In equations (6.5)-(6.8) the inlet coordinate a, the speed-load parameter V, and the viscosity pressure coefficient Q are considered to be known while the exit coordinate c and the lubrication film thickness H_0 together with the functions of pressure $p(x)$ and gap $h(x)$ in the contact region $[a, c]$ have to be determined from the solution of the problem.

6.2.2 Numerical Procedure for Moderately to Heavily Loaded EHL Contacts

We will consider the simplest case of a Newtonian fluid. In this case will be able to illustrate all significant features of an EHL problem solution for a moderately to heavily loaded contact. In dimensionless variables (6.3) and (6.4), the problem is reduced to the system of equations (6.5)-(6.8). We will assume that the lubricant viscosity is controlled by Barus exponential relationship (6.9). Therefore, for the given inlet coordinate a and parameters V and Q we need to determine the functions of pressure $p(x)$ and gap $h(x)$ as well as the exit coordinate c and lubrication film thickness H_0. It is easier to numerically solve the problem if the contact region $[a, c]$ with the unknown boundary c can be replaced by the interval $[-1, 1]$. It is achieved by introducing the substitution

$$x = \frac{c+a}{2} + \frac{c-a}{2}\xi, \tag{6.10}$$

where ξ is the new independent variable. Using (6.10) reduces the system to

$$\frac{2H_0^2}{V(c-a)} \frac{d}{d\xi} [h^3 \exp(-Qp)\frac{dp}{d\xi} - h] = 0, \tag{6.11}$$

$$p(-1) = p(1) = \frac{dp(1)}{d\xi} = 0, \tag{6.12}$$

$$H_0(h-1) = (\frac{c+a}{2} + \frac{c-a}{2}\xi)^2 - c^2 + \frac{c-a}{\pi} \int\limits_{-1}^{1} p(t)\ln \mid \frac{1-t}{\xi-t} \mid dt, \tag{6.13}$$

$$\int\limits_{-1}^{1} p(t)dt = \frac{\pi}{c-a}. \tag{6.14}$$

For heavily loaded contacts $V \ll 1$ and/or $Q \gg 1$ (see the preceding two sections). It was shown that the pressure is close to the Hertzian one for purely dry contacts away from the inlet $\xi = -1$ and exit $\xi = 1$ points and the gap is approximately constant and equal to 1. At the same time within the inlet and exit zones the pressure and gap are equally governed by both the lubricant viscous flow and elastic displacements of the solid surfaces. Moreover, as it was shown by asymptotic methods in heavily loaded contacts the lubrication film thickness H_0 and exit coordinate c are predominantly determined by the problem solution within the inlet and exit zones, respectively. The iterative method for solution of the problem has to take into account the fact that in the inlet and exit zones the viscous and elastic effects are of the same order.

To simplify the problem and to reduce it to a system of linear algebraic equations, we will use the method of quasi-linearization [16] which represents linearization of the problem about its k-th iteration p_k, h_k, H_{0k}, c_k (k is the iteration number). Application of this method leads to the system

$$\frac{2H_{0k}^3}{V(c_k-a)} \frac{d}{d\xi} \left(h_k^3 e^{-Qp_k} \frac{d\triangle p_{k+1}}{d\xi}\right) - \frac{c_k-a}{\pi} \int\limits_{-1}^{1} \frac{\triangle p_{k+1}(t)dt}{t-\xi}$$

$$+ \frac{6H_{0k}^3}{V(c_k-a)} \frac{d}{d\xi} \left(h_k^2 e^{-Qp_k} \frac{\partial h}{\partial p} \mid_k \triangle p_{k+1} \frac{dp_k}{d\xi}\right)$$

$$- \frac{2H_{0k}^3 Q}{V(c_k-a)} \frac{d}{d\xi} \left(h_k^3 e^{-Qp_k} \triangle p_{k+1} \frac{dp_k}{d\xi}\right) + \triangle H_{0k+1}\{ \frac{6H_{0k}^2}{V(c_k-a)} \frac{d}{d\xi} \left(h_k^3 e^{-Qp_k} \frac{dp_k}{d\xi}\right)$$

$$+ \frac{6H_{0k}^3}{V(c_k-a)} \frac{d}{d\xi} \left(h_k^2 e^{-Qp_k} \frac{\partial h}{\partial H_0} \mid_k \frac{dp_k}{d\xi}\right)\}$$

$$+\triangle c_{k+1}\{-\frac{2H_{0k}^3}{V(c_k-a)}\frac{d}{d\xi}(h_k^3 e^{-Qp_k}\frac{dp_k}{d\xi})$$

$$+\frac{6H_{0k}^3}{V(c_k-a)}\frac{d}{d\xi}(h_k^2 e^{-Qp_k}\frac{\partial h}{\partial c}\mid_k \frac{dp_k}{d\xi})-c_k-(c_k-a)\xi \qquad (6.15)$$

$$-\frac{1}{\pi}\int_{-1}^{1}\frac{p_k(t)dt}{t-\xi}\} = H_{0k}\{\frac{dh_k}{d\xi}-\frac{2H_{0k}^3}{V(c_k-a)}\frac{d}{d\xi}(h_k^3 e^{-Qp_k}\frac{dp_k}{d\xi})\},$$

$$\triangle p_{k+1}(-1) = \triangle p_{k+1}(1) = 0, \qquad (6.16)$$

$$\int_{-1}^{1}\triangle p_{k+1}(t)dt + \frac{\pi}{(c_k-a)^2}\triangle c_{k+1} = \frac{\pi}{c_k-a}-\int_{-1}^{1}p_k(t)dt. \qquad (6.17)$$

The equation for constant c follows from the requirement that the lubricant flux is conserved, i.e., from equation (6.11) integrated with respect to ξ from ξ to 1 with the help of the boundary condition in (6.12) and equation (6.13)

$$\frac{2H_0^2}{V(c-a)}h^3 e^{-Qp}\frac{dp}{d\xi} = h-1.$$

Application of quasi-linearization to the latter equation gives

$$\frac{2H_{0k}^3}{V(c_k-a)}h_k^3 e^{-Qp_k}\frac{d\triangle p_{k+1}}{d\xi}-\frac{c_k-a}{\pi}\int_{-1}^{1}\triangle p_{k+1}\ln\mid\frac{1-t}{t-\xi}\mid dt$$

$$+\frac{6H_{0k}^3}{V(c_k-a)}h_k^2 e^{-Qp_k}\frac{\partial h}{\partial p}\mid_k \triangle p_{k+1}\frac{dp_k}{d\xi}-\frac{2H_{0k}^3 Q}{V(c_k-a)}h_k^3 e^{-Qp_k}\triangle p_{k+1}\frac{dp_k}{d\xi}$$

$$+\triangle H_{0k+1}\{\frac{6H_{0k}^2}{V(c_k-a)}h_k^3 e^{-Qp_k}\frac{dp_k}{d\xi}+\frac{6H_{0k}^3}{V(c_k-a)}h_k^2 e^{-Qp_k}\frac{\partial h}{\partial H_0}\mid_k \frac{dp_k}{d\xi})\} \qquad (6.18)$$

$$+\triangle c_{k+1}\{-\frac{2H_{0k}^3}{V(c_k-a)}h_k^3 e^{-Qp_k}\frac{dp_k}{d\xi})+\frac{6H_{0k}^3}{V(c_k-a)}\frac{d}{d\xi}(h_k^2 e^{-Qp_k}\frac{\partial h}{\partial c}\mid_k \frac{dp_k}{d\xi})$$

$$-(1+\xi)(\frac{c_k+a}{2}+\frac{c_k-a}{2}\xi)+2c_k-\frac{1}{\pi}\int_{-1}^{1}p_k(t)\ln\mid\frac{1-t}{t-\xi}\mid dt\}$$

$$= H_{0k}\{dh_k-1-\frac{2H_{0k}^2}{V(c_k-a)}h_k^3 e^{-Qp_k}\frac{dp_k}{d\xi}\}.$$

In equations (6.15)-(6.18) function $\triangle p_{k+1} = p_{k+1}-p_k$ and constants $\triangle H_{0k+1} = H_{0k+1}-H_{0k}$ and $\triangle c_{k+1} = c_{k+1}-c_k$ are incremental changes in the corresponding quantities from iteration to iteration, $\frac{\partial h}{\partial p}\mid_k \triangle p_{k+1}$ is the result of application of the linear operator (the derivative of h_k with respect to p_k) to function $\triangle p_{k+1}$, $\frac{\partial h}{\partial H_0}\mid_k$ and $\frac{\partial h}{\partial c}\mid_k$ are the partial derivatives of h_k with respect to H_0 and c, respectively, calculated on the k-th iteration.

Further on, we introduce two sets of nodes on $[-1,1]$: integral nodes $\{\xi_i\}$, $i = 1,\dots,I$, $\xi_1 = -1$, $\xi_I = 1$ and semi-integral nodes $\{\xi_{i+1/2} = (\xi_i+\xi_{i+1})/2\}$, $i = 1,\dots,I-1$. After that we integrate equations (6.15) and (6.18) with respect to ξ over the intervals $[\xi_{j-1/2},\xi_{j+1/2}]$ and $[\xi_{I-1},\xi_I]$, respectively, and use the approximation

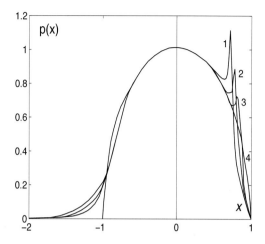

FIGURE 6.3
The profiles of the pressure distribution $p(x)$ obtained for $Q = 7$ and $V = 0.2$ (curve 1), $V = 0.1$ (curve 2), $V = 0.05$ (curve 3), and Hertzian pressure distribution (curve 4) (after Airapetov, Kudish, and Panovko [14]). Reprinted with permission from Allerton Press.

$$\tfrac{dp}{d\xi}(\xi_{i+1/2}) \approx \tfrac{p(\xi_{i+1})-p(\xi_i)}{\xi_{i+1}-\xi_i}$$

and the rectangle rule [15]

$$\int_{-1}^{1} \tfrac{p(t)dt}{t-\xi_i} \approx \sum_{j=1}^{I-1} \tfrac{p(\xi_j)(\xi_{j+1}-\xi_j)}{\xi_{j+1/2}-\xi_i}, \quad i = 1,\dots,I,$$

to obtain a system of linear algebraic equations for $\{\triangle p_{k+1}(\xi_i)\}$, $i = 1,\dots,I$, and constants $\triangle H_{0k+1}$ and $\triangle c_{k+1}$.

As soon as values $\{\triangle p_{k+1}(\xi_i)\}$, $i = 1,\dots,I$, $\triangle H_{0k+1}$, and $\triangle c_{k+1}$ are found from the solution of the above–mentioned system of linear algebraic equations we can determine the new iterates of our solution

$$p_{k+1} = p_k + \triangle p_{k+1}, \quad H_{0k+1} = H_{0k} + \triangle H_{0k+1}, \quad c_{k+1} = c_k + \triangle c_{k+1}.$$

After that the new function of gap h_{k+1} can be easily calculated using usual quadrature formulas, for example, the trapezoidal rule. To initiate this iterative process, we have to provide the initial approximations for p, H_0, and c. For moderately to heavily loaded regimes that can be done as follows

$$c_0 = 1, \quad H_{00} = 0.272(VQ)^{3/4}, \quad p_0(\xi) = \sqrt{1 - (\tfrac{1+a}{2} + \tfrac{1-a}{2}\xi)^2}$$

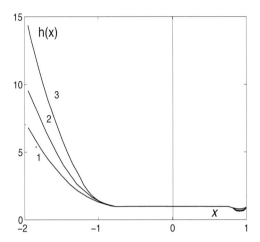

FIGURE 6.4
The profiles of the gap distribution $h(x)$ obtained for $Q = 7$ and $V = 0.2$ (curve 1), $V = 0.1$ (curve 2), and $V = 0.05$ (curve 3) (after Airapetov, Kudish, and Panovko [14]). Reprinted with permission from Allerton Press.

or by taking the values of p, H_0, and c from the known solution with the closest set of input parameters a, V, and Q. The iterative process runs until the solution converges with the desired precision.

It is important to mention that in equations (6.15) and (6.18) the terms proportional to $\triangle H_{0k+1}$ and $\triangle c_{k+1}$ represent the naturally occurring regularization terms for solution of this ill-posed lubrication problem. In [15] it is rigorously proven that a similarly designed numerical method for solution of one-dimensional linear first kind singular integral equations converges to the exact solution of the problem. In this sense the mathematical features of the EHL problem for heavily loaded regimes is very close to the first kind singular integral equations considered in [15].

The proposed numerical method possesses some advantages in comparison with other methods. Namely, it allows simultaneous determination of constants H_0 and c along with pressure $p(\xi)$, automatic regularization procedure by using the terms proportional to $\triangle H_{0k+1}$ and $\triangle c_{k+1}$, and the guarantee of the constant lubricant flux through the gap between the lubricated solids. The method converges well for lightly and moderately to heavily loaded regimes. However, it still does not converge for severely heavily loaded regimes. Some remedies related to stability of heavily to severely heavily loaded EHL contacts are proposed in the following sections.

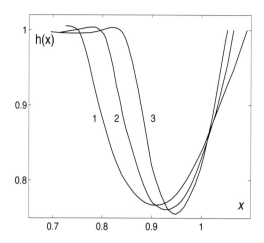

FIGURE 6.5
Magnification of the profiles of the gap distribution $h(x)$ in the vicinity of the exit point obtained for $Q = 7$ and $V = 0.2$ (curve 1), $V = 0.1$ (curve 2), and $V = 0.05$ (curve 3) (after Airapetov, Kudish, and Panovko [14]). Reprinted with permission from Allerton Press.

6.2.3 Some Numerical Results for Heavily Loaded EHL Contacts

The numerical results described below are obtained for the fixed inlet coordinate $a = -2$, pressure coefficient $Q = 7$, and slide-to-roll ratio $s_0 = 2(u_2 - u_1)/(u_2 + u_1) = 0.01$. The results were obtained on a nonuniform set of nodes $\{\xi_i\}$ more dense near the exit point 1. The total number of nodes used was $I = 100$. For moderate values of parameters V and Q, the increase of the total number of nodes I from 100 to 200 does not noticeably change the solution of the problem. The relative error was maintained at the level of 10^{-4}. The results are presented in Fig. 6.3-6.7, where curves marked with 1 correspond to $V = 0.2$, curves marked with 2 correspond to $V = 0.1$, and curves marked with 3 correspond to $V = 0.05$. Figure 6.3 shows pressure distributions for different values of parameter V. Also, for comparison in Fig. 6.3 is shown the Hertzian distribution of pressure $\sqrt{1 - x^2}$ marked by 4. It is clear that the pressure distribution has a narrow spike close to the exit point $x = c$. In [17] by solving asymptotically valid equations in the exit zone it was determined that this narrow spike exists only when the lubricant viscosity μ is strongly dependent on pressure p. Specifically, in the latter study it was found that if the lubricant viscosity $\mu(p) = \exp(Qp^m), Q \gg 1$, then the narrow spike does not exist when $m \leq 0.25$ and it does exist when $m \geq 0.75$. The pres-

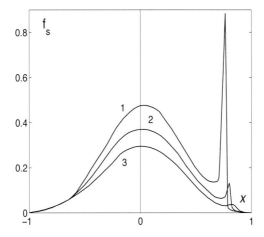

FIGURE 6.6
The profiles of the sliding frictional stress distribution $f_s(x)$ obtained for
$Q = 7$ and $V = 0.2$ (curve 1), $V = 0.1$ (curve 2), and $V = 0.05$ (curve
3) (after Airapetov, Kudish, and Panovko [14]). Reprinted with permission
from Allerton Press.

sure distributions in Fig. 6.3 indicate that as V increases the pressure spike
increases in height and moves toward the center of the contact. The pressure
distribution exhibits a similar behavior when $|\,a\,|$ $(a < 0)$ increases but to a
much lesser extent. Moreover, when varying the inlet coordinate a within the
region $a < -2$ it practically does not affect the numerical results. For a significantly closer to the beginning of the Hertzian region $x = -1$ (for example,
$a = -1.25$) there is a small difference in the pressure distribution in the inlet
and exit zones, especially near the spike. However, this difference increases as
a approaches -1. These changes in pressure in the inlet zone affect the value
of the film thickness H_0 and the exit coordinate c.

The graphs of the gap distributions $h(x)$ are presented in Fig. 6.4. It is clear
that in the cental part of the contact (Hertzian region) we have a practically
constant gap $h(x) = 1$. Moreover, the minimum of the film thickness is by no
more than 15% smaller than the gap value in the center of the contact. In more
detail it can be seen from Fig. 6.5 where a blow up of the gap distributions
are shown in the exit zone.

Variations of the dimensionless sliding $f_s(x) = \frac{V}{12H_0}\frac{\mu s_0}{h}$ and rolling $f_r(x)$
$= \frac{H_0 h}{2}\frac{dp}{dx}$ frictional stresses are presented in Fig. 6.6 and 6.7, respectively.
Obviously, the shape of the distributions of the sliding frictional stress $f_s(x)$
resembles the shape of pressure $p(x)$ distributions. The rolling frictional stress
$f_r(x)$ is mainly concentrated (different from zero) in the inlet and exit zones,

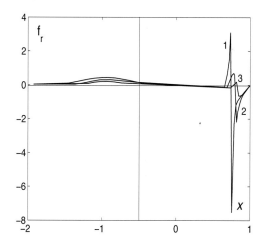

FIGURE 6.7

The profiles of the rolling frictional stress distribution $f_r(x)$ obtained for $Q = 7$ and $V = 0.2$ (curve 1), $V = 0.1$ (curve 2), and $V = 0.05$ (curve 3) (after Airapetov, Kudish, and Panovko [14]). Reprinted with permission from Allerton Press.

and it is proportional to and, to a certain extent, resembles $\frac{1}{\mu}\frac{dp}{dx}$. For fixed values of the slide-to-roll ratio s_0, sufficiently small parameter V and/or sufficiently large parameter Q the ratio of the rolling and sliding frictional stresses $f_r(x)/f_s(x)$ is small in the entire contact.

For the values of parameter $V = 0.2$, 0.1, 0.005, the following results for the film thickness $H_0 = 0.3637$, $0.2342, 0.1479$ and the exit coordinate $c = 1.078$, 1.059, 1.044 were obtained. Based on the corrected Ertel-Grubin formula (see Section 6.6) $H_0 = 0.272(VQ)^{3/4}$ for the used values of V and Q we get $H_0 = 0.3501$, 0.2082, 0.1238, respectively. The comparison of the two sets of H_0 values reveals no surprise, i.e., $H_0^{Ertel-Grubin} <$ $H_0^{EHLnumerical}$. The explanation of this relationship between the Ertel-Grubin $H_0^{Ertel-Grubin}$ and the numerically obtained $H_0^{EHLnumerical}$ is based on the results of the asymptotic analysis of the over-critical regimes in Section 6.6. In these cases the dimensionless minimal film thickness is equal to $h_{min} = 0.7653$, 0.7587, 0.7545, respectively. Therefore, based on this we get $H_{min} = h_{min}H_0 = 0.2806$, 0.1777, 0.1116.

The results for a lightly loaded contact differ significantly from the ones for heavily loaded conditions. It can clearly be seen from Fig. 6.8 the results of which (pressure and gap) are obtained for $V = 20$ and $Q = 4$. In particular, the pressure distribution $p(x)$ (curve 1) possesses a clear maximum at $x \approx 0.2$ and it differs significantly from the Hertzian pressure distribution (curve 3).

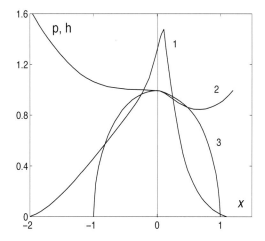

FIGURE 6.8
The profiles of the pressure distribution $p(x)$ (curve 1) and gap $h(x)$ (curve 2) obtained for $V = 20$ and $Q = 4$. Curve 3 is the Hertzian pressure distribution (after Airapetov, Kudish, and Panovko [14]). Reprinted with permission from Allerton Press.

This pattern of pressure behavior is the continuation of the trend for the pressure spike behavior indicated above in Fig. 6.3. At the same time, the gap distribution $h(x)$ (curve 2) is no longer constant in the central part of the contact region but rather resembles the gap determined by function $1 + \frac{x^2}{H_0}$.

Similar results were obtained for $Q = 5$ and $Q = 6$ and same other parameters. The pressure spike, exit coordinate c, and film thickness H_0 decrease with decreasing value of Q.

6.3 Asymptotic Analysis of a Heavily Loaded Lubricated Contact of Elastic Solids. Pre-Critical Lubrication Regimes for Newtonian Liquids

In the classic formulation the EHL problem has been studied in numerous papers and monographs by numerical and approximate analytical methods since the end of the 1940. Some references to these papers and monographs can be found in Dowson et al. [5], Ertel [6], Grubin [7], and Kudish et al. [8], the latest achievements in numerical methods of solving the considered problem are described by Houpert et al. [9], Bissett [10], Hamrock [11] (also see the bibliography in these papers), Lubrecht and Venner [12], Evans and Hughes [18], and others.

The lubrication regimes studied in this section correspond to the case of a relatively weak ("non-prevailing" in any zone of a contact region) dependence of lubricant viscosity μ on pressure p. These regimes were studied earlier by Dowson et al. [5], Houpert et al. [9], Bissett [10], Hamrock [11], Lubrecht and Venner [12], and Evans and Hughes [18] using numerical methods. Ertel [6], Grubin [7], Archard et al. [19], and Crook [20] considered cases of rapidly growing with pressure lubricant viscosity using approximate analytical methods. The method used by Crook [20] differs from the methods employed by Ertel [6] and Grubin [7] only by more precise techniques under the same prior assumptions. The method proposed by Archard et al. [19] is the extension of the methods published by Ertel [6], Grubin [7], and Crook [20] for the case of a weak relationship between the viscosity μ and pressure p. The main difference between the methods published by Archard et al. [19] and Ertel [6], Grubin [7], and Crook [20] is in approximations of the gap function $h(x)$ and in a parabolic approximation for pressure $p(x)$ in the zone of large pressure. The purpose of the first modification is to take into account the pressure gradient along the lubricant flow and of the second one is to simplify the approximate calculations. All in all, these approximate analytical studies of the considered problem are based on certain prior assumptions which should be checked and are vital and necessary for application of the aforementioned analytical methods.

The main mathematical difficulties in the analytical analysis of the classic EHL problem are

1. The essential non-linearity of the problem causing the possibility of existence solutions with qualitatively different structure depending on the values of the problem input parameters.

2. The integro-differential form of the problem equations causing the existence of a small parameter ω (possible definitions of the small parameter ω will be considered in Section 6.3.2) and boundary layers for heavily loaded lubrication regimes.

3. The proximity of the considered problem to the classic contact problem of elasticity described by an integral equation of the first kind numerical solutions of which are generally unstable with respect to small perturbations. (The latter problem is the limiting case for the EHL problem for the small parameter $\omega = 0$.)

4. The presence of the unknown dimensionless free exit boundary $x = c$ (exit point) and the dimensionless exit lubrication film thickness H_0.

5. The possibility to discriminate and to differentiate between the influence of different zones (inlet and exit zones) and regions (Hertzian region) of the lubricated contact on the EHL problem solution.

The purpose of this analysis is not to receive one more approximate formula for lubrication film thickness (which already exist in abundance) but to give a complete analysis of the aforementioned classic problem without any contradictions and prior assumptions about its solution. In addition to that, we will understand better the structure of the EHL problem solution and the reason of why there exist so many similar formulas for the lubrication film thickness.

The approach employed in the present and the following sections differs from the previously published ones in that it does not use any prior assumptions about the solution of the problem. A method of matched asymptotic expansions (see Van-Dyke [21] and Kevorkian and Cole [22]) is applied to studying the equations of the EHL problem. The early history of applying the methods of regular and matched asymptotic expansions to various EHL problems can be traced through papers of one of the authors (Kudish [23]-[26]). Kudish [17] investigated the lubrication problem for non-Newtonian fluids under iso- and non-isothermal conditions using the method of regular asymptotic expansions (Kevorkian and Cole [22]). For the first time the method of matched asymptotic expansions was applied to the simplest lubrication problem with Newtonian fluid in an isothermal heavily loaded contact by Kudish [23, 24]. In the next papers (Kudish [17, 25]) this method has been further developed for the case of Newtonian fluids. Also, Kudish [17, 25] has established the conditions for the existence of the second pressure peak and for the dependence of its magnitude, width, and location on the inlet coordinate, temperature, etc. (compare with similar results by Kudish [17]). In papers [24, 26], Kudish investigated a lubrication problem for rough heavily loaded contact lubricated by a non-Newtonian fluid with general rheology. A combination of methods of regular and matched asymptotic expansions was used. In all these cases, the structure of the contact region and the solution behavior were studied and asymptotically valid formulas for the film thickness were obtained.

The method allows to determine the structure of the solution, to study the boundary layers (inlet and exit zones), to obtain asymptotically valid equations, and, most importantly, to get asymptotically accurate estimates for the lubrication film thickness and friction force. Furthermore, several difficulties of the numerical solution of the problem are avoided, such as the presence of a small parameter at the higher derivative in the problem equations and potential instabilities of the problem solution.

As it is our first encounter with matched asymptotic expansions in application to EHL problems we will start with the simplest possible problem. We will consider a plane problem of isothermal lubrication for heavily loaded elastic solids. The lubricant will be considered to be a Newtonian fluid. We will start by studying the so called pre-critical lubrication regimes, the so-called over-critical lubrication regimes are considered in Section 6.6. The structure of the contact area and of the solution will be determined. In the boundary layers (inlet and exit zones), two types of equivalent systems of equations for the major terms of asymptotic expansions for unknown quantities will be obtained. Furthermore, some asymptotic formulas for lubrication film thickness will be derived. The regimes of starved and fully flooded lubrication will be analyzed.

6.3.1 Equivalent Problem Formulations

The problem equations (6.5)-(6.8) can be presented in other equivalent forms. The first way of doing that is by solving equation (6.7) for pressure $p(x)$ and taking into account boundary conditions (6.6) which would produce the equivalent form (Vorovich et al. [27]) of our system expressed in terms of pressure $p(x)$

$$p(x) = R(x)\left[1 - \frac{1}{2\pi} \int\limits_a^c \frac{dM(p,h)}{dt} \frac{dt}{R(t)(t-x)}\right], \quad R(x) = \sqrt{(x-a)(c-x)}, \quad (6.19)$$

$$\int\limits_a^c \frac{dM(p,h)}{dt} \frac{dt}{R(t)} = \pi(a+c), \quad \int\limits_a^c \frac{dM(p,h)}{dt} \frac{t\,dt}{R(t)} = \pi[(\frac{c-a}{2})^2 + \frac{(a+c)^2}{2} - 1], \quad (6.20)$$

$$M(p,h) = \frac{H_0^3}{V} \frac{h^3}{\mu} \frac{dp}{dx}, \quad (6.21)$$

$$H_0(h-1) = x^2 - c^2 + \frac{2}{\pi} \int\limits_a^c p(t) \ln \mid \frac{c-t}{x-t} \mid dt. \quad (6.22)$$

It is obtained by inverting the singular Cauchy integral in equations (6.5)-(6.7). The equivalence of the systems of equations (6.5)-(6.8) and (6.19)-(6.22) takes place under the following condition (Vorovich et al. [27])

$$\frac{1}{\pi} \int\limits_a^c M(p,h)\frac{dt}{R(t)} + \frac{2}{\pi} \int\limits_a^c p(t) \ln \frac{1}{|c-t|} dt + c^2 - \frac{1}{2}(\frac{c-a}{2})^2 - (\frac{a+c}{2})^2$$

$$(6.23)$$

$$= \ln \mid \frac{4}{c-a} \mid.$$

Equation (6.23) describes the condition necessary for the existence of a bounded solution for $p(x)$ (Vorovich et al. [27]), i.e., the solution that is equal to zero at $x = a$ and $x = c$.

Moreover, because of the equivalence of systems (6.5)-(6.7) and (6.19)-(6.22) in the latter system equation (6.22) for gap $h(x)$ can be replaced by (the Reynolds) equation

$$M(p,h) = H_0(h-1), \quad (6.24)$$

which follows from integration of equation (6.5) with the last boundary condition from (6.6) (also see (6.21)). Therefore, system of equations (6.19)-(6.22) is equivalent to the system of equations (6.19)-(6.21) and (6.25).

Another equivalent form of our original system (6.5)-(6.8) can be obtained by noticing that equation (6.5) is equivalent to (see (6.21))

$$\frac{dM(p,h)}{dx} = H_0 \frac{dh}{dx}, \quad (6.25)$$

and if the function of gap $h(x)$ is known then pressure $p(x)$ can be expressed in the form (see (6.5), (6.6), and (6.21))

$$p(x) = H_0 \int\limits_a^x \frac{h(t)-1}{W(p(t),h(t))} dt, \ W(p,h) = M(p,h)/\frac{dp}{dx}. \tag{6.26}$$

Substituting equations (6.25) and (6.26) into equations (6.19) and (6.20) we obtain the representation of the problem in99999 terms of gap $h(x)$ in the form

$$H_0 \int\limits_a^x \frac{h(t)-1}{W(p(t),h(t))} dt = R(x)\left[1 - \frac{H_0}{2\pi} \int\limits_a^c \frac{dh}{dt} \frac{dt}{R(t)(t-x)}\right],$$
$$\tag{6.27}$$

$$R(x) = \sqrt{(x-a)(c-x)},$$

$$H_0 \int\limits_a^c \frac{dh}{dt} \frac{dt}{R(t)} = \pi(a+c), \ H_0 \int\limits_a^c \frac{dh}{dt} \frac{tdt}{R(t)} = \pi[(\frac{c-a}{2})^2 + \frac{(a+c)^2}{2} - 1], \tag{6.28}$$

$$p(x) = H_0 \int\limits_a^x \frac{h(t)-1}{W(p(t),h(t))} dt, \ W(p,h) = M(p,h)/\frac{dp}{dx}, \tag{6.29}$$

where $M(p,h)$ is determined by equation (6.21).

Two equivalent systems of equations (6.19)-(6.21), (6.25) and (6.27)-(6.29), (6.21) expressed in terms of pressure $p(x)$ and gap $h(x)$, respectively, in some cases are useful for numerical solution of EHL problems for heavily loaded contacts. However, the direct numerical solution of the system of equations (6.27)-(6.29), (6.21) exhibits some patterns of instability.

6.3.2   Asymptotic Analysis of the Problem for Heavily Loaded Lubricated Contact

We will consider equations (6.5)-(6.8) together with the equivalence condition (6.23). Let us analyze a heavily loaded contact for which the presence of a small parameter ω in the problem equations is typical, for example, $\omega = V \ll 1$. In this case the deformation effects in the elastic solids almost everywhere in the contact prevail over the lubrication effects. That results in the smallness of the first term of equation (6.5) compared to the second term everywhere in the contact except for narrow boundary layers that are next to the inlet $x = a$ and exit $x = c$ points, respectively.

Thus, the contact will be called heavily loaded if

$$H_0(h-1) \ll 1 \ and \ \frac{d}{dx}\left[\frac{H_0^3}{V} \frac{h^3}{\mu} \frac{dp}{dx}\right] \ll 1$$
$$\tag{6.30}$$

$$for \ x - a \gg \epsilon_q \ and \ c - x \gg \epsilon_g,$$

where $\epsilon_q = \epsilon_q(\omega) \ll 1$ and $\epsilon_g = \epsilon_g(\omega) \ll 1$ are characteristic sizes of the inlet and exit zones, which are boundary layers located next to points $x = a$ and $x = c$ (Van-Dyke [21] and Kevorkian and Cole [22]). The characteristic size of the inlet zone $\epsilon_q(\omega)$ is determined by the given inlet coordinate a (see (6.34)) and depends on ω. The characteristic size of the exit zone $\epsilon_g(\omega)$ is unknown and is determined by the exit coordinate c (see (6.35)). The region determined by inequalities $x - a \gg \epsilon_q$ and $c - x \gg \epsilon_g$ will be called external (Van-Dyke [21] and Kevorkian and Cole [22]) or Hertzian region (Vorovich et al. [27]).

We will begin the analysis of the problem with investigation of the external region temporarily assuming that the boundaries of the contact region $x = a$ and $x = c$ are known and fixed. Then with the help of estimate (6.30) equations (6.5)-(6.8) yield

$$x^2 - c^2 + \tfrac{2}{\pi} \int\limits_a^c p(t) \ln \left| \tfrac{c-t}{x-t} \right| \, dt = o(1),$$

$$(6.31)$$

$$x - a \gg \epsilon_q \ and \ c - x \gg \epsilon_g.$$

Equations (6.31) and (6.8) with the accuracy of up to $o(1)$ coincide with the equations of a classic contact problem of elasticity (Vorovich et al. [27]) and have the well–known solution

$$p_0(x) = \sqrt{(x-a)(c-x)} + \tfrac{1+2ac+(c-a)^2/4-(a+c)x}{2\sqrt{(x-a)(c-x)}},$$

$$(6.32)$$

$$x - a \gg \epsilon_q \ and \ c - x \gg \epsilon_g.$$

Thus, in the external region the problem solution has the form

$$p(x) = p_0(x) + o(1), \quad x - a \gg \epsilon_q \ and \ c - x \gg \epsilon_g. \tag{6.33}$$

Let us assume that for $\omega \ll 1$ the coordinate of the inlet point a is given by

$$a = -1 + \alpha_1 \epsilon_q, \quad \alpha_1 = O(1), \quad \omega \ll 1, \tag{6.34}$$

where α_1 is a given non-positive constant. Then the exit coordinate c may be found in the form

$$c = 1 + \beta_1 \epsilon_g + o(\epsilon_g), \quad \beta_1 = O(1), \quad \omega \ll 1, \tag{6.35}$$

where β_1 is an unknown constant and it is subject to calculation during the solution process.

It follows from equations (6.32)-(6.35) that in the external region $p_0(x) = O(1)$ for $\omega \ll 1$. Besides, under the assumption

$$\epsilon_g = O(\epsilon_q) \ for \ \omega \ll 1 \tag{6.36}$$

the estimate $p_0(x) = O(\epsilon_q^{1/2})$, $\omega \ll 1$, is valid for in the inlet and exit zones, i.e., for $r = (x - a)/\epsilon_q = O(1)$ and $s = (x - c)/\epsilon_g = O(1)$, where r and s are local coordinates in the inlet and exit zones. According to the principle of matched asymptotic expansions (Van-Dyke [21]), one can obtain the following estimate for $p(x)$ in the inlet and exit zones, i.e.,

$$p(x) = O(\epsilon_q^{1/2}) \ for \ r = O(1) \ and \ s = O(1), \ \omega \ll 1. \tag{6.37}$$

Based on estimate (6.37) the problem solution in the inlet and exit zones will be searched in the form

$$p(x) = \epsilon_q^{1/2} q(r) + o(\epsilon_q^{1/2}), \ q(r) = O(1) \ r = O(1), \ \omega \ll 1, \tag{6.38}$$

$$p(x) = \epsilon_g^{1/2} g(s) + o(\epsilon_g^{1/2}), \ g(s) = O(1) \ s = O(1), \ \omega \ll 1, \tag{6.39}$$

where $q(r)$ and $g(s)$ are the major terms of the pressure $p(x)$ asymptotic expansions in the inlet and exit zones.

Using (6.32) equations (6.7), (6.8), and (6.23) can be transformed into

$$H_0(h-1) = \frac{2}{\pi} \int_a^c [p(t) - p_0(t)] \ln \left| \frac{c-t}{x-t} \right| \, dt, \tag{6.40}$$

$$\int_a^c [p(t) - p_0(t)] dt = 0, \tag{6.41}$$

$$\int_a^c M(p,h) \frac{dt}{R(t)} + 2 \int_a^c [p(t) - p_0(t)] \ln \frac{1}{|c-t|} dt = 0. \tag{6.42}$$

Integrating equation (6.5) with respect to x from x to c and using the last boundary condition in (6.6) equations (6.21) and (6.40) yield

$$M(p,h) = \frac{2}{\pi} \int_a^c [p(t) - p_0(t)] \ln \left| \frac{c-t}{x-t} \right| \, dt. \tag{6.43}$$

Resolving (6.43) for $p(x) - p_0(x)$ leads to the estimate (Vorovich et al. [27])

$$p(x) - p_0(x) = O(M(p,h))$$

$$for \ x - a \gg \epsilon_q \ and \ c - x \gg \epsilon_g, \ \omega \ll 1, \tag{6.44}$$

For further analysis it is necessary to make the following temporary assumptions

$$M(p,h) \ll \epsilon_q^{3/2} \ for \ x - a \gg \epsilon_q \ and \ c - x \gg \epsilon_g, \ \omega \ll 1,$$

$$\int_a^c M(p,h) \frac{dt}{R(t)} \ll \epsilon_q^{3/2}, \ \omega \ll 1. \tag{6.45}$$

Obviously, estimate (6.30) and the second estimate in (6.45) are independent. The validity of the above assumptions will be examined after the analysis of the problem is completed.

Estimates (6.44) and (6.45) in the external region lead to

$$p(x) - p_0(x) \ll \epsilon_q^{3/2} \ for \ x - a \gg \epsilon_q \ and \ c - x \gg \epsilon_g, \ \omega \ll 1. \qquad (6.46)$$

The integrals of function $p(x) - p_0(x)$ in (6.41) and (6.42) can be expressed by a sum of three integrals: over the inlet and exit zones and over the external region. Therefore, using estimates (6.45), (6.46), and expressions (6.34), (6.35), (6.38), and (6.39) equations (6.41) and (6.42) can be reduced to the form

$$\epsilon_q^{3/2} \int\limits_0^\infty [q(t) - q_a(t)]dt + \epsilon_g^{3/2} \int\limits_{-\infty}^0 [g(t) - g_a(t)]dt + \ldots = 0, \qquad (6.47)$$

$$\epsilon_q^{3/2} \ln \tfrac{1}{2} \int\limits_0^\infty [q(t) - q_a(t)]dt + \epsilon_g^{3/2} \int\limits_{-\infty}^0 [g(t) - g_a(t)] \ln \tfrac{1}{|t|} dt$$

$$+\epsilon_g^{3/2} \ln \tfrac{1}{\epsilon_g} \int\limits_{-\infty}^0 [g(t) - g_a(t)]dt \ldots = 0, \qquad (6.48)$$

where functions $q_a(r)$ and $g_a(s)$ are the major terms of the inner asymptotic expansions of the external asymptotic (6.32), (6.33) in the inlet and exit zones, which are determined by the equalities

$$q_a(r) = \sqrt{2r} + \tfrac{\alpha_1}{\sqrt{2r}}, \quad g_a(s) = \sqrt{-2s} - \tfrac{\beta_1}{\sqrt{-2s}}. \qquad (6.49)$$

Using the fact $\epsilon_g^{3/2} \ln \tfrac{1}{\epsilon_g} \gg \epsilon_q^{3/2}$ for $\omega \ll 1$ that follows from the assumption (6.36), from (6.47), and from (6.48), we obtained

$$\int\limits_{-\infty}^0 [g(t) - g_a(t)]dt = 0, \qquad (6.50)$$

$$\int\limits_0^\infty [q(t) - q_a(t)]dt = 0, \qquad (6.51)$$

$$\int\limits_{-\infty}^0 [g(t) - g_a(t)] \ln \tfrac{1}{|t|} dt = 0. \qquad (6.52)$$

To get the asymptotic expansions of the gap function h in the inlet and exit zones, the integral in (6.40) can be expressed by the sum of three integrals over the inlet and exit zones and over the external region. Thus, following the described estimating procedure and using equations (6.38), (6.39), (6.46), (6.49)-(6.52) the equations for the major terms of the asymptotic expansions $h_q(r)$ and $h_g(s)$ of the gap function h in the inlet and exit zones are obtained in the form

$$H_0[h_q(r) - 1] = \epsilon_q^{3/2} \tfrac{2}{\pi} \int\limits_0^\infty [q(t) - q_a(t)] \ln \tfrac{1}{|r-t|} dt + \ldots, \qquad (6.53)$$

$$H_0[h_g(s) - 1] = \epsilon_g^{3/2} \tfrac{2}{\pi} \int\limits_{-\infty}^0 [g(t) - g_a(t)] \ln \tfrac{1}{|s-t|} dt + \ldots.$$

It can be shown that the contributions of the effects of elasticity and lubrication to the solution of the problem in the inlet and exit zones are of the same order of magnitude. The commensurability of the aforementioned effects leads to the commensurability of the major terms of the asymptotic expansions of the terms of equation (6.5) in the inlet zone. Therefore, using expressions (6.38), (6.39), (6.53) and estimates

$$\mu(p, Q) = O(1) \; for \; r = O(1) \; and \; s = O(1), \; \omega \ll 1 \qquad (6.54)$$

following from the accepted class of relations for $\mu(p, Q)$, the major term of the asymptotic for the lubrication film thickness H_0 becomes (Kudish [23])

$$H_0 = A(V \epsilon_q^2)^{1/3} + \ldots, \; \omega \ll 1, \qquad (6.55)$$

where $A(\alpha_1) = O(1)$ for $\omega \ll 1$ and $A(\alpha_1)$ is an unknown nonnegative constant independent from ω and ϵ_q. Taking into account estimate (6.55) a similar analysis of the exit zone confirms the validity of the assumption $\epsilon_g^{3/2} \ln \frac{1}{\epsilon_g} \gg \epsilon_q^{3/2}$ for $\omega \ll 1$ and allows to set (see (6.36))

$$\epsilon_g = \epsilon_q. \qquad (6.56)$$

Now, it is useful to introduce the definitions of the regimes of starved and fully flooded lubrication, which are determined by the lubricant flux entering the contact or the position of the inlet coordinate a. These lubrication regimes can be easily expressed by means of the order of proximity of the inlet coordinate a to the left boundary $x = -1$ of the Hertzian dry contact. The conditions when

$$h(x) - 1 \ll 1 \; for \; all \; x \in [a, c], \; \omega \ll 1 \qquad (6.57)$$

are called regimes of starved lubrication, and the conditions when

$$h(x) - 1 = O(1) \; for \; x - a = O(\epsilon_q) \; and \; c - x = O(\epsilon_g), \; \omega \ll 1 \qquad (6.58)$$

are called regimes of fully flooded lubrication. From (6.53) and (6.56) follows that starved lubrication regimes (6.57) take place under the condition

$$\epsilon_q^{3/2} \ll H_0, \; \omega \ll 1, \qquad (6.59)$$

and fully flooded lubrication regimes (6.58) – under the condition

$$\epsilon_q^{3/2} = O(H_0), \; \omega \ll 1. \qquad (6.60)$$

Thus, using formula (6.55) for the exit film thickness H_0 we obtain the conditions for starved

$$\epsilon_q \ll V^{2/5}, \; \omega \ll 1, \qquad (6.61)$$

and fully flooded

$$\epsilon_q = V^{2/5}, \quad \omega \ll 1, \tag{6.62}$$

lubrication regimes. For fully flooded lubrication regimes equations (6.55) and (6.62) lead to the formula for the lubrication film thickness

$$H_0 = AV^{3/5}, \quad \omega \ll 1. \tag{6.63}$$

Finally, using equations (6.53), (6.55), (6.61), and (6.62), the final equations for starved and fully flooded lubrication regimes can be written in the form, respectively,

$$h_q(r) = 1, \tag{6.64}$$

$$h_g(s) = 1; \tag{6.65}$$

$$A[h_q(r) - 1] = \frac{2}{\pi} \int_0^\infty [q(t) - q_a(t)] \ln \frac{1}{|r-t|} dt, \tag{6.66}$$

$$A[h_g(s) - 1] = \frac{2}{\pi} \int_{-\infty}^0 [g(t) - g_a(t)] \ln \frac{1}{|s-t|} dt. \tag{6.67}$$

The asymptotic analysis of equation (6.5) in the inlet zone and of the first boundary condition in (6.6) based on expressions (6.38), (6.53), and (6.55) yields the asymptotically valid equations in the inlet zone

$$\frac{dM_0(q, h_q, \mu_q, r)}{dr} = \frac{2}{\pi} \int_0^\infty \frac{q(t) - q_a(t)}{t - r} dt, \quad q(0) = 0, \tag{6.68}$$

$$M_0(q, h, \mu, r) = A^3 \frac{h^3}{\mu} \frac{dq}{dr}. \tag{6.69}$$

Using (6.39), (6.53), (6.55), and (6.56), a similar analysis of equation (6.5) in the exit zone and of the two last boundary conditions in (6.6) leads to the equations

$$\frac{dM_0(g, h_g, \mu_g, s)}{ds} = \frac{2}{\pi} \int_{-\infty}^0 \frac{g(t) - g_a(t)}{t - s} dt, \quad g(0) = \frac{dg(0)}{ds} = 0. \tag{6.70}$$

In equations (6.68) and (6.70) functions $\mu_q(r)$ and $\mu_g(s)$ are the major terms of the asymptotic expansions of $\mu(p, Q)$ in the inlet and exit zones.

To close the systems of obtained equations, it is necessary to add to them the conditions

$$q(r) \to q_a(r), \quad r \to \infty, \tag{6.71}$$

$$g(s) \to g_a(s), \quad s \to -\infty, \tag{6.72}$$

following from the principle of asymptotic matching (Van-Dyke [21]).

Thus, using the proposed asymptotic analysis of the EHL problem formula (6.55) for the film thickness is obtained, as well as two closed systems of integro-differential equations: in the inlet zone - equations (6.68), (6.69),

(6.71), (6.49), (6.50), and (6.64) for the starved lubrication regimes (or (6.66) for fully flooded lubrication regimes) for functions $q(r)$, $h_q(r)$, and constant A, and in the exit zone - equations (6.70), (6.69), (6.72), (6.49), (6.49), and (6.65) for the starved lubrication regimes (or (6.67) for fully flooded lubrication regimes) for functions $g(s)$, $h_g(s)$, and constant β_1 (see equation (6.35)).

Finally, it should be noted that estimate (6.30) involved in the definition of heavily loaded lubrication regimes follows from (6.32)-(6.35), and (6.55). Besides, estimates (6.45) follow from (6.30), (6.38), (6.39), and (6.55).

In some cases estimate (6.54) playing the validation role of the entire asymptotic analysis in the inlet and exit zones imposes some restrictions. For example, if the lubricant viscosity $\mu(p, Q)$ is determined by the generalized exponential function

$$\mu = \exp(Qp^m), \ Q \gg 1 \ for \ \omega \ll 1 \ (Q, \ m \geq 0), \tag{6.73}$$

then the restrictions on ϵ_q follow from (6.54) and take the form

$$\epsilon_q \ll Q^{-2/m} \ or \ \epsilon_q = O(Q^{-2/m}), \ \omega \ll 1. \tag{6.74}$$

These estimates should be taken into account together with estimates (6.61) and (6.62).

Estimates (6.61), (6.62), and (6.74) define, the so–called, pre-critical lubrication regimes. In general, for a given $\epsilon_q = \epsilon_q(V, Q)$ the pre-critical regimes are characterized by the following condition

$$\mu(\epsilon_q^{1/2}, Q) = O(1), \ \epsilon_q \ll 1, \ \omega \ll 1, \tag{6.75}$$

while the over-critical lubrication regimes are determined by the condition

$$\mu(\epsilon_q^{1/2}, Q) \gg 1, \ \epsilon_q \ll 1, \ \omega \ll 1. \tag{6.76}$$

Now, the detailed definitions for pre-critical starved and fully flooded lubrication regimes can be given. The pre-critical starved lubrication regimes are those for which the relationships (6.57) and (6.75) are valid. The fully flooded pre-critical lubrication regimes are those for which relations (6.58) and (6.75) are valid.

It should be noted that the described asymptotic analysis is valid if the small parameter $\omega = Q^{-1} \ll 1$ or $\omega = V \ll 1$ and conditions (6.75) (which in the case of viscosity μ from (6.73) coincides with conditions (6.74)) are valid.

After the asymptotic solutions of the problem in the inlet and exit zones are obtained, we can determine the uniformly valid approximate solution of the problem for pressure $p_u(x)$ and gap $h_u(x)$ in the form [21]

$$p_u(x) = \frac{\epsilon_q}{2} q(\frac{x-a}{\epsilon_q}) g(\frac{x-c}{\epsilon_q}), \ h_u(x) = h_q(\frac{x-a}{\epsilon_q}) h_g(\frac{x-c}{\epsilon_q}). \tag{6.77}$$

where a and c are determined by formulas (6.34) and (6.35). It can be clearly seen that in the Hertzian region $p_u(x)$ and $h_u(x)$ are practically equal to the

Hertzian pressure $\sqrt{1-x^2}$ and 1, respectively, while in the inlet and exit zones they practically coincide with $\epsilon_q^{1/2}q(r)$, $h_q(r)$ and $\epsilon_q^{1/2}g(s)$, $h_g(s)$, respectively. For starved lubrication with high precision $h_u(x) = 1$ in the entire contact region. For pre-critical regimes, these uniformly valid in the entire contact region functions can be used anywhere the numerical (i.e., also approximate) solutions for pressure $p(x)$ and gap $h(x)$ are used.

6.3.3 Asymptotic Analysis of the System of Equations (6.19)-(6.23)

For $\omega \ll 1$ in the inlet and exit zones, let us obtain the systems of asymptotically valid equations equivalent to those derived in Section 6.3.2. Suppose we have a heavily loaded lubricated contact and, therefore, estimate (6.30) is true and the inlet coordinate a satisfies equation (6.34).

Using (6.21) and the definition of a heavily loaded contact (6.30), from (6.19) in the external region, we can find

$$p(x) = p_0(x) + o(1), \quad p_0(x) = \sqrt{(x-a)(c-x)},$$

$$x - a \gg \epsilon_q \text{ and } c - x \gg \epsilon_g. \tag{6.78}$$

Taking the coordinate of the exit point c according to (6.35) and using assumption (6.36), we obtain that $p_0(x) = O(\epsilon_q^{1/2})$ for $r = O(1)$ and $s = O(1)$, $\omega \ll 1$. As a result, in the inlet and exit zones, we will search the solution in the form (6.38) and (6.39).

Let us consider equation (6.19) in the inlet zone. The integral in (6.19) can be expressed as a sum of three integrals: over the inlet and exit zones and over the external region. Formula (6.55) for H_0 and the following asymptotically valid equation for $q(r)$ in the inlet zone

$$q(r) = \sqrt{2r}\left[1 - \frac{1}{2\pi}\int\limits_{0}^{\infty} \frac{d}{dt}M_0(q, h_q, \mu_q, t)\frac{dt}{\sqrt{2t}(t-r)}\right] \tag{6.79}$$

can be obtained by estimating each of these integrals using expressions (6.34), (6.35), (6.38), and the first estimate in (6.45). A similar analysis in the exit zone leads to equality (6.56) and the following equation for $g(s)$

$$g(s) = \sqrt{-2s}\left[1 - \frac{1}{2\pi}\int\limits_{-\infty}^{0} \frac{d}{dt}M_0(g, h_g, \mu_g, t)\frac{dt}{\sqrt{-2t}(t-s)}\right], \tag{6.80}$$

where the functions $h_q(r)$, $h_g(s)$, $\mu_q(q)$, and $\mu_g(g)$ have the same meaning as earlier.

Using (6.30), (6.34), (6.35), (6.38), (6.39), (6.55), and (6.56) the asymptotic analysis of equations (6.20) and (6.21) leads to a system of linear algebraic equations for A^3 and β_1. The solution of this system has the form

$$1 = \pi\alpha_1/\int\limits_{0}^{\infty} \frac{d}{dt}M_0(q, h_q, \mu_q, t)\frac{dt}{\sqrt{2t}}, \tag{6.81}$$

$$\beta_1 = \frac{1}{\pi} \int_{-\infty}^{0} \frac{d}{dt} M_0(g, h_g, \mu_g, t) \frac{dt}{\sqrt{-2t}}. \tag{6.82}$$

Also, equation (6.81) can be represented in the form

$$A^3 = \pi \alpha_1 / \int_{0}^{\infty} \frac{d}{dt} \left(\frac{h_q^3}{\mu_q} \frac{dq}{dt} \right) \frac{dt}{\sqrt{2t}}. \tag{6.83}$$

Therefore, in the inlet zone the problem is reduced to the system of equations (6.69), (6.79), (6.81), and equations (6.64) or (6.66) (depending on whether the lubrication regime is starved or fully flooded) for functions $q(r)$, $h_q(r)$, and constant A. In the exit zone the problem is reduced to the system of equations (6.69), (6.80), (6.82) and equations (6.65) or (6.67) (depending on whether the lubrication regime is starved or fully flooded) for functions $g(s)$, $h_g(s)$, and constant β_1. Note that equality (6.52) is the condition of equivalence of the latter systems and the corresponding systems from Section 6.3.2.

It should be pointed out that asymptotic relationships (6.71) and (6.72) can be obtained from equations (6.69), (6.79)-(6.81). Obviously, the accepted assumption (6.30) is valid.

It can be shown that for $\mu = 1$ and regimes of starved lubrication the solutions of the asymptotically valid systems of equations of Sections 6.3.2 and 6.3.3 in the inlet and exit zones have the following properties

$$q(r, \alpha_1) = |\alpha_1|^{1/2} q(r/|\alpha_1|, -1),$$

$$g(s, \alpha_1) = |\alpha_1|^{1/2} g(s/|\alpha_1|, -1), \tag{6.84}$$

$$A(\alpha_1) = |\alpha_1|^{2/3} \theta(-\alpha_1) A(-1), \quad \beta_1(\alpha_1) = |\alpha_1| \beta_1(-1),$$

where $\theta(x)$ is a step function, $\theta(x) = 0$, $x \leq 0$ and $\theta(x) = 1$, $x > 0$. It follows from (6.82) that the formulation of the problem makes sense only for $a \leq -1$ ($\alpha_1 \leq 0$), which means that $H_0 \geq 0$ ($A \geq 0$). Obviously, $H_0 = A = 0$ for $a = -1$ ($\alpha_1 = 0$).

Let us finalize the results of the section. A detailed analysis of the structure of the solution of the problem for heavily loaded lubricated contacts in pre-critical regimes is proposed. It is shown that the contact region consists of three characteristic zones: the zone with predominating "elastic solution" - the Hertzian region and the narrow inlet and exit zones (boundary layers) compared to the size of the Hertzian region. In the inlet and exit zones, the influences of the solids' elasticity and viscous flow of lubricant are of the same order of magnitude. In the inlet and exit zones two equivalent asymptotically valid systems of equations describing the solution of the problem are derived.

The solutions of these systems depend on a smaller number of initial problem parameters than the solution of the original problem (6.5)-(6.9). It was shown that regimes of starved and fully flooded lubrication can be realized in the framework of pre-critical regimes. In all cases the characteristic size of the inlet and exit zones in a heavily loaded lubricated contact can be considered small compared to the size of the Hertzian region. The asymptotic formula for the lubrication film thickness is obtained using the fact that the terms of the equation describing lubrication flow and elastic effects are of the same order of magnitude. The lubrication film thickness depends on the particular regime of lubrication and the solution behavior in the inlet zone. The definition of pre-critical regimes is given. Based on that definition the limitations of the applied method were derived. The explicit expressions for coefficient A in the film thickness formula and the exit coordinate β_1 in terms of pressure and gap are derived. This analysis of the problem provides a better understanding of the structure of the solution of the problem in pre-critical lubrication regimes.

The asymptotic method presented in this section will be extended to non-Newtonian lubricants in Chapter 9.

6.4 How to Use the Numerical Solutions of the Asymptotic Equations in the Inlet and Exit Zones?

Now, let us consider how to practically use the numerical solutions of the asymptotic equations for pre-critical regimes in the inlet and exit zones and the analytical formulas for film thickness H_0. Suppose the lubricant viscosity is described by exponential Barus relationship $\mu(p) = \exp(Qp)$. The original EHL problem involves three parameters a, V, and Q. As we already know for heavily loaded EHL contacts $V \ll 1$ and/or $Q \gg 1$.

First, we need to understand whether we are dealing with a pre-critical or over-critical heavily loaded lubrication regime. Let us keep in mind that in a fully flooded pre-critical lubrication regime the characteristic size of the inlet zone is either $\epsilon_q = V^{2/5}$ if $Q_0 = QV^{1/5} = O(1)$ or $\epsilon_q = Q^{-2}$ otherwise while in a fully flooded over-critical lubrication regime the characteristic sizes of the inlet ϵ_q-zone and ϵ_0-zone are $\epsilon_q = (VQ)^{1/2}$ and $\epsilon_0 = Q^{-2}$ ($\epsilon_q \gg \epsilon_0$), respectively. By calculating the value of $Q_0 = QV^{1/5}$ we can determine the type of regime we have at hand, i.e., (A) if $Q_0 \ll 1$ or $Q_0 = O(1)$ then the regime is pre-critical or (B) if $Q_0 \gg 1$ then the regime is over-critical.

In case (A) (pre-critical regimes) when $V^{2/5} \ll Q^{-2}$ we can accept $\epsilon_q = Q^{-2}$, $\alpha_1 = (a+1)Q^2$, $Q_0 = V^{1/5}Q \ll 1$ while when $V^{2/5} = O(Q^{-2})$ we can accept either $\epsilon_q = Q^{-2}$, $\alpha_1 = (a+1)Q^2$, $Q_0 = 1$ or $\epsilon_q = V^{2/5}$, $\alpha_1 = (a+1)V^{-2/5}$, $Q_0 = V^{1/5}Q = O(1)$. Besides that $r = \frac{x-a}{\epsilon_q}$ and $s = \frac{x-c}{\epsilon_q}$. After that we can solve numerically the asymptotic equations for a pre-critical lubrication regime and obtain two sets of solutions: α_1, A, $q(r)$, $h_q(r)$ in the inlet zone and β_1, $g(s)$, $h_g(s)$ in the exit zone. Based on these solutions we determine $H_0 = AV^{3/5}$ if $\epsilon_q = V^{2/5}$ and $H_0 = A(VQ^{-4})^{1/3}$ if $\epsilon_q = Q^{-2}$. The uniformly valid asymptotic approximations of pressure and gap distributions are determined by formulas (6.77), i.e.,

$$p_u(x) = \tfrac{\epsilon_q}{2} q\left(\tfrac{x-a}{\epsilon_q}\right) g\left(\tfrac{x-c}{\epsilon_q}\right), \ \ h_u(x) = h_q\left(\tfrac{x-a}{\epsilon_q}\right) h_g\left(\tfrac{x-c}{\epsilon_q}\right), \qquad (6.85)$$

where a and c are determined by formulas (6.34) and (6.35), i.e., $a = -1 + \alpha_1 \epsilon_q$ and $c = 1 + \beta_1 \epsilon_q$. In cases when $\alpha_1 \ll 1$ we have starved lubrication regimes $(h_q(r) = h_g(s) = 1)$ and we can take $h_u(x) = 1$.

In case (B) (over-critical regimes) when $V^{2/5} \gg Q^{-2}$ we can expect to get $H_0 = A(VQ)^{3/4}$. For practical purposes we still can use solutions of the asymptotic equations obtained for the case of pre-critical lubrication regimes. At the same time we can expect that the lubrication film is formed in the inlet ϵ_q- and ϵ_0-zones of sizes $\epsilon_q = (VQ)^{1/2}$ and $\epsilon_0 = Q^{-2}$. The exit zones have the same respective sizes. The numerical step size should be chosen accordingly.

To summarize, it is important that as a byproduct of the asymptotic analysis right away we get a structural formula for the film thickness H_0. Only the coefficient of proportionality A is unknown in this formula. To determine this coefficient one needs to solve the asymptotic equations in the inlet zone or fit the structural formula for H_0 to the available experimental data.

6.5 Asymptotic Analysis of Heavily Loaded Contacts of Elastic Solids Lubricated by Compressible Fluids

Let us consider a steady isothermal EHL problem for a heavily loaded contact of two infinite parallel cylinders lubricated by a compressible Newtonian fluid. Under the assumptions of this chapter in dimensionless variables, the problem can be reduced to the Reynolds equation

$$\tfrac{d}{dx}\{\rho[M(p,h) - H_0 h]\} = 0, \ \ M(p,h) = \tfrac{H_0^3}{V}\tfrac{h^3}{\mu}\tfrac{dp}{dx}, \qquad (6.86)$$

which expresses the fact of fluid flux preservation through the gap between two solids. The rest of the problem equations coincide with equations (6.6)-(6.8). In equation (6.88) $\rho = \rho(p)$ is the fluid density expressed by a monotonically increasing function of pressure such that $\rho(0) = 1$ (here the density is scaled by the density ρ_0 at the atmospheric pressure).

By integrating equation (6.88) with boundary conditions (6.6), it can be rewritten in the form

$$\rho[M(p,h) - H_0 h] = -H_0.$$

Solving the latter equation for $H_0(h-1)$ and differentiating it the Reynolds equation can be rewritten in the form

$$\tfrac{d}{dx}[M_\rho(p,h) - H_0 h] = 0, \ \ M_\rho(p,h) = M(p,h) + H_0[\tfrac{1}{\rho(p)} - 1]. \qquad (6.87)$$

Therefore, the EHL problem for compressible lubricants can be reduced to equations (6.87) and (6.6)-(6.8). The latter system of equations is equivalent to the system of equations (6.19)-(6.23) in which function $M(p, h)$ is replaced by function $M_\rho(p, h)$ from (6.87).

After that the asymptotic analysis conducted in Section 6.3 can be applied in its entirety to the EHL problem in case of a compressible Newtonian lubricant by replacing $M(p, h)$ by $M_\rho(p, h)$ from (6.87). As a result of this analysis, we obtain two systems: (6.68), (6.69), (6.71), (6.49), (6.50), and (6.64) for the starved lubrication regimes (or (6.66) for fully flooded lubrication regimes) for functions $q(r)$, $h_q(r)$, and constant A in the inlet zone and (6.70), (6.69), (6.72), (6.49), (6.49), and (6.65) for the starved lubrication regimes (or (6.67) for fully flooded lubrication regimes) for functions $g(s)$, $h_g(s)$, and constant β_1 as well as the asymptotic formula (6.55) for the film thickness H_0. In addition to that we obtain two system of equations (6.69), (6.79), (6.81), and equations (6.64) or (6.66) (depending on whether the lubrication regime is starved or fully flooded) for functions $q(r)$, $h_q(r)$, and constant A in the inlet zone and (6.69), (6.80), (6.82) and equations (6.65) or (6.67) (depending on whether the lubrication regime is starved or fully flooded) for functions $g(s)$, $h_g(s)$, and constant β_1 in the exit zone. The latter pairs of systems of equations are equivalent. In these systems function M_0 from (6.69) must be replaced by function $M_{\rho 0}$ as follows

$$M_{\rho 0}(p, h, \mu, x) = M_0(p, h, \mu, x) + (\tfrac{V^{2/5}}{\epsilon_q})^{5/6} A[\tfrac{1}{\rho(p)} - 1]$$

$$= A^3 \tfrac{h^3}{\mu} \tfrac{dp}{dx} + (\tfrac{V^{2/5}}{\epsilon_q})^{5/6} A[\tfrac{1}{\rho(p)} - 1]. \tag{6.88}$$

In formula (6.55) for the the film thickness H_0 constant A besides other parameters depends on the behavior of lubricant density $\rho(p)$ in the inlet zone.

This asymptotic analysis is valid as long as in the Hertzian region

$$(\tfrac{V^{2/5}}{\epsilon_q})^{5/6}[\tfrac{1}{\rho(p)} - 1] \ll 1 \ for \ x - a \gg \epsilon_q \ and \ c - x \gg \epsilon_g, \tag{6.89}$$

and in the inlet and exit zones

$$(\tfrac{V^{2/5}}{\epsilon_q})^{5/6}[\tfrac{1}{\rho(\epsilon_q^{1/2})} - 1] = O(1), \ \omega \ll 1. \tag{6.90}$$

Numerical solutions of the asymptotic equations for incompressible fluids show (see Section 6.9) that for $\alpha_1 < 0$ we have $A > 0$. Therefore, based on formula (6.81), the fact that $\rho(p)$ is a monotonically increasing function of p and that $dq(r)/dr > 0$, we can expect that the value of constant A for the case of a compressible fluid is smaller than the one in the corresponding case of an incompressible fluid.

6.6 Asymptotic Analysis of a Heavily Loaded Lubricated Contact of Elastic Solids. Over-Critical Lubrication Regimes for Newtonian Liquids

In this section we will consider the same problem as in the previous one, i.e., the plane problem of isothermal lubrication for heavily loaded elastic bodies with Newtonian fluid. However, this time we will consider the over-critical lubrication regimes. The problem will be analyzed using the methods of matched asymptotic expansions. The structure of the contact area will be investigated. It will be ascertained that there exist two boundary layers (two inlet zones) with different characteristic sizes at one side of the external (Hertzian) region and two boundary layers (two exit zones) at its opposite side. In each of the larger inlet and exit zones as well as in the Hertzian region, the asymptotically valid equations for the major terms of asymptotic expansions of the unknown quantities will be considered and some solved for the case of a Barus lubricant viscosity. Some structural asymptotic formulas for lubrication film thickness will be obtained. The approximate Ertel-Grubin's and Greenwood's methods and their prior assumptions will be analyzed in detail.

6.6.1 Problem Formulation

Let us consider a classic plane isothermal problem for an incompressible Newtonian lubricant in a heavily loaded contact of smooth elastic cylinders. Under these conditions in terms of the dimensionless variables (6.3) and parameters (6.4), the problem is reduced to the classic system of integro-differential equations (6.5)-(6.8). In Section 6.3 the problem formulation is described in more detail as well as the main mathematical difficulties in analysis of this problem are given. For over-critical lubrication regimes the possibility to separate the influence of different regions of a lubricated contact on the solution of the problem is even more crucial than for pre-critical regimes considered in Section 6.3. The reason for that is the integral nature of the gap dependence on pressure as well as the fact that in over-critical lubrication regimes the solution structure is far from obvious.

The purpose of this study is to progress in the understanding of the considered classic problem for over-critical regimes without using any prior assumptions. We will apply the methods of matched asymptotic expansions (Kevorkian and Cole [21]) to analyzing problem (6.5)-(6.8), which will allow to take into account the mutual influence of elastic and hydrodynamic effects properly. The method allows analyzing the structure of the solution, obtaining asymptotically valid analytical solutions in the larger inlet and exit zones, getting analytical approximate solutions in the Hertzian region, and for different regimes obtaining an asymptotically valid estimate for the lubrication film thickness H_0. However, this analysis stops short of establishing the asymptotically valid equations in smaller inlet and exit zones. Nonetheless, the analysis will provide a solid basis for a number of important conclusions. Moreover, in Section 6.10 we will see that the asymptotically valid equations derived for pre-critical lubrication regimes under certain conditions also describe over-critical lubrication regimes.

It is necessary to note that in papers by Kudish [23] - [26], [28] on heavily loaded lubrication contacts mostly pre-critical regimes were discussed for which the estimate (6.75) (Kudish [24, 26, 28]) holds. The maximum characteristic size of the inlet zone for which the lubrication regime is still a pre-critical one we will call the critical characteristic size of the inlet zone $\epsilon_q = \epsilon_0$. To determine ϵ_0 we need to consider the definitions of the pre- and over-critical regimes (6.75) and (6.76), respectively. The value of ϵ_0 is such that for any $\epsilon_q \gg \epsilon_0$ estimate (6.76) is valid and for $\epsilon_q = \epsilon_0$ estimate (6.75) is valid, respectively. If it appears that $\epsilon_0 \ll 1$ for the given problem parameters, then it means that the viscosity μ depends on a large parameter Q. Thus, if $\epsilon_0 \ll 1$, then for $\epsilon_q = O(\epsilon_0)$ we have pre-critical regimes and for $\epsilon_q \gg \epsilon_0$ – over-critical regimes. For example, let $\mu = \exp(Qp^m)$, where m is a positive constant. Then $\epsilon_0 = Q^{-2/m} \ll 1$ for $\omega = Q^{-1} \ll 1$ (see also Kudish [23] - [26], [28]). Obviously, if $\epsilon_0 \gg 1$ or $\epsilon_0 = O(1)$ for $\omega \ll 1$ then the over-critical regimes cannot be realized and pre-critical lubrication regimes occur.

It can be shown that the conditions that indicate the occurrence of over-critical regimes are identical to those that are considered by Ertel [6], Grubin [7], Crook [20], and Greenwood [29]. Some classifications of lubrication regimes are given by Greenwood [29] and Johnson [30].

6.6.2   Structure of the Solution. Asymptotic Solutions in the External Region

Let us assume that $\mu = \mu(p, Q)$ and for $p = O(1)$ the viscosity is very high, i.e., $\mu(p, Q) \gg 1$ for some small parameter $\omega \ll 1$. The specific ways the small parameter can be introduced will be considered later. One of these ways is to define it as $\omega = Q^{-1} \ll 1$.

Let us consider a heavily loaded contact for which almost everywhere in the contact region the elastic deformation effects prevail over the effects caused by the lubrication viscous flow. That results in the smallness of the first term of equation (6.5) compared to 1 everywhere in the contact except for the narrow inlet and exit zones (which are located next to the inlet $x = a$ and exit $x = c$ points).

As in the preceding section a contact will be called heavily loaded if far away from the inlet and exit points estimates (6.30) are valid. We will take the inlet coordinate in the form similar to (6.34)

$$a = a_0 + \alpha_1 \epsilon_q + \dots, \quad \alpha_1, \ a_0 = O(1), \ \omega \ll 1, \qquad (6.91)$$

where a_0 is unknown and is defined by equation (6.95), α_1 is a given non-positive constant, and ϵ_q is a given characteristic size of the inlet zone, $\epsilon_q \ll 1$ for $\omega \ll 1$ (ϵ_q may depend on the problem parameters Q and V differently). If $\epsilon_q = O(min(V^{2/5}, \epsilon_0))$ then this is a pre-critical regime, which can be studied by the method described in the preceding section. Further, we will assume that

$$\epsilon_q(\omega) \gg \epsilon_0(\omega), \ \omega \ll 1. \tag{6.92}$$

That means that the lubrication regime is over-critical and the solution structure is different from the one analyzed in the preceding section. Thus, in order to analyze the problem, we should develop another asymptotic approach different from the one presented in Section 6.6.

The exit coordinate c will be searched in the form

$$c = c_0 + \beta_1 \epsilon_q + \ldots, \ \beta_1, \ c_0 = O(1), \ \omega \ll 1, \tag{6.93}$$

where constant c_0 is unknown and is defined by equation (6.96) while constant β_1 is unknown and should be determined from the solution of the problem. The points with $x = a_0$ and $x = c_0$ are close to -1 and 1, respectively, and they represent the left and right end-points of the Hertzian region where the external bounded solution $p_{ext}(x)$ of the problem is asymptotically valid.

According to the principle of matched asymptotic expansions (Kevorkian and Cole [21]) for $\epsilon_q = O(\epsilon_0)$, $\omega \ll 1$, in the inlet and exit zones the estimate $p(x) = O(\epsilon_q^{1/2})$ can be obtained. Therefore, it is natural to assume that for $\epsilon_q \gg \epsilon_0$ in the inlet and exit zones the estimate

$$p(x) = O(\epsilon_0^{1/2}), \ x - a_0 = O(\epsilon_q) \ and \ x - c_0 = O(\epsilon_q) \ for \ \omega \ll 1 \tag{6.94}$$

also holds.

Also, let us determine the left and right boundaries of the external (Hertzian) region $x = a_0$ and $x = c_0$

$$a_0 = -1 + \alpha_0 \epsilon_0^{1/2} \epsilon_q^{1/2} + \ldots, \ \alpha_0 = O(1), \ \omega \ll 1, \tag{6.95}$$

$$c_0 = 1 + \beta_0 \epsilon_0^{1/2} \epsilon_q^{1/2} + \ldots, \ \beta_0 = O(1), \ \omega \ll 1, \tag{6.96}$$

by satisfying the conditions of no singularity in the solution $p_{ext}(x)$ of the external problem for all $x \in [a_0, c_0]$, i.e., in particular, for $x = a_0$ and $x = c_0$. The values of constants α_0 and β_0 are unknown and must be determined in accordance with the mentioned conditions of no singularity at $x = a_0$ and $x = c_0$.

Using the pressure estimates (6.94) in all inlet and exit zones the solution of the problem in the inlet and exit ϵ_q-zones will be searched in the form

$$p(x) = \epsilon_0^{1/2} q(r) + o(\epsilon_0^{1/2}), \ q(r) = O(1),$$

$$r = \frac{x - a_0}{\epsilon_q} = O(1), \ \omega \ll 1, \tag{6.97}$$

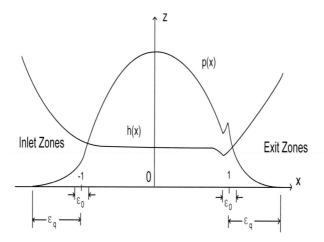

FIGURE 6.9
The schematic view of a heavily loaded contact in an over-critical lubrication regime.

$$p(x) = \epsilon_0^{1/2} g(s) + o(\epsilon_0^{1/2}), \ g(s) = O(1),$$

$$s = \tfrac{x - c_0}{\epsilon_q} = O(1), \ \omega \ll 1,$$

$$(6.98)$$

where r and s are the local independent variables in the inlet and exit ϵ_q-zones, which are adjacent to the inlet $x = a$ and exit $x = c$ points and have the characteristic size of ϵ_q, $q(r)$ and $g(s)$, are the unknown functions representing the leading terms of the pressure asymptotic expansions in the inlet and exit ϵ_q-zones.

Further, it can be shown that asymptotic expansions (6.97) and (6.98) cannot be matched with the asymptotic expansions of the external solution $p_{ext}(x)$. It is caused by ignoring the presence of small inlet and exit zones (ϵ_0-zones) with characteristic size of ϵ_0 located around points $x = a_0$ and $x = c_0$. In the inlet and exit ϵ_0-zones, the effects of elasticity and lubrication flow are of the same order of magnitude. The solutions in the inlet ϵ_0- and ϵ_q-zones must match as well as the inner asymptotic of the external solution $p_{ext}(x)$ and the solution in the inlet ϵ_0-zone (see Fig. 6.9). Similarly, the solutions in the exit ϵ_0- and ϵ_q-zones must match as well as the inner asymptotic of the external solution $p_{ext}(x)$ and the solution in the exit ϵ_0-zone. Note that the absence of singularities in $p_{ext}(x)$ at $x = a_0$ and $x = c_0$ leads to the estimate $p_{ext}(x) = O(\epsilon_0^{1/2})$ for $x - a_0 = O(\epsilon_0)$ and $x - c_0 = O(\epsilon_0)$, $\omega \ll 1$. That allows for matching $p_{ext}(x)$ with the solutions in the inlet and exit ϵ_0-zones.

Based on the described behavior of the solution $p_{ext}(x)$ in the inlet and exit ϵ_0-zones, the solution of the problem can be searched in the form

$$p(x) = \epsilon_0^{1/2} q_0(r_0) + o(\epsilon_0^{1/2}), \quad q_0(r_0) = O(1),$$

$$r_0 = \frac{x-a_0}{\epsilon_0} = O(1), \quad \omega \ll 1, \tag{6.99}$$

$$p(x) = \epsilon_0^{1/2} g_0(s_0) + o(\epsilon_0^{1/2}), \quad g_0(s_0) = O(1),$$

$$s_0 = \frac{x-c_0}{\epsilon_0} = O(1), \quad \omega \ll 1, \tag{6.100}$$

where $q_0(r_0)$ and $g_0(s_0)$ are unknown functions, which should be determined, and r_0 and s_0 are the local point coordinates in the inlet and exit ϵ_0-zones, respectively.

For further analysis it is more convenient to re-scale the problem equations by using as the characteristic dimensional film thickness the central film thickness $h_0 = h(x = 0.5(a_0 + c_0))$ instead of the exit film thickness h_e. That will cause only one of the dimensionless problem equations to change as follows

$$H_0(h-1) = x^2 - (\tfrac{a_0+c_0}{2})^2 + \frac{2}{\pi} \int\limits_a^c p(t) \ln \left| \frac{\frac{a_0+c_0}{2}-t}{x-t} \right| dt, \tag{6.101}$$

where the dimensionless central film thickness H_0 is given by the formula

$$H_0 = \frac{2R' h_0}{a_H^2}. \tag{6.102}$$

Based on asymptotic representations (6.93)-(6.100) for some $\eta(\omega)$ such that $\eta/\epsilon_0 \gg 1$ and $\eta/\epsilon_q \ll 1$ for $\omega \ll 1$, we have

$$\int\limits_a^c p(t) \ln \left| \frac{\frac{a_0+c_0}{2}-t}{x-t} \right| dt$$

$$= \left\{ \int\limits_a^{a_0-\eta} + \int\limits_{a_0-\eta}^{a_0+\eta} + \int\limits_{a_0+\eta}^{c_0-\eta} + \int\limits_{c_0-\eta}^{c_0+\eta} + \int\limits_{c_0+\eta}^c \right\} p(t) \ln \left| \frac{\frac{a_0+c_0}{2}-t}{x-t} \right| dt$$

$$= \epsilon_0^{1/2} \epsilon_q \int\limits_{\alpha_1-\alpha_0(\epsilon_0/\epsilon_q)^{1/2}-\alpha_2\epsilon_0/\epsilon_q+\ldots}^{-\eta/\epsilon_q} q(r) \ln \left| \frac{\frac{c_0-a_0}{2}-\epsilon_q r}{x-a_0-\epsilon_q r} \right| dr$$

$$+\epsilon_0^{3/2} \int\limits_{-\eta/\epsilon_0}^{\eta/\epsilon_0} q_0(r_0) \ln \left| \frac{\frac{c_0-a_0}{2}-\epsilon_0 r_0}{x-a_0-\epsilon_0 r_0} \right| dr_0 \tag{6.103}$$

$$+ \int\limits_{a_0+\eta}^{c_0-\eta} p_{ext}(t) \ln \left| \frac{\frac{a_0+c_0}{2}-t}{x-t} \right| dt$$

$$+\epsilon_0^{3/2} \int\limits_{-\eta/\epsilon_0}^{\eta/\epsilon_0} g_0(s_0) \ln \left| \frac{\frac{a_0-c_0}{2}-\epsilon_0 s_0}{x-c_0-\epsilon_0 s_0} \right| ds_0$$

$$+\epsilon_0^{1/2}\epsilon_q \int\limits_{\eta/\epsilon_q}^{\beta_1+\cdots} g(s)\ln\left|\frac{\frac{a_0-c_0}{2}-\epsilon_q s}{x-c_0-\epsilon_q s}\right|ds+\cdots$$

$$=\epsilon_0^{1/2}\epsilon_q \int\limits_{\alpha_1-\alpha_0(\epsilon_0/\epsilon_q)^{1/2}-\alpha_2\epsilon_0/\epsilon_q+\cdots}^{0} q(r)\ln\left|\frac{\frac{c_0-a_0}{2}-\epsilon_q r}{x-a_0-\epsilon_q r}\right|dr$$

$$+\epsilon_0^{3/2}\int\limits_{-\eta/\epsilon_0}^{\eta/\epsilon_0}[q_0(r_0)-q(0)\theta(-r_0)-\epsilon_0^{-1/2}p_{ext}(a_0+\epsilon_0 r_0)\theta(r_0)]$$

$$\times\ln\left|\frac{\frac{c_0-a_0}{2}-\epsilon_0 r_0}{x-a_0-\epsilon_0 r_0}\right|dr_0+\int\limits_{a_0}^{c_0}p_{ext}(t)\ln\left|\frac{\frac{a_0+c_0}{2}-t}{x-t}\right|dt$$

$$+\epsilon_0^{3/2}\int\limits_{-\eta/\epsilon_0}^{\eta/\epsilon_0}[g_0(s_0)-g(0)\theta(s_0)-\epsilon_0^{-1/2}p_{ext}(c_0+\epsilon_0 s_0)\theta(-s_0)]$$

$$\times\ln\left|\frac{\frac{a_0-c_0}{2}-\epsilon_0 s_0}{x-c_0-\epsilon_0 s_0}\right|ds_0+\epsilon_0^{1/2}\epsilon_q \int\limits_{0}^{\beta_1+\cdots} g(s)\ln\left|\frac{\frac{a_0-c_0}{2}-\epsilon_q s}{x-c_0-\epsilon_q s}\right|ds+\cdots,$$

where $\theta(x)$ is a step-function such that $\theta(x)=0$ for $x\le 0$ and $\theta(x)=1$ for $x>0$. Here we made a natural assumption (which is substantiated in Section 6.6.4) that in the inlet and exit ϵ_0-zones $q(r)=q(0)+O(\epsilon_0)$ and $g(s)=g(0)+O(\epsilon_0)$, respectively.

Now, let us obtain the asymptotic of the bounded solution of the problem in the external region $[a_0,c_0]$. However, first we need to find a singular solution of the problem. Suppose that in the latter case the boundaries of the external contact region $x=a_0$ and $x=c_0$ are fixed. Let us assume that in the external region the estimate

$$M(p,h)\ll\epsilon_0^{3/2}\ for\ x-a_0\gg\epsilon_0,\ c_0-x\gg\epsilon_0,\ \omega\ll 1,\qquad(6.104)$$

holds (see a similar estimate in (6.30), Kudish [28]), where $M(p,h)=\frac{H_0^3}{V}\frac{h^3}{\mu}\frac{dp}{dx}$. Estimating the integrals in (6.7) and (6.8), and using (6.34), (6.92), (6.93), (6.95)-(6.104) for $\omega\ll 1$ in the external region we obtain the equations

$$x^2-(\tfrac{a_0+c_0}{2})^2+\tfrac{2}{\pi}\int\limits_{a_0}^{c_0}p_{ext}(t)\ln\left|\frac{\frac{a_0+c_0}{2}-t}{x-t}\right|dt$$

$$+\epsilon_0^{1/2}\epsilon_q\tfrac{2}{\pi}\int\limits_{\alpha_1-\alpha_0(\epsilon_0/\epsilon_q)^{1/2}+\cdots}^{0} q(r)\ln\left|\frac{\frac{c_0-a_0}{2}-\epsilon_q r}{x-a_0-\epsilon_q r}\right|dr$$

$$+\epsilon_0^{1/2}\epsilon_q\tfrac{2}{\pi}\int\limits_{0}^{\beta_1+\cdots} g(s)\ln\left|\frac{\frac{a_0-c_0}{2}-\epsilon_q s}{x-c_0-\epsilon_q s}\right|ds+\cdots=0,$$

(6.105)

$$\int\limits_{a_0}^{c_0}p_{ext}(t)dt$$

$$+\epsilon_0^{1/2}\epsilon_q\left\{\int_{\alpha_1-\alpha_0(\epsilon_0/\epsilon_q)^{1/2}+\dots}^{0} q(r)dr + \int_0^{\beta_1+\dots} g(s)ds\right\} + \dots = \frac{\pi}{2}.$$

It can be shown that equations (6.105) approximate the original problem (6.5)-(6.8) in the external region with the accuracy of $O(\epsilon_0^{3/2})$.

The solution of equations (6.105) will be searched by regular perturbation methods in the form

$$p_{ext}(x) = p_0(x) + \epsilon_0^{1/2}\epsilon_q p_1(x) + o(\epsilon_0^{1/2}\epsilon_q), \tag{6.106}$$

$$x - a_0 \gg \epsilon_0, \quad c_0 - x \gg \epsilon_0.$$

Substituting (6.106) in (6.105) and equating coefficients of the same powers of ϵ_0 and ϵ_q (which are considered to be independent, $\epsilon_q \gg \epsilon_0$) for $p_0(x)$ and $p_1(x)$, we obtain the following sets of equations:

$$x^2 - (\tfrac{a_0+c_0}{2})^2 + \tfrac{2}{\pi}\int_{a_0}^{c_0} p_0(t)\ln\left|\frac{\frac{a_0+c_0}{2}-t}{x-t}\right|dt = 0, \quad \int_{a_0}^{c_0} p_0(t)dt = \tfrac{\pi}{2}, \tag{6.107}$$

$$\int_{a_0}^{c_0} p_1(t)\ln\left|\frac{\frac{a_0+c_0}{2}-t}{x-t}\right|dt$$

$$= -\int_{\alpha_1-\alpha_0(\epsilon_0/\epsilon_q)^{1/2}+\dots}^{0} q(r)\ln\left|\frac{\frac{c_0-a_0}{2}-\epsilon_q r}{x-a_0-\epsilon_q r}\right|dr$$

$$\tag{6.108}$$

$$- \int_0^{\beta_1+\dots} g(s)\ln\left|\frac{\frac{a_0-c_0}{2}-\epsilon_q s}{x-c_0-\epsilon_q s}\right|ds,$$

$$\int_{a_0}^{c_0} p_1(t)dt = -\int_{\alpha_1-\alpha_0(\epsilon_0/\epsilon_q)^{1/2}+\dots}^{0} q(r)dr - \int_0^{\beta_1+\dots} g(s)ds.$$

The singular solutions of systems (6.107) and (6.108) have the form [27] (also see Section 6.3.2)

$$p_0(x) = \sqrt{(x-a_0)(c_0-x)} + \frac{1+2a_0c_0+(c_0-a_0)^2/4-(c_0+a_0)x}{2\sqrt{(x-a_0)(c_0-x)}}, \tag{6.109}$$

$$p_1(x) = \frac{1}{\pi\sqrt{1-y^2}}\left\{-\frac{2}{c_0-a_0}\left[\int_{\alpha_1-\alpha_0(\epsilon_0/\epsilon_q)^{1/2}+\dots}^{0} q(r)dr\right.\right.$$

$$\left.+\int_0^{\beta_1+\dots} g(s)ds\right] - \frac{1}{\pi}\int_{-1}^{1}\frac{\sqrt{1-t^2}dt}{t-y}\int_{\alpha_1-\alpha_0(\epsilon_0/\epsilon_q)^{1/2}+\dots}^{0}\frac{q(r)dr}{\frac{c_0-a_0}{2}(1+t)-\epsilon_q r} \tag{6.110}$$

$$-\frac{1}{\pi}\int_{-1}^{1}\frac{\sqrt{1-t^2}dt}{t-y}\int_0^{\beta_1+\dots}\frac{g(s)ds}{\frac{c_0-a_0}{2}(t-1)-\epsilon_q s}\right\}, \quad y = \frac{2x-a_0-c_0}{c_0-a_0}.$$

In the last two integrals in (6.110), the integrands have singularities at points $t = y$ and at two discrete points $r = 0$, $t = -1$ ($t \geq -1$, $r \leq 0$) and $s = 0$, $t = 1$ ($t \leq 1$, $s \geq 0$), respectively. We will show that it is possible to change the order of integration in the repeated integrals in (6.110) as in integrals of regular functions

$$
\int_{-1}^{1} \frac{\sqrt{1-t^2}\,dt}{t-y} \int_{\alpha_1 - \alpha_0(\epsilon_0/\epsilon_q)^{1/2}+\ldots}^{0} \frac{q(r)dr}{\frac{c_0 - a_0}{2}(1+t) - \epsilon_q r}
$$

$$
= \int_{\alpha_1 - \alpha_0(\epsilon_0/\epsilon_q)^{1/2}+\ldots}^{0} q(r)dr \int_{-1}^{1} \frac{\sqrt{1-t^2}\,dt}{(t-y)[\frac{c_0 - a_0}{2}(1+t) - \epsilon_q r]},
$$

(6.111)

$$
\int_{-1}^{1} \frac{\sqrt{1-t^2}\,dt}{t-y} \int_{0}^{\beta_1 + \ldots} \frac{g(s)ds}{\frac{c_0 - a_0}{2}(t-1) - \epsilon_q s}
$$

$$
= \int_{0}^{\beta_1 + \ldots} g(s)ds \int_{-1}^{1} \frac{\sqrt{1-t^2}\,dt}{(t-y)[\frac{c_0 - a_0}{2}(t-1) - \epsilon_q s]}.
$$

Let us first consider the integral involving $q(r)$. By introducing a small enough value $\epsilon < 0$ this integral can be transformed into a sum of three integrals

$$
\int_{-1}^{1} \frac{\sqrt{1-t^2}\,dt}{t-y} \int_{\alpha_1 - \alpha_0(\epsilon_0/\epsilon_q)^{1/2}+\ldots}^{0} \frac{q(r)dr}{\frac{c_0 - a_0}{2}(1+t) - \epsilon_q r}
$$

$$
= \int_{\alpha_1 - \alpha_0(\epsilon_0/\epsilon_q)^{1/2}+\ldots}^{\epsilon} q(r)dr \int_{-1}^{1} \frac{\sqrt{1-t^2}\,dt}{(t-y)[\frac{c_0 - a_0}{2}(1+t) - \epsilon_q r]}
$$

(6.112)

$$
- \frac{q(0)}{\epsilon_q} \int_{-1}^{1} \frac{\sqrt{1-t^2}}{t-y} \ln \left| \frac{t+1}{t+1 - \frac{2\epsilon\epsilon_q}{c_0 - a_0}} \right| \, dt
$$

$$
+ \int_{\epsilon}^{0} [q(r) - q(0)]dr \int_{-1}^{1} \frac{\sqrt{1-t^2}\,dt}{(t-y)[\frac{c_0 - a_0}{2}(1+t) - \epsilon_q r]}.
$$

Here in the first and the third integrals the order of integration is changed (Gakhov [31]). Similarly, we obtain

$$
\int_{-1}^{1} \frac{\sqrt{1-t^2}\,dt}{t-y} \int_{0}^{\beta_1 + \ldots} \frac{g(s)ds}{\frac{c_0 - a_0}{2}(t-1) - \epsilon_q s}
$$

$$
= \int_{-\epsilon}^{\beta_1 + \ldots} g(s)ds \int_{-1}^{1} \frac{\sqrt{1-t^2}\,dt}{(t-y)[\frac{c_0 - a_0}{2}(t-1) - \epsilon_q s]}
$$

(6.113)

$$
- \frac{g(0)}{\epsilon_q} \int_{-1}^{1} \frac{\sqrt{1-t^2}}{t-y} \ln \left| \frac{t-1 + \frac{2\epsilon\epsilon_q}{c_0 - a_0}}{t-1} \right| \, dt
$$

$$+ \int\limits_{0}^{-\epsilon} [g(s) - g(0)]ds \int\limits_{-1}^{1} \frac{\sqrt{1-t^2}\,dt}{(t-y)[\frac{c_0-a_0}{2}(t-1)-\epsilon_q s]}.$$

Using the expressions for the integrals

$$\int\limits_{-1}^{1} \frac{\sqrt{1-t^2}\,dt}{(t-y)[\frac{c_0-a_0}{2}(t+1)-\epsilon_q r]} = -\frac{2\pi}{c_0-a_0}\left\{1 - \frac{\sqrt{t_0^2-1}}{y-t_0}\right\},$$

$$\int\limits_{-1}^{1} \frac{\sqrt{1-t^2}\,dt}{(t-y)[\frac{c_0-a_0}{2}(t-1)-\epsilon_q s]} = -\frac{2\pi}{c_0-a_0}\left\{1 + \frac{\sqrt{t_1^2-1}}{y-t_1}\right\}, \qquad (6.114)$$

$$t_0 = -1 + \frac{2\epsilon_q r}{c_0-a_0} \le -1, \; t_1 = 1 + \frac{2\epsilon_q s}{c_0-a_0} \ge 1,$$

and calculating the inner integrals in (6.112) and (6.113) by using substitution $t = (1-\tau^2)/(1+\tau^2)$ and the integral Bierens De Haan [32] (p. 194, No. 15)

$$\int\limits_{0}^{\infty} \frac{\ln(p^2+\tau^2)d\tau}{q^2-\tau^2} = -\frac{\pi}{q}\arctan\frac{q}{p}$$

we obtain [31, 33]

$$\int\limits_{-1}^{1} \frac{\sqrt{1-t^2}}{t-y}\ln\left|\frac{t+1}{t-\tau_0}\right|dt = 2\pi\sqrt{1-y^2}\{\tfrac{1}{2}\arccos y$$

$$+ \arctan\sqrt{\frac{1-y}{1+y}\frac{\tau_0+1}{\tau_0-1}} - \arctan\sqrt{\frac{1-y}{1+y}}\} - \pi(1+\tau_0+\sqrt{\tau_0^2-1}) \qquad (6.115)$$

$$+\pi y\ln|\tau_0 - \sqrt{\tau_0^2-1}|, \; \tau_0 = -1 + \frac{2\epsilon\epsilon_q}{c_0-a_0}, \; |y|<1, \; \epsilon<0,$$

$$\int\limits_{-1}^{1} \frac{\sqrt{1-t^2}}{t-y}\ln\left|\frac{t-\tau_1}{t-1}\right|dt = 2\pi\sqrt{1-y^2}\{\arctan\sqrt{\frac{1-y}{1+y}}$$

$$- \arctan\sqrt{\frac{1-y}{1+y}\frac{\tau_1+1}{\tau_1-1}} + \frac{\pi}{2} - \tfrac{1}{2}\arccos y\} - \pi(1-\tau_1+\sqrt{\tau_1^2-1}) \qquad (6.116)$$

$$-\pi y\ln|\tau_1 + \sqrt{\tau_1^2-1}|, \; \tau_1 = 1 - \frac{2\epsilon\epsilon_q}{c_0-a_0}, \; |y|<1, \; \epsilon<0.$$

Using formulas (6.114), (6.115), and (6.116) in expression (6.110) for function $p_1(x)$ and taking the limit as $\varepsilon \to -0$ for any fixed value of y ($|y|<1$), we arrive at

$$p_1(x) = \frac{2}{\pi(c_0-a_0)\sqrt{1-y^2}}\left\{-\int\limits_{\alpha_1-\alpha_0(\epsilon_0/\epsilon_q)^{1/2}+\dots}^{0} q(r)\frac{\sqrt{t_0^2(r)-1}}{y-t_0(r)}dr\right.$$

$$\left.+ \int\limits_{0}^{\beta_1+\dots} g(s)\frac{\sqrt{t_1^2(s)-1}}{y-t_1(s)}ds\right\} + \frac{2}{\pi\varepsilon_q}[q(0)-g(0)]\{\tfrac{1}{2}\arccos y \qquad (6.117)$$

$$- \arctan\sqrt{\frac{1-y}{1+y}}\}.$$

However, taking into account the identity $\frac{1}{2}\arccos y - \arctan\sqrt{\frac{1-y}{1+y}} = 0$, for any y ($|\,y\,| < 1$), we find the final expression for $p_1(x)$ in the form

$$p_1(x) = \frac{2}{\pi(c_0-a_0)\sqrt{1-y^2}} \left\{ - \int\limits_{\alpha_1-\alpha_0(\epsilon_0/\epsilon_q)^{1/2}+\dots}^{0} q(r)\frac{\sqrt{t_0^2(r)-1}}{y-t_0(r)}dr \right.$$

$$\left. + \int\limits_{0}^{\beta_1+\dots} g(s)\frac{\sqrt{t_1^2(s)-1}}{y-t_1(s)}ds \right\}, \tag{6.118}$$

$$t_0 = -1 + \frac{2\epsilon_q r}{c_0-a_0} \leq -1, \; t_1 = 1 + \frac{2\epsilon_q s}{c_0-a_0} \geq 1.$$

It can be shown by the direct substitution that expression (6.118) for $p_1(x)$ satisfies equations (6.108). To do that and some other calculations, we will use the following integrals:

$$\int\limits_{-1}^{1} \frac{\ln|y-t|dt}{\sqrt{1-t^2}(t-x)} = -\frac{\pi\,sign(x)}{\sqrt{x^2-1}} \ln\frac{x-y}{x+sign(x)\sqrt{x^2-1}}, \; |\,x\,| > 1, \; |\,y\,| < 1,$$

$$\int\limits_{-1}^{1} \frac{\ln|y-t|dt}{\sqrt{1-t^2}(t-x)} = -\frac{\pi\,sign(x)}{\sqrt{x^2-1}} \ln\frac{xy-1+\sqrt{(y^2-1)(x^2-1)}}{|x|+\sqrt{x^2-1}}, \tag{6.119}$$

$$|\,x\,| > 1, \; |\,y\,| > 1,$$

the expressions for which are obtained by using substitution $t = (1-\tau^2)/(1+\tau^2)$ and the integral Bierens De Haan [32] (p. 194, No. 15) $\int\limits_{0}^{\infty} \frac{\ln(p^2+\tau^2)d\tau}{q^2-\tau^2}$ presented earlier as well as the other two integrals Bierens De Haan [32] (p. 193, No. 13 and p. 194, No. 16)

$$\int\limits_{0}^{\infty} \frac{\ln(p^2+\tau^2)d\tau}{q^2+\tau^2} = \frac{\pi}{q}\ln(p+q), \; \int\limits_{0}^{\infty} \frac{\ln(p^2-\tau^2)^2 d\tau}{q^2+\tau^2} = \frac{\pi}{q}\ln(p^2+q^2).$$

Furthermore, it is interesting to note that the outlined procedure of finding $p_1(x)$ is equivalent but different from the one used by Kuznetsov in [34] for solution of a dry (without lubrication) contact problem with a known additional load applied outside of a contact.

Let us obtain a bounded solution $p_{ext}(x)$ of system (6.105) with the precision of $o(\epsilon_0^{1/2}\epsilon_q)$ using formula (6.106) and solutions (6.109) and (6.118). It can be done by choosing the proper values for constants α_0 and β_0 involved in expressions (6.95) and (6.96) for a_0 and c_0, respectively. It is obvious that for $p_{ext}(x)$ to be bounded for any y ($|\,y\,| < 1$) it is necessary to require that the sum of the second terms in $p_0(x)$ and $\epsilon_0^{1/2}\epsilon_q p_1(x)$ be a regular function at $y = \pm 1$ with the precision of $o(\epsilon_0^{1/2}\epsilon_q)$. That leads to the requirement that in $p_{ext}(x) = p_0(x) + \epsilon_0^{1/2}\epsilon_q p_1(x) + \dots$ the expression representing the coefficient of $1/\sqrt{1-y^2}$ is equal to zero at $y = \pm 1$ with the precision of $o(\epsilon_0^{1/2}\epsilon_q)$. The latter leads to

$$1 - (\tfrac{c_0-a_0}{2})^2 + \tfrac{c_0^2-a_0^2}{2} - \epsilon_0^{1/2}\epsilon_q\tfrac{2}{\pi}\Big\{ - \int\limits_{\alpha_1+\dots}^{0} q(r)\tfrac{\sqrt{t_0^2(r)-1}}{1+t_0(r)}dr$$

$$+ \int\limits_{0}^{\beta_1+\dots} g(s)\tfrac{\sqrt{t_1^2(s)-1}}{1+t_1(s)}ds\Big\} = 0,\ t_0 = -1 + \tfrac{2\epsilon_q r}{c_0-a_0} \le -1,$$

$$1 - (\tfrac{c_0-a_0}{2})^2 - \tfrac{c_0^2-a_0^2}{2} - \epsilon_0^{1/2}\epsilon_q\tfrac{2}{\pi}\Big\{ \int\limits_{\alpha_1+\dots}^{0} q(r)\tfrac{\sqrt{t_0^2(r)-1}}{1-t_0(r)}dr$$

$$- \int\limits_{0}^{\beta_1+\dots} g(s)\tfrac{\sqrt{t_1^2(s)-1}}{1-t_1(s)}ds\Big\} = 0,\ t_1 = 1 + \tfrac{2\epsilon_q s}{c_0-a_0} \ge 1.$$

Using expressions (6.95) and (6.96) for a_0 and c_0, respectively, and estimating the integrals involved, we obtain the asymptotic formulas for α_0 and β_0 in the form

$$\alpha_0 = \tfrac{2}{\pi}\int\limits_{\alpha_1}^{0}\tfrac{q(r)dr}{\sqrt{-2r}} + \dots,\quad \beta_0 = -\tfrac{2}{\pi}\int\limits_{0}^{\beta_1}\tfrac{g(s)ds}{\sqrt{2s}} + \dots. \tag{6.120}$$

Because $q(r) \ge 0$, $\alpha_1 < 0$, and $g(s) \ge 0$, $\beta_1 > 0$, we get $\alpha_0 \ge 0$ and $\beta_0 \le 0$. Therefore, formulas (6.95), (6.96), and (6.120) for the boundary points a_0 and c_0 indicate that the external region $[a_0, c_0]$ is slightly narrower than the classic Hertzian region $[-1, 1]$ and belongs to the interior of the latter one.

Formula (6.118) for $p_1(x)$ can be rewritten further as follows

$$p_1(x) = \tfrac{2}{\pi(c_0-a_0)\sqrt{1-y^2}}\Big\{ - \int\limits_{\alpha_1-\alpha_0(\epsilon_0/\epsilon_q)^{1/2}+\dots}^{0}\tfrac{q(r)t_0(r)dr}{\sqrt{t_0^2(r)-1}}$$

$$+ \int\limits_{0}^{\beta_1+\dots}\tfrac{g(s)t_1(s)ds}{\sqrt{t_1^2(s)-1}} + y\Big[- \int\limits_{\alpha_1-\alpha_0(\epsilon_0/\epsilon_q)^{1/2}+\dots}^{0}\tfrac{q(r)dr}{\sqrt{t_0^2(r)-1}}$$

$$+ \int\limits_{0}^{\beta_1+\dots}\tfrac{g(s)ds}{\sqrt{t_1^2(s)-1}}\Big] \tag{6.121}$$

$$+(y^2-1)\Big[\int\limits_{\alpha_1-\alpha_0(\epsilon_0/\epsilon_q)^{1/2}-\alpha_2\epsilon_0/\epsilon_q+\dots}^{0}\tfrac{q(r)dr}{\sqrt{t_0^2(r)-1}(y-t_0(r))}$$

$$- \int\limits_{0}^{\beta_1+\dots}\tfrac{g(s)ds}{\sqrt{t_1^2(s)-1}(y-t_1(s))}\Big]\Big\}.$$

Making use of (6.121) provides us with the asymptotic of $p_1(x)$ in the form

$$p_1(x) = \epsilon_q^{-1/2} \frac{(\alpha_0 + \beta_0)y - \alpha_0 + \beta_0}{2\sqrt{1-y^2}} + \frac{\sqrt{1-y^2}}{\pi} \left\{ \int\limits_{\alpha_1 + \ldots}^{0} \frac{q(r)dr}{\sqrt{t_0^2(r)-1}(y-t_0(r))} \right.$$

$$\left. - \int\limits_{0}^{\beta_1 + \ldots} \frac{g(s)ds}{\sqrt{t_1^2(s)-1}(y-t_1(s))} \right\} + \ldots, \tag{6.122}$$

$$t_0 = -1 + \frac{2\epsilon_q r}{c_0 - a_0}, \quad t_1 = 1 + \frac{2\epsilon_q s}{c_0 - a_0}.$$

We need to determine a more detailed asymptotic behavior of $p_1(x)$ in the vicinity of points $x = a_0$, $x = c_0$, i.e., in the vicinity of $y = \pm 1$. To achieve that we need to analyze the behavior of the integrals in (6.122). Considering the first integral in (6.122), we have

$$\int\limits_{\alpha_1 + \ldots}^{0} \frac{q(r)dr}{\sqrt{t_0^2(r)-1}(y-t_0(r))} = \int\limits_{\alpha_1 + \ldots}^{0} \frac{[q(r)-q(0)]dr}{\sqrt{t_0^2(r)-1}(y-t_0(r))}$$

$$\tag{6.123}$$

$$+ q(0) \left\{ \int\limits_{-\infty}^{0} \frac{dr}{\sqrt{t_0^2(r)-1}(y-t_0(r))} - \int\limits_{-\infty}^{\alpha_1 + \ldots} \frac{dr}{\sqrt{t_0^2(r)-1}(y-t_0(r))} \right\}.$$

In (6.123), the first integral in the right-hand side of the expression is a bounded regular function for any $|y| < 1$ because $q(r)$ is a continuously differentiable function, and, therefore, the integrand is a regular function for $r \in [\alpha_1 - \alpha_0(\epsilon_0/\epsilon_q)^{1/2} + \ldots, 0]$. The third term in the right-hand side of (6.123) is also a bounded regular function for any $|y| < 1$ because $q(0)$ is finite and the integrand is a regular function for any $|y| < 1$ and $r \le \alpha_1 - \alpha_0(\epsilon_0/\epsilon_q)^{1/2} + \ldots$ while as $r \to -\infty$ the integrand vanishes as $(-r)^{-3/2}$. In addition, we have

$$\int\limits_{-\infty}^{0} \frac{dr}{\sqrt{t_0^2(r)-1}(y-t_0(r))} = \frac{c_0 - a_0}{2} \frac{\epsilon_q^{-1}}{\sqrt{1-y^2}} \left\{ \frac{\pi}{2} - \arctan\sqrt{\frac{1+y}{1-y}} \right\}.$$

Using the expression for $t_0(r)$, as a result of the above considerations we obtain that

$$\int\limits_{\alpha_1 + \ldots}^{0} \frac{q(r)dr}{\sqrt{t_0^2(r)-1}(y-t_0(r))} = F_r(y)$$

$$\tag{6.124}$$

$$+ \frac{2\epsilon_q^{-1}q(0)}{\sqrt{1-y^2}} \left\{ \frac{\pi}{2} - \arctan\sqrt{\frac{1+y}{1-y}} \right\} + \ldots,$$

where $F_r(y)$ is a bounded regular function for all $|y| < 1$. A similar analysis can be done for the second integral in (6.122). Finally, the asymptotic of $p_1(x)$ in the vicinity of $y = \pm 1$ can be represented in the form

$$p_1(x) = \epsilon_q^{-1/2} \frac{(\alpha_0 + \beta_0)y - \alpha_0 + \beta_0}{2\sqrt{1-y^2}} + \frac{q(0)}{\epsilon_q} \left\{ 1 - \frac{2}{\pi} \arctan\sqrt{\frac{1+y}{1-y}} \right\}$$

$$\tag{6.125}$$

$$+ \frac{g(0)}{\epsilon_q} \left\{ 1 - \frac{2}{\pi} \arctan\sqrt{\frac{1-y}{1+y}} \right\} + \frac{\sqrt{1-y^2}}{\pi} G(y) + o(\epsilon_q^{-1}),$$

where $G(y)$ is a bounded regular function for all $\mid y \mid < 1$.

It is important to emphasize that the contributions of all other terms in the latter expression for $p_1(x)$ are of the order of $o(\epsilon_q^{-1})$, which provide a contribution to $p_{ext}(x)$ in the inlet and exit ϵ_0-zones of the order of $o(\epsilon_0^{1/2})$ (see formula (6.106) for $p_{ext}(x)$).

Using the expressions for a_0 and c_0, we find the asymptotic of $p_0(x)$ as follows

$$p_0(x) = \sqrt{1 - y^2} - \epsilon_0^{1/2}\epsilon_q^{1/2}\frac{(\alpha_0+\beta_0)y - \alpha_0+\beta_0}{2\sqrt{1-y^2}} + o(\epsilon_0^{1/2}\epsilon_q^{1/2}). \qquad (6.126)$$

Now, based on functions $p_0(x)$ and $p_1(x)$ we can determine the asymptotic representations of $p_{ext}(x)$ in the inlet and exit ϵ_0-zones as follows

$$p_{ext}(x) = \epsilon_0^{1/2}\{q(0) + \sqrt{2r_0}\theta(r_0)\} + o(\epsilon_0^{1/2}),$$
$$\qquad (6.127)$$
$$p_{ext}(x) = \epsilon_0^{1/2}\{g(0) + \sqrt{-2s_0}\theta(-s_0)\} + o(\epsilon_0^{1/2}).$$

By setting values of α_0 and β_0 to be equal to the expressions in (6.120), we eliminate singularities at $x = a_0$ and $x = c_0$ in the terms of $p_{ext}(x)$ of the orders of magnitude proportional to $\epsilon_q^{1/2}$ and $\epsilon_0^{1/2}$.

Therefore, for $r_0 = O(1)$ and $s_0 = O(1)$, $\omega \ll 1$, the first two leading terms of the asymptotic expansions (6.106) for $p_{ext}(x)$ in the inlet and exit ϵ_0-zones are of order of $\epsilon_0^{1/2}$, i.e., of the same order of magnitude as the problem solution in the inlet and exit ϵ_0-zones. As the result of that the latter asymptotic expansions of the external solution $p_{ext}(x)$ potentially allow for matching $p_{ext}(x)$ with the solutions in the inlet and exit ϵ_0-zones (see estimate (6.94)).

6.6.3 Auxiliary Gap Function $h_H(x)$. Behavior of $h(x)$ in the Inlet and Exit ϵ_q-Zones

Now, let us define an auxiliary gap function $h_H(x)$, which will be used to evaluate the asymptotic behavior of gap $h(x)$ in the inlet and exit ϵ_q-zones by the following equation

$$H_0[h_H(x) - 1] = x^2 - (\tfrac{c_0+a_0}{2})^2 + \tfrac{2}{\pi}\int_{a_0}^{c_0} p_{ext}(t)\ln\mid\frac{\frac{c_0+a_0}{2}-t}{x-t}\mid dt$$

$$+\epsilon_0^{1/2}\epsilon_q\tfrac{2}{\pi}\int_{\alpha_1-\alpha_0(\epsilon_0/\epsilon_q)^{1/2}+\dots}^{0} q(r)\ln\mid\frac{\frac{c_0-a_0}{2}-\epsilon_q r}{x-a_0-\epsilon_q r}\mid dr \qquad (6.128)$$

$$+\epsilon_0^{1/2}\epsilon_q\tfrac{2}{\pi}\int_{0}^{\beta_1+\dots} g(s)\ln\mid\frac{\frac{c_0-a_0}{2}+\epsilon_q s}{x-c_0-\epsilon_q s}\mid ds.$$

It follows from (6.106)-(6.108) that $h_H(x) = 1$, $x \in [a_0, c_0]$. Also, for further analysis we will need the asymptotic behavior of $h_H(x)$ in the inlet and exit ϵ_q-

and ϵ_0-zones. To obtain this behavior we need to consider the corresponding integrals (see (6.128)) of $p_0(x)$ and $p_2(x)$. We will make use of the following integrals

$$\frac{2}{\pi}\int_{-1}^{1}\sqrt{1-t^2}\ln\left|\frac{t}{y-t}\right|dt = -y^2 + |y|\sqrt{y^2-1} - \ln\left||y|+\sqrt{y^2-1}\right|,$$

$$\frac{2}{\pi}\int_{-1}^{1}\frac{1}{\sqrt{1-t^2}}\ln\left|\frac{t}{y-t}\right|dt = -\ln\left||y|+\sqrt{y^2-1}\right|,$$

$$\frac{1}{\pi}\int_{-1}^{1}\frac{t}{\sqrt{1-t^2}}\ln\left|\frac{t}{y-t}\right|dt = y - sign(y)\sqrt{y^2-1} \ \ for \ \ |y|\geq 1.$$

Based on the above integrals and (6.109), we get

$$x^2 - \left(\frac{c_0+a_0}{2}\right)^2 + \frac{2}{\pi}\int_{a_0}^{c_0}p_0(t)\ln\left|\frac{\frac{c_0+a_0}{2}-t}{x-t}\right|dt$$

$$= \left(\frac{c_0-a_0}{2}\right)^2|y|\sqrt{y^2-1} - \ln\left||y|+\sqrt{y^2-1}\right| \qquad (6.129)$$

$$+\frac{c_0^2-a_0^2}{2}sign(y)\sqrt{y^2-1} \ \ for \ \ |y|\geq 1.$$

Using the fact that the integrands in the integrals in expression (6.118) for $p_1(x)$ are regular functions and using the integrals in (6.119), we obtain

$$\int_{a_0}^{c_0}p_1(t)\ln\left|\frac{\frac{c_0+a_0}{2}-t}{x-t}\right|dt =$$

$$\int_{\alpha_1-\alpha_0(\epsilon_0/\epsilon_q)^{1/2}+\ldots}^{0}q(r)\ln\left|\frac{[yt_0(r)-1]sign(y)-\sqrt{(y^2-1)(t_0^2(r)-1)}}{t_0(r)}\right|dr$$

$$+\int_{0}^{\beta_1+\ldots}g(s)\ln\left|\frac{[yt_1(s)-1]sign(y)+\sqrt{(y^2-1)(t_1^2(s)-1)}}{t_1(s)}\right|ds$$

$$(6.130)$$

$$for \ \ |y|\geq 1, \ t_0(r) = -1+\frac{2\epsilon_q r}{c_0-a_0}, \ t_1(s) = 1+\frac{2\epsilon_q s}{c_0-a_0}.$$

From (6.128) with the help of (6.93), (6.94), (6.97)-(6.104), (6.106), (6.109), and (6.118), we readily obtain the asymptotic expansions of $h_H(x)$ for $r = O(1)$ and $s = O(1)$, $\omega \ll 1$,

$$H_0[h_H(x)-1] = -\epsilon_q^{3/2}\frac{4}{3}r\sqrt{-2r}\theta(-r) + o(\epsilon_q^{3/2}),$$

$$r = O(1), \ \omega \ll 1, \qquad (6.131)$$

$$H_0[h_H(x)-1] = \epsilon_q^{3/2}\frac{4}{3}s\sqrt{2s}\theta(s) + o(\epsilon_q^{3/2}), \ s = O(1), \ \omega \ll 1. \qquad (6.132)$$

Now, let us get the asymptotic behavior of $h_H(x)$ in the inlet and exit ϵ_0-zones, i.e., for $r_0 = O(1)$ and $s_0 = O(1)$, $\omega \ll 1$. In the inlet ϵ_0-zone $y = -1 + 2\epsilon_0 r_0/(c_0 - a_0)$ and the two–term asymptotic expansion of the expression in the left-hand side of (6.129) has the form

$$x^2 - (\tfrac{c_0+a_0}{2})^2 + \tfrac{2}{\pi} \int\limits_{a_0}^{c_0} p_0(t) \ln \mid \tfrac{\frac{c_0+a_0}{2}-t}{x-t} \mid dt$$

$$= -2\alpha_0 \epsilon_0 \epsilon_q^{1/2} \sqrt{-2r_0} + \tfrac{2}{3}(-2r_0)^{3/2} \epsilon_0^{3/2} + o(\epsilon_0^{3/2})$$

(6.133)

while the asymptotic of the integral of $p_1(x)$ from (6.130) is given by

$$\int\limits_{a_0}^{c_0} p_1(t) \ln \mid \tfrac{\frac{c_0+a_0}{2}-t}{x-t} \mid dt = 2\alpha_0 \epsilon_0 \epsilon_q^{1/2} \sqrt{-2r_0} + o(\epsilon_0^{3/2}).$$

(6.134)

Therefore, from expression (6.128) for $h_H(x)$ and the obtained asymptotic expressions (6.133) and (6.134) in the inlet ϵ_0-zone, we find that

$$H_0[h_H(x) - 1] = -\epsilon_0^{3/2} \tfrac{4}{3} r_0 \sqrt{-2r_0}\, \theta(-r_0) + O(\epsilon_0^{3/2}),$$

$$r_0 = O(1), \quad \omega \ll 1.$$

(6.135)

Similarly, in the exit ϵ_0-zone we have the asymptotic estimate

$$H_0[h_H(x) - 1] = \epsilon_0^{3/2} \tfrac{4}{3} s_0 \sqrt{2s_0}\, \theta(s_0) + O(\epsilon_0^{3/2}),$$

$$s_0 = O(1), \quad \omega \ll 1,$$

(6.136)

where the terms of the order of $O(\epsilon_0^{3/2})$ are produced by the last two integrals in expression (6.128) for $h_H(x)$. The fact that it is very difficult to obtain the exact analytical expressions for these terms prevents us from deriving the asymptotically valid equations in the inlet and exit ϵ_0-zones.

Let us consider the behavior of gap $h(x)$ in the inlet and exit ϵ_q-zone. By transforming equation (6.7) with the help of equation (6.128), we find

$$H_0[h(x) - 1] = H_0[h_H(x) - 1]$$

$$+ \tfrac{2}{\pi} \int\limits_{a}^{c} p(t) \ln \mid \tfrac{\frac{c_0+a_0}{2}-t}{x-t} \mid dt - \tfrac{2}{\pi} \int\limits_{a_0}^{c_0} p_{ext}(t) \ln \mid \tfrac{\frac{c_0+a_0}{2}-t}{x-t} \mid dt$$

$$- \epsilon_0^{1/2} \epsilon_q \tfrac{2}{\pi} \int\limits_{\alpha_1 - \alpha_0(\epsilon_0/\epsilon_q)^{1/2}+\ldots}^{0} q(r) \ln \mid \tfrac{\frac{c_0-a_0}{2}-\epsilon_q r}{x-a_0-\epsilon_q r} \mid dr$$

$$- \epsilon_0^{1/2} \epsilon_q \tfrac{2}{\pi} \int\limits_{0}^{\beta_1 + \ldots} g(s) \ln \mid \tfrac{\frac{c_0-a_0}{2}+\epsilon_q s}{x-c_0-\epsilon_q s} \mid ds.$$

(6.137)

Using expressions (6.34), (6.93), (6.97)-(6.104), (6.106), estimating the integrals over all zones of the contact, and taking into account estimates from (6.131) and (6.137), we obtain the asymptotic estimates of gap $h(x)$ in the inlet and exit ϵ_q-zones in the form

$$H_0[h(x) - 1] = -\epsilon_q^{3/2}\tfrac{4}{3}r\sqrt{-2r}\theta(-r) + o(\epsilon_q^{3/2}), \ r = O(1), \ \omega \ll 1, \quad (6.138)$$

$$H_0[h(x) - 1] = \epsilon_q^{3/2}\tfrac{4}{3}s\sqrt{2s}\theta(s) + o(\epsilon_q^{3/2}), \ s = O(1), \ \omega \ll 1. \quad (6.139)$$

As a matter of fact, asymptotic estimates (6.138) and (6.139) mean that gap $h(x)$ is a given function in the inlet and exit ϵ_q-zones if the film thickness H_0 is known. Essentially, it is determined by the external pressure $p_{ext}(x)$ in the Hertzian region. This pressure is close to the Hertzian one $\sqrt{1 - x^2}\theta(1 - x^2)$ in a dry contact of elastic cylinders.

In the inlet and exit ϵ_0-zones, we can obtain the estimate

$$H_0[h(x) - 1] = O(\epsilon_0^{3/2}) \ for \ r_0 = O(1) \ and \ s_0 = O(1), \ \omega \ll 1. \quad (6.140)$$

6.6.4   Asymptotic Solutions for $q(r)$ and $g(s)$ in the Inlet and Exit ϵ_q-Zones

Based on estimates (6.138) and (6.139) one can conclude that in the inlet and exit ϵ_q-zones with a small error the lubricant flow is realized in the gap with almost given configuration. Using estimates (6.97) and (6.138), requiring that the two terms of equation (6.5) are of the same order of magnitude, and estimating these terms for $r = O(1)$, $\omega \ll 1$, we find the estimate for the central film thickness H_0 in the form

$$H_0 = A(V\epsilon_0^{-1/2}\epsilon_q^{5/2})^{1/3} + \dots, \ A = O(1), \ \omega \ll 1, \quad (6.141)$$

where $A = A(\alpha_1) \geq 0$ is a constant independent from ω, ϵ_0, and ϵ_q.

Further, based on (6.97), (6.98), (6.138), (6.139), and (6.141) from equation (6.5) and the first and third conditions in (6.6) in the inlet and exit ϵ_q-zones, we obtain the asymptotically valid systems of equations for function $q(r)$

$$\frac{A^3 h_q^3}{\mu_q(q)}\frac{dq}{dr} = -\tfrac{4}{3}r\sqrt{-2r}\theta(-r), \ q(\alpha_1) = 0, \quad (6.142)$$

$$h_q(r) = 1 \ for \ \epsilon_q \ll (V\epsilon_0^{-1/2})^{1/2}, \ \omega \ll 1, \quad (6.143)$$

or

$$A[h_q(r) - 1] = -\tfrac{4}{3}r\sqrt{-2r}\theta(-r) \ for \ \epsilon_q = (V\epsilon_0^{-1/2})^{1/2}, \ \omega \ll 1, \quad (6.144)$$

and for $g(s)$

$$\frac{A^3 h_g^3}{\mu_g(g)}\frac{dg}{ds} = \tfrac{4}{3}s\sqrt{2s}\theta(s) - \tfrac{4}{3}\beta_1\sqrt{2\beta_1}, \ g(\beta_1) = 0, \quad (6.145)$$

$$h_g(s) = 1 \; for \; \epsilon_q \ll (V\epsilon_0^{-1/2})^{1/2}, \; \omega \ll 1, \tag{6.146}$$

or

$$A[h_g(s) - 1] = \tfrac{4}{3}s\sqrt{2s}\theta(s) \; for \; \epsilon_q = (V\epsilon_0^{-1/2})^{1/2}, \; \omega \ll 1, \tag{6.147}$$

where $h_q(x)$, $\mu_q(q)$ and $h_g(x)$, $\mu_g(q)$ are the major terms of the asymptotic expansions of gap $h(x)$ and viscosity $\mu(p)$ in the inlet and exit ϵ_q-zones, respectively. Relations (6.143), (6.146) and (6.144), (6.147) are valid for starved and fully flooded lubrication regimes, respectively.

It is important to note that the value of constant A is not determined and cannot be determined by solving the system of equations (6.142)-(6.144) in the inlet ϵ_q-zone as well as the value of β_1 is not determined by equations (6.145)-(6.147) in the exit ϵ_q-zone. These constants are determined by the inlet and exit ϵ_0-zones, respectively.

Let us consider a two-term asymptotic approximations of pressure $p(x)$ in the inlet and exit ϵ_q-zones. The next to the Hertzian contribution of the order of magnitude $O(\epsilon_q^{3/2})$ to the function of gap $h(x)$ in the inlet and exit ϵ_q-zones is the contribution of the displacement produced by pressure $\epsilon_0^{1/2}\epsilon_q p_1(x)$. The order of magnitude of this contribution is $\epsilon_0^{1/2}\epsilon_q$. Therefore, by estimating the terms of the Reynolds equation and by taking into account that the latter contribution is of the order of magnitude of $\epsilon_0^{1/2}\epsilon_q$, we obtain that such a two-term approximation of pressure should be searched as follows $p(x) = \epsilon_0^{1/2}q(r) + \epsilon_0\epsilon_q^{-1/2}q_1(r) + \ldots$ and $p(x) = \epsilon_0^{1/2}g(s) + \epsilon_0\epsilon_q^{-1/2}g_1(s) + \ldots$ (where $q_1(r)$ and $g_1(s)$ are new unknown functions of order 1) in the inlet and exit ϵ_q-zones, respectively. We will not engage in this process because $\epsilon_0 \ll \epsilon_q$ and the contribution of the second terms in the pressure representation in the inlet and exit ϵ_q-zones to the solution of the problem in the inlet and exit ϵ_0-zones is negligibly small.

Therefore, in the inlet and exit ϵ_q-zones the problem is reduced to integration of nonlinear first-order differential equations for the leading terms of the asymptotic representations. Let us consider an example of solutions of systems (6.142)-(6.144) and (6.145)-(6.147) for the case of $\mu = exp(Qp)$. We will have $\epsilon_0 = Q^{-2}$, $\omega = Q^{-1} \ll 1$, $\mu_q(q) = \exp(q)$, $\mu_g(g) = \exp(g)$, and

$$H_0 = A(VQ\epsilon_q^{5/2})^{1/3} + \ldots, \; A = O(1), \; \epsilon_q \ll (VQ)^{1/2}, \; \omega \ll 1, \tag{6.148}$$

$$H_0 = A(VQ)^{3/4} + \ldots, \; A = O(1), \; \epsilon_q = (VQ)^{1/2}, \; \omega \ll 1. \tag{6.149}$$

Then for starved and fully flooded lubrication regimes we, respectively, obtain

$$q(r) = -\ln\left\{1 + \tfrac{8\sqrt{2}}{15A^3}[(-r)^{5/2}\theta(-r) - (-\alpha_1)^{5/2}\theta(-\alpha_1)]\right\}, \tag{6.150}$$

$$q(r) = -\ln\left\{1 + \tfrac{1}{2A}(\tfrac{9}{4A})^{1/3}J_4[\tau(\alpha_1), \tau(r)]\right\},$$

$$\tau(r) = (\tfrac{4\sqrt{2}}{3A})^{1/3}\sqrt{-r}\theta(-r),   \tag{6.151}$$

$$J_4(\alpha, y) = \tfrac{1}{3}\left[\tfrac{2t^2}{3(1+t^3)} - \tfrac{t^2}{(1+t^3)^2} - \tfrac{1}{9}\ln\tfrac{(t+1)^2}{t^2-t+1} + \tfrac{2\sqrt{3}}{9}\arctan\tfrac{2t-1}{\sqrt{3}}\right]\Big|_\alpha^y,$$

where $\theta(r)$ is a step function, $\theta(r) = 0$, $r \leq 0$ and $\theta(r) = 1$, $r > 0$. Obvious mechanical considerations lead to the fact that the solution $q(r)$ for $r \geq \alpha_1$ of equations (6.142)-(6.144) is bounded. Moreover, function $q(r)$ is a nondecreasing function of r and $q(r) = q(0)$ for $r > 0$. Therefore, the direct matching of function $q(r) = q(0)$ as $r \to \infty$ with the major term of the inner expansion of the external solution (6.145) is impossible without considering the asymptotic of the solution in the inlet ϵ_0-zone. From the explanation presented it follows that the major term of the two asymptotic expansions that match solution $\epsilon_0^{1/2}q_0(r_0)$ in the inlet ϵ_0-zone as $r_0 \to \pm\infty$ takes the form

$$q_a(r_0) = q(0) + \sqrt{2r_0}\theta(r_0).   \tag{6.152}$$

In the particular case of $\mu = \exp(Qp)$ from the fact that $q(0)$ is bounded (i.e., the logarithms are determined) and from expressions (6.150) and (6.151) for different lubrication regimes, we obtain the low bounds for constant A as follows

$$A > (\tfrac{8\sqrt{2}}{15})^{1/3} \mid \alpha_1 \mid^{5/6} \theta(-\alpha_1)$$

$$= 0.91 \mid \alpha_1 \mid^{5/6} \theta(-\alpha_1), \;\; \epsilon_q \ll (VQ)^{1/2},   \tag{6.153}$$

$$A > (\tfrac{3}{4\sqrt{2}})^{1/2}\{-J_4[\tau(\alpha_1), 0]\}^{3/4}, \;\; \epsilon_q = (VQ)^{1/2}.$$

For $\alpha_1 = -\infty$ the last inequality in (6.153) is reduced to

$$A > \tfrac{(54\sqrt{3}\pi^3)^{1/4}}{27} = 0.272.   \tag{6.154}$$

Furthermore, note that for vanishing r the estimate $dq/dr = O((-r)^{3/2})$ follows from (6.142). In view of $r = r_0\epsilon_0/\epsilon_q$ and from the last estimate for dq/dr as r vanishes, we obtain

$$\tfrac{dp}{dx} = O(\epsilon_0^2\epsilon_q^{-5/2}), \;\; r_0 = O(1), \;\; \omega \ll 1.   \tag{6.155}$$

For the same case of $\mu = \exp(Qp)$ in the exit ϵ_q-zone from equations (6.145)-(6.147) for $g(s)$, we find that for starved lubrication regimes

$$g(s) = -\ln\left\{1 + \tfrac{8\sqrt{2}}{15A^3}[\beta_1^{5/2}\theta(\beta_1) - s^{5/2}\theta(s)\right.$$

$$\left. -\tfrac{5}{2}\beta_1^{3/2}\theta(\beta_1)[\beta_1 - s\theta(s)]]\right\},   \tag{6.156}$$

while for fully flooded lubrication regimes

$$g(s) = -\ln\left\{1 + \frac{1}{2A}\left(\frac{9}{4A}\right)^{1/3}[J_4[\tau(s), \tau(\beta_1)]\right.$$

$$\left. -\frac{4\sqrt{2}}{3A}\beta_1^{3/2}\theta(\beta_1)J_1[\tau(s), \tau(\beta_1)]]\right\}, \quad \tau(s) = \left(\frac{4\sqrt{2}}{3A}\right)^{1/3}\sqrt{s}\theta(s), \tag{6.157}$$

$$J_1(\alpha, y) = \left[\frac{4t^2}{9(1+t^3)} + \frac{t^2}{3(1+t^3)^2} - \frac{2}{27}\ln\frac{(t+1)^2}{t^2-t+1} + \frac{4\sqrt{3}}{27}\arctan\frac{2t-1}{\sqrt{3}}\right]\Big|_\alpha^y,$$

where $J_4(\alpha, y)$ is defined in (6.151).

Estimating $\frac{d^2 g(s)}{ds^2}$ for vanishing s such that $s = \frac{\epsilon_0}{\epsilon_q}s_0$, $s_0 = O(1)$, $\omega \ll 1$, based on (6.145)-(6.147) we find that

$$\frac{d^2 g(s)}{ds^2} = O(\epsilon_0^{1/2}\epsilon_q^{-1/2}), \quad s_0 = O(1), \quad \omega \ll 1. \tag{6.158}$$

Therefore, we can conclude that in the exit ϵ_0-zone

$$\frac{d^2 p(x)}{dx^2} = O(\epsilon_0\epsilon_q^{-5/2}), \quad s_0 = O(1), \quad \omega \ll 1. \tag{6.159}$$

This estimate coincides with the estimate for d^2p/dx^2 in the ϵ_0-inlet zone, which follows from (6.155).

The above analysis of the behavior of $p_{ext}(x)$ and $g(s)$ in the ϵ_0-exit zone indicates that the major term of the two asymptotic expansions that match the solution $\epsilon_0^{1/2}g_0(s_0)$ in the exit ϵ_0-zone as $s_0 \to \pm\infty$ has the form

$$g_a(s_0) = g(0) + \sqrt{-2s_0}\theta(-s_0). \tag{6.160}$$

6.6.5   Asymptotic Analysis of the Inlet and Exit ϵ_0-Zones

Let us consider the asymptotic relationships in the inlet and exit ϵ_0-zones in more detail. For this purpose it is necessary to estimate the integrals in equations (6.8) and (6.137). Let us introduce a function $\eta(\omega)$ so that $\epsilon_0 \ll \eta(\omega) \ll \epsilon_q$ for $\omega \ll 1$. We have

$$\int_a^c p(t)\ln\left|\frac{\frac{c_0+a_0}{2}-t}{x-t}\right| \, dt = \left[\int_a^{a_0-\eta} + \int_{a_0-\eta}^{a_0+\eta} + \int_{a_0+\eta}^{c_0-\eta} + \int_{c_0-\eta}^{c_0+\eta} + \int_{c_0+\eta}^c\right]p(t)$$

$$\times \ln\left|\frac{\frac{c_0+a_0}{2}-t}{x-t}\right| \, dt.$$

From the last equality using expressions (6.97)-(6.104), (6.106), (6.152), and (6.160), we find

$$\int\limits_{a}^{c} p(t) \ln \left| \frac{\frac{c_0+a_0}{2}-t}{x-t} \right| dt$$

$$= \epsilon_0^{1/2}\epsilon_q \int\limits_{\alpha_1-\alpha_0(\epsilon_0/\epsilon_q)^{1/2}-\alpha_2\epsilon_0/\epsilon_q+\dots}^{0} q(r) \ln \left| \frac{\frac{c_0-a_0}{2}-\epsilon_q r}{x-a_0-\epsilon_q r} \right| dr$$

$$+\epsilon_0^{3/2} \int\limits_{-\infty}^{\infty} \widetilde{q}_0(r_1) \ln \left| \frac{\frac{c_0-a_0}{2}-\epsilon_0 r_1}{x-a_0-\epsilon_0 r_1} \right| dr_1 + \int\limits_{a_0}^{c_0} p_{ext}(t) \ln \left| \frac{\frac{c_0+a_0}{2}-t}{x-t} \right| dt \qquad (6.161)$$

$$+\epsilon_0^{3/2} \int\limits_{-\infty}^{\infty} \widetilde{g}_0(s_1) \ln \left| \frac{\frac{c_0-a_0}{2}+\epsilon_0 s_1}{c_0-x+\epsilon_0 s_1} \right| ds_1$$

$$+\epsilon_0^{1/2}\epsilon_q \int\limits_{0}^{\beta_1+\dots} g(s) \ln \left| \frac{\frac{c_0-a_0}{2}+\epsilon_q s}{x-c_0-\epsilon_q s} \right| ds + o(\epsilon_0^{3/2}), \; \omega \ll 1,$$

where $\widetilde{q}_0(r_0) = q_0(r_0) - q_a(r_0)$ and $\widetilde{g}_0(s_0) = g_0(s_0) - g_a(s_0)$. The estimates $\epsilon_0 \ll \eta \ll \epsilon_q$ are used for deriving equation (6.161). In a similar way we obtain the asymptotic expression for the integral in equation (6.8).

Using this procedure of estimating the integral in equation (6.8) and using formulas (6.161), we can transform equations (6.137) and (6.8) as follows

$$H_0[h(x) - 1] = \epsilon_0^{3/2}\frac{2}{\pi} \int\limits_{-\infty}^{\infty} \widetilde{q}_0(r_1) \ln \left| \frac{\frac{c_0-a_0}{2}-\epsilon_0 r_1}{x-a_0-\epsilon_0 r_1} \right| dr_1$$

$$\qquad (6.162)$$

$$+\epsilon_0^{3/2}\frac{2}{\pi} \int\limits_{-\infty}^{\infty} \widetilde{g}_0(s_1) \ln \left| \frac{\frac{c_0-a_0}{2}+\epsilon_0 s_1}{c_0-x+\epsilon_0 s_1} \right| ds_1 + o(\epsilon_0^{3/2}),$$

$$\epsilon_0^{3/2} \int\limits_{-\infty}^{\infty} \widetilde{q}_0(r_1)dr_1 + \epsilon_0^{3/2} \int\limits_{-\infty}^{\infty} \widetilde{g}_0(s_1)ds_1 + o(\epsilon_0^{3/2}) = 0.$$

The matching conditions on the boundaries of the external region and the inlet and exit ϵ_q-zones, obviously, have the form (see (6.105))

$$\widetilde{q}_0(r_0) \to 0, \; r_0 \to \pm\infty, \qquad (6.163)$$

$$\widetilde{g}_0(s_0) \to 0, \; s_0 \to \pm\infty. \qquad (6.164)$$

Now, let us consider the inlet ϵ_0-zone. Using the asymptotic of $h_H(x)$ for $r_0 = O(1)$, $\omega \ll 1$, from (6.135) and (6.162) with the precision of $o(\epsilon_0^{3/2})$, $\omega \ll 1$, we find the expression for $h(x)$ in the form

$$H_0[h(x) - 1] = \epsilon_0^{3/2} \ln \frac{c_0-a_0}{2\epsilon_0}\frac{2}{\pi} \int\limits_{-\infty}^{\infty} \widetilde{q}_0(r_1)dr_1 + O(\epsilon_0^{3/2}),$$

$$\qquad (6.165)$$

$$r_0 = O(1), \; \omega \ll 1.$$

Obviously, this expression for $h(x)$ does not allow for matching $q_0(r_0)$ with $q_a(r_0)$ as $r_0 \to \infty$. Thus, in order to match these solutions it is necessary for the right side in (6.165) to be equal to zero. In turn, from the second equation in (6.162) and the mentioned conclusion we obtain

$$\int_{-\infty}^{\infty} [q_0(r_0) - q_a(r_0)]dr_0 = 0, \tag{6.166}$$

$$\int_{-\infty}^{\infty} [g_0(s_0) - g_a(s_0)]ds_0 = 0. \tag{6.167}$$

With the help of (6.166) and (6.167) for the regimes of starved and fully flooded lubrication in the inlet and exit ϵ_0-zones from the first equation in (6.162), we find that in the inlet and exit ϵ_0-zones $h(x) - 1 \ll 1$ and

$$H_0[h(x) - 1] = \epsilon_0^{3/2} \Big\{ -\tfrac{4}{3}r_0\sqrt{-2r_0}\theta(-r_0)$$

$$+\tfrac{2}{\pi} \int_{-\infty}^{\infty} [q_0(t) - q_a(t)] \ln \tfrac{1}{|r_0 - t|}dt \Big\} + O(\epsilon_0^{3/2}), \quad r_0 = O(1), \ \omega \ll 1, \tag{6.168}$$

$$H_0[h(x) - 1] = \epsilon_0^{3/2} \Big\{ \tfrac{4}{3}s_0\sqrt{2s_0}\theta(s_0)$$

$$+\tfrac{2}{\pi} \int_{-\infty}^{\infty} [g_0(t) - g_a(t)] \ln \tfrac{1}{|s_0 - t|}dt \Big\} + O(\epsilon_0^{3/2}), \quad s_0 = O(1), \ \omega \ll 1, \tag{6.169}$$

respectively. Unfortunately, it is very difficult if not impossible to specify analytically in more detail the terms $O(\epsilon_0^{3/2})$ in equations (6.168) and (6.169). This represents the main obstacle in obtaining the asymptotically valid equations for $q_0(r_0)$ and $g_0(s_0)$ in the inlet and exit ϵ_0-zones, respectively.

From equations (6.168), (6.169), and formula (6.141) for H_0, we obtain that in the inlet and exit ϵ_0-zones

$$h(x) - 1 \ll 1 \ for \ r_0 = O(1) \ and \ s_0 = O(1), \quad \omega \ll 1. \tag{6.170}$$

This is true as long as $\epsilon_0^{3/2} \ll H_0$ (see the formula for H_0 (6.141) and the size of $\epsilon_q = (V\epsilon_0^{-1/2})^{1/2}$ for fully flooded over-critical lubrication regimes from (6.144)), i.e.,

$$\epsilon_0 \ll V^{2/5}, \ \omega \ll 1. \tag{6.171}$$

The latter inequality represents the condition that discriminates between the pre- and over-critical lubrication regimes. Namely, if

$$\epsilon_0 = O(V^{2/5}) \ or \ \epsilon_0 \gg V^{2/5}, \ \omega \ll 1, \tag{6.172}$$

then the lubrication regime is pre-critical (see estimate (6.62) for $\epsilon_q = V^{2/5}$ for fully flooded pre-critical lubrication regimes and the definitions of pre-critical

(6.75) and over-critical (6.76) regimes) while if estimate (6.171) is satisfied then the lubrication regime is over-critical.

Moreover, it is necessary to demand matching the asymptotic expansions of the first and second derivatives of pressure p on the boundaries of the inlet and exit ϵ_q- and ϵ_0-zones, i.e., the validity of estimates (6.155) and (6.158) on the boundaries of the inlet and exit ϵ_q- and ϵ_0-zones, respectively. With this in mind we can say that in the inlet and exit zones the distribution of pressure p depends on the independent variables $r = \frac{x-a_0}{\epsilon_q}$, $r_0 = \frac{x-a_0}{\epsilon_0}$ and $s = \frac{x-c_0}{\epsilon_q}$, $s_0 = \frac{x-c_0}{\epsilon_0}$, respectively. Therefore, in the inlet and exit zones the operator of differentiation d/dx can be represented as follows

$$\frac{d}{dx} = \frac{1}{\epsilon_q}\frac{d}{dr} + \frac{1}{\epsilon_0}\frac{d}{dr_0}, \quad \frac{d}{dx} = \frac{1}{\epsilon_q}\frac{d}{ds} + \frac{1}{\epsilon_0}\frac{d}{ds_0}.$$

Taking into account the estimates for dp/dx in the ϵ_0-inlet zone and for d^2p/dx^2 in the ϵ_0-exit zone as well as estimates (6.138), (6.139) for gap h in the ϵ_q-inlet and exit zones and estimates (6.170) in the ϵ_0-inlet and exit zones, we obtain in the inlet zones

$$\frac{d}{dx}\left(\frac{h^3}{\mu}\frac{dp}{dx}\right) = \frac{\epsilon_0^{1/2}}{\epsilon_q^2}\frac{d}{dr}\left(\frac{h_q^3}{\mu_q}\frac{dq}{dr}\right) + \frac{\epsilon_0}{\epsilon_q^{5/2}}\frac{d}{dr_0}\left(\frac{1}{\mu_q(q_0)}\frac{dq_0}{dr_0}\right) + \dots, \tag{6.173}$$

and in the exit zones

$$\frac{d}{dx}\left(\frac{h^3}{\mu}\frac{dp}{dx}\right) = \frac{\epsilon_0^{1/2}}{\epsilon_q^2}\frac{d}{ds}\left(\frac{h_g^3}{\mu_g}\frac{dg}{ds}\right) + \frac{\epsilon_0}{\epsilon_q^{5/2}}\frac{d}{ds_0}\left(\frac{1}{\mu_g(g_0)}\frac{dg_0}{ds_0}\right) + \dots. \tag{6.174}$$

In addition, in the ϵ_0-inlet and exit zones in expressions (6.173) and (6.174) we need to take into account that $r = \epsilon_0\epsilon_q^{-1}r_0$ and $s = \epsilon_0\epsilon_q^{-1}s_0$, respectively. Therefore, based on equations (6.142)-(6.144) and (6.145)-(6.147) for $q(r)$ and $g(s)$, we conclude that

$$\frac{H_0^3\epsilon_0^{1/2}}{V\epsilon_q^2}\frac{d}{dr}\left(\frac{h_q^3}{\mu_q(q)}\frac{dq}{dr}\right) = -\epsilon_0^{1/2}2\sqrt{-2r_0}\theta(-r_0),$$

$$\frac{H_0^3\epsilon_0^{1/2}}{V\epsilon_q^2}\frac{d}{ds}\left(\frac{h_g^3}{\mu_g(g)}\frac{dg}{ds}\right) = \epsilon_0^{1/2}2\sqrt{2s_0}\theta(s_0). \tag{6.175}$$

At last, we substitute (6.168), (6.169), (6.179), and (6.174) in equation (6.5) and in boundary conditions (6.6), and take into account equations (6.175) and in the inlet and exit ϵ_0-zones. That allows us to derive the asymptotically valid equations

$$\frac{d}{dr_0}M_0(q_0, 1, \mu_q(q_0), r_0) = \frac{2}{\pi}\int\limits_{-\infty}^{\infty}\frac{q_0(t)-q_a(t)}{t-r_0}dt + O(1),$$

$$r_0 = O(1), \quad \omega \ll 1, \tag{6.176}$$

$$\frac{d}{ds_0}M_0(g_0, 1, \mu_g(g_0), s_0) = \frac{2}{\pi}\int\limits_{-\infty}^{\infty}\frac{g_0(t)-g_a(t)}{t-s_0}dt + O(1),$$

(6.177)

$$s_0 = O(1), \quad \omega \ll 1,$$

where function M_0 is defined by the equality

$$M_0(p, h, \mu, x) = A^3\frac{h^3}{\mu}\frac{dp}{dx},$$

(6.178)

functions μ_q and μ_g are the leading terms of the asymptotic expansions of viscosity $\mu(p, Q)$ in the inlet and exit ϵ_0-zones. In these equations terms $O(1)$ are certain functions of the order of 1 that vanish as $r_0 \to \pm\infty$ and $s_0 \to \pm\infty$.

Thus, the considered problem for $\omega \ll 1$ is reduced to a set of problems in different zones of the contact area. In the inlet ϵ_q-zone the asymptotic analogue of the original problem equations (6.5)-(6.8) is the system of equations (6.142)-(6.144) for function $q(r)$, in the inlet ϵ_0-zone the asymptotic analogue of equations (6.5)-(6.8) is the system of equations (6.152), (6.163), (6.166), (6.176), and (6.178) for function $q_0(r_0)$ and constant A

$$\frac{d}{dr_0}M_0(q_0, 1, \mu_q(q_0), r_0) = \frac{2}{\pi}\int\limits_{-\infty}^{\infty}\frac{q_0(t)-q_a(t)}{t-r_0}dt + O(1),$$

$$q_0(r_0) \to q_a(r_0) = q(0) + \sqrt{2r_0}\theta(r_0), \quad r_0 \to \pm\infty,$$

(6.179)

$$\int\limits_{-\infty}^{\infty}[q_0(t) - q_a(t)]dt = 0,$$

in the Hertzian region the asymptotic analogue is represented by equations (6.105) whose solution is represented by formulas (6.106), (6.109), and (6.118), in the exit ϵ_0-zone the analogue of the original system is given by equations (6.160), (6.164), (6.167), (6.177), and (6.178) for function $g_0(s_0)$ and constant β_1

$$\frac{d}{ds_0}M_0(g_0, 1, \mu_g(g_0), s_0) = \frac{2}{\pi}\int\limits_{-\infty}^{\infty}\frac{g_0(t)-g_a(t)}{t-s_0}dt + O(1),$$

$$g_0(s_0) \to g_a(s_0) = g(0) + \sqrt{-2s_0}\theta(-s_0), \quad s_0 \to \pm\infty,$$

(6.180)

$$\int\limits_{-\infty}^{\infty}[g_0(t) - g_a(t)]dt = 0,$$

and in the exit ϵ_q-zone the asymptotic analogue of the original problem equations (6.5)-(6.8) is the system of equations (6.145)-(6.147) for function $g(s)$.

It is important to mention that the presence of terms $O(1)$ in equations (6.179) and (6.180) is essential. For example, if $\mu_q(q) = \mu_g(q) = \exp(q)$ and these terms are removed from the latter equations then their only solutions are $q_0(r_0) = q_a(r_0)$, $A^3\exp(-q(0)) = 0$ and $g_0(s_0) = g_a(s_0)$, A^3

$\times \exp(-g(0)) = 0$, respectively. As our analysis showed the values of constant A in the lubrication film thickness formulas (6.148) and (6.149) are not determined by the inlet ϵ_q-zone, Hertzian region, or the exit zones but determined by the problem solution in the inlet ϵ_0-zone. On the other hand, lack of knowledge about the specifics of terms $O(1)$ in equations (6.179) and (6.180) does not allow for actual numerical calculation of constant A involved in the lubrication film thickness formulas (6.148) and (6.149) based on these equations.

In Section 6.10 we will see that the asymptotic equations derived for pre-critical regimes are actually valid for over-critical regimes assuming that in this case $\alpha_1 \gg 1$.

Note, that in the external region the estimate $\mu \gg 1$ leads to the validity of estimates (6.104) and $H_0(h-1) \ll 1$ in the external region. That validates the entire analysis of the problem.

Obviously, the described analysis is valid if the small parameter of the problem $\omega = Q^{-1} \ll 1$ or $\omega = V \ll 1$. The method is also applicable to the problems under non-isothermal conditions and the problems with non-Newtonian fluids. In these cases only solutions in the inlet and exit ϵ_q zones and function M_0 from (6.178) must be adjusted for thermal conditions or fluid non-Newtonian rheology. The rest of the analysis and formulas remain intact.

The described analysis allows for several important conclusions.

1. Under various conditions when $\mu(p) = \exp(Qp)$ the EHL problem solution exhibits a sharp narrow (pin–like) spike located close to the exit point $x = c$. The above analysis shows that this spike cannot be realized in the exit ϵ_q-zone where the pressure distribution described by function $g(s)$ is a monotonically decreasing function of s as well as it is not realized in the Hertzian region. Therefore, it can be realized only in the exit ϵ_0-zone. Theoretically, the pressure spike width and its height is supposed to decrease as the value of Q increases. It is clear from the fact that for $Q \gg 1$ the characteristic size of the exit ϵ_0-zone is $Q^{-2} \ll 1$ and the order of magnitude of pressure $p(x)$ in the exit zones is $\epsilon_0^{1/2} = Q^{-1} \ll 1$. However, in many numerical solutions the pressure spike height increases significantly as Q grows. Primarily, it is caused by pressure instability in the exit ϵ_0-zone. To describe this pressure spike numerically in a sufficiently accurate manner, the step size of the numerical calculations in the exit ϵ_0-zone has to be significantly smaller than its characteristic size ϵ_0. A similar conclusion can be made about the precision of the value of constant A and the step size in the inlet ϵ_0-zone. For example, if $\mu(p) = \exp(Qp)$ and $Q = 25$ (which is a pretty typical practical situation) we have $\epsilon_0 = Q^{-2} = 0.0016$. If we consider that the inlet and exit ϵ_0-zones should be covered by at least 20 step sizes, then the step size should about 0.00008. This requirement places a significant demand on computer memory and clock speed as the solution should be determined at more than 2.1/0.00008=26,250 nodes. Obviously, calculations with the step size (in the inlet and exit ϵ_0-zones) greater than 0.00008 will produce results with an error greater than

5%. The usually used step sizes for calculations in such cases are at least an order of magnitude larger. In more details the subject of precision and stability of numerical solutions of EHL problems is considered in the following sections.

2. The situation with solution precision is better in cases of pre-critical lubrication regimes but, still, for desired solution precision a proper care should be taken of the step size in numerical calculations. For example, for fully flooded pre-critical regimes with $V = 0.001$, the characteristic size of the inlet zone is $\epsilon_q = V^{2/5} = 0.063$. To have at least 20 nodes in the inlet zone the step size is supposed to be below 0.00315. Therefore, if the original EHL problem needs to be solved with such precision and a constant step size throughout the contact region then it would require to consider problem solution at more than 650 nodes.

3. Depending on the values of the parameters of a lubricated contact, the lubrication film thickness may be described by different formulas. Examples of that are the film thickness formulas (6.63) and (6.149) for fully flooded pre- and over-critical lubrication regimes, respectively. We learned that constant A in the film thickness formula (6.149) for over-critical lubrication regimes is determined by the problem solution in the inlet ϵ_0-zone. However, practically it cannot be determined from the problem solution in the inlet ϵ_0-zone as the actual equations are not completely specified. In spite of that formula (6.149) as well as formula (6.63) are well suited for use in curve fitting of experimentally obtained data. Also, they can be used in curve fitting of sufficiently accurate numerically obtained data (see comment 1).

4. The other conclusion derived from this analysis concerns a large number of existing formulas for film thickness. These formulas are obtained by curve fitting using either numerical or experimental data. The existence of numerous formulas for the film thickness of Newtonian lubricants can be mainly explained by several reasons: (a) by some inaccuracies in the input parameters of experimentally obtained data on film thickness, (b) by using numerically obtained film data obtained for different pre- and over-critical lubrication regimes and treating this data as a homogeneous pool of data for curve fitting while from the above analysis it is obvious that it is not (as the governing relationships are not the same), (c) the sets of data obtained by different numerical methods with various step sizes have very different precision.

6.7 Choosing Pre- or Over-Critical Lubrication Regimes and Small Parameter ω

In this section we consider two very different regimes of lubrication: pre- and over-critical lubrication regimes. Because the solutions for these two regimes are different and, in particular, the formulas for the lubrication film thickness H_0 are different, it is important to learn how to recognize which regime of lubrication is realized in each particular case.

Let us assume that the function of lubricant viscosity $\mu(p)$ is known. There are only two choices for the problem small parameter: $\omega = V$ or ω equal to some small parameter involved in the relationship for viscosity $\mu(p)$. Whatever choice of the small parameter ω is made the characteristic size of the inlet zone in fully flooded pre-critical lubrication regimes is equal to $\epsilon_q = V^{2/5}$ (see the expression for ϵ_q in (6.62)).

Now, we need to determine the critical size ϵ_0 of the inlet zone. It is determined as the maximum solution of equation (6.75). For practical purposes we will replace equation (6.75) by

$$\mu(\epsilon_0^{1/2}) = C, \qquad (6.181)$$

where C is a constant of the order of magnitude of 1, for example, $C = e$. Therefore, the critical size ϵ_0 of the inlet zone is the maximum among all solutions of equation (6.181). If $\mu(p)$ is a monotonically increasing function of p then ϵ_0 is a unique solution of the above equation.

After that, the value of ϵ_0 must be compared to the characteristic size $\epsilon_q = V^{2/5}$ of the inlet region in a pre-critical lubrication regime. If $\epsilon_0 \gg \epsilon_q$, then the lubrication regime is pre-critical, the problem small parameter ω can be taken equal to $V \ll 1$, and in the inlet and exit zones the lubricant viscosity can be assumed to be $\mu(p) = 1$. Moreover, the lubrication film thickness H_0 should be calculated according to the formula $H_0 = A(V\epsilon_q^2)^{1/3}$ for $\epsilon_q \ll V^{2/5}$ or $\epsilon_q = V^{2/5}$ (see (6.55)).

If $\epsilon_0 = O(\epsilon_q) = O(V^{2/5})$, then the problem small parameter ω again can be taken equal to $V \ll 1$ or to the parameter involved in the relationship of $\mu(p)$, which causes $\mu(p)$ to increase significantly as pressure p increases from 0 to 1. In this case the lubrication regime is still pre-critical and the above formula should be used for calculating the values of the film thickness H_0. However, the lubricant viscosity $\mu(p)$ can no longer be assumed to be equal to 1 in the inlet and exit zones.

The above situation can be easily illustrated for the case of an exponential viscosity $\mu(p) = exp(Qp)$. In this case $\epsilon_0 = Q^{-2}$ and the estimate $\epsilon_0 \gg \epsilon_q$ (we need to remember that in a pre-critical regime $max(\epsilon_q) = V^{2/5}$) or $\epsilon_0 = O(\epsilon_q) = O(V^{2/5})$ means that a pre-critical regime is realized if $Q \ll V^{-1/5}$ or $Q = O(V^{-1/5})$ while $V \ll 1$. In other words, a pre-critical regime can occur only when the pressure viscosity coefficient Q is relatively small or moderate in value.

If $\epsilon_0 \ll \epsilon_q = V^{2/5}$, then the lubrication regime is over-critical, the small parameter ω can be chosen as a specific small parameter involved in the expression for the lubricant viscosity $\mu(p)$ that causes its significant growth from $\mu(p) = 1$ for $p = 0$ to $\mu(p) \gg 1$ for $p = 1$. In this case the value of the lubrication film thickness H_0 should be calculated for the specific values of ϵ_0

and ϵ_q based on formula $H_0 = A(V\epsilon_0^{-1/2}\epsilon_q^{5/2})^{1/3}$ (see formula (6.141)) for over-critical regimes (also, see criteria (6.171) and (6.172)).

The latter conditions also can be easily illustrated for the case of an exponential viscosity $\mu(p) = exp(Qp)$ for which $\epsilon_0 = Q^{-2}$ and estimate $\epsilon_0 \ll \epsilon_q = V^{2/5}$ means that an over-critical regime is realized if $Q \gg V^{-1/5}$ while $V \ll 1$. In other words, an over-critical regime can occur only when the pressure viscosity coefficient Q is sufficiently large.

The approaches for replacing the actual viscosity dependence $\mu(p)$ by a standardized exponential like dependence are discussed in Section 9.5.

6.8 Analysis of the Ertel-Grubin Method

The Ertel-Grubin method is an approximate engineering method developed in the forties by Ertel [6] and Grubin [7]. It was used for analysis of heavily loaded lubricated contacts. Initially, this method was developed for the case of an isothermal problem for Newtonian fluids. A more careful analysis of this problem under the same prior assumptions was done by Crook [20]. Later, the Ertel-Grubin method was generalized and applied to studying of a wide class of lubrication problems, including non-isothermal problems, the problems for non-Newtonian lubricants, etc. Some related studies are cited in [29].

Let us analyze the essence of the Ertel-Grubin method in the simplest case of an isothermal problem for smooth elastic cylinders and an incompressible Newtonian lubricant under heavily loaded conditions. Then the problem can be reduced to equations (6.5)-(6.8).

In the preceding sections it is shown that the conditions of heavily loaded contact occur for $V \ll 1$ or $Q \gg 1$. Further, we assume that $Q \gg 1$.

Let us formulate the basic assumptions used in the Ertel-Grubin method.

1. It is assumed that the fluid viscosity is determined by the relation $\mu = exp(Qp)$.

2. At the left boundary of the Hertzian zone (represented by the interval $[-1, 1]$), i.e., in the inlet zone, it is assumed that the following estimate is valid

$$Qp(-1) \gg 1 \; for \; Q \gg 1. \qquad (6.182)$$

Let us reproduce the formal scheme of the Ertel-Grubin method. From (6.5)-(6.7) after one integration with respect to x, we obtain

$$M(p, h) = H_0(h - 1). \qquad (6.183)$$

Taking into account the exponential smallness of $M(p, h)$ in the Hertzian (external) region we find that in the Hertzian region $H_0(h - 1) \ll 1$. Let us take in (6.7) and (6.8) $a + 1 \ll 1$ and $c - 1 \ll 1$. From the equation

$H_0(h - 1) = 0$ in the Hertzian region for $Q \gg 1$, we obtain the function of pressure $p(x)$, which with high accuracy is close to the Hertzian pressure $p(x) = \sqrt{1 - x^2}\theta(1 - x^2)$. Hence, with a corresponding accuracy from (6.7), we get $h(x) = h_0(x)$, where

$$H_0(h_0 - 1) = [|\,x\,|\,\sqrt{x^2 - 1} - \ln(|\,x\,| + \sqrt{x^2 - 1})]\theta(x^2 - 1). \qquad (6.184)$$

Integrating equation (6.183) with respect to x from a to x and taking into account the accepted relationship for $\mu = \exp(Qp)$ and equation (6.184), we obtain

$$1 - \exp[-Qp(x)] = \frac{VQ}{H_0^2} \int\limits_{a}^{x} \frac{h_0(t) - 1}{h_0^3(t)} dt. \qquad (6.185)$$

Let us set $x = -1$ in equation (6.186) and use assumption (6.182). It results in an approximate equation

$$1 = \frac{VQ}{H_{0EG}^2} \int\limits_{a}^{-1} \frac{h_0(t) - 1}{h_0^3(t)} dt \qquad (6.186)$$

for H_{0EG}, which is the Ertel-Grubin lubrication film thickness. Ertel [6] and Grubin [7], by numerically integrating equation (6.186) for the inlet boundary $a = -\infty$ obtained the formula for the film thickness H_{0EG}

$$H_{0EG} = 0.254(VQ)^{8/11}. \qquad (6.187)$$

A more careful asymptotic analysis of the integral in (6.186) with the help of estimate (6.138) produces the formula

$$H_{0EG} = A_0(VQ)^{3/4}, \quad A_0 = (\tfrac{2\sqrt{3}\pi^3}{19683})^{1/4} = 0.272. \qquad (6.188)$$

The same result is received by Crook [20].

In a similar way for starved lubrication regimes, we obtain

$$H_{0EG} = A_1(-\alpha_1)^{5/6}\theta(-\alpha_1)(VQ\epsilon_q^{5/2})^{1/3}, \quad A_1 = (\tfrac{8\sqrt{2}}{15})^{1/3} = 0.91. \qquad (6.189)$$

A little more careful analysis leads to the inequality

$$1 > \frac{VQ}{H_0^2} \int\limits_{a}^{-1} \frac{h(t) - 1}{h^3(t)} dt. \qquad (6.190)$$

Using the assumptions of this section and previously obtained asymptotic estimates for $h(x)$ in the inlet ϵ_q- and ϵ_0-zones for the film thickness H_0 in a fully flooded over-critical lubrication regime, we obtain

$$H_0 > H_{0EG} = A_0(VQ)^{3/4}, \qquad (6.191)$$

where coefficient A_0 is determined from solution of the equation obtained from inequality (6.153) in which the sign of inequality is replaced by the sign of equality.

Although Greenwood's [29] method represents a certain improvement of Ertel-Grubin's method, it is still based on Ertel-Grubin's assumptions. The results arising from the replacement of assumption (6.182) by $Qp(-1) = 0.2$ are discussed by Greenwood [29] where he mentioned several defects of the Ertel-Grubin method.

Obviously, the intended area of applicability of the Ertel-Grubin method is limited to only over-critical lubrication regimes. That can be seen from the agreement of the orders of magnitude of H_0 in (6.149), (6.148) and (6.188), (6.189) for $\omega = Q^{-1} \ll 1$, respectively. For pre-critical lubrication regimes the Ertel-Grubin method is not applicable at all because everywhere in the inlet zone we have $\mu = O(1)$, $\omega \ll 1$, and, hence, assumption (6.182) is not valid. Besides that, in the inlet zone the function of gap $h(x)$ is significantly different from $h_0(x)$ in (6.184).

The great merit of the Ertel-Grubin method was in its simplicity and ability to predict the value of lubrication film thickness (see estimate (6.141) for $\epsilon_0 = Q^{-2} \ll 1$ and (6.189)) of a correct order.

Nevertheless, the Ertel-Grubin method has significant defects. Among them, we can classify the assumption that the approximate expression (6.184) is valid for gap $h(x)$ everywhere in the inlet zone up to point $x = -1$, which is the beginning of the Hertzian region. That leads to ignoring the very existence of the inlet ϵ_0-zone. In other words, this assumption leads to a non-controllable error in calculation of constant A in formula (6.141) for the lubrication film thickness H_0.

Strictly speaking, assumption (6.182) makes the Ertel-Grubin method contradictory and flawed. Actually, according to (6.97), (6.150), and (6.151), we have $Qp(-1) = q(-\alpha_0(\epsilon_0/\epsilon_q)^{1/2}) + \ldots = q(0) + \ldots = O(1)$ for $Q \gg 1$. The latter contradicts assumption (6.182) that $Qp(-1) \gg 1$ for $Q \gg 1$, which can be reformulated as $q(0) \gg 1$ for $Q \gg 1$. In particular, the assumption represented by (6.182) results in an artificially lowered value of the lubrication film thickness calculated by the Ertel-Grubin method compared to the actual values. Therefore, it is not surprising that coefficient A_0 in the formula for film thickness in fully flooded lubrication regime in the Ertel-Grubin equation (6.188) coincides with the low bound for A presented in (6.154).

The fact that the lubrication film thickness values calculated using (6.188) and measured experimentally are close (in spite of the aforementioned artificially lowered value) can be explained by reduction of the actual film thickness due to heat generation, lubricant compressibility, and surface roughness. However, in cases of high temperatures, high rolling and sliding speeds, etc., the deviation between the these theoretical and experimental values of film thickness is significant.

Let us summarize the results of the section. The over-critical lubrication regimes are defined and studied. The detailed consideration is given to the structure of the solution of the problem for heavily loaded lubricated contact in over-critical regimes. It is shown that the contact region consists of five characteristic zones: the zone with prevailing "elastic solution" - the Hertzian region, two different small inlet and small exit zones compared to the size of the Hertzian region.

A two-term asymptotic approximation of the solution is found in the Hertzian region. The expression for the major term of the latter solution depends only on the total applied load and the coordinates of the beginning and end of this region. The expression for the second term is determined by the viscous lubricant flow in the inlet and exit gap between "non-deformable" solids the shapes of which are determined by the major asymptotic term of pressure in the Hertzian region. The unknown coordinates of the beginning and end of the Hertzian region are found as certain functions of pressure in the inlet and exit ϵ_q-zones that are located next to the inlet and exit coordinates of the lubricated contact.

The size of the ϵ_0-inlet zone is defined by the parameter controlling the growth rate of the lubricant viscosity with pressure. In the inlet and exit ϵ_0-zones, the influences of solids' elasticity and the viscous lubrication flow are of the same order of magnitude. The characteristic sizes of the inlet and exit ϵ_0-zones are equal. The second inlet zone (ϵ_q-inlet zone), which is located next to the ϵ_0-inlet zone (ϵ_0-inlet zone is located between the Hertzian region and the ϵ_q-inlet zone) has a characteristic size ϵ_q determined by the position of the inlet coordinate a. The size of the inlet ϵ_q-zone is much larger than the size of the inlet ϵ_0-zone. In the inlet ϵ_q-zone the solution of the problem is determined by the viscous lubricant flow in the inlet gap between almost non-deformable solids whose shape is determined by the first two terms of the pressure approximation in the Hertzian region.

In the inlet and exit ϵ_q-zones the asymptotically valid systems of equations describing the solution of the problem are derived and solved. The solutions of these systems depend on a smaller number of problem parameters than the solution of the original problem. It is shown that regimes of starved and fully flooded lubrication can be realized in the frame work of over-critical regimes. In all cases the characteristic size of the inlet ϵ_q-zone can be considered small compared to the size of the Hertzian region. The lubrication film thickness formula obtained in the analysis of the system depends on the particular regime of lubrication. The limitations of the applied method are derived.

A simple set of rules for determining whether the regime is pre- or over-critical is devised (see estimates (6.171) and (6.172)).

The position and nature of often observed pressure spike in the vicinity of the exit point $x = c$ are discussed. It is stated that the usually abnormal behavior of the numerically obtained pressure spike for large viscosity pressure coefficients Q is caused by solution instability. Certain requirements on the step size for numerical solution of the original EHL problem for pre- and over-critical lubrication regimes are established.

Formulas (6.63) and (6.149) for the lubrication film thickness in pre- and over-critical regimes are well suited for use in curve fitting of experimentally and numerically obtained data.

The existence of numerous film thickness formulas for Newtonian lubricants obtained theoretically and/or experimentally is explained, in particular, based on comparison of the governing relations for the film thickness for different pre- and over-critical regimes.

The described method can be easily extended on the case of non-Newtonian fluids. The proposed analysis reveals the structure of the solution of the problem in over-critical regimes. This analysis provides some aid in numerical solution of the problem.

The approximate Ertel-Grubin and Greenwood methods are analyzed, and the contradictions of their original assumptions and, hence, of the final results are demonstrated.

6.9   Numerical Solution of Asymptotically Valid Equations for Pre-Critical Regimes in the Inlet and Exit Zones for Newtonian Liquids

In the preceding section it was mentioned that the major difficulty encountered in numerical solution of EHL problems for heavily loaded contacts is instability of pressure $p(x)$ in the inlet and mostly exit zones. There are two interconnected sources of this instability in the original equations (6.5)-(6.9). These are due to fast growth of viscosity μ with pressure p (i.e., the presence of a large parameter in μ) and for high μ due to the proximity of these equations to a problem for an integral equation of the first kind

$$x^2 - c^2 + \tfrac{2}{\pi} \int\limits_a^c p(t) \ln \left| \tfrac{c-t}{x-t} \right| dt = 0, \ \int\limits_a^c p(t) dt = \tfrac{\pi}{2},$$

the numerical solutions of which are often unstable. In case of an exponential viscosity $\mu(p) = \exp(Qp)$ the solution instability usually reveals itself for sufficiently large values of Q. For the same reason as the original equations of the EHL problem, the asymptotically valid in the inlet zone equations (6.68), (6.69), (6.71), (6.49), (6.50), and (6.64) for the starved lubrication regimes (or (6.66) for fully flooded lubrication regimes) and in the exit zone equations (6.70), (6.69), (6.72), (6.49), (6.49), and (6.65) for the starved lubrication regimes (or (6.67) for fully flooded lubrication regimes) are also susceptible to numerical instability. In fact, in the above equations the large parameter Q is removed but now functions $q(r)$ and $g(s)$ approach infinity as $r \to \infty$ and $s \to -\infty$, respectively. Therefore, the numerical solutions of these equations may still be unstable due to the necessity to solve them for large r and s for which the equations are still close to integral equations of the first kind.

The numerical instability can be avoided if instead of solving the above equations we solve the pair of equivalent systems (6.69), (6.79), (6.81), and (6.64) for the starved lubrication regimes (or (6.66) for fully flooded lubrication regimes) and (6.69), (6.80), (6.82), and (6.65) for the starved lubrication regimes (or (6.67) for fully flooded lubrication regimes) in the inlet and exit zones, respectively. Theoretically, the above two types of systems of asymptotically valid equations are equivalent. However, numerically the latter asymptotically valid equations in the inlet and exit zones provide the opportunity to get stable converging solutions. It is due to the fact that equations (6.79) and (6.80) are analytically (not numerically) resolved with respect to the major term of the solution and provide stable asymptotic pressure behavior as $r \to \infty$ and $s \to -\infty$, respectively. In this form the equations guarantee the correct solution behavior at infinity (see relationships in (6.49), (6.71), and (6.72)) while for small and moderate r and s the equations are no longer close to integral equations of the first kind. Moreover, numerical solutions (iterations) of the latter systems converge faster for faster–growing lubricant viscosity.

6.9.1   Numerical Solution in the Inlet Zone

Let us consider the inlet region. Function M_0 we will represent in the form

$$M_0(q, h_q, \mu_q, r) = W(q, h_q, \mu_q, r)\frac{dq}{dr}, \ W(q, h_q, \mu_q, r) = \frac{A^3 h_q^3}{\mu_q(q)}. \qquad (6.192)$$

By integrating equations (6.66) and (6.68) with respect to r and using the fact that $h_q(r) \to 1$ and $M_0(q, h_q, \mu_q, r) \to 0$ as $r \to \infty$, we obtain

$$h_q(r) = 1 + \frac{W(q, h_q, \mu_q, r)}{A}\frac{dq}{dr}. \qquad (6.193)$$

Now, we introduce two sets of nodes $r_k = k\triangle r$ and $r_{k+1/2} = 1/2(r_k + r_{k+1})$, $k = 0, \ldots$, where $\triangle r$ is a positive step size. The equations will be solved by iterations. It is clear that constant A is a monotonically increasing function of $\mid \alpha_1 \mid$, $\alpha_1 \leq 0$. That enables us to consider two slightly different approaches to numerical solution of the problem. The first is to solve the problem and determine A and $q(r)$ for the given value of $\alpha_1 < 0$. The other one is based on determining α_1 and $q(r)$ for the given value of A. The second approach allows us to solve the problems in the inlet and exit zones independently of each other.

Let us consider the second approach first. Assuming that $r_{N+1/2}$ is sufficiently large, we will approximate the integral in equation (6.79) at $r = r_k$ as follows

$$\int\limits_0^\infty \frac{d}{dt} M_0(q, h_q, \mu_q, t)\frac{dt}{\sqrt{2t}(t - r_k)}$$

$$\approx \int\limits_0^{r_{1/2}} \frac{d}{dt}[W\frac{dq}{dt}]\frac{dt}{\sqrt{2t}(t - r_k)} + \sum_{j=1}^{N}\int\limits_{r_{i-1/2}}^{r_{i+1/2}} \frac{d}{dt}[W\frac{dq}{dt}]\frac{dt}{\sqrt{2t}(t - r_k)}$$

$$(6.194)$$

$$\approx \frac{d}{dr}[W\frac{dq}{dr}]\mid_{r_1}\int\limits_0^{r_{1/2}}\frac{dt}{\sqrt{2t}(t-r_k)} + \sum_{j=1}^{N}\frac{d}{dr}[W\frac{dq}{dr}]\mid_{r_i}\int\limits_{r_{i-1/2}}^{r_{i+1/2}}\frac{dt}{\sqrt{2t}(t-r_k)}$$

$$\approx \frac{1}{\triangle r^2}[W_{3/2}(q_2-q_1) - W_{1/2}(q_1-q_0)]\frac{1}{\sqrt{2r_k}}\ln\mid\frac{\sqrt{r_{1/2}}-\sqrt{r_k}}{\sqrt{r_{1/2}}+\sqrt{r_k}}\mid +$$

$$\frac{1}{\triangle r^2}\sum_{j=1}^{N}[W_{j+1/2}(q_{j+1}-q_j) - W_{j-1/2}(q_j-q_{j-1})]\frac{1}{\sqrt{2r_k}}$$

$$\times \ln\mid\frac{\sqrt{r_{j+1/2}}-\sqrt{r_k}}{\sqrt{r_{j-1/2}}-\sqrt{r_k}}\frac{\sqrt{r_{j-1/2}}+\sqrt{r_k}}{\sqrt{r_{j+1/2}}+\sqrt{r_k}}\mid.$$

Therefore, by satisfying equation (6.79) at nodes r_k and approximating the integral according to (6.194), we obtain the system of nonlinear equations for $i+1$-st iterates q_k^{i+1} in the form

$$q_k^{i+1} = \sqrt{2r_k} - \frac{1}{2\pi\triangle r^2}\sum_{j=1}^{N}[W_{j+1/2}^{i+1}(q_{j+1}^{i+1}-q_j^{i+1})$$

$$-W_{j-1/2}^{i+1}(q_j^{i+1}-q_{j-1}^{i+1})]\gamma_{jk},\ k=0,\dots,N+1,$$
(6.195)

$$\gamma_{jk} = \ln\mid\frac{\sqrt{r_{3/2}}-\sqrt{r_k}}{\sqrt{r_{3/2}}+\sqrt{r_k}}\mid,\ j=1,$$

$$\gamma_{jk} = \ln\mid\frac{\sqrt{r_{j+1/2}}-\sqrt{r_k}}{\sqrt{r_{j-1/2}}-\sqrt{r_k}}\frac{\sqrt{r_{j-1/2}}+\sqrt{r_k}}{\sqrt{r_{j+1/2}}+\sqrt{r_k}}\mid,\ j>1,$$
(6.196)

where i is the iteration number. By introducing a new set of unknowns $v_k = q_k^{i+1} - q_k^i$, we can propose the following approximation of equations (6.195)

$$q_k^i + v_k + \frac{1}{2\pi\triangle r^2}\sum_{j=1}^{N}[W_{j+1/2}^i(q_{j+1}^i-q_j^i) - W_{j-1/2}^i(q_j^i-q_{j-1}^i)$$

$$+W_{j+1/2}^i(v_{j+1}-v_j) - W_{j-1/2}^i(v_j-v_{j-1})]\gamma_{jk} = \sqrt{2r_k},$$
(6.197)

$$k=0,\dots,N+1.$$

Therefore, we obtained a finite system of $N+2$ linear algebraic equations in $N+2$ unknowns v_k, $k=0,\dots,N+1$. Here the influence factors γ_{jk} are determined from (6.196), N is a sufficiently large integer. Obviously, the approximation error gets smaller as number N increases and $\triangle r$ decreases.

After solution of system (6.197) for values v_k, $k=0,\dots,N+1$, is done the new iterates of pressure q_k^{i+1}, $k=0,\dots,N+1$, are calculated from the formula $q_k^{i+1} = q_k^i + v_k$.

The next step of the process after the new iterates of q_k^{i+1} are obtained is calculation of the new iterates of gap $h_q(r_{k+1/2})$. For regimes of starved lubrication $h_q(r_{k+1/2}) = 1$, $k=0,\dots,N$. For fully flooded lubrication regimes the values of gap $h_q(r_{k+1/2})$, $k = 0, \ldots, N$, can be found by solving the nonlinear equation (6.193) at nodes $r_{k+1/2}$, $k = 0, \ldots, N$, assuming that the values of pressure q_k, $k = 0, \ldots, N+1$, are known. Solution of equation (6.193) is done iteratively based on a modified Newton's method

$$h_q^{i+1}(r_{k+1/2}) = h_q^i(r_{k+1/2}) - D/D_{h_q}, \quad D = h_q^i(r_{k+1/2}) - 1$$

$$-\frac{W(q_{k+1/2}^{i+1}, h_q^i(r_{k+1/2}), \mu_q(q_{k+1/2}^{i+1}), r_{k+1/2})}{A} \frac{q_{k+1}^{i+1} - q_k^{i+1}}{\triangle r},$$

$$D_{h_q} = 1 - \alpha_* \frac{W_{h_q}(q_{k+1/2}^{i+1}, h_q^i(r_{k+1/2}), \mu_q(q_{k+1/2}^{i+1}), r_{k+1/2})}{A} \frac{q_{k+1}^{i+1} - q_k^{i+1}}{\triangle r},$$

$$q_{k+1/2}^{i+1} = \frac{q_{k+1}^{i+1} + q_k^{i+1}}{2}, \quad k = 0, \ldots, N,$$

(6.198)

where α_* is a sufficiently small positive number (for example, $\alpha_* = 0.05$). The introduction of small parameter α_* is due to the fact that the value of derivative D_{h_q} changes its sign (may become equal or close to zero) at some point $r_{k+1/2} > 0$ while for large $r_{k+1/2}$ derivative D_{h_q} is close to 1. The iteration of $h_q(r_{k+1/2})$ should be done in the order from $k = N - 1$ to $k = 0$. The initial approximation for gap $h_q^0(r_{k+1/2})$ can be obtained from the equation

$$h_q^0(r_{k+1/2}) = 1 + \frac{W(q_{k+1/2}^1, 1, \mu_q(q_{k+1/2}^1), r_{k+1/2})}{A} \frac{q_{k+1}^1 - q_k^1}{\triangle r},$$

$$k = 0, \ldots, N.$$

(6.199)

Notice, that the iteration process based on

$$h_q^{i+1}(r_{k+1/2}) = 1$$

$$+\frac{W(q_{k+1/2}^{i+1}, h_q^i(r_{k+1/2}), \mu_q(q_{k+1/2}^{i+1}), r_{k+1/2})}{A} \frac{q_{k+1}^{i+1} - q_k^{i+1}}{\triangle r}, \quad k = 0, \ldots, N$$

(6.200)

instead of (6.198), which also converges but at a slightly slower pace.

After the new iterates of q_k^{i+1} and $h_q^{i+1}(r_{k+1/2})$ are obtained the value of α_1 is determined from equation (6.81). Using the approximation of the integral in (6.81) similar to the one used in (6.194) we obtain the formula

$$\alpha_1 = \frac{1}{\pi \triangle r} \sum_{j=1}^{N} \delta_j \frac{W_{j+1/2}^{i+1}(q_{j+1}^{i+1} - q_j^{i+1}) - W_{j-1/2}^{i+1}(q_j^{i+1} - q_{j-1}^{i+1})}{\sqrt{2r_j}},$$

$$\delta_j = 2, \ j = 1; \ \delta_j = 1, \ j > 1.$$

(6.201)

The initial approximation of q_k can be taken based on a modified asymptotic of the Hertzian pressure such as

$$q_k^0 = \gamma r_k, \ 0 \le r_k \le r_*; \ q_k^0 = \sqrt{2r_k} + \frac{\alpha_1}{\sqrt{2r_k}}, \ r_k > r_*,$$

$$r_* = 0.79A, \ \gamma = 1.06A^{-1/2}, \ \alpha_1 = -0.53A. \tag{6.202}$$

For larger values of A it may be beneficial to take the solution of the problem obtained for A closest to the one at hand as the initial approximation for q_k. That would accelerate the iteration process convergence. The iteration process stops when the desired precision ε is reached.

Now, let us consider the first approach. The only change in the second approach is that coefficient A is no longer constant but changes from iteration to iteration as follows

$$\frac{A^{i+1}}{A^i} = \left\{ \frac{\pi \triangle r \alpha_1}{\sum\limits_{j=1}^{N} \delta_j [W_{j+1/2}^{i+1}(q_{j+1}^{i+1} - q_j^{i+1}) - W_{j-1/2}^{i+1}(q_j^{i+1} - q_{j-1}^{i+1})](2r_j)^{-1/2}} \right\}^{1/3}, \tag{6.203}$$

$$\delta_j = 2, \ j = 1; \ \delta_j = 1, \ j > 1,$$

where A^i is the i-th iterate of coefficient A. All other formulas derived for the first approach remain in force except for the fact that in equations (6.197)-(6.202) A is replaced by A^i, A^{i+1}, and A^0, respectively. As we know, in this approach the value of α_1 can be used to get a better initial approximation for gap $h_q^0(r_{k+1/2})$ than the one determined by equation (6.199). It can be obtained from the equation (see formula (6.144))

$$h_q^0(r_{k+1/2}) = 1 - \frac{4}{3A} \eta_{k+1/2} \sqrt{-2\eta_{k+1/2}} \theta(-\eta_{k+1/2}),$$

$$\eta_{k+1/2} = r_{k+1/2} + \alpha_1, \ k = 0, \dots, N. \tag{6.204}$$

The convergence of the schemes based on the first and second approaches is about the same for small and moderate values of a. Moreover, for starved lubrication regimes they converge very fast. Usually, it takes about 10 iterations to get a solution with the absolute precision of 10^{-4}. For fully flooded lubrication regimes and large $|a|$, $(a < 0)$ the convergence is much slower due to slow convergence of the gap function $h_q(r_{k+1/2})$. For the same solution precision it may take more than 200 iterations, to reach the solution. However, the scheme based on the first approach converges faster than the one based on the second approach. This is due to the fact that usually the initial approximation $h_q^0(r)$ is better and for large $|a|$, $(a < 0)$ coefficient A is very insensitive to even large variations in α_1.

6.9.2 Numerical Solution in the Exit Zone

In a similar fashion a numerical scheme can be proposed in the exit zone. For convenience, in the exit zone we will introduce the new variable $r = -s$. Then, at $r = r_k$ using boundary condition $\frac{dg(0)}{dr} = 0$, the integral in equation (6.80) can be approximated as follows (for comparison see (6.194)):

$$\int\limits_0^\infty \frac{d}{dt} M_0(g, h_g, \mu_g, t) \frac{dt}{\sqrt{2t(t-r_k)}}$$

$$\approx \int\limits_0^{r_{1/2}} \frac{d}{dt}[W\frac{dg}{dt}] \frac{dt}{\sqrt{2t(t-r_k)}} + \sum_{j=1}^N \int\limits_{r_{i-1/2}}^{r_{i+1/2}} \frac{d}{dt}[W\frac{dg}{dt}] \frac{dt}{\sqrt{2t(t-r_k)}}$$

$$\approx \frac{d}{dr}[W\frac{dg}{dr}] \mid_{0.5r_{1/2}} \int\limits_0^{r_{1/2}} \frac{dt}{\sqrt{2t(t-r_k)}}$$

$$+ \sum_{j=1}^N \frac{d}{dr}[W\frac{dg}{dr}] \mid_{r_i} \int\limits_{r_{i-1/2}}^{r_{i+1/2}} \frac{dt}{\sqrt{2t(t-r_k)}} \tag{6.205}$$

$$\approx \frac{2}{\triangle r^2} W_{1/2}(g_1 - g_0) \frac{1}{\sqrt{2r_k}} \ln \mid \frac{\sqrt{r_{1/2}} - \sqrt{r_k}}{\sqrt{r_{1/2}} + \sqrt{r_k}} \mid +$$

$$\frac{1}{\triangle r^2} \sum_{j=1}^N [W_{j+1/2}(g_{j+1} - g_j) - W_{j-1/2}(g_j - g_{j-1})] \frac{1}{\sqrt{2r_k}}$$

$$\times \ln \mid \frac{\sqrt{r_{j+1/2}} - \sqrt{r_k}}{\sqrt{r_{j-1/2}} - \sqrt{r_k}} \frac{\sqrt{r_{j-1/2}} + \sqrt{r_k}}{\sqrt{r_{j+1/2}} + \sqrt{r_k}} \mid .$$

By satisfying equation (6.80) at nodes r_k and approximating the integral according to (6.205), we obtain the system of nonlinear equations for $i + 1$-st iterates g_k^{i+1} in the form

$$g_k^{i+1} + \frac{1}{\pi \triangle r^2} W_{1/2}^{i+1}(g_1^{i+1} - g_0^{i+1}) \ln \mid \frac{\sqrt{r_{1/2}} - \sqrt{r_k}}{\sqrt{r_{1/2}} + \sqrt{r_k}} \mid$$

$$+ \frac{1}{2\pi \triangle r^2} \sum_{j=1}^N [W_{j+1/2}^{i+1}(g_{j+1}^{i+1} - g_j^{i+1}) - W_{j-1/2}^{i+1}(g_j^{i+1} - g_{j-1}^{i+1})] \tag{6.206}$$

$$\times \ln \mid \frac{\sqrt{r_{j+1/2}} - \sqrt{r_k}}{\sqrt{r_{j-1/2}} - \sqrt{r_k}} \frac{\sqrt{r_{j-1/2}} + \sqrt{r_k}}{\sqrt{r_{j+1/2}} + \sqrt{r_k}} \mid = \sqrt{2r_k}, \ k = 0, \dots, N + 1,$$

where i is the iteration number. By introducing a new set of unknowns $v_k = g_k^{i+1} - g_k^i$ and linearizing the equations in the vicinity of g_k^i, we obtain a finite system of $N + 2$ linear algebraic equations in $N + 2$ unknowns v_k, $k = 0, \dots, N + 1$:

$$g_k^i + v_k + \frac{1}{\pi \triangle r^2} [W_{1/2}^i(g_1^i - g_0^i) + W_{1/2}^i(v_1 - v_0)] \ln \mid \frac{\sqrt{r_{1/2}} - \sqrt{r_k}}{\sqrt{r_{1/2}} + \sqrt{r_k}} \mid$$

$$+ \frac{1}{2\pi \triangle r^2} \sum_{j=1}^N [W_{j+1/2}^i(g_{j+1}^i - g_j^i) - W_{j-1/2}^i(g_j^i - g_{j-1}^i) \tag{6.207}$$

$$+ W_{j+1/2}^i(v_{j+1} - v_j) - W_{j-1/2}^i(v_j - v_{j-1})] \gamma_{jk} = \sqrt{2r_k},$$

$$k = 0, \ldots, N + 1,$$

where N is a sufficiently large integer and constants γ_{jk} are determined in (6.195). Obviously, the approximation error gets smaller as number N increases and $\triangle r$ decreases.

After solution of system (6.207) for values v_k, $k = 0, \ldots, N + 1$, is done the new iterates of pressure g_k^{i+1}, $k = 0, \ldots, N + 1$, are calculated from the formula $g_k^{i+1} = g_k^i + v_k$.

Having the new iterates of g_k^{i+1} we can calculate the new iterates of gap $h_g(r_{k+1/2})$ in the fashion similar to the one used in the inlet zone. For regimes of starved lubrication $h_g(r_{k+1/2}) = 1$, $k = 0, \ldots, N$. For fully flooded lubrication regimes the values of the gap $h_g(r_k)$, $k = 0, \ldots, N$, can be found by solving the nonlinear (Reynolds) equation

$$h_g(r) = 1 - \frac{W(g, h_g, \mu_g, r)}{A} \frac{dg}{dr} \qquad (6.208)$$

at nodes $r_{k+1/2}$, $k = 0, \ldots, N$. As in the inlet zone, solution of equation (6.208) can be done iteratively based on a modified Newton's method

$$h_g^{i+1}(r_{k+1/2}) = h_g^i(r_{k+1/2}) - D/D_{h_g}, \ D = h_g^i(r_{k+1/2}) - 1$$

$$+ \frac{W(g_{k+1/2}^{i+1}, h_g^i(r_{k+1/2}), \mu_g(g_{k+1/2}^{i+1}), r_{k+1/2})}{A} \frac{g_{k+1}^{i+1} - g_k^{i+1}}{\triangle r},$$

$$D_{h_g} = 1 + \alpha_* \frac{W_{h_g}(g_{k+1/2}^{i+1}, h_g^i(r_{k+1/2}), \mu_g(g_{k+1/2}^{i+1}), r_{k+1/2})}{A} \frac{g_{k+1}^{i+1} - g_k^{i+1}}{\triangle r}, \qquad (6.209)$$

$$g_{k+1/2}^{i+1} = \frac{g_{k+1}^{i+1} + g_k^{i+1}}{2}, \ k = 0, \ldots, N,$$

(α_* is a sufficiently small positive number) or based on the iteration process

$$h_g^{i+1}(r_{k+1/2}) = 1$$

$$- \frac{W(g_{k+1/2}^{i+1}, h_g^i(r_{k+1/2}), \mu_g(g_{k+1/2}^{i+1}), r_{k+1/2})}{A} \frac{g_{k+1}^{i+1} - g_k^{i+1}}{\triangle r}, \ k = 0, \ldots, N. \qquad (6.210)$$

Opposite to the case of the inlet zone the iteration process (6.210) converges as fast as the one presented in (6.209).

The initial approximation for gap $h_g^0(r_{k+1/2})$ can be taken as $h_g^0(r_{k+1/2}) = 1$, $k = 0, \ldots, N$, or it can be obtained from the equation

$$h_g^0(r_{k+1/2}) = 1$$

$$- \frac{W(g_{k+1/2}^1, 1, \mu_g(g_{k+1/2}^1), r_{k+1/2})}{A} \frac{g_{k+1}^1 - g_k^1}{\triangle r}, \ k = 0, \ldots, N. \qquad (6.211)$$

After the new iterates of g_k^{i+1} and $h_g^{i+1}(r_{k+1/2})$ are obtained, the value of β_1 is determined from equation (6.82) in which s is replaced by $-r$. Using the approximation of the integral in (6.82) similar to the one used in the inlet zone, we obtain the formula

$$\beta_1 = 2\frac{W_{1/2}(g_1^{i+1}-g_0^{i+1})}{\pi\triangle r\sqrt{2\triangle r}}$$

$$+\frac{1}{\pi\triangle r}\sum_{j=1}^{N}\frac{W_{j+1/2}^{i+1}(g_{j+1}^{i+1}-g_j^{i+1})-W_{j-1/2}^{i+1}(g_j^{i+1}-g_{j-1}^{i+1})}{\sqrt{2r_j}}.$$

(6.212)

The initial approximation of g_k can also be taken as a modified asymptotic of the Hertzian pressure, however, different from the one used in the inlet zone, i.e.,

$$g_k^0 = \gamma r_k^2, \ 0 \le r_k \le r_*; \ g_k^0 = \sqrt{2r_k} - \frac{\beta_1}{\sqrt{2r_k}}, \ r_k > r_*,$$

(6.213)

$$r_* = 0.3A, \ \gamma = 3.37A^{-3/2}, \ \beta_1 = 0.36A.$$

In the exit zone the patterns of convergence for starved and fully flooded lubrication regimes are similar to the ones in the inlet region. However, for fully flooded lubrication regimes the convergence of this scheme with the absolute precision of 10^{-4} is much faster than in the inlet region. This is mainly due to the fact that the function of gap $h_g(r_{k+1/2})$ is everywhere relatively close to 1. Therefore, the convergence process usually takes less than 20 iterations.

The other major difference between the solution convergence in the inlet and exit zones is the fact that for elevated values of Q_0 (for $\mu_q(q) = e^{Q_0 q}$ and $\mu_g(g) = e^{Q_0 g}$) and A in the inlet zone the solutions remain stable while in the exit zone they become unstable. This instability manifests itself in two different ways: the elevated sensitivity of $\max g(s)$ as a function of N and $\triangle r$ as well as in small oscillations of the derivative of pressure $dg(s)/ds$ and gap $h_g(s)$ between the position of the pressure spike and the exit point $s = 0$. The latter type of instability in the region adjacent to the exit point, when it manifests itself, is easily corrected by using simple averaging of the gap values, i.e., $h_g(r_{k-1/2}) = 0.5[h_g(r_{k-1/2}) + h_g(r_{k+1/2})]$. That slightly increases the discrepancy in the Reynolds equation. However, it still remains within acceptable limits. Because of this regularization treatment of gap $h_g(s)$, we need to take a special care of the values of $h_g(\triangle r/2)$ and $h_g(3\triangle r/2)$. That is done by using the fact that $h_g(0) = 1$ and by employing a cubic interpolation as follows

$$h_g(r) = \frac{1}{\triangle r^3}\{-\frac{8}{315}(r - r_{5/2})(r - r_{7/2})(r - r_{9/2})$$

$$+\frac{1}{5}h_g(r_{5/2})r(r - r_{7/2})(r - r_{9/2})$$

$$-\frac{2}{7}h_g(r_{7/2})r(r - r_{5/2})(r - r_{9/2})$$

$$+\frac{1}{9}h_g(r_{9/2})r(r - r_{5/2})(r - r_{7/2})\}, \ r = r_{1/2}, \ r_{3/2}.$$

(6.214)

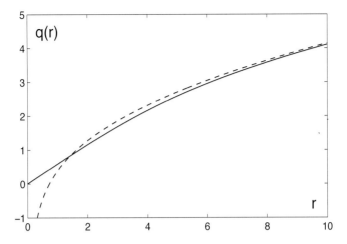

FIGURE 6.10
Main term of the asymptotic of the pressure distribution $q(r)$ (solid curve) and the asymptote of the Hertzian pressure $q_a(r)$ (dashed curve) in the inlet zone of a starved lubricated contact for $Q_0 = 1$ and $A = 2$. (after Kudish and Covitch [3]). Reprinted with permission from CRC Press.

The sensitivity of $\max g(s)$ as a function of N and $\triangle r$ is much harder to treat. Unfortunately, the usual regularization techniques do not work. A simple and very effective method of regularization for numerical solution of asymptotically valid problems in the inlet and exit zones will be proposed in Section 8.2. A practically identical regularization approach will be employed for solution of the original EHL problem.

6.9.3 Some Numerical Results for Pre-Critical Regimes in the Inlet and Exit Zones

The above–described numerical methods converge well to stable solutions in the inlet zone for starved and fully flooded lubrication regimes. In the exit zone they also converge well for sufficiently small values of A and Q_0 (see the discussion below). Obviously, for starved lubrication regimes in the inlet and exit zones the equations are nonlinear if μ_q and μ_g depend on $q(r)$ and $g(s)$, respectively. Otherwise, the equations are linear and their solution is achieved in just one iteration.

Let us consider the case of an exponential viscosity for which $\mu(p) = \exp(Qp)$. First, we will consider regimes of starved lubrication for which $\epsilon_q^{1/2} = Q^{-1} \ll V^{1/5}$, $Q \gg 1$ and, therefore, we have $\mu_q(q) = e^q$ and

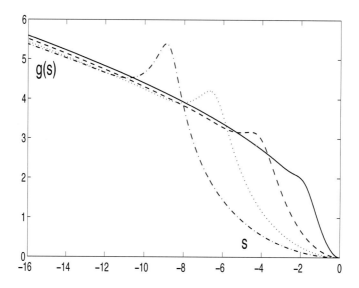

FIGURE 6.11

Main term of the asymptotic of the pressure distribution $g(s)$ in the exit zone of a starved lubricated contact for $Q_0 = 1$, $A = 1$ (solid curve), $A = 2$ (dashed curve), $A = 3$ (dotted curve), and $A = 4$ (dash-dotted curve). (after Kudish and Covitch [3]). Reprinted with permission from CRC Press.

$\mu_g(g) = e^g$ (i.e., $Q_0 = 1$). The absolute error of calculations was chosen to be not higher than $\varepsilon = 10^{-4}$. To check the convergence of the numerical scheme for $A = 2$, three series of calculations were done for $N = 400$, $\triangle r = 0.0625$, $N = 800$, $\triangle r = 0.03125$, and $N = 1,600$, $\triangle r = 0.015625$ ($N\triangle r = 25$). The solution precision was reached after 8-13 iterations. The maximum relative errors of the solutions obtained for $N = 400$, $N = 800$ as well as for $N = 1,600$ in the values of α_1, β_1, $q(r)$, and $g(s)$ were found to be not greater than 0.58% and 0.33%. Therefore, the rest of calculations was done for $N = 800$, $\triangle r = 0.03125$. The relative error in the integral conditions $\int\limits_0^\infty [q(r) - q_a(r)]dr = 0$ and $\int\limits_{-\infty}^0 [g(s) - g_a(s)]ds = 0$ did not exceed 0.7%. The proximity of these integrals to zero serves as a gauge whether the product $N\triangle r$ is large enough to provide sufficient solution precision. The graphs of $q(r)$ for $A = 2$ and $g(s)$ for four values of the parameter $A = 1$, 2, 3, and 4 are given in Fig. 6.10 and 6.11. For these values of A the corresponding values of α_1 are equal to -0.5016, -1.4045, -2.5244, and -3.9783 while the values of β_1 are equal to 0.3646, 0.8966, 1.5017, and 2.1552. Obviously, the values of $\mid \alpha_1 \mid$ and β_1 increase as the value of A does. For the above values of A

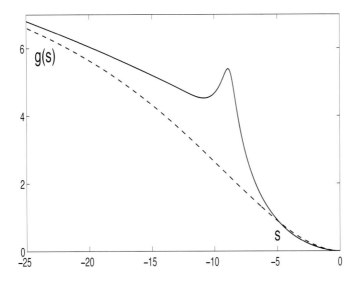

FIGURE 6.12
Main term of the asymptotic of the pressure distribution $g(s)$ in the exit zone of a starved lubricated contact for $A = 4$ and viscosity $\mu_g(g) = 1$ (dashed curve) and $\mu_g(g) = e^g$ (solid curve). (after Kudish and Covitch [3]). Reprinted with permission from CRC Press.

the curves of $q(r)$ resemble each other. Therefore, just one curve of $q(r)$ (solid curve) and for comparison the curve of the Hertzian pressure asymptote $q_a(r)$ from (6.49) (dashed curve) for $A = 2$ are given in Fig. 6.10. It can be clearly seen from Fig. 6.10 that $q(r)$ is a monotonically increasing function of r which approaches its asymptote $q_a(r)$ and does not exhibit any signs of instability or oscillations, in particular, for large r. Figure 6.11 demonstrates the behavior of all four curves of $g(s)$ (solid, dashed, dotted, and dash-dotted curves correspond to $A = 1$, 2, 3, and 4, respectively). Each of the curves of $g(s)$ possesses a local maximum (pressure spike) that shifts closer to the center of the contact (to $s = -\infty$) and increases in value as the values of $\mid \alpha_1 \mid$ and A increase. As in the case of $q(r)$ for large s the behavior of $g(s)$ practically coincides with the behavior of its asymptote $g_a(s)$ and does not exhibit any oscillations or signs of instability.

Let us consider some starved lubrication regimes for which the viscosity in the inlet and exit zones can be considered constant. In these cases the problem equations are linear and their solution does not require iterations. For an exponential viscosity with $\mu(p) = \exp(Qp)$ and for regimes of starved lubrication with $\epsilon_q^{1/2} \ll Q^{-1} \ll V^{1/5}$, $Q \gg 1$, we have $\mu_q(q) = \mu_g(g) =$

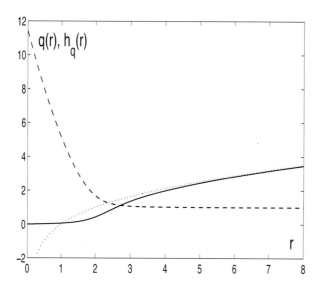

FIGURE 6.13

Main terms of the asymptotic distributions of pressure $q(r)$ (solid curve), gap $h_q(r)$ (dashed curve), and Hertzian pressure asymptote $q_a(r)$ (dotted curve) in the inlet zone of a fully flooded lubricated contact for $A = 0.525$ and $Q_0 = 1$. (after Kudish and Covitch [3]). Reprinted with permission from CRC Press.

1 and $Q_0 = 0$. Because of solutions in the inlet and exit zones approach their asymptotes slower than for the case of exponential viscosity, we will do calculations for $N = 1,600$ and $\triangle r = 0.03125$. We will consider solutions of the problem in the inlet and exit zones for $A = 1$, 2, 3, and 4 for which the values of α_1 are equal to -0.4759, -1.3905, -2.5721, and -3.9444 while the values of β_1 are equal to 0.5080, 1.4363, 2.6561, and 4.1274, respectively. The numerical solutions for $q(r)$ in the inlet zone qualitatively resemble the one in Fig. 6.10. They are monotonically increasing functions of r which approach $\sqrt{2r}$ as $r \to \infty$, however, at a much slower pace for larger values of A than for the corresponding cases of $\mu_q(q) = e^q$ described above. From the data presented above it follows that for the same values of constant A the values of α_1 for the cases of $\mu_q(q) = 1$ and $\mu_q(q) = e^q$ do not vary much. Effectively, the above behavior indicates that for the same values of ϵ_q and A the inlet zone is wider in the cases of constant viscosity ($\mu_q(q) = 1$) compared to the cases when $\mu_q(q) = e^q$. That requires an increased number of nodes N for numerical solution in the former cases.

For starved lubrication regimes and constant viscosity $\mu_g(g) = 1$ in the exit zone the solutions for $g(s)$ are monotonically decreasing functions of s, which

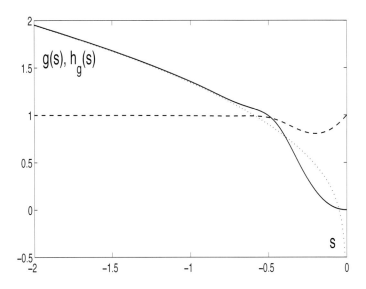

FIGURE 6.14
Main terms of the asymptotic distributions of pressure $g(s)$ (solid curve), gap $h_g(s)$ (dashed curve), and the Hertzian pressure asymptote $g_a(s)$ (dotted curve) in the exit zone of a fully flooded lubricated contact for $A = 0.525$ and $Q_0 = 2.5$. (after Kudish and Covitch [3]). Reprinted with permission from CRC Press.

approach $\sqrt{-2s}$ as $s \to -\infty$. Also, it happens slower than in the corresponding cases of $\mu_g(g) = e^g$. Therefore, for $\epsilon_q^{1/2} \ll Q^{-1} \ll V^{1/5}$, $Q \gg 1$ and $\mu_g(g) = 1$ the solution functions $g(s)$ differ significantly from the solutions for $g(s)$ obtained for $\epsilon_q^{1/2} = Q^{-1} \ll V^{1/5}$, $Q \gg 1$, $\mu_g(g) = e^g$, and depicted in Fig. 6.11. That can be clearly seen from Fig. 6.12 in which for $A = 4$ graphs of two functions $g(s)$ for $\mu_g(g) = 1$ (dashed curve) and $\mu_g(g) = e^g$ (solid curve) are presented. Moreover, from the presented data it is clear that for regimes of starved lubrication for the same values of ϵ_q and constant A the exit zone for constant viscosity is also larger than for exponential viscosity ($\mu_g(g) = e^g$) and the difference in size increases as A increases. Therefore, the numerical solution in the former case requires an increased number of nodes N.

It is important to realize that for starved lubrication regimes two different solutions in both the inlet and exit zones can be converted into each other by a simple transformation as long as for both solutions the value of $Q_0 A^{3/4}$ is the same. Moreover, for small values of Q_0 for starved lubrication regimes it is easy come up with simple analytical formulas for A and β_1 as functions of

α_1. By employing the following transformation

$$(\widetilde{r}, \widetilde{s}, \widetilde{\alpha_1}, \widetilde{\beta_1}) = \gamma^2(r, s, \alpha_1, \beta_1), \quad (\widetilde{q}, \widetilde{g}) = \gamma(q, g), \quad \widetilde{A} = \gamma^{4/3} A \qquad (6.215)$$

the asymptotically valid equations (6.69), (6.79), (6.81), and (6.64) in the inlet zone and equations (6.69), (6.80), (6.82), and (6.65) in the exit zone remain the same if variables $(r, \alpha_1, A, q, h_q, \mu_q, s, \beta_1, g, h_g, \mu_g)$ are replaced by $(\widetilde{r}, \widetilde{\alpha_1}, \widetilde{A}, \widetilde{q}, h_q, \widetilde{\mu_q}, \widetilde{s}, \widetilde{\beta_1}, \widetilde{g}, h_g, \widetilde{\mu_g})$. Here $\mu_q = \exp(Q_0 q)$, $\mu_g = \exp(Q_0 g)$ and $\widetilde{\mu_q} = \exp(Q_* \widetilde{q})$, $\widetilde{\mu_g} = \exp(Q_* \widetilde{g})$, $Q_* = Q_0 \gamma^{-1}$. Therefore, in the inlet zone for the input parameters Q_0 and A the original problem for $q(r)$ and α_1 is equivalent to the modified problem for $\widetilde{q}$ and $\widetilde{\alpha_1}$ for the input parameters $Q_* = Q_0 A^{3/4}$ and $\widetilde{A} = 1$ if $\gamma = A^{-3/4}$, respectively, i.e., these two problems are described by the same equations with the above mentioned replacements. The same situation takes place in the exit zone. Based on the latter and the physics of the lubrication process, we can conclude that parameters $\widetilde{\alpha_1}$ and $\widetilde{\beta_1}$ are certain monotonically increasing functions f_i and f_e of just one parameter Q_*, i.e., $\alpha_1 = A^{3/2} f_i(Q_0 A^{3/4})$ and $\beta_1 = A^{3/2} f_e(Q_0 A^{3/4})$. Therefore, for $Q_0 = 0$ by inverting the latter formulas for α_1 and β_1, we obtain

$$A = A_1 \mid \alpha_1 \mid^{2/3}, \quad \beta_1 = \beta_{10} \mid \alpha_1 \mid, \quad A_1 = 1.64, \quad \beta_{10} = 1.054, \qquad (6.216)$$

where constants A_1 and β_{10} are the values of A and β_1 obtained from the numerical solution of the original systems of asymptotically valid in the inlet and exit zones equations for $Q_0 = 0$ and $\alpha_1 = -1$. Formulas (6.216) can be used for pre-critical starved lubrication regimes for $Q_0 = 0$ in conjunction with the formula for the film thickness H_0 for Newtonian lubricant (6.55), i.e.,

$$H_0 = 1.64(a + 1)^{2/3} V^{1/3}, \qquad (6.217)$$

where a is the inlet coordinate.

Moreover, for $\mu_q = \exp(Q_0 q)$ and $\mu_g = \exp(Q_0 g)$ as the value of the parameter $Q_* = Q_0 A^{3/4}$ increases in the inlet zone the solution approaches its asymptotic a bit slower and in the exit zone the maximum of the pressure spike increases and it moves away from the exit point toward the center of the contact region.

Now, let us consider some results for pre-critical fully flooded lubrication regimes. Let us consider some examples for exponential viscosity $\mu(p) = \exp(Qp)$. Then for $\epsilon_q = V^{2/5} = O(Q^{-2})$, $Q \gg 1$, we have $\mu_q(q) = e^{Q_0 q}$, $\mu_g(g) = e^{Q_0 g}$, and $Q_0 = Q V^{1/5} = O(1)$. For fully flooded lubrication regimes in the inlet zone the iterations starting with the initial approximation from (6.202) also converge to stable solutions, however, it takes more iterations. For example, for $Q_0 = 1$, $A = 0.525$, $N = 480$, $\triangle r = 0.03125$, and the absolute precisions $\varepsilon = 0.01$, $\varepsilon = 0.001$, and $\varepsilon = 0.0001$, the solutions in the inlet zone converged after 156, 333, and 517 number of iterations, respectively. The solutions obtained for these three levels of precision are as follows: $\alpha_1 = -1.8236$, $h_q(\triangle r/2) = 10.6001$, $\alpha_1 = -1.9072$, $h_q(\triangle r/2) = 11.2607$, and

TABLE 6.1

The dependence of coefficient A and the inlet gap $h_q(\triangle r/2)$ on the inlet coordinate α_1. (after Kudish and Covitch [3]). Reprinted with permission from CRC Press.

α_1	A	$h_q(\triangle r/2)$
0	0	1
-0.016	0.100	1.079
-0.059	0.200	1.238
-0.142	0.300	1.527
-0.322	0.400	2.168
-0.992	0.500	5.262
-1.930	0.525	11.342
-4.325	0.535	33.804

$\alpha_1 = -1.9160$, $h_q(\triangle r/2) = 11.3310$, respectively. For $Q_0 = 1$, $A = 0.525$, $N = 800$, and $\triangle r = 0.015625$, the graphs of $q(r)$, $h_q(r)$, and the asymptote $q_a(r)$ of the Hertzian pressure are given in Fig. 6.13.

For $Q_0 = 1$ and several values of coefficient A the values of the inlet coordinate α_1 and gap $h_q(\triangle r/2)$ are presented in Table 6.1. This information gives an idea of how quickly lubricant starvation develops (i.e., the proximity of the inlet coordinate α_1 to zero) and how it affects the lubrication film thickness H_0, which is directly proportional to coefficient A (see formula (6.63)). It is not unexpected that when coefficient A approaches its limiting value the gap at the inlet point increases (without bound). Therefore, for values of A close to its limiting value the number of iterations required to get a converged solution increases. Note, that $A = 0.525$ is relatively close to its limiting value which requires so many iterations. The evidence of that is in the fact that for $Q_0 = 1$, $A = 0.535$, $N = 800$, $\triangle r = 0.015625$, and $\varepsilon = 0.001$ we have $\alpha_1 = -4.325$ and $h_q(\triangle r/2) = 33.804$, i.e., the value of $h_q(\triangle r/2)$ almost tripled while the value of A is increased from 0.525 to 0.535 by just 0.01 (or 1.9%). Also, the data from Table 6.1 shows that the size of the entire inlet zone is usually very small, i.e., for $V \ll 1$ it is about $6V^{2/5} \ll 1$. Based on the numerical analysis of the fully flooded inlet zone and formula (6.63) for $|a| \gg 1$, $a < 0$, we can obtain some formulas for the lubrication film thickness H_0 as follows

$$H_0 = AV^{3/5}, \quad A = 0.535, \quad Q_0 = 1; \quad A = 0.676, \quad Q_0 = 2. \qquad (6.218)$$

In the exit zone, for fully flooded pre-critical regimes for $Q_0 = 1$, $A = 0.525$, $N = 800$, $\triangle r = 0.00390625$, and $\varepsilon = 0.0001$ the solution converges after 17 iterations, and we get $\beta_1 = 0.120$ and $\min h_g(s) = 0.773$. It is inter-

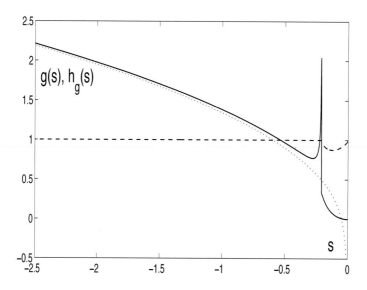

FIGURE 6.15
Main terms of the asymptotic distributions of pressure $g(s)$ (solid curve),
gap $h_g(s)$ (dashed curve), and the Hertzian pressure asymptote $g_a(s)$ (dotted
curve) in the exit zone of a fully flooded lubricated contact for $A = 0.525$
and $Q_0 = 10$. (after Kudish and Covitch [3]). Reprinted with permission from
CRC Press.

esting that for pre-critical fully flooded lubrication regimes in the exit zone
for $Q_0 \leq 2.5$ the pressure distribution $g(s)$ instead of having a pressure spike
behaves monotonically with a mild "hump." However, for larger values of Q_0
we get a pressure spike which increases in height and becomes very thin as
Q_0 increases. This pressure and gap behavior are illustrated in Fig. 6.14 and
6.15 the data for which is obtained for $A = 0.525$, $\varepsilon = 0.001$, $Q_0 = 2.5$, and
$Q_0 = 10$, respectively. The general pressure hump/spike behavior is similar to
the one considered under starved lubrication conditions.

6.10 Numerical Validation of the Asymptotic Analysis. Some Additional Properties of the Original EHL Problem Solutions

To validate the asymptotic approach developed earlier for pre-critical lubrication regimes, let us compare the solutions of the EHL problem in the asymptotic and original formulations. Generally, it can be done in two different ways: on the conceptual and detailed numerical levels.

On the conceptual level it can be done as follows. If the asymptotic approach is valid, then for Newtonian fluids for fully flooded pre-critical lubrication regimes the formula for the film thickness $H_0 = AV^{3/5}$ (see formula (6.63)) should be correct. More specifically, for large enough $\mid a \mid$, $a < 0$, (which can be judged by the value of $h(a)$ in comparison with 1) the values of coefficient A and $\beta_1 = (c-1)/V^{2/5}$ (see formulas (6.35), (6.36), (6.62)) are supposed to be functions of only $Q_0 = QV^{1/5}$. For the asymptotic method to be valid, the characteristic size of the inlet zone $\epsilon_q = V^{2/5}$ (see formula (6.62)) should be small and the lubrication regime is supposed to be pre-critical. Therefore, for sufficiently large $\mid a \mid$, $a < 0$, practically fixed value of the parameter $Q_0 = QV^{1/5}$ (because Q_0 is fixed while V is small) and for different values of parameter V the value of coefficient $A = H_0 V^{-3/5}$ should be constant. Let us consider two examples. For $N = 1,601$, $\triangle y = 0.00125$, $V = 0.05$ and $V = 0.1$ the values of $\epsilon_q = V^{2/5} = 0.302$ and $\epsilon_q = V^{2/5} = 0.398$ are relatively small. For $Q = 0$ we have $\mu(\epsilon_q^{1/2}, Q) = \exp(\epsilon_q^{1/2}Q) = 1$, which means that the lubrication regime is pre-critical (see definition of precritical regimes (6.75)). Notice that the solution of the original EHL problem for $a = -2$, $V = 0.05$, and $Q = 0$ gives $H_0 = 0.066$ and $c = 1.030$ while for $a = -2$, $V = 0.1$ the solution data is presented in Table 8.3. Therefore, for $a = -2$, $Q = 0$, $V = 0.05$, and $V = 0.1$ the values of coefficient $A = H_0 V^{-3/5}$ are equal to 0.401 and 0.404 while the values of $\beta_1 = (c-1)/V^{2/5}$, are equal to 0.098 and 0.091, respectively. The differences between these pairs of values of A and β_1 are of about 0.75% and 7%. For smaller values of V the agreement between the solutions of the asymptotic and original EHL problems is even better. That validates the asymptotic approach on the conceptual level.

Validation of the asymptotic approach on the detailed numerical level involves several steps that include comparison of numerical solutions of asymptotic and original EHL problems for pre-critical lubrication regimes and matching some properties of the asymptotic and original EHL problems for both pre- and over-critical lubrication regimes. Let us start with the comparison of numerical solutions of asymptotic and original EHL problems for Newtonian fluids in pre-critical lubrication regimes. To make this comparison we will have to go through several steps. First, for the given values of V, Q, and a we determine a solution of the original EHL problem. Then we solve the asymptotic equations for $Q_0 = QV^{1/5}$ and $\alpha_1 = (a+1)V^{-2/5}$ for the same values of parameters V, Q, and a. After that we compare the values of the film thickness H_0 from formula (6.63) obtained from numerical solutions of the asymptotic and original EHL problems. Similarly, we compare the values of exit coordinate c and $\min h$ obtained from numerical solution of the asymptotic and original EHL problems. Let us do the comparison for a series of three solutions obtained for $a = -2$, $V = 0.05$ ($\epsilon_q = V^{2/5} = 0.302$), $Q = 0$ ($Q_0 = 0$), $Q = 1.821$ ($Q_0 = 1$), and $Q = 3.641$ ($Q_0 = 2$) (based on the relationship $Q = Q_0 V^{1/5}$) with the corresponding asymptotic solutions. For all these solutions $\epsilon_q \doteq V^{2/5} = 0.302$ and $\alpha_1 = -3.314$ (see formula $a = -1 + \alpha_1 \epsilon_q$). The solutions of the original EHL problem are represented in Table 6.2. In Table 6.3 we present the values of $H_{0(asym)}$, $c_{(asym)}$, and $\min h_g(s)$ obtained from numerical solutions of the asymptotically valid equations in the inlet and exit zones and formulas $H_0 = AV^{3/5}$ and $c = 1 + \beta_1 \epsilon_q$ for the same values of parameters Q, V, and $\alpha_1 = -3.314$. The original EHL problem is solved for $N = 1,325$, $\triangle y = 0.001509434$ while the asymptotic problems are solved for $N = 1000$, $\triangle r = 0.0075$. Notice that the step sizes $\triangle y = 0.001509434$ and $\triangle r = 0.0075$ approximately satisfy the relationship $\triangle y = \frac{2\epsilon_q}{c-a} \triangle r$ obtained in Section 6.11.

TABLE 6.2
Parameters H_0, c, and $\min h(y)$
obtained from solution of the original
EHL problem for $a = -2$, $V = 0.05$,
and different values of Q. (after
Kudish and Covitch [3]). Reprinted
with permission from CRC Press.

Q (Q_0)	H_0	c	$\min h(y)$
0 (0)	0.066	1.030	0.793
1.821 (1)	0.089	1.038	0.771
3.641 (2)	0.110	1.044	0.761

From the data presented in Tables 6.2 and 6.3, it is clear that in spite of the fact that $\epsilon_q = 0.302$ is not very small the agreement between the solutions of the original and asymptotically valid equations of the EHL problem is very good. Moreover, the comparison of the values of the film thickness H_0, minimum gap $\min h$, and exit coordinate c obtained from the numerical solution of the original EHL problem and the asymptotic ones shows that the difference is smaller or equal than 2.7%. The precision of the asymptotic solution becomes even better for smaller values of V.

Now, let us validate the asymptotic approach differently. The asymptotic equations for Newtonian fluids in pre-critical lubrication regimes (6.69),

TABLE 6.3
Parameters $H_{0(asym)}$, $c_{(asym)}$, and
$\min h_g(s)$ obtained from solution of the
asymptotic EHL problem for $\alpha_1 = -3.314$
and different values of $Q(Q_0)$. (after Kudish
and Covitch [3]). Reprinted with permission
from CRC Press.

Q (Q_0)	$H_{0(asym)}$	$c_{(asym)}$	$\min h_g(s)$
0 (0)	0.0642	1.0291	0.791
1.821 (1)	0.0883	1.0347	0.770
3.641 (2)	0.1106	1.0391	0.760

(6.79), (6.81), and (6.64) for the starved lubrication regimes (or (6.66) for
fully flooded lubrication regimes) and (6.69), (6.80), (6.82), and (6.65) for the
starved lubrication regimes (or (6.67) for fully flooded lubrication regimes)
show that their solutions depend only on two parameters α_1 and $Q_0 = QV^{1/5}$.
Therefore, if the asymptotic analysis for pre-critical lubrication regimes is
valid, then we can expect that for different values of parameter V and the
same values of parameters α_1 and Q_0 solutions of the original EHL problem
suppose to exhibit a property that the value of coefficient $A = H_0 V^{-3/5}$ is
constant. Let us examine these statement using two series of solutions of the
original EHL problem obtained for $V = 0.05$, $a = -2$ $(\alpha_1 = -3.314)$ and
$V = 0.01$, $a = -1.525234$ $(\alpha_1 = -3.314)$. In both series of calculations, we
will assume that $N = 1,325$, $\triangle y = 0.001509434$ and $Q_0 = 1$ and $Q_0 = 2$.
These values of Q_0 clearly indicate (see the definition of pre-critical lubrication regimes (6.75)) that the lubrication regime is pre-critical. The results of
these calculations are presented in Table 6.4.

TABLE 6.4
Parameters H_0, c, $h(\triangle y/2)$, and $\min h(y)$ obtained from
solution of the original EHL problem for cases of $V = 0.05$, 0.01,
and $Q_0 = 1$, 2. (after Kudish and Covitch [3]). Reprinted with
permission from CRC Press.

a	V	Q (Q_0)	H_0	c	$h(\triangle y/2)$	$\min h(y)$
-2.000	0.05	1.821 (1)	0.0887	1.0383	24.880	0.7708
-2.000	0.05	3.641 (2)	0.1097	1.0442	20.268	0.7607
-1.525	0.01	2.512 (1)	0.0337	1.0211	23.473	0.7703
-1.525	0.01	5.024 (2)	0.0419	1.0242	19.023	0.7595

From the data of Table 6.4 for $V = 0.05$, 0.01 and $Q_0 = 1$, we have
$A = H_0 V^{-3/5} = 0.5349$ and 0.5336, respectively, and for $V = 0.05$, 0.01 and
$Q_0 = 2$ we have $A = H_0 V^{-3/5} = 0.6617$ and 0.6640, respectively. Therefore,

the difference between the corresponding values of constant A is less than 0.34%, which again validates the asymptotic analysis for pre-critical lubrication regimes.

Finally, let us find out if the asymptotic equations derived for Newtonian fluids in pre-critical regimes can be used for calculations for over-critical lubrication regimes. To do that we will consider the case of fully flooded over-critical lubrication regime (see the definition of over-critical lubrication regimes (6.76)) with viscosity $\mu = \exp(Qp)$ and inlet coordinate $a = -1 + \alpha_{10}(VQ)^{1/2}$. Then the same inlet coordinate in pre-critical lubrication regime would be $a = -1 + \alpha_1 V^{2/5}$, which means that $\alpha_1 = \alpha_{10} Q_0^{1/2}$, $Q_0 = QV^{1/5}$. As it was mentioned earlier, the solution of asymptotic equations for pre-critical lubrication regime depend only on the values of parameters α_{10} and Q_0. It means that we can expect to get film thickness $H_0 = AV^{3/5}$, where $A = A(\alpha_{10}, Q_0)$. Therefore, if over-critical lubrications regimes can be described by the asymptotic equations for pre-critical lubrication regimes, then we can expect that we get a good match of the values of the lubrication film thickness H_0 obtained from solution of the asymptotic equations for pre-critical lubrication regimes with the solution of the original EHL problem. Let us consider the case of $a = -2$, $V = 0.05$, and $Q = 10$ ($Q_0 = 5.4928$). This is clearly an over-critical lubrication regime because $e^{Q_0} = 242.94 \gg 1$ (see (6.76)). In this case $\alpha_1 = -3.314$ and from the solution of the asymptotic equations for pre-critical lubrication regimes obtained for $N = 1,000$ and $\triangle r = 0.0075$ we have $A = 1.0824$ and, therefore, $H_0 = AV^{3/5} = 0.1794$. The solution of the original EHL problem for the same values of parameters a, V, Q, and $N = 1,325$, $\triangle y = 0.001509434$ gives $H_0 = 0.1743$. The difference between these values of H_0 is 2.9%. Therefore, the asymptotic equations derived for pre-critical regimes can be used for calculations for over-critical lubrication regimes as well. Also, this conclusion allows us not to do any further analysis of heavily loaded over-critical lubrication regimes beyond the one conducted in Section 6.6.

Moreover, taking into account the fact that the solution of the asymptotic equations for pre-critical lubrication regimes depends only on parameter α_{10} and Q_0 if we compare formulas for the lubrication film thickness for pre-critical $H_0 = BV^{3/5}$ and over-critical $H_0 = A(VQ)^{3/4}$ lubrication regimes, we obtain that $B = AQ_0^{3/4}$ for $Q_0 \gg 1$. Therefore, for heavily loaded pre- and over-critical lubrication regimes (which are determined by the fact that the characteristic size of the inlet zone $\epsilon_q = V^{2/5} \ll 1$ and $\epsilon_q = (VQ)^{1/2} \ll 1$, respectively), one can create a map of level curves for $H_0 V^{-3/5} = f_*(Q_0)$, i.e., for any values of parameters V and Q for which $Q_0 = QV^{1/5} = const$ we have $H_0 V^{-3/5} = const$.

The analysis of the numerical results in this section validates the asymptotic approaches used for the cases of heavily loaded pre- and over-critical lubrication regimes. Moreover, it reveals that the asymptotic equations obtained for pre-critical lubrication regimes can be used for calculations of over-critical lubrication regimes.

6.11 Numerical Precision, Grid Size, and Stability Considerations

The availability of the asymptotically valid in the inlet and exit zones equations provides a unique opportunity to analyze in detail the sensitivity of the numerical solutions to the value of the step size $\triangle r$. Let us consider an incompressible fluid with Newtonian rheology with $\epsilon_q = Q_0 Q^{-1} \ll V^{2/5}$ or $\epsilon_q = Q_0 Q^{-1} = O(V^{2/5})$, $\mu_q(q) = e^{Q_0 q}$, and $\mu_g(g) = e^{Q_0 g}$. For $A = 1, 2, 3$, $Q_0 = 1$, and absolute error $\varepsilon = 10^{-4}$ for both starved and fully flooded lubrication regimes in the inlet and exit zones the solution stabilizes for $\triangle r \leq 0.00335$ while $N \triangle r$ is kept approximately equal to 6.7. Therefore, the sufficient accuracy of the solution of the original EHL problem in the inlet and exit zones can be achieved if the step size $\triangle y$ would be smaller or equal to $\triangle y = 0.00335 \frac{2}{c-a} \epsilon_q \leq \frac{0.0067}{c-a} V^{2/5}$. For example, for realistic values of $V = 0.01$, $a = -2$, and $c \approx 1$ this requirement translates into $\triangle y \leq 0.00071$. Therefore, if the span of the lubricated contact region is about 3, then the number of evenly spaced nodes covering the region should be about $4,238$ or higher.

TABLE 6.5

Data on the sensitivity of the inlet coordinate α_1 as a function of the step size $\triangle r$ for $A = 2$ and $Q_0 = 2.5$. (after Kudish and Covitch [3]). Reprinted with permission from CRC Press.

$\triangle r$ (N)	α_1
0.0700000 (100)	-1.217
0.0350000 (200)	-1.236
0.0175000 (400)	-1.250
0.0087500 (800)	-1.260
0.0043750 (1,600)	-1.268
0.0021875 (3,200)	-1.273

Let us consider the sensitivity of the asymptotic solutions for starved lubrication regimes (i.e., for $h_q(r) = h_g(s) = 1$) for the case of elevated Q_0. The

TABLE 6.6
Data on the sensitivity of the pressure spike
maximum $\max g(r)$ in the exit zone and the exit
coordinate β_1 as functions of the step size $\triangle r$ for
$A = 2$ and $Q_0 = 2.5$. (after Kudish and Covitch
[3]). Reprinted with permission from CRC Press.

$\triangle r$ (N)	$\max g(s)$	β_1	C.N.
0.00600 (1,200)	14.799	1.015	$1.37 \cdot 10^9$
0.00415 (1,600)	17.746	1.050	$5.06 \cdot 10^9$
0.00335 (2,000)	19.970	1.758	$1.14 \cdot 10^{10}$
0.00280 (2,400)	21.780	1.880	$2.16 \cdot 10^{10}$
0.00240 (2,800)	23.469	2.021	$3.74 \cdot 10^{10}$
0.00210 (3,200)	24.814	1.109	$5.84 \cdot 10^{10}$
0.00187 (3,600)	26.278	1.118	$8.84 \cdot 10^{10}$
0.00168 (4,000)	27.708	1.140	$1.30 \cdot 10^{11}$

values of the inlet coordinate α_1 as a function of the step size $\triangle r$ and number
of nodes N are given in Table 6.5) for $A = 2$, $Q_0 = 2.5$, and the precision
$\varepsilon = 10^{-3}$. The values of the pressure spike maximum $\max g(s)$ in the exit zone,
the exit coordinate β_1, and the system's Jacobian condition number C.N. as
functions of the step size $\triangle r$ and number of nodes N are given in Table 6.6
for $A = 2$, $Q_0 = 2.5$, and $\varepsilon = 10^{-3}$. The relationship between the step size
$\triangle r$ and the number of nodes N is kept in such a way that $N\triangle r \approx 6.7$. It
can be clearly seen that in the inlet zone α_1 as a function of $\triangle r$ reaches its
limiting/constant value for relatively large step sizes $\triangle r \leq 0.0021875$. In the
exit zone for large s pressure $g(s)$ is well behaved for all values of N, i.e., it
approaches its asymptote and it is stable. Typical examples of the behavior of
$g(s)$ for $A = 1$, 1.5, 2, $\mu_g(g) = e^{Q_0 g}$, $Q_0 = 2.5$, $\triangle r = 0.00335$, and $N = 2,000$
are presented in Fig. 6.16. It is clear from Fig. 6.16 that for sufficiently large
values of A and/or Q_0 the pressure spike is extremely thin and high (pin
like) while for relatively small values of A and Q_0 (for example, for $A = 1$
and $Q_0 = 2.5$) the pressure spike is reasonably wide and not excessively high.
For the values of A and Q_0 for which the pressure spike is thin and high
its maximum $\max g(s)$ in the exit zone remains sensitive to the value of the
step size $\triangle r$ even for $\triangle r < 0.00168$ and $N > 4,000$ (see Table 6.6). The
system's Jacobian condition number C.N. is high for all values of N and it
increases from $1.37 \cdot 10^9$ to $1.3 \cdot 10^{11}$ as N increases from 1,200 to 4,000 and
$\triangle r$ decreases from 0.006 to 0.00168. Compare these condition numbers with
the condition number C.N. $= 2.81 \cdot 10^8$ for the case of a stable solution for
$A = 1$, $\mu_g(g) = e^{Q_0 g}$, $Q_0 = 2.5$, $\triangle r = 0.00335$, and $N = 2,000$ (see Fig. 6.16)
which is much lower (about two orders of magnitude) than the ones for the
cases presented in Table 6.6 for which the solutions are unstable. This elevated
solution sensitivity as well as small oscillations of the derivative of pressure
$dg(s)/ds$ (not the distribution of pressure $g(s)$ itself) between the position of

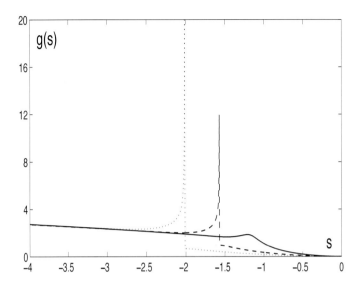

FIGURE 6.16
Main term of the asymptotic of the pressure distribution $g(s)$ in the exit zone of a starved lubricated contact for $A = 1$ (solid curve), $A = 1.5$ (dashed curve), $A = 2$ (dotted curve), and $\mu_g(g) = e^{Q_0 g}$, $Q_0 = 2.5$, $\triangle r = 0.00335$, $N = 2,000$. (after Kudish and Covitch [3]). Reprinted with permission from CRC Press.

the pressure spike and the exit point $s = 0$ are the indications of instability. This instability is caused by a strong dependence of the viscosity $\mu_g(g) = e^{Q_0 g}$ on g or equally by high value of constant A. This is in perfect agreement with the fact that the values of the system's Jacobian condition number *C.N.* are very high (see Table 6.6). The unstable conditions are described by inequality $Q_0 A^{3/4} > 3$ (see transformation (6.215) and its discussion). For values of A and Q_0 such that $Q_0 A^{3/4} \leq 3$, the solution in the exit zone is perfectly stable.

For fully flooded lubrication regimes the typical pressure behavior in the inlet and exit zones is similar to the ones for starved lubrication regimes. The solutions for pressure $q(r)$ and gap $h_q(r)$ in the inlet zone are stable for all combinations of values of Q_0 and A while the solutions for pressure $g(s)$ and gap $h_g(s)$ in the exit zone for elevated values of Q_0 demonstrate instability. The instability presents itself by small oscillations of gap $h_g(s)$ in the area between the pressure spike and the exit point $s = 0$ and by producing an excessively high pressure spike. The former is caused by small oscillations of the derivative $dg(s)/ds$ in the aforementioned area. However, the former and latter manifestations of instability are caused by strong dependence of viscosity $\mu_g(g)$ on pressure g. The only remedy to this instability situation in the exit zone is a proper regularization. The gap oscillations are remedied by a simple averaging regularization procedure described earlier in this section. That causes some loss of precision, i.e., the discrepancy in the Reynolds equation increases but still remains within acceptable range of ± 0.015. A more sophisticated approach to regularization will be described later.

At the same time, in the inlet zone the problem solution is stable for all combinations of values of Q_0 and A. Therefore, film thickness H_0 can be determined in a stable manner in the inlet zone as it depends on the solution for $A = A(\alpha_1)$ only in the inlet zone. It is sufficient to know film thickness H_0 to be able to determine the sliding frictional stress $f(x)$ and the sliding friction force F_S (see Section 10.8) as they mainly depend on pressure $p(x)$ in the Hertzian region which for heavily loaded lubricated contacts is close to the Hertzian pressure $\sqrt{1 - x^2}$. The latter can be considered as one of the advantages of the asymptotic approach in comparison with the solution of the original (non-asymptotic) problem, which involves dealing with a possibly numerically unstable solution in the exit zone.

After analyzing the numerical results obtained from solution of asymptotically valid equations in the inlet and exit zones, we can make a conclusion about the value of the step size $\triangle y$, which would provide the adequate precision for the solution of the original (non-asymptotic) EHL problem. As the previous consideration shows the major limitation to achieving the desired precision of the solution of the original EHL problem in the most of the contact region (excluding the exit zone) is the ability to resolve the inlet zone sufficiently accurately. Suppose the step size $\triangle r_*$ provides the sufficient precision to the solution of the asymptotically valid equations in the inlet zone. Then it is clear that to get the same solution details and precision from a numerical solution of the original EHL problem for pre-critical lubrication regimes it is necessary to use the step size $\triangle y = \frac{2}{c-a}\epsilon_q \triangle r_*$. Similarly, for over-critical lubrication regimes the step size for solution of the original EHL problem should be $\triangle y = \frac{2}{c-a}\epsilon_0 \triangle r_*$ (recall that for over-critical regimes $\epsilon_q \gg \epsilon_0$). Therefore, if $\epsilon_q \ll 1$ and $\epsilon_0 \ll 1$ to get an accurate numerical solution of the original EHL problem, we should require that step size $\triangle y \ll \triangle r_*$. For example, for $V = 10^{-3}$, $Q = 1$, $a = -2$, and $c \approx 1$ for pre-critical lubrication regimes the step size should not be greater than $\triangle y = 0.000092015$ assuming that $\triangle r_* = 0.0021875$. For the case of evenly spaced nodes, the latter step size translates into the necessity to have more than 10,868 nodes in the lubricated contact region $[a, c]$. We have to keep in mind that in the Hertzian region the requirements on the step size can be significantly relaxed. In spite of that it is clear that this requirement on the step size still puts severe demands on computer resources required for solution of EHL problems for line contacts with high precision. Problems for lubricated rough surfaces in line contacts require even more detailed description and that imposes even higher demands on computer memory and speed.

Keeping in mind that the properties of the EHL problems for smooth point (two-dimensional) contacts are similar to the problem for line contacts the absolute precision of $\epsilon = 10^{-4}$ can be guaranteed for $V = 10^{-3}$, $Q = 1$ if the step size is not greater than 0.000092015 and the number of equidistant nodes is not less than $10,868 \times 10,868 = 118,113,424$. Of course we need to realize that in the Hertzian region and in the direction perpendicular to lubricant flow the grid can be more coarse. Nonetheless, that makes it practically impossible to solve such problems adequately accurate. The situation with rough point EHL contacts is even worse.

However, in most numerical studies of the original EHL problem the step size is chosen independently of the above condition and it is much larger than the one which would provide sufficiently high accuracy of the solution. To determine numerical convergence it is customary to use the comparison of numerical solutions obtained for two step sizes $\triangle y_1$ and $\triangle y_2$ such that $\triangle y_2 = 0.5 \triangle y_1$. The above analysis shows that even if they differ by just a decimal of a percent it does not provide the necessary assurance of the solution sufficient precision.

To conclude this section it is important to emphasize the advantages of the asymptotic approach. They are due to the fact that the asymptotic equations contain smaller number of the original problem input parameters and the numerical solutions of these equations are always stable in the inlet zone and easy to obtain. In the exit zone the problem solution is stable in a certain range of the input parameters. Moreover, the presented asymptotic analysis provides simple structural formulas for film thickness. The subsequent numerical analysis of the asymptotic equations in the inlet zone supplies the only coefficient in the film thickness formulas that is not determined analytically but is obtained numerically. The values of this coefficient finalize the film thickness formulas.

6.12   Closure

The classic EHL problem for a heavily loaded line contact with Newtonian incompressible lubricant is studied using methods of matched asymptotic expansions and numerical methods. In heavily loaded lubricated contacts two different regimes of lubrication are recognized: pre-critical and over-critical regimes. Overall, the difference between the two regimes is in how strongly the lubricant viscosity depends on pressure and how much lubricant is available at the inlet in the contact. Two different asymptotic approaches to the analysis of the pre- and over-critical regimes are developed while both are based on matched asymptotic expansions. A detailed structure of the inlet and exit zones and the Hertzian region of a lubricated contact is analyzed. Several asymptotically based formulas for the lubrication film thickness in heavily loaded EHL contacts are derived analytically. Each of these formulas involve just one coefficient of proportionality which can be determined from numerical problem solution in the inlet zone or by curve fitting of independently obtained experimental values for the lubrication film thickness. The classic Ertel-Grubin method is analyzed and its shortcomings are revealed. The numerical methods for asymptotically valid equations in the inlet and exit zones are developed and realized. The issues of numerical precision, proper grid size, and stability of the numerical solutions for heavily loaded lubricated contacts are discussed. Several ways of the validation of the asymptotic analysis by comparison its results with the direct numerical solutions of the EHL problem in the original formulation were presented.

6.13   Exercises and Problems

1. (a) What is the difference between pre- and over-critical lubrication regimes in heavily loaded contacts? What are the magnitudes of the contributions of each of the zones in the contact region to the problem solution? Explain the procedure by which pre- and over-critical regimes are determined for the case of lubricant viscosity $\mu = \exp(Qp)$, $Q \gg 1$ (see equations (6.75), (6.76), and (6.92)).

(b) Is it possible for an over-critical regime to occur for a lubricant with constant viscosity? Explain.

2. Describe in your own words the basis for the derivation of the formula for the film thickness H_0 from (6.55) for pre-critical lubrication regimes.

3. For a Newtonian lubricant with constant viscosity, derive formulas (6.84).

4. For a Newtonian lubricant, provide a detailed derivation of formula (6.141) for the film thickness H_0 under over-critical lubrication regimes.

5. Using the numerical data for $q(0)$ and the asymptotic expansion for $p_{ext}(x)$ from (6.145) show that assumption 2 (expressed by estimate (6.182)) laid in the foundation of the Ertel-Grubin method is violated.

6. Apply the asymptotic procedures employed for pre- and over-critical regimes for the Barus viscosity $\mu = e^{Qp}$ to the case of viscosity determined by the formula $\mu = \exp(Qp^m)$, where m is a positive constant. Derive the structural dimensionless formulas for the exit and central lubrication film thickness for pre- and over-critical lubrication regimes, respectively. Compare this formulas to the ones obtained for the Barus viscosity.

7. Is it possible to employ the Ertel-Grubin procedure for the case of viscosity determined by the formula $\mu = \exp(Qp^m)$ (where m is a positive constant)? Explain the difference in application the asymptotic procedure developed above and the Ertel-Grubin procedure.

8. Describe the differences in the precision of the numerical solutions in the

inlet and exit zones of heavily loaded EHL contact. Describe the symptoms of solution numerical instability in the exit zones of heavily loaded EHL contacts.

References

[1] Bhushan, B. 1999. *Principles and Applications of Tribology*. Toronto: John Wiley & Sons, Inc., 586 - 591.

[2] Cheng, H.S. 1980. Fundamentals of Elastohydrodynamic Contact Phenomena. In *Fundamentals of Tribology*, N.P. Suh and N. Saka, eds. Cambridge, MA: The MIT Press, 1009 - 1048.

[3] Kudish, I.I. and Covitch, M.J. 2010. *Modeling and Analytical Methods in Tribology*. Boca Raton: CRC Press.

[4] Hamrock, B.J. 1994. *Fundamentals of Fluid Film Lubrication*. New York: McGraw-Hill.

[5] Dowson, D. and Higginson, G.R. 1966. *Elastohydrodynamic Lubrication*. London: Pergamon Press.

[6] Ertel, M.A. 1945. Hydrodynamic Calculation of Lubricated Contact for Curvilinear Surfaces. Dissertation, *Proc. of CNIITMASh*, 1-64.

[7] Grubin, A.N. 1949. The Basics of the Hydrodynamic Lubrication Theory for Heavily Loaded Curvilinear Surfaces. *Proc. CNIITMASh* 30:126-184.

[8] Kudish, I.I. and Marchenko, S.M. 1985. Some Problems and Mathematical Methods in the Elastohydrodynamic Lubrication Theory. *Izvestiya Severo−Kavkazskogo Nauchnogo Centra Vysshey Shkoly, Estestwennye Nauki* 3:46-52.

[9] Houpert, L.G. and Hamrock, B.J. 1986. Fast Approach for Calculating Film Thickness and Pressures in Elastohydrodynamically Lubricated Contacts at High Loads. *ASME J. Tribology* 108, No. 3:441–452.

[10] Bissett, E.J. and Glander, D.W. 1988. A Highly Accurate Approach that Resolves the Pressure Spike of Elastohydrodynamic Lubrication. *ASME J. Tribology* 110, No. 2:241-246.

[11] Hamrock, B.J., Ping Pan, and Rong-Tsong Lee. 1988. Pressure Spikes in Elastohydro-dynamically Lubricated Conjunctions. *ASME J. Tribology* 110, No. 2:279-284.

[12] Venner, C.H. and Lubrecht, A.A. 2000. *Multilevel Methods in Lubrication*. Amsterdam: Elsevier.

[13] Kostreva, M.M. 1984. Elastohydrodynamic Lubrication: A Nonlinear Complementarity Problem. *Intern. J. Numerical Methods in Fluids* 4:377-397.

[14] Airapetov, E.L., Kudish, I.I., and Panovko, M.Ya. 1992. Numerical Solution of Heavily Loaded Elastohydrodynamic Contact. *Soviet J. Fric. and Wear* 13, No. 6:1-7.

[15] Belotserkovsky, S.M. and Lifanov, I.K. 2000. *Method of Discrete Vortices*. Boca Raton: CRC Press.

[16] Bellman, R.E. and Kalaba, R.E. 1965. *Quasilinearization and Nonlinear Boundary − Value Problems*. New York: Elsevier.

[17] Kudish, I.I. 1978. Elastohydrodynamic Problem for a Heavily Loaded Rolling Contact. *Proc. Acad. Sci. of Armenia SSR, Mechanics* 31, No. 1:65-78.

[18] Evans, H.P. and Hughes, T.G. 2000. Evaluation of Deflection in Semi-Infinite Bodies by a Differential Method. *Proc. Instn. Mech. Engrs.* 214, Part C :563-584.

[19] Archard, J.F. and Baglin, K.P. 1986. Elastohydrodynamic Lubrication - Improvements in Analytic Solutions. *Proc. Inst. Mech. Eng.* 200, No. C4 :281-291.

[20] Crook, A.W. 1961. The Lubrication of Rollers II. Film Thickness with Relation to Viscosity and Speed. *Philosophical Trans. Royal Soc. of London, Ser. A, Math., Phys. and Eng. Sci.* 254, No. 1040:223-236.

[21] Van Dyke, M. 1964. *Perturbation Methods in Fluid Mechanics*. New-York: Academic Press.

[22] Kevorkian, J. and Cole, J.D. 1985. *Perturbation Methods in Applied Mathematics. Applied Mathematics Series, Vol.* 34. New York: Springer-Verlag.

[23] Kudish, I.I. 1977. Hydrodynamic Lubrication Theory of Rolling Cylindrical Bodies. *Abstracts of the 2nd All − Union Conf. on Elastohydrodynamic Theory of Lubric. and Its Pract. Appl. in Technology* (1976), Kujbyshev, 11; *Proc. 2nd All − Union Conf. on Elastohydrodynamic Theory of Lubrication and Its Practical Applications in Industry*, Kujbyshev, 33-38.

[24] Kudish, I.I. 1982. Asymptotic Methods of Study for Plane Problems of the Elastohydrodynamic Lubrication Theory in Heavy Loaded Regimes. Part 1. Isothermal Problem. *Proc. Acad. Sci. of Armenia SSR, Mechanics* 35, No. 5:46-64.

[25] Kudish, I.I. 1978. Asymptotic Analysis of a Plane Non-isothermal Elastohydro-dynamic Problem for a Heavily Loaded Rolling Contact. *Proc. Acad. Sci. of Armenia SSR, Mechanics* 31, No. 6:16-35.

[26] Kudish, I.I. 1983. Asymptotic Method of Study for Plane Problems of the Elastohydrodynamic Lubrication Theory for Heavily Loaded Regimes.

Part 2. Non-isothermal Problem. *Proc. Acad. Sci. Armenia SSR, Mechanics* 36, No. 5:47-59.

[27] Vorovich, I.I., Aleksandrov, V.M., and Babeshko, V.A. 1974. *Non − Classical Mixed Problems of Elasticity*. Moscow: Nauka.

[28] Kudish, I.I. 1996. Asymptotic Analysis of a Problem for a Heavily Loaded Lubricated Contact of Elastic Bodies. Pre- and Over-critical Lubrication Regimes for Newtonian Fluids. *Dynamic Systems and Applications*, Dynamic Publishers, Atlanta, 5, No. 3:451-478.

[29] Greenwood, J.A. 1972. An Extension of the Grubin Theory of Elasto-hydrodynamic Lubrication. *Phys. D. Appl. Phys.* 5:2195-2211.

[30] Johnson, K.L. 1970. Regimes of Elastohydrodynamic Lubrication. *J. Mech. Eng. Sci.* 12, No. 1:9-16.

[31] Gakhov, F.D. 1977. *Boundary Value Problems*. 3rd ed., Moscow: Nauka.

[32] Bierens De Haan, D. 1867. *Nouvelles Tables, D'Integrales Definies*. New York and London: Hafner Publishing Co.

[33] Pykhteev, G.N. 1980. *Exact Methods of Calculation for the Cochy Type Integrals*. Novosibirsk: Nauka.

[34] Kuznetsov, E.A. 1982. Two-Dimensional Contact Problem with an Additional Load Applied Outside a Stamp Taken into Account. *Soviet Appl. Mech. (English translation of Prikladnaya Mechanika)* 18, No. 5:462-468.

Part V

Isothermal and Thermal EHL Problems for Line Contacts and Lubricants with Newtonian and Non-Newtonian Rheologies

In Part V of the monograph thermal problems for line contacts lubricated by Newtonian fluids and isothermal and thermal problems for line contacts lubricated by non-Newtonian fluids are analyzed. The lubricated contacts are considered being heavily loaded. The problems are studied by asymptotic and numerical methods. Based on the solution of thermal EHL problems an efficient way to regularize the generally unstable numerical solutions of isothermal heavily loaded line EHL contacts is proposed and realized.

7

Thermal EHL Problems for Heavily Loaded Line Contacts with Newtonian Lubricant

7.1 Introduction

In this chapter we will consider a special case of a thermal EHL (TEHL) problem for Newtonian lubricant when the contact surface temperatures are known and equal. To simplify and analyze the problem we will develop a powerful perturbation technique. This approach will allow us to reduce the whole thermal EHL problem (which includes the energy equation for lubricant) to an EHL problem for a modified Reynolds equation in a certain sense similar to the classic isothermal one. The latter problem can be solved numerically or studied by the appropriate asymptotic methods. Historically, this approach was first developed by Kudish in [1] - [5] for analysis of isothermal and thermal EHL problem for Newtonian and non-Newtonian fluid lubricants and greases. Later, it was employed for fluid lubricants with non-Newtonian rheologies in [6].

The general case of a thermal problem for a non-Newtonian lubricant with heat dissipation in contact solids is considered in the next chapter.

7.2 Formulation and Analysis of a TEHL Problem for Newtonian Fluid in Heavily Loaded Contacts

Let us consider the behavior of a Newtonian fluid in a TEHL contact of two elastic cylinders separated by a lubrication film and loaded with a given normal force acting along their center line. The details of the problem set up are given in Section 5.2. We will assume that contact surface temperatures $T_{w1}(x)$ and $T_{w2}(x)$ are known and equal, i.e., $T_{w1}(x) = T_{w2}(x) = T_{w0}(x)$. It is a well-known experimental fact that the lubricant viscosity μ increases with lubricant pressure p and decreases with temperature T. To model this dependence of lubricant viscosity μ on temperature T, we will take*

$$\mu(p, T) = \mu^0(p) \exp[-\alpha_T(T - T_a)], \qquad (7.1)$$

where $\mu^0(p)$ is independent from lubricant temperature T and is a monotonically increasing with pressure p lubricant viscosity at the ambient temperature T_a and α_T is the temperature viscosity coefficient which may depend only on pressure p. With respect to lubricant heat conductivity λ for simplicity we will assume that it may depend on pressure p but is independent from temperature T, i.e., $\lambda = \lambda(p)$. Therefore, both μ^0 and λ are independent from the coordinate z across the lubrication layer.

To model this problem we will accept all assumptions concerning the relative sizes of the film thickness, contact, and solid radii as well as the direction and nature of the heat flow in lubricated contacts made in Section 5.2. In the case of Newtonian fluid in dimensional variables $F(x) = x$ and, therefore, from equations (5.2) and boundary conditions (5.3), we derive an equation for the sliding frictional stress $f(x)$ in the form

$$f(x) = \left\{ u_2 - u_1 - \frac{dp}{dx} \int_{-h/2}^{h/2} \frac{zdz}{\mu} \right\} / \int_{-h/2}^{h/2} \frac{dz}{\mu}. \qquad (7.2)$$

Integrating the third equation in (5.2) with respect to z from $z = -h/2$ to $z = h/2$ with the boundary conditions for the velocity component w from (5.3) and using the expression for $f(x)$ from (7.2), we obtain the modified Reynolds equation and boundary conditions (see (5.9))

$$\frac{d}{dx}\left\{ \left[\int_{-h/2}^{h/2} \frac{z^2 dz}{\mu} - \left(\int_{-h/2}^{h/2} \frac{zdz}{\mu} \right)^2 / \int_{-h/2}^{h/2} \frac{dz}{\mu} \right] \frac{dp}{dx} \right.$$

$$\left. + (u_2 - u_1) \int_{-h/2}^{h/2} \frac{zdz}{\mu} / \int_{-h/2}^{h/2} \frac{dz}{\mu} - \frac{u_1+u_2}{2} h \right\} = 0, \qquad (7.3)$$

$$p(x_i) = p(x_e) = \frac{dp(x_e)}{dx} = 0.$$

Using the reduced energy equation (5.47), the expressions for $F(x)$, and $f(x)$, we obtain the equations for the lubricant temperature T (see equations (5.47))

$$\frac{\partial^2 T}{\partial z^2} = -\frac{1}{\lambda \mu}\left\{ u_2 - u_1 + \frac{dp}{dx}\left[z \int_{-h/2}^{h/2} \frac{ds}{\mu} - \int_{-h/2}^{h/2} \frac{sds}{\mu} \right] \right\}^2 / \left\{ \int_{-h/2}^{h/2} \frac{ds}{\mu} \right\}^2, \qquad (7.4)$$

$$T(x, -\tfrac{h}{2}) = T_{w1}(x), \quad T(x, \tfrac{h}{2}) = T_{w2}(x).$$

---

*Similarly, we can assume that $\mu(p, T) = \frac{\mu^0(p)}{1+\delta(T-T_a)}$. For this lubricant viscosity, the further analysis of the problem is similar but simpler.

For heavily loaded TEHL contact we introduce dimensionless variables (6.3), (5.48), $\{T', T'_{wi}\} = \{T, T_{wi}\}/T_a - 1$ and parameters (6.4) as well as two additional dimensionless parameters

$$\delta_T = \alpha_T T_a, \quad \kappa_N = \frac{\mu_0 (u_1 + u_2)^2}{4\lambda_a T_a}. \tag{7.5}$$

In these dimensionless variables the problem is reduced to the system (primes at the dimensionless variables are omitted)

$$\frac{d}{dx}\left\{ \frac{12 H_0^2}{V} \left[\int_{-h/2}^{h/2} \frac{z^2 dz}{\mu} - \left(\int_{-h/2}^{h/2} \frac{z dz}{\mu} \right)^2 / \int_{-h/2}^{h/2} \frac{dz}{\mu} \right] \frac{dp}{dx} \right.$$

$$\left. + s_0 \int_{-h/2}^{h/2} \frac{z dz}{\mu} / \int_{-h/2}^{h/2} \frac{dz}{\mu} - h \right\} = 0, \quad p(a) = p(c) = \frac{dp(c)}{dx} = 0, \tag{7.6}$$

$$\frac{\partial^2 T}{\partial z^2} = -\frac{\kappa_N}{\lambda \mu} \left\{ s_0 + \frac{12 H_0^2}{V} \frac{dp}{dx} \left[z \int_{-h/2}^{h/2} \frac{ds}{\mu} - \int_{-h/2}^{h/2} \frac{s ds}{\mu} \right] \right\}^2 / \left\{ \int_{-h/2}^{h/2} \frac{ds}{\mu} \right\}^2, \tag{7.7}$$

$$T(x, -\tfrac{h}{2}) = T_{w1}(x), \quad T(x, \tfrac{h}{2}) = T_{w2}(x),$$

$$H_0(h - 1) = x^2 - c^2 + \frac{2}{\pi} \int_a^c p(t) \ln \left| \frac{c - t}{x - t} \right| dt, \tag{7.8}$$

$$\int_a^c p(t) dt = \frac{\pi}{2}. \tag{7.9}$$

In equations (7.6)-(7.9) the dimensionless parameter V is determined by the formula

$$V = \frac{24 \mu_a (u_1 + u_2) R'^2}{a_H^3 p_H} \tag{7.10}$$

coincides with parameter V determined by (6.4).

In the introduced dimensionless variables, we have

$$\mu(p, T) = \mu^0(p) \exp(-\delta_T T). \tag{7.11}$$

Therefore, the system can be rewritten in the final form

$$\frac{d}{dx}\{M(p, h) - H_0 h\} = 0, \quad p(a) = p(c) = \frac{dp(c)}{dx} = 0, \tag{7.12}$$

$$M(p, h) = \frac{12 H_0^3}{V} \left[\int_{-h/2}^{h/2} z^2 e^{\delta_T T} dz - \left(\int_{-h/2}^{h/2} z e^{\delta_T T} dz \right)^2 \right.$$

$$\left. / \int_{-h/2}^{h/2} e^{\delta_T T} dz \right] \frac{1}{\mu^0} \frac{dp}{dx} + H_0 s_0 \int_{-h/2}^{h/2} z e^{\delta_T T} dz / \int_{-h/2}^{h/2} e^{\delta_T T} dz, \tag{7.13}$$

$$\frac{\partial^2 T}{\partial z^2} = -\kappa_N \frac{\mu^0}{\lambda} e^{\delta_T T} \left\{ s_0 + \frac{12H^2}{V} \frac{1}{\mu^0} \frac{dp}{dx} \left[z \int_{-h/2}^{h/2} e^{\delta_T T} ds \right. \right.$$

$$\left. \left. - \int_{-h/2}^{h/2} e^{\delta_T T} s ds \right] \right\}^2 / \left(\int_{-h/2}^{h/2} e^{\delta_T T} ds \right)^2, \tag{7.14}$$

$$T(x, -\tfrac{h}{2}) = T_{w1}(x), \quad T(x, \tfrac{h}{2}) = T_{w2}(x),$$

$$H_0(h - 1) = x^2 - c^2 + \frac{2}{\pi} \int_a^c p(t) \ln \left| \frac{c-t}{x-t} \right| dt, \tag{7.15}$$

$$\int_a^c p(t) dt = \frac{\pi}{2}. \tag{7.16}$$

7.3 Some Analytical Approximations of TEHL Problems for Newtonian Fluids

Obviously, this is a highly complex nonlinear system of integro-differential equations. The goal of our analytical approach based on the regular perturbation method is for a sufficiently high slide-to-roll ratio s_0 to simplify the problem. To achieve this goal we will try to find a perturbation series solution for lubricant temperature T and use it in the generalized Reynolds equation to simplify it. Instead of using a small parameter for this purpose we will be using a small function of x which represents the ratio of the rolling and sliding frictional stresses

$$\nu(x) = \frac{H_0 h}{2f} \frac{dp}{dx}. \tag{7.17}$$

More specifically, we will assume that in the inlet and exit zones of the lubricated contact

$$\nu(x) = \frac{H_0 h}{2f} \frac{dp}{dx} \ll 1, \quad \omega \ll 1, \tag{7.18}$$

where ω is a small parameter characterizing heavily loaded lubrication regimes. It can be either $\omega = V \ll 1$ or $\omega = Q^{-1} \ll 1$. It will be shown that if condition (7.18) holds in the inlet and exit zones then it holds in the central (Hertzian) region of the contact as well.

Let us expand $T(x, z)$, $T_{w1}(x)$, and $T_{w2}(x)$ in asymptotic power series in $\nu(x) \ll 1$ as follows

$$T(x, z) = T_0(x, z) + \nu(x) T_1(x, z) + O(\nu^2(x)),$$

$$T_{w1}(x) = T_{w10}(x) + \nu(x) T_{w11}(x) + O(\nu^2(x)), \tag{7.19}$$

$$T_{w2}(x) = T_{w20}(x) + \nu(x) T_{w21}(x) + O(\nu^2(x)),$$

$$T_0(x, z),\ T_1(x, z),\ T_{w10}(x),\ T_{w11}(x), T_{w20}(x),\ T_{w21}(x) = O(1),\ \omega \ll 1,$$

where $T_0(x, z)$ and $T_1(x, z)$ are the unknown functions that need to be determined while $T_{w10}(x)$, $T_{w20}(x)$, $T_{w11}(x)$, and $T_{w21}(x)$ are given functions.

Substituting expansions (7.19) in equations (7.14), we obtain a sequence of boundary-value problems

$$\frac{\partial^2 T_0}{\partial z^2} = -\kappa_N s_0^2 \frac{\mu^0}{\lambda} e^{\delta_T T_0} / \left(\int_{-h/2}^{h/2} e^{\delta_T T_0} ds \right)^2,$$

$$(7.20)$$

$$T_0(x, -\tfrac{h}{2}) = T_{w10}(x),\ T_0(x, \tfrac{h}{2}) = T_{w20}(x),$$

$$\frac{\partial^2 T_1}{\partial z^2} = -\kappa_N s_0^2 \frac{\mu^0}{\lambda} e^{\delta_T T_0} \left[\frac{4z}{h} + \delta_T T_1 - 2\delta_T \int_{-h/2}^{h/2} T_1 e^{\delta_T T_0} ds \right.$$

$$\left. / \int_{-h/2}^{h/2} e^{\delta_T T_0} ds - \frac{4}{h} \int_{-h/2}^{h/2} s e^{\delta_T T_0} ds / \int_{-h/2}^{h/2} e^{\delta_T T_0} ds \right] / \left(\int_{-h/2}^{h/2} e^{\delta_T T_0} ds \right)^2,$$

$$(7.21)$$

$$T_1(x, -\tfrac{h}{2}) = T_{w11}(x),\ T_1(x, \tfrac{h}{2}) = T_{w21}(x).$$

Here we used the fact that

$$\nu(x) = \frac{6H_0^2}{V s_0} \frac{h}{\mu^0} \frac{dp}{dx} \int_{-h/2}^{h/2} e^{\delta_T T_0} ds \{1 + o(1)\},\ \omega \ll 1. \qquad (7.22)$$

Using the conditions for our special case we obtain $T_{w10} = T_{w20} = T_{w0}$ and $T_{w11} = T_{w21} = 0$, where $T_{w0}(x)$ is a given function. It is easy to see that the solution of problem (7.20) has the form

$$T_0 = T_{w0} + \frac{2}{\delta_T} \ln\{\cosh(\tfrac{Bh}{2}) / \cosh(Bz)\},$$

$$(7.23)$$

$$B = \frac{2}{h} \ln\{\eta + \sqrt{\eta^2 + 1}\},\ \eta = \frac{|s_0|}{2} \sqrt{\frac{\delta_T \kappa_N \mu^0}{2\lambda}} e^{-\delta_T T_{w0}}.$$

Function T_0 is even with respect to variable z. Therefore,

$$\int_{-h/2}^{h/2} e^{\delta_T T_0} ds = e^{\delta_T T_{w0}} \frac{\sinh(Bh)}{B},\ \int_{-h/2}^{h/2} s e^{\delta_T T_0} ds = 0. \qquad (7.24)$$

Formulas (7.24) allow to simplify problem (7.21) for T_1 and reduce it to the form

$$\frac{\partial^2 T_1}{\partial z^2} = -\kappa_N s_0^2 \frac{\mu^0}{\lambda} e^{\delta_T T_0} \left[\frac{4z}{h} + \delta_T T_1 \right.$$

$$\left. -2\delta_T \int_{-h/2}^{h/2} T_1 e^{\delta_T T_0} ds / \int_{-h/2}^{h/2} e^{\delta_T T_0} ds \right] / \left(\int_{-h/2}^{h/2} e^{\delta_T T_0} ds \right)^2, \qquad (7.25)$$

$$T_1(x, -\tfrac{h}{2}) = 0,\ T_1(x, \tfrac{h}{2}) = 0.$$

It is easy to see that the solution of problem (7.25) is an odd function of z, which has the form

$$T_1 = \frac{2}{\delta_T}\{\frac{\tanh(Bz)}{\tanh(Bh/2)} - \frac{2z}{h}\}. \tag{7.26}$$

It is worth mentioning that in a similar fashion can be obtained the expressions for the further terms $T_k(x, z)$, $k \geq 2$. Below it is shown that to analyze the problem for $p(x)$, $h(x)$, H_0, and c it is not sufficient to approximate $T(x, z)$ by just $T_0(x, z)$ because by doing so some terms in the generalized Reynolds equation of the same order of magnitude as the retained ones will be lost. On the other hand, retaining both terms $T_0(x, z)$ and $T_1(x, z)$ in the approximation of $T(x, z)$ supplies a correct approximation for the generalized Reynolds equation with function $M(p, h)$ as follows (see (7.13))

$$M(p, h) = \frac{H_0^3}{V}\frac{h^3}{\mu^0}\frac{dp}{dx}R_T(x),$$

$$R_T(x) = e^{\delta_T T_{w0}}\frac{3(1+\eta^2)}{\ln^2(\eta+\sqrt{\eta^2+1})}\left\{1 + \ln(1 + \eta^2)\right.$$

$$\left.- \frac{\sqrt{\eta^2+1}}{\eta}\ln(\eta + \sqrt{\eta^2 + 1}) - 2\frac{\int_0^{\ln(\eta+\sqrt{\eta^2+1})}\ln(\cosh(t))dt}{\ln(\eta+\sqrt{\eta^2+1})}\right\} \tag{7.27}$$

$$+O(\nu^2), \; \omega \ll 1, \; \beta = \frac{|s_0|}{2}\sqrt{\frac{\delta_T \kappa_N}{2\lambda}}, \; \eta = \beta\sqrt{\mu^0 e^{-\delta_T T_{w0}}},$$

where

$$\int_0^z \ln(\cosh(t))dt = \frac{z^2}{2} - z\ln 2 + \frac{\pi^2}{24} - \frac{1}{2}\sum_{k=1}^{\infty}\frac{(-1)^{k+1}}{k^2}e^{-2kz},$$

$$\int_0^z \ln(\cosh(t))dt = \frac{z^3}{6} + O(z^4), \; z \ll 1. \tag{7.28}$$

Here it is taken into account that $\int_{-h/2}^{h/2} z e^{\delta_T T}dz = O(\nu) \ll 1$ and the expression for $M(p, h)$ from formula (7.27) is based on just the first and the last terms in equation (7.13). In formula (7.27) function $R_T(x)$ represents the ratio of two functions $M(p, h)$ determined for non-isothermal and isothermal conditions (see formula (6.21)). A typical graph of function $R_T(x)$ obtained based on the Hertzian pressure $\sqrt{1 - x^2}$ and $T_{w0}(x) = 0$ is given in Fig. 7.1. Obviously, the proximity of the solutions of the isothermal and non-isothermal lubrication problems is determined by the behavior of function $R_T(x)$. The closer $R_T(x)$ is to unity the closer are these solutions to each other and vice versa the greater the value of $R_T(x)$ the farther these solutions are apart.

It can be shown that if $T_{w0} = 0$ and $\eta \to 0$ then $M(p, h) \to \frac{H_0^3}{V}\frac{h^3}{\mu^0}\frac{dp}{dx}$, which corresponds to the case of isothermal regime for Newtonian lubricant.

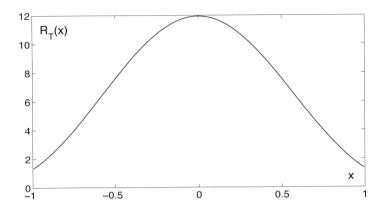

FIGURE 7.1
A typical graph of $R_T(x)$ as a function of x. (after Kudish and Covitch [7]).
Reprinted with permission from CRC Press.

On the other hand, for $\eta \to \infty$ (which represents the conditions of high heat generation in the lubrication layer), we have

$$M(p,h) \to \frac{H_0^3}{V} \frac{h^3}{\mu^0} \frac{dp}{dx} R_\infty(x), \;\; R_\infty(x) = \frac{3(1+\eta^2)}{\ln^2(\eta+\sqrt{\eta^2+1})}(1-2\ln 2),$$

$$\beta = \frac{|s_0|}{2}\sqrt{\frac{\delta_T \kappa_N}{2\lambda}}, \;\; \eta = \beta\sqrt{\mu^0 e^{-\delta_T T_{w0}}}. \tag{7.29}$$

The comparison of formula (7.29) with the one for the isothermal case (6.21) shows that for $\eta \gg 1$ the effective viscosity is the function $\frac{\mu^0}{R_\infty(x)}$. Usually, surface temperature $T_{w0}(x)$ of the contacting solids experiences an increase as x varies from the inlet to the exit point. In most cases this increase in $T_{w0}(x)$ is modest and depends on the material properties of contacting solids and the amount of heat generated in the lubrication layer. Therefore, if for $\eta \gg 1$ the above–mentioned effective viscosity grows with η, then it grows slowly not in a usual exponential manner as the lubricant viscosity μ^0 at the ambient temperature does. That makes the case of $\eta \gg 1$ more similar to the case of constant viscosity.

The above properties of function $M(p,h)$ hold a promise of solution regularization in cases of fast growing with pressure isothermal viscosity (see Chapter 8).

However, one has to remember that for sufficiently large values of $R_T(x)$ the approximation of pressure by the Hertzian one (as it is done in the asymptotic analysis) in the region away from the contact inlet and exit points might no longer be valid due to the thermal elastic deformations of the solids in contact.

We succeeded in reducing the non-isothermal problem to the approximate one which resembles the isothermal EHL problem. The point made about the necessity of taking into account not only T_0 but also T_1 for correct determination of the film thickness H_0 can be clearly seen from the expression for $M(p,h)$ in (7.27) if $\nu(x) \ll 1$ in the inlet zone.

Let us analyze the equations of the approximate problem (7.12), (7.15), (7.16), and (7.27) in more detail in accordance with the asymptotic methods developed for heavily loaded EHL contacts in Sections 6.3 and 6.6. To do that we define a new function

$$G(p,h) = \frac{H_0^3}{V} \frac{h^3}{\mu^0} e^{\delta_T T_{w0}} \frac{dp}{dx} / M(p,h), \tag{7.30}$$

which allows to consider and compare various lubrication regimes. Now, we will consider the behavior of function $G(p,h)$ in three cases: $\eta \ll 1$, $\eta \sim 1$, and $\eta \gg 1$. Using the above expression for $M(p,h)$, we have

$$G(p,h) = 1 + O(\eta), \ \eta \ll 1, \tag{7.31}$$

$$G(p,h) = O(1) + O(\eta), \ \eta \sim 1, \tag{7.32}$$

$$G(p,h) = \frac{\ln^2(\eta + \sqrt{\eta^2+1})}{3\eta^2} [1 + O(\frac{1}{\ln \eta})], \ \eta \gg 1. \tag{7.33}$$

Based on the analysis of Section 6.3 and estimates (7.31)-(7.33) we conclude that the pre-critical lubrications regimes are realized when in the inlet zone $x + 1 = O(\epsilon_q)$ we have

$$\mu^0(\epsilon_q^{1/2}) = O(1), \ \eta(\epsilon_q^{1/2}) \ll 1, \ \omega \ll 1, \tag{7.34}$$

$$\mu^0(\epsilon_q^{1/2}) = O(1), \ \eta(\epsilon_q^{1/2}) = O(1), \ \omega \ll 1, \tag{7.35}$$

$$\frac{\ln^2[\eta(\epsilon_q^{1/2}) + \sqrt{\eta^2(\epsilon_q^{1/2})+1}]}{\eta^2(\epsilon_q^{1/2})} \mu^0(\epsilon_q^{1/2}) = O(1), \ \eta(\epsilon_q^{1/2}) \gg 1, \ \omega \ll 1, \tag{7.36}$$

while the critical size of the inlet zone ϵ_0 is determined as before as $\epsilon_0 = \max(\epsilon_q)$, where values of ϵ_q satisfy one of the estimates (7.34)-(7.36).

On the other hand, for pre-critical regimes of lubrication ($\epsilon_q \ll 1$, $\omega \ll 1$) it can be shown that if $\nu(x) \ll 1$, $\omega \ll 1$, in the inlet zone, then the same estimate holds in the exit zone and the Hertzian region. For pre-critical regimes the condition $\nu(x) \ll 1$, $\omega \ll 1$, in the inlet zone is equivalent to

$$\epsilon_q \ll |s_0|^{6/5} V^{2/5} \{\mu^0(\epsilon_q^{1/2}) e^{-\delta_T T_{w0}(\epsilon_q^{1/2})} \frac{B}{\sinh(Bh)}\}^{6/5}, \ \omega \ll 1, \tag{7.37}$$

where function B is determined by equation (7.23) and in this case it is calculated based on $\mu^0(\epsilon_q^{1/2})$ and $T_{w0}(\epsilon_q^{1/2})$. Using the asymptotic representations (6.53) for gap $h(x)$ in the inlet and exit zones, it is easy to analyze the pre-critical regimes of starved lubrication. In particular, for the film thickness H_0 we get the formula in (6.55) where constant A in addition to dependence on

α_1 (see the relationship for a in (6.34)) also is influenced by μ^0, δ_T, λ, κ_N, and T_{w0}.

Let us introduce the following functions

$$M_0(q, h, \mu^0, \delta_T, \eta, x) = A^3 \frac{h^3}{\mu^0} \frac{dp}{dx} R_T(x), \qquad (7.38)$$

where $R_T(x)$ is determined by formula (7.27). Then the proposed asymptotic formula (6.55) for the film thickness allows to obtain two closed systems of integro-differential equations: in the inlet zone - equations (6.68), (7.38), (6.71), (6.49), (6.50), (6.64) for the starved lubrication regimes for functions $q(r)$, $h_q(r)$, and constant A, and in the exit zone - equations (6.70), (7.38), (6.72), (6.49), (6.49), (6.65) for the starved lubrication regimes for functions $g(s)$, $h_g(s)$, and constant β_1 (see equation (6.35)). It is important to remember that in the asymptotically valid equations in the inlet and exit zones function M_0 needs to be replaced by $M_0(q, h_q, \mu_q^0 e^{-\delta_{Tq} T_{w0q}}, \delta_{Tq}, \eta_q, r)$ and $M_0(g, h_q, \mu_g^0 e^{-\delta_{Tg} T_{w0g}}, \delta_{Tg}, \eta_g, s)$, respectively. Indexes q and g indicate the main terms of asymptotic expansions of the corresponding functions. Similarly, in the inlet and exit zones we can obtain the asymptotically valid equations (6.69), (6.79), and (6.81) for functions $q(r)$, $h_q(r)$, and constant A and equations (6.69), (6.80), and (6.82) for functions $g(s)$, $h_g(s)$, and constant β_1 with the above–mentioned replacements for functions M_0 and W_0.

Now, let us consider the over-critical lubrication regimes for which $\epsilon_q \gg \epsilon_0$. As before, the validity of the estimate $\nu(x) \ll 1$ in the inlet ϵ_q-zone guarantees its validity in the entire contact (see estimate (6.155) in the ϵ_0-inlet zone). It can be shown that the estimate $\nu(x) \ll 1$ in the inlet ϵ_q-zone is equivalent to

$$\epsilon_q \ll | s_0 |^{3/2} \left(\frac{V}{\epsilon_0^{1/2}}\right)^{1/2} \{\mu^0(\epsilon_0^{1/2}) e^{-\delta_T T_{w0}(\epsilon_q^{1/2})} \frac{B}{\sinh(Bh)}\}^{3/2}, \quad \omega \ll 1, \qquad (7.39)$$

where function B is determined by equation (7.23) and in this case it is calculated based on $\mu^0(\epsilon_0^{1/2})$ and $T_{w0}(\epsilon_q^{1/2})$. The further analysis of the over-critical lubrication regimes is done in accordance with the asymptotic method described in Section 6.6. In particular, the lubrication film thickness can still be determined from formula (6.141), where constant A depends on α_1, κ_N, δ_T, μ^0, λ, and T_{w0}.

For fully flooded lubrication regimes in the inlet zone the estimate $\nu(x) \ll 1$, $\omega \ll 1$, is no longer valid. For the cases of pre-critical regimes when in the inlet zone $\nu(x) \sim 1$, $\omega \ll 1$, or in the cases of over-critical regimes when in the inlet ϵ_q-zone $\nu(x) \sim 1$, $\omega \ll 1$, a qualitative analysis can still be conducted and structurally the same formulas for the film thickness H_0 as in the cases when $\nu(x) \ll 1$, $\omega \ll 1$, can be obtained.

7.4　Numerical Solutions of Asymptotic TEHL Problems for Newtonian Lubricants

The general approaches to solution of asymptotically valid equations in the inlet and exit zones of non-isothermal lubricated contacts under pre-critical lubrications conditions coincide with the ones employed for solution of similar isothermal problems, which are described in Sections 6.9.1 and 6.9.2. The main difference in these approaches is due to different calculation of functions M and W, which involves temperature effects. Besides that in the exit zone for $\mu_g = e^{Q_0 g}$, large values of Q_0, and small values of β the consequent iterates may experience oscillations about the exact problem solutions (these are not instability oscillations but oscillations from one iteration to another). To dampen these oscillations one may use the following approach $g_n^{k+1} = 0.5(g_n^k + g_n^{k+1})$, $n = 1, \ldots, N + 1$, where k is the iteration number. However, it is not necessary. Therefore, we will just illustrate on several examples of starved lubrication regimes the influence of non-isothermal conditions on the problem solution in the inlet and exit zones.

TABLE 7.1

Comparison of the inlet coordinates for isothermal α_1^0 and non-isothermal α_1^t solutions for different values of coefficient A and $Q_0 = 1$, $\beta = 0.1$, and $T_{w0} = 0$. (after Kudish and Covitch [7]). Reprinted with permission from CRC Press.

A	1	2	3
α_1^0	-0.5016	-1.4045	-2.5244
α_1^t	-0.5178	-1.4423	-2.5894

Let us consider starved lubrication conditions, i.e., $h_q(r) = h_g(s) = 1$ for lubricant viscosity, which in the inlet and exit zones is described by equations $\mu_q^0 = e^{Q_0 q}$ and $\mu_g^0 = e^{Q_0 g}$, respectively. In practice, the surface temperature $T_{w0}(x) \geq 0$ tends to grow with x. For simplicity, we will consider only the case of $T_{w0} = 0$. Therefore, in our further analysis δ_T will not play an independent role besides its influence on the value of parameter β (see formula (7.27)).

Qualitatively, in the inlet zone of a non-isothermal lubricated contact pressure $q(r)$ behaves exactly the same way as in an isothermal contact. Quantitatively, these solutions are slightly different, which accounts for the presence of

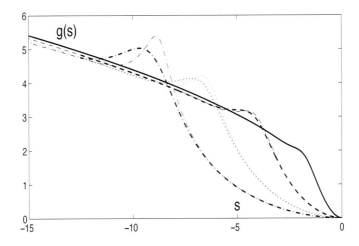

FIGURE 7.2
Graphs of $g(s)$ for isothermal (thin curves) and non-isothermal (thick curves) problems obtained for $A = 1$ (solid curves), $A = 2$ (dashed curves), $A = 3$ (dotted curves), $A = 4$ (dash-dotted curves) and $Q_0 = 1$, $\beta = 0.1581$, $N = 1000$, and $\triangle r = 0.015625$. (after Kudish and Covitch [7]). Reprinted with permission from CRC Press.

temperature effects. The numerical solutions are stable. We will not present graphs of such solutions as the similar ones were considered in detail in Section 6.9.3. Instead, for comparison the values of inlet coordinate α_1 obtained under isothermal conditions (labeled by α_1^0) and non-isothermal conditions (labeled by α_1^t) are presented in Table 7.1 for $Q_0 = 1$, $\beta = 0.1$ and several values of the coefficient A. The variations of α_1 as a function of parameter β are presented in Table 7.2 for $Q_0 = 1$ and $A = 2$.

TABLE 7.2
The dependence of α_1 on parameter β for $Q_0 = 1$,
$A = 2$, and $T_{w0} = 0$. (after Kudish and Covitch [7]).
Reprinted with permission from CRC Press.

β	0.01	0.03162	0.0707	0.1	0.3162
α_1	-1.4078	-1.4155	-1.4305	-1.4423	-1.5331

Data in Tables 7.1 and 7.2 show that as parameters β and/or A increase the inlet coordinate α_1 moves farther away from the contact center. Also, it

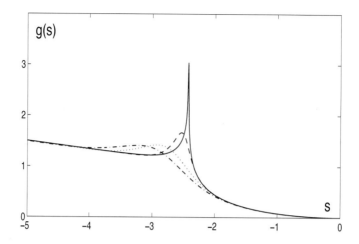

FIGURE 7.3

Graphs of $g(s)$ for non-isothermal problem obtained for $A = 2$, $Q_0 = 2.5$, $N = 1600$, $\triangle r = 0.0078125$, and $\beta = 0.0122$ (solid curve), $\beta = 0.0707$ (dashed curve), $\beta = 0.2236$ (dotted curve), $\beta = 0.3162$ (dash-dotted curve). (after Kudish and Covitch [7]). Reprinted with permission from CRC Press.

can be viewed as follows: as the amount of heat generated in the lubrication layer (represented by parameter β) increases lubricant viscosity decreases and it takes more lubricant in the inlet zone to maintain the same film thickness H_0 (proportional to coefficient A) as in the case of a corresponding isothermal contact.

In the exit zone, qualitatively and quantitatively solutions of the non-isothermal and isothermal problems are similar. However, the pressure spike in non-isothermal solutions (if exists) is lower and smoother than in the corresponding isothermal cases. That can be seen in Fig. 7.2 where for $Q_0 = 1$, $\beta = 0.1581$, $N = 1000$, $\triangle r = 0.015625$, and $A = 1$, 2, 3, and 4 the comparison of the isothermal and non-isothermal solutions is presented. It is clear from Fig. 7.2 that for these data pressure distributions $g(s)$ in non-isothermal contacts are very close to the corresponding ones in isothermal contacts. Moreover, as parameter β increases the pressure spike gets relatively smaller and it tends to move away from the exit point $s = 0$. An example of such a behavior of $g(s)$ is shown in Fig. 7.3 obtained for $A = 2$, $Q_0 = 2.5$, $N = 1600$, $\triangle r = 0.00415$ and several values of parameter β from $\beta = 0.004472$ through $\beta = 0.3162$. For significantly larger values of β the solutions for the exit zone pressure $g(s)$ become monotonic.

In cases of fully flooded lubrication regimes, we cannot use the above obtained approximation for function $M(p, h)$. However, from the numerical point of view it is still useful as we can rearrange function $M(p, h)$ as the one obtained asymptotically multiplied by the ratio of the original function $M(p, h)$ and its asymptotic representation. After that the same numerical iterative procedure as for the cases of starved lubrication regimes can be used while the values of the above ratio should be taken from the previous iteration and temperature T should be determined from the numerical solution of the original problem (7.14) in the inlet and exit zones. We will not pursue this any further as the behavior of the solution in the exit zone under fully flooded lubrication conditions is clear from the features the solution exhibits under starved lubrication conditions in non-isothermal contacts and under fully flooded lubrication conditions in isothermal contacts (see Section 6.9.3).

7.5   Closure

A basic EHL problem for a thermal heavily loaded line contact with Newtonian lubricant is studied using methods of matched asymptotic expansions and numerical methods. The asymptotic solution for the lubricant temperature T is obtained analytically. The solution for T was substituted into the Reynolds equation which allowed to simplify it significantly and to reduce it to an analog of the Reynolds equation for isothermal EHL problems. The rest of the asymptotic analysis was done in a manner very similar to the one used in the case of isothermal conditions. In the inlet and exit zones numerical methods for asymptotically valid problem solution are obtained and realized by a simple modification of the numerical methods presented in the preceding chapter. A formula for the film thickness is obtained analytically in the form identical to the case of isothermal conditions. The only difference in this formula is the value of the coefficient of proportionality A. The value of this coefficient of proportionality A can be determined from numerical problem solution in the inlet zone or by curve fitting of independently obtained experimental values for the lubrication film thickness.

7.6   Exercises and Problems

1. Explain why the assumption that the ratio $\nu(x)$ of the rolling and sliding frictional stresses is small in the inlet zone of a heavily loaded contact guarantees its smallness in the entire contact.

2. Show that in pre-critical lubrication regimes in the inlet and exit zones of heavily loaded thermal EHL contact with constant viscosity by a proper

scaling the asymptotic equations can be reduced to the equations for the corresponding isothermal problem. Find that scaling.

3. Describe the differences and similarities in the expressions for the film thickness H_0 for pre-critical lubrication regimes for thermal and isothermal heavily loaded EHL contacts.

4. Determine the two-term asymptotic solution for the lubricant temperature $T(x, z)$ in case when $\nu(x) \ll 1$ and the surface temperatures $T_{w1}(x)$ and $T_{w2}(x)$ are not equal.

5. Under the conditions of Problem 4 derive the approximate Reynolds equation.

References

[1] Kudish, I.I. 1978. Asymptotic Analysis of a Plane Non-isothermal Elastohydro-dynamic Problem for a Heavily Loaded Rolling Contact. *Proc. Acad. Sci. of Armenia SSR, Mechanics* 31, No. 6:16-35.

[2] Alexandrov, V.M. and Kudish, I.I. 1980. Problem of Elastohydrodynamic Theory of Lubrication for a Viscous Fluid with Complex Rheology. *J. Mech. Solids* 15, No. 4:57-68.

[3] Kudish, I.I. 1981. On Solution of some Contact and Elastohydrodynamic Problems. 3*rd Intern. Tribology Congress EUROTRIB* 81, *Warsaw, Tribological Processes in Contact Areas of Lubricated Solid Bodies* 2:251-271.

[4] Kudish, I.I. 1983. Asymptotic Method of Study for Plane Problems of the Elastohydrodynamic Lubrication Theory for Heavily Loaded Regimes. Part 2. Non-isothermal Problem. *Proc. Acad. Sci. Armenia SSR, Mechanics* 36, No. 5:47-59.

[5] Kudish, I.I. 1996. Asymptotic Analysis of a Problem for a Heavily Loaded Lubricated Contact of Elastic Bodies. Pre- and Over-critical Lubrication Regimes for Newtonian Fluids. *Dynamic Systems and Applications*, Dynamic Publishers, Atlanta, 5, No. 3:451-478.

[6] Myllerup, C.M. and Hamrock, B.J. 1994. Perturbation Approach to Hydrodynamic Lubrication Theory. *ASME J. Tribology* 116, No. 1:110-118.

[7] Kudish, I.I. and Covitch, M.J. 2010. *Modeling and Analytical Methods in Tribology*. Boca Raton: CRC Press.

8

Numerical Solution of the Original Isothermal EHL Problem for Newtonian Lubricant Revisited. Regularization Approach and Stable Numerical Method

8.1 Introduction

In Section 6.3.1 it has been shown that systems (6.5)-(6.8) and (6.19)-(6.22) are equivalent. In Section 6.2 it has been mentioned that for truly heavily loaded contacts the numerical procedure described in that section based on the former system of equations still tends to be unstable. In this chapter we will use the ideas developed for obtaining stable and accurate solutions of asymptotically valid equations to propose an improved (compared to Section 6.2) numerical scheme for heavily loaded EHL contacts with Newtonian lubricants in the original (non-asymptotic) formulation. These regularization approaches stem from the solution of the TEHL problem from the preceding chapter.

8.2 Numerical Solution of Asymptotic Isothermal EHL Problems for Newtonian Lubricant Revisited. Regularization Approach

The history of numerical methods in application to solution of EHL problems for heavily loaded lubricated contacts started with the revolutionary work of A.I. Petrusevich [1] in the end of 1940s in which, for the first time, he demonstrated the existence of a pressure spike in the exit zone. Since that time a series of different numerical methods has been developed. Among them are various modifications of Newton's and gradient methods, multi-level multi-grid methods, methods involving usage of fast Fourier transform, etc. Each of these methods has its own advantages and shortcomings.

All of these methods have one common drawback: they produce inaccurate and unstable solutions for heavily loaded EHL contacts lubricated by fluids with high pressure viscosity coefficients, which was demonstrated in Section 6.11. This numerical instability is primarily caused by the proximity of the EHL problems for heavily loaded contacts to an integral equation of the first kind and by extremely strong nonlinearity of the problem resulting from high values of the pressure viscosity coefficient. The numerical instability manifests itself in oscillating either pressure or its derivative as well as gap in the vicinity of a pressure spike located in the exit zone of a lubricated contact, in strong dependence/sensitivity of the pressure spike maximum on the numerical step size/number of nodes used, high condition number of the approximating algebraic system of equations and its proximity to a system with singular matrix, etc. This is a typical signature of an ill-posed problem.

The extensive numerical experiments showed that just decreasing the numerical step size does not alleviate the solution instability as it occurs in most ill-posed problems. A detailed numerical analysis of asymptotically valid equations approximating the original EHL problem in the inlet and exit zones of heavily loaded lubricated contact allows to point out several facts. It has been observed that in the inlet zone of such a contact (primarily responsible for lubrication film formation) an accurate stable solution can be obtained for a sufficiently small step size. That has been achieved by solving numerically the problem equations resolved for pressure involved in the expression for the gap (see Section 6.9.1). Moreover, that means that the equally accurate solution of the original (non-asymptotic) EHL problem can be obtained only for sufficiently small step sizes in the inlet zone. In particular, to get an equally accurate solution of the original EHL problem its step size is supposed to be equal to the characteristic size of the inlet zone (which depends on the problem input parameters) times the step size of the asymptotically valid problem sufficient for obtaining solution with desired precision (see Section 6.11).

In the exit zone of a heavily loaded EHL contact for sufficiently high pressure viscosity coefficients Q (proportional to constant Q_0), even for very small step sizes, the problem solution does not possess adequate precision and it remains unstable (see Section 6.11). Unfortunately, the usually used regularization methods do not improve solution stability. Luckily, the solution can be regularized by employing the procedure based on the naturally occurring process of heat generation in a lubrication film. A properly chosen sufficiently small heat generation in combination with a properly designed numerical scheme may lead to accurate and stable solutions in the exit zone (see Section 6.11).

Mathematically, for $T_{w0} = 0$ the above regularization is described by function $M(p, h)$ from equations (7.27) and (7.28)

$$M(p, h) = \frac{H_0^3}{V} \frac{h^3}{\mu} \frac{dp}{dx} R_T(x), \quad R_T(x) = \frac{3(1+\eta^2)}{\ln^2(\eta+\sqrt{\eta^2+1})} \left\{ 1 + \ln(1 + \eta^2) \right.$$

$$\left. -\frac{\sqrt{\eta^2+1}}{\eta} \ln(\eta + \sqrt{\eta^2 + 1}) - 2 \frac{\int_0^{\ln(\eta+\sqrt{\eta^2+1})} \ln(\cosh(t)) dt}{\ln(\eta+\sqrt{\eta^2+1})} \right\}, \quad \eta = \beta\sqrt{\mu},$$

$$(8.1)$$

where integral in (8.1) is calculated based on formulas (7.28), μ is the lubricant viscosity, and β is a certain nonnegative parameter. In all other respects the numerical schemes in the inlet and exit zones are identical to the ones presented in Sections 6.9.1 and 6.9.2.

Such characteristics of a heavily loaded lubricated contact as lubrication film thickness, exit coordinate of a contact, and maximum of pressure spike are determined predominantly by the inlet and exit zones. Moreover, the main contributions to these parameters come from the portions of the inlet and exit zones where pressure is relatively low. The regions where pressure is high are practically irrelevant for determining film thickness and exit coordinate of a contact. That explains the success of employing the regularization procedure based on introduction of heat generation. In fact, for nominally small heat generation in the regions where pressure is small the effect of heat generation on film thickness is small. At the same time, in high pressure zones heat generation is significantly higher, which dampens pressure growth and, therefore, the unstable growth of pressure in the pressure spike area. This is the conceptual foundation of this regularization approach.

TABLE 8.1
The illustration of convergence/stability of non-isothermal solution in the exit zone for $\beta = 0.012247$. (after Kudish and Covitch [2]). Reprinted with permission from CRC Press.

$\triangle r$ (N)	$C.N.$	β_1	$\max g(s)$ @ s_{max}
0.015625 (800)	$4.63 \cdot 10^7$	0.6317	4.5559 @ -2.4375
0.006000 (1,200)	$1.96 \cdot 10^8$	0.6562	4.7372 @ -2.4376
0.004150 (1,600)	$5.48 \cdot 10^8$	0.6394	4.7661 @ -2.4375

The level of heat generation (the value of constant β) should be chosen in such a way that the solution obtained based on it would be stable and, at the same time, close to the solution of the corresponding non-regularized (original) EHL problem. That can be achieved by comparing relatively stable solution characteristics such as the exit coordinates and positions of the pressure spike (see Table 6.6 in Section 6.11 and Table 8.1) obtained based on regularized and non-regularized solutions. The comparison of the maxima of pressure spikes has no part in this process because the maximum of pressure spike is an unstable characteristic of the original EHL problem.

The above regularization approach leads to good results in both the inlet and exit zones. Let us consider some examples in the exit zone for lubricant viscosity described by equation $\mu_g^0 = e^{Q_0 g}$. Small values of β provide a natural regularization of the problem. For $A = 2$ and $Q_0 = 2.5$ for which the isothermal non-regularized solution exhibits some instability signs (see Fig. 6.16 and

TABLE 8.2

The illustration of convergence/stability of non-isothermal solution in the exit zone for $\beta = 0.014142$. (after Kudish and Covitch [2]). Reprinted with permission from CRC Press.

$\triangle r$ (N)	$C.N.$	β_1	$\max g(s)$ @ s_{max}
0.015625 (800)	$4.62 \cdot 10^7$	0.6308	4.5269 @ -2.4375
0.006000 (1,200)	$1.25 \cdot 10^9$	0.6421	4.6294 @ -2.4420
0.004150 (1,600)	$4.56 \cdot 10^9$	0.6455	4.6381 @ -2.4443

Table 6.6) while for $A = 2$, $Q_0 = 2.5$ and $\beta = 0.012247$ and $\beta = 0.014142$ the solutions in general and the pressure spike, in particular, are stable. That can be seen from Tables 8.1 and 8.2, where the values of the condition number $C.N.$, the exit coordinate β_1, the maximum of pressure spike $\max g(s)$ and its location s_{max} are presented for several values of parameters N (number of nodes) and $\triangle r$ (numerical step size). Also, it can be seen from the comparison of the graphs of $g(s)$ in Fig. 8.1 obtained for isothermal non-regularized and regularized/non-isothermal cases for $A = 2$, $Q_0 = 2.5$, $\beta = 0.012247$, $N = 1600$, and $\triangle r = 0.0078125$. Obviously, for small β the position of the pressure spike practically does not change and almost coincides with the position of the pressure spike following from the non-regularized problem.

For small values of β all parameters of the regularized/non-isothermal problem solution are only slightly different from the ones for the isothermal problem except for the fact that the pressure spike behaves in a stable manner and there are no signs of solution instability in the region between the pressure spike and the contact exit point (see discussions in Section 6.9.3 and Section 6.11). We can get an idea of what changes in the isothermal solution can be caused by the introduction of small heat generation. In particular, for $A = 2$, $Q_0 = 1$, $\beta = 0.012247$, $N = 1600$, and $\triangle r = 0.0078125$ the stable isothermal and regularized/non-isothermal solutions have the following parameters $\beta_1 = 0.8966$, $\max g(s) = 3.2809$ reached at $s_{max} = -4.5802$ and $\beta_1 = 0.8635$, $\max g(s) = 3.1714$ reached at $s_{max} = -4.6059$, respectively. Also, it can be seen from the comparison of the values of the condition number $C.N.$ for isothermal (see Table 6.6 in Section 6.11) and regularized/non-isothermal problems (see Table 8.1). In the latter case these values are by at least 10% lower which promotes stability of numerical solutions. Therefore, the introduction of thermal effects provides a natural way of numerical solution regularization. It is important to realize that to regularize the exit solution for larger values of Q_0 it is necessary to use a greater value of parameter β.

Also, it is important to point out that the pressure viscosity coefficient Q is equal to $Q_0 \epsilon_q^{-1/2}$. It means that, for example, for fully flooded pre-critical lubrication regimes $\epsilon_q = V^{2/5}$ and $Q = Q_0 V^{-1/5}$. Therefore, if $V = 0.001$ and for $Q_0 = 10$ we have a stable solution it means that a stable solution

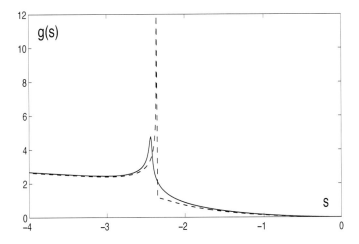

FIGURE 8.1
Graphs of $g(s)$ for non-isothermal problem (solid curve) and for isothermal problem (dashed curve) obtained for $A = 2$, $Q_0 = 2.5$, $\beta = 0.012247$, $N = 1600$, and $\triangle r = 0.0078125$. (after Kudish and Covitch [2]). Reprinted with permission from CRC Press.

can be obtained for $Q = 39.8$. These kind of values of the pressure viscosity coefficient Q represent the upper bound of the range of Q in practice.

To conclude this section it is worth to make the following remark. It is important to keep in mind that from the physical point of view for large values of Q_0 the pressure spike itself does not have much of an influence on the rest of the solution as it practically does not affect any of the vital EHL parameters possibly with just one exception - the rolling frictional stress, which is small in most cases. It is caused by the fact that the pressure spike is narrow. Therefore, the choice of the value of parameter β for regularization of isothermal solutions can be made without much influence of the resulting from it size of the pressure spike. We can also stress that due to solution stability in the inlet zone to determine the lubrication film thickness H_0 and sliding frictional stress $f(x)$ and force F_S (see Section 10.8) it is sufficient to solve the EHL problem in just the inlet zone and to use the Hertzian pressure outside of it. It is a distinct advantage of the asymptotic approach.

8.3 Numerical Method for the Original Isothermal EHL Problem Revisited. Regularization Approach

To improve numerical stability of problem solutions for Newtonian lubricants in heavily loaded contacts, we will employ equations (6.19)-(6.22)

$$p(x) = R(x)\left[1 - \frac{1}{2\pi}\int\limits_a^c \frac{dM(p,h)}{dt}\frac{dt}{R(t)(t-x)}\right], \quad R(x) = \sqrt{(x-a)(c-x)}, \quad (8.2)$$

$$\int\limits_a^c \frac{dM(p,h)}{dt}\frac{dt}{R(t)} = \pi(a+c), \quad \int\limits_a^c \frac{dM(p,h)}{dt}\frac{tdt}{R(t)} = \pi[(\tfrac{c-a}{2})^2 + \tfrac{(a+c)^2}{2} - 1], \quad (8.3)$$

$$M(p,h) = \frac{H_0^3}{V}\frac{h^3}{\mu}\frac{dp}{dx}, \quad (8.4)$$

$$H_0(h-1) = x^2 - c^2 + \frac{2}{\pi}\int\limits_a^c p(t)\ln\mid\tfrac{c-t}{x-t}\mid dt, \quad (8.5)$$

resolved for pressure $p(x)$ involved in the expression for gap $h(x)$ and the regularization approach based on the introduction of heat generation in the contact (i.e., on modification of function $M(p,h)$). The latter means that for large values of pressure viscosity coefficient Q (where $\mu = \exp(Qp)$) instead of using function $M(p,h)$ determined by (8.4) we will be using function $M(p,h)$ calculated as follows (see equations (7.27) and (7.28) for $T_{w0} = 0$)

$$M(p,h) = \frac{H_0^3}{V}\frac{h^3}{\mu}\frac{dp}{dx}R_T, \quad R_T = \frac{3(1+\eta^2)}{\ln^2(\eta+\sqrt{\eta^2+1})}\left\{1 + \ln(1+\eta^2)\right.$$

$$\left. -\frac{\sqrt{\eta^2+1}}{\eta}\ln(\eta+\sqrt{\eta^2+1}) - 2\frac{\int\limits_0^{\ln(\eta+\sqrt{\eta^2+1})}\ln(\cosh(t))dt}{\ln(\eta+\sqrt{\eta^2+1})}\right\}, \quad \eta = \beta\sqrt{\mu}, \quad (8.6)$$

where the integral in (8.6) is calculated based on formulas (7.28) and the details of the dependence of β on s_0 and other parameters can be found in (7.27). The suitability of the above modification of function $M(p,h)$ is based on two facts: (a) for $\beta \to 0$ function $M(p,h)$ from (8.6) approaches function $M(p,h)$ from (8.4), which corresponds to the isothermal case and (b) $M(p,h)$ from (8.6) provides sufficient regularization of solutions of the original isothermal EHL problem. The latter has been demonstrated in Section 8.2 for equations asymptotically valid in the inlet and exit zones.

By introducing the substitution $x = \frac{a+c}{2} + \frac{c-a}{2}y$ equations (8.2)-(8.6) can be reduced to equations within fixed boundaries $[-1,1]$ as follows

$$p(y) = \frac{c-a}{2}\sqrt{1-y^2}\left[1 - \frac{1}{2\pi}(\tfrac{2}{c-a})^3\int\limits_{-1}^1\frac{d}{d\tau}(W\frac{dp}{d\tau})\frac{d\tau}{\sqrt{1-\tau^2}(\tau-y)}\right], \quad (8.7)$$

$$\int_{-1}^{1} \frac{d}{d\tau}\left(W\frac{dp}{d\tau}\right)\frac{d\tau}{\sqrt{1-\tau^2}} = \pi(a+c)\left(\frac{c-a}{2}\right)^2,$$

(8.8)

$$\int_{-1}^{1} \frac{d}{d\tau}\left(W\frac{dp}{d\tau}\right)\frac{\tau d\tau}{\sqrt{1-\tau^2}} = \pi\frac{c-a}{2}\left[\left(\frac{c-a}{2}\right)^2 - 1\right],$$

$$H_0(h-1) = \frac{a^2+2ac-3c^2}{4} + \frac{c^2-a^2}{2}y + \left(\frac{c-a}{2}\right)^2 y^2$$

(8.9)

$$+ \frac{c-a}{\pi}\int_{-1}^{1} p(\tau)\ln\left|\frac{1-\tau}{y-\tau}\right| d\tau,$$

where $M(p,h)$ is replaced by function $W(p,h)$ as follows

$$W(p,h) = \frac{H_0^3}{V}\frac{h^3}{\mu}$$

(8.10)

for small to moderate values of Q, and it is replaced by

$$W(p,h) = \frac{H_0^3}{V}\frac{h^3}{\mu}R_T, \quad R_T = \frac{3(1+\eta^2)}{\ln^2(\eta+\sqrt{\eta^2+1})}\left\{1 + \ln(1+\eta^2)\right.$$

(8.11)

$$\left. -\frac{\sqrt{\eta^2+1}}{\eta}\ln(\eta+\sqrt{\eta^2+1}) - 2\frac{\int_{0}^{\ln(\eta+\sqrt{\eta^2+1})}\ln(\cosh(t))dt}{\ln(\eta+\sqrt{\eta^2+1})}\right\}, \quad \eta = \beta\sqrt{\mu},$$

for large values of Q. The integral in (8.11) is defined in formulas (7.28).

Let us introduce two sets of nodes: $y_k = -1 + (k-1)\triangle y$, $k = 1, \ldots, N$, and $y_{k+1/2} = 0.5(y_k + y_{k+1})$, where $\triangle y$ is the step size. Now, we can approximate the integral

$$\int_{-1}^{1} \frac{d}{d\tau}\left(W\frac{dp}{d\tau}\right)\frac{d\tau}{\sqrt{1-\tau^2}(\tau-y_k)} \approx \int_{-1}^{y_{3/2}} \frac{d}{d\tau}\left(W\frac{dp}{d\tau}\right)\frac{d\tau}{\sqrt{1-\tau^2}(\tau-y_k)}$$

$$+ \sum_{i=2}^{N-1}\int_{y_{i-1/2}}^{y_{i+1/2}} \frac{d}{d\tau}\left(W\frac{dp}{d\tau}\right)\frac{d\tau}{\sqrt{1-\tau^2}(\tau-y_k)} + \int_{y_{N-1/2}}^{1} \frac{d}{d\tau}\left(W\frac{dp}{d\tau}\right)\frac{d\tau}{\sqrt{1-\tau^2}(\tau-y_k)}$$

$$\approx \frac{d}{dy}\left(W\frac{dp}{dy}\right)\Big|_{y_{5/4}}\int_{-1}^{y_{3/2}} \frac{d\tau}{\sqrt{1-\tau^2}(\tau-y_k)} + \sum_{i=2}^{N-1}\frac{d}{dy}\left(W\frac{dp}{dy}\right)\Big|_{y_i}$$

(8.12)

$$\times \int_{y_{i-1/2}}^{y_{i+1/2}} \frac{d\tau}{\sqrt{1-\tau^2}(\tau-y_k)} + \frac{d}{dy}\left(W\frac{dp}{dy}\right)\Big|_{y_{N-1/4}}\int_{y_{N-1/2}}^{1} \frac{d\tau}{\sqrt{1-\tau^2}(\tau-y_k)},$$

$$y_{5/4} = -1 + \frac{\triangle y}{4}, \quad y_{N-1/4} = 1 - \frac{\triangle y}{4}.$$

Taking into account that $\frac{dp(1)}{dy} = 0$ the second derivatives in (8.12) can be approximated as follows

$$\frac{d}{dy}\left(W\frac{dp}{dy}\right)\big|_{y_{5/4}} \approx \frac{1}{\triangle y^2}[W_{5/2}(p_3-p_2)-W_{3/2}(p_2-p_1)],$$

$$\frac{d}{dy}\left(W\frac{dp}{dy}\right)\big|_{y_i} \approx \frac{1}{\triangle y^2}[W_{i+1/2}(p_{i+1}-p_i)-W_{i-1/2}(p_i-p_{i-1})], \qquad (8.13)$$

$$\frac{d}{dy}\left(W\frac{dp}{dy}\right)\big|_{y_{N-1/4}} \approx -\frac{2}{\triangle y^2}W_{N-1/2}(p_N-p_{N-1}).$$

Integrals involved in (8.12) are calculated by employing the substitution $\tau = \frac{2t}{1+t^2}$ and have the form

$$I(a,b,y_k) = \int_a^b \frac{d\tau}{\sqrt{1-\tau^2}(\tau-y_k)} = \frac{1}{\sqrt{1-y_k^2}}J(a,b,y_k),$$

$$J(a,b,y_k) = \ln\left|\,\frac{t_b-t_{1k}}{t_a-t_{1k}}\frac{t_a-t_{2k}}{t_b-t_{2k}}\,\right|,$$

$$(8.14)$$

$$t_a = \frac{1-\sqrt{1-a^2}}{a},\ \ t_b = \frac{1-\sqrt{1-b^2}}{b},\ \ t_{1k} = \frac{1-\sqrt{1-y_k^2}}{y_k},\ \ t_{2k} = \frac{1+\sqrt{1-y_k^2}}{y_k}.$$

In (8.14) t_a and t_b are solutions of the equations $\frac{2t}{1+t^2} = a$ and $\frac{2t}{1+t^2} = b$ chosen in such a way that both $|\,t_a\,|$ and $|\,t_b\,|$ are not greater than 1.

By satisfying equation (8.7) at nodes $y_k,\ k=1,\ldots,N$, we obtain a system of N nonlinear algebraic equations

$$p_k + \frac{1}{2\pi\triangle y^2}\left(\frac{2}{c-a}\right)^2[W_{5/2}(p_3-p_2)-W_{3/2}(p_2-p_1)]J(-1,y_{3/2},y_k)$$

$$+\frac{1}{2\pi\triangle y^2}\left(\frac{2}{c-a}\right)^2\sum_{i=2}^{N-1}[W_{i+1/2}(p_{i+1}-p_i)-W_{i-1/2}(p_i-p_{i-1})]$$

$$(8.15)$$

$$\times J(y_{i-1/2},y_{i+1/2},y_k) - \frac{1}{\pi\triangle y^2}\left(\frac{2}{c-a}\right)^2W_{N-1/2}(p_N-p_{N-1})$$

$$\times J(y_{N-1/2},1,y_k) = \frac{c-a}{2}\sqrt{1-y_k^2},\ k=1,\ldots,N.$$

This kind of integral approximation was tested on dry contact problems (see Section 4.3.2) with the punch bottom described by function Cy^{2m} (C and m are positive constants). These contact problems possess well–known simple exact analytical solutions. The numerical solutions of these problems obtained using the integral approximation similar to (8.12)-(8.14) provide excellent approximations for the exact ones. The absolute error of the numerical solution depends on the values of constants C and m as well as the step size $\triangle y$.

To iteratively solve this system we introduce a set of new unknowns $z_k = p_k^{n+1} - p_k^n,\ k=1,\cdots,N$ (n is the iteration number) and reduce the system of nonlinear equations (8.15) to a system of N linear algebraic equations

$$\frac{1}{2\pi\triangle y^2}\left(\frac{2}{c^n-a}\right)^2 \sum_{i=2}^{N-1} [W_{i+1/2}^n(z_{i+1}-z_i) - W_{i-1/2}^n(z_i-z_{i-1})]\tilde{J}_{ik}$$

$$-\frac{1}{\pi\triangle y^2}\left(\frac{2}{c^n-a}\right)^2 W_{N-1/2}^n(z_N-z_{N-1})\tilde{J}_{Nk} + z_k$$

$$+\frac{1}{2\pi\triangle y^2}\left(\frac{2}{c^n-a}\right)^2 \sum_{i=2}^{N-1} [W_{i+1/2}^n(p_{i+1}^n-p_i^n) - W_{i-1/2}^n(p_i^n-p_{i-1}^n)]\tilde{J}_{ik} \qquad (8.16)$$

$$-\frac{1}{\pi\triangle y^2}\left(\frac{2}{c^n-a}\right)^2 W_{N-1/2}^n(p_N^n-p_{N-1}^n)\tilde{J}_{Nk} + p_k^n$$

$$= \frac{c^n-a}{2}\sqrt{1-y_k^2}, \ k=1,\ldots,N,$$

$$\tilde{J}_{2k} = J(-1, y_{3/2}, y_k) + J(y_{3/2}, y_{5/2}, y_k),$$

$$\tilde{J}_{ik} = J(y_{i-1/2}, y_{i+1/2}, y_k), \ 2 < i < N, \ \tilde{J}_{Nk} = J(y_{N-1/2}, 1, y_k).$$

Now, let us derive formulas for calculation of film thickness H_0 and exit coordinate c. To do that we need to approximate the integrals involved in equations (8.8). Following the same approach as was employed for evaluation of the integral in (8.12), we obtain

$$\int_{-1}^{1} \frac{d}{d\tau}\left(W\frac{dp}{d\tau}\right)\frac{d\tau}{\sqrt{1-\tau^2}} \approx \frac{1}{\triangle y} \sum_{i=2}^{N-1} [W_{i+1/2}(p_{i+1}-p_i)$$

$$-W_{i-1/2}(p_i-p_{i-1})]L_i - \frac{2}{\triangle y}W_{N-1/2}(p_N-p_{N-1})L_N,$$

$$\tag{8.17}$$

$$\int_{-1}^{1} \frac{d}{d\tau}\left(W\frac{dp}{d\tau}\right)\frac{\tau d\tau}{\sqrt{1-\tau^2}} \approx \frac{1}{\triangle y} \sum_{i=2}^{N-1} [W_{i+1/2}(p_{i+1}-p_i)$$

$$-W_{i-1/2}(p_i-p_{i-1})]M_i - \frac{2}{\triangle y}W_{N-1/2}(p_N-p_{N-1})M_N,$$

$$L_2 = \frac{1}{\sqrt{1-y_{5/4}^2}} + \frac{1}{\sqrt{1-y_2^2}}, \ L_N = \frac{1}{\sqrt{1-y_{N-1/4}^2}},$$

$$M_2 = \frac{y_{5/4}}{\sqrt{1-y_{5/4}^2}} + \frac{y_2}{\sqrt{1-y_2^2}}, \ M_N = \frac{1}{\sqrt{1-y_{N-1/4}^2}}, \qquad (8.18)$$

$$L_i = \frac{1}{\sqrt{1-y_i^2}}, \ M_i = \frac{y_i}{\sqrt{1-y_i^2}}, \ i=3,\ldots,N-1.$$

These integral approximations are consistent with the ones used for approximation of integrals in formulas (6.201) and (6.212) and employed for solution of asymptotically valid equations.

Knowing the values of these integrals we divide the second equation in (8.8) by the first equation, which eliminates H_0 from the ratio and produces the equation

$$\gamma = \int\limits_{-1}^{1} \frac{d}{d\tau}(W\frac{dp}{d\tau})\frac{\tau d\tau}{\sqrt{1-\tau^2}} / \int\limits_{-1}^{1} \frac{d}{d\tau}(W\frac{dp}{d\tau})\frac{d\tau}{\sqrt{1-\tau^2}}, \qquad (8.19)$$

$$\gamma = \frac{\frac{c-a}{2} - \frac{2}{c-a}}{c+a}. \qquad (8.20)$$

The integrals in (8.19) are calculated based on formulas (8.17) and (8.18). Solving equation (8.20) for c as a quadratic equation provides two solutions

$$c = a\frac{1 \pm 2\sqrt{\gamma^2 + \frac{1-2\gamma}{a^2}}}{1-2\gamma}. \qquad (8.21)$$

To choose the correct sign in (8.21), we notice that when the problem parameters approach the ones for a dry (not lubricated) contact then $\gamma \to 0$ and we should have $a \to -1$ and $c \to -a$. Expanding formula (8.21) for small γ we easily understand that to satisfy the above conditions we must choose sign minus. Therefore, for the exit coordinate we get the following formula

$$c = a\frac{1 - 2\sqrt{\gamma^2 + \frac{1-2\gamma}{a^2}}}{1-2\gamma}. \qquad (8.22)$$

After the value of c is determined we calculate film thickness H_0 from the first equation in (8.8) assuming that the integral in the left–hand side is approximated based on formulas (8.17) and (8.18). As a result of that we get a formula

$$H_0^3 = \pi(c+a)(\frac{c-a}{2})^2 \triangle y / \Big\{ \sum_{i=2}^{N-1} \Big[\frac{W_{i+1/2}}{H_0^3}(p_{i+1} - p_i)$$

$$-\frac{W_{i-1/2}}{H_0^3}(p_i - p_{i-1})\Big]L_i - 2\frac{W_{N-1/2}}{H_0^3}(p_N - p_{N-1})L_N\Big\}. \qquad (8.23)$$

Finally, to calculate the gap between the lubricated solids $h(y)$, we can employ a discrete analog of equation (8.9) in the form

$$H_0^{n+1}(h_{k+1/2}^{n+1} - 1) = \frac{a^2 + 2ac^{n+1} - 3(c^{n+1})^2}{4} + \frac{(c^{n+1})^2 - a^2}{2}y_{k+1/2}$$

$$+(\frac{c^{n+1}-a}{2})^2 y_{k+1/2}^2 + \frac{c^{n+1}-a}{\pi} \sum_{i=1}^{n-1} N_i(y_{k+1/2}), \ k = 1, \dots, N-1,$$

$$N_i(y) = [p_i^{n+1} - y_i\frac{p_{i+1}^{n+1} - p_i^{n+1}}{y_{i+1} - y_i}][K_1(y_{i+1}) - K_1(y_i) - K_2(y - y_{i+1}) \qquad (8.24)$$

$$+K_2(y - y_i)] + \frac{p_{i+1}^{n+1} - p_i^{n+1}}{y_{i+1} - y_i}[K_3(y_{i+1}) - K_3(y_i) - K_4(y - y_{i+1}, y)$$

$$+K_4(y - y_i, y)], \ K_1(y) = (y - 1)[\ln(1 - y) - 1],$$

$$K_2(y) = y[1 - \ln | y |], \ K_3(y) = \frac{(1-y)^2}{2}[\ln(1 - y) - \tfrac{1}{2}] + K_1(y),$$

$$K_4(y, \zeta) = \frac{y^2}{2}[\ln | y | -\tfrac{1}{2}] + \zeta K_2(y).$$

Also, we can use iteration based approach. Instead of using the discrete analogue of equation (8.9), we can use the equation

$$h = 1 + \frac{1}{H_0} \frac{2}{c-a} W \frac{dp}{dy}, \tag{8.25}$$

which is obtained from the original Reynolds equation (6.5) rewritten in variable y and integrated once with respect to y with the use of the boundary condition $\frac{dp(1)}{dy} = 0$ (see boundary conditions (6.6)). Equation (8.25) can be satisfied at nodes $y_{k+1/2}$, $k = 1, \ldots, N - 1$, and solved using a modified Newton's method similar to the one proposed in Section 6.9.1 (see iterative procedure described in (6.198)). However, it is easier to use a simple iteration process (for comparison see (6.200)), which leads to the following

$$h_{k+1/2}^{n+1} = 1 + \frac{1}{H_0^{n+1}} \frac{2}{c^{n+1}-a} W(p_{k+1/2}^{n+1}, h_{k+1/2}^n) \frac{p_{k+1}^{n+1}-p_k^{n+1}}{\triangle y},$$

$$p_{k+1/2}^{n+1} = 0.5(p_k^{n+1} + p_{k+1}^{n+1}), \ k = 0, \ldots, N - 1, \tag{8.26}$$

where n is the iteration number.

Now, let us describe the steps of the iteration process as a whole. For certainty, we will assume that $\mu = \exp(Qp)$. The iteration process is designed in such a way that for the given values of parameters a, V, and Q we are searching for the sets of p_k, $k = 1, \ldots, N$, $h_{k+1/2}$, $k = 1, \ldots, N - 1$, and constants H_0 and c. First we take some initial approximations for values of p_k^0, $h_{k+1/2}^0$, H_0^0, and c^0. For example, these can be taken as follows

$$H_0^0 = 0.272(VQ)^{3/4}, \ Q > 0 \ and \ H_0 = 0.2, \ Q = 0, \ c^0 = 1,$$

$$p_k^0 = 0 \ if \ | \tfrac{a+c^0}{2} + \tfrac{c^0-a}{2} y_k | > 1,$$

$$p_k^0 = \sqrt{1 - y_k^2} \ otherwise, \ k = 1, \ldots, N,$$

$$h_{k+1/2}^0 = [| x_{k+1/2} | \sqrt{x_{k+1/2}^2 - 1} - \ln(| x_{k+1/2} |$$

$$+ \sqrt{x_{k+1/2}^2 - 1})]\theta(x_{k+1/2}^2 - 1), \ x_{k+1/2} = \tfrac{a+c^0}{2} + \tfrac{c^0-a}{2} y_{k+1/2}, \tag{8.27}$$

$$k = 1, \ldots, N - 1,$$

or from the earlier obtained solution with the closest set of parameters a, V, and Q. Suppose we have the n-th iterates of p_k^n, $k = 1, \ldots, N$, $h_{k+1/2}^n$, $k = 1, \ldots, N - 1$, H_0^n and c^n. Then, the new iterates of p_k^{n+1}, $k = 1, \ldots, N$, are determined from the formula $p_k^{n+1} = z_k + p_k^{n+1}$, where values of z_k, $k = 1, \ldots, N$, are obtained from solution of a system of linear algebraic equations (8.16). After that the new iterate c^{n+1} is determined from the equation (see equations (8.17)-(8.19) and (8.22))

$$c^{n+1} = a \frac{1 - 2\sqrt{(\gamma^{n+1})^2 + \frac{1 - 2\gamma^{n+1}}{a^2}}}{1 - 2\gamma^{n+1}}, \tag{8.28}$$

where the value of γ^{n+1}, based on formula (8.19), is calculated using approximations (8.17), (8.18), and the new iterates p_k^{n+1}, $k = 1, \ldots, N$. That follows by calculation of the new iterate H_0^{n+1} based on formula (8.23)

$$(H_0^{n+1})^3 = (H_0^n)^3 \pi (c^{n+1} + a) \left(\frac{c^{n+1} - a}{2} \right)^2 \triangle y$$

$$\Big/ \Big\{ \sum_{i=2}^{N-1} [W_{i+1/2}^{n+1} (p_{i+1}^{n+1} - p_i^{n+1}) - W_{i-1/2}^{n+1} (p_i^{n+1} - p_{i-1}^{n+1})] L_i \tag{8.29}$$

$$-2 W_{N-1/2}^{n+1} (p_N^{n+1} - p_{N-1}^{n+1}) L_N \Big\},$$

in which c, p_i, and $W_{i+1/2}$ must be replaced by c^{n+1}, p_i^{n+1}, and $W_{i+1/2}^{n+1}$, respectively. After that the new set of iterates $h_{k+1/2}^{n+1}$, $k = 1, \ldots, N - 1$, is obtained from equation (8.26). The iteration process continues until the desired precision is reached.

For large number of nodes N a multi-grid technique in combination with interpolation can be employed to accelerate the solution process by reducing the average order of the systems of linear equations solved. That would require solving the problem for a cascade of denser grids. The interpolation is used to obtain the initial approximation for the solution on a denser grid using the data from the solution on a less dense grid.

The necessity of regularization is determined based on the properties of the obtained solution and, in particular, its sensitivity to the number of nodes N, presence of pressure p or gap h oscillations close to the exit point, etc. In case the regularization is needed, one has to replace function W from formula (8.10) by the one described in formula (8.11). The rest of the iteration procedure remains the same. The value of parameter β in (8.11) is adjusted empirically to provide stability and, at the same time, proximity to the undisturbed solution.

For very large values of Q (such as over 25-30) at the early stages of the iteration process, it is prudent to slow down the iterations due to the fact that the initial approximation used is usually very far from the problem solution and the pressure distribution depends strongly on even small variations in the gap distribution (because the problem is stiff). That can be accomplished by introducing interpolation/relaxation as follows

$$h_{k+1/2}^{n+1} = (1 - \alpha_*) h_{k+1/2}^n + \alpha_* h_{k+1/2}^{n+1}, \quad k = 0, \ldots, N - 1,$$

$$H_0^{n+1} = (1 - \alpha_*) H_0^n + \alpha_* H_0^{n+1}, \quad c^{n+1} = (1 - \alpha_*) c^n + \alpha_* c^{n+1}, \tag{8.30}$$

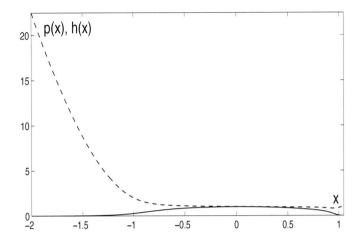

FIGURE 8.2
Graphs of pressure $p(x)$ (solid curve) and gap $h(x)$ (dashed curve) obtained for $V = 0.1$, $Q = 0$, and $\beta = 0$. (after Kudish and Covitch [2]). Reprinted with permission from CRC Press.

where α_* is a sufficiently small positive constant. For example, α_* can be taken equal to 0.1 or 0.15 at the initial stages of the iteration process. At the later stages of the iteration process, the value of this constant α_* can be increased and even taken equal to 1. It is customary that for fully flooded lubrication regimes at the later stages the rate of process convergence decreases in the inlet zone where $h(y) \gg 1$ due to significant nonlinearity of Reynolds equation in h.

Obviously, if formula (8.11) for function W is used the described method provides the solution to the thermal EHL problem in the regimes of starved lubrication.

The described regularized iteration approach works equally well for isothermal and non-isothermal EHL problems for lubricants with non-Newtonian rheology. The only adjustment that is required is the replacement of function W by the one corresponding to the problem at hand. The above–described regularization without any changes can be extended on spatial EHL problems by replacing terms $\partial p / \partial x$ and $\partial p / \partial y$ by terms (see (8.11)) $R_T \partial p / \partial x$ and $R_T \partial p / \partial y$, respectively.

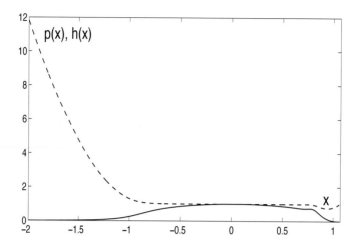

FIGURE 8.3

Graphs of pressure $p(x)$ (solid curve) and gap $h(x)$ (dashed curve) obtained
for $V = 0.1$, $Q = 5$, and $\beta = 0$. (after Kudish and Covitch [2]). Reprinted
with permission from CRC Press.

8.4   Some Numerical Solutions of the Original Regularized Isothermal EHL Problem

The described regularization procedure allows to solve the EHL problems for a
wide range of pressure viscosity coefficient Q (from 0 to 40), which covers the
range of its variations in practice. In the further numerical analysis, we will be
using $a = -2$, $V = 0.1$, $N = 1601$, $\triangle y = 2/(N-1)$, and absolute precision
$\varepsilon = 0.001$. We will demonstrate the behavior of the solution of the EHL
problem on five examples of heavily loaded contacts: for (a) $Q = 0$, (b) $Q = 5$,
(c) $Q = 10$, (d) $Q = 20$, and (e) $Q = 35$ obtained for $\beta = 0$, 0, 0.0325, 0.045,
and 0.85, respectively. For these cases the values of the solution parameters
H_0, c, $h(\triangle y/2)$, $\min h(y)$, and maximum of the pressure spike $\max p(y)$ for
the original EHL problem are gathered in Table 8.3 The graphs of pressure
$p(x)$ and gap $h(x)$ for cases (a)-(e) are presented in Figs. 8.2-8.6, respectively.

It is obvious from these graphs that as Q increases inlet gap $h(\triangle y/2)$ monotonically decreases while film thickness H_0, exit coordinate c, and the maximum of pressure spike $\max p(y)$ monotonically increase. That becomes clear
if you take into account two facts that as Q increases (i) a lubricated contact transitions from a fully flooded pre-critical to an over-critical lubrication
regime in which the characteristic size of the inlet and exit zones transitions

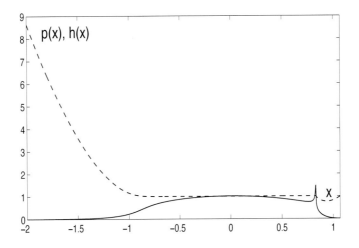

FIGURE 8.4
Graphs of pressure $p(x)$ (solid curve) and gap $h(x)$ (dashed curve) for regularized problem obtained for $V = 0.1$, $Q = 10$, and $\beta = 0.0325$. (after Kudish and Covitch [2]). Reprinted with permission from CRC Press.

from the value proportional to $\epsilon_q = V^{2/5}$ to $\epsilon_q = (VQ)^{1/2}$ and (ii) the inlet coordinate $\alpha_1 = (a + 1)/\epsilon_q$ decreases. Moreover, as Q increases the pressure spike shifts toward the center of the contact. There are no signs of solution instability, the graphs of pressure $p(x)$ and gap $h(x)$ are smooth. All these results are repeatable with high precision for larger number of nodes N (smaller step sizes $\triangle y$).

TABLE 8.3
Solution parameters H_0, c, $h(\triangle y/2)$, $\min h(y)$, and $\max p(y)$ for $V = 0.1$ and different values of Q and β. (after Kudish and Covitch [2]). Reprinted with permission from CRC Press.

Q	β	H_0	c	$h(\triangle y/2)$	$\min h(y)$	$\max p(y)$
0	0	0.102	1.036	22.187	0.792	–
5	0	0.197	1.060	11.827	0.760	0.752
10	0.0325	0.278	1.075	8.624	0.762	1.462
20	0.4500	0.408	1.094	6.153	0.769	1.983
35	0.8500	0.558	1.113	4.761	0.777	2.096

It is important to remember that for fixed values of V the minimum value of parameter β which provides a stable problem solution is a monotonically

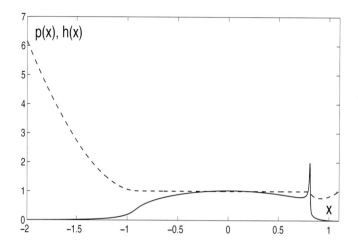

FIGURE 8.5

Graphs of pressure $p(x)$ (solid curve) and gap $h(x)$ (dashed curve) for regularized problem obtained for $V = 0.1$, $Q = 20$, and $\beta = 0.45$. (after Kudish and Covitch [2]). Reprinted with permission from CRC Press.

increasing function of Q. Usually, even significant increase of the value of β (for example, doubling of β) changes the solution parameters by less then 0.75% while the pressure spike maximum $\max p(x)$ may change by couple percents. At the same time, the film thickness H_0 behaves very conservatively by changing by less than 0.2%. In cases when the film thickness H_0 changes more significantly to preserve the value of H_0 it is advisable to use parameter β as a monotonically increasing function from zero in the inlet zone to a certain value in the exit zone where solution instability may occur. This is one of the reasons why it is preferable to solve separately the non-regularized asymptotic problem in the inlet zone and a regularized asymptotic problem in the exit zone.

Let us stress the fact that in the original non-regularized formulation numerical solutions of isothermal EHL problem for heavily loaded contacts tend to be unstable while being regularized in the described manner the solutions become stable. The conclusion that can be drawn from this fact is that for heavily loaded lubricated contacts the isothermal formulation of the EHL problem is inadequate and should be replaced by the thermal formulation of the EHL problem with minimum heat generation. The level of heat generations should be chosen based on numerical results.

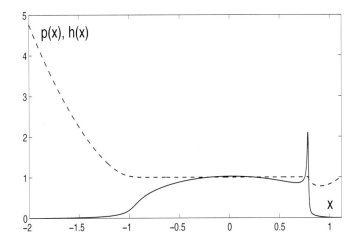

FIGURE 8.6
Graphs of pressure $p(x)$ (solid curve) and gap $h(x)$ (dashed curve) for regularized problem obtained for $V = 0.1$, $Q = 35$, and $\beta = 0.85$. (after Kudish and Covitch [2]). Reprinted with permission from CRC Press.

8.5   Closure

A simple regularization approach to regularization of numerical solution of asymptotically valid equations in the exit zone (as well as in the inlet zone) of an isothermal heavily loaded line EHL contact is proposed. The approach is based on solution of TEHL problem with relatively small heat generation in the lubrication layer. The application of the regularization method produced stable numerical solutions in the inlet and exit zones of isothermal heavily loaded line EHL contacts. This regularization approach is extended on the original (not asymptotically derived) equations of the isothermal EHL problem for heavily loaded line contacts. The regularized method allowed to obtain stable highly precise numerical solutions for the values of the dimensionless pressure viscosity coefficient Q at least up to $Q = 35$.

8.6   Exercises and Problems

1. Describe the idea behind the proposed regularization procedure.

2. In case of a constant viscosity and small β estimate the difference between regularized and non-regularized solutions in the inlet and exit zones of a heavily loaded EHL contact. (Hint: Use a proper scaling to reduce the regularized problems to their non-regularized analogs and, then, estimate the difference between these solutions using the Taylor formula.)

3. For large pressure viscosity coefficients $Q \gg 1$ compare and describe the similarities and differences of the regularized and non-regularized numerical solutions of the EHL problem for heavily loaded contact in its asymptotic and original formulations.

References

[1] Petrusevich A.I. 1951. Fundamental Conclusions from the Contact-Hydrodynamic Theory of Lubrication. *Izv. Acad. Nauk, SSSR (OTN)*, 2, 209.

[2] Kudish, I.I. and Covitch, M.J. 2010. *Modeling and Analytical Methods in Tribology*. Boca Raton: CRC Press.

9

Some Analytical Approximations and Numerical Solutions of EHL Problems for Non-Newtonian Fluids

9.1 Introduction

The chapter is dedicated to the asymptotic analysis of steady heavily loaded isothermal lubricated line contacts under the no-slip condition and two opposite limiting cases of pure rolling and relatively high slide-to-roll ratio. The problem is considered for two general classes of non-Newtonian lubricant rheologies when the shear strain and stress can be expressed as certain explicit functions of shear stress and strain, respectively. Some approximations of the generalized Reynolds equation for non-Newtonian fluids that resemble the Reynolds equation for a Newtonian fluid are obtained. The main idea of the method in the isothermal case is analytical solution of the problem for the sliding shear stress and the consequent reduction the problem to asymptotic and/or numerical solution just for the pressure and gap functions. The procedure for deriving formulas for the isothermal lubrication film thickness is based on the methods outlined in Sections 6.3, 6.6, and Chapter 7 and it is presented for the cases when the influence of the lubrication shear stresses on surface normal and tangential displacements can be neglected. A number of examples illustrating application of the described technique is given. Some general issues as to what is the range of parameters when the proposed approximation provides asymptotically correct solutions and how to define appropriately the pressure viscosity coefficient are sorted out in the case of an isothermal EHL problem for a non-Newtonian lubricant. It is shown that in certain cases the asymptotic procedure described below provides asymptotically correct solutions only in regimes of starved lubrication while in other cases it is valid for both starved and fully flooded lubrication regimes.

Let us consider a line lubricated contact for cylindrical solids of radii R_1 and R_2 made of elastic materials with Young's moduli E_1 and E_2 and Poisson's ratios ν_1 and ν_2, respectively. Far from the contact the cylinders' surfaces are steadily moving with linear velocities u_1 and u_2 and pressed against each other with a normal force P. The lubricant separating the solids is an incompressible non-Newtonian fluid. Under the classical assumptions [1] of slow motion and

narrow gap between the surfaces the rheology of the lubricant can be described
by the equations

$$\mu\frac{\partial u}{\partial z} = F(\tau) \ or \ \tau = \Phi(\mu\frac{\partial u}{\partial z}),$$ (9.1)

where u is the lubricant velocity along the x-axis, $u = u(x, z)$, τ is the shear
stress, $\tau = \tau(x, z)$, and μ is the lubricant viscosity that depends on lubricant
pressure p and temperature T, $\mu = \mu(p, T)$, F and Φ are functions describing
the lubricant rheology. The coordinate system is introduced in such a manner
that the x-axis is directed along the contact in the direction of motion, the y-
axis is directed along cylinders' axes, and the z-axis is directed along the line
connecting the cylinders' centers (see Fig. 5.1). In equations (9.1) functions F
and Φ are given odd smooth functions, which are inverses of each other and
$F(0) = \Phi(0) = 0$.

9.2 Formulation of Isothermal Line EHL Problem for Non-Newtonian Fluids in Heavily Loaded Contacts

First, we will consider an isothermal EHL problem for non-Newtonian lubricants in the case when the tangential surface displacements can be neglected.
Under isothermal conditions the lubricant viscosity μ is independent of the
lubricant temperature T, i.e., $\mu = \mu(p)$ and, therefore, μ is independent of z.
Let us introduce the dimensionless variables for heavily loaded contact

$$\{x', a, c\} = \frac{1}{a_H}\{x, x_i, x_e\}, \ p' = \frac{p}{p_H}, \ \{h', z'\} = \frac{1}{h_e}\{h, z\},$$

$$\mu' = \frac{\mu}{\mu_a}, \ \{\tau', f', \Phi'\} = \frac{\pi R'}{P}\{\tau, f, \Phi\}, \ F' = \frac{2h_e}{\mu_a(u_1+u_2)}F,$$ (9.2)

and parameters

$$s_0 = 2\frac{u_2-u_1}{u_2+u_1}, \ V = \frac{24\mu_a(u_1+u_2)R'^2}{a_H^3 p_H}, \ H_0 = \frac{2R'h_e}{a_H^2},$$ (9.3)

Based on the above assumptions and using the no-slip boundary conditions for
the lubricant velocity u at the contact surfaces $z = \pm h/2$ in the dimensionless
variables (9.2) and (9.3), the EHL problem can be reduced to the following
system of equations [2] (further primes at the dimensionless variables are
omitted)

$$\frac{d}{dx}\left[\frac{1}{\mu}\int\limits_{-h/2}^{h/2} zF(f + H_0z\frac{dp}{dx})dz - h\right] = 0,$$ (9.4)

$$\frac{1}{\mu}\int\limits_{-h/2}^{h/2} F(f + H_0z\frac{dp}{dx})dz = s_0,$$ (9.5)

$$p(a) = p(c) = \frac{dp(c)}{dx} = 0, \tag{9.6}$$

$$H_0(h-1) = x^2 - c^2 + \frac{2}{\pi} \int_a^c p(t) \ln \left| \frac{c-t}{x-t} \right| dt, \tag{9.7}$$

$$\int_a^c p(t)dt = \frac{\pi}{2}. \tag{9.8}$$

In equations (9.2)-(9.8) function f is the sliding frictional stress, $f = f(x)$, h is the gap between the contact surfaces, $h = h(x)$, s_0 is the slide-to-roll ratio based on surface rigid velocities u_1 and u_2, a and c are the x- coordinates of the inlet and exit points of the contact, H_0 is the dimensionless exit film thickness. In equations (9.2) x_i and x_e are the dimensional x-coordinates of the inlet and exit points of the contact, p_H and a_H are the Hertzian maximum pressure and the Hertzian half-width of the contact, $p_H = \sqrt{\frac{E'P}{\pi R'}}$, $a_H = 2\sqrt{\frac{R'P}{\pi E'}}$, $\frac{1}{R'} = \frac{1}{R_1} \pm \frac{1}{R_2}$, $\frac{1}{E'} = \frac{1-\nu_1^2}{E_1} + \frac{1-\nu_2^2}{E_2}$, h_e is the dimensional lubrication film thickness at the exit from the contact, $h_e = h(x_e)$, μ_a is the lubricant viscosity at ambient temperature T_a. In most cases the rheology functions F and Φ and the lubricant viscosity μ also involve parameter V as well as some other parameters.

Therefore, for the given values of parameters a, V, s_0, and other parameters involved as well as for the given functions $\mu(p)$, $F(x)$, and $\Phi(x)$ the solution of the EHL problem is represented by parameters c, H_0 and functions $f(x)$, $p(x)$, and $h(x)$.

9.3 Isothermal Line EHL Problem for Pure Rolling Conditions. Pre- and Over-Critical Lubrication Regimes. Some Numerical Examples

One of the most often encountered in practice condition is lubrication under almost purely rolling conditions. Because of that we will first analyze the problem in the cases of pure rolling conditions in heavily loaded EHL contacts. For the slide-to-roll ratio $s_0 = 0$ due to the fact that $F(x)$ is an odd function of x and $F(0) = 0$ the solution of equation (9.5) is obviously $f(x) = 0$. In this case the generalized Reynolds equation (9.5) is reduced to

$$\frac{d}{dx}\{M(\mu, p, h, \frac{dp}{dx}, V, H_0) - H_0 h\} = 0, \tag{9.9}$$

$$M(\mu, p, h, \frac{dp}{dx}, V, H_0) = \frac{H_0}{\mu} \int_{-h/2}^{h/2} zF(H_0 z \frac{dp}{dx})dz. \tag{9.10}$$

Therefore, the lubrication problem is reduced to solution of the system of equations (9.9), (9.10), (9.6)-(9.8).

Heavily loaded EHL contact conditions are usually caused by low surface velocities or high load applied to the contact and/or the kind of lubricant viscosity that experiences steep increase with pressure. In any of these cases the EHL problem contains a small parameter $\omega \ll 1$. Depending on what causes the lubricated contact to be heavily loaded (including properties of the lubricant viscosity μ) this small parameter can be defined differently. For example, when the lubricant viscosity μ increases with pressure p just moderately, then the heavy loaded conditions of the contact are caused by the smallness of parameter V. Therefore, in this case the small parameter ω can be taken equal to V, i.e., $\omega = V \ll 1$. In the cases when heavy loading is caused by fast increase of the lubricant viscosity μ with pressure p from 1 to $\mu(1) \gg 1$ the small parameter ω can be taken, for example, equal to the reciprocal of $ln(\mu(1))$, i.e., $\omega = 1/ln(\mu(1)) \ll 1$. In particular, for the exponential law $\mu = \exp(Qp)$ (where $Q = \alpha p_H$, α is the pressure viscosity coefficient), the latter definition of ω leads to $\omega = Q^{-1} \ll 1$.

Let us consider some examples of lubricants with Newtonian and non-Newtonian rheologies.

Example 1. For a lubricant of Newtonian rheology we have (see equation (9.10))

$$F(x) = \frac{12H_0}{V}x, \; M = \frac{H_0^3}{V}\frac{h^3}{\mu}\frac{dp}{dx}. \tag{9.11}$$

Example 2. Let us consider a lubricant with Ostwald-de Waele (power law) rheology [3]. Then we have

$$F(x) = \frac{12H_0}{V_n} \mid x \mid^{(1-n)/n} x, \; V_n = V(\frac{P}{\pi R'G})^{(n-1)/n}, \tag{9.12}$$

where G is the characteristic shear modulus of the fluid. It is easy to see that in this case equation (9.10) leads to

$$M = \frac{24n}{2n+1}\frac{H_0^{\frac{2n+1}{n}}}{V_n}\frac{1}{\mu}(\frac{h}{2})^{\frac{2n+1}{n}} \mid \frac{dp}{dx} \mid^{\frac{1-n}{n}}\frac{dp}{dx}. \tag{9.13}$$

Interestingly enough, for $n = 1$ the Ostwald-de Waele rheology is reduced to the Newtonian one and equation (9.13) for M is reduced to equation (9.11) for a lubricant with Newtonian rheology.

Example 3. Let us consider a lubricant with Reiner-Philippoff-Carreau rheology [3] described by the function

$$\Phi(x) = \frac{V}{12H_0}x\{\eta + (1 - \eta)[1 + (\frac{V}{12H_0G_0} \mid x \mid)^m]^{\frac{n-1}{m}}\}, \tag{9.14}$$

$$\eta = \frac{\mu_\infty}{\mu_0}, \; G_0 = \frac{\pi R'G}{P},$$

where μ_0 and μ_∞ are the lubricant viscosities at the shear rates $\partial u/\partial z = 0$ and $\partial u/\partial z = \infty$, respectively, which are scaled according to formulas (9.2), and G is the characteristic shear stress of the fluid, $G > 0$, n and m are constants, $0 \leq n \leq 1$, $m > 0$.

Let us consider two limiting cases for equation (9.14). If the shear stress τ is relatively small, then the lubricant behaves like a Newtonian fluid, i.e.,

$$\Phi(x) = \frac{V}{12H_0}x \ if \ \frac{V}{12H_0G_0}\mu\frac{\partial u}{\partial z} \ll 1. \tag{9.15}$$

In this case $F(x)$ is determined by equation (9.11) and, therefore, the Reynolds equation coincides with the one for Newtonian fluid. In the opposite case of relatively large shear stress τ, the lubricant behaves according to power law

$$\Phi(x) = (1 - \eta)(\frac{V}{12H_0})^n(\frac{|x|}{G_0})^{n-1}x \ if \ \frac{V}{12H_0G_0}\mu\frac{\partial u}{\partial z} \gg 1. \tag{9.16}$$

Solving equation (9.16) for x we get the expression for $F(x)$ in the form similar to (9.12)

$$F(x) = \frac{12H_0}{V(1-\eta)^{1/n}}[\frac{|x|}{G_0}]^{(1-n)/n}x, \ if \ \frac{V}{12H_0G_0}\mu\frac{\partial u}{\partial z} \gg 1. \tag{9.17}$$

It this case equation (9.10) leads to

$$M = \frac{24n}{2n+1}\frac{H_0^{\frac{2n+1}{n}}}{V(1-\eta)^{1/n}G_0^{(1-n)/n}}\frac{1}{\mu}(\frac{h}{2})^{\frac{2n+1}{n}} \mid \frac{dp}{dx}\mid^{\frac{1-n}{n}}\frac{dp}{dx}. \tag{9.18}$$

Now, let us consider the structure of a heavily loaded lubricated contact. In the central part of a heavily loaded lubricated contact, the Hertzian region, the pressure distribution $p(x)$ is well approximated by the Hertzian one equal to $\sqrt{1 - x^2}\theta(1-x^2)$ while the gap $h(x)$ is practically equal to 1 (see the preceding sections). The behavior of the film thickness H_0 is predominantly determined by the EHL problem solution in the inlet zone that is a small vicinity of the inlet point $x = a$ while the location of the exit point $x = c$ is predominantly determined by the behavior of this solution in the exit zone - small vicinity of $x = c$. In preceding sections it was established that depending on the behavior of the lubricant viscosity and contact operating parameters there are possible two very different mechanisms of lubrication: pre- and over-critical lubrications regimes. These regimes reflect different mechanical processes that cause the contact to be heavily loaded. At the same time these regimes correspond to the different definitions of the small parameter ω discussed earlier (also see [4]). It was established earlier that the sizes of the inlet and exit zones are small in comparison with the size of the Hertzian region which is approximately equal to 2. Let us assume that the characteristic size of the inlet zone is $\epsilon_q = \epsilon_q(\omega) \ll 1$ for $\omega \ll 1$. Based on the principle of asymptotic matching of the solutions in the Hertzian region and inlet zone, in the inlet zone where $r = O(1)$, $r = (x - a)/\epsilon_q$, $\epsilon_q(\omega) \ll 1$ for $\omega \ll 1$, both the lubricant viscosity μ and gap h are of the order of one while $p(x) = O(\epsilon_q^{1/2})$, $\omega \ll 1$. We will assume that the lubricant viscosity μ is a non-decreasing function of pressure p. To characterize the above–mentioned lubrication regimes, let us introduce a critical size $\epsilon_0(\omega)$ of the inlet zone as the maximum value of $\epsilon_q(\omega)$ for which the following estimate holds (see Section 6.6)

$$\mu(\epsilon_q^{1/2}(\omega)) = O(1), \ \omega \ll 1. \tag{9.19}$$

Therefore, $\mu(\epsilon_q^{1/2}(\omega)) = O(1)$ for $\epsilon_q = O(\epsilon_0(\omega))$ and $\mu(\epsilon_q^{1/2}(\omega)) \gg 1$ for $\epsilon_q \gg \epsilon_0(\omega)$, $\omega \ll 1$. Now, it is possible to define the necessary conditions for pre- and over-critical lubrication regimes. A pre-critical lubrication regime is such a regime that the size of the inlet zone ϵ_q satisfies the estimate

$$\epsilon_q(\omega) = O(\min(1, \epsilon_0(\omega))), \ \omega \ll 1, \tag{9.20}$$

while an over-critical lubrication regime is a regime for which the size of the inlet zone ϵ_q satisfies the estimate

$$\epsilon_0(\omega) \ll \epsilon_q(\omega) \ll 1, \ \omega \ll 1. \tag{9.21}$$

Keeping in mind that $\epsilon_q(\omega) \ll 1$ for $\omega \ll 1$ from estimates (9.19)-(9.21), we obtain that over-critical regimes cannot be realized if $\epsilon_0(\omega) \gg 1$ for $\omega \ll 1$.

Clearly, the existence of the critical size $\epsilon_0(\omega)$ of the inlet zone depends on the particular behavior of the lubricant viscosity μ and the chosen small parameter ω. For example, if $\mu = \exp[Qp/(1+Q_1p)]$, $Q, Q_1 \geq 0$, $Q \geq Q_1$, $Q \gg 1$ and $\omega = Q^{-1} \ll 1$, then $\epsilon_0(\omega) = Q^{-2}$. On the other hand, for the same as in the above example function μ, $Q \gg 1$, and $\omega = V \ll 1$ pre-critical regimes can be realized only for $\epsilon_q(V) = O(Q^{-2})$ while over-critical regimes are possible for $\epsilon_q(V) \gg Q^{-2}$, $Q \gg 1$.

9.3.1 Pre-Critical Lubrication Regimes

In Section 6.3 it was shown that for pre-critical lubrication regimes it is beneficial to represent our system of equations (9.9), (9.10), (9.6)-(9.8) in the equivalent form (Vorovich et al. [5], see also (6.19)-(6.23))

$$p(x) = R(x)\left[1 - \frac{1}{2\pi}\int_a^c \frac{dM(p,h)}{dt}\frac{dt}{R(t)(t-x)}\right], \ R(x) = \sqrt{(x-a)(c-x)}, \tag{9.22}$$

$$\int_a^c \frac{dM(p,h)}{dt}\frac{dt}{R(t)} = \pi(a+c), \ \int_a^c \frac{dM(p,h)}{dt}\frac{tdt}{R(t)} = \pi[(\tfrac{c-a}{2})^2 + \tfrac{(a+c)^2}{2} - 1], \tag{9.23}$$

$$M(p,h) = \frac{H_0}{\mu}\int_{-h/2}^{h/2} zF(H_0z\tfrac{dp}{dx})dz, \tag{9.24}$$

$$H_0(h-1) = x^2 - c^2 + \frac{2}{\pi}\int_a^c p(t)\ln\mid\tfrac{c-t}{x-t}\mid dt, \tag{9.25}$$

$$\frac{1}{\pi}\int_a^c M(p,h)\frac{dt}{R(t)} + \frac{2}{\pi}\int_a^c p(t)\ln\frac{1}{|c-t|}dt$$

$$+c^2 - \tfrac{1}{2}(\tfrac{c-a}{2})^2 - (\tfrac{a+c}{2})^2 = \ln\mid\tfrac{4}{c-a}\mid. \tag{9.26}$$

We will make the same assumption as in Section 6.3 that in a heavily loaded lubricated contact away from the contact boundaries $x = a$ and $x = c$ the first term W in equation (9.9) is negligibly small in comparison with the second one - $H_0 h$ (see similar assumption in (6.30)). Moreover, it was shown that in a pre-critical lubrication regime the inlet zone is represented by just one boundary layer [6] that has a homogeneous structure. It means that in the entire inlet zone the effects of the viscous fluid flow and elastic displacements are of the same order of magnitude and none of them prevails over another. In such cases in the inlet and exit zones, we have (see Section 6.3)

$$p = O(\epsilon_q^{1/2}), \ \tfrac{dp}{dx} = O(\epsilon_q^{-1/2}), \ h - 1 = O(\tfrac{\epsilon_q^{3/2}}{H_0}), \ r = O(1),$$

$$p = O(\epsilon_g^{1/2}), \ \tfrac{dp}{dx} = O(\epsilon_g^{-1/2}), \ h - 1 = O(\tfrac{\epsilon_g^{3/2}}{H_0}), \ s = O(1),$$

(9.27)

where $\epsilon_q = \epsilon_q(\omega) \ll 1$ and $\epsilon_g = \epsilon_g(\omega) \ll 1$ are characteristic sizes of the inlet and exit zones, which are boundary layers located next to points $x = a$ and $x = c$ (Van Dyke [6]). The characteristic size of the inlet zone $\epsilon_q(\omega)$ is determined by the given inlet coordinate (see (6.34))

$$a = -1 + \alpha_1 \epsilon_q, \ \alpha_1 = O(1) \ for \ \omega \ll 1,$$

(9.28)

and depends on ω. The characteristic size of the exit zone $\epsilon_g(\omega)$ is unknown and is determined by the exit coordinate (see (6.35))

$$c = 1 + \beta_1 \epsilon_g, \ \beta_1 = O(1) \ for \ \omega \ll 1.$$

(9.29)

In equations (9.28) and (9.29) α_1 is a given non-positive constant while β_1 is an unknown constant and is subject to calculation during the solution process. The region determined by the inequalities $x - a \gg \epsilon_q$ and $c - x \gg \epsilon_g$ will be called external (Van Dyke [6]) or Hertzian region (Vorovich et al. [5]).

Using the assumption that in a heavily loaded lubricated contact away from the contact boundaries $x = a$ and $x = c$ the first term W in equation (9.9) is negligibly small in comparison with the second one and temporarily assuming that the boundaries of the contact region $x = a$ and $x = c$ are known and fixed in the external region, we obtain (see equation 6.31))

$$x^2 - c^2 + \tfrac{2}{\pi} \int_a^c p(t) \ln | \tfrac{c-t}{x-t} | \ dt = o(1)$$

(9.30)

$$for \ x - a \gg \epsilon_q \ and \ c - x \gg \epsilon_g.$$

Equations (9.30) and (9.8) with the accuracy of up to $o(1)$ provide us with the solution of a classic contact problem of elasticity (Vorovich et al. [5]), which has the form

$$p_0(x) = \sqrt{(x - a)(c - x)} + \tfrac{1 + 2ac + (c-a)^2/4 - (a+c)x}{2\sqrt{(x-a)(c-x)}}$$

(9.31)

$$for \ x - a \gg \epsilon_q \ and \ c - x \gg \epsilon_g.$$

Thus, in the external region, the problem solution has the form

$$p(x) = p_0(x) + o(1) \; for \; x - a \gg \epsilon_q \; and \; c - x \gg \epsilon_g. \tag{9.32}$$

Based on estimates (9.27) we will try to find the solution of the problem in the inlet and exit zones in the form (see (6.38) and (6.39))

$$p(x) = \epsilon_q^{1/2} q(r) + o(\epsilon_q^{1/2}), \; q(r) = O(1) \; r = O(1), \; \omega \ll 1, \tag{9.33}$$

$$p(x) = \epsilon_g^{1/2} g(s) + o(\epsilon_g^{1/2}), \; g(s) = O(1) \; s = O(1), \; \omega \ll 1, \tag{9.34}$$

where $q(r)$ and $g(s)$ are the major terms of the pressure $p(x)$ asymptotic expansions in the inlet and exit zones. Obviously, the solutions from (9.33) and (9.34) must match the asymptotic representations of $p_0(x)$ at the boundaries of the inlet and exit zones with the Hertzian region (for details see Section 6.3).

To proceed further along the lines of the analysis outlined in the preceding sections and to obtain formulas for the lubrication film thickness H_0, we have to make an assumption about the behavior of the rheology function F or Φ. Therefore, we can assume that the rheology of the lubricant is such that

$$F(H_0 \epsilon_q^{-1/2} y(t)) = V^{-k}(\omega^{-l}\epsilon_q^{-1/2} H_0^{n+1})^{1/m} F_0(y(t)) + \dots,$$
$$\tag{9.35}$$
$$F_0(y(t)) = O(1) \; for \; y(t) = O(1) \; and \; H_0 \epsilon_q^{-1/2} \ll 1,$$

where k, l, m, and n are constants, $m > 0$. Now, by making the necessary estimates and following the analysis of Section 6.3, we will show that we can take $\epsilon_g = \epsilon_q$, which will allow us to arrive at the formula for the film thickness

$$H_0 = A(V^{km}\omega^l \epsilon_q^{\frac{3m+1}{2}})^{\frac{1}{m+n+1}} + \dots, \; A(\alpha_1) = O(1), \; \omega \ll 1, \tag{9.36}$$

where $A(\alpha_1)$ is an unknown nonnegative constant independent from ω and ϵ_q, which is determined by the solution of the problem in the inlet zone. A further analysis along the lines of Section 6.3 in the inlet and exit zones lead to the following asymptotically valid equations:

$$\frac{dM_0(q,h_q,\mu_q,r)}{dr} = \frac{2}{\pi} \int\limits_0^\infty \frac{q(t)-q_a(t)}{t-r} dt,$$
$$\tag{9.37}$$

$$q(0) = 0, \; q(r) \to q_a(r) \; as \; r \to \infty,$$

$$\int\limits_0^\infty [q(t) - q_a(t)]dt = 0, \tag{9.38}$$

$$\frac{dM_0(g,h_g,\mu_g,s)}{ds} = \frac{2}{\pi} \int\limits_{-\infty}^0 \frac{g(t)-g_a(t)}{t-s} dt,$$
$$\tag{9.39}$$

$$g(0) = \frac{dg(0)}{ds} = 0, \; g(s) \to g_a(s) \; as \; s \to -\infty,$$

$$\int\limits_{-\infty}^{0} [g(t) - g_a(t)]dt = 0, \tag{9.40}$$

where

$$M_0(p, h, \mu, x) = \frac{A^{\frac{m+n+1}{m}}}{\mu} \int\limits_{-h/2}^{h/2} z F_0(z\tfrac{dp}{dx}), \tag{9.41}$$

functions $\mu_q(r)$ and $\mu_g(s)$ are the major terms of the asymptotic expansions of $\mu(p, Q)$ in the inlet and exit zones and functions $q_a(r)$ and $g_a(s)$ are the major terms of the inner asymptotic expansions of the external asymptotic (9.31), (9.32) in the inlet and exit zones which are determined by the equalities (see (6.49))

$$q_a(r) = \sqrt{2r} + \frac{\alpha_1}{\sqrt{2r}}, \quad g_a(s) = \sqrt{-2s} - \frac{\beta_1}{\sqrt{-2s}}. \tag{9.42}$$

The expressions for functions $h_q(r)$ and $h_g(s)$ depend on the lubrication conditions. In case of starved lubrication $\epsilon_q^{3/2} \ll H_0$, we have (see Section 6.3)

$$\epsilon_q \ll \omega^{\frac{2l}{3n+2}}, \quad \omega \ll 1, \tag{9.43}$$

and the expressions for the gap in the inlet and exit zones are

$$h_q(r) = 1, \tag{9.44}$$

$$h_g(s) = 1. \tag{9.45}$$

In case of fully flooded lubrication regimes, we can take

$$\epsilon_q = \omega^{\frac{2l}{3n+2}}, \quad \omega \ll 1. \tag{9.46}$$

That leads to the following formula for the film thickness in a fully flooded lubricated contact

$$H_0 = A(V^{km}\omega^l)^{\frac{3}{3n+2}} + \dots, \quad A(\alpha_1) = O(1), \quad \omega \ll 1, \tag{9.47}$$

where $A(\alpha_1)$ is an unknown nonnegative constant independent from ω which is determined by the solution of the problem in the inlet zone. In this case the expressions for $h_q(r)$ and $h_g(s)$ have the form

$$A[h_q(r) - 1] = \frac{2}{\pi} \int\limits_{0}^{\infty} [q(t) - q_a(t)] \ln \tfrac{1}{|r-t|} dt, \tag{9.48}$$

$$A[h_g(s) - 1] = \frac{2}{\pi} \int\limits_{-\infty}^{0} [g(t) - g_a(t)] \ln \tfrac{1}{|s-t|} dt. \tag{9.49}$$

The above analysis is valid for the small parameter of the problem $\omega = Q^{-1} \ll 1$ or $\omega = V \ll 1$ as long as $\mu(p, Q) = O(1)$ in the inlet and exit zones. More details on such restrictions can be found in Section 6.3.

Now, let us consider equation (9.22) and (9.23). In the inlet and exit zones the integrals in these equations can be expressed as a sum of three integrals:

over the inlet and exit zones, and over the external region. Making the necessary estimates we obtain the formula (9.36) for the film thickness H_0 and the following asymptotically valid equations for $q(r)$ and A in the inlet zone

$$q(r) = \sqrt{2r}\left[1 - \frac{1}{2\pi}\int_0^\infty \frac{d}{dt}M_0(q, h_q, \mu_q, t)\frac{dt}{\sqrt{2t}(t-r)}\right],$$

$$\pi\alpha_1 = \int_0^\infty \frac{d}{dt}M_0(q, h_q, \mu_q, t)\frac{dt}{\sqrt{2t}},$$

(9.50)

as well as asymptotically valid equations for $g(s)$ and β_1 in the exit zone

$$g(s) = \sqrt{-2s}\left[1 - \frac{1}{2\pi}\int_{-\infty}^0 \frac{d}{dt}M_0(g, h_g, \mu_g, t)\frac{dt}{\sqrt{-2t}(t-s)}\right],$$

$$\beta_1 = \frac{1}{\pi}\int_{-\infty}^0 \frac{d}{dt}M_0(g, h_g, \mu_g, t)\frac{dt}{\sqrt{-2t}},$$

(9.51)

where the functions $h_q(r)$, $h_g(s)$, $\mu_q(q)$, and $\mu_g(g)$ have the same meaning as earlier.

It is important to remember the general conditions of applicability of the proposed method. Therefore, we need to remember that (see Section 6.3) for a given $\epsilon_q = \epsilon_q(V, Q)$ the pre-critical regimes are characterized by the following condition

$$\mu(\epsilon_q^{1/2}, Q) = O(1), \quad \epsilon_q \ll 1, \quad \omega \ll 1,$$

(9.52)

while the over-critical regimes are determined by the condition

$$\mu(\epsilon_q^{1/2}, Q) \gg 1, \quad \epsilon_q \ll 1, \quad \omega \ll 1.$$

(9.53)

Let us consider pre-critical lubrication regimes for Examples 1-3 presented earlier when the small parameter $\omega = V \ll 1$. The other choice of the small parameter can be $\omega = Q^{-1} \ll 1$.

Example 1. For a Newtonian fluid in equation (9.35), we have $k = 0$, $l = m = n = 1$ or $k = 1$, $l = 0$, $n = 1$ for $\omega = V \ll 1$ and $k = 1$, $l = 0$, $m = n = 1$ for $\omega = Q^{-1} \ll 1$. Therefore, equations (9.36) and (9.47) are reduced to

$$H_0 = A_s(V\epsilon_q^2)^{1/3}, \quad A_s = O(1), \quad \epsilon_q \ll \epsilon_f = V^{2/5}, \quad \omega \ll 1,$$

(9.54)

for starved lubrication and

$$H_0 = A_f V^{3/5}, \quad A_f = O(1), \quad \epsilon_q = O(V^{2/5}), \quad \omega \ll 1,$$

(9.55)

for fully flooded lubrication. These formulas coincide with the ones derived in Section 6.3 for Newtonian fluids.

Example 2. For a fluid with the Ostwald-de Waele (power law) rheology described by equations (9.12) (see also (9.108)), we have $k = 0$, $l = m = n$

for $\omega = V_n \ll 1$ and $k = 1$, $l = 0$, $m = n$ for $\omega = Q^{-1} \ll 1$. Therefore, we obtain [7]

$$H_0 = A_s (V_n^n \epsilon_q^{\frac{3n+1}{2}})^{\frac{1}{2n+1}}, \quad A_s = O(1), \quad \epsilon_q \ll \epsilon_f = V_n^{\frac{2n}{3n+2}}, \quad \omega \ll 1, \qquad (9.56)$$

for starved lubrication and

$$H_0 = A_f V_n^{\frac{3n}{3n+2}}, \quad A_f = O(1), \quad \epsilon_q = O(\epsilon_f), \quad \omega \ll 1, \qquad (9.57)$$

for fully flooded lubrication.

For starved lubrication and $Q = 0$ we can obtain a simple formula for coefficient A_s, (see the procedure outlined in Section 6.11)

$$A_s = A_1 \mid \alpha_1 \mid^{\frac{3n+1}{2(2n+1)}}, \qquad (9.58)$$

where $A_1 = A_1(n)$ is the value of A_s which should be obtained numerically for $Q = 0$ and $\alpha_1 = -1$.

Example 3. The case of a lubricant with Reiner-Philippoff-Carreau rheology described by equation (9.111) contains three distinct cases. In both cases when $\frac{V}{12H_0 G_0} \mid \mu \frac{\partial u}{\partial z} \mid \ll 1$ or $\frac{V}{12H_0 G_0} \mid \mu \frac{\partial u}{\partial z} \mid = O(1)$ for $\omega \ll 1$ the formulas for the film thickness H_0 coincides with the ones for a Newtonian fluid (see Example 1). However, the coefficients of proportionality A_s and A_f in the latter case differ from the Newtonian case and depends on the values of parameters n, m and G_0. In case when $V/(12H_0 G_0) \mid \mu \partial u/\partial z \mid \gg 1$ for $\omega \ll 1$ (here the small parameter ω and the corresponding powers k, l, m, and n are determined as in Example 2) the rheological function Φ is well approximated by the power law represented by formulas (9.108), where parameter V_n must be replaced by $V_n = V[G_0(1 - \eta)]^{\frac{1}{n}}/G_0 \ll 1$. In this case by solving equation (9.108) for x it is easy to determine function

$$F(x) = \frac{1}{(1 - \eta)^{1/n}} \frac{12H_0}{V} \left(\frac{\mid x \mid}{G_0} \right)^{\frac{1-n}{n}} x.$$

Therefore, the formulas for the film thickness H_0 coincide with the ones for a fluid with the power law rheology (see Example 2). The above conditions on $\frac{V}{12H_0 G_0} \mid \mu \frac{\partial u}{\partial z} \mid$ impose some limitations on the problem parameters which are not restrictive in the limiting case of Newtonian behavior and pretty restrictive in the limiting case of "power law" behavior. The latter restrictions have the form of $\epsilon_q \gg (1 - \eta)^4 G_0^{4(1-n)} V^{-2}$ for starved lubrication conditions and $V \gg [(1-\eta)G_0^{1-n}]^{3/2}$. Practically these restrictions mean that the pure rolling operating conditions for which a fluid with Reiner-Philippoff-Carreau rheology demonstrates non-Newtonian behavior hardly exist. It is clear from the mechanical point of view: under conditions of pure rolling the shear rates and shear stresses remain relatively small, which prevents the fluid from exhibiting its non-Newtonian behavior.

Let us summarize the results of the section. A detailed analysis of the structure of the problem solution for heavily loaded contacts lubricated by non-Newtonian fluid in pre-critical regimes is proposed. It is shown that the contact region consists of three characteristic zones: the zone with predominately "elastic solution" - the Hertzian region and the narrow inlet and exit zones (boundary layers) compared to the size of the Hertzian zone. In the inlet and exit zones the influences of the solids' elasticity and the viscous lubricant flow are of the same order of magnitude. In the inlet and exit zones two equivalent asymptotically valid systems of equations describing the solution of the problem are derived.

The solutions of these systems depend on a smaller number of initial problem parameters than the solution of the original problem. In all cases the characteristic size of the inlet and exit zones can be considered small compared to the size of the Hertzian region. The asymptotic formulas for the lubrication film thickness are derived. The lubrication film thickness depends on a particular regime of lubrication and the solution behavior in the inlet zone. The definition of pre-critical regimes is given. Based on that definition the limitations of the applied method were derived.

9.3.2   Numerical Solution of Asymptotically Valid Equations for Non-Newtonian Fluids in the Inlet and Exit Zones in Pre-Critical Regimes

It was mentioned earlier for the case of Newtonian fluids that one of the major difficulties encountered in numerical solution of EHL problems for heavily loaded contacts is instability of pressure $p(x)$ in the inlet and exit zones. As for Newtonian fluids for non-Newtonian fluids, there are two interconnected sources of this instability in the original equations (9.9), (9.10), (9.6)-(9.8) that are due to fast growth of viscosity μ with pressure p (i.e., the presence of a large parameter in μ) and for high μ due to the proximity of these equations to an integral equation of the first kind numerical solutions of which are inherently unstable. In case of an exponential viscosity $\mu(p) = \exp(Qp)$, the solution instability usually reveals itself for sufficiently large values of Q. For the same reason as the original equations of the EHL problem the asymptotically valid in the inlet zone equations (9.37), (9.38), (9.41), (9.42), and (9.44) for starved lubrication regimes or (9.48) for fully flooded lubrication regimes and in the exit zone equations (9.39)-(9.42), and (9.45) for starved lubrication regimes or (9.49) for fully flooded lubrication regimes are also at risk of being numerically unstable. The detailed explanation of why that is so you can can find in Section 6.9 on numerical solution of asymptotically valid equations for Newtonian fluids. To make the numerical solution of asymptotically valid equations stable, we will instead solve equations (9.50) and (9.51) for $q(r)$ and $g(s)$, respectively.

Let us introduce two sets of nodes $r_k = k \triangle r$, $k = 0, \ldots, N + 2$, and $r_{k+1/2} = 1/2(r_k + r_{k+1})$, $k = 0, \ldots, N + 1$, where $\triangle r$ is a positive step size and N is a sufficiently large positive integer. The equations will be solved by iteratively. It is clear that coefficient A is a monotonically increasing function of $|\alpha_1|$, $\alpha_1 \leq 0$. That enables us instead of determining the value of A for the given value of α_1 to determine the value of α_1 for the given value of A. In addition, such an approach allows us to solve the problem in the inlet and exit zones independently of each other. The numerical scheme for equation (9.50) and other equations related to it has the same form as the scheme in equations (6.197)-(6.201), where function $W(q, h_q, \mu_q, r)$ is determined by equations (9.41) and

$$M_0(q, h_q, \mu_q, r) = W(q, h_q, \mu_q, r)\frac{dq}{dr}. \qquad (9.59)$$

Therefore, function $W(q, h_q, \mu_q, r)$ is defined by the equation

$$W(p, h, \mu, x) = \frac{A^{\frac{m+n+1}{m}}}{\mu \frac{dp}{dx}} \int\limits_{-h/2}^{h/2} z F_0(z \frac{dp}{dx}). \qquad (9.60)$$

The initial approximation of q_k can be taken as the asymptotic of the Hertzian pressure, i.e., $q_k^0 = \sqrt{2r_k}$, $k = 0, \ldots, N+2$. The iteration process stops when the desired precision ε is reached.

In the exit zone the numerical scheme is developed in a similar fashion.

Let us consider the example of power law fluid under pure rolling conditions (see Example 2 above). In this case the expression for function W has the form

$$W(p, h, \mu, x) = \frac{24n}{2n+1} \frac{A^{\frac{2n+1}{n}}}{\mu} \left(\frac{h}{2}\right)^{\frac{2n+1}{n}} \left| \frac{dp}{dx} \right|^{\frac{1-n}{n}}. \qquad (9.61)$$

We will discuss some results for regimes of starved lubrication for $\epsilon_q^{1/2} = Q^{-1} \ll V^{1/5}$, $Q \gg 1$ and exponential viscosity $\mu = \exp(Qp)$. In such a case the viscosity in the inlet and exit zones has the form $\mu_q(q) = e^{Q_0 q}$ and $\mu_g(g) = e^{Q_0 g}$ for $Q_0 = 1$. The absolute error of calculations was chosen to be $\varepsilon = 0.0001$. The convergence of this numerical schemes was established in the case of Newtonian fluid in Section 6.9. All of the following calculations were done for $N = 800$, $\triangle r = 0.03125$ and $N = 2,000$, $\triangle r = 0.0078125$ in the inlet and exit zones, respectively. For example, for $A = 3$ and $n = 0.75$, $n = 1$, and $n = 1.25$ the values of α_1 are equal to -2.2491, -2.5244, -2.7490, and the values of β_1 are equal to 1.2908, 1.5017, 1.5950, respectively. For the above values of n functions $q(r)$ are monotonically increasing functions of r, which approach their asymptotes $q_a(r)$, the latter, in turn, approach $\sqrt{2r}$ as $r \to \infty$. These curves do not exhibit any signs of instability or oscillations. Qualitatively and quantitatively the behavior of pressure distribution $q(r)$ for non-Newtonian fluids $(n \neq 1)$ is very similar to the one for Newtonian lubricants $(n = 1)$, which is illustrated in Fig. 6.10 and, therefore, the graphs of $q(r)$ for power law fluid are not presented here. For comparison three graphs of functions $g(s)$ are given for $A = 3$ in Fig. 9.1, two of which are determined for non-Newtonian fluid with $n = 0.75$ (solid curve) and $n = 1.25$ (dotted

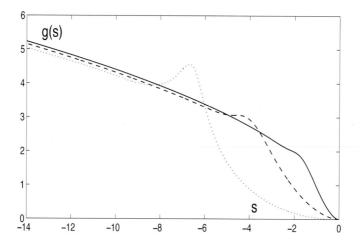

FIGURE 9.1

Main term of the asymptotic of the pressure distribution $g(s)$ in the exit zone of a starved lubricated contact for $Q_0 = 1$, $A = 3$, $n = 0.75$ (solid curve), $n = 1$ (dashed curve), and $n = 1.25$ (dotted curve). (after Kudish and Covitch [8]). Reprinted with permission from CRC Press.

curve) while the third one is determined for the case of Newtonian fluid with $n = 1$ (dashed curve). It is clear from this figure that the pressure spike increases with n. Note that β_1 increases monotonically with n and the pressure spike increases in value and moves toward the contact center as n increases. As in the case of $q(r)$ for large r the behavior of $g(s)$ for large s resembles the behavior of its asymptote $g_a(s)$ and does not exhibit any oscillations or signs of instability.

Figures 9.2 and 9.3 demonstrate the behavior of three curves of $g(s)$ for $Q_0 = 1$, $n = 0.75$ and $n = 1.25$ (solid, dashed, and dotted curves correspond to $A = 1$, 2, 3, respectively). In addition to that, in Fig. 9.2 for $n = 0.75$ and $A = 4$ the pressure distribution $g(s)$ in the exit zone is given by a dash-dotted curve. For $n = 1.25$ and $A = 4$ the pressure distribution $g(s)$ exhibits the same signs of instability as in the cases of Newtonian fluid described in detail in Section 6.9.3, which can be cured by the regularization described in Section 8.2. In particular, for $Q_0 = 1$, $n = 0.75$, $A = 1$, 2, 3, and 4, we have $\alpha_1 = -0.4366$, -1.2408, -2.2491, -3.4009 and $\beta_1 = 0.2879$, 0.7509, 1.2908, 1.8803, respectively, while for $Q_0 = 1$, $n = 1.25$, $A = 1$, 2, 3, and 4, we have $\alpha_1 = -0.5526$, -1.5357, -2.7490, -4.1261 and $\beta_1 = 0.3850$, 0.9510, 1.5950, 2.6671, respectively. For $n = 0.75$, $A = 1$, 2, and $n = 1.25$, $A = 1$, the pressure distributions $g(s)$ in the exit zone are monotonic, i.e., they possess no local maxima. Contrary to that, for

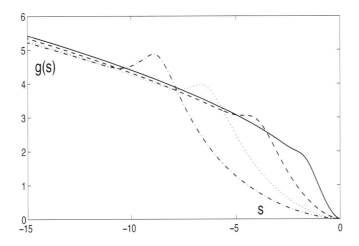

FIGURE 9.2
Main term of the asymptotic of the pressure distribution $g(s)$ in the exit zone of a starved lubricated contact for $Q_0 = 1$, $n = 0.75$, $A = 1$ (solid curve), $A = 2$ (dashed curve), $A = 3$ (dotted curve), and $A = 4$ (dash-dotted curve). (after Kudish and Covitch [8]). Reprinted with permission from CRC Press.

$n = 0.75$, $A = 3$, 4 and $n = 1.25$, $A = 2$, 3, 4, the pressure distributions $g(s)$ do possess local maxima, which shift closer to the center of the lubricated contact (to $s = -\infty$) and increase in value as $|\alpha_1|$ and A increase. As parameter n increases the height of the pressure spike increases and it shifts closer to the contact center. As in the case of $q(r)$ for large s, the behavior of $g(s)$ resembles the behavior of its asymptote $g_a(s)$ and does not exhibit any oscillations or signs of instability. Also, for starved lubrication regimes, we can get a transformation similar to (6.215) presented for the case of Newtonian fluid.

Let us consider some results for pre-critical fully flooded lubrication regimes. Qualitatively, the behavior of pressure $q(r)$ and $g(s)$ and gap $h_q(r)$ and $h_g(s)$ is similar to the one for Newtonian fluids described in Section 6.9.3 and the behavior of $q(r)$ and $g(s)$ considered above for starved lubrication regimes. For $A = 0.4$, $Q_0 = 1$, and $n = 1.25$ ($\alpha_1 = -0.8045$, $h_q(\triangle r/2) = 4.9248$, $\beta_1 = 0.0858$, $\min h_g(s) = 0.7860$) and $A = 0.4$, $Q_0 = 5$, and $n = 1.25$ ($\alpha_1 = -0.2234$, $h_q(\triangle r/2) = 1.6192$, $\beta_1 = 0.0610$, $\min h_g(s) = 0.9859$) examples of such solution behavior in the inlet and exit zones are given in Fig. 9.4, 9.5 and 9.6, 9.7, respectively. In the inlet zone the solutions are obtained for $N = 1,200$ and $\triangle r = 0.015625$ while in the exit zones we used $N = 480$ and $\triangle r = 0.0078125$. The absolute precision of both inlet and exit zones solutions was $\varepsilon = 10^{-4}$. To get a better understanding of the solution behavior in the exit

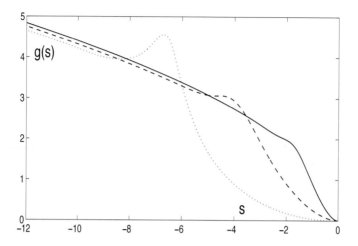

FIGURE 9.3
Main term of the asymptotic of the pressure distribution $g(s)$ in the exit zone
of a starved lubricated contact for $Q_0 = 1$, $n = 1.25$, $A = 1$ (solid curve),
$A = 2$ (dashed curve), and $A = 3$ (dotted curve). (after Kudish and Covitch
[8]). Reprinted with permission from CRC Press.

zone in Fig. 9.8 and 9.9 are presented graphs of pressure $g(s)$ and gap $h_g(s)$ for
four series of input parameters: $A = 0.4$, $n = 0.75$, $Q_0 = 5$, and $Q_0 = 10$ ($\beta_1 =$
0.0496, $\max g(s) = 0.7821$, $\min h_g(s) = 0.8967$, and $\beta_1 = 0.0397$, $\max g(s) =$
0.8537, $\min h_g(s) = 0.9222$, respectively) and $A = 0.4$, $n = 1.25$, $Q_0 = 5$,
and $Q_0 = 10$ ($\beta_1 = 0.0647$, $\max g(s) = 0.8850$, $\min h_g(s) = 0.8487$, and
$\beta_1 = 0.08209$, $\max g(s) = 2.2897$, $\min h_g(s) = 0.877$, respectively). Obviously,
pressure $g(s)$ and gap $h_g(s)$ distributions are smoother for $n = 0.75$ than for
$n = 1.25$. The presence of a pin-like pressure spike for $n = 1.25$ and $Q_0 = 10$
is a manifestation of mild instability. A similar behavior of pressure $g(s)$ and
gap $h_g(s)$ can be observed for $A = 0.4$, $n = 1.25$, $Q_0 = 20$, $N = 1,200$,
$\triangle r = 0.00168$ ($\beta_1 = 0.043$, $\max g(s) = 2.236$, $\min h_g(s) = 0.92$) the graphs
of which are depicted in Fig. 9.10.

Qualitatively, there are significant differences between these two series of
solutions. Obviously, for smaller values of Q_0 the sizes of the inlet and exit
zones as well as the inlet gap $h_q(\triangle r/2)$ are much wider than for larger values
of Q_0. It can be clearly seen from the comparison of graphs and effective zone
sizes in Fig. 9.4 and 9.6 in the inlet zones and in Fig. 9.5 and 9.7 in the exit
zones, respectively. In the exit zone, $\min h_g(s)$ is smaller and located farther
from the exit point for smaller values of Q_0. Moreover, for smaller values of
Q_0 the pressure distribution $g(s)$ in the exit zone is monotonic while for larger
Q_0 it has a spike/local maximum. The value of this pressure spike increases

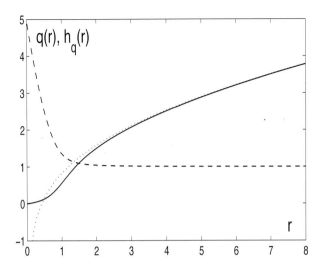

FIGURE 9.4
Main terms of the asymptotic distributions of pressure $q(r)$ (solid curve), gap $h_q(r)$ (dashed curve), and the Hertzian pressure asymptote $q_a(r)$ (dotted curve) in the inlet zone of a fully flooded lubricated contact for $A = 0.4$, $Q_0 = 1$, and $n = 1.25$. (after Kudish and Covitch [8]). Reprinted with permission from CRC Press.

and it shifts toward the center of the lubricated contact as Q_0 increases.

In the inlet zone the iteration process converges to a stable solution within a tried range of the input parameters. In the exit zone the process converges to a stable solution for, in a sense, relatively small to moderate values of A, Q_0, and n. To improve convergence of the iteration process, one can modify calculation of the first several iterations of the methods used in Sections 6.9.1 and 6.9.2 as follows: $q_k^{i+1} = q_k^i + \gamma v_k$ and $g_k^{i+1} = g_k^i + \gamma v_k$, where values v_k are determined by solution of systems of linear equations presented in these sections. The relaxation parameter γ can, for example, be taken equal to 0.5. However, the ultimate way to improve convergence and to get rid of solution instability is to use regularization described in Section 8.2.

Here, we need to reiterate (see Section 6.11) that to obtain sufficiently accurate solution the step size $\triangle x$ for numerical solution of the original EHL problem for lubricants with non-Newtonian rheology should be be determined as $\epsilon_q \triangle r$, where $\triangle r$ is the step size which provides the adequate precision for numerical solution of asymptotically valid equations in the inlet and exit zones.

To conclude this section it is important to emphasize that the advantages

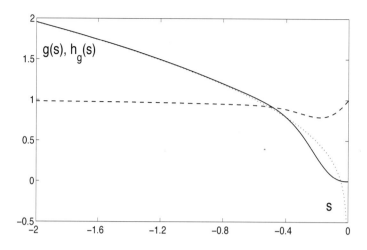

FIGURE 9.5

Main terms of the asymptotic distributions of pressure $g(s)$ (solid curve), gap $h_g(s)$ (dashed curve), and the Hertzian pressure asymptote $g_a(s)$ (dotted curve) in the exit zone of a fully flooded lubricated contact for $A = 0.4$, $Q_0 = 1$, and $n = 1.25$. (after Kudish and Covitch [8]). Reprinted with permission from CRC Press.

of the asymptotic approach to lubricated contacts with non-Newtonian fluids are the same as for the cases with Newtonian fluids (see Section 6.11).

9.3.3 Over-Critical Lubrication Regimes

To develop the asymptotic approach to solution of the EHL problem for non-Newtonian fluid involved in over-critical lubrication regime, we will closely follow the procedure of Section 6.6. In particular, we will use the following system of equation (see equations (9.9), (9.10), and (6.101))

$$\frac{d}{dx}\{M(\mu, p, h, \tfrac{dp}{dx}, V, H_0) - H_0 h\} = 0, \quad p(a) = p(c) = \tfrac{dp(c)}{dx} = 0, \quad (9.62)$$

$$M(\mu, p, h, \tfrac{dp}{dx}, V, H_0) = \tfrac{H_0}{\mu} \int\limits_{-h/2}^{h/2} z F(H_0 z \tfrac{dp}{dx}) dz, \quad (9.63)$$

$$H_0(h-1) = x^2 - (\tfrac{a_0 + c_0}{2})^2 + \tfrac{2}{\pi} \int\limits_{a}^{c} p(t) \ln \left| \tfrac{\frac{a_0 + c_0}{2} - t}{x - t} \right| dt, \quad (9.64)$$

$$\int\limits_{a}^{c} p(t) dt = \tfrac{\pi}{2} \quad (9.65)$$

to describe the problem.

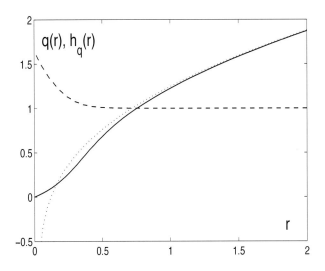

FIGURE 9.6
Main terms of the asymptotic distributions of pressure $q(r)$ (solid curve), gap $h_q(r)$ (dashed curve), and the Hertzian pressure asymptote $q_a(r)$ (dotted curve) in the inlet zone of a fully flooded lubricated contact for $A = 0.4$, $Q_0 = 5$, and $n = 1.25$. (after Kudish and Covitch [8]). Reprinted with permission from CRC Press.

In Section 6.6 it was shown that in over-critical lubrication regimes the inlet zone is represented by not one but two boundary layers [6] of different sizes. Moreover, the inlet zone has a non-homogeneous structure. It means that in the inlet sub-zone most distant from the Hertzian region (called the inlet ϵ_q-zone) the effects of the viscous fluid flow through a gap are mainly created by the Hertzian pressure, which provides a dominating contribution to the elastic surface displacements. In the inlet sub-zone next to the Hertzian region (called the inlet ϵ_0-zone), the effects of the viscous fluid flow and elastic surface displacements are of the same order of magnitude. The size of the ϵ_0-zone is the critical size of the inlet region ϵ_0 and it is much smaller than ϵ_q - the size of the ϵ_q-zone. The same is true about the two exit ϵ_0- and ϵ_q-zones. In such cases in the ϵ_0-zones we have (see Section 6.6 and equations (6.99) and (6.100), in particular)

$$p(x) = O(\epsilon_0^{1/2}), \ \frac{dp(x)}{dx} = O(\epsilon_0^{-1/2}), \ h(x) - 1 = O(\frac{\epsilon_0^{3/2}}{H_0}), \ \omega \ll 1, \quad (9.66)$$

while in the ϵ_q-zones we have

$$p = O(\epsilon_0^{1/2}), \ \frac{dp}{dx} = O(\epsilon_0^{1/2}\epsilon_q^{-1}), \ h - 1 = O(\frac{\epsilon_q^{3/2}}{H_0}), \ \omega \ll 1. \quad (9.67)$$

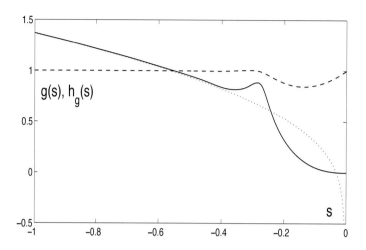

FIGURE 9.7
Main terms of the asymptotic distributions of pressure $g(s)$ (solid curve), gap $h_g(s)$ (dashed curve), and the Hertzian pressure asymptote $g_a(s)$ (dotted curve) in the exit zone of a fully flooded lubricated contact for $A = 0.4$, $Q_0 = 5$, and $n = 1.25$. (after Kudish and Covitch [8]). Reprinted with permission from CRC Press.

Most of the analysis of Section 6.6 deals with the properties of the solutions of the integral equations related to the equation for gap h. This analysis remains exactly the same for the problem at hand. However, some changes have to be made in the analysis in the inlet and exit ϵ_0- and ϵ_q-zones. To proceed further along the lines of the analysis of over-critical lubrication regimes outlined in Section 6.6 and to obtain formulas for the lubrication film thickness H_0, we need to make two assumptions about the behavior of the rheology function F in the inlet and exit zones. These assumptions can be derived from the assumption (9.35) by replacing in it $\epsilon_q^{-1/2}$ by $\epsilon_0^{1/2}\epsilon_q^{-1}$ in the inlet and exit ϵ_q-zones and by replacing $\epsilon_q^{-1/2}$ by $\epsilon_0^{-1/2}$ in the inlet and exit ϵ_0-zones. As a result of that we get

$$F(H_0\epsilon_0^{1/2}\epsilon_q^{-1}y(t)) = V^{-k}(\omega^{-l}\epsilon_0^{1/2}\epsilon_q^{-1}H_0^{n+1})^{1/m}F_0(y(t)) + \dots,$$

$$F_0(y(t)) = O(1) \ for \ y(t) = O(1) \ and \ H_0\epsilon_0^{1/2}\epsilon_q^{-1} \ll 1,$$

(9.68)

$$F(H_0\epsilon_0^{-1/2}y(t)) = V^{-k}(\omega^{-l}\epsilon_0^{-1/2}H_0^{n+1})^{1/m}F_0(y(t)) + \dots,$$

$$F_0(y(t)) = O(1) \ for \ y(t) = O(1) \ and \ H_0\epsilon_0^{-1/2} \ll 1,$$

(9.69)

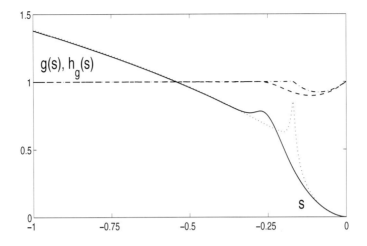

FIGURE 9.8
Main terms of the asymptotic distributions of pressure $g(s)$ (solid curve), gap $h_g(s)$ (dashed curve) for $Q_0 = 5$, and of pressure $g(s)$ (dotted curve), gap $h_g(s)$ (dash-dotted curve) for $Q_0 = 10$, in the exit zone of a fully flooded lubricated contact, $A = 0.4$, $n = 0.75$. (after Kudish and Covitch [8]). Reprinted with permission from CRC Press.

where k, l, m, and n are certain constants. The solution of the problem in the inlet and exit ϵ_q- and ϵ_0-zones will be searched in the form of asymptotic representations (see Section 6.6)

$$p = \epsilon_0^{1/2} q(r) + o(\epsilon_0^{1/2}), \quad q(r) = O(1), \quad r = \tfrac{x - a_0}{\epsilon_q} = O(1), \quad \omega \ll 1, \qquad (9.70)$$

$$p = \epsilon_0^{1/2} g(s) + o(\epsilon_0^{1/2}), \quad g(s) = O(1), \quad s = \tfrac{x - c_0}{\epsilon_q} = O(1), \quad \omega \ll 1, \qquad (9.71)$$

$$p = \epsilon_0^{1/2} q_0 + o(\epsilon_0^{1/2}), \quad q_0(r_0) = O(1), \quad r_0 = \tfrac{x - a_0}{\epsilon_0} = O(1), \quad \omega \ll 1, \qquad (9.72)$$

$$p = \epsilon_0^{1/2} g_0 + o(\epsilon_0^{1/2}), \quad g_0(s_0) = O(1), \quad s_0 = \tfrac{x - c_0}{\epsilon_0} = O(1), \quad \omega \ll 1, \qquad (9.73)$$

where the pair r and s and r_0 and s_0 are the local independent variables in the inlet and exit ϵ_q- and ϵ_0-zones, respectively, $q(r)$ and $g(s)$ are unknown functions representing the major terms of the pressure asymptotic expansions in the inlet and exit ϵ_q-zones while $q_0(r_0)$ and $g_0(s_0)$ are unknown functions representing the major terms of the pressure asymptotic expansions in the inlet and exit ϵ_0-zones. Constants a_0 and c_0 are determined in (6.95) and (6.96), respectively, as follows

$$a_0 = -1 + \alpha_0 \epsilon_0^{1/2} \epsilon_q^{1/2} + \ldots; \quad \alpha_0 = O(1), \quad \omega \ll 1, \qquad (9.74)$$

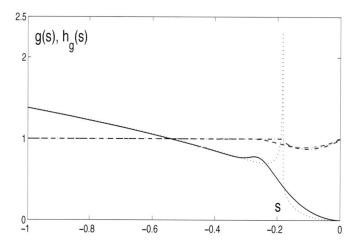

FIGURE 9.9
Main terms of the asymptotic distributions of pressure $g(s)$ (solid curve), gap $h_g(s)$ (dashed curve) for $Q_0 = 5$, and of pressure $g(s)$ (dotted curve), gap $h_g(s)$ (dash-dotted curve) for $Q_0 = 10$, in the exit zone of a fully flooded lubricated contact, $A = 0.4$, $n = 1.25$. (after Kudish and Covitch [8]). Reprinted with permission from CRC Press.

$$c_0 = 1 + \beta_0 \epsilon_0^{1/2} \epsilon_q^{1/2} + \ldots; \ \beta_0 = O(1), \ \omega \ll 1, \tag{9.75}$$

where constants α_0 and β_0 must be determined from the conditions that the external solution (see (6.106))

$$p_{ext}(x) = p_0(x) + \epsilon_0^{1/2} \epsilon_q p_1(x) + \ldots, \ x - a_0 \gg \epsilon_0, \ c_0 - x \gg \epsilon_0, \tag{9.76}$$

is finite at points a_0 and c_0 (for details see Section 6.6).

As before (see (6.104)) we assume that in the external region the estimate

$$M(p, h) \ll \epsilon_0^{3/2} \ for \ x - a_0 \gg \epsilon_0, \ c_0 - x \gg \epsilon_0, \ \omega \ll 1, \tag{9.77}$$

holds. That provides the solutions for $p_0(x)$ and $p_1(x)$ in the external region in the form of equations (6.109) and (6.118) while constants α_0 and β_0 are determined by equations (6.120).

After that we estimate the behavior of gap h in the inlet and exit zones (see Section 6.6). Using (9.68) and estimating the terms of equations (9.62)-(9.65) in the inlet and exit ϵ_q-zones in the manner similar to Section 6.6 we get the formula for the central film thickness

$$H_0 = A(V^{km}\omega^l \epsilon_0^{-1/2} \epsilon_q^{\frac{3m+2}{2}})^{\frac{1}{m+n+1}}, \ A = O(1), \tag{9.78}$$

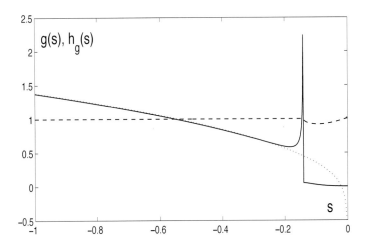

FIGURE 9.10
Main terms of the asymptotic distributions of pressure $g(s)$ (solid curve), gap $h_g(s)$ (dashed curve), and the Hertzian pressure asymptote $g_a(s)$ (dotted curve) in the exit zone of a fully flooded lubricated contact for $A = 0.4$, $Q_0 = 20$, and $n = 1.25$. (after Kudish and Covitch [8]). Reprinted with permission from CRC Press.

and equations for $q(r)$

$$M_0(q, h_q, \mu_q, r) = -\tfrac{4}{3}r\sqrt{-2r}\theta(-r), \quad q(\alpha_1) = 0, \tag{9.79}$$

and for $h_q(r)$ for starved lubrication regimes

$$h_q(r) = 1, \quad \epsilon_q \ll (\omega^l \epsilon_0^{-1/2})^{\frac{2}{3n+1}}, \quad \omega \ll 1, \tag{9.80}$$

or for $h_q(r)$ for fully flooded lubrication regimes

$$A[h_q(r) - 1] = -\tfrac{4}{3}r\sqrt{-2r}\theta(-r), \quad \epsilon_q = (\omega^l \epsilon_0^{-1/2})^{\frac{2}{3n+1}}, \quad \omega \ll 1. \tag{9.81}$$

By using the definition of fully flooded lubrication regimes $H_0 = O(\epsilon_q^{3/2})$ and formula (9.78), we can obtain the central film thickness formula and the characteristic size of the inlet and exit ϵ_q-zones for fully flooded lubrication regimes in the form (see Section 6.6)

$$H_0 = A(V^{km}\omega^l \epsilon_0^{-1/2})^{\frac{3}{3n+1}}, \quad A = O(1), \quad \epsilon_q = (V^{km}\omega^l \epsilon_0^{-1/2})^{\frac{2}{3n+1}}. \tag{9.82}$$

Similarly, we derive equations for $g(s)$

$$M_0(g, h_g, \mu_g, s) = \tfrac{4}{3}s\sqrt{2s}\theta(s) - \tfrac{4}{3}\beta_1\sqrt{2\beta_1}, \quad g(\beta_1) = 0, \tag{9.83}$$

and for $h_g(s)$ for starved lubrication regimes

$$h_g(s) = 1, \quad \epsilon_q \ll (\omega^l \epsilon_0^{-1/2})^{\frac{2}{3n+1}}, \quad \omega \ll 1, \tag{9.84}$$

or for $h_g(s)$ for fully flooded lubrication regimes

$$A[h_g(s) - 1] = \tfrac{4}{3} s \sqrt{2s} \theta(s), \quad \epsilon_q = (\omega^l \epsilon_0^{-1/2})^{\frac{2}{3n+1}}, \quad \omega \ll 1, \tag{9.85}$$

$$M_0(p, h, \mu, x) = \frac{A^{\frac{m+n+1}{m}}}{\mu} \int\limits_{-h/2}^{h/2} F_0(z \tfrac{dp}{dx}) dz, \tag{9.86}$$

where A is a constant independent of ω, ϵ_0, and ϵ_q, functions $h_q(x)$, $\mu_q(q)$ and $h_q(x)$, $\mu_q(q)$ are the major terms of the asymptotic expansions of gap $h(x)$ and viscosity $\mu(p)$ in the inlet and exit ϵ_q-zones, respectively.

Note, that the value of constant A is not determined and cannot be determined by solving the system of equations (9.79)-(9.81) in the inlet ϵ_q-zone as well as the value of β_1 is not determined by equations (9.83)-(9.85) in the exit ϵ_q-zone. These constants are determined by the inlet and exit ϵ_0-zones, respectively. Thus, in the inlet and exit ϵ_q-zones the problem is reduced to integration of nonlinear first-order differential equations.

Obvious mechanical considerations lead to the conclusion that the solution $q(r)$ for $r \geq \alpha_1$ of equations (9.79)-(9.81) is bounded. Moreover, because of monotonicity of the rheological function F, the main term of the asymptotic expansion $q(r)$ is a nondecreasing function of r and $q(r) = q(0)$ for $r > 0$. Therefore, the direct matching of function $q(r) = q(0)$ as $r \to \infty$ without considering the asymptotic in the inlet ϵ_0-zone with the major term of the inner expansion of the external solution is impossible. Therefore, the major term of the two asymptotic expansions that match the solution $\epsilon_0^{1/2} q_0(r_0)$ in the inlet ϵ_0-zone as $r_0 \to \pm\infty$ take the form (for details see Section 6.6)

$$q_a(r_0) = q(0) + \sqrt{2r_0}\theta(r_0). \tag{9.87}$$

A similar asymptotic in the exit zones has the form

$$g_a(s_0) = g(0) + \sqrt{-2s_0}\theta(-s_0). \tag{9.88}$$

The further analysis of the inlet and exit ϵ_0-zones follows exactly the same procedure outlined in Section 6.6.

It is important to remember that from equations (6.168), (6.169), and formula (9.82) for H_0 we obtain that in the inlet and exit ϵ_0-zones

$$h(x) - 1 \ll 1 \ for \ r_0 = O(1) \ and \ s_0 = O(1), \quad \omega \ll 1. \tag{9.89}$$

This is true as long as $\epsilon_0^{3/2} \ll H_0$ (see the formula for H_0 (9.82) and the size of the inlet and exit ϵ_0-zones is much smaller than the size of the inlet and exit ϵ_q-zones for the fully flooded over-critical lubrication regimes from (9.82)), i.e.,

$$\epsilon_0 \ll (\omega^l \epsilon_0^{-1/2})^{\frac{2}{3n+1}}, \ \omega \ll 1. \tag{9.90}$$

The latter inequality represents the condition that discriminates between the pre- and over-critical lubrication regimes. Namely, if

$$\epsilon_0 = O((\omega^l \epsilon_0^{-1/2})^{\frac{2}{3n+1}}) \text{ or } \epsilon_0 \gg (\omega^l \epsilon_0^{-1/2})^{\frac{2}{3n+1}}, \ \omega \ll 1, \tag{9.91}$$

then the lubrication regime is pre-critical. Otherwise, the lubrication regime is over-critical.

Therefore, the considered problem for $\omega \ll 1$ is reduced to a set of problems in different zones of the contact area. In particular, in the inlet ϵ_q-zone the asymptotic analogue of the original problem is the system of equations (9.79), (9.86), and (9.80) or (9.81) for function $q(r)$, in the external region the asymptotic solution of the problem is represented by formulas for $p_0(x)$ and $p_1(x)$ in the form of equations (6.109) and (6.118) while constants α_0 and β_0 are determined by equations (6.120), and in the exit ϵ_q-zone the asymptotic analogue of the original problem is the system of equations (9.83), (9.86), and (9.84) or (9.85) for function $g(s)$.

Note, that in the external region the estimate $\mu \gg 1$ leads to the validity of estimates (9.77) and $H_0(h-1) \ll 1$ in the external region. That validates the entire analysis of the problem. Obviously, the described analysis is valid if parameter of the problem $\omega = Q^{-1} \ll 1$ or $\omega = V \ll 1$. The method is also applicable to the problems with non-Newtonian fluids under non-isothermal conditions. In these cases only solutions in the inlet and exit ϵ_q-zones and function M_0 from (6.178) must be adjusted for thermal conditions or fluid non-Newtonian rheology. The rest of the analysis and formulas remains intact.

Let us consider over-critical lubrication regimes for Examples 1-3 presented earlier when the lubricant viscosity is given by the exponential equation $\mu = \exp(Qp)$, $Q \gg 1$, and the small parameter $\omega = V \ll 1$. Here parameter Q may be a certain function of parameter V. In this case the critical inlet zone size is $\epsilon_0 = Q^{-2}$. Similarly, if $\mu = \exp(Qp^m)$, $Q \gg 1$ then $\epsilon_0 = Q^{-\frac{2}{m}} \ll 1$. Let us consider the first case or the second case for $m = 1$. Also, we can take the small parameter $\omega = V \ll 1$ (see the definitions of k, l, m, and n for different definitions of ω in the preceding subsection of this section).

Example 1. For a Newtonian fluid in equations (9.68) and (9.69), we have $k = 0$, $l = m = n = 1$ for $\omega = V \ll 1$. Therefore, equations (9.78) and (9.82) are reduced to formulas (6.148) and (6.149) for starved and fully flooded lubrication regimes, respectively.

Example 2. For a fluid with the Ostwald-de Waele (power law) rheology described by equations (9.12), we have $k = 0$, $l = m = n$ for $\omega = V \ll 1$. Based on (9.78) and (9.82) we obtain

$$H_0 = A_s(V_n^n Q \epsilon_q^{\frac{3n+2}{2}})^{\frac{1}{2n+1}}, \ A_s = O(1), \ \epsilon_q \ll \epsilon_f = (V_n^n Q)^{\frac{2}{3n+1}}, \tag{9.92}$$

for starved lubrication and

$$H_0 = A_f (V_n^n Q)^{\frac{3}{3n+1}}, \ A_f = O(1), \ \epsilon_q = \epsilon_f, \tag{9.93}$$

for fully flooded lubrication.

Example 3. The case of a lubricant with Reiner-Philippoff-Carreau rheology described by equation (9.14) contains three distinct cases. In both cases when $\frac{V}{12H_0G_0} \mid \mu \frac{\partial u}{\partial z} \mid \ll 1$ or $\frac{V}{12H_0G_0} \mid \mu \frac{\partial u}{\partial z} \mid = O(1)$ for $\omega \ll 1$, the formulas for the film thickness H_0 coincide with the one for a Newtonian fluid (see Example 1). In case when $V/(12H_0G_0) \mid \mu \partial u / \partial z \mid \gg 1$ for $\omega \ll 1$, the rheological function Φ is well approximated by the power law from equations (9.12), where parameter V_n must be replaced by $V_n = V[G_0(1-\eta)]^{\frac{1}{n}}/G_0 \ll 1$. In this case the formulas for the film thickness H_0 coincide with the ones for a fluid with the power law rheology (see Example 2, equations (9.92) and (9.93)). However, the condition $V/(12H_0G_0) \mid \mu \partial u / \partial z \mid \gg 1$ for $\omega \ll 1$ can hardly be realized in practice. The reason for that is for pure sliding the shear stress and shear sliding are relatively small and, therefore, the lubrication fluid does not develop non-Newtonian behavior.

9.4 Isothermal Line EHL Problem for Relatively Large Sliding Conditions. Pre- and Over-Critical Lubrication Regimes

In general, heavily loaded EHL contact conditions are usually caused by low surface velocities or high load applied to the contact and/or the kind of lubricant viscosity that experiences steep increase with pressure. In any of these cases the EHL problem contains a small parameter $\omega \ll 1$. Depending on what causes the lubricated contact to be heavily loaded (including properties of the lubricant viscosity μ), this small parameter can be defined differently. For example, when the lubricant viscosity μ increases with pressure p in a lubricated contact moderately then the heavily loaded conditions of the contact are caused by the smallness of parameter V. Therefore, in this case the small parameter ω can be taken equal to V, i.e., $\omega = V \ll 1$. In the cases when heavy loading is caused by fast increase of the lubricant viscosity μ with pressure p from 1 to $\mu(1) \gg 1$, the small parameter ω can be taken, for example, equal to the reciprocal of $ln(\mu(1))$, i.e., $\omega = 1/ln(\mu(1)) \ll 1$. In particular, for the exponential law $\mu = \exp(Qp)$ (where $Q = \alpha_p p_H$, α_p is the pressure viscosity coefficient) and the latter definition of ω leads to $\omega = Q^{-1} \ll 1$.

We will analyze the EHL problems with non-Newtonian lubricants in cases of heavily loaded contacts with relatively large slide-to-roll ratios s_0. To make these conditions more precise let us introduce a small function (for comparison see function $\nu(x)$ in (7.17))

$$\nu(x) = \frac{H_0 h}{2 f_0} \frac{dp}{dx} \ll 1, \ \omega \ll 1, \tag{9.94}$$

representing the ratio of rolling $0.5 H_0 h dp/dx$ and the main term of asymptotic expansion of sliding f frictional stresses. Let us assume that this function is small in the entire contact or just in the inlet zone (see below) of the contact, i.e., $\nu(x) \ll 1$, $\omega \ll 1$, regardless of a particular definition of parameter ω.

We will try to find the asymptotic representation for $f(x)$ in the form

$$f(x) = f_0(x) + \nu(x) f_1(x) + \nu^2(x) f_2(x) + \nu^3(x) f_3(x)$$
$$+ O(\nu^4(x)), \ \nu(x) \ll 1, \tag{9.95}$$

where functions $f_0(x)$, $f_1(x)$, $f_2(x)$, and $f_3(x)$ are the consecutive terms of the asymptotic of $f(x)$, which have to be determined. Originally, this technique was introduced in [2, 9] and then used later by the author in [7, 10] and by other researchers, in particular, in [11].

By substituting the representation for $f(x)$ from equation (9.95) into equation (9.5), expanding it for $\nu(x) \ll 1$, and equating the terms with the same powers of $\nu(x)$, we obtain

$$f_0(x) = \Phi(\frac{\mu s_0}{h}), \ f_1(x) = 0, \ f_2(x) = -\frac{f_0^2 F''(f_0)}{6 F'(f_0)}, \ f_3(x) = 0, \ \dots. \tag{9.96}$$

The obtained solution for $f(x)$ allows for its elimination from the set of the problem unknowns and reducing the problem to determining just two functions: pressure p and gap h. In fact, by substituting equations (9.94)-(9.96) into equation (9.4), we obtain the approximate generalized Reynolds equation

$$\frac{d}{dx}\{M(\mu, p, h, \frac{dp}{dx}, V, s_0, H_0) - H_0 h\} = 0, \tag{9.97}$$

where the expressions for function M depend on the number of terms retained in equations (9.95) and (9.96). If the first one or two terms of the expansion for $f(x)$ from (9.95) are retained (i.e., $f(x) = f_0(x) + O(\nu^2(x))$, $\nu(x) \ll 1$), then

$$M = \frac{H_0^2 h^3 F'(f_0)}{12\mu} \frac{dp}{dx}, \tag{9.98}$$

while if the first three or four terms of the expansion for $f(x)$ from (9.95) are retained (i.e., $f(x) = f_0(x) + \nu(x) f_2(x) + O(\nu^4(x))$, $\nu(x) \ll 1$), then

$$M = \frac{H_0^2 h^3 F'(f_0)}{12\mu} \frac{dp}{dx}\{1 + \frac{H_0^2 h^2}{8}[\frac{F'''(f_0)}{5 F'(f_0)} - \frac{1}{3}(\frac{F''(f_0)}{F'(f_0)})^2](\frac{dp}{dx})^2\}. \tag{9.99}$$

In cases when the lubricant rheology is given by the second equation in (9.1), the derivatives of function $F(f)$ with respect to f involved in equations (9.98) and (9.99) can be expressed the following way:

$$F'(f_0) = \frac{1}{\Phi'(\lambda)}, \ F''(f_0) = -\frac{\Phi''(\lambda)}{[\Phi'(\lambda)]^3},$$

$$F'''(f_0) = \frac{3[\Phi''(\lambda)]^2 - \Phi'''(\lambda)\Phi'(\lambda)}{[\Phi'(\lambda)]^5}, \ \lambda = \frac{\mu s_0}{h}, \tag{9.100}$$

where function $\Phi(\lambda)$ is differentiated with respect to λ. Therefore, in these cases the expressions for function M which correspond to equations (9.98) and (9.99) can be represented in the forms

$$M = \frac{H_0^2 h^3}{12\mu\Phi'(\lambda)} \frac{dp}{dx}, \quad \lambda = \frac{\mu s_0}{h}, \tag{9.101}$$

$$M = \frac{H_0^2 h^3}{12\mu\Phi'(\lambda)} \frac{dp}{dx} \{ 1 + \frac{H_0^2 h^2}{120} \frac{4[\Phi''(\lambda)]^2 - 3\Phi'''(\lambda)\Phi'(\lambda)}{[\Phi'(\lambda)]^4} (\frac{dp}{dx})^2 \}, \tag{9.102}$$

respectively.

As a result of this analysis for the regimes for which $\nu(x) \ll 1$ for $a \leq x \leq c$, we obtain the following simplified approximate formulations of the isothermal EHL problem for the generalized Reynolds equation

$$\frac{d}{dx} \{ M(\mu, p, h, \frac{dp}{dx}, V, s_0, H_0) - H_0 h \} = 0, \tag{9.103}$$

$$p(a) = p(c) = \frac{dp(c)}{dx} = 0, \tag{9.104}$$

$$H_0(h - 1) = x^2 - c^2 + \frac{2}{\pi} \int\limits_a^c p(t) \ln | \frac{c-t}{x-t} | \, dt, \tag{9.105}$$

$$\int\limits_a^c p(t)dt = \frac{\pi}{2}, \tag{9.106}$$

where the expressions for function M are determined by formulas (9.98) and (9.99) or (9.101) and (9.102).

Let us consider some examples of lubricants with Newtonian and non-Newtonian rheologies.

Example 1. For a lubricant of Newtonian rheology, we have (see equations (9.1) and (9.98)):

$$F(x) = \frac{12H_0}{V}x, \quad \Phi(x) = \frac{V}{12H_0}x, \quad M = \frac{H_0^3}{V} \frac{h^3}{\mu} \frac{dp}{dx}. \tag{9.107}$$

It is important to realize that the generalized Reynolds equation (9.103) coincides with the exact Reynolds equation for a lubricant of Newtonian rheology.

Example 2. Let us consider a lubricant with Ostwald-de Waele (power law) rheology [3]. Then we have

$$F(x) = \frac{12H_0}{V_n} | x |^{(1-n)/n} x, \quad \Phi(x) = (\frac{V_n}{12H_0})^n | x |^{n-1} x,$$

$$V_n = V(\frac{P}{\pi R'G})^{(n-1)/n}, \tag{9.108}$$

where G is the fluid characteristic shear modulus. It is easy to see that equation (9.98) leads to

$$M = \frac{H_0^2}{12n} (\frac{12H_0}{V_n})^n \frac{h^3}{\mu} (\frac{\mu|s_0|}{h})^{1-n} \frac{dp}{dx}, \tag{9.109}$$

while equation (9.99) leads to the formula

$$M = \frac{H_0^2}{12n}\left(\frac{12H_0}{V_n}\right)^n\frac{h^3}{\mu}\left(\frac{\mu|s_0|}{h}\right)^{1-n}\frac{dp}{dx}\left[1 + \frac{(n-1)(n+2)H_0^2}{120n^2}\left(\frac{12H_0}{V_n}\right)^{2n}\right.$$

$$\left. \times\left(\frac{\mu|s_0|}{h}\right)^{-2n}\left(h\frac{dp}{dx}\right)^2\right].$$

(9.110)

Interestingly enough, for $n = 1$ the Ostwald-de Waele rheology is reduced to the Newtonian one and equations (9.108)-(9.110) are reduced to equations (9.107) for a lubricant with Newtonian rheology.

   Example 3. Let us consider a lubricant with Reiner-Philippoff-Carreau rheology [3] described by the function

$$\Phi(x) = \frac{V}{12H_0}x\{\eta + (1-\eta)[1 + (\frac{V}{12H_0G_0}\mid x\mid)^m]^{\frac{n-1}{m}}\},$$

(9.111)

$$\eta = \frac{\mu_\infty}{\mu_0}, \quad G_0 = \frac{\pi R'G}{P},$$

where μ_0 and μ_∞ are the lubricant viscosities at the shear rate $\partial u/\partial z = 0$ and $\partial u/\partial z = \infty$, respectively, which are scaled according to formulas (9.2), and G is the characteristic shear stress of the fluid, $G > 0$, n and m are constants, $0 \leq n \leq 1$, $m > 0$. Based on formula (9.101) we conclude that

$$M = \frac{H_0^2 h^3}{12\mu_0\Phi'(\lambda)}\frac{dp}{dx},$$

$$\Phi'(\lambda) = \frac{V}{12H_0}\{\eta + (1-\eta)[1 + \mid\lambda\mid^m]^{\frac{n-1}{m}}[1 + (n-1)\frac{|\lambda|^m}{1+|\lambda|^m}]\}, \qquad (9.112)$$

$$\lambda = \frac{V}{12H_0}\frac{\mu_0 s_0}{hG_0}.$$

For the case of formula (9.102) the expression for function M can be represented in a similar form, however, it is much more complex than equation (9.112) and is not presented here. For $n = 1$ expressions (9.112) are reduced to the ones for the case of Newtonian fluids.

   Now, let us consider the structure of a heavily loaded lubricated contact. In the central part of a heavily loaded lubricated contact, the Hertzian region, the pressure distribution $p(x)$ is well approximated by the Hertzian one equal to $\sqrt{1-x^2}\theta(1-x^2)$ while gap $h(x)$ is practically equal to 1 (see the preceding sections). The behavior of the film thickness H_0 is predominantly determined by the EHL problem solution in the inlet zone that is a small vicinity of the inlet point $x = a$ while the location of the exit point $x = c$ is predominantly determined by the behavior of this solution in the exit zone - small vicinity of $x = c$. In preceding sections it was established that depending on the behavior of the lubricant viscosity and contact operating parameters there are possible two very different mechanisms of lubrication: pre- and over-critical lubrications regimes. These regimes reflect different mechanical processes that cause the contact to be heavily loaded. At the same time these regimes correspond to the different definitions of the small parameter ω discussed earlier (also see

[4]). It was established earlier that the sizes of the inlet and exit zones are small in comparison with the size of the Hertzian region, which is approximately equal to 2. Let us assume that the characteristic size of the inlet zone is $\epsilon_q = \epsilon_q(\omega) \ll 1$ for $\omega \ll 1$. Based on the principle of asymptotic matching of the solutions in the Hertzian region and inlet zone, in the inlet zone where $r = O(1)$, $r = (x-a)/\epsilon_q$, $\epsilon_q(\omega) \ll 1$ for $\omega \ll 1$ both the lubricant viscosity μ and gap h are of the order of one while $p(x) = O(\epsilon_q^{1/2})$, $\omega \ll 1$. We will assume that the lubricant viscosity μ is a non-decreasing function of pressure p. To characterize the above–mentioned lubrication regimes, let us introduce a critical size $\epsilon_0(\omega)$ of the inlet zone as the maximum value of $\epsilon_q(\omega)$ for which the following estimate holds (see Section 6.6)

$$\mu(\epsilon_q^{1/2}(\omega)) = O(1), \ \omega \ll 1. \tag{9.113}$$

Therefore, $\mu(\epsilon_q^{1/2}(\omega)) = O(1)$ for $\epsilon_q = O(\epsilon_0(\omega))$ and $\mu(\epsilon_q^{1/2}(\omega)) \gg 1$ for $\epsilon_q \gg \epsilon_0(\omega)$, $\omega \ll 1$. Now, it is possible to define the necessary conditions for pre- and over-critical lubrication regimes. A pre-critical lubrication regime is such a regime that the size of the inlet zone ϵ_q satisfies the estimate

$$\epsilon_q(\omega) = O(\min(1, \epsilon_0(\omega))), \ \omega \ll 1, \tag{9.114}$$

while an over-critical lubrication regime is a regime for which the size of the inlet zone ϵ_q satisfies the estimate

$$\epsilon_0(\omega) \ll \epsilon_q(\omega) \ll 1, \ \omega \ll 1. \tag{9.115}$$

Keeping in mind that $\epsilon_q(\omega) \ll 1$ for $\omega \ll 1$ from estimates (9.113)-(9.115) we obtain that over-critical regimes cannot be realized if $\epsilon_0(\omega) \gg 1$ for $\omega \ll 1$.

Clearly, the existence of the critical size $\epsilon_0(\omega)$ of the inlet zone depends on the particular behavior of the lubricant viscosity μ and the chosen small parameter ω. For example, if $\mu = \exp[Qp/(1+Q_1 p)]$, $Q, Q_1 \geq 0$, $Q \geq Q_1$, $Q \gg 1$ and $\omega = Q^{-1} \ll 1$, then $\epsilon_0(\omega) = Q^{-2}$. On the other hand, for the same as in the above example function μ, $Q \gg 1$, and $\omega = V \ll 1$ pre-critical regimes can be realized only for $\epsilon_q(V) = O(Q^{-2})$ while over-critical regimes are possible for $\epsilon_q(V) \gg Q^{-2}$, $Q \gg 1$.

9.4.1　Pre-Critical Lubrication Regimes

In Section 6.3 it was shown that in a pre-critical lubrication regime the inlet zone is represented by just one boundary layer [6] that has a homogeneous structure. It means that in the entire inlet zone the effects of the viscous fluid flow and elastic displacements are of the same order of magnitude and none of them prevails over another. In such cases in the inlet region, we have (see Section 6.6)

$$p = O(\epsilon_q^{1/2}), \ \tfrac{dp}{dx} = O(\epsilon_q^{-1/2}), \ h - 1 = O(\tfrac{\epsilon_q^{3/2}}{H_0}), \ r = O(1). \tag{9.116}$$

To proceed further along the lines of the analysis outlined in the preceding sections and to obtain formulas for the lubrication film thickness H_0, we have to make an assumption about the behavior of the rheology function Φ. Therefore, we can assume that the rheology of the lubricant is such that

$$\Phi\left(\frac{\mu s_0}{h}\right) = V^k \omega^l \mid s_0 \mid^m sign(s_0) H_0^{-n} \Phi_0\left(\frac{\mu}{h}\right),$$

$$\Phi_0\left(\frac{\mu}{h}\right) = O(1), \ r = O(1), \ \omega \ll 1,$$

(9.117)

where $\Phi_0(r)$ is a certain function of r while l, m, k, and n are certain constants. To estimate functions involved in equations (9.95) and (9.96), we obtain the asymptotic behavior of Φ', Φ'', and Φ''' in the inlet zone as follows

$$\Phi'\left(\frac{\mu s_0}{h}\right) = \frac{V^k \omega^l |s_0|^{m-1}}{H_0^n} \Phi_1\left(\frac{\mu}{h}\right), \ \Phi_1\left(\frac{\mu}{h}\right) = O(1), \ r = O(1),$$

(9.118)

$$\Phi''\left(\frac{\mu s_0}{h}\right) = \frac{V^k \omega^l |s_0|^{m-2}}{H_0^n} \Phi_2\left(\frac{\mu}{h}\right), \ \Phi_2\left(\frac{\mu}{h}\right) = O(1), \ r = O(1),$$

(9.119)

$$\Phi'''\left(\frac{\mu s_0}{h}\right) = \frac{V^k \omega^l |s_0|^{m-3}}{H_0^n} \Phi_3\left(\frac{\mu}{h}\right), \ \Phi_3\left(\frac{\mu}{h}\right) = O(1), \ r = O(1),$$

(9.120)

where $\Phi_1(\mu/h)$, $\Phi_2(\mu/h)$, and $\Phi_3(\mu/h)$ are certain functions of r. Similar relationships for Φ, Φ' Φ'', and Φ''' hold in the exit zone. Let us consider the generalized Reynolds equation following from equations (9.98), (9.101), and (9.103). Taking into account equations (9.100), (9.116), and (9.118) and comparing the orders of magnitude of the terms in equation (9.103) in the inlet zone, we obtain a formula for the film thickness

$$H_0 = A(V^k \omega^l \mid s_0 \mid^{m-1} \epsilon_q^2)^{\frac{1}{n+2}}, \ A = O(1), \ \omega \ll 1.$$

(9.121)

In the latter equation A is a coefficient of proportionality, which depends only on the specifics of the rheology function F (and/or Φ) and the lubricant viscosity μ, and it is independent of ω, s_0, and ϵ_q. The value of coefficient A can be obtained experimentally or by numerical solution of the system of asymptotically valid in the inlet zone equations similar to the ones derived in Section 6.3 for a Newtonian lubricant. The actual derivation of the asymptotically valid equations in the inlet and exit zones is left as an exercise for the reader. Equation (9.121) is valid for both starved and fully flooded lubrication regimes. The definitions of the starved and fully flooded lubrication regimes are given in (6.59) and (6.60), respectively. Using equation (9.121) and the definitions of starved and fully flooded regimes, we obtain the formulas

$$H_0 = A_s(V^k \omega^l \mid s_0 \mid^{m-1} \epsilon_q^2)^{\frac{1}{n+2}},$$

$$A_s = O(1), \ \epsilon_q \ll \epsilon_f = (V^k \omega^l \mid s_0 \mid^{m-1})^{\frac{2}{3n+2}}, \ \omega \ll 1,$$

(9.122)

for starved lubrication and

$$H_0 = A_f(V^k \omega^l \mid s_0 \mid^{m-1})^{\frac{3}{3n+2}}, \ A_f = O(1), \ \epsilon_q = O(\epsilon_f), \ \omega \ll 1,$$

(9.123)

for fully flooded lubrication. The value of ϵ_f (see equation (9.122)) represents the characteristic size of the inlet zone in the case of fully flooded pre-critical lubrication regimes. In equations (9.122) and (9.123) coefficients A_s and A_f differ from each other and can be determined experimentally or numerically. It should be stressed that coefficients A_s and A_f are independent of ω, s_0, V, ϵ_0, and ϵ_q.

Now, let us find the conditions under which the applied perturbation analysis for pre-critical regimes is valid, i.e., let us determine when $\nu(x) \ll 1$, $\omega \ll 1$, in the inlet zone. Using (9.94)-(9.96), and (9.121), we obtain the condition

$$\epsilon_q \ll \epsilon_\nu = (V^k \omega^l \mid s_0 \mid^{m+n+1})^{\frac{2}{3n+2}}, \ \omega \ll 1, \qquad (9.124)$$

that guarantees that $\nu(x) \ll 1$, $\omega \ll 1$ in the inlet zone.

By comparing the magnitudes of the values of ϵ_f and ϵ_ν, we easily find that the above analysis is valid for both starved and fully flooded regimes if $\epsilon_f = O(\epsilon_\nu)$, $\omega \ll 1$, which for $n + 2 > 0$ is equivalent to the estimates

$$\mid s_0 \mid \gg 1 \ or \ \mid s_0 \mid = O(1), \ \omega \ll 1. \qquad (9.125)$$

Let us consider pre-critical lubrication regimes for Examples 1-3 presented earlier when the small parameter $\omega = V \ll 1$. A similar analysis can be done for the small parameter $\omega = Q \gg 1$.

Example 1. For a Newtonian fluid (see equations (9.107)) in equation (9.117), we have $k = 0$, $l = m = n = 1$. Therefore, equations (9.122) and (9.123) are reduced to (see Section 6.3)

$$H_0 = A_s(V\epsilon_q^2)^{1/3}, \ A_s = O(1), \ \epsilon_q \ll \epsilon_f = V^{2/5}, \ V \ll 1, \qquad (9.126)$$

for starved lubrication and

$$H_0 = A_f V^{3/5}, \ A_f = O(1), \ \epsilon_q = O(V^{2/5}), \ V \ll 1, \qquad (9.127)$$

for fully flooded lubrication.

Example 2. For a fluid with the Ostwald-de Waele (power law) rheology described by equations (9.108), we have $k = 0$, $l = m = n$. Therefore, we obtain [7]

$$H_0 = A_s(V_n^n \mid s_0 \mid^{n-1} \epsilon_q^2)^{\frac{1}{n+2}},$$
$$(9.128)$$
$$A_s = O(1), \ \epsilon_q \ll \epsilon_f = (V_n^n \mid s_0 \mid^{n-1})^{\frac{2}{3n+2}}, \ V_n \ll 1,$$

for starved lubrication and

$$H_0 = A_f(V_n^n \mid s_0 \mid^{n-1})^{\frac{3}{3n+2}}, \ A_f = O(1), \ \epsilon_q = O(\epsilon_f), \ V_n \ll 1, \qquad (9.129)$$

for fully flooded lubrication.

Example 3. The case of a lubricant with Reiner-Philippoff-Carreau rheology described by equation (9.111) contains three distinct cases. In both cases when

$V \mid s_0 \mid /(12H_0G_0) \ll 1$ or $V \mid s_0 \mid /(12H_0G_0) = O(1)$ for $V \ll 1$ the formulas for the film thickness H_0 coincides with the ones for a Newtonian fluid (see Example 1). However, the coefficients of proportionality A_s and A_f in the latter case differ from the Newtonian case and depend on the values of parameters n and m. In case when $V \mid s_0 \mid /(12H_0G_0) \gg 1$ for $V \ll 1$ the rheological function Φ is well approximated by the power law represented by formulas (9.108), where parameter V_n must be replaced by $V_n = V[G_0(1 - \eta)]^{\frac{1}{n}}/G_0 \ll 1$. In this case the formulas for the film thickness H_0 coincide with the ones for a fluid with the power law rheology (see Example 2).

Now, let us consider the cases when function M is determined by equations (9.99) or (9.102). The second terms in brackets are supposed to be of the order of magnitude of $\nu^2(x) \ll 1$, $V \ll 1$, in the inlet zone. Using equations (9.102), (9.118)-(9.121) it can be shown that the second terms in brackets of equations (9.99) and (9.102) are small if $\epsilon_q \ll \epsilon_\nu$, $\epsilon_\nu = (V^k\omega^l \mid s_0 \mid^{n+m+1})^{\frac{2}{3n+2}}$, $\omega \ll 1$ (see estimate (9.124)). Therefore, all derived formulas for the lubrication film thickness are valid in the case of the generalized Reynolds equations based on the expressions for the function M from equations (9.99) or (9.102).

9.4.2 Over-Critical Lubrication Regimes

In Section 6.6 it was shown that in over-critical lubrication regimes the inlet zone is represented by not one but two boundary layers [6] of different sizes. Moreover, the inlet zone has a non-homogeneous structure. It means that in the inlet sub-zone most distant from the Hertzian region (called the inlet ϵ_q-zone) the effects of the viscous fluid flow through a gap mainly created by the Hertzian pressure dominates over the surface elastic displacements. In the inlet sub-zone next to the Hertzian region (called the inlet ϵ_0-zone) the effects of the viscous fluid flow and surface elastic displacements are of the same order of magnitude. The size of the ϵ_0-zone is the critical size of the inlet region ϵ_0, and it is much smaller than the size ϵ_q of the ϵ_q-zone. In such cases in the ϵ_0-zone we have (see Section 6.6 and equation (6.99), in particular)

$$p = O(\epsilon_0^{1/2}), \ \tfrac{dp}{dx} = O(\epsilon_0^{-1/2}), \ h - 1 = O(\tfrac{\epsilon_0^{3/2}}{H_0}), \ r_0 = O(1), \qquad (9.130)$$

while in the ϵ_q-zone we have

$$p = O(\epsilon_0^{1/2}), \ \tfrac{dp}{dx} = O(\epsilon_0^{1/2}\epsilon_q^{-1}), \ h - 1 = O(\tfrac{\epsilon_q^{3/2}}{H_0}), \ r = O(1). \qquad (9.131)$$

To proceed further along the lines of the analysis for over-critical lubrication regimes outlined in Section 6.6 and to obtain formulas for the lubrication film thickness H_0, we will make the assumption about the behavior of the rheology function Φ given in (9.117). Therefore, estimates (9.118)-(9.120) hold. Let us consider the generalized Reynolds equation following from equations (9.98), (9.101), and (9.103). Taking into account equation (9.100) and estimates (9.130) and (9.131) and comparing the orders of magnitude of the terms in equation (9.103) in the inlet ϵ_q-zone, we obtain a formula for the film thickness

$$H_0 = A(V^k \omega^l \mid s_0 \mid^{m-1} \epsilon_0^{-1/2} \epsilon_q^{5/2})^{\frac{1}{n+2}}, \quad A = O(1), \quad \omega \ll 1. \tag{9.132}$$

In equation (9.132) A is a coefficient of proportionality that depends only on the specifics of the rheology function F (and/or Φ) and the lubricant viscosity μ and it is independent of ω, s_0, V, ϵ_0, and ϵ_q. Coefficient A can be obtained experimentally or by numerical solution of the system of asymptotically valid in the inlet and exit ϵ_0-zones equations. In the inlet and exit ϵ_q-zones functions $q(r)$, $g(s)$, $h_q(r)$, and $h_g(s)$ satisfy equations (see preceding section) (9.79), (9.80) or (9.81) and (9.83), (9.84) or (9.85), respectively. In these equations function M_0 can be defined with the help of one of the expressions (9.101) or (9.102) for function M. For example, based on expressions (9.101) and (9.118) the equation for M_0 has the form

$$M_0(p, h, \mu, x) = \frac{A^{n+2}}{12} \frac{h^3}{\mu \Phi_1(\frac{\mu}{h})} \frac{dp}{dx}. \tag{9.133}$$

Using the definitions of starved and fully flooded lubrication regimes given in (6.59) and (6.60), respectively, as well as equation (9.132), we derive the formulas

$$H_0 = A_s(V^k \omega^l \mid s_0 \mid^{m-1} \epsilon_0^{-1/2} \epsilon_q^{5/2})^{\frac{1}{n+2}},$$

$$A_s = O(1), \quad \epsilon_q \ll \epsilon_f = (V^k \omega^l \mid s_0 \mid^{m-1} \epsilon_0^{-1/2})^{\frac{2}{3n+1}}, \quad \omega \ll 1, \tag{9.134}$$

for starved lubrication and

$$H_0 = A_f(V^k \omega^l \mid s_0 \mid^{m-1} \epsilon_0^{-1/2})^{\frac{3}{3n+1}},$$

$$A_f = O(1), \quad \epsilon_q = O(\epsilon_f), \quad \omega \ll 1, \tag{9.135}$$

for fully flooded lubrication. The value of ϵ_f (see equation (9.134)) represents the characteristic size of the ϵ_q-zone of the inlet region in the case of fully flooded over-critical lubrication regimes. In (9.134) and (9.135) coefficients A_s and A_f differ from each other and can be determined experimentally or numerically. It should be stressed that coefficients A_s and A_f are independent of ω, s_0, V, ϵ_0, and ϵ_q.

Now, let us find the conditions under which the applied perturbation analysis for over-critical regimes is valid, i.e., let us determine when $\nu(x) \ll 1$, $\omega \ll 1$, in the ϵ_q-zone of the inlet region. Using equations (9.94)-(9.96), and (9.132) we obtain the condition

$$\epsilon_q \ll \epsilon_{\nu q} = (V^k \omega^l \mid s_0 \mid^{n+m+1} \epsilon_0^{-1/2})^{\frac{2}{3n+1}}, \quad \omega \ll 1, \tag{9.136}$$

that guaranties that $\nu(x) \ll 1$ in the inlet ϵ_q-zone. The condition that guaranties that $\nu(x) \ll 1$ in the ϵ_0-zone is different and has the form

$$\epsilon_q \ll \epsilon_{\nu 0} = (V^k \omega^l \mid s_0 \mid^{n+m+1} \epsilon_0^{\frac{2n+3}{2}})^{\frac{2}{5(n+1)}}, \quad \omega \ll 1. \tag{9.137}$$

We have the relation $\epsilon_{\nu 0} \ll \epsilon_{\nu q}$ if $\epsilon_q \gg \epsilon_0$ that is typical for over-critical regimes. By comparing the magnitudes of the values of ϵ_f and $\epsilon_{\nu q}$, we obtain that the above analysis is valid for both starved and fully flooded regimes if $\epsilon_f = O(\epsilon_{\nu q})$, $\omega \ll 1$, that is represented by estimates (9.125) while the above analysis is valid for just starved lubrication if $\epsilon_{\nu q} \ll \epsilon_f$, $\omega \ll 1$.

Let us consider over-critical lubrication regimes for Examples 1-3 presented earlier when the lubricant viscosity is given by the exponential equation $\mu = \exp(Qp)$, $Q \gg 1$, and the small parameter $\omega = V \ll 1$. Here parameter $Q \gg 1$ may be a certain function of V. In this case the critical inlet zone size is $\epsilon_0 = Q^{-2}$, $Q \gg 1$. The analysis for a different choice of the small parameter ω can be done in a similar fashion.

Example 1. For a Newtonian fluid (see equations (9.107)) in equation (9.117), we have $l = 0$, $m = n = 1$. Therefore, equations (9.134) and (9.135) are reduced to formulas (6.148) and (6.149) for starved lubrication and for fully flooded lubrication regimes, respectively.

Example 2. For a fluid with the Ostwald-de Waele (power law) rheology described by equations (9.108), we have $k = 1$, $l = 0$, $m = n$. We obtain

$$H_0 = A_s(V_n^n Q \mid s_0 \mid^{n-1} \epsilon_q^{5/2})^{\frac{1}{n+2}},$$

$$A_s = O(1), \quad \epsilon_q \ll \epsilon_f = (V_n^n Q \mid s_0 \mid^{n-1})^{\frac{2}{3n+1}}, \quad V \ll 1, \tag{9.138}$$

for starved lubrication and

$$H_0 = A_f(V_n^n Q \mid s_0 \mid^{n-1})^{\frac{3}{3n+1}}, \quad A_f = O(1), \quad \epsilon_q = O(\epsilon_f), \quad V \ll 1, \tag{9.139}$$

for fully flooded lubrication. Here coefficients A_s and A_f are independent of the input parameters V_n, Q, and s_0. Formula (9.139) indicates an interesting property of the film thickness H_0 as a function of the slide-to-roll ratio s_0. Depending on fluid rheology we may have $n < 1$, $n = 1$, or $n > 1$. It follows from formula (9.139) that for $n < 1$ (pseudoplastic fluid) or $n > 1$ (dilatant fluid) the film thickness H_0 is a decreasing or increasing function of s_0, respectively, while for $n = 1$ it is independent of s_0. Therefore, for $n > 1$ and large sliding the film thickness H_0 in a thermal EHL contact may be not as prone to decrease as it usually happens for Newtonian fluids ($n = 1$) and pseudoplastic ones ($n < 1$).

Example 3. The case of a lubricant with Reiner-Philippoff-Carreau rheology described by equation (9.111) contains three distinct cases. In both cases when $\frac{V s_0}{12 H_0 G_0} \mid \mu \frac{\partial u}{\partial z} \mid \ll 1$ or $\frac{V s_0}{12 H_0 G_0} \mid \mu \frac{\partial u}{\partial z} \mid = O(1)$ for $V \ll 1$ the formulas for the film thickness H_0 coincide with the one for a Newtonian fluid (see Example 1). In case when $\frac{V s_0}{12 H_0 G_0} \mid \mu \frac{\partial u}{\partial z} \mid \gg 1$ for $V \ll 1$, the rheological function Φ is well approximated by the power law from equations (9.108), where parameter V_n must be replaced by $V_n = V[G_0(1 - \eta)]^{\frac{1}{n}}/G_0 \ll 1$. In this case the formulas for the film thickness H_0 coincide with the ones for a fluid with the power law rheology (see Example 2, equations (9.138) and (9.139)).

For the case of large sliding the numerical solutions in the inlet and exit zones can be obtained based on an iteration processes practically identical to the ones described in detail and implemented for Newtonian fluids in Section 6.9.3 and Sections 6.11 and 8.2 and for non-Newtonian fluids under pure rolling conditions in Section 9.3. Because of the solution properties are also similar to the ones discussed in these sections, they will not be considered here.

9.5   Choosing Pre- and Over-critical Lubrication Regimes, Small Parameter ω, and Pressure Viscosity Coefficient

Nowadays, there are various ways to define the pressure viscosity coefficient of a lubricant that, obviously, lead to different values of the lubrication film thickness if the same formula for the film thickness is used. It is important to establish the correct practical way of determining the pressure viscosity coefficient in exponential-like law for lubricant viscosity used for standard calculation of the lubrication film thickness. This standardization would allow for the proper comparison of various lubricants with different pressure-viscosity relationships. However, it is important to remember that the shear stress calculated based on such determined exponential–like viscosity equation may produce a significant error in comparison with the shear stress obtained based on the actual lubricant viscosity relationship $\mu(p)$. We will consider the most frequently used regimes of fully flooded lubrication that allow for obtaining practical and consistent results.

Let us assume that for the known rheology function Φ (and/or F) as well as the function of lubricant viscosity $\mu(p)$ the values of the lubrication film thickness H_0 are obtained numerically or experimentally. The question is how to practically describe these numerically or experimentally obtained values of H_0 by formulas (9.123) and (9.135) derived for the film thickness in fully flooded lubrication regimes that are widely used in engineering practice? We need to find a relatively easy and practical way of determining the unknown components involved in formulas (9.123) and (9.135). These components include coefficient A_f, small parameter ω, and the critical size ϵ_0 of the inlet region.

First, we need to realize that the asymptotic expression (9.117) for the rheological function Φ in the inlet region is independent of any parameters involved in the viscosity dependence $\mu(p)$ because the lubricant rheology is independent of a particular viscosity dependence on pressure and, also, $\mu(p)$ is always of the order of 1 in the inlet region. Second, there are basically only two choices for the problem small parameter: $\omega = V$ or ω equal to some small parameter involved in the relationship for viscosity $\mu(p)$. Third, if the problem small parameter ω is chosen to be equal to V, then the value of power l in (9.117) can be taken equal to zero as parameter V is already reflected in the behavior of the rheological function Φ in (9.117). Therefore, whatever choice of the small parameter ω is made we can consider $l = 0$ and the characteristic size of the inlet zone in fully flooded pre-critical regimes be equal to $\epsilon_f = (|s_0|^{m-1} V^k)^{\frac{2}{3n+2}}$ (see the expression for ϵ_f in (9.122)).

Our choice of the proper approximation of the lubricant viscosity relationship is supposed to be based on the lubrication regime a contact is involved in, i.e., either pre- or over-critical fully flooded regime. The actual reason for that is the fact that in these regimes film thickness is determined by different zones of a contact. In particular, for pre-critical lubrication regimes film thickness and coefficient proportionality A_f are determined by the inlet zone of the characteristic size $\epsilon_f = (|s_0|^{m-1} V^k)^{\frac{2}{3n+2}}$ while in over-critical lubrication regimes the film thickness and coefficient proportionality A_f are determined by the behavior of pressure in the inlet ϵ_0-zone. Therefore, the type of the lubrication regime should be established first, and then the approximation for viscosity $\mu(p)$ should be chosen.

Now, we need to determine the critical size ϵ_0 of the inlet zone. It is determined based on the solutions of equation (9.113). For practical use we will replace equation (9.113) by

$$\mu(\epsilon_0^{1/2}) = C, \tag{9.140}$$

where C is a constant of the order of magnitude 1. For certainty let us fix $C = e$. Therefore, the critical size ϵ_0 of the inlet zone is the maximum among all solutions of equation (9.140). If $\mu(p)$ is a monotonically increasing function of p, then ϵ_0 is a unique solution of the above equation.

After that, the value of ϵ_0 must be compared to the characteristic size ϵ_f of the inlet region in a pre-critical lubrication regime. If $\epsilon_0 \gg \epsilon_f$ then the lubrication regime is pre-critical, the problem small parameter ω can be taken equal to $V \ll 1$, and for the purpose of lubrication film calculations $\mu(p) = 1$. If ϵ_0 is of the order of ϵ_f, then the problem small parameter ω again can be taken equal to $V \ll 1$ or to the parameter involved in the relationship of $\mu(p)$, which causes $\mu(p)$ to increase significantly as pressure p increases from 0 to 1. In this case the lubrication regime is still pre-critical and formula (9.122) should be used for calculating the values of coefficient A_f and film thickness H_0 based on the available numerical or experimental data. In the latter case $\mu(p)$ no longer can be considered constant in the inlet and exit zones.

If $\epsilon_0 \ll \epsilon_f$, where $\epsilon_f(\omega)$ is defined by (9.122), then the lubrication regime is over-critical, the small parameter ω can be chosen as a specific small parameter involved in the expression for the lubricant viscosity $\mu(p)$ that causes its significant growth from $\mu(p) = 1$ at $p = 0$ to $\mu(p) \gg 1$ at $p = 1$. In addition to that the values of coefficient A_f and film thickness H_0 should be calculated based on formula (9.135) for over-critical lubrication regimes using the available numerical or experimental data (see also the criteria (6.171) and (6.172) derived for the case of Newtonian lubricant).

Now, we are in a position to be able to replace the actual viscosity dependence on pressure by an appropriate exponential-like dependence $\mu(p) = \exp(Qp^k)$ and determine its viscosity pressure coefficient Q. To do that we need to expand function $\ln[\mu(p)]$ for small p. Suppose the result of this expansion is

$$\ln[\mu(p)] = \sigma p^\alpha + \ldots, \quad p \ll 1, \qquad (9.141)$$

where σ and α are nonnegative constants.

Also, suppose $\epsilon_0 \ll \epsilon_f(\omega) = (\mid s_0 \mid^{m-1} V^k)^{\frac{2}{3n+2}}$ then the lubrication regime is over-critical and the problem small parameter ω can be chosen as $\omega = \sigma^{-1} \ll 1$, i.e., dependent on a specific parameter involved in the relationship $\mu(p)$. Therefore, by retaining the same dependence on pressure p as in equation (9.141) and using the just determined ϵ_0, it follows from (9.141) that the actual relationship for the lubricant viscosity $\mu(p)$ can be replaced by the exponential–like relationship

$$\mu(p) = \exp(\epsilon_0^{-\alpha/2} p^\alpha) \qquad (9.142)$$

that would provide the film thickness values H_0 practically equal to those obtained for the actual relationship for the lubricant viscosity $\mu(p)$. This is due to the fact that the inlet ϵ_0-zone, where the order of magnitude of pressure p is $\epsilon_0^{1/2}$, is mainly responsible for the formation of the film thickness H_0. Therefore, $Q = \epsilon_0^{-\alpha/2} = \sigma$ can be considered as an appropriate approximation of the viscosity pressure coefficient in an exponential–like viscosity dependence $\mu(p) = \exp(Qp^\alpha)$.

Otherwise, if $\epsilon_0 = O(\epsilon_f(\omega))$ or $\epsilon_0 \gg \epsilon_f(\omega)$, $\epsilon_f(\omega) = (\mid s_0 \mid^{m-1} V^k)^{\frac{2}{3n+2}}$, then the lubrication regime is pre-critical and the problem small parameter ω can be chosen as $\omega = V \ll 1$. In this case formula (9.123) for pre-critical lubrication regimes should be used for calculation of coefficient A_f and film thickness H_0 based on the available numerical or experimental data. Using equation (9.141) for the assumed behavior of the lubricant viscosity $\mu(p)$, we obtain that the actual relationship for the lubricant viscosity $\mu(p)$ can be replaced by the exponential–like relationship of the form

$$\mu(p) = \exp(\epsilon_f^{-\alpha/2} p^\alpha) \; if \; \epsilon_0 = O(\epsilon_f(\omega)) \ll 1, \; \omega \ll 1,$$

$$\mu(p) = 1 \; if \; \epsilon_0 \gg \epsilon_f = (\mid s_0 \mid^{m-1} V^k)^{\frac{2}{3n+2}}, \; \omega \ll 1, \qquad (9.143)$$

that will provide the film thickness values practically equal to those obtained for the actual relationship for the lubricant viscosity $\mu(p)$. This is due to the fact that the inlet zone, where the order of magnitude of pressure p is $\epsilon_f^{1/2}$, is mainly responsible for the formation of the film thickness H_0. Therefore, in this case $Q = \epsilon_f^{-\alpha/2}(\omega) \gg 1$ if $\epsilon_0 = O(\epsilon_f(\omega)) \ll 1, \omega \ll 1$ and $Q = 0$ if $\epsilon_0 \gg \epsilon_f = (\mid s_0 \mid^{m-1} V^k)^{\frac{2}{3n+2}}, \omega \ll 1$, can be considered as an appropriate approximation of the viscosity pressure coefficient Q in $\mu(p) = \exp(Qp^\alpha)$ as it provides the correct behavior of μ with respect to p for $p \ll 1$ and provides the correct values for $\mu(p)$ in the most important for determining film thickness H_0 part of the contact - the inlet zone.

It is clear that the above definition of the viscosity pressure coefficient Q is not necessarily equal to $d\mu(p)/dp$ at $p = 0$ as it is assumed by many researchers and users of the formulas for the lubrication film thickness. In fact, the pressure viscosity coefficient Q is determined by the type of lubrication regime (pre- or over-critical) and behavior of the lubricant viscosity $\mu(p)$ somewhere in the middle of the inlet zone but not necessarily at $p = 0$. The value of constant C in equation (9.140) can be adjusted for the specifics of a particular lubricant viscosity to reflect the particular lubrication conditions better.

It is very important to understand that this viscosity approximation can be used only for the lubrication film calculations. However, an attempt to use this viscosity approximation for friction calculations can lead to significant distortions.

Let us consider a couple of more examples.

Example 4. Let us consider the case of a constant viscosity $\mu = 1$. Then the maximum solution of equation (9.140) is $\epsilon_0 = \infty$. The above considerations indicate that only pre-critical regimes can be realized in this situation, and the problem small parameter ω can be chosen as $\omega = V \ll 1$. Therefore, the film thickness is described by formulas (9.122) and (9.123).

Example 5. Let us consider the example of a Vinogradov-Malkin viscosity relationship

$$\mu(p) = \exp[\alpha_1 R_0(\tfrac{1}{R-R_0} - \tfrac{1}{1-R_0})],$$

$$R = 1 - \tfrac{1}{\alpha_2}\ln(1 + \alpha_3 p), \quad \alpha_3 = \alpha_2 \tfrac{p_H}{p_0}, \tag{9.144}$$

where α_1, α_2, α_3, and R_0 are positive dimensionless constants, p_0 is the characteristic pressure. Then for $C = e$ the solution of equation (9.140) has the form

$$\epsilon_0^{1/2} = \tfrac{1}{\alpha_3}\{\exp[\tfrac{\alpha_2(1-R_0)^2}{1-R_0+\alpha_1 R_0}] - 1\}. \tag{9.145}$$

Note, that for real lubricants the value of $\epsilon_0^{1/2}$ determined from (9.145) is small. By expanding the expression for $\mu(p)$ for small p, we obtain

$$\ln[\mu(p)] = \sigma p + \ldots, \quad \sigma = \tfrac{\alpha_1 \alpha_3}{\alpha_2} \tfrac{R_0}{(1-R_0)^2}, \quad p \ll 1. \tag{9.146}$$

It is obvious from (9.146) that $d\mu(p)/dp \mid_{p=0} = \sigma$. However, according to the above outlined procedure the adequate pressure viscosity coefficient Q in an exponential–like viscosity relationship is different from σ. Depending on whether one of the following two estimates $\epsilon_0 = O(\epsilon_f(\omega))$ or $\epsilon_0 \gg \epsilon_f(\omega)$ is satisfied or $\epsilon_0 \ll \epsilon_f(\omega)$, where $\epsilon_f(\omega)$ is defined by (9.122), the film thickness H_0 has to be determined from formulas (9.123) or (9.135), which correspond to pre- and over-critical lubrication regimes, respectively. After the determination of the lubrication regime (pre- or over-critical) is done the pressure viscosity coefficient $Q = \epsilon_0^{-1/2}$ (see formula (9.145)) for over-critical regimes and $Q = \epsilon_f^{-1/2}$ (see formula (9.143)) for pre-critical regimes. The small parameter ω we can take as $\omega = \sigma^{-1} \ll 1$.

9.6 Closure

Isothermal EHL problems for heavily loaded line contacts lubricated by general non-Newtonian lubricant were studied asymptotically and numerically. Cases of pure rolling and high slide-to-roll ration were considered. For the case of high slide-to-roll ratio using regular perturbations method the Reynolds equation for general non-Newtonian lubricant was reduced to an analog of the Reynolds equation for the case of isothermal EHL problem for Newtonian lubricant. After that the problem was studied by the earlier developed methods of matched asymptotic expansions. A number of rheology dependent formulas for the lubrication film thickness were analytically derived. Each of these formulas involve the parameters characterizing lubricant rheology as well as a coefficient of proportionality such as A_s and A_f the values of which can be determined from numerical problem solution in the inlet zone or by curve fitting of independently obtained experimental values for the lubrication film thickness. Some numerical examples of the solutions in the inlet and exit zones of heavily loaded contacts for lubricants with different rheologies are given.

9.7 Exercises and Problems

1. For $\nu(x) \ll 1$ for a lubrication fluid of general non-Newtonian rheology $\mu \frac{\partial u}{\partial z} = F(\tau)$ find a six-term solution of equation (9.5) for $f(x)$ and, then, derive a corresponding to this solution approximate Reynolds equation.

2. For lubricants with general non-Newtonian rheology and $\nu(x) \ll 1$ in the inlet and exit zones using the formula (9.121) for the film thickness H_0 for pre-critical lubrication regimes derived the asymptotically valid equations of the problem in the inlet and exit zones of heavily loaded EHL contact.

3. For $\nu(x) \ll 1$ (i.e., when condition (9.124) is satisfied) using the formula (9.121) for the film thickness H_0 for pre-critical lubrication regimes check whether in the inlet and exit zones of pre-critically heavily loaded EHL contact the term of function M from (9.102) in the approximate Reynolds equation (9.103) proportional to $(\frac{dp}{dx})^3$ is much smaller than the term proportional to

$\frac{dp}{dx}$.

4. Show that a similar statement as in Problem 2 is valid for the case of over-critical lubrication regimes.

5. Describe the process of determining whether the lubrication regime is pre- or over-critical.

References

[1] Hamrock, B.J. 1994. *Fundamentals of Fluid Film Lubrication*. New York: McGraw-Hill.

[2] Kudish, I.I. 1979. Elastohydrodynamic Problems for Rough Bodies with non-Newtonian Lubrication. *Dopovidi Akademii Nauk Ukrains'koi RSR*, Seriya A, No. 11:915-920.

[3] Van Wazen, J.R., Lyons, J.W., Kim, K.Y., and Cowell, R.E. 1963. *Viscosity and Flow Measurment. A Laboratory Handbook of Rheology*. New York-London: Interscience Publishers, John Wiley & Sons.

[4] Kudish, I.I. 1996. Asymptotic Analysis of a Problem for a Heavily Loaded Lubricated Contact of Elastic Bodies. Pre- and Over-critical Lubrication Regimes for Newtonian Fluids. *Dynamic Systems and Applications*, Dynamic Publishers, Atlanta, 5, No. 3:451-478.

[5] Vorovich, I.I., Aleksandrov, V.M., and Babeshko, V.A. 1974. *Non − Classical Mixed Problems of Elasticity*. Moscow: Nauka.

[6] Kevorkian, J. and Cole, J.D. 1985. *Perturbation Methods in Applied Mathematics. Applied Mathematics Series, Vol. 34*. New York: Springer-Verlag.

[7] Alexandrov, V.M. and Kudish, I.I. 1980. Problem of Elastohydrodynamic Theory of Lubrication for a Viscous Fluid with Complex Rheology. *J. Mech. Solids* 15, No. 4:57-68.

[8] Kudish, I.I. and Covitch, M.J. 2010. *Modeling and Analytical Methods in Tribology*. Boca Raton: CRC Press.

[9] Kudish, I.I. 1978. Asymptotic Analysis of a Plane Non-isothermal Elastohydro-dynamic Problem for a Heavily Loaded Rolling Contact. *Proc. Acad. Sci. of Armenia SSR, Mechanics* 31, No. 6:16-35.

[10] Kudish, I.I. 1983. Asymptotic Method of Study for Plane Problems of the Elastohydrodynamic Lubrication Theory for Heavily Loaded Regimes. Part 2. Non-isothermal Problem. *Proc. Acad. Sci. Armenia SSR, Mechanics* 36, No. 5:47-59.

[11] Myllerup, C.M. and Hamrock, B.J. 1994. Perturbation Approach to Hydrodynamic Lubrication Theory. *ASME J. Tribology* 116, No. 1:110-118.

10

TEHL Problems for Lubricants with General Non-Newtonian Rheology in Line Contacts

---

10.1 Introduction

In this chapter we will analyze both thermal and elastic effects in a contact lubricated by a fluid with general non-Newtonian rheology including the effect of heat dissipation in contact solids. The problem will be considered under the classic assumptions (see Sections 4.2 and 5.2).

10.2 TEHL Problem Formulation for Non-Newtonian Fluids

First, let us consider some basic facts related to thermal conductivity of contacting solids. Suppose there is a half-plane that is subjected to the action of a point heat source with the heat flux q directed into the half-plane. The heat source is moving along the half-plane boundary with a steady velocity u. Additionally, the material of the half-plane is characterized by the density ρ, specific heat c, and thermal diffusivity k. At this point we will ignore thermal expansion of the solids. We will use the rectangular coordinate system which is introduced as follows: the x- and z-axes of the coordinate system are directed along the half-plane surface in the direction of the heat source motion and into the half-plane down, respectively (the y-axis is perpendicular to the x- and z-axes). According to [1], a moving point source of heat currently located at the origin of the coordinate system at the point with coordinate x on the half-plane surface causes the following temperature T of the half-plane surface

$$T = \tfrac{q}{\pi k \rho c} e^{-\frac{ux}{2k}} K_0(|\tfrac{ux}{2k}|)\},$$ (10.1)

where K_0 is the modified Bessel functions [2]. Therefore, for a distributed source of heat using superposition/convolution, we obtain

$$T = \tfrac{1}{\pi k \rho c} \int\limits_{x_i}^{x_e} q(\xi) e^{-\frac{u(x-\xi)}{2k}} K_0(|\tfrac{u(x-\xi)}{2k}|) + T_a,$$ (10.2)

where T_a is the ambient surface temperature.

Now, let us consider the TEHL problem for two cylinders with parallel axes made of elastic materials with elastic moduli E_i and Poisson's ratios ν_i ($i = 1, 2$) and moving with surface speeds u_1 and u_2. The cylinders are loaded with normal force P. This problem formulation was discussed in detail in previous sections.

To solve the problem we will be using asymptotic methods. We will assume that the shear stress in lubricant generates heat, i.e.,

$$\tau F(\tau) \geq 0. \tag{10.3}$$

In addition, we will assume that the predominant direction of the heat flow generated in the contact is normal to the contact surfaces and is directed into the solids [3]. As in Chapter 7 we will assume that the lubricant viscosity μ satisfies the relationship (7.1).

The derivation of the TEHL problem equations is practically identical to the equation derivations conducted in Section 5.2 and Chapter 7. Therefore, based on the assumptions made, the problem can be reduced to the following system of non-linear integro-differential equations [1, 3, 4]

$$\frac{d}{dx}\left\{ \int_{-h/2}^{h/2} \frac{z}{\mu} F(f + z\frac{dp}{dx})dz - \frac{u_1+u_2}{2}h \right\} = 0,$$

$$p(x_i) = p(x_e) = \frac{dp(x_e)}{dx} = 0,$$

$$\int_{-h/2}^{h/2} \frac{1}{\mu} F(f + z\frac{dp}{dx})dz = u_2 - u_1,$$

$$h = h_e + \frac{x^2 - x_e^2}{2R'} + \frac{2}{\pi E'} \int_{x_i}^{x_e} p(t) \ln \mid \frac{x_e - t}{x - t} \mid dt, \quad \int_{x_i}^{x_e} p(\xi)d\xi = P,$$

$$\frac{\partial^2 T}{\partial z^2} = -\frac{1}{\lambda\mu}(f + z\frac{dp}{dx})F(f + z\frac{dp}{dx}),$$

$$T(x, -\tfrac{h}{2}) = T_{w1}(x), \quad T(x, \tfrac{h}{2}) = T_{w2}(x),$$

$$q_1 = \lambda\frac{\partial T}{\partial z} \mid_{z=-h/2}, \quad q_2 = -\lambda\frac{\partial T}{\partial z} \mid_{z=h/2},$$

$$(10.4)$$

$$T_{w1}(x) = \frac{1}{\pi k_1 \rho_1 c_1} \int_{x_i}^{x_e} q_1(\xi) \exp[-\frac{u_1(x-\xi)}{2k_1}]K_0(\mid \frac{u_1(x-\xi)}{2k_1} \mid)d\xi + T_{a1},$$

$$T_{w2}(x) = \frac{1}{\pi k_2 \rho_2 c_2} \int_{a}^{c} q_2(\xi) \exp[-\frac{u_2(x-\xi)}{2k_2}]K_0(\mid \frac{u_2(x-\xi)}{2k_2} \mid)d\xi + T_{a2},$$

$$(10.5)$$

where F is the function describing the lubricant rheology (see equation (9.1)), λ is the coefficient of lubricant heat conductivity, T_{a1} and T_{a2} are the temperatures of the lower and upper contact surfaces far away from the contact region, respectively, λ is the coefficient of lubricant heat conductivity, T_{w1} and T_{w2} are the surface temperatures of the lower and upper cylinders, respectively, and q_1 and q_2 are the heat fluxes directed in the lower and upper cylinders, respectively. The contacting solid materials' densities ρ_1, ρ_2, specific heat parameters c_1, c_2, and the thermal diffusivities k_1 and k_2 for lower (marked with index 1) and upper (marked with index 2) are considered to be known. The rest of the parameters and variables are the same as in Chapter 9.

In addition to equations (5.47), (6.3), (9.2), and (7.5), let us introduce some additional dimensionless variables and parameters as follows

$$T'_{ai} = \frac{T_{ai}}{T_a} - 1 = (-1)^i \Delta_\infty, \ q'_i = \frac{h_e}{\lambda_a T_a} q_i, \ \lambda_i = \frac{u_i a_H}{2k_i}, \ (i = 1, 2),$$

$$T' = \frac{T}{T_a} - 1, \ T_a = \frac{T_{a1} + T_{a2}}{2}, \ \Delta_\infty = \frac{T_{a2} - T_{a1}}{T_{a1} + T_{a2}}, \ s_0 = 2\frac{u_2 - u_1}{u_2 + u_1}, \quad (10.6)$$

$$\delta = \alpha_T T_a, \ \kappa = \frac{(u_1 + u_2)P}{4\pi \lambda_a T_a} \left(\frac{a_H}{R'}\right)^2, \ \eta_0 = \frac{k_1 \rho_1 c_1}{k_2 \rho_2 c_2}, \ \Lambda = \frac{4\lambda_a R'}{\pi k_1 \rho_1 c_1 a_H},$$

where λ_a is the coefficient of lubricant heat conductivity at ambient pressure.

We will postulate that the lubricant viscosity μ can be represented as a product of two parts: one that is independent of the lubricant temperature T and the other that is a function of the lubricant temperature T. In particular, we will assume that in dimensionless variables (10.6), the lubricant viscosity can be represented in the form

$$\mu(p, T) = \mu^0(p)e^{-\delta_T T}, \ \frac{\partial \mu^0(p)}{\partial T} = 0, \quad (10.7)$$

where $\mu^0(p)$ is the lubricant viscosity at the ambient temperature, δ is the temperature viscosity coefficient, $\delta_T \geq 0$. Coefficient δ_T may be a constant or a function of pressure p. With respect to the lubricant heat conductivity λ, we will assume that it may depend only on pressure p and is independent of the lubricant temperature T.

By using the aforementioned dimensionless variables and parameters under the above assumptions the TEHL problem can be reduced to the following system of equations (for simplicity primes at the dimensionless variables are omitted)

$$\frac{d}{dx}\left\{ \int_{-h/2}^{h/2} \frac{z}{\mu} F(f + H_0 z \frac{dp}{dx})dz - h \right\} = 0, \ p(a) = p(c) = \frac{dp(c)}{dx} = 0, \quad (10.8)$$

$$\int_{-h/2}^{h/2} \frac{1}{\mu} F(f + H_0 z \frac{dp}{dx})dz = s_0, \quad (10.9)$$

$$H_0(h - 1) = x^2 - c^2 + \frac{2}{\pi} \int_a^c p(t) \ln \left| \frac{c - t}{x - t} \right| dt, \quad (10.10)$$

$$\int_a^c p(\xi)d\xi = \tfrac{\pi}{2}, \tag{10.11}$$

$$\tfrac{\partial^2 T}{\partial z^2} = -\tfrac{\kappa H_0}{\lambda \mu}(f + H_0 z \tfrac{dp}{dx})F(f + H_0 z \tfrac{dp}{dx}), \tag{10.12}$$

$$T(x, -\tfrac{h}{2}) = T_{w1}(x), \ \ T(x, \tfrac{h}{2}) = T_{w2}(x).$$

$$q_1 = \lambda \tfrac{\partial T}{\partial z}\mid_{z=-h/2}, \ \ q_2 = -\lambda \tfrac{\partial T}{\partial z}\mid_{z=h/2}, \tag{10.13}$$

$$T_{w1}(x) = \tfrac{\Lambda}{2H_0}\int_a^c q_1(\xi)e^{-\lambda_1(x-\xi)}K_0(\mid \lambda_1(x-\xi)\mid)d\xi - \Delta_\infty,$$

$$\tag{10.14}$$

$$T_{w2}(x) = \tfrac{\Lambda \eta_0}{2H_0}\int_a^c q_2(\xi)e^{-\lambda_2(x-\xi)}K_0(\mid \lambda_2(x-\xi)\mid)d\xi + \Delta_\infty.$$

Therefore, for the given values of constants a, Q, s_0, δ_T, κ, η_0, Λ, λ_1, λ_2, Δ_∞ and functions F, μ^0, and λ we have to find constants H_0 - exit film thickness, c - exit coordinate, and functions $p(x)$ - pressure, $h(x)$ - gap, $f(x)$ - sliding frictional stress, $T(x, z)$ - lubricant temperature, $T_{wi}(x)$ and $q_i(x)$ - surface temperatures and heat fluxes, $(i = 1, 2)$.

10.3   Asymptotic Analysis of the Problem for Heavily Loaded Line Contacts

The section is dedicated to the asymptotic analysis of steady heavily loaded thermal lubricated line contacts under the condition of relatively high slide-to-roll ratio. Some approximations of the generalized Reynolds equation for non-Newtonian fluids that resemble the Reynolds equation for a Newtonian fluid are obtained. The main idea of the method for the non-isothermal problem is based on analytical solution of the problem for the sliding shear stress, temperature, and heat flux and the reduction of the problem to asymptotic and/or numerical solution for the pressure, and gap. The procedure for deriving formulas for the lubrication film thickness is based on the methods outlined in Sections 6.3, 6.6, and Chapter 7 and it is presented for the cases when the influence of the lubrication shear stresses on surface normal and tangential displacements can be neglected. A number of examples illustrating application of the described technique is given. Some general issues as to what is the range of parameters when the proposed approximation provides asymptotically correct solutions and how to define appropriately the pressure viscosity coefficient are sorted out. It is shown that in certain cases the asymptotic procedure described below provides asymptotically correct solutions only in regimes of starved lubrication while in other cases it is valid for both starved and fully flooded lubrication regimes. A two-term asymptotic approximation for lubricant temperature T is obtained analytically and finalized numerically. One of the applications of the developed technique is the analysis of the dimple phenomenon (see Chapter 18).

The whole analysis proposed here is for the case of heavily loaded lubricated line contacts. Therefore, it is fair to assume that there is a small parameter ω, which makes these contacts heavily loaded. Our main goal is to reduce equations (10.7)-(10.14) to a problem analogous to the isothermal EHL problem presented in Chapter 9. It means that using some perturbation techniques the temperature $T(x, z)$ and sliding frictional stress $f(x)$ will be determined and solution of equations (10.7)-(10.14) will be reduced to solution of equations for just two functions $p(x)$ and $h(x)$ and two parameters H_0 and c. The reduced equations have to be solved numerically. We will consider only the problem for heavily loaded TEHL contacts with relatively large slide-to-roll ratios s_0. As before we introduce function $\nu(x)$ according to (9.94) which is small in the inlet zone. In case of over-critical regimes we need $\nu(x)$ to be small in the inlet ϵ_0-zone which guarantees that $\nu(x)$ is small in the entire contact. If there is a need to obtain the problem solution in the entire contact region not just the formula for the lubrication film thickness H_0, then $\nu(x)$ would be considered small in the entire contact region. Obviously, it may impose additional limitations on the problem input parameters. The whole procedure of TEHL problem solution consists of a number of steps. On Step 1 we will assume that functions $f(x)$ and $T(x, z)$ can be represented by the expansions

$$f(x) = f_0(x) + \nu(x)f_1(x) + O(\nu^2(x)), \quad \nu(x) \ll 1, \tag{10.15}$$

$$T(x, z) = T_0(x, z) + \nu(x)T_1(x, z) + O(\nu^2(x)), \quad \nu(x) \ll 1, \tag{10.16}$$

where functions $f_0(x)$ and $f_1(x)$ are the consecutive terms of the asymptotic of $f(x)$ to be determined from equation (10.9) while functions $T_0(x, z)$ and $T_1(x, z)$ are the consecutive terms of the asymptotic of $T(x, z)$ to be determined from equations (10.7) and (10.12). Based on equations (10.7), (10.15), and (10.16), we can conclude that

$$\frac{1}{\mu(p,T)} = \frac{e^{\delta_T T_0(x,z)}}{\mu^0(p)}[1 + \nu(x)\delta_T T_1(x, z) + O(\nu^2(x))], \quad \nu(x) \ll 1, \tag{10.17}$$

$$F(f + H_0 z \tfrac{dp}{dx}) = F(f_0) + \nu(f_1 + \tfrac{2z}{h}f_0)F'(f_0) + O(\nu^2), \quad \nu \ll 1. \tag{10.18}$$

Similarly, based on representation (10.16) for the lubricant temperature $T(x, z)$, we can conclude that the surface temperatures $T_{w1}(x)$ and $T_{w2}(x)$ can be searched in the form of the following expansions

$$T_{w1}(x) = T_{w10}(x) + \nu(x)T_{w11}(x) + O(\nu^2(x)),$$

$$T_{w2}(x) = T_{w20}(x) + \nu(x)T_{w21}(x) + O(\nu^2(x)), \quad \nu(x) \ll 1. \tag{10.19}$$

We have to keep in mind that q_1 and q_2 from equations (10.13) also can be represented as asymptotic expansions in the form

$$q_1(x) = q_{10}(x) + \nu(x)q_{11}(x) + O(\nu^2(x)),$$

$$q_2(x) = q_{20}(x) + \nu(x)q_{21}(x) + O(\nu^2(x)), \quad \nu(x) \ll 1,$$

(10.20)

where q_{10}, q_{20} and q_{11}, q_{21} are used in equations (10.24) and (10.25).

On Step 2 of the solution process we substitute expansions (10.15)-(10.19) into equations (10.9) and (10.12), expand the necessary terms, and equate the terms with the same powers of $\nu(x)$. As a result of that we obtain the following equations

$$f_0(x) = \Phi(\tfrac{\mu^0 s_0}{I_1}), \quad \tfrac{f_1(x)}{f_0(x)} = -\tfrac{1}{I_1}\big[\tfrac{F(f_0)}{f_0 F'(f_0)}\delta_T I_{T1} + \tfrac{2}{h}I_z\big],$$

$$I_1 = \int\limits_{-h/2}^{h/2} e^{\delta_T T_0}dz, \quad I_{T1} = \int\limits_{-h/2}^{h/2} T_1 e^{\delta_T T_0}dz, \quad I_z = \int\limits_{-h/2}^{h/2} z e^{\delta_T T_0}dz,$$

(10.21)

$$\frac{\partial^2 T_0}{\partial z^2} = -\kappa H_0 \frac{f_0 F(f_0)}{\lambda \mu^0} e^{\delta_T T_0},$$

$$T_0(x, -\tfrac{h}{2}) = T_{w10}(x), \quad T_0(x, \tfrac{h}{2}) = T_{w20}(x),$$

(10.22)

$$\frac{\partial^2 T_1}{\partial z^2} = -\kappa H_0 \frac{f_0 F(f_0)}{\lambda \mu^0} e^{\delta_T T_0}\{\delta_T T_1 + [1 + \tfrac{f_0 F'(f_0)}{F(f_0)}](\tfrac{f_1}{f_0} + \tfrac{2z}{h})\},$$

$$T_1(x, -\tfrac{h}{2}) = T_{w11}(x), \quad T_1(x, \tfrac{h}{2}) = T_{w21}(x).$$

(10.23)

$$T_{w10}(x) = \tfrac{\Lambda}{2H_0} \int\limits_a^c q_{10}(\xi)e^{-\lambda_1(x-\xi)}K_0(|\lambda_1(x-\xi)|)d\xi - \Delta_\infty,$$

$$T_{w20}(x) = \tfrac{\Lambda \eta_0}{2H_0} \int\limits_a^c q_{20}(\xi)e^{-\lambda_2(x-\xi)}K_0(|\lambda_2(x-\xi)|)d\xi + \Delta_\infty,$$

(10.24)

$$\nu(x)T_{w11}(x) = \tfrac{\Lambda}{2H_0} \int\limits_a^c \nu(\xi)q_{11}(\xi)e^{-\lambda_1(x-\xi)}K_0(|\lambda_1(x-\xi)|)d\xi,$$

$$\nu(x)T_{w21}(x) = \tfrac{\Lambda \eta_0}{2H_0} \int\limits_a^c \nu(\xi)q_{21}(\xi)e^{-\lambda_2(x-\xi)}K_0(|\lambda_2(x-\xi)|)d\xi.$$

(10.25)

First, let us try to find the exact solution of the nonlinear boundary-value problem (10.22) in the form

$$T_0(x, z) = \tfrac{T_{w10}+T_{w20}}{2} + \tfrac{1}{\delta_T} \ln \tfrac{R}{\cosh^2[N(z-z_m)]},$$

(10.26)

where R, N, and z_m are new unknown functions of x to be determined from solution of boundary-value problem (10.22). By substituting the expression for T_0 from formula (10.26) into equations (10.22), we obtain three equations

$$\frac{2\lambda N^2}{\kappa \delta_T H_0} = R\frac{f_0 F(f_0)}{\mu^0} \exp[\delta_T \tfrac{T_{w10}+T_{w20}}{2}],$$

(10.27)

$$\frac{R}{\cosh^2[\frac{N}{2}(h+2z_m)]} = e^{-\Delta}, \quad \frac{R}{\cosh^2[\frac{N}{2}(h-2z_m)]} = e^{\Delta}, \quad \Delta = \delta_T \frac{T_{w20}-T_{w10}}{2}. \quad (10.28)$$

Based on the fact that the values of κ, δ_T, and μ^0 are positive from inequality (10.3) and equations (10.28), we get that function R is positive. Therefore, solving equations (10.28) we obtain

$$R = \cosh[\tfrac{N}{2}(h - 2z_m)] \cosh[\tfrac{N}{2}(h + 2z_m)],$$

$$z_m = \tfrac{1}{2N} \ln\{\tfrac{\sinh[\frac{1}{2}(Nh+\Delta)]}{\sinh[\frac{1}{2}(Nh-\Delta)]}\}. \quad (10.29)$$

Thus, solution of the boundary-value problem (10.22) is reduced to solution of an algebraic equation (10.27) for N after substituting into it the expressions for functions R and z_m from formulas (10.29). In most cases this equation for N can be solved only numerically. However, in some special cases such as the Newtonian rheology it can be done analytically. As it follows from formula (10.26) function T_0 attains its maximum at $z = z_m$. Obviously, function T_0 reaches its maximum between the contact surfaces if $\mid z_m \mid \leq h/2$ and it is a monotonic function if $\mid z_m \mid \geq h/2$.

The expressions for functions q_{10} and q_{20} are as follows

$$q_{10} = \tfrac{2\lambda N}{\delta_T} \tanh[\tfrac{N}{2}(h + 2z_m)], \quad q_{20} = \tfrac{2\lambda N}{\delta_T} \tanh[\tfrac{N}{2}(h - 2z_m)], \quad (10.30)$$

while for q_{11} and q_{21} we have equations (10.42). Substituting these expressions in equations (10.24) we obtain a nonlinear system of equations for $T_{w10}(x)$ and $T_{w20}(x)$ in the form

$$T_{w10} = \frac{\Lambda}{H_0} \int_a^c \tfrac{\lambda N}{\delta_T} \tanh[N(\tfrac{h}{2} + z_m)] e^{-\lambda_1(x-\xi)} K_0(\mid \lambda_1(x - \xi) \mid) d\xi$$

$$-\Delta_\infty,$$
$$(10.31)$$

$$T_{w20} = \frac{\Lambda \eta_0}{H_0} \int_a^c \tfrac{\lambda N}{\delta_T} \tanh[N(\tfrac{h}{2} - z_m)] e^{-\lambda_2(x-\xi)} K_0(\mid \lambda_2(x - \xi) \mid) d\xi$$

$$+\Delta_\infty,$$

solution of which can be obtained numerically using iterations. As the initial approximations for these iterations, which are described below, can be taken functions $T_{w10}(x) = -\Delta_\infty$, $T_{w20}(x) = \Delta_\infty$.

Obviously, if $\eta_0 = 1$, $\lambda_1 = \lambda_2$, and $z_m = \Delta_\infty = 0$, then the surface temperatures $T_{w10}(x)$ and $T_{w20}(x)$ are equal. On the other hand, based on the fact that function $F(x)$ is an odd function from equation (10.27) we get that function $N(x)$ is independent of the sign of the slide-to-roll ratio s_0, i.e., it is even with respect to s_0. Therefore, we can conclude that in the above case $T_{w2}(x) - T_{w1}(x) = O(\nu(x)) \ll 1$.

Now, let us find the solution of the linear boundary-value problem (10.23) for $T_1(x, z)$ in the form

$$T_1(x,z) = \tfrac{1}{\delta_T}\{A_1 + B_1(z - z_m) + C_1\tanh[N(z - z_m)]$$

$$+ D_1(z - z_m)\tanh[N(z - z_m)]\},$$

(10.32)

where A_1, B_1, C_1, and D_1 are unknown functions of just x that have to be determined from solution of problem (10.23) that takes into account the expressions for f_1 and I_{T1} from equations (10.21) and functions T_{w11} and T_{w21}. After that it is easy to get

$$I_1 = I_A = \exp[\delta_T \tfrac{T_{w10}+T_{w20}}{2}]\tfrac{\sinh(Nh)}{N},$$

$$I_z = -I_1\tfrac{R}{N\sinh(Nh)}\{\tfrac{Nh}{2}\tfrac{\sinh(2Nz_m)}{R} - \Delta\},$$

$$I_{T1} = \tfrac{1}{\delta_T}\{A_1 I_A + B_1 I_B + C_1 I_C + D_1 I_D\},\ I_B = I_z - z_m I_1,$$

(10.33)

$$I_C = -I_1\tfrac{\sinh(2Nz_m)}{2R},\ \ I_D = I_1\tfrac{\sinh(Nh)}{4RN}\{\tfrac{Nh}{2}[1 + \tfrac{\sinh^2(2Nz_m)}{\sinh^2(Nh)}]$$

$$+ 2Nz_m\tfrac{\sinh(2Nz_m)}{\sinh(Nh)} - \tfrac{2NhR^2}{\sinh^2(Nh)} + \tfrac{2R}{\sinh(Nh)}\},$$

$$\sinh(2Nz_m) = \tfrac{\sinh(Nh)\sinh(\Delta)}{\cosh(Nh)-\cosh(\Delta)}.$$

Equations (10.28) and (10.29) were used to simplify the expressions for I_z and I_{T1} in (10.33).

Taking into account that the boundary-value problem described by equations (10.23) is linear with respect to T_1 it is obvious that the system for functions A_1, B_1, C_1, and D_1 that follows from satisfying (10.23) is a system of linear algebraic equations. Using linear independence of functions 1, $z - z_m$, $\tanh[N(z - z_m)]$, $(z - z_m)\tanh[N(z - z_m)]$ for functions A_1, B_1, C_1, and D_1, we have the system which can be represented in the form

$$B_1 = -\tfrac{2}{h}[1 + \tfrac{f_0 F'(f_0)}{F(f_0)}],$$

$$D_1\tfrac{2\lambda N}{\kappa\delta_T H_0} = -\tfrac{f_0 F(f_0)}{\mu^0}R\exp[\delta_T\tfrac{T_{w10}+T_{w20}}{2}]\{A_1$$

(10.34)

$$+ [1 + \tfrac{f_0 F'(f_0)}{F(f_0)}](\tfrac{f_1}{f_0} + \tfrac{2z_m}{h})\},$$

$$A_1 + C_1\tanh[\tfrac{N}{2}(h - 2z_m)] + D_1(\tfrac{h}{2} - z_m)\tanh[\tfrac{N}{2}(h - 2z_m)]$$

$$= -B_1(\tfrac{h}{2} - z_m) + \delta_T T_{w21}$$

(10.35)

$$A_1 - C_1\tanh[\tfrac{N}{2}(h + 2z_m)] + D_1(\tfrac{h}{2} + z_m)\tanh[\tfrac{N}{2}(h + 2z_m)]$$

$$= B_1(\tfrac{h}{2} + z_m) + \delta_T T_{w11}.$$

Using the expression for f_1 from (10.21) and some manipulations of the system (10.35) it can be reduced to solution of three systems of linear equations with the same coefficient matrix which have to be solved for three sets of unknowns $\{A_0,\ C_0,\ D_0\}$, $\{A_+,\ C_+,\ D_+\}$, and $\{A_-,\ C_-,\ D_-\}$ and three sets of the right-hand sides so that functions A_1, C_1, and D_1 can be represented in the form

$$\{A_1,\ C_1,\ D_1\} = \{A_0,\ C_0,\ D_0\} + \{A_+,\ C_+,\ D_+\}\tfrac{\delta_T(T_{w11}+T_{w21})}{2}$$

$$+\{A_-,\ C_-,\ D_-\}\tfrac{\delta_T(T_{w21}-T_{w11})}{2} \tag{10.36}$$

The above systems of linear algebraic equations are as follows

$$A_0 + a_{12}C_0 + a_{13}D_0 = \tfrac{2}{h}[1 + \tfrac{f_0 F'(f_0)}{F(f_0)}](\tfrac{I_z}{I_1} - z_m),$$

$$A_0 + a_{22}C_0 + a_{23}D_0 = -\tfrac{2z_m}{h}[1 + \tfrac{f_0 F'(f_0)}{F(f_0)}], \tag{10.37}$$

$$a_{32}C_0 + a_{33}D_0 = 1 + \tfrac{f_0 F'(f_0)}{F(f_0)},$$

$$A_+ + a_{12}C_+ + a_{13}D_+ = 0,$$

$$A_+ + a_{22}C_+ + a_{23}D_+ = 1, \tag{10.38}$$

$$a_{32}C_+ + a_{33}D_+ = 0,$$

$$A_- + a_{12}C_- + a_{13}D_- = 0,$$

$$A_- + a_{22}C_- + a_{23}D_- = 0, \tag{10.39}$$

$$a_{32}C_- + a_{33}D_- = 1,$$

respectively, where coefficients are calculated according to the formulas

$$a_{12} = [1 + \tfrac{f_0 F'(f_0)}{F(f_0)}]\tfrac{I_C}{I_1},\ a_{13} = [1 + \tfrac{f_0 F'(f_0)}{F(f_0)}]\tfrac{I_D}{I_1},$$

$$a_{22} = -\tfrac{\sinh(2Nz_m)}{2R},\ a_{23} = \tfrac{h}{4R}[\sinh(Nh) + \tfrac{2z_m}{h}\sinh(2Nz_m)], \tag{10.40}$$

$$a_{32} = \tfrac{\sinh(Nh)}{2R},\ a_{33} = -\tfrac{h}{4R}[\sinh(2Nz_m) + \tfrac{2z_m}{h}\sinh(Nh)].$$

Systems (10.37)-(10.40) have the following solutions

$$A_0 = \tfrac{2}{h}[1 + \tfrac{f_0 F'(f_0)}{F(f_0)}](\tfrac{I_z}{I_1} - z_m) - a_{12}C_0 - a_{13}D_0,$$

$$C_0 = \tfrac{1}{D}[1 + \tfrac{f_0 F'(f_0)}{F(f_0)}]\{\tfrac{2}{h}\tfrac{I_z}{I_1}a_{33} - a_{13} + a_{23}\},$$

$$D_0 = \tfrac{1}{D}[1 + \tfrac{f_0 F'(f_0)}{F(f_0)}]\{a_{12} - a_{22} - \tfrac{2}{h}\tfrac{I_z}{I_1}a_{32}\}, \tag{10.41}$$

$$A_+ = -a_{12}C_+ - a_{13}D_+,\ C_+ = -\tfrac{a_{33}}{D},\ D_+ = \tfrac{a_{32}}{D},$$

$$A_- = -a_{12}C_- - a_{13}D_-, \ C_- = \frac{a_{23} - a_{13}}{D}, \ D_- = \frac{a_{12} - a_{22}}{D},$$

$$D = a_{33}(a_{12} - a_{22}) - a_{32}(a_{13} - a_{23}).$$

It can be shown that for the case of $T_{w10} = T_{w20}$ and $T_{w11} = T_{w22} = 0$ we have $T_0(x, -z) = T_0(x, z)$ and $z_m = I_z = I_C = a_{12} = a_{22} = a_{33} = 0$. That leads to $A_0 = D_0 = 0$ and, therefore, $T_1(x, z) = \frac{1}{\delta_T}[1 + \frac{f_0 F'(f_0)}{F(f_0)}]\{\frac{\tanh(Nz)}{\tanh(\frac{Nh}{2})} - \frac{2z}{h}\}$. For a Newtonian fluid $\frac{f_0 F'(f_0)}{F(f_0)} = 1$ and function $T_1(x, z)$ coincides with the one determined for the same conditions by formula (7.26).

Then, based on equations (10.13), (10.20), (10.36), and (10.41), we obtain the expressions for functions $q_{11}(x)$ and $q_{21}(x)$ in the form

$$q_{11}(x) = L_1(x) + P_1(x)T_{w11}(x) + R_1(x)T_{w21}(x),$$

$$q_{21}(x) = L_2(x) + P_2(x)T_{w11}(x) + R_2(x)T_{w21}(x),$$

$$L_1 = \frac{\lambda}{\delta_T}(B_1 + C_0\phi_1 - D_0\theta_1),$$

$$P_1 = \frac{\lambda}{2}[(C_+ - C_-)\phi_1 + (D_- - D_+)\theta_1],$$

$$R_1 = \frac{\lambda}{2}[(C_+ + C_-)\phi_1 - (D_+ + D_-)\theta_1], \ \phi_1 = \frac{N}{\cosh^2[\frac{N}{2}(h + 2z_m)]},$$

$$L_2 = -\frac{\lambda}{\delta_T}(B_1 + C_0\phi_2 + D_0\theta_2),$$

$$P_2 = \frac{\lambda}{2}[(C_- - C_+)\phi_2 + (D_- - D_+)\theta_2],$$

$$R_2 = -\frac{\lambda}{2}[(C_+ + C_-)\phi_2 + (D_+ + D_-)\theta_2], \ \phi_2 = \frac{N}{\cosh^2[\frac{N}{2}(h - 2z_m)]},$$

$$\theta_1 = \tanh[\frac{N}{2}(h + 2z_m)] + \frac{N(h + 2z_m)}{2\cosh^2[\frac{N}{2}(h + 2z_m)]},$$

$$\theta_2 = \tanh[\frac{N}{2}(h - 2z_m)] + \frac{N(h - 2z_m)}{2\cosh^2[\frac{N}{2}(h - 2z_m)]}.$$

$$(10.42)$$

By substituting the expressions for functions $q_{11}(x)$ and $q_{21}(x)$ from formulas (10.42) into equations (10.25) and by introducing new unknown functions

$$\Theta_{w1}(x) = \nu(x)T_{w11}(x), \ \Theta_{w2}(x) = \nu(x)T_{w21}(x), \qquad (10.43)$$

we obtain a system of linear integral equations for functions $\Theta_{w1}(x)$ and $\Theta_{w2}(x)$

$$\Theta_{w1} = \frac{\Lambda}{2H_0}\int_a^c [P_1(\xi)\Theta_{w1}(\xi) + R_1(\xi)\Theta_{w2}(\xi)]e^{-\psi_1(x,\xi)}$$

$$\times K_0(|\ \psi_1(x,\xi)\ |)d\xi + \frac{\Lambda}{2H_0}\int_a^c \nu(\xi)L_1(\xi)e^{-\psi_1(x,\xi)}$$

$$(10.44)$$

$$\times K_0(|\ \psi_1(x,\xi)\ |)d\xi,\ \ \psi_1(x,\xi) = \lambda_1(x-\xi),$$

$$\Theta_{w2} = \frac{\Lambda\eta_0}{2H_0} \int\limits_a^c [P_2(\xi)\Theta_{w1}(\xi) + R_2(\xi)\Theta_{w2}(\xi)]e^{-\psi_2(x,\xi)}$$

$$\times K_0(|\ \psi_2(x,\xi)\ |)d\xi + \frac{\Lambda\eta_0}{2H_0} \int\limits_a^c \nu(\xi)L_2(\xi)e^{-\psi_2(x,\xi)} \tag{10.45}$$

$$\times K_0(|\ \psi_2(x,\xi)\ |)d\xi,\ \ \psi_2(x,\xi) = \lambda_2(x-\xi).$$

System (10.46), (10.45) can be solved numerically for functions Θ_{w1} and Θ_{w2}. Then the two-term asymptotic expansions for lubricant temperatures (see equations (10.19)) in the form

$$T(x,z) = T_0(x,z) + \Theta_1(x,z) + \dots,\ \ \Theta_1(x,z) = \nu(x)T_1(x,z). \tag{10.46}$$

On the next and last Step 3, we consider the approximation for the generalized Reynolds equation. Using the appropriate expansions of functions involved in equation (10.8) (see above), we arrive at the following approximate generalized Reynolds equation

$$\frac{d}{dx}\left\{ \frac{F(f_0)+\nu f_1 F'(f_0)}{\mu^0}I_z + \nu\frac{2}{h}\frac{f_0 F'(f_0)}{\mu^0}I_{z^2} + \nu\delta_T\frac{F(f_0)}{\mu^0}I_{zT1} - h \right\} = 0,$$

$$I_{z^2} = \int\limits_{-h/2}^{h/2} z^2 e^{\delta_T T_0}dz,\ I_{zT1} = \int\limits_{-h/2}^{h/2} zT_1 e^{\delta_T T_0}dz, \tag{10.47}$$

$$p(a) = p(c) = \frac{dp(c)}{dx} = 0.$$

Substituting the expression for f_1 from formula (10.21) into equation (10.47), we obtain

$$\frac{d}{dx}\left\{ \left[\frac{2}{h}\frac{f_0 F'(f_0)}{F(f_0)}(I_{z^2} - \frac{I_z^2}{I_1}) + \delta_T(I_{zT1} - \frac{I_z I_{T1}}{I_1})\right]\frac{s_0}{I_1}\frac{H_0 h}{2f_0}\frac{dp}{dx} + s_0\frac{I_z}{I_1} \right.$$

$$\left. -h \right\} = 0,\ p(a) = p(c) = \frac{dp(c)}{dx} = 0. \tag{10.48}$$

To bring the generalized Reynolds equation (10.48) to its final form we need the expressions for integrals I_z and I_{T1} from (10.33) which have the form

$$I_{z^2} = \int\limits_{-h/2}^{h/2} z^2 e^{\delta_T T_0}dz = \frac{R}{N^3}\exp[\delta_T\frac{T_{w10}+T_{w20}}{2}]\{\frac{N^2 h^2 \sinh(Nh)}{4R}$$

$$\tag{10.49}$$

$$-Nh\ln(R) + 2\int\limits_{-N(h/2+z_m)}^{N(h/2-z_m)} \ln\cosh(y)dy\},$$

$$I_{zT1} = \int\limits_{-h/2}^{h/2} zT_1 e^{\delta_T T_0} dz = \frac{1}{\delta_T}\{A_1 I_z + B_1(I_{z^2} - z_m I_z)$$

$$+C_1(I_D + z_m I_C) + D_1(I_E + z_m I_D)\},$$

$$I_E = \int\limits_{-h/2}^{h/2} (z - z_m)^2 \tanh[N(z - z_m)]e^{\delta_T T_0} dz \tag{10.50}$$

$$= \frac{R}{N^3} \exp[\delta_T \tfrac{T_{w10}+T_{w20}}{2}]\{-\tfrac{N^2}{2R^2}(\tfrac{h^2}{4} + z_m^2)\sinh(Nh)\sinh(2Nz_m)$$

$$-\tfrac{N^2 h z_m}{4R^2}[\sinh^2(Nh) + \sinh^2(2Nz_m)] + N^2 h z_m$$

$$-\tfrac{Nh}{2R}\sinh(2Nz_m) - \tfrac{Nz_m}{R}\sinh(Nh) + \Delta\},$$

where the expressions for I_C and I_D follow from (10.33). In formula (10.49) the integral of function $ln[cosh(y)]$ can be calculated using any high precision numerical method or using a proper series (see formulas (7.28) and [5]).

It is easy to see that the contribution of the second group of terms within brackets which depend on $T_1(x,z)$ is of the same order of magnitude as the terms in the first group in Reynolds equation (10.48). There are a number of approaches published aiming at simplifying the Reynolds equation at hand which used just functions $f_0(x)$ and $T_0(x)$ while completely ignoring functions from the next order of approximation, i.e., functions $f_1(x)$ and $T_1(x,z)$. In effect, such approaches neglect the second group of terms within the brackets in (10.48) by assuming that $f_1(x) = T_1(x,z) = 0$. That alone may significantly change the lubrication film thickness obtained from the truncated compared to full Reynolds equations.

With the help of equations (10.33), (10.36), (10.49), and (10.50) equation (10.48) can be rewritten in the form

$$\frac{d}{dx}\Big\{\Big[\frac{2}{h}\frac{f_0 F'(f_0)}{F(f_0)}(I_{z^2} - \tfrac{I_z^2}{I_1}) + J_0(p,h)\Big]\frac{s_0}{I_1}\frac{H_0 h}{2f_0}\frac{dp}{dx}$$

$$+[J_1(p,h)(\Theta_{w11} + \Theta_{w21}) + J_2(p,h)(\Theta_{w21} - \Theta_{w11})]\frac{s_0}{I_1}$$

$$+s_0\tfrac{I_z}{I_1} - h\Big\} = 0, \quad p(a) = p(c) = \frac{dp(c)}{dx} = 0, \tag{10.51}$$

$$J_0(p,h) = B_1[I_{z^2} - \tfrac{I_z^2}{I_1} - z_m(I_z - I_1)] + C_0(I_D + z_m I_C - \tfrac{I_z}{I_1}I_C)$$

$$+D_0(I_E + z_m I_D - \tfrac{I_z}{I_1}I_D),$$

$$J_1(p,h) = \tfrac{\delta_T}{2}\{C_+(I_D + z_m I_C - \tfrac{I_z}{I_1}I_C)$$

$$+ D_+(I_E + z_m I_D - \tfrac{I_z}{I_1}I_D)\},$$

$$J_2(p,h) = \tfrac{\delta_T}{2}\{C_-(I_D + z_m I_C - \tfrac{I_z}{I_1}I_C)$$

$$+ D_-(I_E + z_m I_D - \tfrac{I_z}{I_1}I_D)\},$$

(10.52)

where Θ_{w11} and Θ_{w21} are the boundary values (10.43) of function $\Theta_1(x,z)$ from (10.46) at $z = \mp\tfrac{h}{2}$, respectively.

Above, we presented a relatively simple and well structured way to determine functions f_0, f_1, T_0, T_1, T_{w10}, T_{w20}, T_{w11}, and T_{w21} assuming that functions $p(x)$ and $h(x)$ are known. Therefore, the original problem described by equations (10.8)-(10.14) is reduced to problem of finding functions $p(x)$ and $h(x)$ from equations (10.48) and (10.11) while function $h(x)$ is determined from equation (10.10).

For heavily loaded lubricated contact this TEHL problem can be analyzed asymptotically based on the developed earlier techniques. This analysis can be performed if

$$\left[\frac{2}{h}\frac{f_0 F'(f_0)}{F(f_0)}(I_{z^2} - \tfrac{I_z^2}{I_1}) + J_0(p,h)\right]\frac{s_0}{I_1}\frac{H_0^2 h}{2f_0}\frac{dp}{dx}$$

$$+[J_1(p,h)(\Theta_{w11} + \Theta_{w21}) + J_2(p,h)(\Theta_{w21} - \Theta_{w11})]\frac{s_0 H_0}{I_1}$$

(10.53)

$$+ s_0 H_0 \tfrac{I_z}{I_1} \ll 1, \; x - a \gg \epsilon_q, \; c - x \gg \epsilon_q, \; \omega \ll 1,$$

where ω is a small parameter involved in the problem while ϵ_q is the characteristic size of the inlet and exit zones of the lubricated contact (see sections on asymptotic analysis of EHL problems). The results of such an asymptotic analysis would be similar to the ones obtained for the corresponding isothermal case of EHL problem for non-Newtonian fluids considered in Section 9.4.

Interestingly enough, in the special case of an isothermal problem we have $\delta_T = 0$ and $T_0 = T_1 = 0$. Therefore, $I_z = I_{T1} = I_{zT1} = 0$ and the approximation of the generalized Reynolds equation given by (10.51), (10.52) is reduced to the isothermal case described by equations (9.97) and (9.98), which become exact in the case of Newtonian lubricant.

The above problem can be solved numerically by employing iterative approaches similar to the ones used in Chapter 8.

The problem analyzed above can be generalized for the cases in which it is necessary to consider normal and tangential surface displacements caused by temperature and tangential stresses. That would require to consider the average speed of the surface $0.5(v_1 + v_2)$ and relative sliding speed $s = v_2 - v_1$ as functions of x. Here velocities v_1 and v_2 can be determined from the formula

$$v_i = u_i\left(1 + \frac{du_{xi}}{dx}\right), \; i = 1, 2,$$

where u_{xi} are the tangential displacements of the lower ($i = 1$) and upper ($i = 2$) surfaces assuming that terms with powers of du_{xi}/dx higher than one can be neglected. A simplifying assumption, which may be appropriate for using formulas from [4, 6] for determining the surface displacements u_{xi} is to replace the actual surface velocities $v_1(x)$ and $v_2(x)$ (which vary from point to point) by their constant rigid counterparts u_1 and u_2, respectively. Most of the problem asymptotic analysis coincides with the one presented above, the analysis of the expressions for $0.5(v_1 + v_2)$, $s = v_2 - v_1$, and $h(x)$ which include tangential stresses and heat fluxes also can be done in a similar manner.

Now, let us make a general conclusion about some properties of the solution of the problem for soft solids considered in Section 18.2, which is related to the one we considered in this section. In a heavily loaded lubricated contact of one hard and another soft elastic solids, we still can expect that outside of the inlet and exit zones the pressure will be close to the Hertzian one $p(x) = \sqrt{1 - x^2}$. Then for high slide-to-roll ratio $s_0 \gg 1$ the surface temperatures $T_{w1}(x)$ and $T_{w2}(x)$ will rise due to heat generation in the contact. Moreover, this rise in $T_{w1}(x)$ and $T_{w2}(x)$ will be very similar to the increase of functions $T_{w10}(x)$ and $T_{w20}(x)$ (as they differ from $T_{w1}(x)$ and $T_{w2}(x)$ by $O(\nu(x)) \ll 1$). This temperature rise will likely produce an addition surface displacement which may result in a creation of a "dimple." Now, the question is may such a "dimple" appearance serve as an explanation of the experimentally observed dimple effect [7] which will be considered in Section 18.2.

Let us consider this question in more detail. As it follows from equation (10.27) for $N(x)$ and the fact that $F(x)$ is an odd function functions $T_{w10}(x)$ and $T_{w20}(x)$ are even functions with respect to s_0. Therefore, with accuracy of $O(\nu(x)) \ll 1$ functions $T_{w1}(x)$ and $T_{w2}(x)$ are independent of the sign of the slide-to-roll ratio $s_0 \gg 1$. That means that for large s_0 in both cases of $s_0 < 0$ and $s_0 > 0$ we either get or do not get a dimple. Therefore, if we try to explain the development of dimples in heavily loaded lubricated contacts of hard and soft materials using just the thermal effects in lubricated contacts as it was done by Kaneta and Young [8, 9] we will predict that in both cases of $s_0 < 0$ and $s_0 > 0$ we either have a dimple of almost the same size or not have a dimple at all. However, according to the experimental data of Kaneta [7] a dimple was observed in the contact region between hard and soft materials for $s_0 = 2$ and it was not observed for the same materials for $s_0 = -2$. Therefore, for high slide-to-roll ratio s_0 the attempt to explain the development of a dimple for $s_0 = 2$ and its absence for $s_0 = -2$ in a heavily loaded lubricated contact of hard and soft materials by just considering normal thermal displacement of surfaces fails. It means that to explain the dimple phenomenon it is necessary to consider the normal and tangential surface displacements. The tangential displacements cause variations in surface speeds. The influence of the thermal effects in heavily loaded lubricated contacts with $s_0 \gg 1$ would just make the the surface normal displacements larger, i.e., thermal effects would make the development of dimples more pronounced. For more detail on the "dimple" phenomena see Chapter 18.

10.4 Numerical Analysis of the Heat Transfer in the Contact Solids

Based on the solutions and data presented in Chapters 7 and 9 for thermal and isothermal EHL problems for non-Newtonian lubricant rheology it is clear what would be the behavior of most of the contact characteristics (such as H_0, c, $p(x)$, $h(x)$, $T(x,z)$, etc.) if the surfaces temperatures $T_{w1}(x)$ and $T_{w2}(x)$ are given. Therefore, in this section we will focus on the analysis of the behavior of surface temperatures $T_{w1}(x)$ and $T_{w2}(x)$. Being concerned only with surface temperatures, we will not determine the actual pressure $p(x)$ and gap $h(x)$ behavior but replace them by the Hertzian ones $p(x) = \sqrt{1 - x^2}$ and $h(x) = 1$ which provide good approximations outside of the inlet and exit zones of heavily loaded lubricated contacts. This choice of $p(x)$ limits our analysis because functions Θ_{w11} and Θ_{w21} are proportional to $\frac{dp}{dx}$ (see (10.43)) which approaches infinity as $x \to \pm 1$. It means that proper analysis of Θ_{w11} and Θ_{w21} requires the actual distributions of $p(x)$ and $h(x)$ realized in a lubricated contact while under the above approximations this analysis would be not reliable. However, these assumptions still allow for adequate analysis of surface temperatures based on the behavior of T_{w10} and T_{w20} (see expansions (10.19)).

For simplicity, let us consider the case of a Newtonian lubricant. Then the rheological function $F(x) = \frac{12H_0}{V}x$ and $\frac{f_0 F'(f_0)}{F(f_0)} = 1$. Therefore, the solution of equations (10.27)-(10.29) is

$$N = \frac{2}{h}\ln(\eta + \sqrt{\eta^2 + 1}),$$

$$\eta = \{\sinh^2(\tfrac{\Delta}{2}) + \tfrac{\kappa \delta_T V s_0^2}{96\lambda}\mu^0 e^{-\delta_T \frac{T_{w10} + T_{w20}}{2}}\}^{1/2}, \quad \Delta = \delta_T \frac{T_{w20} - T_{w10}}{2}, \tag{10.54}$$

while the main term of the sliding frictional stress is

$$f_0 = \frac{V}{12H_0} \frac{\mu^0 s N}{\sinh Nh} e^{-\delta_T \frac{T_{w10} + T_{w20}}{2}}. \tag{10.55}$$

For $\Delta = 0$, $s = s_0$, $\mu^0 = \mu$, and $T_{w10} = T_{w20} = T_{w0}$ function N coincides with the expression for B obtained independently for the case of Newtonian fluid in Chapter 7 if you take into account that $\kappa_N = \frac{\kappa V}{12}$, which follows from the definitions of parameters κ_N in Chapter 7 (see formulas (7.5)) and κ in this section (see formulas (10.6)). Moreover, for $\delta_T \to 0$ we have $N \to 0$ and the expression for f_0 approaches to the isothermal one $\frac{\mu^0 s_0}{h}$.

To obtain solutions $T_{w10}(x)$ and $T_{w20}(x)$ of the system of nonlinear equations (10.31), we will use the following iteration process

$$T_{w10}^{n+1} = \frac{\Lambda}{H_0} \int_a^c \frac{\lambda N^{n+1,n}}{\delta_T} \tanh[N^{n+1,n}(\tfrac{h}{2} + z_m^{n+1,n})]e^{-\lambda_1(x-\xi)}$$

$$\times K_0(\mid \lambda_1(x - \xi) \mid)d\xi - \Delta_\infty, \tag{10.56}$$

$$T_{w20}^{n+1} = \frac{\Lambda \eta_0}{H_0} \int_a^c \frac{\lambda N^{n+1,n+1}}{\delta_T} \tanh[N^{n+1,n+1}(\tfrac{h}{2} - z_m^{n+1,n+1})]$$

$$\text{(10.57)}$$

$$\times e^{-\lambda_2(x-\xi)} K_0(|\,\lambda_2(x-\xi)\,|)d\xi + \Delta_\infty,$$

where $N^{n+1,n}$ and $z_m^{n+1,n}$ are calculated based on T_{w10}^{n+1} and T_{w20}^n while $N^{n+1,n+1}$ and $z_m^{n+1,n+1}$ are determined based on previously determined T_{w10}^{n+1} and still unknown T_{w20}^{n+1} (n is the iteration index). The system is solved using Newton's method sequentially, i.e., the iteration cycle is based on solution of decoupled equations and it starts with determining one iteration of equation (10.56) and it continues with determining one iteration of equation (10.58). After that the calculation cycle is repeated until the solution converges to the desired precision. It is assumed here that $a,\, ,\, H_0,\, \delta_T,\, \lambda,\, \lambda_1,\, \lambda_2,\, \Lambda$, and $p(x),\, h(x)$ are known. As the initial approximations for these iterations can be taken functions $T_{w10}(x) = -\Delta_\infty$, $T_{w20}(x) = \Delta_\infty$ of the closest available solution of the problem.

System (10.31) also can be solved by application of Newton's method simultaneously to both equations. However, this approach will require handling solution of algebraic systems of double order compared to the previous approach. This approach might be more appropriate for large pressure viscosity coefficients Q in Barrus viscosity $\mu^0(p) = e^{Qp}$ relationship.

Solution of (10.31) involves the following integral approximations

$$\int_a^c g(\xi)e^{-\lambda_1(x_k-\xi)}K_0(|\,\lambda_1(x_k-\xi)\,|)d\xi$$

$$\text{(10.58)}$$

$$\approx \sum_{i=1}^{N-1} \frac{g(\xi_i)+g(\xi_{i+1})}{2}e^{-\lambda_2(x_k-\xi_{i+1/2})}K_0(|\,\lambda_2(x_k-\xi_{i+1/2})\,|)\Delta x,$$

where $\{x_k\}$ ($k = 1, \ldots, N_x$) is the set of integer nodes, $x_{k+1} = x_k + \Delta x$, $\{x_{i+1/2}\}$ ($i = 1, \ldots, N_x - 1$) is the set of midpoints $x_{i+1/2} = \frac{x_i + x_{i+1}}{2}$, Δx is the step size.

After the sets of $T_{w10}(x_k)$ and $T_{w20}(x_k)$ ($k = 1, \ldots, N_x$) are determined the surface temperatures outside of the lubricated contact (i.e., for $x < a$ and $x > c$) can be obtained from equations (10.31) by calculating the integrals of the already determined functions.

To illustrate the behavior of the surface temperatures, let us consider some examples. Let us assume that $a = -1$, $c = 1$, $V = 0.15$, $s_0 = 2$, $H_0 = 0.075$, $\delta_T = 1$, $\frac{\Lambda}{H_0} = 13.333$, $p(x) = \sqrt{1 - x^2}$, and $h(x) = 1$. For simplicity we will assume that λ and δ_T are constants and $\lambda = 1$. We will consider some numerical results for different combinations of parameters κ, Q, λ_1, λ_2, Δ_∞, and η_0. As Λ increases (H_0 decreases) the surface temperatures $T_{w10}(x)$ and $T_{w20}(x)$ increase as long as $|\,z_m\,| \le \frac{h}{2}$. That follows from equations (10.31) and the fact that $\tanh[N(\frac{h}{2} \pm z_m)] \ge 0$ for all $|\,x\,| \le 1$. As the value of $\frac{\kappa \delta_T V s_0^2}{96\lambda}$ increases the amount of heat generated in the contact increases which causes surface temperatures $T_{w10}(x)$ and $T_{w20}(x)$ to increase (see Fig. 10.1).

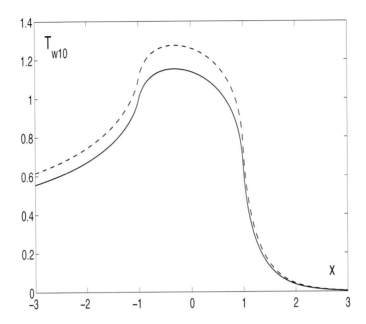

FIGURE 10.1

Dependence of the surface temperatures $T_{w10}(x) = T_{w20}(x)$ on $\frac{\kappa \delta_T V s_0^2}{96\lambda}$ for $\kappa = 100$ (solid curve) and $\kappa = 200$ (dashed curve), where $Q = 1.2$, $\lambda_1 = \lambda_2 = 1$, $\Delta_\infty = 0$, and $\eta_0 = 1$. (after Kudish and Covitch [10]). Reprinted with permission from CRC Press.

The numerical results show that if the values of λ_1 and λ_2 are close to each other and Δ_∞ is close to 0 while η_0 is not very far from 1 then the graphs of surface temperatures $T_{w10}(x)$ and $T_{w20}(x)$ are pretty close to each other (see Fig. 10.2). However, the heat fluxes in the contact surfaces $q_{10}(x)$ and $q_{20}(x)$ may differ significantly in these cases. Two examples of the heat fluxes $q_{10}(x)$ and $q_{20}(x)$ and sliding frictional stress f_0 behavior are given in Fig. 10.3 and 10.4. In cases when λ_1 is significantly different from λ_2, Δ_∞ is different from 0 (i.e., surface temperatures of the solids are different far away from the contact) while η_0 is significantly different from 1 we can expect the surface temperatures $T_{w10}(x)$ and $T_{w20}(x)$ of the solids to be different. Some examples illustrating this behavior are shown in Fig. 10.2.

A very interesting dynamics of heat and temperature distributions can be seen in Fig. 10.5 - 10.7 for the case of different surface temperatures far away from the contact (i.e., for $\Delta_\infty = 1$). Obviously, the level of surface temperature rises much more for the originally (far from the contact) cooler surface. In this

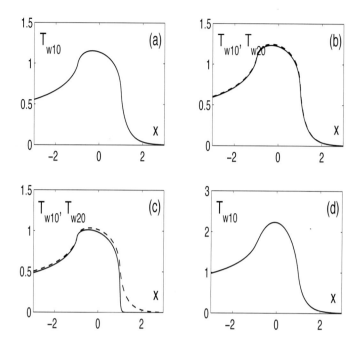

FIGURE 10.2
Dependence of the surface temperatures $T_{w10}(x)$ (solid curves) and $T_{w20}(x)$ (dashed curves) on parameters Q, λ_1, λ_2, and η_0: $Q = 1.2$, $\lambda_1 = \lambda_2 = 1$, $\eta_0 = 1$ - case (a), $Q = 1.2$, $\lambda_1 = \lambda_2 = 1$, $\eta_0 = 5$ - case (b), $Q = 1.2$, $\lambda_1 = 10$, $\lambda_2 = 1$, $\eta_0 = 1$ - case (c), and $Q = 7.5$, $\lambda_1 = \lambda_2 = 1$, $\eta_0 = 1$ - case (d), where $\Delta_\infty = 0$ and $\kappa = 100$. (after Kudish and Covitch [10]). Reprinted with permission from CRC Press.

case the heat fluxes $q_{10}(x)$ and $q_{20}(x)$ only in the center of the contact region directed in both solid surfaces (Fig. 10.6) while at the ends of the contact the heat flows from one contact surface into another (i.e., $q_{10}(x)$ and $q_{20}(x)$ have opposite signs). That finds its reflection in the behavior of functions $N(x)$ and $z_m(x)$ (see Fig. 10.7). It can be seen in the behavior of function $z_m(x)$, which represents the position of the vertex of the lubricant temperature $T(x, z)$ distribution with respect to z. In cases when $| z_m |\geq \frac{h}{2} = \frac{1}{2}$ the distribution of temperature $T(x, z)$ between the contact surfaces is monotonic and, therefore, the heat flux is directed from one contact surface to another. That, obviously, is represented by the behavior of heat fluxes $q_{10}(x)$ and $q_{20}(x)$ as well as by surface temperatures $T_{w10}(x)$ and $T_{w20}(x)$ (see Fig. 10.5 and 10.6).

If λ_1 is greater than λ_2 (i.e., the lower surface moves faster than the upper

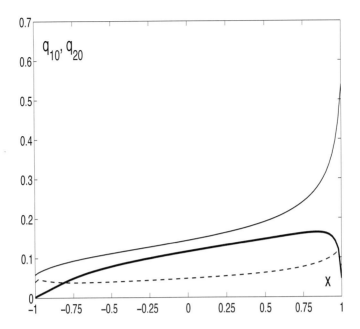

FIGURE 10.3
Dependence of the heat fluxes $q_{10}(x)$ and $q_{20}(x)$ at the contact surfaces obtained for $Q = 1.2$, $\lambda_1 = 10$, $\lambda_2 = 1$ (solid and dashed line, respectively) and for $Q = 7.5$, $\lambda_1 = \lambda_2 = 1$ for which $q_{10}(x) = q_{20}(x)$ (double solid line), where $\Delta_\infty = 0$, $\eta_0 = 1$, and $\kappa = 100$. (after Kudish and Covitch [10]). Reprinted with permission from CRC Press.

one), then the temperature $T_{w10}(x)$ of the lower surface rises slower and to a slightly lower lever than the temperature $T_{w20}(x)$ of the upper surface due to a lesser exposure to heat generated in the contact. Moreover, for faster–moving surface (assuming that $\lambda_1 > 0$ and $\lambda_2 > 0$, i.e., both surfaces are moving from left to right) the effect of heat generation in the contact extends to a lesser degree ahead of the contact, i.e., for $x > 1$ (see Fig. 10.2, case (c)).

If, for example, the lower surface has larger product $k_1\rho_1 c_1$, then the upper surface (i.e., $\eta_0 > 1$) then temperature $T_{w20}(x)$ rises in the contact a little higher than $T_{w10}(x)$ (see Fig. 10.2, case (b)). This is due to slower heat dissipation in the upper solid.

For $\lambda_1 > 0$ and $\lambda_2 > 0$, both solids are moving in the positive direction (from left to right). For small values of the pressure viscosity coefficient Q the maximum of the surface temperatures $T_{w10}(x)$ and $T_{w20}(x)$ occurs to the left of $x = 0$ (i.e., in a lubricated contact the surface temperature maximum

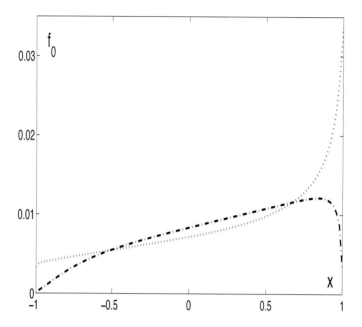

FIGURE 10.4
Two examples of the distribution of the sliding frictional stress $f_0(x)$ obtained for $Q = 1.2$, $\lambda_1 = 10$ (dotted curve) and $Q = 7.5$, $\lambda_1 = 1$ (dashed curve), where $\lambda_2 = 1$, $\Delta_\infty = 0$, $\eta_0 = 1$, and $\kappa = 100$. (after Kudish and Covitch [10]). Reprinted with permission from CRC Press.

is shifted from $x = 0$ toward the contact inlet). As the pressure viscosity coefficient Q increases the temperature distributions lean forward and the temperature maximum point approaches $x = 0$ (see Fig. 10.2, cases (a)-(d)). At the same time the level of the surface temperatures rises due to the rise of lubricant viscosity μ^0 and, therefore, increased heat generation in the contact.

As the value of δ_T increases the surface temperatures in the contact increase to lower levels than for smaller δ_T. This is due to fast decrease of lubricant viscosity $\mu = \mu^0 e^{-\delta_T T}$ and heat generation in the contact with increase of lubricant temperature T.

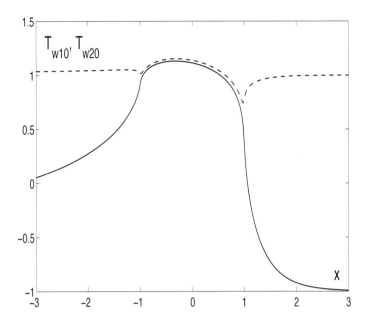

FIGURE 10.5
Distribution of surface temperatures $T_{w10}(x)$ (solid curve) and $T_{w20}(x)$ (dashed curve) obtained for $Q = 1.2$, $\lambda_1 = \lambda_2 = 1$, $\Delta_\infty = 1$, and $\eta_0 = 1$. (after Kudish and Covitch [10]). Reprinted with permission from CRC Press.

10.5 Regularization for Isothermal Heavily Loaded Line EHL Contacts with Non-Newtonian Fluids

There are two major wide classes of non-Newtonian fluids: pseudo-plastic and dilatant ones. Numerical solution of heavily loaded isothermal EHL problems for pseudo-plastic lubricant fluids is less and for dilatant lubricant fluids is more prone to solution instability compared to the case of Newtonian fluids. Therefore, in many cases there is still need for regularization of such isothermal numerical solutions. In regularization of heavily loaded isothermal EHL contacts with non-Newtonian fluid, we will follow the same idea as in the case of a Newtonian fluid. However, in this case we need to use the main term of the asymptotic solution for temperature T_0 obtained for a fluid with a particular non-Newtonian rheology. It is easy to do by using the special case of the solution from preceding section. Namely, assuming that the dimensionless

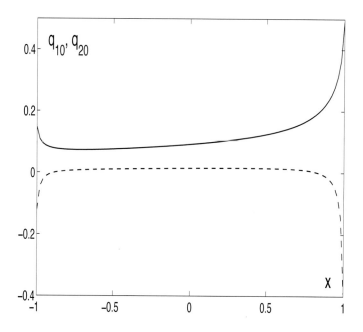

FIGURE 10.6

Distribution of heat fluxes $q_{10}(x)$ (solid curve) and $q_{20}(x)$ (dashed curve) at contact surfaces obtained for $Q = 1.2$, $\lambda_1 = \lambda_2 = 1$, $\Delta_\infty = 1$, and $\eta_0 = 1$. (after Kudish and Covitch [10]). Reprinted with permission from CRC Press.

$T_{w1} = T_{w2} = 0$ we obtain $z_m = I_z = 0$ (see equations (10.26)-(10.32))

$$T_0(x, z) = \frac{1}{\delta_T} \ln \frac{R}{\cosh^2(Nz)}, \quad R = \cosh^2(\tfrac{Nh}{2}), \tag{10.59}$$

where values of N are determined from solution of the equation

$$\frac{4\lambda}{\kappa \delta_T H_0 s_0} N \tanh(\tfrac{Nh}{2}) = \Phi(\tfrac{\mu^0 s_0}{h} \tfrac{Nh}{\sinh(Nh)}). \tag{10.60}$$

Similarly, from the above conditions and formulas (10.32)-(10.35) follows that

$$T_1(x, z) = \frac{1}{\delta_T}\Big[1 + \frac{f_0 F'(f_0)}{F(f_0)}\Big]\Big\{\frac{\tanh(Nz)}{\tanh(\frac{Nh}{2})} - \frac{2z}{h}\Big\},$$

$$\tag{10.61}$$

$$f_0 = \Phi(\tfrac{\mu^0 s_0}{h} \tfrac{Nh}{\sinh(Nh)}).$$

Besides that we obtain $f_1 = I_{T1} = 0$. Therefore, the Reynolds equation is reduced to the equation (see (10.48))

$$\frac{d}{dx}\Big\{\Big[\frac{2}{h}\frac{f_0 F'(f_0)}{F(f_0)}I_{z^2} + \delta_T I_{zT1}\Big]\frac{Ns_0}{\sinh(Nh)}\frac{H_0 h}{2f_0}\frac{dp}{dx} - h\Big\} = 0. \tag{10.62}$$

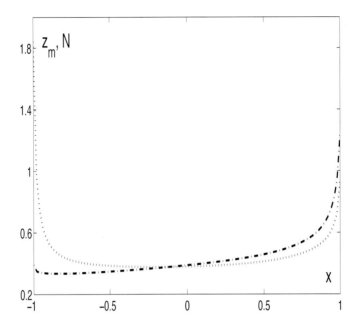

FIGURE 10.7
Distribution of functions $z_m(x)$ (dotted curve) and $N(x)$ (dash-dotted curve) obtained for $Q = 1.2$, $\lambda_1 = \lambda_2 = 1$, $\Delta_\infty = 1$, and $\eta_0 = 1$. (after Kudish and Covitch [10]). Reprinted with permission from CRC Press.

To finalize the generalized Reynolds equation (10.62), we need the particular form of expressions for integrals (see (10.49) and (10.50))

$$I_{z^2} = \frac{\sinh(Nh)}{N^3} \left\{ \frac{N^2 h^2}{4} - \frac{1}{\tanh(\frac{Nh}{2})} \left[Nh \ln \cosh(\frac{Nh}{2}) \right. \right.$$

$$\left. \left. - 2 \int_0^{Nh/2} \ln \cosh(y) dy \right] \right\}, \tag{10.63}$$

$$I_{zT1} = \frac{1}{\delta_T} \left[1 + \frac{f_0 F'(f_0)}{F(f_0)} \right] \left\{ -\frac{2}{h} I_{z^2} + \frac{I_D}{\tanh(\frac{Nh}{2})} \right\},$$

$$I_D = \frac{\cosh^2(\frac{Nh}{2})}{N^2} \left[\frac{Nh}{2} \tanh^2(\frac{Nh}{2}) - \frac{Nh}{2} + \tanh(\frac{Nh}{2}) \right], \tag{10.64}$$

where the expression for I_D follows from (10.33).

Finally, recalling the Reynolds equation (9.97) for the isothermal EHL problem with function M from (9.98), we can conclude that the regularized problem is reduced to solution of the following equations

$$\frac{d}{dx}\{M(\mu,p,h,\frac{dp}{dx},V,s_0,H_0) - H_0 h\} = 0, \quad p(a) = p(c) = \frac{dp(c)}{dx} = 0, \quad (10.65)$$

$$M = \frac{H_0^2 h^3 F'(f_0)}{12\mu}\frac{dp}{dx}R_T, \tag{10.66}$$

$$R_T = \frac{6}{h^2}[\frac{2}{h}I_{z^2} + \frac{F(f_0)}{f_0 F'(f_0)}\delta_T I_z T1], \quad f_0 = \Phi(\frac{\mu s_0}{h}\frac{Nh}{\sinh(Nh)}),$$

$$H_0(h-1) = x^2 - c^2 + \frac{2}{\pi}\int\limits_a^c p(t)\ln\mid\frac{c-t}{x-t}\mid dt, \tag{10.67}$$

$$\int\limits_a^c p(t)dt = \frac{\pi}{2}, \tag{10.68}$$

where I_{z^2} and $\delta_T I_z T1$ are determined according to (10.63) and (10.64) while function N is obtained by solving equation

$$\frac{V s_0}{24\beta^2 H_0}N\tanh(\frac{Nh}{2}) = \Phi(\frac{\mu s_0}{h}\frac{Nh}{\sinh(Nh)}), \tag{10.69}$$

where β is a fictitious sufficiently small positive constant. This parameter β is chosen in such a way that the numerical solution is stable and, at the same time, the regularized solution is sufficiently close to the non-regularized one. In equations (10.65)-(10.69) it is assumed that $\mu^0 = \mu$, where μ is the isothermal lubricant viscosity at the ambient temperature. To construct a regularized numerical algorithm equations (10.65)-(10.69) have to be resolved for $p(x)$ by inverting equation (10.67) and one has to follow the procedure of numerical calculations outlined in Chapter 8.

For the case of a Newtonian fluid, equations (10.65)-(10.69) coincide with the regularized problem proposed in Chapter 8.

10.6　Friction in Heavily Loaded Lubricated Contacts

After the solution for a heavily loaded EHL contact is obtained, we can determine friction forces at the contact surfaces from the formulas

$$F_T = F_S \pm F_R, \quad F_S = \int\limits_a^c f(x)dx, \quad F_R = \frac{H_0}{2}\int\limits_a^c h(x)\frac{dp(x)}{dx}dx, \tag{10.70}$$

where F_S and F_R are the sliding and rolling friction forces, respectively, and $f(x)$ is the sliding frictional stress determined from the problem solution. Formulas for $f(x)$ for various cases can be found in the preceding sections.

For sufficiently hard elastic materials for both pre- and over-critical lubrication regimes in the cental region of the contact the pressure is close to the

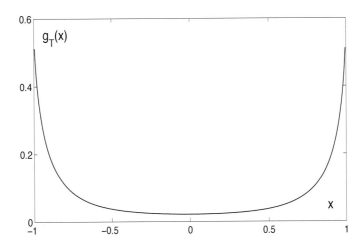

FIGURE 10.8

A typical functional dependence of $g_T(\eta)$ on η in the Hertzian region. (after Kudish and Covitch [10]). Reprinted with permission from CRC Press.

Hertzian pressure $\sqrt{1-x^2}$ and the gap is practically equal to 1 while outside of the Hertzian region pressure is small and gap is greater than 1 in the inlet zone and it is about 1 in the exit zone. Because of that in these cases calculation of the sliding frictional stress $f(x)$ and sliding friction force F_S pressure $p(x)$ and gap $h(x)$ can be replaced by $\sqrt{1-x^2}$ and 1, respectively. Moreover, due to the symmetry of the Hertzian pressure about $x = 0$ and because of the proximity of $h(x)$ to 1 in the central region of the contact, the main contributions to the rolling friction stress F_R come from the inlet and exit zones.

For example, for Newtonian fluid in an isothermal contact the sliding frictional stress $f(x)$ is given by formula

$$f(x) = \frac{V}{12H_0}\frac{\mu s_0}{h}. \qquad (10.71)$$

Therefore, for $\mu(p) = \exp(Qp)$ and $Q \gg 1$ using Watson Lemma (see [11]) for the dimensionless sliding friction force, we obtain an analytical formula

$$F_S = \int\limits_a^c f(x)dx \approx \frac{V s_0}{12H_0}\sqrt{\frac{2\pi}{Q}}e^Q + \dots \qquad (10.72)$$

Similar, analytical formulas for F_S can be obtained for other lubricant rheologies.

In case of a Newtonian fluid involved in a thermal EHL contact with prevailing sliding frictional stress (i.e., $\nu(x) \ll 1$), the dimensionless viscosity described by $\mu(p, T) = \mu^0(p) \exp(-\delta_T T)$, and equal surface temperatures $T_{w1} = T_{w2} = T_{w0}$ the sliding frictional stress $f(x)$ is given by the approximate formulas (see Chapter 7)

$$f(x) \approx \frac{V s_0}{12 H_0} \frac{\mu^0 e^{-\delta_T T_{w0}}}{h} g_T(\eta), \ g_T(\eta) = \frac{4(\eta + \sqrt{\eta^2 + 1})^2 \ln(\eta + \sqrt{\eta^2 + 1})}{(\eta + \sqrt{\eta^2 + 1})^4 - 1}, \quad (10.73)$$

$$\eta = \frac{|s_0|}{2} \sqrt{\frac{\delta_T \kappa}{2\lambda} \mu^0 e^{-\delta_T T_{w0}}}.$$

It is important to recognize the fact that the difference in formulas (10.71) and (10.73) for the sliding frictional stress $f(x)$ is in just the multiplier $e^{-\delta_T T_{w0}} g_T(\eta)$. It can be shown that $g_T(\eta) \to 1$ as $\eta \to 0$. Therefore, for $T_{w0} = 0$ and $\eta \to 0$ the expression for the sliding frictional stress $f(x)$ in a non-isothermal lubricated contact converges to the one in an isothermal contact (see formulas (10.71) and (10.73)). If $\mu^0(p) = \exp(Qp)$, then for sufficiently large values of Q parameter $\eta \gg 1$, which leads to the estimate $g_T(\eta) = \ln(2\eta)/\eta^2$. In turn, in this case it means that the sliding frictional stress $f(x)$ and the sliding friction force F_S are approximately equal to

$$f(x) \approx \frac{VQ}{3 H_0} \frac{\lambda}{\delta_T \kappa s_0} \frac{p}{h}, \ F_S \approx \frac{\pi VQ}{6 H_0} \frac{\lambda}{\delta_T \kappa s_0}. \quad (10.74)$$

In the expression for F_S it is assumed that λ/δ_T is a constant. For $T_{w0}(x) = 0$ a typical graph of $g_T(\eta)$ in the Hertzian region $[-1, 1]$ calculated based on the Hertzian pressure $p = \sqrt{1 - x^2}$ is shown in Fig. 10.8. Obviously, for large values of η the sliding frictional stress $f(x)$ in the most of the Hertzian region and friction force F_S are significantly smaller than the corresponding values in the case of isothermal lubrication.

Formulas (10.73) and (10.74) can be easily generalized for the case of a Newtonian fluid and different surface temperatures $T_{w1} \neq T_{w2}$ (see Section 10.4 for specific formulas).

In case of pure rolling the slide-to-roll ratio $s_0 = 0$ and $f(x) = 0$. Therefore, under pure rolling conditions $F_S = 0$ and the frictional stresses $\pm \frac{H_0 h}{2} \frac{dp(x)}{dx}$ and friction force F_R are completely determined by the solution behavior in the inlet and exit zones. Usually, under pure rolling conditions friction is low.

Under most lubrication conditions with mixed rolling and sliding, the sliding frictional stress $f(x)$ dominates the rolling one $\pm \frac{H_0 h}{2} \frac{dp(x)}{dx}$, which is equivalent to the case of $\nu(x) \ll 1$ (see preceding sections). Under such conditions the friction force F_T is practically completely represented by its sliding part F_S. For the known lubrication film thickness H_0, the sliding friction force F_S can be determined based on the Hertzian pressure $\sqrt{1 - x^2}$ and, if necessary, the lubricant temperature $T(x, z)$, which, in turn, can be calculated based on the Hertzian pressure. In other words, under such condition the knowledge of the solution behavior in the inlet and exit zones of a lubricated contact is needed only for determining the film thickness H_0. For a particular non-Newtonian lubricant rheology, the specific formulas for the frictional sliding stress $f(x)$ and sliding friction force F_S can be obtained in a similar fashion based on the results of preceding sections.

10.7   Closure

Some modern thermal EHL problems for lubricants with non-Newtonian rheology were analyzed. The analysis was done using regular and matched asymptotic expansions as well as numerical methods. An approximate Reynolds equation which takes into account heat generated in lubricated contact was derived. The numerical method for solution of the problem for heat dissipation in contact solids was proposed and realized. Several numerical solutions for contact surface temperatures resulted from heat produced in the lubricated contact and transferred to contact surfaces were presented. It was shown that the behavior of surface temperatures in lubricated contacts significantly depends on the lubricated contact parameters and surface temperatures far away from the lubricated contact.

10.8   Exercises and Problems

1. For a non-Newtonian lubricant with constant viscosity, derive formulas (9.122) and (9.123) for film thickness H_0 for starved and fully flooded lubrication regimes.

2. For a non-Newtonian lubricant, provide a detailed derivation of formula (9.134) and (9.135) for the film thickness H_0 under over-critical lubrication regimes.

3. What is the physical meaning of the expression for $\nu(x)$ from formulas (7.17) and (9.94)?

4. Why is it necessary for proper approximation of the Reynolds equation to determine the two-term asymptotic solution for the lubricant temperature T and why is it not sufficient to use just the main term for T? (Hint: See the expression for function M from (7.27).)

5. For starved lubrication regimes for power law fluid derive the formulas of transformation similar to (6.215).

References

[1] Carslaw, H.S. and Jaeger, J.C. 1959. *Conduction of Heat in Solids.* 2nd ed. Oxford: Clarendon Press.

[2] *Handbook on Mathematical Functions with Formulas, Graphs and Mathematical Tables*, Eds. Abramowitz, M. and Stegun, I.A., National Bureau of Standards, 55, 1964.

[3] Hamrock, B.J. 1994. *Fundamentals of Fluid Film Lubrication.* New York: McGraw-Hill.

[4] Barber, J. 1984. Thermoelastic Displacements and Stresses due to a Heat Source Moving Over the Surface of a Half-Plane. *ASME J. Appl. Mech.* 51:636-640.

[5] Kudish, I.I. 1978. Asymptotic Analysis of a Plane Non-isothermal Elastohydro-dynamic Problem for a Heavily Loaded Rolling Contact. *Proc. Acad. Sci. of Armenia SSR, Mechanics* 31, No. 6:16-35.

[6] Galin, L.A. 1980. *Contact Problems in the Theory of Elasticity and Viscoelasticity.* Moscow: Nauka.

[7] Kaneta M., Nishikawa H., Kanada T., and Matsuda K. 1996. Abnormal Phenomena Appearing in EHL Contacts. *ASME J. Tribology.* 118:886-892.

[8] Wang, J., Yang, P., Kaneta, M., and Nishikawa, H. 2003. On the Surface Dimple Phenomena in Elliptical TEHL Contacts with Arbitrary Entrainment. *ASME J. Tribology* 125, No. 1:102-109.

[9] Kaneta, M., Shigeta, T., and Young, P. 2006. Film Pressure Distributions in Point Contacts Predicted by Thermal EHL Analysis. *Tribology Intern.* 39:812-819.

[10] Kudish, I.I. and Covitch, M.J. 2010. *Modeling and Analytical Methods in Tribology.* Boca Raton: CRC Press.

[11] De Bruijn, N.G. 1958. *Asymptotic Methods in Analysis.* Groningen: North Holland Publishing Co.

Part VI

Stress-Induced Lubricant Degradation in Line EHL Contacts

This part of the monograph is devoted to investigating the effect of stress-induced lubricant degradation of the parameters of lubricated contacts. It is assumed that the lubricant is a Newtonian or non-Newtonian fluid. The stress-induced lubricant degradation is the process of breaking polymer molecules dissolved in lubricant which serve as additives to oil base stock and are called viscosity modifiers (VM). There are several different structures of such polymer molecules: polymer molecules with linear structure, star polymers, etc. We will consider here the simplest case of the effect of degradation of polymer molecules with linear structure on the parameters of line EHL contacts. The specific models of linear and star polymer molecule degradation are based on solution certain kinetic equations describing variations in polymer molecule density distribution along lubricant flow streamlines which are presented in detail in [1]. Here we will use the kinetic model for degradation of polymer molecules with linear structure to analyze the influence of this process on the behavior of lubrication film thickness, lubricant viscosity, friction stresses, etc. Also, we will study the effect of lubricant degradation on contact fatigue of materials involved in a lubricated contact. We will use the model of contact fatigue developed and described in [1]. The analysis of the effects of lubricant degradation will be done numerically. The numerical procedure used will be described in detail.

11

Elastohydrodynamic Lubrication by Formulated Lubricants That Undergo Stress-Induced Degradation

---

11.1   Introduction

Modern lubricating oils are formulated with a variety of additives designed to (1) provide beneficial rheological characteristics to lubricants, (2) to stabilize their physical and chemical properties, and (3) to protect lubricated equipment against wear, fatigue, and corrosion. Under the influence of chemical and mechanical stresses and elevated temperatures lubricants tend to undergo certain reversible and irreversible changes. The reversible changes are caused by temporary alignment of polymeric additives in the direction of flow, resulting in an apparent drop in viscosity. When the liquid returns to a state of rest, the viscosity returns to its initial value. This is known as non-Newtonian rheology. The irreversible changes are due to a number of ongoing processes such as stress-induced scission of polymeric additives, oxidation, contamination, etc. The latter detrimental processes limit the useful life of lubricants and can lead to costly repairs and down time if a lubricated system is not properly maintained. In this chapter we focus on the combined effects of the lubricants' non-Newtonian rheology and stress-induced polymer molecule scission and on changes in lubricant contact parameters.

It is established experimentally that many lubricants exhibit some reversible non-Newtonian rheological properties (see, for example, Bair and Winer [2] and Hoglund and Jacobson [3]). Some rheological models of lubricant behavior are given in Bair and Winer [2] and in Eyring [4]. Also, there exist a number of theoretical studies of the effect of non-Newtonian behavior on various parameters of lubricated contacts such as lubrication film thickness, frictional stress, etc. Examples include studies by Houpert and Hamrock [5] and Kudish [6].

Several experimental studies of non-reversible stress-induced degradation of lubricants are presented in [7, 9] - [12] while the theoretical studies of lubricant degradation with few exceptions such as [13, 14] dealt only with the processes of homogeneous degradation caused by temperature [15] - [18] and radiation [19] effects. A semi-deterministic attempt of a theoretical study of stress-induced degradation was made by Kudish and Ben-Amotz in [20]. In later papers [21] - [23] and in monograph [1] a kinetic probabilistic approach to stress-induced degradation was developed. The approach is based on the derivation and usage of a probabilistic kinetic equation(s) describing the process of stress-induced polymer scission. These kinetic equations have been successfully applied to practical cases. In particular, the numerically simulated data are in excellent agreement with experimental data as it is shown in [1].

In this chapter we consider the application of the developed kinetics approach to the phenomenon of elastohydrodynamic lubrication of surfaces lubricated with non-Newtonian fluids that undergo some changes caused by lubricant degradation. The problem is reduced to a coupled system of the generalized Reynolds equation for non-Newtonian lubricant flow, the equation for the gap between the surfaces of the elastic solids, the equations for the lubricant flow streamlines, the kinetic equation describing the changes in the polymer molecular weight distribution, and the equations for the lubricant viscosity. The generalized isothermal EHL equations are coupled with the kinetic equation through the lubricant viscosity, which depends not only on lubricant pressure but also on the concentration of polymer molecules and the distribution of their chain lengths. The solution of the problem is obtained using numerical methods similar to the ones described by Kudish and Airapetyan in [24] and [25]. The kinetic equation is solved along the lubricant flow streamlines. The solution of the kinetic equation predicts the density of the probabilistic distribution of the polymer molecule chain lengths. The shear stress and the changes in the distribution of polymer molecular weight caused by lubricant degradation affect local lubricant properties. In particular, the lubricant viscosity experiences reversible and irreversible losses and, in general, is a discontinuous function of spacial variables. The changes in the lubricant viscosity alter virtually all parameters of the lubricated contact such as film thickness, friction stresses, pressure, and gap. Several comparisons of lubricants with Newtonian and non-Newtonian rheologies with and without lubricant degradation are considered.

The material presented in this chapter is mainly based on papers [24] - [27].

11.2 Formulation of the EHL Problems for Lubricants with Newtonian and Non-Newtonian Rheologies That Undergo Stress-Induced Degradation

Let us consider a plane EHL problem. Suppose two infinite parallel cylinders steadily move with linear surface speeds u_1 and u_2. The cylinders have radii R_1 and R_2 and are made of elastic materials with Young's moduli E_1 and

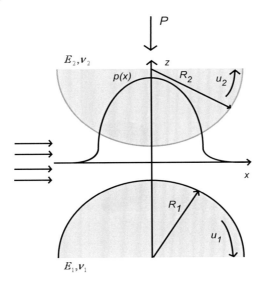

FIGURE 11.1

The general view of a lubricated contact.

E_2 and Poisson's ratios ν_1 and ν_2, respectively. The cylinders are separated by a thin lubrication layer and are loaded by the force P directed along their centers and normal to their axes (see Fig. 11.1).

Let us consider steady isothermal lubrication conditions. We will accept the classic EHL assumptions usually used for non-conformal contacts such as (a) the motion is slow so that the inertia forces can be neglected in comparison with the viscous ones, (b) the lubrication film thickness is much smaller than the size of the contact region, which, in turn, is much smaller than the curvature radii of the contact surfaces, (c) the variation rate of the gap between the cylindrical surfaces in the contact is small, (d) the contact of actual surfaces can be replaced by a contact of two elastic half-planes.

Under the above conditions the lubricant is assumed to be an incompressible fluid with a non-Newtonian rheology described by the equation

$$\frac{\partial u}{\partial z} = \frac{\tau_L}{\mu} G\left(\frac{\tau}{\tau_L}\right),$$
(11.1)

where z is the coordinate of a lubricant particle along the coordinate axis directed across the gap between the cylinders, u is the x-component of the lubricant velocity along the x-axis directed along the direction of the cylinders' motion (and the lubrication layer), τ and τ_L are the local and limiting shear stresses in the lubricant, μ is the lubricant viscosity, $\mu = \mu(x, z)$, G is a given function that determines the rheology of the lubricant fluid. The lubricant viscosity may be shear rate-dependent. In pseudoplastic/dilatant fluids increase in shear rate leads to increase/decrease in fluid viscosity. The shear thinning and thickening represent the reversible loss and gain of fluid viscosity, respectively. Most fluids are pseudoplastic. The extreme example of pseudoplastic fluids is the fluids with the limiting shear stress. The viscosity of the latter fluids vanishes as the shear rate exceeds a certain level. Therefore, the particular rheology considered here is given by the formula

$$G(x) = \tanh^{-1} x, \tag{11.2}$$

proposed by Bair and Winer in [2] and based on a series of experimental studies of lubricant rheological behavior.

According to [2, 3] τ_L is usually a linear function of pressure p, i.e.,

$$\tau_L = \tau_{L0} + \tau_{L1} p, \tag{11.3}$$

where parameters τ_{L0} and τ_{L1} are certain functions of the lubricant temperature.

Then the equations describing the process of lubrication are [24, 25, 28]

$$\frac{\partial \tau}{\partial z} = \frac{dp}{dx}, \ p = p(x), \ \tau = \mu \frac{\partial u}{\partial z}, \ \frac{\partial u}{\partial x} + \frac{\partial w}{\partial z} = 0,$$

$$u(x, -\tfrac{h}{2}) = \frac{u_1}{\sqrt{1+\frac{1}{4}(\frac{dh}{dx})^2}}, \ u(x, \tfrac{h}{2}) = \frac{u_2}{\sqrt{1+\frac{1}{4}(\frac{dh}{dx})^2}}, \tag{11.4}$$

$$w(x, -\tfrac{h}{2}) = -u_1 \frac{dh}{dx}, \ w(x, \tfrac{h}{2}) = u_2 \frac{dh}{dx},$$

where the last two pairs of conditions in (11.4) represent the boundary conditions imposed on the horizontal u and vertical w components of the lubricant velocity.

Integrating the first relationship in (11.4) with respect to z, we obtain

$$\tau = f + z \frac{dp}{dx}, \tag{11.5}$$

where f is the local unknown frictional stress due to sliding, $f = f(x)$. Consequently, integrating equations (11.1), (11.2) and the third of the relationships in (11.4) with respect to z from $-h/2$ to z we receive the expression for the horizontal component of the lubricant velocity u in the form

$$u(x, z) = \frac{u_1}{\sqrt{1+\frac{1}{4}(\frac{dh}{dx})^2}} + \int_{-h/2}^{z} \frac{\tau_L}{\mu} G(\frac{1}{\tau_L}(f + \zeta \frac{dp}{dx})) d\zeta. \tag{11.6}$$

Setting z equal to $h/2$ in equation (11.5) provides us with the equation for the local sliding frictional stress f as follows [6]

$$\int_{-h/2}^{h/2} \frac{\tau_L}{\mu} G(\frac{1}{\tau_L}(f + z \frac{dp}{dx})) dz = \frac{u_2 - u_1}{\sqrt{1+\frac{1}{4}(\frac{dh}{dx})^2}}. \tag{11.7}$$

Then integrating the continuity equation in (11.4) with respect to z from $-h/2$ to $h/2$ and taking into account the boundary conditions from (11.4) imposed on the components u and w of the lubricant velocity we arrive at the equations governing the isothermal EHL problem based on the generalized Reynolds and elasticity theory equations in the form [6, 25, 28]

$$\int_{-h/2}^{h/2} z \frac{\dot{\tau_L}}{\mu} G(\frac{1}{\tau_L}(f + z\frac{dp}{dx}))dz = \frac{u_1+u_2}{2}\left[\frac{h}{\sqrt{1+\frac{1}{4}(\frac{dh}{dx})^2}} - \frac{h_e}{\sqrt{1+\frac{1}{4}(\frac{dh(x_e)}{dx})^2}}\right]$$

$$+ \int_{-h_e/2}^{h_e/2} z \frac{\tau_L(x_e)}{\mu(x_e,z)} G(\frac{f(x_e)}{\tau_L(x_e)})dz, \; p(x_i) = p(x_e) = \frac{p(x_e)}{dx} = 0, \quad (11.8)$$

$$h = h_e + \frac{x^2-x_e^2}{2R'} + \frac{2}{\pi E'}\int_{x_i}^{x_e} p(x')\ln\frac{x_e-x'}{|x-x'|}dx', \; \int_{x_i}^{x_e} p(x')dx' = P,$$

where x_i and x_e are the coordinates of the lubricant inlet and exit points, respectively, $h(x)$ is the gap between the cylinders, h_e is the exit film thickness, $h_e = h(x_e)$, R' is the effective radius of the cylinders, $1/R' = 1/R_1+1/R_2$, E' is the effective elasticity modulus, $1/E' = (1-\nu_1^2)/E_1 + (1-\nu_2^2)/E_2$, P is the load per unit length applied to cylinders. The boundary conditions on pressure p at $x = x_i$ and $x = x_e$ reflect the natural assumption that the pressure at the boundary of the contact region is equal to the ambient atmospheric pressure that is much smaller than the pressure inside of the contact region.

The process of polymer additive degradation caused by chain scission, occurs while the additive dissolved in the lubricant moves along the lubricant flow streamlines. Therefore, we need to formulate the equations for the lubricant flow streamlines $z = z(x)$. Similar to [24, 25] these equations can be represented in differential and integral forms as follows:

$$\frac{dz}{dx} = \frac{w}{u},$$

$$\frac{d}{dx}\int_{-h/2}^{z(x)} u(x,\zeta)d\zeta = 0,$$

$$\int_{-h/2}^{z(x)} u(x,\zeta)d\zeta = \frac{h}{2}\frac{u_1}{\sqrt{1+\frac{1}{4}(\frac{dh}{dx})^2}} + zu(x,z) \quad (11.9)$$

$$- \int_{-h/2}^{z(x)} \zeta\frac{\tau_L}{\mu} G(\frac{1}{\tau_L}(f + \zeta\frac{dp}{dx}))d\zeta.$$

The vertical component of the lubricant velocity $w(x,z)$ theoretically can be determined from the continuity equation and the corresponding boundary conditions [24]. However, the numerical realization of such a process is unstable and, therefore, not recommended for practical calculations.

The stress-induced process of lubricant degradation (i.e., the process of scission of additive polymer molecules with linear structure dissolved in the base stock) is described by the kinetic equation derived and analyzed in [1, 21, 22]. The kinetic equation of lubricant degradation is written for the probabilistic density distribution $W(x, z, l)$ of polymer molecular weight, where l is the polymer molecule chain length. Function $W(x, z, l)$ is introduced in such a way that $W(x, z, l)\Delta l$ is the polymer weight in a unit fluid volume centered at the point with coordinates (x, z) with the polymer molecule chain lengths in the range from l to $l + \Delta l$. In case of steady two-dimensional lubricant motion, the kinetic equation can be represented as follows (see [1])

$$\tau_f \{u\tfrac{\partial W}{\partial x} + w\tfrac{\partial W}{\partial z}\} = 2l \int\limits_{l}^{\infty} R(x, z, L)p_c(l, L)W(x, z, L)\tfrac{dL}{L}$$

$$-R(x, z, l)W(x, z, l),$$

$$(11.10)$$

where R is the probability of polymer molecule scission, $p_c(l, L)$ is the density of the conditional probability of a polymer molecule with the chain length L to break into two pieces with lengths l and $L - l$, τ_f is the time required for one polymer chain to undergo one act of fragmentation. The expressions for functions R and p_c have been derived in [1] and have the form

$$R(x, z, l) = 0 \ if \ l \leq L_*,$$

$$R(x, z, l) = 1 - (\tfrac{l}{L_*})^{\tfrac{2\alpha U_A}{kT}} \exp\left[-\tfrac{\alpha U_A}{kT}(\tfrac{l^2}{L_*^2} - 1)\right] \ if \ l \geq L_*, \qquad (11.11)$$

$$p_c(l, L) = ln(2)\tfrac{4|L-2l|}{L^2} \exp\left[-ln(2)\tfrac{4l(L-l)}{L^2}\right],$$

where L_* is calculated according to the formulas

$$L_* = \sqrt{\tfrac{U_A}{Ca_* l_*^2 |\tau|}}, \ U_A = \tfrac{U}{N_A}. \qquad (11.12)$$

In the above formulas L_* is the characteristic polymer chain length, k is Boltzmann's constant $(1.381 \cdot 10^{-23} \ J/^\circ K)$ and N_A is Avogadros number, $(6.022 \cdot 10^{23} \ mole^{-1})$. In equation (11.12) parameter C is the dimensionless shield constant, a_* and l_* are the polymer bead radius and bond length, and U is the bond dissociation energy per mole. The above functions R and p_c were validated by the comparison of numerically simulated and experimentally obtained data in [1].

We will assume that the distribution of polymer molecules entering the contact region is uniform across the lubrication layer, i.e.,

$$W(x_i, z, l) = W_a(l) \ if \ u(x_i, z) > 0,$$

$$W(x_e, z, l) = W_a(l) \ if \ u(x_e, z) < 0. \qquad (11.13)$$

It should be noted that at the inlet to the contact region, $x = x_i$, some of the lubricant enters the contact, some may turn around and exit the contact at $x = x_i$. A similar situation is possible at the exit point $x = x_e$ from the contact if the contact surfaces are moving in the opposite directions. In each case it is assumed that when the lubricant enters the contact region (whether through the inlet $x = x_i$ or the exit $x = x_e$ cross sections) the distribution of the polymer molecules is described by $W_a(l)$ from (11.13).

It is important to keep in mind that equations (11.10)-(11.13) need to be solved along the lubricant streamlines $z = z(x)$ described by equations (11.6) and (11.9). Some properties of the kinetic equation (11.10) are established in [1]. Among these properties is the fact that the mass of the polymer additive is conserved along the lubricant flow streamlines while the number of polymer molecules tends to increase due to polymer molecule scission.

Due to the fact that the lubricant contains some linear polymer additive which may degrade while passing the lubricated contact, the viscosity of the lubricant μ varies as a function of pressure p, concentration c_p, and the density distribution $W(x, z, l)$ of the polymer additive. In particular, we will assume that the lubricant viscosity depends on the polymer additive distribution according to the Huggins and Mark Houwink equations (see [1, 7, 8])

$$\mu = \mu_a e^{\alpha_p p} \frac{1 + c_p[\eta] + k_H(c_p[\eta])^2}{1 + c_p[\eta]_a + k_H(c_p[\eta]_a)^2}, \quad [\eta] = k' M_W^\beta,$$

$$M_W = \left\{ \int_0^\infty w_*^\beta W(x, z, w_*) dw_* / \int_0^\infty W(x, z, w_*) dw_* \right\}^{1/\beta}, \tag{11.14}$$

where μ_a is the ambient lubricant viscosity, α_p is the pressure coefficient of viscosity, c_p and $[\eta]$ are the polymer concentration and the intrinsic viscosity, respectively, $[\eta]_a$ is the ambient intrinsic viscosity (at the inlet point x_i), k_H is the Huggins constant ($k_H = 0.25$, see [7, 8]), k' and β are the Mark Houwink constants ($k' = 2.7 \cdot 10^{-4} \ dL/g(g/mole)^{-\beta}$, $\beta = 0.74$, see [7, 8]), w_* is the polymer molecule weight, $w_* = w_m l$, and w_m is the monomer molecular weight.

Therefore, for the given values of x_i, R', E', u_1, u_2, P, μ_a, α_p, α, β, τ_{L0}, τ_{L1}, c_p, w_m, T, U, C, a_*, l_*, k', k_H, and the given ambient distribution of polymer additive $W_a(l)$ the problem is reduced to solution of equations (11.2), (11.3), (11.6)-(11.14) with respect to constants x_e, h_e and functions $p(x)$, $h(x)$, $f(x)$, $\mu(x, z)$, and $W(x, z, l)$.

After the solution of the problem is complete one can determine the frictional stresses τ_1 and τ_2 applied to the contact surfaces according to the formulas

$$\tau_i = (-1)^{i-1} f - \frac{h}{2} \frac{dp}{dx}, \tag{11.15}$$

where index $i = 1$ is for the lower contact surface and $i = 2$ is for the upper contact surface, respectively.

In case of an isothermal problem τ_L may be considered independent of z and τ_L can be pulled out of the integrals in equations (11.6)-(11.9). In addition, in case of no lubricant degradation the latter equations can be simplified further as the lubricant viscosity μ is independent of z. Finally, under the above isothermal conditions, the rheological model described by equations (11.2), (11.3), (11.6)-(11.14) allows for analytical calculation of integrals in equations (11.6)-(11.9) that may simplify and speed up the whole process of numerical solution.

Let us introduce the following dimensionless variables:

$$(x', a, c) = \tfrac{1}{a_H}(x, x_i, x_e), \ (z', h') = \tfrac{1}{h_e}(z, h), \ p' = \tfrac{p}{p_H}, \ \mu' = \tfrac{\mu}{\mu_a},$$

$$u' = \tfrac{2u}{u_1+u_2}, \ w' = \tfrac{2a_H w}{(u_1+u_2)h_e}, \ (f', \tau_i') = \tfrac{2h_e(f,\tau)}{\mu_a(u_1+u_2)}, \ \tau_{L0}' = \tfrac{\tau_{L0}}{p_H}, \qquad (11.16)$$

$$(W', W_a') = \tfrac{1}{W_0}(W, W_a),$$

where a_H and p_H are the half-width of and the maximum pressure in a dry Hertzian contact, W_0 is the characteristic value of the density of molecular weight distribution.

The values of parameters x_i, R', E', u_1, u_2, P, μ_a, α_p, α, β, τ_{L0}, τ_{L1}, c_p, w_m, T, U, C, a_*, l_*, k', and k_H determine the specific values of a number of dimensionless parameters that uniquely identify the solution of the problem. These dimensionless parameters are τ_{L0}/p_H, τ_{L1} and

$$a = \tfrac{x_i}{a_H}, \ V = \tfrac{3\pi^2 \mu_a(u_1+u_2)R'E'}{P^2}, \ Q = \alpha p_H, \ s_0 = \tfrac{2(u_2-u_1)}{u_1+u_2},$$

$$\varepsilon = (\tfrac{a_H}{2R'})^2, \ \gamma = \alpha\tfrac{U_A}{kT}, \ \delta = \tfrac{U_A a_H^2}{Ca_* l_*^2 \mu_a(u_1+u_2)R'}, \ \theta = k'c_p w_m^\beta, \qquad (11.17)$$

$$\kappa_f = \tau_f \tfrac{u_1+u_2}{2a_H}.$$

For simplicity, in the further discussion primes at the dimensionless variables are omitted.

Therefore, the solution of the problem is determined by the values of the dimensionless parameters τ_{L0}, τ_{L1}, the parameters from equations (11.17) and by the function $W_a(l)$. The solution of the problem is represented by the dimensionless functions: pressure $p(x)$, gap $h(x)$, sliding frictional stress $f(x)$, lubricant viscosity $\mu(x, z)$, distribution of molecular weight $W(x, z, l)$ and by two dimensionless constants: the exit coordinate c and the exit film thickness

$$H_0 = \tfrac{2h_e R'}{a_H^2}. \qquad (11.18)$$

In the dimensionless variables, the equations of the problem are

$$\int_{-h/2}^{h/2} \tfrac{\tau_L}{\mu} G(\tfrac{1}{\tau_L}(f + z\tfrac{12H_0^2}{V}\tfrac{dp}{dx}))dz = \tfrac{s_0}{\sqrt{1+\tfrac{\varepsilon H_0^2}{4}(\tfrac{dh}{dx})^2}},$$

$$\tau_L = \tfrac{12H_0}{V\sqrt{\varepsilon}}(\tau_{L0} + \tau_{L1}p),$$

$$u(x,z) = \frac{1-s_0/2}{\sqrt{1+\frac{\varepsilon H_0^2}{4}(\frac{dh}{dx})^2}} + \int\limits_{-h/2}^{z(x)} \frac{\tau_L}{\mu} G(\frac{1}{\tau_L}(f + \zeta \frac{12H_0^2}{V}\frac{dp}{dx}))d\zeta,$$

$$\frac{d}{dx}\int\limits_{-h/2}^{z(x)} u(x,\zeta)d\zeta = 0, \quad \int\limits_{-h/2}^{z(x)} u(x,\zeta)d\zeta = \frac{h}{2}\frac{1-s_0/2}{\sqrt{1+\frac{\varepsilon H_0^2}{4}(\frac{dh}{dx})^2}} + zu(x,z)$$

$$(11.19)$$

$$-\int\limits_{-h/2}^{z(x)} \zeta \frac{\tau_L}{\mu} G(\frac{1}{\tau_L}(f + \zeta \frac{12H_0^2}{V}\frac{dp}{dx}))d\zeta, \quad p(a) = p(c) = \frac{dp(c)}{dx} = 0,$$

$$H_0(h-1) = x^2 - c^2 + \frac{2}{\pi}\int\limits_a^c p(x')\ln\frac{c-x'}{|x-x'|}dx', \quad \int\limits_a^c p(x')dx' = \frac{\pi}{2},$$

$$\tau_i = (-1)^i f + \frac{6H_0^2}{V}h\frac{dp}{dx}, \quad i = 1,2,$$

$$\kappa_f(u\frac{\partial W}{\partial x} + w\frac{\partial W}{\partial z}) = 2l\int\limits_l^\infty R(x,z,L)p_c(l,L)W(x,z,L)\frac{dL}{L}$$

$$-R(x,z,l)W(x,z,l),$$

$$(11.20)$$

$$W(a,z,l) = W_a(l) \; if \; u(a,z) > 0,$$

$$W(c,z,l) = W_a(l) \; if \; u(c,z) < 0,$$

$$R(x,z,l) = 0 \; if \; l \leq L_*$$

$$R(x,z,l) = 1 - (\frac{l}{L_*})^{2\gamma}\exp[-\gamma(\frac{l^2}{L_*^2}-1)] \; if \; l \geq L_*,$$

$$L_* = \sqrt{\frac{H_0\delta}{|\tau|}}, \quad \tau = f + z\frac{12H_0^2}{V}\frac{dp}{dx},$$

$$(11.21)$$

$$p_c(l,L) = ln(2)\frac{4|L-2l|}{L^2}\exp[-ln(2)\frac{4l(L-l)}{L^2}],$$

$$\mu(p,W) = e^{Qp}\frac{1+[\eta]+k_H[\eta]^2}{1+[\eta]_a+k_H[\eta]_a^2}, \quad [\eta] = \theta M_W^\beta,$$

$$M_W = \left\{\int\limits_0^\infty l^\beta W(x,z,l)dl / \int\limits_0^\infty W(x,z,l)dl\right\}^{1/\beta}.$$

11.3   Lubricant Flow Topology

Obviously, it is sufficient to only consider cases of non-positive values of the slide-to-roll ratio s_0. Cases with $s_0 > 0$ are identical to the latter ones if the upper and lower solid surfaces are interchanged. Before we consider the numerical method employed for solution of the aforementioned problem, it is necessary to understand the specifics of the topology of the lubricant flow.

In case of pure rolling, $s_0 = 0$ and the flow is symmetric about the x-axis. It has two sets of flow streamlines running through the whole contact and two sets of flow streamlines that enter the contact region through the inlet gap, turn around, and exit the contact region through the same inlet gap. The flow streamlines running through the whole contact region are adjacent to the contact surfaces while the flow streamlines that turn around are adjacent to the x-axis in the inlet zone of the contact. It is important to mention that in the flow there is a special point where both the horizontal u and vertical w components of lubricant velocity are equal to zero. This point is the bifurcation point. The bifurcation point is located in the inlet zone of the lubrication film and outside of the Hertzian region. Because of continuity of the components of the lubricant velocity in the vicinity of the bifurcation point both u and v are small. Moreover, the bifurcation point is the point of intersection of the flow separatrices the curves that separate different zones of the flow, i.e., zones occupied by sets of streamlines running through the whole contact and the ones that turn around and exit through the inlet gap. In the case of pure rolling, there are two symmetric separatrices between the running through the whole contact and turning around streamlines, one separatrix dividing two sets of turning around streamlines, and one separatrix dividing the running through the whole contact streamlines. The latter two coincide with the x-axis.

For cases with $-2 < s_0 < 0$, there are still two pairs of flow streamline sets: two sets of running through the whole contact streamlines and two sets of turning around and exiting through the inlet gap ones. However, they are no longer symmetric about the x-axis. In comparison with the case of pure rolling the entire flow, including the bifurcation point and the separatrices, are shifted toward the upper surface $z = h/2$. For the case of $s_0 = -2$, there is just one set of turning around flow streamlines adjacent to the upper contact surface and one set of flow streamlines running through the whole contact that is adjacent to the lower contact surface everywhere in the contact. Outside of the inlet region the latter set of running through the whole contact flow streamlines is also adjacent to the upper contact surface. For the cases with $-\infty < s_0 < -2$, the lubricant flow is asymmetric. It is represented by two sets of flow streamlines adjacent to the lower and upper contact surfaces and running through the whole contact in opposite directions as well as by two sets of turning around flow streamlines one of which enters the contact region and

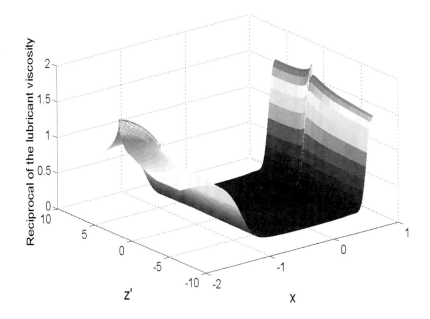

FIGURE 11.2 (See color insert)

The reciprocal of the viscosity μ in the lubricants with Newtonian and non-Newtonian rheologies under mixed rolling and sliding conditions and Series I input data ($s_0 = -0.5$). The variable z' is an artificially stretched z-coordinate across the film thickness (namely, $z' = zh(a)/h(x)$) to make the relationship more transparent (after Kudish and Airapetyan [25]). Reprinted with permission from the ASME.

exits it through the inlet gap and the other that enters and exits the contact through the exit gap. For $-2 < s_0 \leq 0$, there is only one bifurcation point located between the contact surfaces in the inlet zone while for $s_0 = -2$ the bifurcation point belongs to the upper contact surface in the inlet zone. For $-\infty < s_0 < -2$ there are two bifurcation points between the contact surfaces, one in the inlet and the other in the exit zones of the contact.

Under isothermal conditions in case of no degradation, the lubricant viscosity depends only on pressure p, which, in turn, is a function of x. Therefore, the lubricant viscosity depends only on the distance of a fluid particle from the inlet point $x = a$, i.e., the lubricant viscosity μ depends on x and is independent from z. That makes the function of lubricant viscosity μ a continuous function of x and z because the function of pressure $p(x)$ is a differentiable function.

In case of lubricant degradation, the situation changes completely. Fluid particles that approach the separatrices between two sets of turning around and running through streamlines from different sides follow different flow streamlines and are subjected to different shear stresses over different time periods. Therefore, these fluid particles degrade differently that, in turn, leads to different viscosities of these fluid particles. Thus, we can conclude that degrading lubricants possess a remarkable property. Namely, for all slide-to-roll ratios s_0 (except for $s_0 = 0$ and $s_0 = \pm 2$), the separatrices to which fluid particles approach from both sides in the process of their motion are the curves of the lubricant viscosity $\mu(x, z)$ discontinuity (see Fig. 11.2). In spite of that, the shear stress $\tau(x, z)$ is continuous in the entire lubricated region. This is due to the fact that the sliding frictional stress $f(x)$ and the derivative of pressure dp/dx are continuous functions of x. Taking into account these properties of $\mu(x, z)$ and $\tau(x, z)$ we can conclude that $\frac{\partial u}{\partial z}$ as a function of x and z is also discontinuous across the aforementioned separatrices.

The above–mentioned discontinuities of $\mu(x, z)$ and $\frac{\partial u}{\partial z}$ cause some additional complications in numerical solution of an EHL problem for degrading lubricants with Newtonian and non-Newtonian rheologies. One can expect a much more complex problem formulation for a non-isothermal EHL case due to the fact that conservation of heat flux across of the above separatrices where μ and $\frac{\partial u}{\partial z}$ are discontinuous. It is expected that the lubricant temperature T is also discontinuous across the same separatrices as the lubricant viscosity μ is discontinuous.

11.4 Numerical Method for Solution of EHL Problems for Degrading Lubricants in Line Contacts

Certain difficulties in the solution process are caused by the fact that the solution of the above problem is determined within the interval $[a, c]$ where the inlet coordinate a is given while the exit coordinate c is unknown. To simplify the solution process one needs to convert the problem from the interval with a priori unknown boundary $x = c$ into the problem within a region with fixed boundaries, say $[-1, 1]$. That can be achieved by the substitution

$$x = \frac{c+a}{2} + \frac{c-a}{2} y, \tag{11.22}$$

that makes $-1 \leq y \leq 1$. However, unless it is indicated otherwise, we will discuss the numerical method in the x variable.

The very first step toward the problem solution involves choosing an initial approximation. After that the general iterative process is organized as follows. For the known values of pressure $p(x)$, gap $h(x)$, lubricant viscosity $\mu(x, z)$, film thickness H_0, and exit coordinate c, the numerical procedure for solution of the considered problem consists of several steps in the following order: (a) evaluating the sliding frictional stress $f(x)$, (b) determining the horizontal component of the fluid velocity $u(x, z)$ and fluid flux, (c) calculating the flow streamlines $z(x)$, (d) finding separatrices of the lubricant flow some of which may be the curves of discontinuity of the lubricant viscosity, (e) evaluating the lubricant viscosity $\mu(x, z)$ at every point of the flow by solving the kinetic equation along the flow streamlines, (f) calculating the new approximation of the sliding frictional stress $f(x)$, and (g) solution of the modified Reynolds and gap equations for $p(x)$ and $h(x)$. Each of the above steps of the numerical solution of the problem is considered below in detail. To speed up the solution process, we use the multigrid approach along the x-axis, i.e., we use a series of grids on the interval $[a, c]$ with equidistant nodes x_k, $k = 1, \ldots, N_x$, $x_1 = a$ with increasing sequence of total number of nodes N_x. The grid along the z-axis is introduced by using the convenient basis created by so–called pseudo-streamlines, which can be determined from simple equations $z_j^p(x) = h(x)(2j/N_z^p - 1)/2$, $j = 1, \ldots, N_z^p$, where N_z^p is the number of pseudo-streamlines, including the contact surfaces. Along the l-axis, the axis of polymer molecule chain lengths, the chain nodes l_m, $m = 1, \ldots, N_l$, are introduced in a non-uniform manner in such a way that $l_1 > 0$ and $l_m < l_{m+1}$.

11.4.1 Initial Approximation

The initial approximation of the solution of the formulated EHL problem for degrading lubricant is taken as a solution of the corresponding EHL problem for non-degrading lubricant. To determine this solution we have to perform two tasks: to calculate the sliding frictional stress and to solve the modified Reynolds and gap equations. The latter problems are solved using Newtons method with a damping coefficient. As these procedures are essentially the same as for consequent iterations for the case of degrading lubricant, we will describe them in detail later.

To reach the desired input values of some parameters (such as Q, τ_{L0}, etc.), the process of homotopy is employed. That means that the solution is first obtained for the closest in some sense set of values of the input parameters for which it is known a priori that iterations converge fast. After that the solution is found for a sequence of sets of input parameters (leading to the desired set of values of the input parameters) the values of which are just slightly different from each other. For each new set of values of the input parameters the initial approximation for the problem solution is taken from the solution of the problem for the preceding set of values of the input parameters.

11.4.2 Sliding Frictional Stress $f(x)$

The distribution of the sliding frictional stress $f(x)$ is controlled by the first two equations in (11.19). It is obvious that for any lubricant rheology $f(x) = 0$ for $s_0 = 0$. For $s_0 \neq 0$ the aforementioned equations have an analytical exact solution only in the case of lubricants of Newtonian rheology. Therefore, for $s_0 \neq 0$ the aforementioned equations must be solved numerically. At this stage we assume that the distributions of $p(x)$, $h(x)$, $\mu(x,z)$, and constants H_0 and c are known. To solve the above equations for $f(x_k)$, we use a modified Newtons method:

$$f^{i+1} = f^i - \frac{\alpha_f}{D_f}\left[\int_{-h/2}^{h/2} \frac{\tau_L}{\mu} G(\frac{1}{\tau_L}(f^i + z\frac{12H_0^2}{V}\frac{dp}{dx}))dz - \frac{s_0}{\sqrt{1+\frac{\varepsilon H_0^2}{4}(\frac{dh}{dx})^2}}\right], \quad (11.23)$$

$$D_f = \frac{V}{12H_0^2}\frac{\tau_L^2}{\mu_{avg}\frac{dp}{dx}}[G(\frac{1}{\tau_L}(f^i + \frac{6H_0^2}{V}h\frac{dp}{dx})) - G(\frac{1}{\tau_L}(f^i - \frac{6H_0^2}{V}h\frac{dp}{dx}))],$$

$$\mu_{avg} = \frac{h}{\int_{-h/2}^{h/2}\frac{dz}{\mu}}. \quad (11.24)$$

In equation (11.23) α_f is an empirically chosen positive coefficient the value of which is adjusted in the process of iterations. Namely, the initial value of α_f is relatively small and it increases with the number of iterations i until it reaches 1. To calculate the above and similar to it integrals, we use the trapezoidal quadrature formula and the grid created by the pseudo-streamlines. To evaluate just one integral with respect to z and to avoid calculating the second similar integral for the derivative of the first one with respect to f, we use an assumption that for the purpose of calculating the above–mentioned derivative of the first integral the lubricant viscosity $\mu(x,z)$ can be approximated and replaced by its average $\mu_{avg}(x)$ across the gap $h(x)$. That allows to calculate the latter integral analytically and, therefore, to simplify and to speed up the process of calculations of the sliding frictional stress $f(x)$. At nodes x_k at which $dp(x_k)/dx = 0$, the expression for D_f can be easily obtained from equations (11.19) by using the limit as $dp/dx \to 0$

$$D_f = \frac{h\tau_L}{\mu_{avg}}G'(\frac{f^i}{\tau_L}) \; if \; \frac{dp(x_k)}{dx} = 0. \quad (11.25)$$

The initial approximation for $f(x)$ we can take as the one-term asymptotic solution of the first equation in (11.19)

$$f = \tau_L G^{-1}\left[\frac{s_0}{\sqrt{1+\frac{\varepsilon H_0^2}{4}(\frac{dh}{dx})^2}}\frac{1}{\tau_L\int_{-h/2}^{h/2}\frac{dz}{\mu}}\right], \quad (11.26)$$

where G^{-1} is the inverse function of the rheological function G. It is obtained under the condition that the sliding frictional stress $f(x)$ is much higher than the rolling frictional stress $6H_0^2/V(hdp/dx)$ (see Chapter 6 and Kudish [6]).

The numerical results show that the iterative method expressed by equations (11.23)-(11.25) converges as well as the modified Newtons method with the same damping coefficient α_f and the value of D_f calculated using the integral of the derivative of the function G. For the present method the required CPU time is almost twice smaller than for the modified Newtons method.

In the process of solution of the Reynolds equation, we will use not only the solution $f(x)$ of the first two equations in (11.19) obtained by using equations

(11.24)-(11.26) but also function $f_{avg}(x)$ that satisfies similar equations in which the local lubricant viscosity $\mu(x, z)$ is replaced by the lubricant viscosity averaged over the film thickness $\mu_{avg}(x)$. The averaged sliding frictional stress $f_{avg}(x)$ is also obtained using equations (11.24)-(11.26) by replacing $\mu(x, z)$ by $\mu_{avg}(x)$ in equation (11.24).

The described numerical method converges very fast for moderate values of the slide-to-roll ratio s_0, pressure coefficient of viscosity Q, speed-load parameter V, and the parameters τ_{L0} and τ_{L1} controlling the limiting shear stress τ_L. However, in cases of the presence of a limiting stress τ_L ($|\tau| \leq \tau_L$) and for high values of s_0, Q, V, and low values of τ_{L0} and τ_{L1} the frictional stress τ may be very close to τ_L (see (11.21)). It means that even very small variations in the sliding frictional stress $f(x)$ may destabilize the iteration process. Such a situation is typical for ill-conditioned problems. Obviously, such variations of $f(x)$ are inevitable during any iterative solution of the first equation in (11.19). One of the ways to deal with such a problem is to dampen the incremental changes of $f(x)$ during iterations based on Newtons method. For higher values of s_0, Q, V, and smaller values of τ_{L0} and τ_{L1}, this dampening must be stronger. Therefore, under the above conditions in cases of strong dampening the iteration process converges relatively slow.

11.4.3 Horizontal Component of the Lubricant Velocity and Flux

After the initial approximation for the solution is chosen or one full iteration of the problem solution is completed, we know the approximate values of pressure $p(x)$, gap $h(x)$, lubricant viscosity $\mu(x, z)$, film thickness H_0, and the exit coordinate c. Based on the above values of $p(x)$, $h(x)$, $\mu(x, z)$, H_0, and c, the formula for $u(x, z)$ from equations (11.19), and using the trapezoidal quadrature formula for integration with respect to ζ on the grid of pseudo-streamlines $z_j^p(x)$, we obtain the distribution of the lubricant velocity $u(x_k, z_j^p(x_k))$, where k is the number of the node x_k. The lubricant flux described by the integral of u with respect to z from equations (11.19) is also calculated on the pseudo-streamlines $z_j^p(x)$ by the trapezoidal quadrature formula for integration with respect to z.

11.4.4 Lubricant Flow Streamlines

Because polymer scission occurs while lubricant particles move along the flow streamlines $z_j(x)$, $j = 1, \ldots, N_z$, there is a necessity to determine the flow streamlines. It seems that an efficient way to do that is to solve a series of initial-value problems

$$\frac{dz_j}{dx} = \frac{w(x, z_j)}{u(x, z_j)}, \ z_j(x_0) = z_j^p(x_0), \ j = 1, \ldots, N_z,$$

$$x_0 = a \ if \ u(a, z_j^p(a)) > 0, \ x_0 = c \ if \ u(c, z_j^p(c)) < 0,$$

$$(11.27)$$

using one of the methods such as the modified Euler method (N_z is the number of flow streamlines). In equations (11.27) $w(x,z)$ is the z-component of the lubricant velocity. The way the x-coordinate of the initial point of the streamline x_0 is determined in (11.27) depends on the topology of the lubricant flow. The only way to find the z-component of the lubricant velocity $w(x,z)$ is to solve the continuity equation

$$\frac{\partial u}{\partial x} + \frac{\partial w}{\partial z} = 0.$$

It can be done by integrating this equation over small rectangular cells with vertices $(x_k, z_j^p(x_k))$, $(x_{k+1}, z_j^p(x_{k+1}))$, $(x_k, z_{j+1}^p(x_k))$, and $(x_{k+1}, z_{j+1}^p(x_{k+1}))$ and using the corresponding boundary conditions [24, 28]. However, our practice showed that the numerical realization of the above process does not provide sufficient precision and stability in determination of the z-component of the lubricant velocity $w(x,z)$ and flow streamlines $z_j(x)$. Therefore, we do not recommend it for practical calculations. In fact, it causes instability in $u(x,z)$.

The calculation of the lubricant flow streamlines is done based on the integrated form of the equation for the x-component of the lubricant flux following from equations (11.19):

$$\int\limits_{-h(x_k)/2}^{z_j(x_k)} u(x_k, z)dz = \int\limits_{-h(x_0)/2}^{z_j(x_0)} u(x_0, z)dz,$$

$$k = 1, \ldots, N_x, \; j = 1, \ldots, N_1,$$

$$x_0 = a \; if \; u(a, z_j^p(a)) > 0, \; x_0 = c \; if \; u(c, z_j^p(c)) < 0,$$

(11.28)

and linear interpolation with respect to z (N_1 is the number of running through and turning around flow streamlines adjacent to the lower contact surface). This is exactly the procedure used to determine the flow streamlines passing through the whole lubricated contact (see Section 11.5). For turning around flow streamlines, the procedure is more complex.

Let us consider the general procedure of the flow streamline computation for the case when $-2 < s_0 < 0$, and there are still two pairs of flow streamline sets: two sets of running through the whole contact streamlines and two sets of turning around and exiting through the inlet gap ones. Prior to calculating the flow streamlines, we determine the so–called zero-curves in the flow, $z_j^0(x)$, $j = 1, 2$, along which the x-component of the lubricant velocity is equal to zero. It is done using the known distribution of the x-component of the lubricant velocity $u(x,z)$ and linear interpolation with respect to z. After the zero-curves $z_j^0(x)$ are determined, we first calculate the sets of lower flow streamlines $z_j(x)$ that are closer to the lower surface $z = -h/2$ (both running through the whole contact and turning around ones) and then the same kind ones that are closer to the upper contact surface $z = h/2$. We determine the initial points of the flow streamlines $z_j(x)$ starting from the lower surface that corresponds to $j =$

1. If $u(a, z_j^P(a)) > 0$, then the point $(a, z_j^P(a))$ is considered to be the starting point for the flow streamline $z_j(x)$, otherwise it is skipped and we go to the next j. Starting from the initial point we calculate the coordinates of the points on the flow streamline based on equation (11.28) until the flow streamline intersects the lower zero-curve or exits through the exit cross section of the contact at $x = c$. If it exits through the exit cross section of the contact at $x = c$, the determination of the flow streamline is complete. Otherwise, using linear interpolation, we find the point $(a, z_j^e(a))$ in the inlet cross section of the contact closest to the initial point $(a, z_j^P(a))$ at which the lubricant flux through the cross section determined by the points $(a, -h(a)/2)$ and $(a, z_j^e(a))$ is equal to the flux through the cross section determined by the points $(a, -h(a)/2)$ and $(a, z_j^P(a))$. Starting from the point $(a, z_j^e(a))$ we solve the equation

$$\int_{-h(x_k)/2}^{z_j(x_k)} u(x_k, z)dz = \int_{-h(x_0)/2}^{z_j(x_0)} u(x_0, z)dz,$$

$$k = 1, \ldots, N_x, \ j = N_{t1}, \ldots, N_1,$$

(11.29)

$$x_0 = a \ if \ u(a, z_j^e(a)) < 0, \ x_0 = c \ if \ u(c, z_j^e(c)) > 0,$$

to find the exiting branches of the turning around flow streamlines (N_{t1} is the number of the turning around flow streamline closest to the lower contact surface). This process continues until the flow streamline in question reaches the lower zero-curve.

For the upper sets of the flow streamlines, the algorithm is similar. However, it is more convenient to solve similar equation instead of solving equations (11.28) and (11.29)

$$\int_{z_j(x_k)}^{h(x_k)/2} u(x_k, z)dz = \int_{z_j(x_0)}^{h(x_0)/2} u(x_0, z)dz,$$

$$k = 1, \ldots, N_x, \ j = N_1 + 1, \ldots, N_z,$$

$$x_0 = a \ if \ u(a, z_j^P(a)) > 0, \ x_0 = c \ if \ u(c, z_j^P(c)) < 0,$$

(11.30)

$$\int_{z_j(x_k)}^{h(x_k)/2} u(x_k, z)dz = \int_{z_j(x_0)}^{h(x_0)/2} u(x_0, z)dz,$$

$$k = 1, \ldots, N_x, \ j = N_1 + 1, \ldots, N_2,$$

$$x_0 = a \ if \ u(a, z_j^e(a)) < 0, \ x_0 = c \ if \ u(c, z_j^e(c)) > 0,$$

where N_2 is the number of the turning around flow streamline closest to the upper contact surface. In all other respects the solution procedure is identical to that described above.

For different flow topologies, the process of calculation of the flow streamlines is similar to that described above but somewhat different. For example, for $s_0 < -2$ or $s_0 > 2$ we still have two sets of both kind of running through the whole contact and turning around flow streamlines. Therefore, after the computation of the flow streamlines that start at the inlet cross section is complete (see equations (11.28), (11.29), or (11.30)) for this topology of the lubricant flow, we employ a similar process of determining of the flow streamlines that start at the exit cross section of the lubricated contact.

In cases of $s_0 = -2$ or $s_0 = 2$, we have just one set of running through the whole contact and turning around flow streamlines and just two sets of equations (11.28), (11.29), or (11.30) are used, respectively.

11.4.5 Separatrices of the Lubricant Flow

As soon as all flow streamlines are found depending on the flow topology we can determine the separatrices in the flow which are needed for the proper determination of the viscosity of a degrading lubricant. The idea behind the approach that allows determining the separatrices is based on conservation of flux. To find the separatrices, we determine the lubricant fluxes U_1, $U_2, \ldots$ through the cross sections between the contact surfaces and the bifurcation points. After that we find curves (separatrices) that represent the boundaries of the flow regions the flux through any cross section of which is preserved, i.e., is equal to U_1, U_2, and so on.

Let us consider the case of $0 \leq |s_0| < 2$. First, we find the bifurcation point as a point of intersection of the zero-curves

$$z_1^0(x) = z_2^0(x). \tag{11.31}$$

Suppose the x-coordinate of the bifurcation point that satisfies equation (11.31) is x_b. To find the separatrices, we need to solve equations

$$\int_{-h(x_k)/2}^{z_1^s(x_k)} u(x_k, z)dz = \int_{-h(x_b)/2}^{z_1^0(x_b)} u(x_b, z)dz, \ z_1^s \leq z_1^0,$$

$$x_k \leq x_b \ if \ s_0 \leq 0 \ or \ x_k \geq x_b \ if \ s_0 \geq 0,$$

$$\int_{z_2^s(x_k)}^{h(x_k)/2} u(x_k, z)dz = \int_{z_1^0(x_b)}^{h(x_b)/2} u(x_b, z)dz, \ z_2^s \geq z_2^0, \tag{11.32}$$

$$x_k \leq x_b \ if \ s_0 \leq 0 \ or \ x_k \geq x_b \ if \ s_0 \geq 0,$$

$$\int_{-h(x_k)/2}^{z_3^s(x_k)} u(x_k, z)dz = \int_{-h(x_b)/2}^{z_1^0(x_b)} u(x_b, z)dz, \ z_1^0 < z_3^s < z_2^0,$$

$$x_k \leq x_b \ if \ s_0 \leq 0 \ or \ x_k \geq x_b \ if \ s_0 \geq 0,$$

$$\int_{-h(x_k)/2}^{z_4^s(x_k)} u(x_k, z)dz = \int_{-h(x_b)/2}^{z_1^0(x_b)} u(x_b, z)dz,$$

$$x_k \geq x_b \ if \ s_0 \leq 0 \ or \ x_k \leq x_b \ if \ s_0 \geq 0.$$

For $\mid s_0 \mid = 2$ there is just one bifurcation point either on the upper $(s_0 = -2)$ or the lower $(s_0 = 2)$ contact surfaces. It is important to remember that for $s_0 = -2$ and $s_0 = 2$ we have $z_2^0(x) = h(x)/2$ and $z_1^0(x) = -h(x)/2$, respectively. Therefore, we need to solve one of the equations

$$\int_{-h(x_k)/2}^{z_1^s(x_k)} u(x_k, z)dz = \int_{-h(x_b)/2}^{z_1^0(x_b)} u(x_b, z)dz,$$

$$x_k \leq x_b, \ z_1^s(x_k) \leq z_1^0(x_k) \ if \ s_0 = -2,$$

$$\int_{z_2^s(x_k)}^{h(x_k)/2} u(x_k, z)dz = \int_{z_1^0(x_b)}^{h(x_b)/2} u(x_b, z)dz,$$

$$x_k \leq x_b, \ z_2^s(x_k) \geq z_2^0(x_k) \ if \ s_0 = 2.$$

(11.33)

Now, let us consider the case of $\mid s_0 \mid > 2$. First, we need to find the bifurcation points, i.e., the right most point x^L on the zero-curve $z_1^0(x)$ that starts in the inlet zone and the left most point x^R on the zero-curve $z_2^0(x)$ that starts in the exit zone of the contact. After that we need to solve the following equations:

$$\int_{-h(x_k)/2}^{z_1^s(x_k)} u(x_k, z)dz = \int_{-h(x^L)/2}^{z_1^0(x^L)} u(x^L, z)dz,$$

$$x_k \leq x^L, \ z_1^s(x_k) \leq z_1^0(x_k),$$

$$\int_{z_2^s(x_k)}^{h(x_k)/2} u(x_k, z)dz = \int_{z_2^0(x^R)}^{h(x^R)/2} u(x^R, z)dz,$$

$$x_k \leq x^R, \ z_2^s(x_k) \geq z_2^0(x_k),$$

(11.34)

$$\int_{-h(x_k)/2}^{z_3^s(x_k)} u(x_k, z)dz = \int_{-h(x^L)/2}^{z_1^0(x^L)} u(x^L, z)dz,$$

$$x^L \leq x_k \leq x^R.$$

Equations (11.32)-(11.34) are easily solved by employing linear interpolation and using the earlier determined lubricant flux.

11.4.6 Solution of the Kinetic Equation and Evaluating the Lubricant Viscosity

First, let us consider the case of $0 \leq |s_0| < 2$ for which there are two zero-curves that start at the inlet cross section and end (intersect) at the bifurcation point in the inlet zone and there is no zero-curve that passes through the whole contact from $x = a$ to $x = c$. To calculate the lubricant viscosity, we have to determine the distribution of the molecular weight $W(x, z, l)$ along the flow streamlines $z = z_j(x)$. That means we have to integrate the kinetic equation (see equations (11.20)) along all flow streamlines $z_j(x)$, $j = 1, \ldots, N_z$. Therefore, along each of the streamlines $z_j(x)$, $j = 1, \ldots, N_z$, it is convenient to introduce a parameter t (time) according to the formulas

$$dt = \frac{dx}{\kappa_f u(x, z_j(x))}. \tag{11.35}$$

By integrating equation (11.35), we can introduce a partition of the t-axis as follows

$$t_{k+1} = t_k + \frac{1}{\kappa_f} \int\limits_{x_k}^{x_{k+1}} \frac{dx}{u(x, z_j(x))}, \quad k = 1, \ldots, N_x, \ t_1 = 0.$$

Discretizing the latter equation gives the set of time nodes t_k

$$t_{k+1} = t_k + \frac{2(x_{k+1} - x_k)}{\kappa_f [u(x_k, z_j(x_k)) + u(x_{k+1}, z_j(x_{k+1}))]}, \quad k = 1, \ldots, N_x, \ t_1 = 0. \tag{11.36}$$

For simplicity let us introduce the notations

$$R_j(t, l) = R(x_j(t), z_j(t), l), \quad W_j(t, l) = W(x_j(t), z_j(t), l). \tag{11.37}$$

Then the kinetic equation can be reduced to

$$\frac{dW_j}{dt} = 2l \int\limits_l^\infty R_j(t, L) p_c(l, L) W_j(t, L) \frac{dL}{L} - R_j(t, l) W_j(t, l), \tag{11.38}$$

$$j = 1, \ldots, N_z.$$

Using the trapezoidal quadrature formula for approximation of the integral in equation (11.38), that further is denoted by $I(t_n, l_m)$, we get

$$I(t_k^R, t_k, l_m) = I_1(t_k^R, t_k, l_m) + I_2(t_k^R, t_k, l_m),$$

$$I_1(t_k^R, t_k, l_m) = \frac{l_{m+1} - l_m}{2l_m} = R_j(t_k^R, l_m) p_c(l_m, l_m) W_j(t_k, l_m),$$

$$I_2(t_k^R, t_k, l_m) = \frac{1}{2} \sum_{n=m}^{N_l - 2} \frac{l_{n+2} - l_n}{l_{n+1}} R_j(t_k^R, l_{n+1}) p_c(l_m, l_{n+1}) \tag{11.39}$$

$$\times W_j(t_k, l_{n+1}) + \frac{l_{N_l} - l_{N_l-1}}{2l_{N_l}} R_j(t_k^R, l_{N_l}) W_j(t_k, l_{N_l}), \quad m = 1, \ldots, N_l,$$

where t_k^R is an approximation of t_k and $t_k^R = t_k$ or $t_k^R = t_{k-1}$.

By employing the forward difference approximation for the derivative of W_j and satisfying equation (11.38) at $t_{k+1/2} = (t_k + t_{k+1})/2$ and $l = l_m$, we obtain

$$\frac{W_j(t_{k+1}, l_m) - W_j(t_k, l_m)}{t_{k+1} - t_k} = l_m[I(t_k, t_k, l_m) + I(t_k, t_{k+1}, l_m)]$$

$$- \frac{R_j(t_k, l_m)}{2}[W_j(t_k, l_m) + W_j(t_{k+1}, l_m)]. \tag{11.40}$$

The specific approximation of the integral in equation (11.38) used in equation (11.40) is caused by the fact that the probability of scission R depends on the lubricant viscosity μ through the sliding frictional stress f (see equations (11.19) and (11.21)) and that the lubricant viscosity $\mu(x_{k+1}, z_j(x_{k+1}))$ (that corresponds to the viscosity μ of the lubricant particle moving along the flow streamline $z_j(x)$ at the time moment t_{k+1}) is unknown while all values of $W_j(t_{k+1}, l_m)$, $m = 1, \ldots, N_l$, are not determined yet. That follows from equations (11.20) and (11.21) for the viscosity μ that represents the integral dependence between μ and $W(t, l)$. Solving equations (11.39) and (11.40) for $W_j(t_{k+1}, l_m)$, we find the final scheme

$$W_j(t_{k+1}, l_m) = \frac{2 - R_j(t_k, l_m)\Delta t_k}{2 + R_j(t_k, l_m)\Delta t_k[1 - (l_{m+1} - l_m)p_c(l_m, l_m)]} W_j(t_k, l_m)$$

$$+ \frac{2l_m \Delta t_k[I(t_k, t_k, l_m) + I_2(t_k, t_{k+1}, l_m)]}{2 + R_j(t_k, l_m)\Delta t_k[1 - (l_{m+1} - l_m)p_c(l_m, l_m)]}, \tag{11.41}$$

$$j = 1, 2, \ldots, \quad k = 1, 2, \ldots, \quad m = N_l, \ldots, 1,$$

$$W_j(0, l_m) = W_a(l_m), \quad \Delta t_k = t_{k+1} - t_k.$$

It is important to emphasize that at any given point x_{k+1} (time moment t_{k+1}) on any given flow streamline $z_j(x)$ computation of $W_j(t_{k+1}, l_m)$ with respect to l_m starts with $m = N_l$ and ends with $m = 1$, i.e., it is done in the direction from longer molecular chains to shorter ones.

As soon as all $W_j(t_{k+1}, l_m)$, $m = 1, \ldots, N_l$, for the fixed time moment t_{k+1} are computed from equations (11.20) by simple integration using the trapezoidal rule we can determine the viscosity $\mu_j(t_{k+1}) = \mu(x_{k+1}, z_j(x_{k+1}))$ of a lubricant particle that in the process of its motion underwent degradation. That process gives us the distribution of the lubricant viscosity along the flow streamlines.

In the process of solution of the equations for the sliding frictional stress $f(x)$ and the Reynolds equation, we need to perform integration with respect to z. It is almost impossible to do by using the flow parameters distributed along the flow streamlines and, at the same time, it is easy to do knowing the flow parameters on the grid of pseudo-streamlines $z_j^p(x)$, $j = 1, \ldots, N_z^p$. Therefore, there is a need to consider in more detail the way the lubricant viscosity $\mu(x_{k+1}, z_j(x_{k+1}))$ is evaluated along the flow streamlines $z_j(x)$, $j = 1, \ldots, N_z$, is mapped onto the grid of flow pseudo-streamlines. It is done using linear interpolation with respect to z. In particular, for $s_0 = 0$ the lubricant viscosity is continuous everywhere, and the standard linear interpolation provides its values at any point of interest. In the case of $0 <| s_0 |< 2$, there is a necessity to carefully evaluate the lubricant viscosity above and below the separatrix, the lubricant viscosity discontinuity curve, running from the inlet through the whole contact to the exit. To evaluate the lubricant viscosity in the vicinity of this separatrix, we linearly extrapolate the lubricant viscosity from the two closest adjacent to the separatrix flow streamlines that are located above and below the separatrix while preserving the discontinuity of the lubricant viscosity.

For $| s_0 |= 2$ one of the zero-curves that coincides with one of the contact surfaces, the motionless one, that is at the same time a flow streamline and a pseudo-streamline. To determine the lubricant viscosity along such a motionless contact surface we, obviously, cannot use parameter t introduced by equation (11.35). In this case to evaluate the viscosity of a lubricant particle adjacent to the motionless contact surface, we use linear extrapolation of the lubricant viscosity from the two closest adjacent to the motionless contact surface flow streamlines.

For the case of $| s_0 |> 2$, the only zero-curve in the lubricant flow is a separatrix that separates lubricant particles moving in opposite directions. At the same time, this separatrix is a curve of discontinuity of the lubricant flow. To evaluate the lubricant viscosity in the vicinity of this separatrix, we use the same approach as in the case of $0 <| s_0 |< 2$.

11.4.7   Solution of the Reynolds Equation

To solve the modified Reynolds and gap equations, we use the modified Newton's method that is applied to fourth through eighth equations in (11.19) for $z(x) = h(x)/2$. For this purpose the fourth equation in (11.19) is integrated with respect to z and is represented in the form

$$\int_{-h(x)/2}^{h(x)/2} u(x, z)dz = \int_{-1/2}^{1/2} u(c, z)dz. \qquad (11.42)$$

The modified Reynolds equation from 11.19 is solved on a sequence of grids with decreasing step sizes along the x-axis while the grid created by the pseudo-streamlines along the z-axis remains the same. The further calculations are based on the modified Newton's method. Let us introduce the incremental variations of the major variables: $\triangle p(x_k) = \triangle p_k$, $\triangle f(x_k) = \triangle f_k$, $\triangle h(x_k) = \triangle h_k$, $\triangle \mu(x_k, z)$, $k = 1, \ldots, N_x$, $\triangle H_0$, and $\triangle c$. The unknowns in this iterative process are the incremental variations in pressure $\triangle p(x_k)$, $k = 1, \ldots, N_x$, film thickness $\triangle H_0$, and exit coordinate $\triangle c$ while the variations of $\triangle f(x_k)$ and $\triangle h(x_k)$, $k = 1, \ldots, N_x$, are found in terms of the former ones.

Let us assume that the i-th iterates of all contact parameters are known. The main assumptions used in the following numerical procedure are: (a) the term $\varepsilon H_0^2 (dh/dx)^2$ is small and its variation can be neglected, (b) we will assume that the lubricant viscosity variations $\triangle \mu(x_k, z)$, $k = 1, \ldots, N_x$, are mostly caused by variations of pressure $\triangle p(x_k)$, $k = 1, \ldots, N_x$, and the variations of lubricant viscosity due to variations of the molecular weight W can be neglected, (c) the concept of averaged viscosity introduced in calculation of the sliding frictional stress $f(x)$ will be used for approximation of the Jacobian of the system of linearized equations. Then using equations (11.19) and the two-term Taylor formula, we can linearize the modified Reynolds equation (11.42) in the vicinity of its i-th iterates as well as its boundary conditions for p and the integral condition in the form

$$\frac{\triangle h_k}{\sqrt{1 + \frac{\varepsilon H_0^2}{4}(\frac{dh}{dx})^2}} - \frac{h \triangle h_k}{4}\left[\frac{\tau_L}{\mu(x,h/2)}G(\frac{\tau_+}{\tau_L}) - \frac{\tau_L}{\mu(x,-h/2)}G(\frac{\tau_-}{\tau_L})\right]$$

$$-\triangle p_k \int_{-h/2}^{h/2} \frac{\partial \tau_L}{\partial p}\frac{1}{\mu}G(\frac{\tau}{\tau_L})\zeta d\zeta + \triangle p_k \int_{-h/2}^{h/2} \frac{\tau_L}{\mu^2}\frac{\partial \mu}{\partial p}G(\frac{\tau}{\tau_L})\zeta d\zeta$$

$$-\int_{-h/2}^{h/2} \frac{\tau_L}{\mu}G'(\frac{\tau}{\tau_L})\left[-\frac{1}{\tau_L^2}\frac{\partial \tau_L}{\partial p}\tau \triangle p_k + \frac{1}{\tau_L}(\triangle f_k + \frac{24 H_0}{V}\zeta \frac{dp}{dx}\triangle H_0\right.$$

$$\left.+\frac{12 H_0^2}{V}\zeta \triangle(\frac{dp}{dx}) - \frac{12 H_0^2}{V}\zeta \frac{dp}{dx}\frac{\triangle c}{c-a})\right]\zeta d\zeta = \int_{-h/2}^{h/2} \frac{\tau_L}{\mu}G(\frac{\tau}{\tau_L})\zeta d\zeta$$

$$-\frac{h}{\sqrt{1 + \frac{\varepsilon H_0^2}{4}(\frac{dh}{dx})^2}}, \quad k = 2, \ldots, N_x - 1, \quad \triangle p_1 = \triangle p_{N_x} = 0, \qquad (11.43)$$

$$\frac{1}{2}\sum_{k=1}^{N_x - 1}[\triangle p_k + \triangle p_{k+1}](x_{k+1} - x_k)$$

$$+\frac{\triangle c}{2(c-a)}\sum_{k=1}^{N_x - 1}[\triangle p_k + \triangle p_{k+1}](x_{k+1} - x_k) = \frac{\pi}{2}$$

$$-\frac{1}{2}\sum_{k=1}^{N_x - 1}[p_k + p_{k+1}](x_{k+1} - x_k),$$

$$\tau = f + \frac{12 H_0^2}{V}\zeta \frac{dp}{dx}, \quad \tau_+ = f + \frac{6 H_0^2}{V}h\frac{dp}{dx}, \quad \tau_- = f - \frac{6 H_0^2}{V}h\frac{dp}{dx}.$$

To determine the values of $\triangle f(x_k)$ and $\triangle h(x_k)$, $k = 1, \ldots, N_x$, we linearize the equations for $f(x)$ and $h(x)$ from (11.19)

$$\frac{\triangle h_k}{2}\left[\frac{\tau_L}{\mu(x,h/2)}G(\frac{\tau_+}{\tau_L}) + \frac{\tau_L}{\mu(x,-h/2)}G(\frac{\tau_-}{\tau_L})\right]$$

$$+\triangle p_k\Big[\int\limits_{-h/2}^{h/2}\frac{\partial\tau_L}{\partial p}\frac{1}{\mu}G(\frac{\tau}{\tau_L})d\zeta - \int\limits_{-h/2}^{h/2}\frac{\tau_L}{\mu^2}\frac{\partial\mu}{\partial p}G(\frac{\tau}{\tau_L})d\zeta\Big]$$

$$+ \int\limits_{-h/2}^{h/2}\frac{\tau_L}{\mu}G'(\frac{\tau}{\tau_L})[-\frac{1}{\tau_L^2}\frac{\partial\tau_L}{\partial p}\tau\triangle p_k + \frac{1}{\tau_L}(\triangle f_k + \frac{24H_0}{V}\zeta\frac{dp}{dx}\triangle H_0$$

$$+\frac{12H_0^2}{V}\zeta\triangle(\frac{dp}{dx}) - \frac{12H_0^2}{V}\zeta\frac{dp}{dx}\frac{\triangle c}{c-a})]d\zeta = \frac{s_0}{\sqrt{1+\frac{\varepsilon H_0^2}{4}(\frac{dh}{dx})^2}}$$

$$- \int\limits_{-h/2}^{h/2}\frac{\tau_L}{\mu}G(\frac{\tau}{\tau_L})d\zeta, \quad k=1,\dots,N_x, \tag{11.44}$$

$$H_0\triangle h_k = \frac{1}{\pi}\sum_{k=1}^{N_x-1}[\triangle p_k + \triangle p_{k+1}]I(x_k,x_{k+1},c) - \triangle H_0(h-1)$$

$$+[(x_k-c)(x_k+c-2a) - H_0(h-1)]\frac{\triangle c}{c-a}, \tag{11.45}$$

$$I(x_k,x_{k+1},c) = \int\limits_{x_k}^{x_{k+1}}\ln\frac{|c-x|}{|x_k-x|}dx.$$

In the above equations all variables whose indexes are not shown are calculated at x_k based on the values of ith iterates of the problem solution. Equations (11.43)-(11.45) are obtained by making the transition from the variable x to the variable y according the substitution from equation (11.22), then linearizing equations and, finally, returning back to the variable x.

Taking into account equation (11.44) for $\triangle f(x_k)$ and the fact that $\triangle f(x_k)$ are independent from z, we can eliminate it from equation (11.43). After that with the help of the expression for $\triangle h(x_k)$, we can reduce solution of the modified Reynolds equation to a system of $N_x + 2$ linear algebraic equations for the increments of pressure $\triangle p(x_k)$, $k = 1,\dots,N_x$, film thickness $\triangle H_0$, and exit coordinate $\triangle c$. Solution of this system is an extremely long process as coefficients of the system depend on the values of a number of integrals in equations (11.43)-(11.45). In equations (11.43) and (11.44), the necessity of numerical evaluation of a number of integrals with respect to z is the main cause of the significant slow down of solution of the above system. It is obvious, that to solve the equation for $f(x)$ and the modified Reynolds and gap equations we must evaluate the integrals in the right-hand sides of equations (11.43) and (11.44). To significantly simplify and to evaluate analytically the rest of the integrals in equations (11.43), (11.44) that control the Jacobian of the system, we will use instead of the local lubricant viscosity $\mu(x,z)$ and the sliding frictional stress $f(x)$ the averaged viscosity $\mu_{avg}(x)$ and the averaged sliding stress $f_{avg}(x)$, respectively. This assumption allows us to do all integrations analytically using integration by parts which significantly speeds up the calculations of the approximate Jacobian of the system. Integrals in equation (11.45) are evaluated analytically.

After the above system is solved for the incremental variations in pressure, film thickness, and exit coordinate the incremental variations of pressure undergo the process of filtering out high frequencies. It is done by applying the Fast Fourier transform to the set of $\triangle p(x_k)$, $k = 1, \ldots, N_x$, cutting off the tail of high frequencies, and inverting the filtered values of $\triangle p(x_k)$, $k = 1, \ldots, N_x$. The degree of filtering decreases with the increase of the iteration number and stops after a certain number of iterations. As soon as filtering is done the new iterates of pressure, film thickness, and exit coordinate are determined by the formulas

$$p_k^{i+1} = p_k^i + \alpha_R \triangle p_k, \ k = 1, \ldots, N_x, \ H_0^{i+1} = H_0^i + \alpha_R \triangle H_0,$$

$$c^{i+1} = c^i + \alpha_R \triangle c,$$

(11.46)

where α_R is a positive damping coefficient the value of which is changed depending on the iteration number i. For a certain number of first iterations α_R is kept small and as the iteration process progresses, and the approximate solution gets sufficiently close to the exact one the value of α_R gradually increases.

As we move to a grid with smaller step sizes the Jacobian of the system is calculated on just first few iterations after which it is not changed from iteration to iteration until the solution converges to the desired precision. That significantly accelerates the iterative process without hampering its convergence.

The derivative of the gap h is used in a number of steps of the iteration process. To make the derivative of the gap dh/dx consistent with the Reynolds equation after each iteration dh/dx is determined from the Reynolds equation according to the formula

$$\frac{dh}{dx} = -\sqrt{1 + \frac{\varepsilon H_0^2}{4}\left(\frac{dh}{dx}\right)^2} \int\limits_{-h/2}^{h/2} \frac{\partial u(x,z)}{\partial x} dz.$$

(11.47)

The partial derivative $\partial u/\partial x$ is calculated from the expression for the directional derivative du/ds

$$\frac{du}{ds} = \frac{\partial u}{\partial x}\frac{\triangle x}{\sqrt{\triangle x^2 + \varepsilon H_0^2 \triangle z^2}} + \frac{\partial u}{\partial z}\frac{\triangle z}{\sqrt{\triangle x^2 + \varepsilon H_0^2 \triangle z^2}},$$

(11.48)

which is obtained numerically using the expression for $\partial u/\partial z$ that is analytically determined from the equation for $u(x, z)$ (see equations (11.19)).

In case of an isothermal problem τ_L may be considered independent of z, and τ_L can be pulled out of the integrals in equations (11.19). In addition to that, in case of no lubricant degradation the latter equations can be simplified further as the lubricant viscosity μ is independent of z. Finally, under the above isothermal conditions the rheological model described by equation (11.1) allows for analytical calculation of integrals in equations (11.19) that may simplify and speed up the whole process of numerical solution. For a case of non-isothermal lubrication, the numerical method can be design in a similar fashion by averaging not only μ but also τ_L as it may depend on the lubricant temperature and, as a result of that, on variable z.

11.5   Solutions for Lubricants with Newtonian and Non-Newtonian Rheologies without Degradation

In this section we will consider the EHL solutions in cases of lubricants with Newtonian and non-Newtonian rheologies without degradation. That allows determining the reversible loss of lubricant viscosity.

One of the challenges in application of the above approach to practical EHL problems is the necessity to calibrate the kinetics part of the model, i.e., to be able to choose the proper values for constants C and α. The actual values of these parameters must be determined from detailed experimental data with all necessary input and output data available. Below we will consider two series of results obtained for different combinations of parameters C and α: Series I for

$$C = 15.5, \ \alpha = 0.008, \tag{11.49}$$

and Series II for

$$C = 0.055, \ \alpha = 0.008. \tag{11.50}$$

In Series II simulation of only nominally Newtonian lubricant will be considered. In both series of simulations, we used the following dimensional

$$U = 347kJ/mole, \ T = 310°K, \ a_* = 0.374nm, \ l_* = 0.154nm,$$

$$w_m = 35.1g/mole, \ a_H = 10^{-3}m, \ p_H = 1.284GPa, \tag{11.51}$$

and dimensionless parameters

$$a = -2, \ V = 0.1, \ \varepsilon = 10^{-4}, \ \gamma = 1.0774, \ \kappa = 1, \tag{11.52}$$

where a_H and p_H are the Hertzian half-length of the contact and the Hertzian maximum pressure, respectively.

For Series I simulations the following additional values of dimensional parameters [29, 30]

$$\mu_a = 0.00125Pa \cdot s, \ c_p = 1.204g/dL \tag{11.53}$$

and dimensionless parameters

$$a = -2, \ V = 0.1, \ Q = 5, \ \tau_{L0} = 0.002, \ \tau_{L1} = 0.046,$$

$$\delta = 0.671 \cdot 10^8, \ \theta = 4.524 \cdot 10^{-3}, \tag{11.54}$$

are used while for Series II simulations the values of additional dimensional parameters [24, 29, 30]

$$\mu_a = 0.00924 Pa \cdot s, \ c_p = 0.86 g/dL \tag{11.55}$$

and dimensionless parameters

$$Q = 11, \ \delta = 0.256 \cdot 10^{10}, \ \theta = 3.231 \cdot 10^{-3}, \tag{11.56}$$

are used.

In particular simulations we employed: $M = 3001$ is the number of x_m nodes along the x-axis used for calculation of the flow streamlines and the solution of the kinetic equation, $M_R = 501$ and $M_R = 1501$ are the numbers of nodes used for sequential solution of the Reynolds equation on crude and fine grids, respectively, $J = 641$ is the number of nodes across the gap, the number of nodes with respect to the polymer chain length l was 232. For non-degrading lubricants solution of Reynolds equation was done only on the crude grid with the number of x_m nodes equal to $M_R = 501$. Both grids along and across the film layer were uniform while the grid with respect to the polymer chain length was non-uniform. To speed up the solution process on the fine grid, the Jacobian for Newton's method was calculated just during the first few iterations after which it was kept unchanged until the process converged. The intermediate tolerance levels for Newton's method used for calculation of $f(x)$, $p(x)$, H_0, and c were set equal to 10^{-7}. However, the actual solution precision is controlled by the grid step sizes, and in the discussed cases it was about 10^{-4}. Because the solution depends on three variables (x, z, and l) the numerical solution of the problem requires extremely intensive CPU calculations and high computer memory. For the Newtonian lubricant the predominant fraction of the time required for problem solution is associated with solution of the kinetic equation while for the non-Newtonian lubricant the predominant fraction of the time required for problem solution is associated with solution of the equation for $f(x)$, the Reynolds and gap equations, and the kinetic equation. The initial distribution of the polymer molecular weight $W_a(l)$ is taken from test measurements done by Covitch [12].

First, let us consider the non-Newtonian behavior of the lubricant in the contact. In the inlet zone of a heavily loaded lubricated contact, the pressure is small and the lubricant viscosity is close to one while the gap is large. Therefore, in the inlet zone the local sliding frictional stress $f(x)$ is small. Because of that and the fact that the derivative of pressure dp/dx is relatively small the rheological function G behaves like a linear function. It means that the behavior of G resembles the one of a fluid with Newtonian rheology. Therefore, taking into account that in heavily loaded lubricated contacts the film

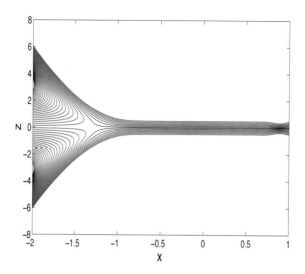

FIGURE 11.3

Flow streamlines $z(x)$ for non-degrading lubricant with Newtonian rheology and Series I input data, $s_0 = 0$ (after Kudish and Airapetyan [25]). Reprinted with permission from the ASME.

thickness H_0 is primarily determined by the inlet zone [31] we should expect that the difference between the lubrication film thicknesses in the contacts with the Newtonian and non-Newtonian lubricants is relatively small. The situation is different in case of a lightly loaded lubricated contact (see [32]) where the film thickness is determined by the solution of the problem in the whole contact. On the other hand, in the above cases for high slide-to-roll ratios s_0 the sliding frictional stress $f(x)$ for a lubricant with non-Newtonian rheology $(G(x) = tanh^{-1}x)$ is noticeably lower than that for a lubricant with Newtonian rheology $(G(x) = x)$ given the same ambient lubricant viscosity.

For a non-degrading lubricant with Newtonian rheology, the solution of the isothermal EHL problem is independent of the slide-to-roll ratio s_0. In particular, for the Newtonian lubricant with the input data for Series I we have $H_0 = 0.1966$ and $c = 1.0513$ while for the input data for Series II we have $H_0 = 0.339$ and $c = 1.052$. For lubricants with non-Newtonian rheology, the solution of the isothermal EHL problem depends on s_0. For the non-Newtonian lubricant (i.e., for Series I input data), $s_0 = 0$, and τ_{L0} decreasing from 0.01 to 0.002 the film thickness monotonically decreases from $H_0 = 0.1967$ to $H_0 = 0.1963$, respectively, while for $s_0 = -0.5$ the film thickness monotonically decreases from $H_0 = 0.1967$ to $H_0 = 0.1961$, respectively. Within the same range of variations of s_0 and τ_{L0}, the value of c varies by no more than seven units in the fifth place after the decimal point.

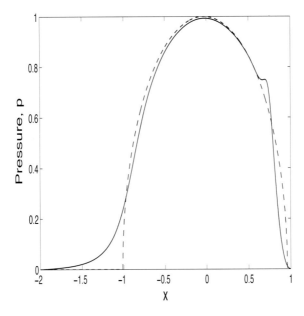

FIGURE 11.4
Pressure distribution $p(x)$ for lubricant with Newtonian (solid line) rheology
and Series I input data, $s_0 = 0$. The Hertzian pressure distribution is represented by a dashed line (after Kudish and Airapetyan [25]). Reprinted with
permission from the ASME.

The above data indicate that the film thickness H_0 for a lubricant with non-Newtonian rheology is practically identical to the one for the lubricant with
the Newtonian rheology. Also, the value of H_0 is almost independent of the
slide-to-roll ratio s_0 and depends significantly on the value of the dimensionless
pressure coefficient Q. At the same time, the sliding frictional stress $f(x)$ is
noticeably affected by the lubricant rheology, pressure coefficient of viscosity
Q, speed-load parameter V, limiting stress τ_L, and the slide-to-roll ratio s_0.

Therefore, the behavior of pressure $p(x)$, gap $h(x)$, film thickness H_0, coordinate of the exit point c, and the lubricant flow streamlines $z(x)$ in the
cases of Newtonian and non-Newtonian lubricants are very similar. For Series
I input data and $s_0 = 0$ the graphs of the flow streamlines $z(x)$, pressure $p(x)$,
and gap $h(x)$ for the lubricant with Newtonian rheology are given in Figs.
11.3-11.5, respectively. For the lubricant with the non-Newtonian rheology
the graphs of p and h are very close to the corresponding ones for the Newtonian fluid. In case of $s_0 = 0$ for any rheological function G the sliding frictional
stress $f(x) = 0$ because $G(x)$ is an odd function of x. For $s_0 = -0.5$ the graphs
of $f(x)$ for the non-degrading Newtonian and non-Newtonian lubricants

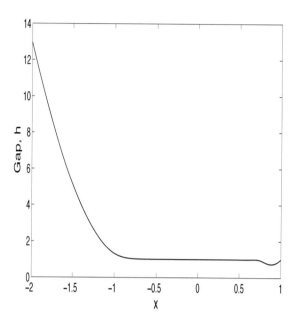

FIGURE 11.5

Gap distribution $h(x)$ for lubricant with Newtonian rheology and Series I input data, $s_0 = 0$ (after Kudish and Airapetyan [25]). Reprinted with permission from the ASME.

are presented in Fig. 11.6. The behavior of $f(x)$ is distinctly different from the one of pressure $p(x)$: the magnitude of the values of $f(x)$ is greater in the case of the Newtonian lubricant than in the case of the non-Newtonian one. This effect is due to the reversible viscosity loss of the non-Newtonian lubricant in comparison with the Newtonian one. A similar behavior of the surface frictional stresses τ_1 and τ_2 can be seen in Fig. 11.7, where τ_1 and τ_2 are given for the lubricants with Newtonian and non-Newtonian rheologies under mixed rolling and sliding conditions ($s_0 = -0.5$) as well as the surface shear stress $\tau_1 = -\tau_2$ for the Newtonian lubricant (dashed-dotted line) under pure rolling conditions ($s_0 = 0$). For the case of the pure rolling ($s_0 = 0$) for the non-Newtonian lubricant, the surface shear stresses $\tau_1 = -\tau_2$ are very close to the ones for the case of the Newtonian lubricant.

For Series II input data and $s_0 = 0$ the behavior of pressure $p(x)$, gap $h(x)$, and the lubricant flow streamlines $z(x)$ resembles the one shown for the case of Newtonian lubricant for Series I input data in Figs. 11.3-11.5. For Series II input data in the case of pure sliding ($s_0 = -2$), the behavior of pressure $p(x)$ and gap $h(x)$ is still very close to the one for the case of pure rolling ($s_0 = 0$). However, the behavior of the streamlines in case of pure sliding ($s_0 = -2$) is

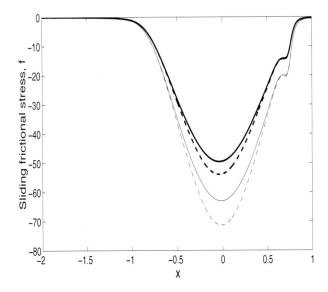

FIGURE 11.6
Sliding frictional stress $f(x)$ distributions for non-degrading lubricants with
Newtonian (dashed line) and non-Newtonian (solid line) rheologies and for
degrading lubricants with Newtonian (thick dashed line) and non-Newtonian
(thick solid line) rheologies for Series I input data, $s_0 = -0.5$ (after Kudish
and Airapetyan [25]). Reprinted with permission from the ASME.

different and it is shown in Fig. 11.8. There is only one set of flow streamlines
entering the contact region and turning around and one set of streamlines
running through the the entire contact.

The reversible loss of viscosity of the non-Newtonian lubricant can be determined as a ratio of the sliding frictional stresses $f(x)$ for non-Newtonian
and Newtonian lubricants in the Hertzian region. It is based on the fact that
in the Hertzian region of a heavily loaded lubricated contact $h(x)-1 \ll 1$ [28].
Therefore, from Fig. 11.6 follows that for $s_0 = -0.5$ the reversible lubricant
viscosity loss of the non-Newtonian lubricant reaches 12% of its original value.

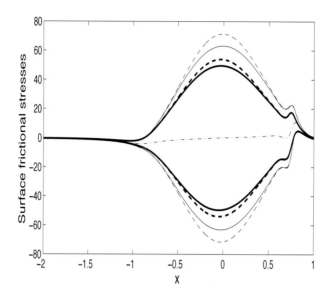

FIGURE 11.7

Frictional stresses τ_1 and τ_2 applied to the contact surfaces for non-degrading (dashed lines) and degrading (thick dashed lines) Newtonian lubricants and for non-degrading (solid lines) and degrading (thick solid lines) non-Newtonian lubricants under mixed rolling and sliding conditions ($s_0 = -0.5$) and surface frictional stresses $\tau_2 = -\tau_1 = (6H_0^2/V)hdp/dx$ for non-degrading Newtonian lubricant (dash-dotted line) under pure rolling conditions ($s_0 = 0$) and Series I input data (after Kudish and Airapetyan [25]). Reprinted with permission from the ASME.

11.6   EHL Solutions for Lubricants with Newtonian and Non-Newtonian Rheologies with Degradation

In this section we will compare the EHL solutions for a lubricant with non-Newtonian rheology experiencing the stress-induced degradation with a solution for a lubricant with a similar rheology but without degradation. That allows determining the irreversible loss of lubricant viscosity. Moreover, we will compare the EHL solutions for degrading lubricants with Newtonian and non-Newtonian rheologies. The simulation input data are taken identical to those in Section 11.5.

Let us consider some general properties of a solution of the isothermal EHL problem for a degrading lubricant with non-Newtonian rheology. The behavior of pressure $p(x)$ and gap $h(x)$ distributions as well as of the film thickness H_0

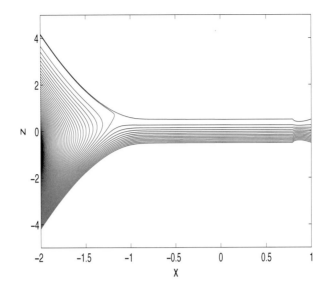

FIGURE 11.8
Flow streamlines $z(x)$ for non-degrading lubricant with Newtonian rheology and Series II input data, $s_0 = -2$ (after Kudish and Airapetyan [24]). Reprinted with permission from the ASME.

and exit coordinate c of the latter problem with respect to parameters a, V, and Q is similar to that described in the preceding chapter for lubricants with Newtonian and non-Newtonian rheologies (also see Kudish [6, 31, 32]). For a lubricant with Newtonian rheology increase in load P leads to increase in the Hertzian pressure p_H, which, in turn, increases Q and causes a relatively slow increase in H_0 and a rapid increase in $f(x)$ and, therefore, in $\tau(x, z)$. That, in turn, leads to a rapid lubricant degradation (see equations (11.20) and (11.21)). For a lubricant with non-Newtonian rheology increase in load P also leads to increase in the sliding frictional stress $f(x)$ and, thus, to increase in the shear stress τ. However, these increases in $f(x)$ and τ are moderated by the lubricant rheology, i.e., these increases are bounded by the limiting stress τ_L. Therefore, for a non-Newtonian lubricant the shear stress τ for high loads P is much lower than for a similar Newtonian lubricant. It means that the degradation process of such a non-Newtonian lubricant runs slower than for a Newtonian counterpart. A usually very small parameter ε almost does not affect the problem solution. For small τ_{L0} and τ_{L1} the shear stress τ in a non-Newtonian lubricant is small, which, in turn, slows down the process of lubricant degradation (see equations (11.21) for R and L_*). As the values of τ_{L0} and τ_{L1} increase, the lubricant shear stress τ and the film thickness H_0

increase. As the slide-to-roll ratio s_0 increases the film thickness H_0 slowly decreases. However, variations in H_0 are insignificant while the increase in τ is directly proportional to the increase in τ_L, i.e., the increase in τ_{L0} and τ_{L1}. The latter causes lubricant degradation to run faster (see equations (11.21) for R and L_*). Moreover, an increase in the value of parameter δ leads to a corresponding increase in the characteristic chain length L_* that causes lubricant degradation to slow down (see equations (11.21) for R and L_*). For smaller values of parameter γ the process of lubricant degradation runs slower than for the larger ones (see equations (11.21) for R). For higher values of θ, the lubricant viscosity responds stronger to changes in the molecular weight distribution, i.e., for higher θ the loss of lubricant viscosity is higher (see equations (11.21) for μ). The value of parameter κ in the left-hand side of the kinetic equation in (11.20) controls the rate of lubricant convection and, therefore, the time a lubricant small volume is present in the contact area. For high values of κ, the time of lubricant presence in the contact area is small and the rate of lubricant degradation is low.

First, let us examine a case of pure rolling ($s_0 = 0$) for Series I input parameters. Under these conditions lubricants with Newtonian and non-Newtonian rheologies degrade to a lesser extent than in cases when $s_0 \neq 0$. The solutions for Newtonian and non-Newtonian lubricant rheologies are qualitatively and quantitatively very close to each other because $|G^{-1}(x)| \leq |x|$ and $G(x) \to x$ as $x \to 0$. The map of flow streamlines for the non-Newtonian lubricant is presented in Fig. 11.9. This map shows that the flow streamlines are symmetric about the line $z = 0$. Because of the symmetry of the problem the lubricant viscosity μ and shear rate $\partial u / \partial z$ are continuous functions of x and z. All of the flow streamlines enter the contact through the inlet cross section of the lubrication film. Some of them that are adjacent to the contact surfaces run through the whole contact. However, the other streamlines turn around and exit the contact region also through the inlet cross section of the film. The pressure and gap distributions are very close to the ones for the non-degrading Newtonian lubricant presented in Figs. 11.4 and 11.5. The maximum difference between the gap distributions for Newtonian and non-Newtonian fluids reaches 8%.

The map of the horizontal component u of the lubricant velocity along the flow streamlines for the non-Newtonian lubricant is given in Fig. 11.10. Because for $s_0 = 0$ the flow is symmetrical about the line $z = 0$, Fig. 11.10 shows the lubricant velocity along four sets of streamlines: two (coinciding) sets of streamlines passing through the whole contact and two (coinciding) sets of streamlines that turn around and exit through the inlet cross section. In this case the exit film thicknesses and exit coordinates for the Newtonian and non-Newtonian degrading lubricants are $H_0 = 0.1829$, $c = 1.0527$ and $H_0 = 0.1834$, $c = 1.0527$, respectively. The comparison of the data for H_0 obtained for degrading and non-degrading lubricants for pure rolling $s_0 = 0$ shows a relatively small affect of lubricant degradation on the film thickness H_0 (only about 8%).

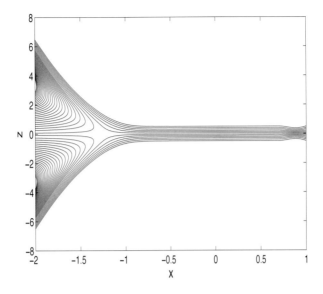

FIGURE 11.9
Flow streamlines for degrading lubricant with non-Newtonian rheology under pure rolling conditions ($s_0 = 0$) and Series I input data (after Kudish and Airapetyan [25]). Reprinted with permission from the ASME.

The distributions of the lubricant viscosity μ and polymer molecular weight W for lubricants with Newtonian and non-Newtonian rheologies are practically identical. The fact that for $s_0 = 0$ lubricants with Newtonian and non-Newtonian rheologies degrade relatively slowly can be also seen from the graphs of the reciprocal of the lubricant viscosity (see Fig. 11.11) and the distribution of the molecular weight W along the flow streamline $z(x)$ that is closest to the line $z = 0$ and running through the whole contact below $z = 0$ (see Figure 11.12). In Fig. 11.5, the variable z' is the artificially stretched z-coordinate across the film thickness (namely, $z' = zh(a)/h(x)$) to make the relationship more transparent. It is clear from the graphs of the reciprocal of the lubricant viscosity (see Fig. 11.11) that for the Newtonian and non-Newtonian lubricants the maximum irreversible viscosity loss reaches about 31%. Lubricant degradation depends on a number of parameters among which the shear stress τ (see equations (11.21)) plays one of the major roles. For $s_0 = 0$ the sliding frictional stress $f(x) = 0$ and, therefore, the shear stress $\tau = (12H_0^2/V)zdp/dx$ reaches its extrema in the inlet and exit zones while in the Hertzian region it is relatively small [31].

That leads to the conclusion that for pure rolling ($s_0 = 0$) practically all lubricant degradation occurs in the inlet zone while the lubricant almost does

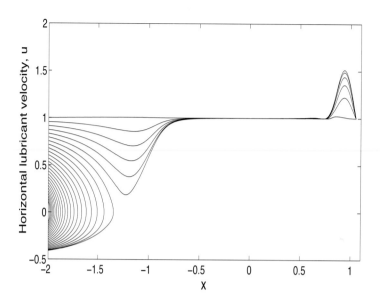

FIGURE 11.10

The horizontal component u of the lubricant velocity along the flow stream-lines for degrading lubricant with non-Newtonian rheology under pure rolling conditions ($s_0 = 0$) and Series I input data (after Kudish and Airapetyan [25]). Reprinted with permission from the ASME.

not degrade in the Hertzian region and the exit zone. The latter also can be clearly seen from the distributions of the molecular weight W in Fig. 11.12.

For the given lubricant flow parameters, the degree of lubricant degradation is higher along the longer streamlines. The length of the flow streamlines adjacent to the contact surfaces is larger and the absolute value of the shear stress along them is slightly higher than for the ones that turn around and exit through the inlet cross section. In spite of that fact lubricant degradation along the turning around streamlines is higher because they are completely located in the inlet of the contact, practically the only contact zone where lubricant degradation takes place. Also, it follows from the expression for the shear stress $\tau = (12H_0^2/V)z\,dp/dx$ that along the streamlines running through the whole contact the degree of lubricant degradation is higher near the contact surfaces. Among the turning around streamlines the degree of lubricant degradation is higher along the longest streamlines that enter the contact region and are closest to the contact surfaces and exit the contact through the inlet cross section close to $z = 0$. Obviously, for higher viscosity μ, the film thickness H_0 is higher while dp/dx changes insignificantly. Therefore, the shear stress τ is higher that promotes faster lubricant degradation and higher viscosity loss.

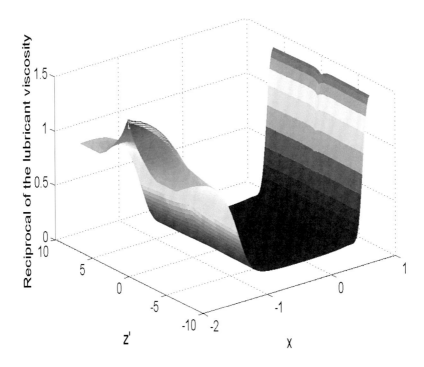

FIGURE 11.11 (See color insert)

The reciprocal of the lubricant viscosity μ in Newtonian and non-Newtonian lubrication film under pure rolling conditions ($s_0 = 0$) and Series I input data. The variable z' is an artificially stretched z-coordinate across the film thickness (namely, $z' = zh(a)/h(x)$) to make the relationship more transparent (after Kudish and Airapetyan [25]). Reprinted with permission from the ASME.

Similarly, for higher pressure viscosity coefficients Q one can expect faster degradation and higher viscosity loss.

Now, let us examine a case of mixed rolling and sliding with $s_0 = -0.5$ for Series II input data. In this case the solutions of the problem for the lubricants with Newtonian and non-Newtonian rheologies are also very close to each other. The maps of the flow streamlines $z(x)$ and the horizontal component u of the lubricant velocity along the flow streamlines for the degrading Newtonian and non-Newtonian lubricants are given in Figs. 11.13 and 11.14, respectively. The lubricant flows are no longer symmetric about $z = 0$ and the maps of their flow streamlines are completely different from the ones for the pure rolling case. There are still two sets of streamlines with flow reversal as well as two sets of flow streamlines adjacent to the lower and upper contact surfaces and running through the whole contact that can be distinctly seen in Fig.

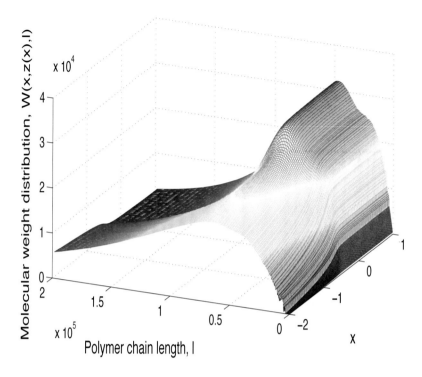

FIGURE 11.12 (See color insert)

The molecular weight distribution W of degrading lubricants with Newtonian and non-Newtonian rheologies along the flow streamline closest to $z = 0$ and running through the whole contact below $z = 0$ under pure rolling conditions ($s_0 = 0$) and Series I input data (after Kudish and Airapetyan [25]). Reprinted with permission from the ASME.

11.13. The exit lubrication film thickness H_0 and the exit coordinate c exhibit behavior very similar to the case of pure rolling. For the non-degrading and degrading Newtonian lubricants, they are equal to $H_0 = 0.1961$, $c = 1.0513$ and $H_0 = 0.1819$, $c = 1.0531$, respectively. For the non-degrading and degrading non-Newtonian lubricants, the film thickness is equal to $H_0 = 0.1961$ and $H_0 = 0.1813$, respectively. In all other respects the distributions of the pressure p and gap h are very similar to the ones for the case of pure rolling. The numerical results show that in the case of mixed rolling and sliding for both lubricants with Newtonian and non-Newtonian rheologies the sliding frictional stress $f(x)$ (see Fig. 11.6) is smaller than in a similar case without degradation but still large enough to cause lubricant degradation to a much greater extent than in the case of pure rolling. In the case of mixed rolling and sliding, the surface frictional stresses τ_1 and τ_2 are much higher (see Fig.

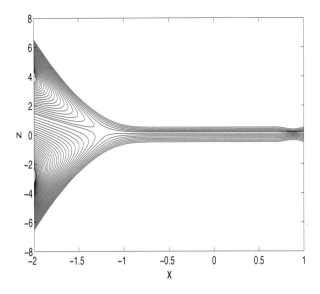

FIGURE 11.13
Flow streamlines $z(x)$ for degrading lubricants with Newtonian and non-Newtonian rheologies under mixed rolling and sliding conditions ($s_0 = -0.5$) and Series I input data (after Kudish and Airapetyan [25]). Reprinted with permission from the ASME.

11.7) than the ones for the case of pure rolling ($\tau_2 = -\tau_1 = (6H_0^2/V)hdp/dx$) and, at the same time, lower than for the case of no degradation. That explains the stronger lubricant degradation in the case of mixed rolling and sliding conditions in comparison with the case of pure rolling conditions (see Figs. 11.11, 11.12, 11.2, and 11.15). For the degrading lubricants, the frictional stresses τ_1 and τ_2 applied to the contact surfaces are on average lower than for the non-degrading lubricant by about 26%. The frictional stress τ in the lubrication film is mostly concentrated near the contact surfaces. Depending on the sign of dp/dx the maximum of the absolute value of the surface frictional stress is reached either on the lower or upper contact surfaces. That leads to faster lubricant degradation near that contact surface and to slower lubricant degradation near the other one. This can be clearly seen from the behavior of the reciprocal of the lubricant viscosity μ in cases of Newtonian and non-Newtonian lubricants (see Fig. 11.2). In the case of $s_0 = -0.5$ the maximum loss of the lubricant viscosity is approximately 41%. That can be seen from the comparison of graphs of the sliding frictional stress $f(x)$ for the non-degrading and degrading lubricants (see Fig. 11.6) as well as from the graph of the distribution of the reciprocal of the lubricant viscosity μ (see

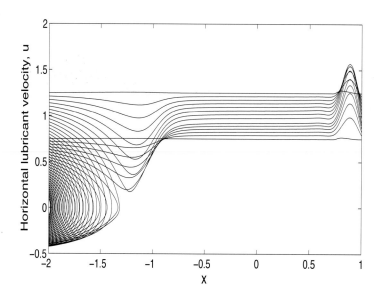

FIGURE 11.14

The horizontal component u of the lubricant velocity along the flow streamlines for degrading lubricants with Newtonian and non-Newtonian rheologies
under mixed rolling and sliding conditions ($s_0 = -0.5$) and Series I input
data (after Kudish and Airapetyan [25]). Reprinted with permission from the
ASME.

Fig. 11.2).

The behavior of the polymer molecule distribution W for the lubricants with
Newtonian and non-Newtonian rheologies along the flow streamline $z_{31}(x)$
running through the whole contact below $z = 0$ and located next to the
first turning around streamline is given in Fig. 11.15. The comparison of the
polymer molecular weight W distributions for the cases of pure rolling and
mixed rolling and sliding (see Fig. 11.12 and 11.15) shows that in the case of
pure rolling the lubricant degrades slower than in the case of mixed rolling
and sliding. In the latter case rapid polymer scission occurs throughout the
entire contact region while in the former case it is mostly concentrated in the
inlet zone of the contact. That can be seen from the values of $W(x, z(x), l)$
for polymer molecules with short chain lengths l at different points along the
flow streamlines. Moreover, from Fig. 11.15, it follows that for Newtonian and
non-Newtonian lubricants under mixed rolling and sliding conditions lubricant
degradation occurs throughout the contact. For non-Newtonian lubricant at
higher values of the slide-to-roll ratio $\mid s_0 \mid$, the relative impact of the rolling
frictional stress $(6H_0^2/V)zdp/dx$ on τ decreases slowly while for the Newtonian

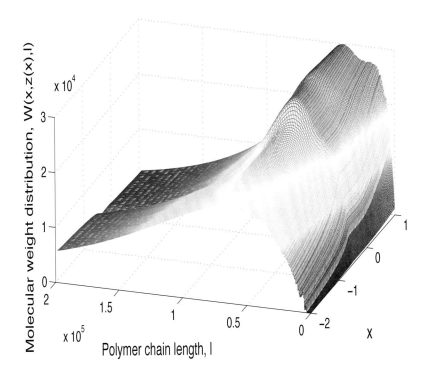

FIGURE 11.15 (See color insert)
The molecular weight distribution W of degrading lubricants with Newtonian and non-Newtonian rheologies along the flow streamline $z_{31}(x)$ closest to $z = 0$ and running through the whole contact below $z = 0$ under mixed rolling and sliding conditions ($s_0 = -0.5$) and Series I input data (after Kudish and Airapetyan [25]). Reprinted with permission from the ASME.

lubricant it decreases significantly. Moreover, the described differences in the degradation behavior of the lubricants with Newtonian and non-Newtonian rheologies may be the key to understanding why synthetic lubricants, which usually exhibit pseudoplastic behavior, degrade at a much slower pace.

Now, let us turn to the case of Series II input data. Under conditions of pure sliding, the behavior of pressure $p(x)$ and gap $h(x)$ in a lubricated contact is very similar to the one described for Series I input data. Even under conditions of pure sliding ($s_0 = -2$) the behavior of pressure $p(x)$ and gap $h(x)$ in a lubricated contact is very similar to the one described for the cases of pure rolling and mixed rolling and sliding. The reduction in the film thickness H_0 due to polymer degradation is slightly higher than for the cases of $s_0 = 0$ and $s_0 = -0.5$, i.e., $H_0 = 0.295$ and $H_0 = 0.339$ with and without lubricant degradation, respectively. For the case of $s_0 = -2$, there is just one set of

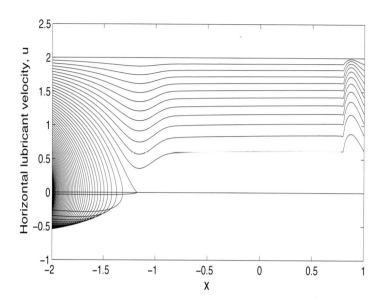

FIGURE 11.16

The horizontal component u of the lubricant velocity along the flow stream-lines for degrading lubricants with Newtonian rheology under pure sliding conditions ($s_0 = -2$) and Series II input data (after Kudish and Airapetyan [24]). Reprinted with permission from the ASME.

turning around flow streamlines adjacent to the motionless upper contact surface and one set of flow streamlines running through the whole contact that is adjacent to the moving lower contact surface everywhere in the contact (see Fig. 11.8). The map of the horizontal component u of the lubricant velocity along these flow streamlines is shown in Fig. 11.16. The frictional stresses τ_1 and τ_2 applied to the surfaces are much higher than in the case of pure rolling $s_0 = 0$. The frictional stress τ in the lubrication layer is mostly concentrated near the motionless upper contact surface ($z = h/2$), and it is relatively low near the moving lower contact surface ($z = -h/2$). That leads to significant lubricant degradation near the upper contact surface and to low lubricant degradation near the moving lower contact surface. The numerical results show that on average in case of pure sliding a lubricant undergoes degradation to a much greater extent than in the case of pure rolling. This can be clearly seen from the behavior of the lubricant viscosity μ in the lubrication layer (see Fig. 11.17). In this case the average loss of the lubricant viscosity reaches about 25%. The local lubricant viscosity losses near the moving lower and motionless upper surfaces reach about 10% and 60%, respectively. For the degrading lubricant, the frictional stresses applied to the contact surfaces are generally

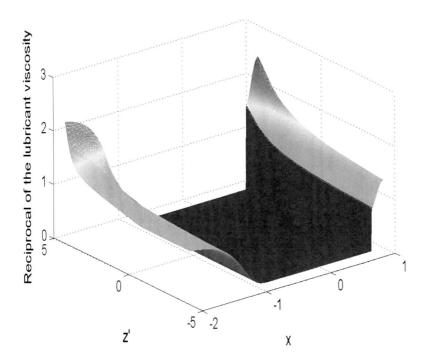

FIGURE 11.17 (See color insert)
The reciprocal of the lubricant viscosity μ in a Newtonian lubrication film under pure sliding conditions ($s_0 = -2$) and Series II input data. The variable z' is an artificially stretched z-coordinate across the film thickness (namely, $z' = zh(a)/h(x)$) to make the relationship more transparent (after Kudish and Airapetyan [24]). Reprinted with permission from the ASME.

lower than for the non-degrading lubricant by about 20%-30%. The general behavior of the polymer molecule distribution $W(x, z(x), l)$ to a certain extent is similar to the one obtained under pure rolling and mixed rolling and sliding conditions, and it is given along the flow streamline $z_{48}(x)$ running through the entire contact and located next to the first turning around streamline (see Fig. 11.18). Moreover, in case of pure sliding the lubricant degrades along the flow streamlines throughout the entire contact (see Fig. 11.18) while in the considered cases of pure rolling and mixed rolling and sliding the degradation occurs mostly in the inlet and zone (see Fig. 11.12 and 11.15).

It is convenient to analyze the process of polymer degradation based on the behavior of the characteristic polymer chain length L_* (see equations (11.21)), which represents the position of the center of the profile of the probability of polymer scission $R(x, z, l)$. In fact, the probability of polymer scission R is

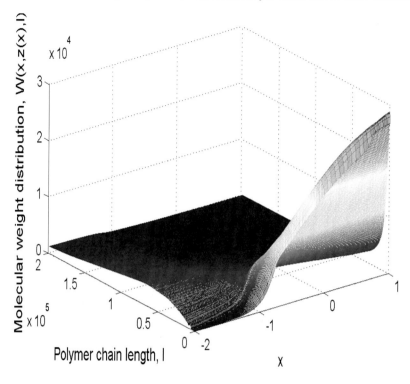

FIGURE 11.18 (See color insert)
The molecular weight distribution W of the degrading lubricant with Newtonian rheology along the flow streamline $z_{48}(x)$ running through the entire contact closest to the first turning around streamline under pure sliding conditions ($s_0 = -2$) and Series II input data (after Kudish and Airapetyan [24]). Reprinted with permission from the ASME.

the main key to the control over the lubricant degradation. As the zone of the chain lengths l, where $R(x, z, l)$ varies from 0 to 1 is quite narrow, one can conclude that the polymer molecules with chain lengths l shorter than L_* practically do not undergo scission while most of the molecules longer that L_* get degraded. It is important to emphasize that the degree of polymer degradation varies from point to point together with the shear stress τ, which, in turn, is a certain function of x and z. In the Hertzian region the shear stress τ is relatively large due to large lubricant viscosity μ while in the inlet and exit zones the values of the pressure derivative dp/dx and, therefore, the rolling frictional stress $(6H_0^2/V)z\,dp/dx$, are higher than the ones in the Hertzian zone. That explains the above-mentioned fact of lubricant degradation throughout the contact.

The effect of lubricant viscosity loss of up to 70% was observed experimentally and documented by Walker, Sanborn, and Winer [9]. In cases discussed in this chapter, the loss of lubricant viscosity reached about 41%. One has to keep in mind that in reality the value of the pressure coefficient of viscosity Q can be much higher than $Q = 5$, which may lead to a much higher viscosity loss.

It is important to realize that the lubricant viscosity loss caused by degradation is controlled by the contact operating conditions as well as by the lubricant rheology, the nature, concentration, and molecular weight distribution of the polymer additive in the lubricant supplied to the contact. Moreover, significant changes in the lubricant viscosity μ and film thickness H_0 may also lead to noticeable variations in frictional stresses τ_1 and τ_2 applied to the contact surfaces. That can also affect fatigue life of solids involved in a contact with degrading lubricant (see Kudish [33] and Section 11.7). Therefore, for any given operating conditions, it is desirable to select a lubricant with certain rheology and concentration and molecular weight distribution of a polymer additive that will provide sufficiently good lubrication conditions (high lubrication film thickness and low frictional stress) while the viscosity loss is reduced to minimum.

In case of a thermal EHL problem due to heat generation in the lubrication film and elevated lubricant temperature T, the lubricant viscosity (in case of no lubricant degradation) is lower than in a comparable isothermal case. That leads to higher values of not only T but also L_*, which slows down the degradation process. The extent to which the polymer degradation process slows down depends on the lubricant nature, nature and distribution of the molecular weight of polymer molecules entering the contact, and on the contact operating conditions.

The results based on such a model can be used for the evaluation of useful lubricant life and for more realistic estimation of the lubrication film thickness and frictional stresses in lubricated contacts. That may provide a solid basis for maintenance scheduling of lubricated machinery operating under various conditions. Moreover, it may enable researchers to create new and more durable lubricant additives for general and specific applications.

To conclude the section we can state that a plane isothermal elastohydrodynamic problem for a line contact lubricated by fluids with Newtonian and non-Newtonian rheologies with degrading polymer additives is modeled. The polymer additive in the lubricant, the viscosity modifier (VM), is considered to be of linear structure. The lubricant degradation is caused by stress-induced scission of polymer molecules while they move through the contact along the flow streamlines. The polymer degradation process is controlled by the probabilistic kinetic equation for the polymer molecular weight distribution. The lubricant flow is described by the generalized Reynolds equation. The lubrication and lubricant degradation processes are coupled through the lubricant viscosity that depends on pressure and temperature as well as on the concentration and distribution of the polymer molecular weight. The problem is solved numerically. In particular, the lubricant flow streamlines, distribution of the polymer molecular weight, reversible and irreversible lubricant viscosity losses, pressure, gap, and film thickness are determined. The stress-induced scission of polymer molecules leads to irreversible changes in the lubricant viscosity - irreversible viscosity loss. Numerical results show that in the cases of pure rolling, mixed rolling and sliding, and pure sliding the viscosity loss reaches about 31%, 41%, and 60%, respectively. For example, in the cases of mixed rolling and sliding, the reversible and irreversible viscosity losses at the exit from the contact can be as high as 12% and 41%, respectively. Moreover, the new property of a flow of degrading lubricant is discovered, i.e., the viscosity of a degrading lubricant generally is a discontinuous function of the position of a fluid particle. The viscosity loss in the inlet zone of the contact causes all parameters of the lubricated contact to change. In particular, in the case of mixed rolling and sliding conditions the lubrication film thickness of the degrading lubricants is about 8% thinner while in the case of pure sliding the lubrication film thickness of the degrading lubricants is about 13% thinner than the one for similar non-degrading lubricants. The effect of lubricant degradation for non-Newtonian lubricants is noticeably lower than for Newtonian lubricants. Reasonably large changes in frictional stresses applied to the contact surfaces (reached up to 40%) may noticeably impact fatigue life.

11.7 Effect of Lubricant Degradation on Contact Fatigue

In practice, most surfaces in relative motion operate in a lubricated environment. One of the often limiting parameters of such contacts is fatigue life. Fatigue life is also used as one of the constraints in design of joints subjected to cyclic loading. It is well known that contact fatigue is affected by contact pressure, frictional stress, residual stress, initial distribution of material flaws, etc. The behavior of contact pressure and, primarily, of frictional stress are determined by lubricant viscous properties. Motor oils are usually formulated by adding to the base oil stock a number of polymeric additives. In spite of the fact that the concentrations of these additives are always low, the presence of the additives significantly modifies and stabilizes lubricant's viscosity and changes lubricant rheology. It is also recognized that in the process of machinery operation lubricant's composition and rheology undergo certain changes. In fact, high operating temperatures and stresses cause degradation of polymeric additives that lead to changes in lubricants viscosity and rheology. Degradation of lubricants causes significant viscosity loss that, in turn, reduces the frictional stress in contacts with sufficiently smooth surfaces operating in elastohydrodynamic regime of lubrication. The reduced frictional stresses raise contact fatigue life. Therefore, to properly predict contact fatigue life it is not sufficient to know the ambient lubricant viscosity and its dependence on pressure in a non-degraded lubricant. To successfully model contact fatigue of lubricated joints with sufficiently smooth surfaces that are completely separated by lubricant, it is necessary to know not only the fatigue resistance and cleanliness of the materials involved in the contact but also to be able to predict the effect of temperature and stress-induced polymer degradation on lubricant viscosity. The latter leads to correct prediction of contact frictional stresses that strongly affect contact fatigue [27, 33] - [37].

The objective of this section is to explore the extent to which lubricant degradation may change contact fatigue life. The analysis is performed numerically based on the developed models of contact fatigue [1, 27, 33] and lubricant degradation in an EHL contact (see Section 11.2 and [24]). The results show that for solids completely separated by lubricants that have the same ambient viscosity, material contact fatigue may vary significantly because of the specific way these lubricants were formulated. In particular, contact fatigue is strongly affected by the initial distribution of the molecular weight of the polymeric additive (viscosity modifier) and by contact operating conditions that in some cases promote rapid lubricant degradation caused by high shearing stresses.

11.7.1 Model of Contact Fatigue

The contact fatigue model used here is based of the assumption that material contact fatigue failure is caused by fatigue crack propagation. Moreover, it is assumed that fatigue crack growth stage in material represents practically the entire material fatigue life. It is also assumed that material operates in an elastic region of parameters and that only small local plastic zones may occur in the vicinity of fatigue crack tips. Fatigue cracks are considered distributed over the entire material with a certain initial statistical distribution of cracks versus their size. Fatigue cracks are allowed to grow along certain directions which are determined by the material stress field. The probability of material failure is determined by the material point at which there is a maximum number of fatigue cracks which reached the critical size at this point relative to the initial number of cracks at the point. The detailed description of the contact fatigue model which will be used here, its validation, and the set of all necessary assumptions laid in the foundation of the model are given in [1]. Here we will present only the necessary formulas for calculation of the probability of failure $P_{glob}(N)$ after N loading cycles for the case of the initial distribution of defects $f(0, x, y, z, l)$ versus defect size l is the same throughout the material, i.e., at every point (x, y, z) except for a thin surface layer where it is assumed that there are no defects. We will assume that $f(0, x, y, z, l)$ versus defect semi-length l is described by a log-normal distribution with constant mean μ_{ln} and standard deviation σ_{ln} . Then [1]

$$P_{glob}(N) = p_m(N),$$

$$p_m(N) = \min_V p(N, x, y, z) = \tfrac{1}{2}\{1 + erf[\tfrac{\ln \min_V l_{oc} - \mu_{ln}}{\sqrt{2}\sigma_{ln}}]\}, \tag{11.57}$$

$$l_{oc} = \left\{ l_c^{\frac{2-n}{2}} + N(\tfrac{n}{2} - 1)g_0 \left[\max_{-\infty < x < \infty} k_{10} \right]^n \right\}^{\frac{2}{2-n}}, \quad l_c = (\tfrac{K_f}{k_{10}})^2,$$

where $erf(x)$ is the error integral, V is the solid volume, K_f is the material fracture toughness, and the normal stress intensity factor k_1 at the tips of small subsurface cracks can be approximated by asymptotic formulas [1]

$$k_1 = \sqrt{l}[Y^r + q^0 \sin^2 \alpha]\theta[Y^r + q^0 \sin^2 \alpha], \quad k_{10} = k_1 l^{-1/2},$$

$$Y^r = Re(Y), \quad Y = \tfrac{1}{\pi} \int\limits_a^c [p(t)\overline{D}_0(t) + \tau(t)\overline{G}_0(t)]dt, \quad \tau = -\lambda p,$$

$$D_0(t) = \tfrac{i}{2}\left[-\tfrac{1}{t-X} + \tfrac{1}{t-\overline{X}} - \tfrac{e^{-2i\alpha}(\overline{X}-X)}{(t-\overline{X})^2} \right], \tag{11.58}$$

$$G_0(t) = \tfrac{1}{2}\left[\tfrac{1}{t-X} + \tfrac{1-e^{-2i\alpha}}{t-\overline{X}} - \tfrac{e^{-2i\alpha}(t-X)}{(t-\overline{X})^2} \right], \quad X = x + iy,$$

where l is the crack semi-length, q^0 is the residual stress in material, λ is the friction coefficient, g_0 and n are the constant parameters of the Paris equation controlling fatigue crack growth, i is the imaginary unit ($i^2 = -1$), $\theta(x)$ is a step function: $\theta(x) = 0$, $x \leq 0$ and $\theta(x) = 1$, $x > 0$.

The angle of fatigue crack propagation α is determined by one of the two angles from formula [1]

$$\tan 2\alpha = -\frac{2y \int\limits_a^c (t-x)T(t,x,y)dt}{\tfrac{\pi}{2}q^0 + \int\limits_a^c [(t-x)^2 - y^2]T(t,x,y)dt}, \tag{11.59}$$

$$T(t, x, y) = \frac{yp(t) + (t-x)\tau(t)}{[(t-x)^2 + y^2]^2},$$

which provides maximum for the crack normal stress intensity factor k_1.

11.7.2   Elastohydrodynamic Model for a Lubricant That Undergoes Stress-Induced Scission

In modeling of line contacts with degrading lubricant, we will follow the model from Section 11.2 developed for steady isothermal line elastohydrodynamic contacts. Namely, two cylinders with parallel axes and elastic constants E_1, ν_1 and E_2, ν_2 move with linear velocities u_1 and u_2, respectively, and subjected to a normal compressive force P. The cylinders are completely separated by a lubrication layer. The lubricant is considered to be a Newtonian fluid (base oil) formulated with a polymeric additive of a linear structure. The lubricant viscosity $\mu(p, W)$ depends not only on lubricant pressure p but also on the density of the polymer molecular weight distribution $W(x, z, l)$, where l is the polymer molecule chain length. The density $W(x, z, l)$ is introduced in such a way that $W(x, z, l)dl$ is the molecular weight of the polymer molecules in a unit volume centered at (x, z) with the chain lengths from the interval $[l, l + dl]$.

In the dimensionless form the equations describing the EHL problem for degrading lubricant are presented in Section 11.2 (see equations (11.16)-(11.21)). For a Newtonian lubricating fluid $G(x) = x$ (see (11.1)) in dimensionless variables in (11.16)-(11.18) such equations are given below (see Section 11.2)

$$f = \left\{ \frac{s_0}{\sqrt{1 + \frac{\varepsilon H_0^2}{4}(\frac{dh}{dx})^2}} - \frac{12 H_0^2}{V} \frac{dp}{dx} \int\limits_{-h/2}^{h/2} \frac{\zeta d\zeta}{\mu} \right\} / \left(\int\limits_{-h/2}^{h/2} \frac{d\zeta}{\mu} \right)$$

$$u(x, z) = \frac{1 - s_0/2}{\sqrt{1 + \frac{\varepsilon H_0^2}{4}(\frac{dh}{dx})^2}} \left\{ 1 + \int\limits_{-h/2}^{z} \frac{d\zeta}{\mu} \Big/ \int\limits_{-h/2}^{h/2} \frac{d\zeta}{\mu} \right\}$$

$$\tag{11.60}$$

$$+ \frac{12 H_0^2}{V} \frac{dp}{dx} \left\{ \int\limits_{-h/2}^{z} \frac{\zeta d\zeta}{\mu} - \int\limits_{-h/2}^{h/2} \frac{\zeta d\zeta}{\mu} \int\limits_{-h/2}^{z} \frac{d\zeta}{\mu} \Big/ \int\limits_{-h/2}^{h/2} \frac{d\zeta}{\mu} \right\},$$

$$\frac{d}{dx} \int\limits_{-h/2}^{z(x)} u(x, \zeta) d\zeta = 0, \ p(a) = p(c) = \frac{dp(c)}{dx} = 0,$$

$$\int\limits_{-h/2}^{z(x)} u(x, \zeta) d\zeta = \frac{1}{\sqrt{1 + \frac{\varepsilon H_0^2}{4}(\frac{dh}{dx})^2}} \left\{ \frac{h}{2} - s_0 \left[\frac{1}{2} + \int\limits_{-h/2}^{z} \frac{\zeta d\zeta}{\mu} \Big/ \int\limits_{-h/2}^{h/2} \frac{d\zeta}{\mu} \right] \right\}$$

$$\tag{11.61}$$

$$+ z u(x, z) - \frac{12 H_0^2}{V} \frac{dp}{dx} \left\{ \int\limits_{-h/2}^{z} \frac{\zeta^2 d\zeta}{\mu} - \int\limits_{-h/2}^{h/2} \frac{\zeta d\zeta}{\mu} \int\limits_{-h/2}^{z} \frac{\zeta d\zeta}{\mu} \Big/ \int\limits_{-h/2}^{h/2} \frac{d\zeta}{\mu} \right\},$$

$$H_0(h - 1) = x^2 - c^2 + \frac{2}{\pi} \int\limits_a^c p(x') \ln \frac{c - x'}{|x - x'|} dx', \ \int\limits_a^c p(x') dx' = \frac{\pi}{2},$$

$$\tag{11.62}$$

$$\tau_i = (-1)^i f + \frac{6 H_0^2}{V} h \frac{dp}{dx}, \ i = 1, 2,$$

$$\kappa_f(u \frac{\partial W}{\partial x} + w \frac{\partial W}{\partial z}) = 2l \int\limits_l^{\infty} R(x, z, L) p_c(l, L) W(x, z, L) \frac{dL}{L}$$

$$- R(x, z, l) W(x, z, l),$$

$$W(a,z,l) = W_a(l) \ if \ u(a,z) > 0, \ W(c,z,l) = W_a(l) \ if \ u(c,z) < 0,$$

$$R(x,z,l) = 0 \ if \ l \leq L_*,$$

$$R(x,z,l) = 1 - (\tfrac{l}{L_*})^{2\gamma} \exp[-\gamma(\tfrac{l^2}{L_*^2} - 1)] \ if \ l \geq L_*,$$

$$L_* = \sqrt{\tfrac{H_0\delta}{|\tau|}}, \ \tau = f + y\tfrac{12H_0^2}{V}\tfrac{dp}{dx}, \tag{11.63}$$

$$p_c(l,L) - ln(2)\tfrac{4|L-2l|}{L^2} \exp[-ln(2)\tfrac{4l(L-l)}{L^2}],$$

$$\mu(p,W) = e^{Qp}\tfrac{1+[\eta]+k_H[\eta]^2}{1+[\eta]_a+k_H[\eta]_a^2}, \ [\eta] = \theta M_W^\beta,$$

$$M_W = \left\{ \int_0^\infty l^\beta W(x,z,l)dl / \int_0^\infty W(x,z,l)dl \right\}^{1/\beta}.$$

Three types of calculations are required for solution of elastohydrodynamic problem for degrading lubricant (see equations (11.60)-(11.63)). The lubricant flow streamlines need to be determined from equations (11.62) and the variations of the density of the molecular weight distribution W along the streamlines have to be calculated from equations (11.63). After the above two calculations are carried out, the EHL problem itself (described by equations (11.62) for $z = h/2$) can be solved. The above process is iterative, and more detail is presented in Section 11.4.

11.7.3   Combined Model for Contact Fatigue and Lubricant That Undergoes Degradation

To combine the two above models, we need to make a number of natural and reasonable assumptions. For heavily loaded smooth contacts in which surfaces are completely separated by a lubricant, the actual pressure p is very close to the Hertzian pressure distribution. The situation with the frictional stress τ applied to the lubricated surfaces is different. There are still no good and reliable methods for determining lubricant rheology and frictional stresses (see [38]). Therefore, the assumptions that the lubricant rheology is Newtonian and the lubricant viscosity changes with pressure according to an exponential law may be crude. Nevertheless, they serve sufficiently well to reveal the main mechanisms of lubricant degradation and provide at least qualitatively correct results (see [1, 12, 33]). A similar analysis for a lubricant with one kind of non-Newtonian rheology was presented in Sections 11.2, 11.5, and 11.6. On the other hand, the results of simulations based on the contact fatigue model show that fatigue life is insensitive to even large local variations in the behavior of the frictional stress applied to the surfaces as long as the frictional force remains the same. It is due to the fact that the maximum of the normal stress intensity factor k_1 at crack tips is reached relatively far from the center of a contact and it is determined by integral expressions given in [1], which to some extent averages the local variations of frictional stress τ. In particular, at a depth of about half-width of a Hertzian contact beneath the surface, the maximum of the normal stress intensity factor k_1 is reached at a distance of at least $1.5a_H - 2a_H$ behind the contact. The goal of this analysis is mainly to qualitatively demonstrate the effect of lubricant degradation on contact fatigue of completely separated by lubricant smooth surfaces. Therefore, based on the above considerations in fatigue modeling we will accept the frictional stress applied to the surfaces to be determined by Coulomb's law $\tau = -\lambda p$ where λ is the coefficient of friction. For one particular set of the problem, parameters λ will be assigned a reasonable value of the friction coefficient. At the same time, the friction coefficient λ for other cases will be determined as follows:

$$\lambda = \lambda_0 \int_a^c \tau(x)dx / \int_{a_0}^{c_0} \tau_0(x)dx, \qquad (11.64)$$

where λ_0 is the assigned value of the friction coefficient for the frictional stress τ_0 and contact boundaries a_0 and c_0 while λ is the coefficient of friction for the frictional stress τ and contact boundaries a and c. Obviously, the two models of contact fatigue and lubricant degradation are coupled together by equation (11.64). In the future, coupling between the contact fatigue and lubricant degradation parts of the model can be improved as more detailed and accurate information on lubricant rheology and calculation of frictional stresses becomes available.

11.7.4 Numerical Results and Discussion

The fatigue crack resistance parameters of the solid material as well as crack parameters used for particular calculations were as follows $g_0 = 8.863 \ MPa^{-n} \cdot m^{1-n/2}$, $n = 6.67$, $\mu_r = 49.62 \ \mu m$, and $\sigma_r = 13.9 \ \mu m$. The residual stress varied from large compressive $q^0 = -237.9 \ MPa$ on the surface to small tensile $q^0 = 0.035 \ MPa$ at the depth of 400 μm while the fracture toughness K_f varied between 15 and 95 $MPa \cdot m^{1/2}$. The other parameter values related to lubricant and polymeric additive properties were $U = 347 \ kJ/mole$, $T = 310°K$, $a_* = 0.374 \ nm$, $l_* = 0.154 \ nm$, $w_m = 35.1 \ g/mole$, $\mu_a = 0.00125 \ Pa \cdot s$, $c_p = 1.204 \ g/dL$. The values for the shield constants were chosen as follows $C = 15.5$ and $\alpha = 0.008$. The dimensionless operating parameters $a = -2$, $V = 0.1$, $Q = 5$, $s_0 = -0.5$, $\varepsilon = 10^{-4}$, $\gamma = 1.0774$, $\delta = 0.671 \cdot 10^8$, $\theta = 4.524 \cdot 10^{-3}$, and $\kappa = 1$ correspond to the Hertzian semi-length of the contact $a_H = 10^{-3}m$ and the Hertzian maximum pressure $p_H = 1.284GPa$.

In this section we compare the global survival probabilities $P_{glob}(N)$ of the lower solid calculated for four cases. The four cases differ from each other only by the initial density of polymer molecular weight distribution $W_a(l)$. It is important to remember that the ambient lubricant viscosity is the same for all four cases. For the base case the initial distribution of the polymer molecular weight $W_{a0}(l)$, $0 \leq l \leq l_{max}$, $l_{max} > 0$, is taken from the experimental study by Covitch [12]. For the other three cases, the initial densities of polymer molecular weight distributions were

$$W_{a1}(l) = W_{a0}(1.25l), \ 0 \leq l \leq 1.25l_{max},$$

$$W_{a2}(l) = W_{a0}(0.5l), \ 0 \leq l \leq 0.5l_{max}, \tag{11.65}$$

$$W_{a3}(l) = W_{a0}(0.25l), \ 0 \leq l \leq 0.25l_{max}.$$

The mean polymer chain lengths l_m that correspond to the initial molecular weight distributions from equations (11.65) were as follows

$$l_{m1} = 1.25l_{m0}, \ l_{m2} = 0.5l_{m0}, \ l_{m3} = 0.25l_{m0}, \tag{11.66}$$

where $w_m s_m = w_m l_{m0} = 1.3196 \cdot 10^5$ is the initial mean chain length determined by the initial molecular weight distribution $W_a(l) = W_{a0}(l)$. For the base case a typical value for the friction coefficient in bearings is taken equal to $\lambda_0 = 0.02$ while for the other three cases according to equations (11.64) and (11.65) we obtained $\lambda_1 = 0.01967$, $\lambda_2 = 0.02109$, and $\lambda_3 = 0.02193$ for $s_0 = 0$ and $\lambda_1 = 0.01925$, $\lambda_2 = 0.02231$, and $\lambda_3 = 0.02427$ for $s_0 = -0.5$. To determine the above values of λ, we used $a = a_0 = -2$ and $c = c_0, c_1, c_2, c_3$ obtained from numerical solutions of the elastohydrodynamic lubrication problem.

The main properties of solutions of the isothermal EHL problem for a degrading lubricant with Newtonian and non-Newtonian rheologies are presented in Section 11.6. The behavior of the pressure $p(x)$ and gap $h(x)$ distributions as well as of the film thickness H_0 and exit coordinate c of the EHL problem with respect to parameters a, V, and Q is similar to the one described in Sections 11.5 and 11.6 for non-degrading lubricants with Newtonian and non-Newtonian rheologies. For a lubricant with Newtonian rheology increase in Q causes a relatively slow increase in H_0 and a rapid increase in $f(x)$ and $\tau(x, z)$. That, in turn, leads to rapid lubricant degradation. A usually very small parameter ε almost does not affect the problem solution.

Lubricant degradation depends on a number of parameters among which the shear stress τ (see equations (11.63)) plays the major role. For $s_0 = 0$ the sliding frictional stress $f(x) = 0$ and, therefore, the shear stress $\tau = (12H_0^2/V)zdp/dx$ reaches its extrema in the inlet and exit zones while in the Hertzian region it is relatively small. That leads to the situation when lubricant degradation is relatively slow and practically all lubricant degradation occurs in the inlet zone while the lubricant almost does not degrade in the Hertzian region and the exit zone.

Now, let us examine a case of mixed rolling and sliding with $s_0 = -0.5$. The exit lubrication film thickness H_0, exit coordinate c, the distributions of pressure p and gap h exhibit behavior very similar to the case of pure rolling. From the comparison of the values of the friction coefficient λ (see

TABLE 11.1

Fatigue life N versus the survival probability P_{glob} for $s_0 = 0$ (after Kudish [27]). Reprinted with permission from the STLE.

$P_{glob}(N)$	N	λ	l_m
0.9	$0.149 \cdot 10^{10}$		
0.75	$0.1845 \cdot 10^{10}$	$\lambda = 0.02$	$l_m = l_{m0}$
0.5	$0.2345 \cdot 10^{10}$		
0.9	$0.1789 \cdot 10^{10}$		
0.75	$0.2218 \cdot 10^{10}$	$\lambda = 0.01967$	$l_m = 1.25 l_{m0}$
0.5	$0.2619 \cdot 10^{10}$		
0.9	$0.820 \cdot 10^{9}$		
0.75	$0.1018 \cdot 10^{10}$	$\lambda = 0.02109$	$l_m = 0.5 l_{m0}$
0.5	$0.1293 \cdot 10^{10}$		
0.9	$0.527 \cdot 10^{9}$		
0.75	$0.652 \cdot 10^{9}$	$\lambda = 0.02193$	$l_m = 0.25 l_{m0}$
0.5	$0.828 \cdot 10^{9}$		

Tables 11.1 and 11.2) it is obvious that variations in λ due to variations of the initial polymer molecular weight and lubricant degradation are smaller in the case of pure rolling ($s_0 = 0$) than in the case of mixed rolling and sliding ($s_0 = -0.5$). The numerical results from Sections 11.5 and 11.6 show that in the case of mixed rolling and sliding, the surface frictional stresses τ_1 and τ_2 are smaller than in a similar case without degradation but still large enough to cause lubricant degradation to a much greater extent than in the case of pure rolling. On the other hand, in the case of mixed rolling and sliding the surface frictional stresses τ_1 and τ_2 are much higher than the ones for the case of pure rolling ($\tau_2 = -\tau_1 = (6H_0^2/V)hdp/dx$) and, at the same time, lower than for the case of no degradation. That explains the stronger lubricant degradation in the case of mixed rolling and sliding conditions in comparison with the case of pure rolling conditions (see Sections 11.5 and 11.6). For the base case of degrading lubricant, the frictional stresses τ_1 and τ_2 applied to the contact surfaces are on average lower than for the non-degrading lubricant by about 26% (see Sections 11.5 and 11.6). In the base case for $s_0 = -0.5$, the maximum loss of the lubricant viscosity reaches about 41% (see Sections 11.5 and 11.6). In many respects the process of lubricant degradation is governed by the characteristic polymer chain length L_* (see equations (11.63)), which represents the position of the beginning of the profile of the probability of polymer scission $R(x, z, l)$. In fact, the probability of polymer scission R is the main key to achieve control over lubricant degradation. A more detailed discussion of the role of R in lubricant degradation can be found in [1] and Sections 11.5 and 11.6.

Now, let us consider the behavior of the global survival probability P_{glob}

TABLE 11.2
Fatigue life N versus the survival probability P_{glob} for $s_0 = -0.5$ (after Kudish [27]). Reprinted with permission from the STLE.

$P_{glob}(N)$	N	λ	l_m
0.9	$0.149 \cdot 10^{10}$		
0.75	$0.1845 \cdot 10^{10}$	$\lambda = 0.02$	$l_m = l_{m0}$
0.5	$0.2345 \cdot 10^{10}$		
0.9	$0.227 \cdot 10^{10}$		
0.75	$0.2805 \cdot 10^{10}$	$\lambda = 0.01925$	$l_m = 1.25 l_{m0}$
0.5	$0.357 \cdot 10^{10}$		
0.9	$0.432 \cdot 10^{9}$		
0.75	$0.535 \cdot 10^{9}$	$\lambda = 0.02231$	$l_m = 0.5 l_{m0}$
0.5	$0.680 \cdot 10^{9}$		
0.9	$0.161 \cdot 10^{9}$		
0.75	$0.200 \cdot 10^{9}$	$\lambda = 0.02427$	$l_m = 0.25 l_{m0}$
0.5	$0.254 \cdot 10^{9}$		

(N_{life}) in all four cases for slide-to-roll ratios $s_0 = 0$ and $s_0 = -0.5$. In the cases when the initial mean polymer chain length l_m varies between $0.25 l_{m0}$ and $1.25 l_{m0}$ the contact fatigue life N_{life} that corresponds to the survival probability $P_{glob}(N_{life}) = 0.9$ for $s_0 = 0$ changes about 3.4 times, i.e., N_{life} varies between $0.354 N_{life0}$ and $1.201 N_{life0}$ cycles, respectively (see Table 11.1 and Fig. 11.19) while for $s_0 = -0.5$ the contact fatigue life N_{life} changes more than 14–fold, i.e., N_{life} varies between $0.108 N_{life0}$ and $1.523 N_{life0}$ cycles, respectively (see Table 11.2 and Fig. 11.20). Here $N_{life0} = 0.149 \cdot 10^{10}$ for $s_0 = 0$ and $s_0 = -0.5$ is the contact fatigue life (in cycles) for the base case when $W_a(l) = W_{a0}(l)$, $l_m = l_{m0}$, and $\lambda = 0.02$. In Tables 11.1 and 11.2 series of data for the number of cycles N that correspond to the survival probability $P_{glob}(N)$ equal to 0.9, 0.75, and 0.5 for $s_0 = 0$ and $s_0 = -0.5$ are presented for the base case and the other three cases from equations (11.64) and (11.65), respectively.

To understand how polymer concentration c_p affects contact fatigue, we need to consider the dependence of the lubricant viscosity $\mu(p, W)$ on c_p. Numerical results show that the distribution of pressure p is almost independent of c_p. Therefore, in dimensional variables the derivative of $\mu(p, W)$ with respect to c_p can be expressed as follows

$$\frac{\partial \mu}{\partial c_p} \approx e^{Qp}([\eta] - [\eta]_a) \frac{1 + 2k_H([\eta] + [\eta]_a) + k_H c_p^2 [\eta][\eta]_a}{(1 + c_p[\eta]_a + k_H c_p^2 [\eta]_a^2)^2}$$

$$+ e^{Qp} c_p \frac{\partial [\eta]}{\partial c_p} \frac{1 + 2k_H c_p[\eta]}{1 + c_p[\eta]_a + k_H c_p^2 [\eta]_a^2}. \tag{11.67}$$

Numerical results show that the first term in equation (11.67) dominates the

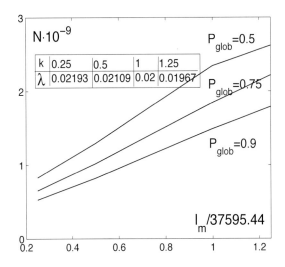

FIGURE 11.19
Graphs of fatigue life N as a function of the average length of polymer additive molecules l_m obtained for pure rolling $s_0 = 0$. (after Kudish and Covitch [1]). Reprinted with permission from CRC Press.

second one and as lubricant moves along the contact the intrinsic viscosity $[\eta]$ monotonically decreases starting with $[\eta]_a$ at the inlet point $x = a$, i.e., $[\eta] - [\eta]_a \leq 0$. Therefore, we can expect that the loss of lubricant viscosity caused by polymer degradation increases with increasing polymer additive concentration c_p. That, in turn, leads to reduced frictional stresses and higher fatigue life.

The described above effects of significant increased fatigue life due to reduced lubricant viscosity and frictional stress are valid only for contacts with sufficiently smooth surfaces so that the direct contact of asperities does not occur even in case of degraded lubricant. In cases of mixed friction, the dry friction created by direct contact of asperities usually dominates the friction realized in lubricated zones where contact surfaces are completely separated by lubricant. Because the coefficient of dry friction is by about one or two orders of magnitude higher than that of a well–lubricated contact one can expect that for contacts with relatively large portion of direct asperity contact the effect of lubricant viscosity loss may not be that significant. Moreover, in such cases the increase in the ambient viscosity may lead to a reduced overall frictional stress due to increased film thickness that separates asperities better and, thus, reduces the portion of the frictional stress due to direct asperity contact more than it raises the frictional stress due to   increased

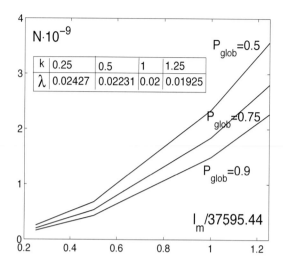

FIGURE 11.20
Graphs of fatigue life N as a function of the average length of polymer additive molecules l_m obtained for mixed rolling and sliding $s_0 = -0.5$. (after Kudish and Covitch [1]). Reprinted with permission from CRC Press.

lubricant viscosity. The illustration of this effect can be found in the results of experimental study by Krantz and Kahraman [39].

The effect of lubricant viscosity loss of up to 70% was observed experimentally and documented by Walker, Sanborn, and Winer [9]. Therefore, in real lubricated contacts the viscosity loss may reach high values, which would lead to lower frictional stresses and much wider margins of fatigue life variations.

In case of a thermal EHL problem due to heat generation in the lubrication film and elevated lubricant temperature T, the lubricant viscosity (in case of no lubricant degradation) is lower than in a comparable isothermal case. That leads to higher value of L_*, which slows down the degradation process. The extent to which the polymer degradation process slows down depends on the lubricant nature, on the nature and distribution of the polymer molecular weight entering the contact, and on the contact operating conditions. Therefore, under higher temperature conditions, the margins of fatigue life variations are narrower than for lower temperatures but are still significant.

The results based on such a model can be used for a more realistic estimation of contact fatigue life as a function of not only the parameters of contact materials but also of lubricant parameters such as lubricant viscosity, chemistry, concentration, and details of the polymeric additive initial molecular weight distribution. That may provide a better understanding of contact fatigue performance of lubricated joints working in different applications and under various conditions. Moreover, better understanding of lubricant influence on contact fatigue may lead to better design of bearing and gear testing procedures. In particular, a special attention should be paid to lubricant to keep it identical in different tests used for determination and comparison of fatigue life in a homogeneous sample of tested bearings or gears.

11.8   Closure

The effect of stress-induced polymeric additive degradation in a lubricated contact and its effect on contact fatigue are studied. The analysis is performed numerically based on the developed models of lubricant degradation and contact fatigue. The results show that degradation of the polymeric additive in a lubricated contact may be significant and, therefore, leads to reduction of lubricant viscosity and frictional stresses. Moreover, the results show that for lubricants with the same ambient viscosity contact fatigue of sufficiently smooth solids completely separated by lubricants may vary significantly due to the specific way these lubricants were formulated. In such cases it is established that contact fatigue is strongly affected by the initial distribution of the molecular weight of the polymeric additive (viscosity modifier) to lubricant and by contact operating conditions that in some cases promote rapid lubricant degradation caused by high lubricant shearing stresses. In particular, for sufficiently smooth surfaces completely separated by lubricant and for lubricants with higher viscosity loss caused by additive degradation, contact fatigue life is higher given same other contact parameters. In cases where the average molecular weight of polymer molecules varied between $3.2989 \cdot 10^4$ and $1.6495 \cdot 10^5$ the contact fatigue life that corresponds to the global survival probability equal to 0.9 for pure rolling ($s_0 = 0$) varied between $0.179 \cdot 10^{10}$ and $0.527 \cdot 10^9$ and for mixed rolling and sliding ($s_0 = -0.5$) varied between $0.227 \cdot 10^{10}$ and $0.161 \cdot 10^9$, i.e., contact fatigue life changed more than 3– and 14–fold, respectively, due to different lubricant degradation patterns. Therefore, to properly determine contact fatigue life it is not sufficient to know the ambient lubricant viscosity and the dependence on pressure of the viscosity of the non-degraded lubricant. Moreover, to successfully predict contact fatigue of lubricated joints, it is necessary to know not only the fatigue resistance and cleanliness of the contact materials but also to be able to take into account the effect of temperature and stress-induced polymer degradation on lubricant viscosity in joints. Better understanding of lubricant influence on contact fatigue may lead to better design of bearing and gear testing procedures.

11.9   Exercises and Problems

1. What are the two general causes of lubricant non-Newtonian behavior in lubricated contacts? List the most important manifestations of lubricant non-Newtonian behavior. What is the usual response of lubricant viscosity to increase in the shear rate? In which cases lubricant viscosity loss is reversible and irreversible?

2. The term $\sqrt{1 + \frac{1}{4}(\frac{dh}{dx})^2}$ is usually very close to 1 and almost never used in formulations of the EHL problems for non-degrading lubricants. What is the primary reason for the inclusion of this term in equations (11.8) for degrading lubricants?

3. Explain why is it necessary to know lubricant flow streamlines to analyze stress-induced lubricant degradation.

4. Describe and discuss the topology of the lubricant flow between the surfaces of contacting solids. What is a possible implication of the topology of flow streamlines in the vicinity of a separatrix running through the entire contact of degrading lubricant for the lubricant viscosity μ in case of some sliding (i.e., $s_0 \neq 0$)?

5. Describe the general approach to the numerical method used for solution of the EHL problem for degrading lubricant. Discuss in detail how: (a) the sliding frictional stress $f(x)$ is determined, (b) the horizontal component $u(x, z)$ of the lubricant velocity and lubricant flux are calculated, (c) the lubricant flow streamlines are obtained, (d) the kinetic equation is solved and lubricant viscosity μ is updated, and (e) the Reynolds equation is solved.

6. Describe the similarities and differences in the behavior of degrading and non-degrading lubricants in heavily loaded contacts. Describe in detail the behavior of the polymer additive molecular weight $W(x, z, l)$ and lubricant viscosity $\mu(x, z)$ in lubricated contacts.

7. Describe the effect the distribution of molecular weight $W(a, z, l)$ or $W(c, z, l)$ of a polymeric additive entering the lubricated contact has on fatigue of a lubricated contact.

References

[1] Kudish, I.I. and Covitch, M.J. 2010. *Modeling and Analytical Methods in Tribology*. Boca Raton: CRC Press.

[2] Bair, S. and Winer, W.O. 1979. Shear Strength Measurements of Lubricants at High Pressure. *J. Lubr. Techn.* 101, No. 3:251-257.

[3] Hoglund, E. and Jacobson, B. 1986. Experimental Investigations of the Shear Strength of Lubricants Subjected to High Pressure and Temperature. *ASME J. Tribology* 108, No. 4:571-578.

[4] Eyring, H. 1936. Viscosity, Plasticity, and Diffusion as Examples of Absolute Reaction Rates. *J. Chem. Phys.* 4, No. 4:283-291.

[5] Houpert, L.G. and Hamrock, B.J. 1985. Elastohydrodynamic Lubrication Calculations Used as a Tool to Study Scuffing. In *Mechanisms and Surface Distress : Global Studies of Mechanisms and Local Analyses of Surface Distress Phenomena*, Eds. D. Dowson, et al. Butterworths, England, 146-162.

[6] Kudish, I.I. 1982. Asymptotic Methods for Studying Plane Problems of the Elastohydrodynamic Lubrication Theory in Heavily Loaded Regimes. Part 1. Isothermal Problem. *Izvestija Akademii Nauk Arm. SSR, Mekhanika* 35, No. 5:46-64.

[7] Crail, I.R.H. and Neville, A.L. 1969. The Mechanical Shear Stability of Polymeric VI Improvers. *J. Inst. Petrol.* 55, No. 542:100-108.

[8] Casale, A. and Porter, R.S. 1971. The Mechanochemistry of High Polymers. *J. Rubber Chem. and Techn.* 44, No.2:534-577.

[9] Walker, D.L., Sanborn, D.M., and Winer, W.O. 1975. Molecular Degradation of Lubricants in Sliding Elastohydrodynamic Contacts. *ASME J. Lubr. Techn.* 97, No. 3:390-397.

[10] Yu, J.F.S, Zakin, J.L., and Patterson, G.K. 1979. Mechanical Degradation of High Molecular Weight Polymers in Dilute Solution. *J. Appl. Polymer Sci.* 23:2493-2512.

[11] Odell, J.A., Keller, A., and Rabin, Y. 1988. Flow-Induced Scission of Isolated Macromolecules. *J. Chem. Phys.* 88, No. 6:4022-4028.

[12] Covitch, M.J. 1998. How Polymer Architecture Affects Permanent Viscosity Loss of Multigrade Lubricants. *SAE Technical Paper* No. 982638.

[13] Herbeaux, J.-L., Flamberg, A., Koller, R.D., and Van Arsdale, W.E. 1998. Assesment of Shear Degradation Simulators. *SAE Technical Paper* No. 982637.

[14] Herbeaux, J.-L. 1996. *Mechanochemical Reactions in Polymer Solutions* Ph.D. Dissertation, University of Houston, Houston (and references therein).

[15] Ziff, R.M. and McGrady, E.D. 1985. The Kinetics of Cluster Fragmentation and Depolymerization. *J. Phys. A : Math. Gen.* 18:3027-3037.

[16] Ziff, R.M. and McGrady, E.D. 1986. Kinetics of Polymer Degradation. *AchS, Macromolecules* 19:2513-2519.

[17] McGrady, E.D. and Ziff, R.M. 1988. Analytical Solutions to Fragmentation Equations with Flow. *AIChE* 34, No. 12:2073-2076.

[18] Montroll, E.W. and Simha, R. 1940. Theory of Depolymerization of Long Chain Molecules. *J. Chem. Phys.* 8:721-727.

[19] Saito, O. 1958. On the Effect of High Energy Radiation to Polymers. I, Cross-Linking and Degradation. *J. Phys. Soc. Jpn.* 13:198-206.

[20] Kudish, I.I. and Ben-Amotz, D. 1999. Modeling Polymer Molecule Scission in EHL Contacts. In *The Advancing Frontier of Engineering Tribology, Proc. 1999 STLE/ASME H.S. Cheng Tribology Surveillance*, Eds.: Q. Wang, J. Netzel, and F. Sadeghi, 176-182.

[21] Kudish, I.I., Airapetyan, R.G., and Covitch, M.J. 2002. Modeling of Kinetics of Strain-Induced Degradation of Polymer Additives in Lubricants. *J. Math. Models and Methods Appl. Sci.* 12, No. 6:1-22.

[22] Kudish, I.I., Airapetyan, R.G., and Covitch, M.J. 2003. Modeling of Kinetics of Stress-Induced Degradation of Polymer Additives in Lubricants and Viscosity Loss. *STLE Tribology Trans.* 46, No. 1:1-11.

[23] Kudish, I.I., Airapetyan, R.G., Hayrapetyan, G.R., and Covitch, M.J. 2005. Kinetics Approach to Modeling of Stress Induced Degradation of Lubricants Formulated with Star Polymer Additives. *STLE Tribology Trans.* 48, No. 2:176-189.

[24] Kudish, I.I., and Airapetyan, R.G. 2003. Modeling of Line Contacts with Degrading Lubricant. *ASME J. Tribology* 125, No. 3:513-522.

[25] Kudish, I.I., and Airapetyan, R.G. 2004. Lubricants with Newtonian and Non-Newtonian Rheologies and Their Degradation in Line Contacts. *ASME J. Tribology* 126, No. 1:112-124.

[26] Kudish, I.I., and Airapetyan, R.G. 2003. A New Approach to Modeling of Stress-Induced Degradation of Formulated Lubricants and Their Behavior in Lubricated Contacts. *Proc. 2nd Tribology in Environmental Design Intern. Conf.* September 8-10, Bournemouth, UK.

[27] Kudish, I.I. 2005. Effect of Lubricant Degradation on Contact Fatigue. *STLE Tribology Trans.* 48, No. 1:100-107.

[28] Hamrock, B.J. 1991. *Fundamentals of Fluid Film Lubrication*. Cleveland: NASA Reference Publication 1255.

[29] Billmeyer, F.W., Jr. 1966. *Textbook of Polymer Science*. New York: John Wiley & Sons.

[30] Crespi, G., Valvassori, A., and Slisi, U. 1977. Olefin Copolymers. In *The Stereo Rubbers*. Ed. W.M. Saltman. New York: John Wiley & Sons, 365-431.

[31] Kudish, I.I. 1996. Asymptotic Analysis of a Problem for a Heavily Loaded Lubricated Contact of Elastic Bodies. Pre- and Over-critical Lubrication Regimes for Newtonian Fluids. *Dynamic Systems and Applications*, Dynamic Publishers, Atlanta, 5, No. 3:451-476.

[32] Kudish, I.I. 1981. Some Problems of Elastohydrodynamic Theory of Lubrication for a Lightly Loaded Contact. *J. Mech. Solids* 16, No. 3:75-88.

[33] Kudish, I.I. 2000. A New Statistical Model of Contact Fatigue. *STLE Tribology Trans.* 43, No. 4:711-721.

[34] Kudish, I.I. and Burris, K.W. 2000. Modern State of Experimentation and Modeling in Contact Fatigue Phenomenon. Part I. Contact Fatigue versus Normal and Tangential Contact and Residual Stresses. Nonmetallic Inclusions and Lubricant Contamination. Crack Initiation and Crack Propagation. Surface and Subsurface Cracks. *STLE Tribology Trans.* 43, No. 2:187-196.

[35] Tallian, T., Hoeprich, M., and Kudish, I.I. 2001. Author's Closure. *STLE Tribology Trans.* 44, No. 2:153-155.

[36] Kudish, I.I. 2002. Lubricant-Crack Interaction, Origin of Pitting, and Fatigue of Drivers and Followers. *STLE Tribolology Trans.* 45, No. 4:583-594.

[37] Kudish, I.I. 1987. Contact Problems of The Theory of Elasticity for Prestressed Bodies with Cracks. *J. Appl. Mech. and Tech. Phys.* 28, No. 2:295-303.

[38] Jacod, B., Venner, C.H., and Lugt, P.M. 2003. Extension of the Friction Mastercurve to Limiting Shear Stress Models. *ASME J. Tribology* 125, No. 3:739-746.

[39] Krantz, T.L. and Kahraman, A. 2004. An Experimental Investigation of the Influence of the Lubricant Viscosity and Additives on Gear Wear. *STLE Tribology Trans.* 47, No. 1:138-148.

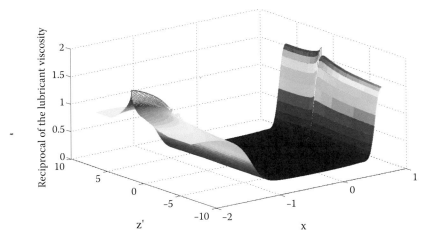

FIGURE 11.2
The reciprocal of the viscosity μ in the lubricants with Newtonian and non-Newtonian rheologies under mixed rolling and sliding conditions and Series I input data ($s_0 = -0.5$). The variable z' is an artificially stretched z-coordinate across the film thickness (namely, $z' = zh(a)/h(x)$) to make the relationship more transparent (after Kudish and Airapetyan [25]). Reprinted with permission from the ASME.

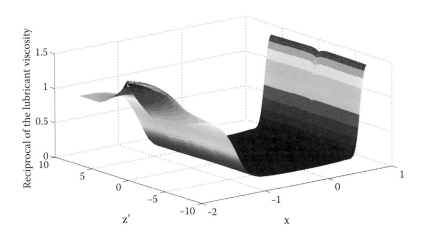

FIGURE 11.11
The reciprocal of the lubricant viscosity μ in Newtonian and non-Newtonian lubrication film under pure rolling conditions ($s_0 = 0$) and Series I input data. The variable z' is an artificially stretched z-coordinate across the film thickness (namely, $z' = zh(a)/h(x)$) to make the relationship more transparent (after Kudish and Airapetyan [25]). Reprinted with permission from the ASME.

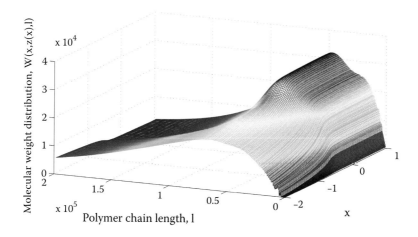

FIGURE 11.12
The molecular weight distribution W of degrading lubricants with Newtonian and non-Newtonian rheologies along the flow streamline closest to $z = 0$ and running through the whole contact below $z = 0$ under pure rolling conditions ($s_0 = 0$) and Series I input data (after Kudish and Airapetyan [25]). Reprinted with permission from the ASME.

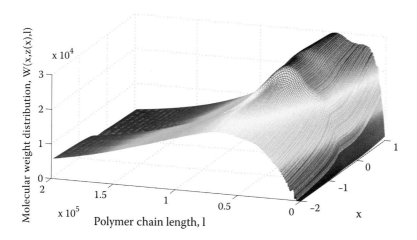

FIGURE 11.15
The molecular weight distribution W of degrading lubricants with Newtonian and non-Newtonian rheologies along the flow streamline $z_{31}(x)$ closest to $z = 0$ and running through the whole contact below $z = 0$ under mixed rolling and sliding conditions ($s_0 = -0.5$) and Series I input data (after Kudish and Airapetyan [25]). Reprinted with permission from the ASME.

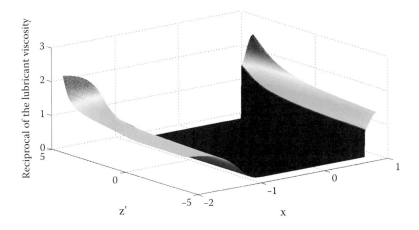

FIGURE 11.17
The reciprocal of the lubricant viscosity μ in a Newtonian lubrication film under pure sliding conditions ($s_0 = -2$) and Series II input data. The variable z' is an artificially stretched z-coordinate across the film thickness (namely, $z' = zh(a)/h(x)$) to make the relationship more transparent (after Kudish and Airapetyan [24]). Reprinted with permission from the ASME.

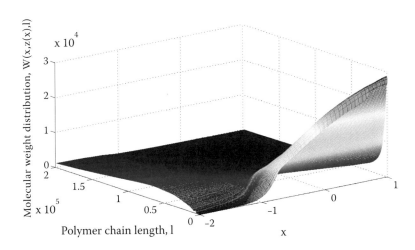

FIGURE 11.18
The molecular weight distribution W of the degrading lubricant with Newtonian rheology along the flow streamline $z_{48}(x)$ running through the entire contact closest to the first turning around streamline under pure sliding conditions ($s_0 = -2$) and Series II input data (after Kudish and Airapetyan [24]). Reprinted with permission from the ASME.

Part VII

Isothermal and Thermal EHL Problems for Point Contacts and Lubricants with Different Rheologies

In this part of the monograph the asymptotic technique developed for heavily loaded line EHL contacts has been extended on analysis of heavily loaded point EHL problems. It is assumed that the lubricant is a Newtonian or non-Newtonian fluid. The technique is effective along the "central" lubricant flow streamlines. The notion of "central" lubricant flow streamlines depends on the geometry of the solids in contact and the type of motion (translational and/or rotational) they are involved in. Along these "central" flow streamlines solutions of the point EHL problems are controlled by the corresponding solutions of the line EHL problems. The types of point lubricated contacts studied here are the contacts where (a) the lubricant is entrained along the smaller Hertzian semi-axis, (b) the direction of the lubricant entrainment is skewed with respect to the smaller Hertzian semi-axis, and (c) the lubricant entrainment is skewed and, also, the lubricant is involved in a spinning motion. For these problems the structures of the lubricated contacts are different and they are revealed in the proposed analysis. The asymptotically valid equations are derived in the inlet and exit zones of heavily loaded point EHL contacts which depend on lubricant rheology but are independent from the specific motion the lubricant is involved in. Lubricant motion dependent analytical formulas for the lubrication film thickness are derived.

12

Isothermal EHL Problems for Heavily Loaded Point Contacts with Newtonian Lubricants

---

12.1 Introduction

EHL problems for line and point contact even in the simplest case of Newtonian lubricant are reduced to boundary-value problems for very complex systems of nonlinear integro-differential equations. One of the very first approaches developed for solution of EHL problems for heavily loaded line contacts was the analytical method of Ertel-Grubin [1, 2]. The approximate result of the method is a formula for the lubrication film thickness which, under certain conditions, is in a relatively good agreement with experimentally obtained data. Later, this method was advanced in several studies [3]-[5] to improve the agreement between the analytical and experimental results. For elliptical heavily loaded EHL contacts a semi-analytical approach similar to [1, 2] was developed by Cheng [6]. These studies practically exhausted the advancements in analytical approaches to solution of EHL problems. All of these approaches are based on certain strong assumptions about the solution of the EHL problem validation of which raises a number of questions (see [7]). The major advantage of these methods is that they provide specific approximate analytical formulas for the lubrication film thickness without much of numerical calculations.

A different analytical approach to EHL problems for heavily loaded line contacts based on asymptotic methods was developed in a number of studies presented in the preceding chapters and collected in [7]. These studies are based on perturbation techniques and some asymptotic methods, in particular, on the method of matched asymptotic expansions [8, 9]. These methods do not use any prior assumptions about the problem solution. Some of the results of the asymptotic analysis are a solid understanding of the solution structure, establishing possible lubrication regimes, and analytical structural formulas for the lubrication film thickness. Among other benefits of the asymptotic methods are the following: reduction of the number of the input EHL problem parameters which determine the numerical solution, development of effective numerical approaches for solution of asymptotically valid as well as the original EHL problems for high pressure viscosity coefficients which usually cause numerical solutions to be unstable, and discovering a simple approach to regularization of generally unstable solutions in heavily loaded isothermal EHL contacts [7]. It was determined that formulas for the lubrication film thickness are lubrication regime dependent, i.e., contrary to a widespread belief, there are series of formulas for the lubrication film thickness different for distinctly different heavily loaded lubrication regimes. The above analytical analysis is augmented by a stable numerical approach to solution of asymptotically valid equations. The success in getting stable numerical solutions for asymptotically valid equations allowed for the development of a stable numerical approach to solution of the EHL problem in the original (non-asymptotic) formulation. This numerical method is based on the solution structure and general properties of the EHL problem discovered through the EHL problem asymptotic analysis. In particular, the original EHL problem equations for line contacts are transformed to a different much more beneficial for numerical solution form. In addition to that, the grid size necessary for sufficient solution precision is determined based on the EHL problem input parameters (i.e., based on the characteristic size of the inlet zone) and solution structure analyzed asymptotically [7].

Obviously, the initial application of the asymptotic approach requires a deeper analytical analysis compared to the existing direct numerical methods in application to solution of EHL problems. However, this asymptotic analysis is worth the additional effort and, in essence, is just the extension of the classic asymptotic analysis proposed by Reynolds which led to significant simplification of the full Navier-Stokes equations and derivation of the Reynolds equation. It is clear that the usefulness of the latter analysis is hard to deny. In preceding chapters it became clear that asymptotic methods have some advantages compared to the direct numerical solution of the EHL problems in their original formulations. Some of these specific reasons why the asymptotic methodology applied to heavily loaded line EHL contacts followed by numerical methods applied to asymptotic equations is attractive are collected in Table 12.1. In Table 12.1, the term "direct numerical methods" is related to the solution methods which make use of only numerical approximation formulas such as various quadrature and finite difference formulas while the essence of the term "asymptotic methods" is analytical analysis of the problem which is presented below.

Advancements in computers led to the development of a variety of numerical methods such as methods based on Newton-Raphson's method [11, 12], the multilevel multi-grid methods [13, 14], and methods which use the fast Fourier transform [15, 16] as well as some others. The accessability of computers and the relative simplicity of their use compared to analytical studies led to their domination in practical applications.

Let us consider the typical steps involved in numerically based study of EHL problems with the goal to determine a formula for the lubrication film thickness.

TABLE 12.1
Comparison of the features of the direct numerical methods with the asymptotic methods followed by based on them numerical methods.

	Direct numerical methods	Asymptotic and numerical methods
Produce an approximate solution	Yes	Yes
Produce a clear structure of the solution and the characteristic sizes of the inlet and exit zones of EHL contacts as certain functions of the problem input parameters	No	Yes
Produce the proper numerical grid which is directly related to the specific values of the EHL problem input parameters (through the characteristic size of the inlet and exit zones) and allow for its optimization	No	Yes
Using self-similarity allow for reduction of the number of EHL problem input dimensionless parameters and, as a result of that, allow for simpler parametric numerical analysis of the problem	No	Yes
Produce clear understanding of distinct lubrication regimes and distinct formulas for the film thickness for these regimes as functions of the problem input parameters	No	Yes
Provide simple analytical formulas for the sliding frictional stress in contacts lubricated by fluids with Newtonian and non-Newtonian rheologies	No	Yes
For determining just the film thickness (a) calculations in the inlet zone separate from the rest of the contact,	No	Yes
(b) produce significant (about 3 times) reduction of the number of grid nodes,	No	Yes
(c) easy increase of lubrication film precision by decreasing the grid size	No	Yes
Produce detailed description of the exit zone including the sharp pressure spike	No	Yes
Provide a simple and seamless extension of numerical schemes and results for isothermal EHL contacts on the cases of thermal EHL contacts with lubricants of various rheologies	No	Yes
Propose regularization resulting in stable numerical solutions of isothermal EHL problems for heavily loaded contacts with lubricant viscosity strongly dependent on pressure	No, except for [10]	Yes

(i) First, a researcher picks a series of sets of problem input parameters. Usually, these sets of parameters are picked without any regard to the fact that they may belong to distinctly different lubrication regimes.

(ii) The EHL problem is solved numerically and the values of the lubrication film thicknesses for each set of the input parameters are obtained. The grid size for numerical calculations is chosen without the specific knowledge of the characteristic sizes of the inlet and exit zones responsible for the deviation of the solution from the Hertzian solution for a dry contact of the same solids as well as for producing the lubrication film and pressure spike in the exit zone.

(iii) The series of the obtained values for the lubrication film thickness is curve fitted to a power (sometimes a little more complex) expression of the corresponding input parameters.

There are some pitfalls connected with this approach such as

(a) the numerical solutions are not necessarily obtained with sufficient precision due to the grid choice not related to the characteristic size of the inlet and exit zones (see comparison in [17]) and

(b) the formulas obtained represent a "hybrid" of formulas for different lubrication regimes [7] which, in reality, do not represent any of the actual relationships between the input parameters and the film thickness.

The asymptotically based solutions do not suffer from drawbacks (a) and (b) and provide analytically derived and asymptotically valid formulas for the film thickness for each specific regime of lubrication. In the end, both approaches produce similarly looking results. However, the asymptotically obtained formulas are more precise and correspond directly to the specific lubrication regimes at hand.

The precision and stability of numerical solutions for EHL problems is paramount. Unfortunately, after about 60 years of predominantly numerical research efforts the situation with numerical stability of solutions for heavily loaded EHL contacts is still not resolved satisfactorily while most researchers assume that this is normal that the existing numerical solutions for heavily loaded line and point EHL contacts are unstable, nonconvergent, and insufficiently precise. Let us consider an example of one of the most advanced, widely used, and well documented numerical methods in EHL - the Multilevel method [14]. The solutions for heavily loaded lubricated contacts obtained using this method suffer from instability and, therefore, low precision and lack of convergence. That can be clearly seen from the graphs of pressure distributions in Fig. 6.9 and 6.12 [14] exhibiting erratic pressure behavior in the exit part of the contact occupied by horse shoe shaped pressure spike. These graphs also indicate that the fluid mass is not conserved due to discontinuities in pressure gradient. Unfortunately, the situation with other numerical methods is not better. This situation with instability of numerical solutions for heavily loaded EHL contacts requires remedy. For heavily loaded line EHL contacts such a regularization remedy is proposed in [7] based on the asymptotic solution of thermal EHL problems. It is extremely important to realize that this regularization works equally successfully for asymptotically valid equations and the EHL equations in their original form used by all researchers.

The ideas behind the asymptotic methods applied to heavily loaded point EHL contacts are very simple. In each zone of the lubricated contact (inlet and exit zones as well as the Hertzian region) the terms of the equations of the EHL problem are estimated. The formulas for the film thickness and asymptotically valid equations in the inlet and exit zones are derived based on the fact that at the points where the inlet and exit zones border/overlap with the Hertzian region the orders of the magnitudes of the main terms of the equations involved are supposed to be the same [7].

In this chapter it is shown that the properties of the solutions for the lubricated line and point contacts are similar. In particular, (1) for heavily loaded contacts both problems are described by equations which are close to integral equations of the first kind. That as well as the fact that (2) usually the lubricant viscosity is a strongly nonlinear (exponentially fast growing) function of pressure creates a problem with numerically unstable solutions. The way to eliminate solution instability related to reason (1) is to analytically resolve the EHL equations with respect to pressure involved in the integral equation for the gap. Unfortunately, for point contacts it is generally impossible to do due to the geometry of the contact region boundary while for line contacts it can be easily done (see preceding chapters and [7]). For line contacts this approach works well for up to moderately high values of pressure viscosity coefficients [17]. To solve EHL problems for higher pressure viscosity coefficients, i.e. to remedy the situation with solution numerical instability caused by reason (2), a certain regularization technique can be devised which slightly modifies the problem equations while keeping the problem solutions stable and close to solutions of the non-regularized problem.

Due to the well developed understanding of the EHL problem solution behavior for heavily loaded line contacts it would be beneficial to see if EHL problems for heavily loaded point contacts can somehow be reduced/compared to the corresponding problems for heavily loaded line contacts. Therefore, the purpose of this paper is to establish the relationship between the EHL problems and their solutions for heavily loaded line and point contacts. It will be done using the asymptotic methods similar to the ones used in the preceding chapters which will provide the benefits described above.

12.2 Problem Formulation

Let us consider a steady isothermal lubricated contact of two elastic solids. The solids' elastic moduli and Poisson's ratios are E_1, E_2 and ν_1, ν_2, respectively. We will introduce a coordinate system with the xy-plane parallel to the plane of the contact and the z-axis normal to the xy-plane directed from the lower solid upward to the upper solid. In this coordinate system the shapes of the solid surfaces can be approximated by elliptical paraboloids: $z = z_2(x, y) = \frac{x^2}{2R_{2x}} + \frac{y^2}{2R_{2y}}$ and $z = z_1(x, y) = -\frac{x^2}{2R_{1x}} - \frac{y^2}{2R_{1y}}$, where R_{1x}, R_{2x}, R_{1y}, and R_{2y} are the surface curvature radii of the solid surfaces along the x- and y-axes, respectively. The two solids are loaded by a normal force P acting along the z-axis and are moving with the surface velocities (u_1, v_1) and (u_2, v_2), respectively. The solids are separated by a layer of an incompressible viscous lubricant with Newtonian rheology. Using the traditional assumptions that the solids' motions are steady and relatively slow, the size of the contact region is significantly smaller than the solid radii, and the lubrication film thickness is much smaller than the contact size [11] we obtain equations (4.4) and (4.10) (see Chapter 4). Using no slippage boundary conditions for the lubricant velocity components

$$u(x, y, -\tfrac{h}{2}) = u_1, \ v(x, y, -\tfrac{h}{2}) = v_1, \ u(x, y, \tfrac{h}{2}) = u_2, \ v(x, y, \tfrac{h}{2}) = v_2,$$

$$w(x, y, -\tfrac{h}{2}) = -\tfrac{u_1}{2}\tfrac{\partial h}{\partial x} - \tfrac{v_1}{2}\tfrac{\partial h}{\partial y}, \ w(x, y, \tfrac{h}{2}) = \tfrac{u_2}{2}\tfrac{\partial h}{\partial x} + \tfrac{v_2}{2}\tfrac{\partial h}{\partial y},$$

(12.1)

and integrating equations (4.10) with respect to z from $-\frac{h}{2}$ to z we obtain expressions for the lubricant velocities $u(x, y, z)$ and $v(x, y, z)$. By integrating the continuity equation (4.4) with respect to z from $-\frac{h}{2}$ to $\frac{h}{2}$, carefully changing the order of differentiation and integration and using the last pair of boundary conditions from (12.1) as well as the integrals of u and v with respect to z from $-\frac{h}{2}$ to $\frac{h}{2}$ we obtain the classic Reynolds equation which describes variations in pressure $p(x, y)$ as a function of variations in gap $h(x, y)$ between the contact surfaces. The boundary conditions for pressure follow from the fact that it is negligibly small outside of the contact compared to its values in the contact. The other condition on pressure follows from the requirement of the absence of cavitation at the exit boundary from the contact. Finally, the expression for the gap h and the balance conditions follow from the equations (4.14) and the second equation in (4.15). Finally, we derive the equations of the isothermal EHL problem in the form:

$$\frac{\partial}{\partial x}\left(\frac{h^3}{12\mu}\frac{\partial p}{\partial x}\right) + \frac{\partial}{\partial y}\left(\frac{h^3}{12\mu}\frac{\partial p}{\partial y}\right) = \frac{u_1 + u_2}{2}\frac{\partial h}{\partial x} + \frac{v_1 + v_2}{2}\frac{\partial h}{\partial y},$$

$$p(x, y)\mid_\Gamma = 0, \ \frac{dp(x,y)}{d\overrightarrow{n}}\mid_{\Gamma_e} = 0,$$

$$h(x, y) = h_0 + \frac{x^2}{2R_x} + \frac{y^2}{2R_y} + \frac{1}{\pi E'}\{\int\int_\Omega \frac{p(x_1, y_1)dx_1 dy_1}{\sqrt{(x-x_1)^2 + (y-y_1)^2}}$$

(12.2)

$$- \int\int_\Omega \frac{p(x_1, y_1)dx_1 dy_1}{\sqrt{x_1^2 + y_1^2}}\}, \ \int\int_\Omega p(x_1, y_1)dx_1 dy_1 = P,$$

where $p(x, y)$ and $h(x, y)$ are the pressure and gap distributions in the contact, respectively, μ is the lubricant viscosity, $\mu = \mu(p)$, h_0 is the central film thickness at the point with coordinates $(0,0)$, R_x, R_y, and E' are the effective radii along the x- and y-axes and the effective modulus of elasticity, $\frac{1}{R_x} = \frac{1}{R_{1x}} + \frac{1}{R_{2x}}$, $\frac{1}{R_y} = \frac{1}{R_{1y}} + \frac{1}{R_{2y}}$, and $\frac{1}{E'} = \frac{1-\nu_1^2}{E_1} + \frac{1-\nu_2^2}{E_2}$, Ω and Γ are the contact region and its boundary, respectively, and $\overrightarrow{n}$ is the unit external normal vector to the boundary Γ. The inlet Γ_i and exit Γ_e portions of the contact boundary Γ through which the lubricant enters and leaves the contact, respectively, constitute the whole boundary of Ω, i.e. $\Gamma_i \bigcup \Gamma_e = \Gamma$. The inlet Γ_i and exit Γ_e boundaries are determined by the relationships

$$(x,y) \in \Gamma_i \ if \ \overrightarrow{F_f} \cdot \overrightarrow{n} < 0; \ (x,y) \in \Gamma_e \ if \ \overrightarrow{F} \cdot \overrightarrow{n} \geq 0,$$

$$\overrightarrow{F_f} = (u_1 + u_2, v_1 + v_2)\frac{h}{2} - \frac{h^3}{12\mu}(\frac{\partial p}{\partial x}, \frac{\partial p}{\partial y}),$$

(12.3)

where $\overrightarrow{F_f}$ is the fluid flux through the contact boundary Γ and $\overrightarrow{a} \cdot \overrightarrow{b}$ is the dot product of vectors $\overrightarrow{a}$ and $\overrightarrow{b}$.

For simplicity we will assume that the lubricant viscosity μ satisfies the Barus equation

$$\mu(p) = \mu_a \exp(\alpha p),$$

(12.4)

where μ_a and α are the ambient lubricant viscosity and the pressure coefficient of viscosity, respectively.

Let us introduce the dimensionless variables as follows

$$x' = \frac{x}{a_H}, \ y' = \frac{y}{b_H}, \ p' = \frac{p}{p_H}, \ h' = \frac{h}{h_0}, \ \mu' = \frac{\mu}{\mu_a},$$

(12.5)

as well as the following dimensionless parameters

$$V = \frac{24\mu_a(u_1+u_2)R_x^2}{p_H a_H^3}, \ Q = \alpha p_H, \ \delta = \frac{R_x}{R_y}, \ H_{00} = \frac{2R_x h_0}{a_H^2},$$

(12.6)

where a_H, b_H, and p_H are the Hertzian semi-axes (small and large) of the dry elliptical contact and the Hertzian maximum pressure, respectively. It is useful to keep in mind that

$$b_H = [\frac{3}{\pi D(e)} \frac{PR_x}{E'}]^{1/3}, \ a_H = b_H\sqrt{1-e^2}, \ p_H = \frac{3P}{2\pi b_H^2 \sqrt{1-e^2}},$$

$$D(e) = \frac{K(e)-E(e)}{e^2},$$

(12.7)

where e is the dry contact ellipse eccentricity which satisfies the equation

$$\frac{E(e)-(1-e^2)K(e)}{(1-e^2)[K(e)-E(e)]} = \delta,$$

(12.8)

while $K(e)$ and $E(e)$ are full elliptic integrals of the first and second kind, respectively [18].

Using the dimensionless variables from (12.5) and (12.6) for the case of solids motion along the $x-$axis (i.e. for $v_1 + v_2 = 0$) and omitting the primes at the dimensionless variables equations (12.2) and (12.4) assume the form

$$\frac{\partial}{\partial x}\left(\frac{h^3}{\mu}\frac{\partial p}{\partial x}\right) + (1 - e^2)\frac{\partial}{\partial y}\left(\frac{h^3}{\mu}\frac{\partial p}{\partial y}\right) = \frac{V}{H_{00}^2}\frac{\partial h}{\partial x}, \ \ \mu = \exp(Qp),$$

$$p(x, y)\ |_{\Gamma} = 0, \ \ \frac{dp(x,y)}{d\overrightarrow{n}}\ |_{\Gamma_e} = 0,$$

$$H_{00}[h(x, y) - 1] = x^2 + \frac{\delta}{1-e^2}y^2 + \frac{D(e)}{\pi(1-e^2)^{3/2}}\{\int\int_{\Omega} \frac{p(x_1,y_1)dx_1 dy_1}{\sqrt{(x-x_1)^2 + \frac{(y-y_1)^2}{1-e^2}}} \tag{12.9}$$

$$- \int\int_{\Omega} \frac{p(x_1,y_1)dx_1 dy_1}{\sqrt{x_1^2 + \frac{y_1^2}{1-e^2}}}\}, \ \int\int_{\Omega} p(x_1, y_1)dx_1 dy_1 = \frac{2\pi}{3}.$$

In this problem formulation for the given inlet boundary Γ_i and constants V, Q, and δ (and e which can be easily determined from equation (12.8)) we need to find the distributions of pressure $p(x, y)$ and gap $h(x, y)$ as well as the exit boundary Γ_e and the dimensionless central film thickness H_{00}.

12.3   Asymptotic Analysis of a Heavily Loaded Lubricated Contact under Pre-Critical Lubrication Conditions

Let us consider heavily loaded lubrication conditions. They can be caused by low speed $(u_1 + u_2)$, high load P, small radii R_x and R_y, low ambient viscosity μ_a, and high pressure viscosity coefficient α. All these contact parameters are represented by the two dimensionless parameters V and Q. By looking at the expressions for V and Q in (12.6) it becomes clear that heavily loaded conditions can be realized only when V is small and/or Q is large. It is important to keep in mind that it is the most common practical condition in fully lubricated bearing and gear contacts. Therefore, we will assume that system (12.9) contains a small parameter $\omega \ll 1$ which can be equal to $V \ll 1$ or $Q^{-1} \ll 1$.

In heavily loaded lubricated contacts (i.e. for $\omega \ll 1$) at the points away from the contact boundary Γ the gap distribution $h(x, y)$ is very close to being constant along the $x-$axis while the pressure distribution $p(x, y)$ is close to the pressure distribution $p_0(x, y)$ which occurs in a dry contact Ω of a rigid indenter with the radii R_x and R_y and sharp edges with an elastic half-space with the effective elastic modulus E'. Away from the contact boundary Γ this pressure distribution is close to the Hertzian pressure distribution

$$p_{00}(x, y) = \sqrt{1 - x^2 - y^2} \tag{12.10}$$

in a circular (in dimensionless variables) contact

$$x^2 + y^2 \leq 1. \tag{12.11}$$

The region where that occurs we will call the Hertzian region.

In heavily loaded lubricated contacts the film thickness H_{00} separating the solids is small as the situation is close to the case of a dry contact where the film thickness is zero. Due to that and the fact that in the Hertzian region, i.e. at the points where the gap $h(x, y)$ is almost constant along the x-axis, we have $\partial h/\partial x \ll 1$. Therefore, it is clear that in heavily loaded EHL contacts $H_{00}\partial h/\partial x \ll 1$.

The above characterization of heavily loaded EHL contacts is fully supported by direct numerical solutions of equations (12.9) [13].

Based on this characterization of a heavily loaded EHL contact and the fact that the film thickness $H_{00} \ll 1$ we will define a contact to be heavily loaded if

$$H_{00}\frac{\partial h(x,y)}{\partial x} \ll 1, \; (x, y) \in \Omega \setminus \Omega_\epsilon, \tag{12.12}$$

where Ω_ϵ is a narrow ring-like shaped region adjacent to the contact boundary Γ which encompasses the region $\Omega \setminus \Omega_\epsilon$ located in the central part of the contact. Let the characteristic width of Ω_ϵ be $\epsilon_q = \epsilon_q(\omega) \ll 1$ for $\omega \ll 1$.

Condition (12.12) is equivalent to

$$\frac{H_{00}^3}{V}[\frac{\partial}{\partial x}(\frac{h^3}{\mu}\frac{\partial p}{\partial x}) + (1 - e^2)\frac{\partial}{\partial y}(\frac{h^3}{\mu}\frac{\partial p}{\partial y})] \ll 1, \; (x, y) \in \Omega \setminus \Omega_\epsilon, \tag{12.13}$$

as it follows from condition (12.12) and the Reynolds equation (12.9) multiplied by H_{00}^3/V.

Due to the definitions (12.13) and (12.12) and the fact that in the Hertzian region $\partial h/\partial x \ll 1$ it is clear that in the Hertzian region $\Omega \setminus \Omega_\epsilon$ the gap distribution $h(x, y)$ is as follows

$$h(x, y) = h_\infty(y), \; (x, y) \in \Omega \setminus \Omega_\epsilon, \tag{12.14}$$

where $h_\infty(y)$ is the exit film thickness for the given y, i.e. the film thickness at a point of the exit boundary Γ_e with coordinates (x, y). The latter becomes obvious from the further analysis as the derivative of p with respect to y in the Reynolds equation (12.9) in the exit zone can be neglected (see (12.26)) and the obtained equation can be integrated with respect to x from x to $\gamma_e(y)$ ($\gamma_e(y)$ is the x-coordinate of the exit boundary point) with the boundary condition $\partial p(\gamma_e(y), y)/\partial x = 0$ (see (12.17) and (12.26)).

The definitions (12.13) and (12.12) and equations (12.9) indicate that pressure $p_0(x, y)$ (generally different from the Hertzian pressure distribution $p_{00}(x, y)$ from (12.10) due to the shape of the contact region Ω) satisfies the following equations

$$x^2 + \frac{\delta}{1-e^2}y^2 + \frac{D(e)}{\pi(1-e^2)^{3/2}} \{ \int\int_\Omega \frac{p_0(x_1,y_1)dx_1dy_1}{\sqrt{(x-x_1)^2+\frac{(y-y_1)^2}{1-e^2}}}$$

$$- \int\int_\Omega \frac{p_0(x_1,y_1)dx_1dy_1}{\sqrt{x_1^2+\frac{y_1^2}{1-e^2}}} \} = 0, \quad \int\int_\Omega p_0(x_1,y_1)dx_1dy_1 = \frac{2\pi}{3}. \tag{12.15}$$

Equations (12.15) are obtained from the last two equations in (12.9) by setting $H_{00} = 0$. Physically, function $p_0(x, y)$ represents the pressure distribution in a dry fixed contact Ω of a rigid indenter with the radii R_x and R_y and sharp edges with an elastic half-space with the effective elastic modulus E'. In most cases $p_0(x, y)$ cannot be determined analytically due to the complexity of the shape of the boundary Γ. Depending on the fixed shape of Γ the solution $p_0(x, y)$ of equations (12.15) at some points of Γ may vanish or be singular with the singularity of the type $d^{-1/2}$, where d is the normal distance of a point from the region boundary Γ [19]. Therefore, we can conclude that close to the contact boundary Γ pressure $p_0(x, y)$ in the dry contact Ω can be asymptotically represented as follows [19]

$$p_0(x,y) = \epsilon_q^{1/2}q_0 + o(\epsilon_q^{1/2}), \quad q_0(x,y,x_\Gamma,y_\Gamma) = N_0\sqrt{2d_\epsilon} + \frac{N_1}{\sqrt{2d_\epsilon}},$$

$$N_0 = N_0(x_\Gamma,y_\Gamma) = O(1), \quad N_1 = N_1(x_\Gamma,y_\Gamma) = O(1), \tag{12.16}$$

$$d_\epsilon = \frac{\sqrt{(x-x_\Gamma)^2+\frac{(y-y_\Gamma)^2}{1-e^2}}}{\epsilon_q} = O(1), \quad \omega \ll 1, \quad (x,y) \in \Omega_\epsilon, \quad (x_\Gamma,y_\Gamma) \in \Gamma,$$

where the point (x, y) belongs to the the line normal to the boundary Γ through the boundary point (x_Γ, y_Γ). Obviously, the point (x, y) is at a distance of about ϵ_q from the boundary point (x_Γ, y_Γ).

Let us make some simple and, at the same time, natural assumptions about the contact region boundary Γ. We will assume that the inlet and exit boundaries can be described by the equations $x = \gamma_i(y)$ and $x = \gamma_e(y)$ (see Fig. 12.1) and that the inlet Γ_i and exit Γ_e boundaries of the lubricated contact are close to the boundaries of the dry circular contact, i.e.

$$\gamma_i(y) = -\sqrt{1-y^2} + \epsilon_q\alpha_{1p}(y), \quad \alpha_{1p}(y) = O(1), \quad (x,y) \in \Gamma_i,$$

$$\gamma_e(y) = \sqrt{1-y^2} + \epsilon_q\beta_{1p}(y), \quad \beta_{1p}(y) = O(1), \quad (x,y) \in \Gamma_e, \quad \omega \ll 1, \tag{12.17}$$

where $\alpha_{1p}(y)$ is a given function of y while $\beta_{1p}(y)$ has to be determined from the problem solution (For more information on the determination of ϵ_q see equations (12.40)-(12.43) for pre-critical lubrication regimes). Moreover, we will assume that $\gamma_i(y)$ and $\gamma_e(y)$ are differentiable functions and that outside of the ends of HSSPGD zone Ω_{HSSPGD} (which are in the vicinity of the points $(x, y) = (0, -1)$ and $(x, y) = (0, 1)$) we have

$$\frac{d\gamma_i(y)}{dy} = O(1), \quad \frac{d\gamma_e(y)}{dy} = O(1), \quad \omega \ll 1. \tag{12.18}$$

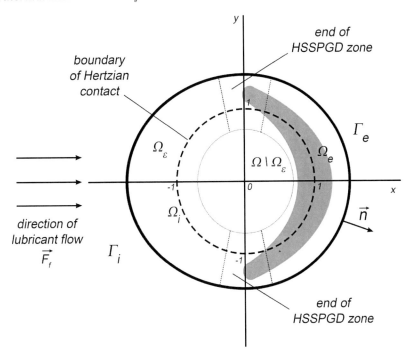

FIGURE 12.1
A schematic view of the lubricated region of a point contact. The area shaded by grey and its immediate vicinity is HSSPGD zone the ends of which are located in the vicinity of the points with coordinates $(0, -1)$ and $(0, 1)$. The points where the contact boundary Γ switches from the inlet Γ_i to the exit Γ_e contact boundary are located within the ends of HSSPGD zone. The dotted circle is the inner boundary of Ω_ϵ and outer boundary of $\Omega \setminus \Omega_\epsilon$.

These assumptions mean that the variations of the contact region boundary Γ are moderate, i.e. the boundary Γ does not experience abrupt step-type variations.

Estimates (12.17) and (12.18), in turn, lead to the following estimates

$$\frac{dN_0(x_\Gamma, y_\Gamma)}{d\overrightarrow{\tau_\Gamma}}, \ \frac{dN_1(x_\Gamma, y_\Gamma)}{d\overrightarrow{\tau_\Gamma}} = O(1), \ \omega \ll 1, \tag{12.19}$$

where $\overrightarrow{\tau_\Gamma}$ is a unit vector tangent to the contact boundary Γ at the point with coordinates $(x_\Gamma, y_\Gamma) \in \Gamma$.

Using (12.15) the last two equations in (12.9) can be rewritten in the form

$$H_{00}[h(x, y) - 1] = \frac{D(e)}{\pi(1-e^2)^{3/2}} \int\int_\Omega [p(x_1, y_1) - p_0(x_1, y_1)] \tag{12.20}$$

$$\times\{\frac{1}{\sqrt{(x-x_1)^2+\frac{(y-y_1)^2}{1-e^2}}}-\frac{1}{\sqrt{x_1^2+\frac{y_1^2}{1-e^2}}}\}dx_1dy_1,$$

$$\underset{\Omega}{\int\int}[p(x_1,y_1)-p_0(x_1,y_1)]dx_1dy_1=0.$$

Now, we are ready for the asymptotic analysis of the EHL problem. It is important to remember that in the region of the contact (inside of the Hertzian circular contact) which is at a distance from the boundary of the Hertzian contact of much more than $O(\epsilon_q)\ll 1$ the pressure distribution $p(x,y)$ is very close to the Hertzian pressure $p_{00}(x,y)$. At the same time, in the inlet and exit zones the actual pressure $p(x,y)$ is significantly different from the Hertzian pressure $p_{00}(x,y)$ as well as from pressure $p_0(x,y)$. Such a situation is a typical set up for application of the matched asymptotic expansions [8, 9]. The application of the matched asymptotic approach to solution of the EHL problems for heavily loaded line contacts is described in detail in [7].

First, we will consider the so called pre-critical heavily loaded lubrication regimes [7] for which the lubricant viscosity varies with pressure moderately. The very basic fact about the matched asymptotic expansions approach is that the solution of the problem $p(x,y)$ in the inlet and exit zones of width $O(\epsilon_q)\ll 1$ is of the order of magnitude of $\epsilon_q^{1/2}$ because otherwise it will not be able to match the external (dry) pressure $p_0(x,y)$ outside of the inlet and exit zones (see (12.16)). Therefore, a heavily loaded lubrication regime is pre-critical if in the inlet zone of the contact Ω_i where $p(x,y)=O(\epsilon_q^{1/2})$, $\epsilon_q\ll 1$, the following estimate holds

$$\mu(p(x,y))=O(1),\ if\ p(x,y)=O(\epsilon_q^{1/2}),\ (x,y)\in\Omega_\epsilon,\ \omega\ll 1. \qquad (12.21)$$

The inlet Ω_i and exit Ω_e zones are narrow zones which are parts of the region Ω_ϵ adjacent to the inlet Γ_i and exit Γ_e boundaries, respectively.

For line and point lubricated contacts the pre-critical lubrication regimes were studied earlier in [11, 14, 20]-[22] using numerical methods.

It is important to realize that according to (12.16) $d_\epsilon=\frac{\sqrt{(x-x_\Gamma)^2+\frac{(y-y_\Gamma)^2}{1-e^2}}}{\epsilon_q}=$ $O(1)$, $\omega\ll 1$, as long as $x-x_\Gamma=O(\epsilon_q)$ and $y-y_\Gamma=O(\epsilon_q)$ for $\omega\ll 1$. Therefore, $p_0(x,y)=O(\epsilon_q^{1/2})$ within the zone Ω_ϵ. In order for the pressure function $p(x,y)$ to match the function $p_0(x,y)$ at the inner boundary of the region Ω_ϵ and the outer boundary of the region $\Omega\setminus\Omega_\epsilon$ it is necessary that they are of the same order of magnitude [8, 9]. Therefore, we have

$$p(x,y)=O(\epsilon_q^{1/2}),\ (x,y)\in\Omega_\epsilon,\ \omega\ll 1. \qquad (12.22)$$

Based on (12.21) and (12.22) the definition of the pre-critical regimes can be presented in the form

$$\mu(\epsilon_q^{1/2})=O(1),\ \omega\ll 1. \qquad (12.23)$$

Let us consider the points in the inlet Ω_i and exit Ω_e zones but outside of the ends of HSSPGD zone Ω_{HSSPGD} which belong to the line $y = y_*$. First, let us consider the inlet zone. We will replace our original coordinate system (x, y) by a new coordinate system (r_p, y), where $r_p = \frac{x - \gamma_i(y_*)}{\epsilon_q}$. Using (12.16)-(12.19), (12.22) and the principle of matching [8, 9] of asymptotic expansions of the problem solutions in the inlet zone $\Omega_i \bigcap \Omega_\epsilon$ and the Hertzian region we obtain that

$$p(x, y_*) = \epsilon_q^{1/2} q_p(r_p, y_*) + o(\epsilon_q^{1/2}), \quad q_p(r_p, y_*) = O(1),$$

$$r_p = \frac{x - \gamma_i(y_*)}{\epsilon_q} = O(1), \quad (x, y_*) \in \Omega_i \bigcap \Omega_\epsilon, \ \omega \ll 1,$$
(12.24)

where $q_p(r_p, y_*)$ is the unknown function which is the main term of the asymptotic expansion of pressure in the inlet zone. In a similar manner we can easily establish that

$$p(x, y_*) = \epsilon_q^{1/2} g_p(s_p, y_*) + o(\epsilon_q^{1/2}), \quad g_p(s_p, y_*) = O(1),$$

$$s_p = \frac{x - \gamma_e(y_*)}{\epsilon_q} = O(1), \quad (x, y_*) \in \Omega_e \bigcap \Omega_\epsilon, \ \omega \ll 1,$$
(12.25)

where $g_p(s_p, y_*)$ is the unknown function which is the main term of the asymptotic expansion of pressure in the exit zone. Functions $q_p(r_p, y_*)$ and $g_p(s_p, y_*)$ further for simplicity will be called asymptotic pressure distributions in the inlet and exit zones.

Based on (12.18) and (12.19) in the inlet Ω_i and exit Ω_e zones outside of the HSSPGD zone Ω_{HSSPGD} we obtain

$$\frac{\partial}{\partial x} = \epsilon_q^{-1} \frac{\partial}{\partial r_p}, \ r_p = O(1); \ \frac{\partial}{\partial x} = \epsilon_q^{-1} \frac{\partial}{\partial s_p}, \ s_p = O(1), \ \omega \ll 1,$$

$$\frac{\partial}{\partial y} = O(1), \ \omega \ll 1.$$
(12.26)

Using (12.18) we can determine that the external normal vector $\overrightarrow{n} = (n_x, n_y)$ has a nonzero x-component $n_x = O(1)$, $\omega \ll 1$. Therefore, based on (12.25) and (12.26) we can conclude that the condition $\frac{dp(x,y)}{d\overrightarrow{n}} \mid_{\Gamma_e} = 0$ at the exit boundary Γ_e (see (12.9)) can be represented in the form $\{ \epsilon_q^{-1/2} \frac{\partial g_p}{\partial s_p} n_x + \epsilon_q^{1/2} \frac{\partial g_p}{\partial y} n_y + \ldots \} \mid_{\Gamma_e} = 0$, $\omega \ll 1$. Obviously, for $\omega \ll 1$ the latter condition leads to the following one $\frac{\partial g_p}{\partial s_p} \mid_{s_p=0} = 0$.

Now, let us consider the asymptotic representations for $\frac{\partial h}{\partial x}$ in the inlet and exit zones. From the first equation in (12.20) we have

$$H_{00} \frac{\partial h}{\partial x} = \frac{D(e)}{\pi(1-e^2)^{3/2}} \int\int_\Omega \frac{(x_1-x)[p(x_1,y_1)-p_0(x_1,y_1)]}{[(x-x_1)^2 + \frac{(y-y_1)^2}{1-e^2}]^{3/2}} dx_1 dy_1.$$
(12.27)

We will consider the portions of the inlet Ω_i and exit Ω_e zones (away from the ends of HSSPGD zone Ω_{HSSPGD}) fully extended along the x-axis (i.e., along the line $y = y_*$) which are practically symmetric about the horizontal line $y = y_*$ (see (12.18)) and with characteristic sizes along the y-axis of about ϵ_q. Due to (12.16)-(12.19) within these zones

$$p(x, y) = p(x, y_*) + O(\epsilon_q),$$

$$p(x, y) - p_0(x, y) = p(x, y_*) - p_0(x, y_*) + O(\epsilon_q), \qquad (12.28)$$

$$(x, y) \in (\Omega_i \bigcap \Omega_\epsilon) \bigcup (\Omega_e \bigcap \Omega_\epsilon), \ w \ll 1.$$

Taking into account the definition (12.12) of a heavily loaded EHL contact and related to it assumption that away from the boundary of the Hertzian contact

$$p(x, y) - p_0(x, y) = O(\epsilon_q^2), \ (x, y) \in \Omega \setminus \Omega_\epsilon, \ w \ll 1, \qquad (12.29)$$

the representations (12.24) and (12.25), as well as the estimate (12.28) and using the local coordinates r_p, s_p, and $\eta = \frac{y - y_*}{\epsilon_q}$ we can estimate that if the point (x, y) belongs to the inlet or exit zones (away from the ends of the HSSPGD zone Ω_{HSSPGD}) then the contribution of the integral of $\frac{(x_1 - x)[p(x_1, y_1) - p_0(x_1, y_1)]}{[(x - x_1)^2 + \frac{(y - y_1)^2}{1 - e^2}]^{3/2}}$ (see equation (12.27)) over the entire contact region Ω excluding the considered inlet or exit zone will be of the order of magnitude $O(\epsilon_q^2)$, $w \ll 1$, while the contribution of the integral of the above function over the inlet or exit zone will be of the order of $O(\epsilon_q^{1/2})$, $w \ll 1$. Therefore, we can conclude that

$$H_{00} \frac{\partial h(x, y_*)}{\partial x} = \frac{\epsilon_q^{1/2}}{\pi} \frac{D(e)}{(1 - e^2)^{3/2}} \int \int_\Omega \frac{(t - r_p)[q_p(t, y_*) - q_{ap}(t, y_*)]}{[(r_p - t)^2 + \frac{\eta^2}{1 - e^2}]^{3/2}} dt d\eta$$

$$+ o(\epsilon_q^{1/2}) = \frac{\epsilon_q^{1/2}}{\pi} \frac{D(e)}{(1 - e^2)^{3/2}} \int_0^\infty (t - r_p)[q_p(t, y_*) - q_{ap}(t, y_*)] dt \qquad (12.30)$$

$$\times \int_{-\infty}^\infty \frac{d\eta}{[(r_p - t)^2 + \frac{\eta^2}{1 - e^2}]^{3/2}} + o(\epsilon_q^{1/2}), \ (x, y_*) \in \Omega_i \bigcap \Omega_\epsilon, \ w \ll 1,$$

where $q_0(x, y, x_\Gamma, y_\Gamma)$ is replaced by $q_{ap}(r_p, y_*) = q_0(\gamma_i(y_*) + r_p \epsilon_q, y_*, \gamma_i(y_*), y_*)$ $= N_{0q}(y_*) \sqrt{2 r_p} + \frac{N_{1q}(y_*)}{\sqrt{2 r_p}}$ due to the fact that the direction of the normal vector $\overrightarrow{n}$ to the contact boundary Γ_i in $\Omega_i \bigcap \Omega_\epsilon$ practically coincides with the x-axis (see (12.18)). Obviously, the last integral in (12.30) is equal to $\frac{2\sqrt{1 - e^2}}{(t - r_p)^2}$. Therefore, from (12.30) we obtain

$$H_{00} \frac{\partial h}{\partial x} = \epsilon_q^{1/2} \frac{2}{\pi} \theta \int_0^\infty \frac{q_p(t, y_*) - q_{ap}(t, y_*)}{t - r_p} dt + o(\epsilon_q^{1/2}), \qquad (12.31)$$

$$(x, y_*) \in \Omega_i \bigcap \Omega_\epsilon, \ w \ll 1,$$

where $\theta = \frac{D(e)}{1-e^2}$. In a similar fashion, in the exit zone $\Omega_e \bigcap \Omega_\epsilon$ we get

$$H_{00}\frac{\partial h}{\partial x} = \epsilon_q^{1/2}\frac{2}{\pi}\theta \int_{-\infty}^{0} \frac{g_p(t,y_*)-g_{ap}(t,y_*)}{t-s_p}dt + o(\epsilon_q^{1/2}),$$

$$(12.32)$$

$$(x,y_*) \in \Omega_e \bigcap \Omega_\epsilon, \ \omega \ll 1,$$

where $g_{ap}(s_p,y_*) = q_0(\gamma_e(y_*) + s_p\epsilon_q, y_*, \gamma_e(y_*), y_*) = N_{0g}(y_*)\sqrt{-2s_p} + \frac{N_{1g}(y_*)}{\sqrt{-2s_p}} + o(1)$, $\epsilon_q \ll 1$.

It is important to realize that the function $N_0(x_\Gamma, y_\Gamma)$ depends only on the behavior of $p_{00}(x,y)$ from (12.10) and it is independent from the shape of the contact boundary Γ which is close to $x^2 + y^2 = 1$ (see (12.17)). It means that at the border between the inlet zone and the Hertzian region, i.e. as $r_p \to \infty$ while $x + \sqrt{1-y^2} = O(\epsilon_q)$ the value of $\epsilon_q^{1/2}N_0\sqrt{2r_p} = \sqrt{1-x^2-y^2} + o(1)$. Substituting in the latter equation $r_p = \frac{x+\sqrt{1-y^2-\epsilon_q\alpha_{1p}}}{\epsilon_q}$ we obtain the equation

$$N_0\sqrt{2}\sqrt{x + \sqrt{1-y^2} - \epsilon_q\alpha_{1p}} = \sqrt{1-x^2-y^2} + o(1), \ \omega \ll 1. \quad (12.33)$$

At the boundary between the inlet zone and the Hertzian region, i.e. for $r_p \to \infty$ and $x + \sqrt{1-y^2} = O(1)$, $\omega \ll 1$, after simple calculations from (12.33) we can conclude that in the inlet zone

$$N_0(y) = \sqrt[4]{1-y^2}. \quad (12.34)$$

In a similar fashion in the exit zone we will get that $N_0(y)$ also satisfies (12.34). Therefore,

$$N_{0q}(y) = N_{0g}(y) = \sqrt[4]{1-y^2}. \quad (12.35)$$

On the other hand, the function $N_1(x_\Gamma, y_\Gamma)$ depends on the distance of the contact boundary Γ from the boundary of the Hertzian contact $x^2 + y^2 = 1$, i.e. functions N_{1q} and N_{1g} are proportional to α_{1p} and β_{1p} (see (12.17)), respectively. In more detail this will be described later (see (12.54)).

The above estimates for the derivatives of p and h are valid for an arbitrary $y = y_*$ as long as the point (x,y) belongs to the inlet or exit zones outside of the ends of HSSPGD zone Ω_{HSSPGD}. Therefore, using the estimates for $H_{00}\frac{\partial h}{\partial x}$ from (12.31), (12.32), and (12.26) in the inlet $\Omega_i \bigcap \Omega_\epsilon$ and exit $\Omega_e \bigcap \Omega_\epsilon$ zones from the requirement that the differential and integral terms in the Reynolds equation in (12.9) are of the same order of magnitude for $\omega \ll 1$ we obtain

$$\frac{\partial}{\partial r_p}M_0(A_p, q_p, h_{qp}, \mu_{qp}, r_p, y) = \frac{2}{\pi}\theta \int_0^{\infty} \frac{q_p(t,y)-q_{ap}(t,y)}{t-r_p}dt, \ q_p(0,y) = 0, \quad (12.36)$$

$$\frac{\partial}{\partial s_p} M_0(A_p, g_p, h_{gp}, \mu_{gp}, s_p, y) = \frac{2}{\pi}\theta \int\limits_{-\infty}^{0} \frac{g_p(t,y) - g_{ap}(t,y)}{t - s_p} dt,$$

(12.37)

$$g_p(0, y) = 0, \quad \frac{\partial g_p(0,y)}{\partial s_p} = 0,$$

where function $M_0(A, p, h, \mu, x, y)$ is determined by the equation

$$M_0(A, p, h, \mu, x, y) = A^3 \frac{h^3(x,y)}{\mu} \frac{\partial p}{\partial x}.$$

(12.38)

and A_p is a dimensionless coefficient of proportionality in the formula for the film thickness

$$H_{00} = A_p (V \epsilon_q^2)^{1/3}, \quad A_p = O(1), \quad \omega \ll 1,$$

(12.39)

where $h_{qp}(r_p, y)$ and $h_{gp}(s_p, y)$ are the gap functions in the inlet $\Omega_i \bigcap \Omega_\epsilon$ and exit $\Omega_e \bigcap \Omega_\epsilon$ zones. In (12.36) and (12.37) functions μ_{qp} and μ_{gp} are the main terms of the asymptotic expansions of $\mu(\epsilon_q^{1/2} q_p)$ and $\mu(\epsilon_q^{1/2} g_p)$, respectively.

The expressions for $h_{qp}(r_p, y)$ and $h_{gp}(s_p, y)$ depend on the lubrication regime. For starved lubrication regimes the gap distribution $h(x, y)$ is very close to $h_\infty(y)$ in the entire contact region. That corresponds to the case of $\epsilon_q^{3/2} \ll H_{00}$ and

$$h_{qp}(r_p, y) = h_\infty(y), \quad \epsilon_q \ll V^{2/5},$$

(12.40)

$$h_{gp}(s_p, y) = h_\infty(y), \quad \epsilon_q \ll V^{2/5},$$

(12.41)

while for fully flooded lubrication regimes $H_{00} = O(\epsilon_q^{3/2})$, $\omega \ll 1$, and in the inlet and exit zones the gap satisfies the equations (see equations (12.31) and (12.32))

$$A_p \frac{\partial h_{qp}}{\partial r_p} = \frac{2}{\pi}\theta \int\limits_{0}^{\infty} \frac{q_p(t,y) - q_{ap}(t,y)}{t - r_p} dt, \quad \epsilon_q = V^{2/5}, \quad \omega \ll 1,$$

(12.42)

$$A_p \frac{\partial h_{gp}}{\partial s_p} = \frac{2}{\pi}\theta \int\limits_{-\infty}^{0} \frac{g_p(t,y) - g_{ap}(t,y)}{t - s_p} dt, \quad \epsilon_q = V^{2/5}, \quad \omega \ll 1.$$

(12.43)

It is obvious from (12.39) that for fully flooded pre-critical lubrication regimes (characterized by the relationship $H_{00} = O(\epsilon_q^{3/2})$, $\omega \ll 1$) we have

$$H_{00} = A_p V^{3/5}, \quad \epsilon_q = \epsilon_f,$$

(12.44)

$$\epsilon_f = V^{2/5}.$$

(12.45)

Here and in (12.39) the value of the coefficient A_p depends on θ, $\alpha_{1p}(0)$, and on $Q\epsilon_q^{1/2}$ for starved lubrication regimes and on $QV^{1/5}$ for fully flooded lubrication regimes, respectively.

It is clear from equations (12.42) and (12.43) that when $r_p \to \infty$ and $s_p \to -\infty$ we get

$$\int\limits_{0}^{\infty} [q_p(t, y) - q_{ap}(t, y)] dt = 0,$$

(12.46)

$$\int\limits_{-\infty}^{0} [g_p(t,y) - g_{ap}(t,y)]dt = 0, \qquad (12.47)$$

because otherwise the functions $h_{qp}(r_p, y)$ and $h_{gp}(s_p, y)$ would be unbounded as $r_p \to \infty$ and $s_p \to -\infty$, respectively. Equations (12.46) and (12.47) are consistent with the second equation in (12.20).

For fully flooded lubrication regimes (i.e., for $\epsilon_q = \epsilon_f$) the values of functions $h_{qp}(r_p, y)$ and $h_{gp}(s_p, y)$ are obtained by integrating equations (12.42) and (12.43) with the conditions $h_{qp}(r_p, y) \to h_\infty(y)$ as $r_p \to \infty$ and $h_{gp}(s_p, y) \to h_\infty(y)$ as $s_p \to -\infty$ as follows

$$A_p[h_{qp}(r_p, y) - h_\infty(y)] = \tfrac{2}{\pi}\theta \int\limits_{0}^{\infty} [q_p(t,y) - q_{ap}(t,y)]\ln\tfrac{1}{|t-r_p|}dt, \qquad (12.48)$$

$$A_p[h_{gp}(s_p, y) - h_\infty(y)] = \tfrac{2}{\pi}\theta \int\limits_{-\infty}^{0} [g_p(t,y) - g_{ap}(t,y)]\ln\tfrac{1}{|t-s_p|}dt. \qquad (12.49)$$

Therefore, in the inlet zone for the given values of θ, $Q\epsilon_q^{1/2}$ (involved in μ_{qp}), $\alpha_{1p}(y)$, and y we derived the system of equations (12.36), (12.46), and (12.40) for starved lubrication regimes or (12.48) for fully flooded lubrication regimes for functions $q_p(r_p, y)$, $h_{qp}(r_p, y)$, $h_\infty(y)$, and constant A_p. In the exit zone for the given values of θ, $Q\epsilon_q^{1/2}$, A_p, $h_\infty(y)$, and y we need to solve the system of equations (12.37), (12.47), and (12.41) for starved lubrication regimes or (12.49) for fully flooded lubrication regimes for functions $g_p(s_p, y)$, $h_{gp}(s_p, y)$, and $\beta_{1p}(y)$. It is important to keep in mind that constant A_p is the coefficient of proportionality in formula (12.39) for the central film thickness and it is determined for $y = 0$. Also, it means that $h_\infty(0) = 1$ while, generally, $h_\infty(y) \neq 1$ for $y \neq 0$. The solution procedure should be implemented according to the following sequence: first, the system of asymptotically valid equations in the inlet zone should be solved for $y = 0$ which determines the value of constant A_p; after that, the values of function $h_\infty(y)$ for $y \neq 0$ are determined from the solution of the asymptotically valid system of equations in the inlet zone and these values play the role similar to the one played by constant A_p for $y = 0$.

The above systems of asymptotic equations in the inlet and exit zones are very similar to the ones derived and analyzed for the cases of line EHL contacts in preceding chapters and in [7, 23, 24]. Namely, by introducing the substitutions

$$\{r, s, \alpha_1, \beta_1\} = \tfrac{1}{r_0}\{r_p, s_p, \alpha_{1p}, \beta_{1p}\}, \quad \{q, g\} = \tfrac{1}{N_0 r_0^{1/2}}\{q_p, g_p\},$$

$$\qquad (12.50)$$

$$\{h_q, h_g\} = \tfrac{1}{h_\infty}\{h_{qp}, h_{gp}\}, \quad \{\mu_q(q), \mu_g(g)\} = \{\mu_{qp}(q_p), \mu_{gp}(g_p)\},$$

and the definitions

$$A = A_p[\theta N_0^4(y)]^{1/5}h_\infty(y), \quad H_0(y) = H_{00}h_\infty(y), \quad r_0 = [\theta^2 N_0^3(y)]^{-2/5} \quad (12.51)$$

we will be able to reduce each of the above systems of asymptotically valid equations to

$$\frac{\partial M_0(A,q,h_q,\mu_q,r,y)}{\partial r} = \frac{2}{\pi} \int\limits_0^\infty \frac{q(t,y)-q_a(t,y)}{t-r}dt, \quad q(0,y) = 0,$$

$$\int\limits_0^\infty [q(t,y) - q_a(t,y)]dt = 0, \quad q_a(r,y) = \sqrt{2r} + \frac{\alpha_1(y)}{\sqrt{2r}}, \tag{12.52}$$

$$h_q(r,y) = 1 \text{ or } A[h_q(r,y) - 1] = \frac{2}{\pi} \int\limits_0^\infty [q(t,y) - q_a(t,y)] \ln \frac{1}{|t-r|} dt,$$

in the inlet zone and to

$$\frac{\partial M_0(A,g,h_g,\mu_g,s,y)}{\partial s} = \frac{2}{\pi} \int\limits_{-\infty}^0 \frac{g(t,y)-g_a(t,y)}{t-s}dt, \quad g(0,y) = \frac{\partial g(0,y)}{\partial s} = 0,$$

$$\int\limits_{-\infty}^0 [g(t,y) - g_a(t,y)]dt = 0, \quad g_a(s,y) = \sqrt{-2s} - \frac{\beta_1(y)}{\sqrt{-2s}}, \tag{12.53}$$

$$h_g(s,y) = 1 \text{ or } A[h_g(s,y) - 1] = \frac{2}{\pi} \int\limits_{-\infty}^0 [g(t,y) - g_a(t,y)] \ln \frac{1}{|t-s|} dt,$$

in the exit zone where function M_0 is determined by formula (12.38).

In equation (12.51) function $H_0(y)$ is the dimensionless analog of the dimensional film thickness h_0 but at the exit from the contact region as well as in the Hertzian region at the points with the fixed ordinate y. In other words, we can expect that there are some variations in the film thickness in the Hertzian region along the y-axis. Besides that, we used the fact that

$$\alpha_{1p} = \frac{N_{1q}}{N_{0q}}, \quad \beta_{1p} = -\frac{N_{1g}}{N_{0g}}. \tag{12.54}$$

We have to keep in mind that due to a particular shape of the contact region boundary Γ the inlet coordinate α_1 is generally a function of y which makes the exit coordinate β_1 also a certain function of y. In addition to that, in the inlet and exit zones the functions of the lubricant viscosity μ_q and μ_g depend on $Q\epsilon_q^{1/2}\theta^{-2/5}(1 - y^2)^{1/10}q(r)$ and $Q\epsilon_q^{1/2}\theta^{-2/5}(1 - y^2)^{1/10}g(s)$, respectively.

Obviously, in case of α_1 independent of y and $\mu_q = \mu_g = 1$ (i.e., in the inlet and exit zones the lubricant viscosity is independent of pressure) we will need to solve systems (12.52) and (12.53) just once and using substitutions (12.50) and (12.51) get the solution of the problem in the inlet and exit zones for all values of y such that $(x,y) \in \Omega_i \bigcap \Omega_e$ and $(x,y) \in \Omega_e \bigcap \Omega_e$.

Systems of equations (12.52) and (12.53) are identical to the corresponding systems of equations (6.68), (6.69), (6.71), (6.49), (6.50), and (6.64) for the starved lubrication regimes (or (6.66) for fully flooded lubrication regimes) in the inlet zone and to equations (6.70), (6.69), (6.72), (6.49), (6.49), and (6.65) for the starved lubrication regimes (or (6.67) for fully flooded lubrication regimes) in the exit zone. These systems of equations are derived in the inlet and exit zones of lubricated heavily loaded line contacts. Therefore, the whole arsenal of analytical and numerical methods developed for the analysis of these equations in Chapters 6 - 10 and the corresponding conclusions made for line contacts are transferable to point contacts. That includes discussions of the numerical methods suitable for solution of asymptotically valid equations in the inlet and exit zones, solution precision, proper grid sizing, and solution stability/instability issues as well as the regularization approach for obtaining stable numerical solutions for high values of the dimensionless pressure viscosity coefficient Q. All this information including a large number of numerical examples can be found in Chapters 6 and 8.

The above substitutions and formulas (12.35), (12.39), (12.51), and (12.54) lead to the conclusion that the exit film thickness H_0 besides being a function of α_1, V, Q, ϵ_q, e is also a function of y and

$$H_0(y) = A[\tfrac{D(e)}{1-e^2}(1 - y^2)]^{-1/5}(V\epsilon_q^2)^{1/3}, \ A = O(1), \ \omega \ll 1, \qquad (12.55)$$

where A is a constant determined from the solution of the asymptotic system of equations (12.52) in the inlet zone for a line contact (which is dependent on α_1, $Q\epsilon_q^{1/2}$, and, also, it is dependent on θ and y through μ_q) while V is determined by the formula (12.6) for a point contact. For pre-critical lubrication regimes the value of $Q\epsilon_q^{1/2}$ can be taken as either 0 for $\epsilon_q \ll Q^{-2}$, $\omega \ll 1$, or equal to $Q\epsilon_q^{1/2} = O(1)$, $\omega \ll 1$, if $\epsilon_q = O(Q^{-2})$, $\omega \ll 1$. In a special case of constant viscosity μ or the case of starved lubrication regime when $\epsilon_q \ll Q^{-2}$, $\omega \ll 1$, the situation is simpler because the value of the constant A is dependent only on $\alpha_1(y)$. In the latter case the dependence of the exit film thickness H_0 on the contact ellipse eccentricity e is limited to just the one shown in (12.55).

The exit film thickness H_0 and, simultaneously, the film thickness in the Hertzian region slowly increases with $\mid y \mid$, i.e. farther away from the contact center $(0,0)$ along the y-axis the film thickness is slightly higher than in the center. It is important to keep in mind that formula (12.55) is valid only outside of the ends of HSSPGD zone Ω_{HSSPGD}, i.e. away from the points with coordinates $(0, -1)$ and $(0, 1)$.

In a similar fashion for fully flooded heavily loaded lubrication regimes when $\epsilon_q = V^{2/5}$ we obtain (see (12.45) and (12.55))

$$H_0(y) = A[\tfrac{D(e)}{1-e^2}(1 - y^2)]^{-1/5}V^{3/5}, \ A = O(1), \ \omega \ll 1. \qquad (12.56)$$

According to Chapter 6 the two systems in (12.52) and (12.53) can be rewritten in the equivalent form as follows

$$q(r,y) = \sqrt{2r}[1 - \tfrac{1}{2\pi}\int\limits_0^\infty \tfrac{\partial}{\partial t} M_0(A,q,h_q,\mu_q,t,y)\tfrac{dt}{\sqrt{2t}(t-r)}],$$

$$A^3 = \pi\alpha_1 / \int\limits_0^\infty \tfrac{\partial}{\partial t}\left(\tfrac{h_q^3}{\mu_q}\tfrac{\partial q(t,y)}{\partial t}\right)\tfrac{dt}{\sqrt{2t}}, \tag{12.57}$$

$$h_q(r,y) = 1 \ \ or \ \ A[h_q(r,y) - 1] = \tfrac{2}{\pi}\int\limits_0^\infty [q(t,y) - q_a(t,y)]\ln\tfrac{1}{|t-r|}dt,$$

$$g(s,y) = \sqrt{-2s}[1 - \tfrac{1}{2\pi}\int\limits_{-\infty}^0 \tfrac{\partial}{\partial t} M_0(A,g,h_g,\mu_g,t,y)\tfrac{dt}{\sqrt{-2t}(t-s)}],$$

$$\beta_1 = \tfrac{1}{\pi}\int\limits_{-\infty}^0 \tfrac{\partial}{\partial t} M_0(A,g,h_g,\mu_g,t,y)\tfrac{dt}{\sqrt{-2t}}, \tag{12.58}$$

$$h_g(s,y) = 1 \ \ or \ \ A[h_g(s,y) - 1] = \tfrac{2}{\pi}\int\limits_{-\infty}^0 [g(t,y) - g_a(t,y)]\ln\tfrac{1}{|t-s|}dt$$

where function $M_0(p,h,\mu,x,y)$ is determined by the equation (12.38).

For fully flooded lubrication regimes by integrating equations (12.52) and (12.53) for q and g the gap functions h_q and h_g from (12.57) and (12.58) can be replaced by equations

$$h_q = 1 + \tfrac{1}{A}M_0(A,q,h_q,\mu_q,r,y), \ \ h_g = 1 - \tfrac{1}{A}M_0(A,g,h_g,\mu_g,s,y). \tag{12.59}$$

To obtain two-dimensional distributions of pressure $p(x,y)$ and gap $h(x,y)$ as well as the distribution of the exit film thickness $H_0(y)$ in the central region away from the ends of the HSSPGD zone Ω_{HSSPGD} one has to numerically solve the asymptotic problems (12.57) and (12.58) for a sequence of y values and create uniformly valid approximations of $p(x,y)$ and $h(x,y)$ along the lines indicated in Section 6.4.

The validity of the definition of heavily loaded lubricated contacts (12.13) and (12.12) easily follows from the these relationships in $\Omega \setminus \Omega_\epsilon$ (i.e., outside of the inlet and exit zones) due to the fact that the film thickness H_{00} is determined by equations (12.39) and (12.44) for starved and fully flooded lubrication conditions, respectively.

In Chapter 6, the extensive research has shown that for numerical analysis of the inlet and exit zones systems (12.57) and (12.58) provide a more stable numerical solution than systems (12.52) and (12.53), respectively. Detailed numerical schemes for solution of the systems in the inlet and exit zones as well as a wide variety of examples of numerical solutions are provided in Chapters 6. Moreover, a very effective regularization approach resulting in stable numerical solutions of original and asymptotically valid equations even for extremely high values of the dimensionless viscosity pressure coefficient Q

such as $Q = 35$ is proposed in Section 8.2 based on a solution regularization technique which takes its roots in the solution of a thermal EHL problem with relatively small heat generation.

The analysis of the zones of the contact region encompassing the ends of HSSPGD zone Ω_{HSSPGD} of a heavily loaded lubricated contact is beyond the scope of this chapter.

12.4 Asymptotic Analysis of a Heavily Loaded Lubricated Contact under Over-Critical Lubrication Conditions

Over-critical lubrication regimes occur when lubricant viscosity rapidly grows with pressure, i.e. $Q \gg 1$. In a heavily loaded lubricated contact the maximum characteristic size of the inlet zone ϵ_0 for which the lubrication regime is still a pre-critical one we will call the critical characteristic size of the inlet zone ϵ_q. To determine ϵ_0 we need to consider the definition (12.23) of the pre-critical lubrication regimes. The value of ϵ_0 is such that for any $\epsilon_q \gg \epsilon_0$ the estimate (12.23) is not valid while for $\epsilon_q = \epsilon_0$ this estimate is still valid. If it appears that $\epsilon_0 \ll 1$ for the given problem parameters then it means that the viscosity μ depends on a large parameter Q. Thus, if $\epsilon_0 \ll 1$ then for $\epsilon_q = O(\epsilon_0)$ we have pre-critical regimes and for $\epsilon_q \gg \epsilon_0$ - over-critical regimes. For example, if $\mu = \exp(Qp)$ then $\epsilon_0 = Q^{-2} \ll 1$ for $\omega = Q^{-1} \ll 1$. Obviously, if $\epsilon_0 \gg 1$ or $\epsilon_0 = O(1)$ for $\omega \ll 1$ then over-critical regimes cannot be realized and pre-critical lubrication regimes occur.

Therefore, the formal definition of over-critical lubrication regimes is as follows

$$\mu(\epsilon_q^{1/2}) \gg 1, \; \epsilon_q(\omega) \gg \epsilon_0(\omega), \; \epsilon_0(\omega) \ll 1, \; \omega \ll 1. \tag{12.60}$$

The over-critical regimes are identical to those that were considered in [1] - [6].

Ertel [1], Grubin [2], Crook [3], and Archard et al. [5], considered cases of line lubricated contacts with rapidly growing with pressure lubricant viscosity using approximate analytical methods. The method used by Crook [3] differs from the methods employed by Ertel [1] and Grubin [2] only by more precise techniques under the same prior assumptions. The method proposed by Archard et al. [5] is the extension of the methods published by Ertel [1], Grubin [2], and Crook [3] for the case of a weak relationship between the viscosity μ and pressure p. The main difference between the methods published by Archard et al. [5] and Ertel [1], Grubin [2], and Crook [3] is in approximations of the gap function $h(x)$ and in a parabolic approximation for pressure $p(x)$ in the region of large pressure. The purpose of the first modification is to take into account the pressure gradient along the lubricant flow and of the second one is to simplify the approximate calculations. For elliptical heavily loaded lubricated contact a procedure similar to Ertel [1] and Grubin [2] was proposed by Cheng [6]. All in all, these approximate analytical and semi-analytical studies of the considered problem are based on certain prior assumptions which should be checked/validated and are vital and necessary for application of the aforementioned analytical methods.

A heavily loaded over-critical lubricated contact has the following structure (see Chapter 6). The inlet zone is represented by two sub-zones: one of a characteristic size $\epsilon_q \gg \epsilon_0$ (inlet ϵ_q-zone) which is adjacent to the the inlet contact boundary Γ_i and a smaller inlet zone of the characteristic size ϵ_0 (inlet ϵ_0-zone) located between the inlet ϵ_q-zone and the Hertzian region where pressure is very close to $p_{00}(x,y) = \sqrt{1 - x^2 - y^2}$. On the exit side of the contact the structure of the exit zone is similar to the one of the inlet zone, i.e. a smaller exit ϵ_0-zone is located between the Hertzian region and the larger exit ϵ_q-zone adjacent to the contact exit boundary Γ_e.

Using the same approach as the one used for derivation of (12.31) and (12.32) outside of the ends of HSSPGD zone Ω_{HSSPGD} we obtain

$$H_{00}\frac{\partial h}{\partial x} = 2x + \frac{2}{\pi}\theta \int\limits_{\gamma_i(y)}^{\gamma_e(y)} \frac{p(t,y)}{t-x}dt + \ldots, \quad \omega \ll 1. \tag{12.61}$$

Conducting the order analysis of the derivatives with respect to x and y in the Reynolds equation in the inlet and exit zones as well as of the derivative $\frac{dp}{d\overrightarrow{n}}$ at the exit from the contact we obtain that in the inlet and exit zones the EHL problem can be reduced to a problem similar to an EHL problem for a line contact

$$\frac{\partial}{\partial x}\left(\frac{h^3}{\mu}\frac{\partial p}{\partial x}\right) = \frac{V}{H_{00}^2}\frac{\partial h}{\partial x}, \quad p(\gamma_i(y),y) = 0, \quad \frac{\partial p(\gamma_e(y),y)}{\partial x} = 0, \quad \mu = \exp(Qp),$$

$$H_{00}\frac{\partial h}{\partial x} = 2x + \frac{2}{\pi}\theta \int\limits_{\gamma_i(y)}^{\gamma_e(y)} \frac{p(t,y)}{t-x}dt + \ldots, \quad \omega \ll 1, \quad h(\gamma_i(y),y) = h_\infty(y),$$

$$\tag{12.62}$$

for points (x,y) outside of the ends of HSSPGD zone and to a balance condition consistent with the last equation in (12.9).

In Chapter 6, it is described in detail how to study over-critical lubrication regimes for line contacts. It is also shown there that systems of asymptotic equations (12.57) and (12.58) (as well as equivalent to them systems (12.52) and (12.53)) are capable of describing the over-critical lubrication regimes. Based on that we can conclude that for over-critical regimes the central film thickness is determined by the formulas

$$H_{00} = A_p(VQ\epsilon_q^{5/2})^{1/3}, \quad \epsilon_q \ll \epsilon_f, \quad A_s = O(1), \quad Q \gg 1, \tag{12.63}$$

$$H_{00} = A_f(VQ)^{3/4}, \quad \epsilon_q = \epsilon_f, \quad A_f = O(1), \quad Q \gg 1, \tag{12.64}$$

$$\epsilon_f = (VQ)^{1/2}, \tag{12.65}$$

where the coefficients A_p and A_f depend on θ and $\alpha_{1p}(0)$. In the manner similar to the one used for pre-critical lubrication regimes from formulas (12.63) we obtain formulas for the exit film thickness as follows

$$H_0 = A_{p1}[\tfrac{D(e)}{1-e^2}(1-y^2)]^{-1/5}(VQ\epsilon_q^{5/2})^{1/3}, \tag{12.66}$$

$$\epsilon_q \ll \epsilon_f, \; A_{p1} = O(1), \; Q \gg 1,$$

for starved over-critical lubrication regimes and

$$H_0 = A_{f1}[\tfrac{D(e)}{1-e^2}(1-y^2)]^{-1/5}(VQ)^{3/4}, \tag{12.67}$$

$$\epsilon_q = \epsilon_f, \; A_{f1} = O(1), \; Q \gg 1,$$

for fully flooded over-critical lubrication regimes.

The values of constants A_p and A_f can be obtained from the solution of the system of asymptotically valid equations in the inlet zone for pre-critical lubrication regimes. For example, for fully flooded over-critical lubrication regime the value of A_f can be obtained the following way. For pre-critical lubrication regimes $\gamma_i = -\sqrt{1-y^2} + \alpha_{1p}\epsilon_f$ (see expression for ϵ_f in (12.65)) while simultaneously for over-critical lubrication regimes $\gamma_i = -\sqrt{1-y^2} + \alpha_{1po}\epsilon_{fo}$ (see expression for ϵ_{fo} in (12.45)). Therefore, we have the corresponding inlet coordinate α_{1p} for pre-critical lubrication regimes $\alpha_{1p} = \alpha_{1po}(QV^{1/5})^{1/2}$. For this inlet coordinate the value of constant $A_f(\alpha_{1p})$ from the solution in the inlet zone of asymptotically valid equations for pre-critical lubrication regimes will produce the value of $A_f(\alpha_{1po})$ in formula (12.64) for over-critical lubrication regimes if the value of $A_f(\alpha_{1p})$ would be substituted in formula (12.44) for H_{00} for pre-critical lubrication regimes and compared with formula (12.64) for H_{00} for over-critical lubrication regimes. To obtain the value of constant $A_p(\alpha_{1po})$ in formula (12.63) we need to solve in the inlet zone the asymptotically valid equations for pre-critical lubrication regimes with the inlet coordinate $\alpha_{1p} = \alpha_{1p0}\frac{\epsilon_{qo}}{\epsilon_q}$ to obtain the value of $A_p(\alpha_{1p})$ where for pre-critical regime calculations $\gamma_i = -\sqrt{1-y^2} + \alpha_{1p}\epsilon_q$ while for over-critical regime calculations $\gamma_i = -\sqrt{1-y^2} + \alpha_{1po}\epsilon_{qo}$ and $\epsilon_q = O(\epsilon_0) \ll \epsilon_{qo}$. After that, the comparison of formulas (12.39) and (12.63) will produce the value of $A_p(\alpha_{1po})$ in equation (12.63) for the film thickness H_{00} for over-critical starved lubrication regimes. For more details see Section 6.10.

The validity of the definition of heavily loaded lubricated contacts (12.13) and (12.12) easily follows from the these relationships in $\Omega \setminus \Omega_\epsilon$ (i.e., outside of the inlet and exit zones) due to the fact that the film thickness H_{00} is determined by equations (12.63) for starved and fully flooded lubrication conditions.

The choice between pre- and over-critical lubrication regimes can be made based on a simple analysis from Section 9.5 while the regularization approaches to numerical solution of isothermal EHL problems are discussed in Section 15.6. Numerical studies of the inlet and exit zones based on the derived asymptotically valid equations can be found in Chapter 6. The numerical results include parametric studies, studies of solution precision and stability, etc.

12.5   Discussion and Validation of Results

The asymptotic methods allow not only to reduce the analysis of the central region of a heavily loaded point EHL contact to the analysis of a series of heavily loaded line EHL contacts but also the asymptotic methods allow to finish this analysis and get results which are very close to the ones obtained by direct numerical methods. Therefore, this sequence of results shows so to speak "an unbroken chain of custody," i.e., the analysis of heavily loaded point EHL contacts can be successfully performed asymptotically by reducing the point EHL problem to the problem for line contacts which can be successfully solved asymptotically with the use of specialized numerical methods only in the inlet and exit zones which are described in the preceding chapters on heavily loaded line EHL contacts.

TABLE 12.2

The dependence of the value of A and the inlet gap $h_q(\Delta r/2)$ on the inlet coordinate α_1.

α_1	A	$h_q(\Delta r/2) - 1$
-0.142	0.300	0.527
-0.322	0.400	1.168
-0.992	0.500	4.262
-1.930	0.525	10.342
-4.325	0.535	32.804

First, let us discuss one of the assumptions of the analysis that the inlet and exit zones are small compared to the size of the Hertzian region. This assumption is reflected in formulas (12.17) where ϵ_q is a small characteristic size/width of the inlet and exit zones. Assuming this assumption about the proximity of the contact inlet boundary to the boundary of the Hertzian contact we learned from the above analysis that in the inlet zone outside of the ends of HSSPGD zone the equations for a heavily loaded point EHL contact are reduced to equivalent sets of equations (12.52), (12.53) and (12.57), (12.58). To establish the validity of the above assumption about small width of the inlet and exit zones it is sufficient to check the variations of the inlet coordinate α_1 (see (12.17) and (12.50)) and the corresponding variations in the coefficient A involved in the formula (12.56) for the film thickness H_0 as well as the gap $h_q(\Delta r/2)$ value at the point next to the inlet boundary (Δr is the numerical step size used in the inlet zone). Some numerical solutions of equations (12.57) for fully flooded pre-critical lubrication regimes for $QV^{1/5} = 1$ and different values of constant coefficient A are presented in Table 12.2.

The data from Table 12.2 shows that while the inlet coordinate α_1 varies from -1.930 to -4.325 (i.e. by 224%) the value of constant A increases from 0.525 to 0.535 (i.e., by just 1.9%). It means that the value of constant A (and, therefore, the film thickness H_0 which is proportional to A) is close to its maximum limit. At the same time, the growth of the near inlet gap $h_q(\Delta r/2)$ accelerates significantly as the inlet boundary moves away from the boundary of the Hertzian region (i.e. while α_1 varies from -0.142 to -4.325). The further movement of the inlet boundary α_1 to the left does not produce any significant changes in the values of constant A and film thickness H_0 while the near inlet gap $h_q(\Delta r/2)$ grows progressively larger. We need to remember that at some sufficiently large $\mid \alpha_1 \mid$, $\alpha_1 < 0$, the gap at the inlet $h_q(0)$ becomes so large that the fundamental assumption that the gap is much smaller than the contact width (i.e. $h \ll 2a_H$) used for the classic derivation of the Reynolds equation gets violated and the full Navier-Stokes equations must be used beyond this value of α_1. In particular, for pre-critical lubrication regimes and $a < -1 - 6V^{2/5}$ the gap at the inlet is greater than 50. The numerical results show that as $\alpha_1 \to -\infty$ the inlet gap $h_q(0) \to +\infty$.

Moreover, using asymptotic methods similar to the ones above it can be shown that if the inlet point a is far from the beginning of the Hertzian region $x = -1$, i.e. if $a = -1-\Sigma$, $\Sigma \gg 1$, then for $x+1 = O(\Sigma)$ the pressure $p(x)$ is so small that its contribution to the film thickness H_0, exit coordinate c as well as to the pressure $p(x)$ and gap $h(x)$ distributions is negligibly small in the inlet zones of size $O(V^{2/5})$ for pre-critical regimes and of size $O((VQ)^{1/2})$ for over-critical regimes as well as in the Hertzian region and the exit zones. Therefore, for pre-critical fully flooded lubrication regimes effectively the whole inlet zone is of a small size of the order of magnitude of $O(V^{2/5})$, i.e. $a+1 = O(V^{2/5}) \ll 1$ for $V \ll 1$. Similarly, for over-critical fully flooded lubrication regimes it can be shown that effectively the whole inlet zone is of a small size of the order of magnitude of $O((VQ)^{1/2})$, i.e. $a + 1 = O((VQ)^{1/2}) \ll 1$ for $Q \gg 1$ and $VQ \ll 1$. It means that the assumption that the inlet zone is small is correct.

As it has been mentioned earlier that numerous solutions of asymptotic equations in the inlet and exit zones of a heavily loaded EHL contact can be found in Chapter 6 as well as the validation of the asymptotic method through the comparison of the asymptotic solutions with the solutions of the EHL problem in the original formulation obtained by direct numerical methods. The excellent agreement between the solutions of asymptotically valid equations and equations of the full original EHL problem provides a solid basis for application of these asymptotic methods to the cases of more complex EHL problem for point heavily loaded EHL contacts. In addition to that, the proposed regularization of the equations of the isothermal EHL problem provides for the way of consistently obtaining stable solutions of EHL problems for heavily loaded contacts.

A direct very convincing verification of the fact that for lubricants (2,3-dimethylpentane, lubricants at high temperatures, and water/glycol solutions) with weak (not exponential) dependence of viscosity μ on pressure p the central film thickness h_0 in a point contact is proportional to $(u_1 + u_2)^{0.6}$ is obtained experimentally in [25] and coincides precisely with the dependence obtained in formulas (12.44) and (12.56) obtained for pre-critical lubrication regimes (also see the expression (12.6) for parameter V which is just proportional to $u_1 + u_2$). This dependence of the film thickness h_0 on $u_1 + u_2$ is significantly different from the one for the Barus-type lubricant viscosity (12.4) for which according to formula (12.67) for over-critical lubrication regimes h_0 is proportional to $(u_1 + u_2)^{0.75}$.

In [26] in the formula for the central film thickness h_0 the degrees of parameters $\mu_a(u_1 + u_2)$, α, E', and P are as follows 0.67, 0.53, -0.073, and -0.067, respectively. The ranges of the degrees of these parameters in the formulas (12.56) and (12.67) for the central film thickness h_0 in pre- and over-critical fully flooded lubrication regimes are $\{0.6, 0.75\}$, $\{0, 0.75\}$, $\{-0.867, 0.083\}$, and $\{-0.133, -0.0833\}$, respectively. Obviously, in [26] the degrees for $\mu_a(u_1 + u_2)$, α, and E' belong to the middle of the ranges obtained above from the asymptotic formulas while the degree of P is outside of the corresponding range but gravitates to its higher borders. The latter situation may be caused by a number of factors among which are the precision of the numerical results in [26] and the error introduced by the least squares approximation of the numerical results used for deriving formulas for the central film thickness in [26]. In addition, this situation suggests that among the numerically obtained data sets used in [26] by the least squares approximation of the formula for the film thickness there were more sets which should be characterized as the data sets obtained for over-critical lubrication regimes.

In general, all powers of such parameters as μ_a, α, $u_1 + u_2$, R_x, P, and E' obtained asymptotically for pre- and over-critical heavily loaded lubrication regimes in point contacts are close to the corresponding powers obtained numerically and experimentally, for example see data in [1] - [6], [11, 14, 25, 26]. Another important conclusion which can be drawn from this comparison is that the corresponding parameter powers for line and point contacts are close to each other. Obviously, that supports the whole premise of the current study which is to show certain similarities between heavily loaded point and line EHL contacts. Some insignificant differences in the values of these powers are caused by the way they were calculated (by curve fitting), precision of numerical calculations and experimental measurements as well as by different mixtures of the lubrication regimes (pre- and over-critical) for which these powers were obtained.

Another opportunity to validate the asymptotic results by the results from the direct numerical solution of the EHL problems obtained in [13] stems from the comparison of the film thickness variations in the Hertzian region along the y−axis. It is important to remember that the film thickness at $(0, y)$ coincides with the exit film thickness $H_0(y)$ ($h_e(y)$ in dimensional variables). Using any of the formulas (12.55), (12.56), (12.66), and (12.67) for pre- and over-critical starved and fully flooded lubrication regimes we obtain

$$\frac{h_e(y)}{h_e(y_*)} = \left(\frac{1-y^2}{1-y_*^2}\right)^{-1/5}, \tag{12.68}$$

where y_* is a fixed value. The comparison of the data obtained based on formula (12.68) and based on a direct numerical solution for a circular lubricated contact (i.e., $b_H = a_H$) from [13] (see Fig. 6.11 [13]) is given in Table 12.3. In particular, it follows from Fig. 6.11 [13] that in the Hertzian region in the dimensionless coordinates the contour plots of $h/2$ are practically flat and are located at $y_* = y_1 = \pm 0.2$, $y_2 = \pm 0.295$, $y_3 = \pm 0.367$, $y_4 = \pm 0.417$, and $y_5 = \pm 0.45$ and separated one from another by the increments equal to $\frac{1}{2}\Delta h = 0.01$. It follows from Fig 6.10 [13] that $\frac{1}{2}h_e(y_*) = 1$, i.e. the inner most set of two symmetric contour plots of $h/2$ in Fig. 6.11 [13] correspond to $\frac{1}{2}h_e(y_*) = 1$. Taking into account the precision of the data obtained from these graphs it is obvious from Table 12.3 that formula (12.68) provides results which are in excellent agreement with the data for the central film thickness obtained based on a direct numerical method and presented in Fig. 6.11 from [13]. The applicability range of the proposed method depends on how heavily loaded an EHL contact is. The heavier it is loaded the wider is the range of the method applicability. If $-R_0 < y < R_0$ is the range of the method applicability then $R_0 \to 1$ as $\omega \to 0$.

TABLE 12.3
Comparison of the asymptotically and numerically obtained values of $\frac{h_e(y)}{h_e(y_*)}$ based on Fig. 6.11 [13].

y	0.295	0.367	0.417	0.450
Asymptotic value of $\frac{h_e(y)}{h_e(y_*)}$	1.010	1.021	1.030	1.038
Numerical value of $\frac{h_e(y)}{h_e(y_*)}$ [13]	1.01	1.02	1.03	1.04

Now, let us analyze the dependence of the central film thickness h_0 (i.e. at $y = 0$) on the dry contact eccentricity e in the case of constant viscosity, i.e. when $Q = 0$ and the lubrication regime is pre-critical. To do that we need to determine from (12.8) the dependence of e on the radii ratio δ. That is easy to achieve by using the asymptotic expansions of the full elliptic integrals $E(e)$ and $K(e)$ for small values of e and values of e close to 1 [18]. This analysis shows that e is an increasing function of δ for $\delta - 1 \ll 1$, i.e. for a contact close to circular. Using formulas (12.6) and (12.56) we obtain that

$$h_0 = \frac{\gamma}{[D(e)(1-e^2)^{3/2}]^{1/5}}, \tag{12.69}$$

where for constant lubricant viscosity the value of constant γ is independent from e assuming that the Hertzian contact ellipse semi-axis a_H along the

$x-$axis as well as such parameters as $u_1 + u_2$, μ_a, E', R_x, and P are fixed. Based on formula (12.69) and the dependence of the eccentricity e on δ it can be easily concluded that for $\delta - 1 \ll 1$ the central as well as the exit film thicknesses are monotonically increasing functions of the ellipse eccentricity e. This conclusion coincides with the behavior of the central film thickness h_0 described in [26]. A similar analysis can be done for significantly elongated elliptical contacts, i.e. for $\delta \ll 1$.

12.6   Closure

An asymptotic approach to isothermal steady heavily loaded point EHL contacts with incompressible Newtonian lubricants is developed. The inlet and exit zones away from the ends of the horse-shoe shaped pressure distribution area are analyzed in detail. Asymptotic analysis of heavily loaded contacts lubricated by a Newtonian fluid recognizes two different regimes of lubrication: pre- and over-critical lubrication regimes. Overall, the difference between these two regimes is in how rapidly the lubricant viscosity grows with pressure and how much lubricant is available at the inlet of the contact. For pre-critical lubrication regimes the asymptotically valid equations for pressure and gap have been derived. It has been shown that by a simple transformation the derived equations can be reduced to equations identical to the corresponding equations for pre-critical lubrication regimes in the inlet and exit zones of line heavily loaded contacts. The latter equations have been validated through the comparison with the direct numerical solutions of the original EHL equations. For over-critical lubrication regimes another approach has been used which led to formulas for the lubrication film thickness for starved and fully flooded conditions. The asymptotic approaches to EHL problems offer certain distinct advantages in comparison with the direct numerical solutions of EHL problem such a reduction of the number of problem input parameters and the approach to regularization of numerical solutions. Some numerical results of asymptotically valid equations and their comparison with the solutions of the full original (non-asymptotic) EHL equations are provided. The analysis is validated by the direct comparison of the asymptotic results with experimental ones as well as obtained by one of the direct numerical methods (the multilevel multi-grid method). Each of the analytically derived formulas for the central H_{00} and exit H_0 lubrication film thicknesses contains a coefficient of proportionality which can be found by numerically solving the corresponding asymptotically valid equations in the inlet zone or by just curve fitting the experimentally obtained values of the lubrication film thickness.

12.7 Exercises and Problems

1. (a) Explain why the asymptotic procedure of the derivation of asymptotically valid equations in the inlet and exit zones of heavily loaded point EHL contact is not applicable in the vicinity of points $(0, -1)$ and $(0, 1)$. Which of the assumptions laid in the foundation of the derivation is no longer valid?

(b) Explain how would you approach the derivation of asymptotic equations for heavily loaded point EHL contact in the vicinity of points $(0, -1)$ and $(0, 1)$.

2. Analyze the dependence of the central lubrication film thickness h_0 on the dry ellipse eccentricity e when $e - 1 \ll 1$.

3. Give several possible reasons why the powers of the parameters V and Q in various formulas for the lubrication film thickness h_0 in fully flooded lubrication regimes derived by different authors by a regression of numerically obtained results are different from the ones presented in formulas (12.44) and (12.63) for pre-and over-critical lubrication regimes, respectively.

4. List all known to you advantages and drawbacks of the asymptotic and direct numerical methods in application to EHL problems for point heavily loaded contacts.

5. Explain why for fully flooded lubrication regimes the size of the inlet and exit zones in a heavily loaded point EHL contact is much smaller than the smaller semi-axis a_H of the Hertzian dry contact.

References

[1] Ertel, M.A. 1945. Hydrodynamic Calculation of Lubricated Contact for Curvilinear Surfaces. *Proc. CNIITMASh* :1-64.

[2] Grubin, A.N. 1949. The Basics of the Hydrodynamic Lubrication Theory for Heavily Loaded Curvilinear Surfaces. *Proc. CNIITMASh* 30:126-184.

[3] Crook, A.W. 1961. The Lubrication of Rollers II. Film Thickness with Relation to Viscosity and Speed. *Philosophical Trans. Royal Soc. of London, Ser. A, Math., Phys. and Eng. Sci.* 254, No. 1040:223-236.

[4] Greenwood, J.A. 1972. An Extension of the Grubin Theory of Elasto-hydrodynamic Lubrication. *Phys. D. Appl. Phys.* 5:2195-2211.

[5] Archard, J.F. and Baglin, K.P. 1986. Elastohydrodynamic Lubrication - Improvements in Analytic Solutions. *Proc. Inst. Mech. Eng.* 200, No. C4 :281-291.

[6] Cheng, H.S. 1970. A Numerical Solution to the Elastohydrodynamic Film Thickness in an Elliptical Contact. *J. Lubr. Technol.* 92, No. 1: 155-162.

[7] Kudish, I.I. and Covitch, M.J. 2010. *Modeling and Analytical Methods in Tribology.* Chapman & Hall/CRC.

[8] Van Dyke, M. 1964. *Perturbation Methods in Fluid Mechanics.* New York: Academic Press.

[9] Kevorkian, J. and Cole, J.D. 1985. *Perturbation Methods in Applied Mathematics. Applied Mathematics Series, Vol.* 34, New York, Springer-Verlag.

[10] Habchi, W. 2008. A Full-System Finite Element Approach to Elastohy-drodynamic Lubrication Problems: Application to Ultra-Low-Viscosity Fluids. Ph.D. Thesis. Institut National des Sciences Appliques de Lyon.

[11] Hamrock, B.J. 1994. *Fundamentals of Fluid Film Lubrication.* New York: McGraw-Hill.

[12] Evans, H.P. and Hughes, T.G. 2000. Evaluation of Deflection in Semi-Infinite Bodies by a Differential Method. *Proc. Instn. Mech. Engrs.* 214, Part C, pp. 563-584.

[13] Lubrecht, A.A. and Venner, C.H., 1999. Elastohydrodynamic lubrication of rough surfaces. *Proc. Instn. Mech. Engrs. J. Eng. Tribology.* V. 213, No. 5, pp. 397-404

[14] Venner, C.H. and Lubrecht, A.A. 2000. *Multilevel Methods in Lubrication*. Amsterdam: Elsevier.

[15] Ai, X. L. and Cheng, H. S. Hydrodynamic lubrication analysis of metallic hip joint. Tribol. Trans., 1996, 39, 103-111.

[16] Liu, S., Hua, D., Chen, W.W., and Wang, Q.J. 2007. Tribological modeling: Application of fast Fourier transform. *Tribology International*, V. 40, Issue 8, pp. 1284-1293.

[17] Kudish, I.I. 2010. Asymptotic Analysis, Regularization and Stable Numerical Solutions for Heavily Loaded Line EHL Contacts. Part 3. Regularization and Stable Numerical Solutions of Asymptotic and Original Equations of Isothermal Problems. *Lubrication Science*, V. 22, issue 6-7, pp. 305-322.

[18] Handbook of Mathematical Functions with Formulas, Graphs and Mathematical Tables. 1964. Eds. M. Abramowitz and I.A. Stegun, National Bureau of Standards, Applied Mathematics Series 55.

[19] Vorovich, I.I., Aleksandrov, V.M., and Babeshko, V.A. 1974. *Non − classical Mixed Problems of Elasticity*. Moscow: Nauka.

[20] Dowson, D. and Higginson, G.R. 1966. *Elastohydrodynamic Lubrication*. London: Pergamon Press.

[21] Houpert, L.G. and Hamrock, B.J. 1986. Fast Approach for Calculating Film Thickness and Pressures in Elastohydrodynamically Lubricated Contacts at High Loads. *ASME J. Tribology* 108, No. 3:441–452.

[22] Bissett, E.J. and Glander, D.W. 1988. A Highly Accurate Approach that Resolves the Pressure Spike of Elastohydrodynamic Lubrication. *ASME J. Tribology* 110, No. 2:241-246.

[23] Kudish, I.I. 2010. Asymptotic Analysis, Regularization, and Stable Numerical Solutions for Heavily Loaded Line EHL Contacts. Part 1. Asymptotic and Numerical Analysis of Isothermal Problems, *Lubrication Science*, Vol. 22, issue 6-7, pp. 251-289.

[24] Kudish, I.I. 2010. Asymptotic Analysis, Regularization, and Stable Numerical Solutions for Heavily Loaded Line EHL Contacts. Part 2. Asymptotic and Numerical Analysis of Non-Isothermal Problems, *Lubrication Science*, V. 22, issue 6-7, pp. 291-303.

[25] Kumar, P., Bair, S., Krupka, I., and Hartl, M. 2010. Newtonian Quantitative Elastohydrodynamic Film Thickness with Linear Piezoviscosity, *Tribology International*, V. 43, pp. 2159-2165.

[26] Hamrock, B.J. and Dowson. D., 1981. *Ball Bearing Lubrication. The Elastohydrodynamics of Elliptical Contacts*. New York: John Wiley & Sons.

13

---

*Asymptotic Analysis of Isothermal
Lubricated Heavily Loaded Point Contacts
with Skewed Direction of Entrained Lubricant*

13.1 Introduction

A detailed review of the current state of the methodologies used in solution of
EHL problems for line and point contacts as well as certain advantages and
drawbacks of numerical and asymptotic methods are presented in preceding
chapters. In addition to studies reviewed earlier it is necessary to mention numerical studies of the point EHL problem with skewed lubricant entrainment
[1]-[5]. In these papers the EHL problem with skewed lubricant entrainment
was solved numerically. The distributions of pressure and gap were obtained
as well as a number of formulas for the film thickness were proposed. Also, in
[3] a case of a ball involved in pure spin was considered but the convergence
of the iteration process at high loads was poor.

   This chapter is devoted to the generalization of the results obtained in
Chapter 12 and application of the developed asymptotic approach to solution
of the steady isothermal EHL problem for heavily loaded point contacts with
skewed direction of entrained lubricant. The problem analysis will be done
along the lines of the study conducted in Chapter 12. It is shown that the
whole contact region can be subdivided into three subregions: the central one
which is far away from the other two regions occupied by the ends of the
horse-shoe shaped pressure/gap distribution (HSSPGD). The central region,
in turn, can be subdivided into the Hertzian region and two adjacent boundary
layers - the inlet and exit zones. Moreover, in the central region in the inlet and
exit zones the EHL problem can be reduced to asymptotically valid equations
identical to the ones obtained in the inlet and exit zones of heavily loaded line
EHL contacts. These equations can be analyzed and numerically solved based
on the stable methods using a specific regularization approach which were
developed for lubricated line contacts. Cases of pre-critical and over-critical
lubrication regimes are considered. The byproduct of this asymptotic analysis
is an easy analytical derivation of formulas for the lubrication film thickness for
pre-critical and over-critical lubrication regimes. The latter allows for simple
analysis of the film thickness as a function of contact eccentricity and the

direction of the entrained lubricant at the inlet in the contact.

13.2   Problem Formulation

Let us consider a steady isothermal lubricated contact of two elastic solids, i.e. a ball moving in a grooved raceway. The problem formulation is identical with the one used in the preceding chapter. Therefore, the problem is reduced to the system of equations (12.2)-(12.4). The difference between the problems considered here and in Chapter 12 is that now the speed $v_1 + v_2 \neq 0$, i.e., the lubricant entrainment is skewed. Therefore, let us introduce the dimensionless variables (12.5) and instead of dimensionless parameters (12.6) we will use

$$V = \frac{24\mu_a \sqrt{(u_1+u_2)^2+(v_1+v_2)^2} R_x^2}{p_H a_H^3}, \quad Q = \alpha p_H, \quad \delta = \frac{R_x}{R_y},$$

$$\gamma = \frac{u_1+u_2}{\sqrt{(u_1+u_2)^2+(v_1+v_2)^2}}, \quad H_{00} = \frac{2R_x h_0}{a_H^2}, \tag{13.1}$$

where a_H, b_H, and p_H are the Hertzian semi-axes (small and large) of the dry elliptical contact and the Hertzian maximum pressure, respectively. The dry contact ellipse eccentricity e and the parameters a_H, b_H, and p_H are determined by equations (12.7) and (12.8). Using the dimensionless variables from (12.5) and (13.1) and omitting the primes at the dimensionless variables equations (12.2) and (12.4) assume the form

$$\frac{\partial}{\partial x}\left(\frac{h^3}{\mu}\frac{\partial p}{\partial x}\right) + (1-e^2)\frac{\partial}{\partial y}\left(\frac{h^3}{\mu}\frac{\partial p}{\partial y}\right) = \frac{V}{H_{00}^2}\left\{\gamma\frac{\partial h}{\partial x} + \sqrt{1-e^2}\sqrt{1-\gamma^2}\frac{\partial h}{\partial y}\right\},$$

$$\mu = \exp(Qp), \quad p(x,y)\mid_\Gamma = 0, \quad \frac{dp(x,y)}{d\overrightarrow{n}}\mid_{\Gamma_e} = 0,$$

$$H_{00}[h(x,y)-1] = x^2 + \frac{\delta}{1-e^2}y^2 + \frac{D(e)}{\pi(1-e^2)^{3/2}}\left\{\int\int_\Omega \frac{p(x_1,y_1)dx_1dy_1}{\sqrt{(x-x_1)^2+\frac{(y-y_1)^2}{1-e^2}}}\right. \tag{13.2}$$

$$\left. - \int\int_\Omega \frac{p(x_1,y_1)dx_1dy_1}{\sqrt{x_1^2+\frac{y_1^2}{1-e^2}}}\right\}, \quad \int\int_\Omega p(x_1,y_1)dx_1dy_1 = \frac{2\pi}{3}.$$

In this problem formulation for the given inlet boundary Γ_i and constants V, Q, δ, and e we need to find the distributions of pressure $p(x,y)$ and gap $h(x,y)$ as well as the exit boundary Γ_e and the dimensionless central film thickness H_{00}.

To simplify the further analysis let us make the following substitution (rotation) of the independent variables x and y

$$\chi = \frac{\gamma x + \sqrt{1-e^2}\sqrt{1-\gamma^2}y}{\sqrt{1-e^2+e^2\gamma^2}}, \quad \xi = \frac{-\sqrt{1-e^2}\sqrt{1-\gamma^2}x+\gamma y}{\sqrt{1-e^2+e^2\gamma^2}}, \quad d\chi d\xi = dxdy, \tag{13.3}$$

where

$$x = \frac{\gamma\chi - \sqrt{1-e^2}\sqrt{1-\gamma^2}\xi}{\sqrt{1-e^2+e^2\gamma^2}}, \quad y = \frac{\sqrt{1-e^2}\sqrt{1-\gamma^2}\chi + \gamma\xi}{\sqrt{1-e^2+e^2\gamma^2}}. \tag{13.4}$$

The reason for such a choice of new independent variables will become clear later when we will be considering the problem solution away from the contact boundaries where with high precision the gap is practically constant along straight lines $\xi(x,y) = const$, i.e., $h = h(\xi)$ (see formula (13.10)).

In variables (13.3) equations (12.9) will take the form

$$\frac{(1-e^2)^2(1-\gamma^2)+\gamma^2}{1-e^2+e^2\gamma^2} \frac{\partial}{\partial\chi}\left(\frac{h^3}{\mu}\frac{\partial p}{\partial\chi}\right) - \frac{e^2\gamma\sqrt{1-e^2}\sqrt{1-\gamma^2}}{1-e^2+e^2\gamma^2}\left[\frac{\partial}{\partial\chi}\left(\frac{h^3}{\mu}\frac{\partial p}{\partial\xi}\right) + \frac{\partial}{\partial\xi}\left(\frac{h^3}{\mu}\frac{\partial p}{\partial\chi}\right)\right]$$

$$+ \frac{1-e^2}{1-e^2+e^2\gamma^2}\frac{\partial}{\partial\xi}\left(\frac{h^3}{\mu}\frac{\partial p}{\partial\xi}\right) = \frac{V}{H_{00}^2}\sqrt{1-e^2+e^2\gamma^2}\frac{\partial h}{\partial\chi}, \quad \mu = \exp(Qp),$$

$$p(\chi,\xi)\mid_\Gamma = 0, \quad \frac{dp(\chi,\xi)}{d\overrightarrow{n}}\mid_{\Gamma_e} = 0,$$

$$H_{00}[h(\chi,\xi)-1] = x^2(\chi,\xi) + \frac{\delta}{1-e^2}y^2(\chi,\xi) \tag{13.5}$$

$$+ \frac{D(e)}{\pi(1-e^2)^{3/2}}\left\{\iint\limits_\Omega \frac{p(\chi_1,\xi_1)d\chi_1 d\xi_1}{\sqrt{(\chi-\chi_1)^2 + \frac{(\xi-\xi_1)^2}{1-e^2}}} - \iint\limits_\Omega \frac{p(\chi_1,\xi_1)d\chi_1 d\xi_1}{\sqrt{\chi_1^2 + \frac{\xi_1^2}{1-e^2}}}\right\},$$

$$\iint\limits_\Omega p(\chi_1,\xi_1)d\chi_1 d\xi_1 = \frac{2\pi}{3},$$

where x, y and x_1, y_1 are functions of χ, ξ and χ_1, ξ_1 determined according to (13.4), respectively.

It is clear that for the case of lubricant entrainment occurs along the x-axis we have $\gamma = 1$ and the EHL problem with skewed lubricant entrainment described by equations (13.2) and (13.5) gets reduced to the problem considered in Chapter 12.

13.3 Asymptotic Analysis of a Heavily Loaded Lubricated Contact under Pre-Critical Lubrication Conditions

Let us consider heavily loaded lubrication conditions. They can be caused by low speed $(\sqrt{(u_1+u_2)^2 + (v_1+v_2)^2})$, high load P, small radii R_x and R_y, low ambient viscosity μ_a, and high pressure viscosity coefficient α. All these contact parameters are represented by the two dimensionless parameters V and Q. From the expressions for V and Q in (13.1) it becomes clear that heavily loaded conditions can be realized only when V is small and/or Q is large. It is important to keep in mind that it is the most common practical condition in fully lubricated bearing and gear contacts. Therefore, we will assume that system (13.5) contains a small parameter $\omega \ll 1$ which can be equal to $V \ll 1$ or $Q^{-1} \ll 1$.

In heavily loaded lubricated contacts (i.e. for $\omega \ll 1$) at the points away from the contact boundary Γ the gap distribution $h(\chi, \xi)$ is very close to being constant along the χ–axis while the pressure distribution $p(\chi, \xi)$ is close to the pressure distribution $p_0(\chi, \xi)$ which occurs in a dry contact Ω of a rigid indenter with the radii R_x and R_y and sharp edges with an elastic half-space with the effective elastic modulus E'. Away from the contact boundary Γ this pressure distribution is close to the Hertzian pressure distribution

$$p_{00}(\chi, \xi) = \sqrt{1 - \chi^2 - \xi^2} \tag{13.6}$$

in a circular (in dimensionless variables) contact

$$\chi^2 + \xi^2 \leq 1. \tag{13.7}$$

The region where that occurs we will call the Hertzian region.

In heavily loaded lubricated contacts the film thickness H_{00} separating the solids is small as the situation is close to the case of a dry contact where the film thickness is zero. Due to that and the fact that in the Hertzian region, i.e. at the points where the gap $h(\chi, \xi)$ is almost constant along the χ-axis, we have $\partial h / \partial \chi \ll 1$. The above characterization of heavily loaded EHL contacts is fully supported by direct numerical solutions of equations (13.5) [6].

Based on this characterization of a heavily loaded EHL contact and the fact that the film thickness $H_{00} \ll 1$ we will define a contact to be heavily loaded if

$$H_{00} \frac{\partial h(\chi, \xi)}{\partial \chi} \ll 1, \quad (\chi, \xi) \in \Omega \setminus \Omega_\epsilon, \tag{13.8}$$

where Ω_ϵ is a narrow ring-like shaped region adjacent to the contact boundary Γ which encompasses the region $\Omega \setminus \Omega_\epsilon$ located in the central (Hertzian) part of the contact. Let the characteristic width of Ω_ϵ be $\epsilon_q = \epsilon_q(\omega) \ll 1$ for $\omega \ll 1$.

Condition (13.8) is equivalent to

$$\frac{H_{00}^3}{V} \left\{ \frac{(1-e^2)^2(1-\gamma^2)+\gamma^2}{1-e^2+e^2\gamma^2} \frac{\partial}{\partial \chi} \left(\frac{h^3}{\mu} \frac{\partial p}{\partial \chi} \right) - \frac{e^2\gamma\sqrt{1-e^2}\sqrt{1-\gamma^2}}{1-e^2+e^2\gamma^2} \left[\frac{\partial}{\partial \chi} \left(\frac{h^3}{\mu} \frac{\partial p}{\partial \xi} \right) \right. \right.$$

$$\left. \left. + \frac{\partial}{\partial \xi} \left(\frac{h^3}{\mu} \frac{\partial p}{\partial \chi} \right) \right] + \frac{1-e^2}{1-e^2+e^2\gamma^2} \frac{\partial}{\partial \xi} \left(\frac{h^3}{\mu} \frac{\partial p}{\partial \xi} \right) \right\} \ll 1, \quad (\chi, \xi) \in \Omega \setminus \Omega_\epsilon, \tag{13.9}$$

as it follows from condition (13.8) and the Reynolds equation (13.5) multiplied by H_{00}^3/V.

Due to the definitions (13.9) and (13.8) and the fact that in the Hertzian region $\partial h / \partial \chi \ll 1$ it is clear that in the Hertzian region $\Omega \setminus \Omega_\epsilon$ the gap distribution $h(\chi, \xi)$ is as follows

$$h(\chi, \xi) = h_\infty(\xi), \quad (\chi, \xi) \in \Omega \setminus \Omega_\epsilon, \tag{13.10}$$

where $h_\infty(\xi)$ is the exit film thickness for the given ξ, i.e. the film thickness at a point of the exit boundary Γ_e with coordinates (χ, ξ). The latter becomes obvious from the further analysis as the derivatives of p with respect to ξ in the Reynolds equation (13.5) in the exit zone can be neglected (see (13.22)) and the obtained equation can be integrated with respect to χ from χ to $\gamma_e(\xi)$ (where $\gamma_e(\xi)$ is the χ-coordinate of the exit boundary point) with the boundary condition $\partial p(\gamma_e(\xi), \xi)/\partial \chi = 0$ (see (13.13) and (13.22)).

The definitions (13.9) and (13.8) and equations (13.5) indicate that pressure $p_0(\chi, \xi)$ (generally different from the Hertzian pressure distribution $p_{00}(\chi, \xi)$ from (13.6) due to the shape of the contact region Ω) satisfies the following equations

$$x^2(\chi, \xi) + \frac{\delta}{1-e^2} y^2(\chi, \xi) + \frac{D(e)}{\pi(1-e^2)^{3/2}} \{ \int\int_\Omega \frac{p_0(\chi_1, \xi_1) d\chi_1 d\xi_1}{\sqrt{(x-x_1)^2 + \frac{(y-y_1)^2}{1-e^2}}}$$

$$\text{(13.11)}$$

$$- \int\int_\Omega \frac{p_0(\chi_1, \xi_1) d\chi_1 d\xi_1}{\sqrt{x_1^2 + \frac{y_1^2}{1-e^2}}} \} = 0, \quad \int\int_\Omega p_0(\chi_1, \xi_1) d\chi_1 d\xi_1 = \frac{2\pi}{3}.$$

Physically, function $p_0(\chi, \xi)$ represents the pressure distribution in a dry fixed contact Ω of a rigid indenter with the radii R_x and R_y and sharp edges with an elastic half-space with the effective elastic modulus E'. In most cases $p_0(\chi, \xi)$ cannot be determined analytically due to the complexity of the shape of the boundary Γ. Depending on the fixed shape of Γ the solution $p_0(\chi, \xi)$ of equations (4.9) at some points of Γ may vanish or be singular with the singularity of the type $d^{-1/2}$, where d is the normal distance of a point from the region boundary Γ [7]. Therefore, we can conclude that close to the contact boundary Γ pressure $p_0(\chi, \xi)$ in the dry contact Ω can be asymptotically represented as follows [7]

$$p_0(\chi, \xi) = \epsilon_q^{1/2} q_0 + o(\epsilon_q^{1/2}), \quad q_0(\chi, \xi, \chi_\Gamma, \xi_\Gamma) = N_0 \sqrt{2 d_\epsilon} + \frac{N_1}{\sqrt{2 d_\epsilon}},$$

$$N_0 = N_0(\chi_\Gamma, \xi_\Gamma) = O(1), \quad N_1 = N_1(\chi_\Gamma, \xi_\Gamma) = O(1),$$

$$\text{(13.12)}$$

$$d_\epsilon = \frac{\sqrt{[x(\chi, \xi) - x_\Gamma(\chi, \xi)]^2 + \frac{[y(\chi, \xi) - y_\Gamma(\chi, \xi)]^2}{1-e^2}}}{\epsilon_q} = O(1), \quad \omega \ll 1,$$

$$(\chi, \xi) \in \Omega_\epsilon, \quad (x_\Gamma(\chi, \xi), y_\Gamma(\chi, \xi)) \in \Gamma,$$

where the point $(x(\chi, \xi), y(\chi, \xi))$ belongs to the line normal to the boundary Γ through the boundary point $(x_\Gamma(\chi, \xi), y_\Gamma(\chi, \xi))$. The direction of the normal practically coincides with the direction of the line $\xi(x, y) = const$ while the distance between these two points is about ϵ_q.

Let us make some simple and, at the same time, natural assumptions about the contact region boundary Γ. We will assume that the inlet and exit boundaries can be described by the equations $\chi = \gamma_i(\xi)$ and $\chi = \gamma_e(\xi)$ and that the inlet Γ_i and exit Γ_e boundaries of the lubricated contact are close to the boundaries of the dry circular contact, i.e.

$$\gamma_i(\xi) = -\sqrt{1-\xi^2} + \epsilon_q \alpha_{1p}(\xi), \ \alpha_{1p}(\xi) = O(1), \ (\chi,\xi) \in \Gamma_i,$$

$$\gamma_e(\xi) = \sqrt{1-\xi^2} + \epsilon_q \beta_{1p}(\xi), \ \beta_{1p}(\xi) = O(1), \ (\chi,\xi) \in \Gamma_e, \ \omega \ll 1,$$

(13.13)

where $\alpha_{1p}(\xi)$ is a given function of ξ while $\beta_{1p}(\xi)$ has to be determined from the problem solution. Moreover, we will assume that $\gamma_i(\xi)$ and $\gamma_e(\xi)$ are differentiable functions and that outside of the vicinity of points $(\chi,\xi) = (0,-1)$ and $(\chi,\xi) = (0,1)$ we have

$$\frac{d\gamma_i(\xi)}{d\xi} = O(1), \ \frac{d\gamma_e(\xi)}{d\xi} = O(1), \ \omega \ll 1. \tag{13.14}$$

These assumptions mean that the variations of the contact region boundary Γ are moderate, i.e., the boundary Γ does not experience abrupt step-type variations.

Relationships (13.13) and (13.14), in turn, lead to the following estimates

$$\frac{dN_0(x_\Gamma(\chi,\xi),y_\Gamma(\chi,\xi))}{d\overrightarrow{T_\Gamma}}, \ \frac{dN_1(x_\Gamma(\chi,\xi),y_\Gamma(\chi,\xi))}{d\overrightarrow{T_\Gamma}} = O(1), \ \omega \ll 1, \tag{13.15}$$

where $\overrightarrow{T_\Gamma}$ is a unit vector tangent to the contact boundary Γ at the point with coordinates $(x_\Gamma(\chi,\xi),y_\Gamma(\chi,\xi)) \in \Gamma$.

Using (13.11) the last two equations in (13.5) can be rewritten in the form

$$H_{00}[h(\chi,\xi)-1] = \frac{D(e)}{\pi(1-e^2)^{3/2}} \int\int_\Omega [p(\chi_1,\xi_1) - p_0(\chi_1,\xi_1)]$$

$$\times \left\{ \frac{1}{\sqrt{(x-x_1)^2 + \frac{(y-y_1)^2}{1-e^2}}} - \frac{1}{\sqrt{x_1^2 + \frac{y_1^2}{1-e^2}}} \right\} d\chi_1 d\xi_1, \tag{13.16}$$

$$\int\int_\Omega [p(\chi_1,\xi_1) - p_0(\chi_1,\xi_1)] d\chi_1 d\xi_1 = 0.$$

The proposed below asymptotic approach is valid for the central region occupied by the "central" flow streamlines which are away from the vicinity of points $(\chi,\xi) = (0,-1)$ and $(\chi,\xi) = (0,1)$. Also, this approach is valid for $0 \leq \gamma \leq 1$. In other words, we are proposing an asymptotic analysis of the central part of the lubricated contact which includes the corresponding inlet and exit zones occupied by "central" lubricant streamlines.

Now, we are ready for the asymptotic analysis of the EHL problem. It is important to remember that in the region of the contact (inside of the Hertzian circular contact) which is at a distance from the boundary of the Hertzian contact of much more than $O(\epsilon_q) \ll 1$ the pressure distribution $p(\chi,\xi)$ is very close to the Hertzian pressure $p_{00}(\chi,\xi)$. At the same time, in the inlet and exit zones the actual pressure $p(\chi,\xi)$ is significantly different from the Hertzian pressure $p_{00}(\chi,\xi)$ as well as from pressure $p_0(\chi,\xi)$. We will apply the matched asymptotic expansions approach developed in the preceding chapter to solution of the EHL problems at hand.

First, we will consider the pre-critical heavily loaded lubrication regimes for which the lubricant viscosity varies with pressure moderately. The very basic fact about the matched asymptotic expansions approach is that the solution of the problem $p(\chi,\xi)$ in the inlet and exit zones of width $O(\epsilon_q) \ll 1$ is of the order of magnitude of $\epsilon_q^{1/2}$ because otherwise it will not be able to match the external (dry) pressure $p_0(\chi,\xi)$ outside of the inlet and exit zones (see (13.12)). Therefore, a heavily loaded lubrication regime is pre-critical if in the inlet zone of the contact Ω_i where $p(\chi,\xi) = O(\epsilon_q^{1/2})$, $\epsilon_q \ll 1$, the following estimate holds

$$\mu(p(\chi,\xi)) = O(1), \ if \ p(\chi,\xi) = O(\epsilon_q^{1/2}), \ (\chi,\xi) \in \Omega_\epsilon, \ \omega \ll 1. \qquad (13.17)$$

It is important to realize that according to (13.12) $d_\epsilon = \epsilon_q^{-1}\{(x(\chi,\xi) - x_\Gamma(\chi,\xi))^2 + \frac{(y(\chi,\xi)-y_\Gamma(\chi,\xi))^2}{1-e^2}\}^{1/2} = O(1)$, $\omega \ll 1$, as long as $x(\chi,\xi)-x_\Gamma(\chi,\xi) = O(\epsilon_q)$ and $y(\chi,\xi) - y_\Gamma(\chi,\xi) = O(\epsilon_q)$ for $\omega \ll 1$. Therefore, $p_0(\chi,\xi) = O(\epsilon_q^{1/2})$ within the zone Ω_ϵ. In order for the pressure function $p(\chi,\xi)$ to match the function $p_0(\chi,\xi)$ at the inner boundary of the region Ω_ϵ and the outer boundary of the region $\Omega \setminus \Omega_\epsilon$ it is necessary that they are of the same order of magnitude. Therefore, we have

$$p(\chi,\xi) = O(\epsilon_q^{1/2}), \ (\chi,\xi) \in \Omega_\epsilon, \ \omega \ll 1. \qquad (13.18)$$

Based on (13.17) and (13.18) the definition of the pre-critical regimes can be presented in the form

$$\mu(\epsilon_q^{1/2}) = O(1), \ \omega \ll 1. \qquad (13.19)$$

Let us consider the points in the inlet Ω_i and exit Ω_e zones but outside of the vicinity of points $(\chi,\xi) = (0,-1)$ and $(\chi,\xi) = (0,1)$ which belong to the line $\xi = \xi_*$. First, let us consider the inlet zone. We will replace our original coordinate system (χ,ξ) by a new coordinate system (r_p,ξ), where $r_p = \frac{\chi-\gamma_i(\xi_*)}{\epsilon_q}$. Using (13.12)-(13.15), (13.18) and the principle of matching [8, 9] of asymptotic expansions of the problem solutions in the inlet zone $\Omega_i \bigcap \Omega_\epsilon$ and the Hertzian region we obtain that

$$p(\chi,\xi_*) = \epsilon_q^{1/2}q_p(r_p,\xi_*) + o(\epsilon_q^{1/2}), \ q_p(r_p,\xi_*) = O(1),$$

$$r_p = \frac{\chi-\gamma_i(\xi_*)}{\epsilon_q} = O(1), \ (\chi,\xi_*) \in \Omega_i \bigcap \Omega_\epsilon, \ \omega \ll 1, \qquad (13.20)$$

where $q_p(r_p,\xi_*)$ is the unknown function which is the main term of the asymptotic expansion of pressure in the inlet zone. In a similar manner we can easily establish that

$$p(\chi,\xi_*) = \epsilon_q^{1/2}g_p(s_p,\xi_*) + o(\epsilon_q^{1/2}), \ g_p(s_p,\xi_*) = O(1),$$

$$s_p = \frac{\chi-\gamma_e(\xi_*)}{\epsilon_q} = O(1), \ (\chi,\xi_*) \in \Omega_e \bigcap \Omega_\epsilon, \ \omega \ll 1, \qquad (13.21)$$

where $g_p(s_p, \xi_*)$ is the unknown function which is the main term of the asymptotic expansion of pressure in the exit zone. Functions $q_p(r_p, \xi_*)$ and $g_p(s_p, \xi_*)$ further for simplicity will be called asymptotic pressure distributions in the inlet and exit zones.

Based on (13.14) and (13.15) in the inlet and exit zones away from the vicinity of points $(\chi, \xi) = (0, -1)$ and $(\chi, \xi) = (0, 1)$ we obtain

$$\frac{\partial}{\partial \chi} = \epsilon_q^{-1} \frac{\partial}{\partial r_p}, \ r_p = O(1); \ \frac{\partial}{\partial \chi} = \epsilon_q^{-1} \frac{\partial}{\partial s_p}, \ s_p = O(1), \ \omega \ll 1,$$

$$\frac{\partial}{\partial \xi} = O(1), \ \omega \ll 1. \tag{13.22}$$

Using (13.14) we can determine that the external normal vector $\vec{n} = (n_\chi, n_\xi)$ has a nonzero χ-component $n_\chi = O(1)$, $\omega \ll 1$. Therefore, based on (13.21) and (13.22) we can conclude that the condition $\frac{dp(\chi, \xi)}{d\vec{n}} \mid_{\Gamma_e} = 0$ at the exit boundary Γ_e (see (13.5)) can be represented in the form $\{\epsilon_q^{-1/2} \frac{\partial g_p}{\partial s_p} n_\chi + \epsilon_q^{1/2} \frac{\partial g_p}{\partial \xi} n_\xi + \ldots\} \mid_{\Gamma_e} = 0$, $\omega \ll 1$. Obviously, for $\omega \ll 1$ the latter condition leads to the following one $\frac{\partial g_p}{\partial s_p} \mid_{s_p=0} = 0$.

Now, let us consider the asymptotic representations for $\frac{\partial h}{\partial \chi}$ in the inlet and exit zones. From the first equation in (13.16) we have

$$H_{00} \frac{\partial h}{\partial \chi} = \frac{D(e)}{\pi(1-e^2)^{3/2}} \int \int_\Omega [p(\chi_1, \xi_1) - p_0(\chi_1, \xi_1)]$$

$$\times \frac{\partial}{\partial \chi} \frac{1}{\sqrt{(x-x_1)^2 + \frac{(y-y_1)^2}{1-e^2}}} d\chi_1 d\xi_1. \tag{13.23}$$

We will consider the inlet and exit zones (away from the the vicinity of points $(\chi, \xi) = (0, -1)$ and $(\chi, \xi) = (0, 1)$) fully extended along the χ-axis (i.e., along the line $\xi = \xi_*$) which are practically symmetric about the line $\xi = \xi_*$ (see (13.14)) and with characteristic sizes along the ξ-axis of about ϵ_q. Due to (13.12)-(13.15) within these zones

$$p(\chi, \xi) = p(\chi, \xi_*) + O(\epsilon_q),$$

$$p(\chi, \xi) - p_0(\chi, \xi) = p(\chi, \xi_*) - p_0(\chi, \xi_*) + O(\epsilon_q), \tag{13.24}$$

$$(\chi, \xi) \in (\Omega_i \cap \Omega_e) \cup (\Omega_e \cap \Omega_e), \ \omega \ll 1.$$

Taking into account the definition (13.8) of a heavily loaded EHL contact and related to it assumption that away from the boundary of the Hertzian contact

$$p(\chi, \xi) - p_0(\chi, \xi) = O(\epsilon_q^2), \ (\chi, \xi) \in \Omega \setminus \Omega_e, \ \omega \ll 1, \tag{13.25}$$

the representations (13.20) and (13.21), as well as the estimate (13.24) and using the local coordinates r_p, s_p, and $\eta = \frac{\xi - \xi_*}{\epsilon_q}$ we can estimate that if the point (χ, ξ) belongs to the inlet or exit zones (away from the vicinity of points $(\chi, \xi) = (0, -1)$ and $(\chi, \xi) = (0, 1)$) then the contribution of the integral of $\frac{(x_1-x)[p(\chi_1,\xi_1)-p_0(\chi_1,\xi_1)]}{[(x-x_1)^2+\frac{(y-y_1)^2}{1-e^2}]^{3/2}}$ (see equation (13.23)) over the entire contact region Ω excluding the considered inlet or exit zone will be of the order of magnitude $O(\epsilon_q^2)$, $\omega \ll 1$, while the contribution of the integral of the above function over the inlet or exit zone will be of the order of $O(\epsilon_q^{1/2})$, $\omega \ll 1$. Therefore, in the inlet zone we can conclude that

$$H_{00}\frac{\partial h(\chi,\xi_*)}{\partial \chi}$$

$$= -\frac{\epsilon_q^{1/2}}{\pi}\frac{D(e)}{(1-e^2)^{3/2}}\sqrt{1-e^2+e^2\gamma^2}\int\int_\Omega [q_p(t,\xi_*) - q_{ap}(t,\xi_*)]$$

$$\times\frac{r_p-t+e^2\gamma\sqrt{\frac{1-\gamma^2}{1-e^2}}\eta}{[(r_p-t)^2+2e^2\gamma\sqrt{\frac{1-\gamma^2}{1-e^2}}(r_p-t)\eta+\frac{(1-e^2)^2(1-\gamma^2)+\gamma^2}{1-e^2}\eta^2]^{3/2}}d\eta dt + o(\epsilon_q^{1/2})$$

$$= -\frac{\epsilon_q^{1/2}}{\pi}\frac{D(e)}{(1-e^2)^{3/2}}\sqrt{1-e^2+e^2\gamma^2}\int_0^\infty [q_p(t,\xi_*) - q_{ap}(t,\xi_*)]dt \qquad (13.26)$$

$$\times\int_{-\infty}^\infty \{r_p - t + e^2\gamma\sqrt{\frac{1-\gamma^2}{1-e^2}}\eta\}\{\frac{(1-e^2)^2(1-\gamma^2)+\gamma^2}{1-e^2}[\eta$$

$$+\frac{e^2\gamma\sqrt{(1-e^2)(1-\gamma^2)}}{(1-e^2)^2(1-\gamma^2)+\gamma^2}(r_p - t)]^2 + \frac{(1-\gamma^2)[(1-e^2)^2-e^4\gamma^2]+\gamma^2}{(1-e^2)^2(1-\gamma^2)+\gamma^2}(r_p - t)^2\}^{-3/2}$$

$$+o(\epsilon_q^{1/2}), \quad (\chi,\xi_*) \in \Omega_i \bigcap \Omega_\epsilon, \quad \omega \ll 1,$$

where $q_0(\chi, \xi, \chi_\Gamma, \xi_\Gamma)$ is replaced by $q_{ap}(r_p, \xi_*) = q_0(\gamma_i(\xi_*)+r_p\epsilon_q, \xi_*, \gamma_i(\xi_*), \xi_*)$ $= N_{0q}(\xi_*)\sqrt{2r_p}+\frac{N_{1q}(\xi_*)}{\sqrt{2r_p}}$ due to the fact that the direction of the normal vector $\vec{n}$ to the contact boundary Γ_i in $\Omega_i \bigcap \Omega_\epsilon$ practically coincides with the χ-axis (see (13.14)). Using the fact that $\int_{-\infty}^\infty \frac{\eta d\eta}{(\eta^2+a^2)^{3/2}} = 0$ and $\int_{-\infty}^\infty \frac{d\eta}{(\eta^2+a^2)^{3/2}} = \frac{2}{a^2}$ the last integral in (13.26) is equal to $2\sqrt{\frac{1-e^2}{(1-e^2)^2(1-\gamma^2)+\gamma^2}}\frac{1}{r_p-t}$. Therefore, from (13.26) in the inlet zone we obtain

$$H_{00}\frac{\partial h}{\partial \chi} = \epsilon_q^{1/2}\frac{2}{\pi}\frac{D(e)}{1-e^2}\sqrt{\frac{1-e^2+e^2\gamma^2}{(1-e^2)^2(1-\gamma^2)+\gamma^2}}\int_0^\infty \frac{q_p(t,\xi_*)-q_{ap}(t,\xi_*)}{t-r_p}dt$$

$$(13.27)$$

$$+o(\epsilon_q^{1/2}), \quad (\chi,\xi_*) \in \Omega_i \bigcap \Omega_\epsilon, \quad \omega \ll 1.$$

In a similar fashion, in the exit zone $\Omega_e \bigcap \Omega_\epsilon$ we get

$$H_{00}\frac{\partial h}{\partial \chi} = \epsilon_q^{1/2}\frac{2}{\pi}\frac{D(e)}{1-e^2}\sqrt{\frac{1-e^2+e^2\gamma^2}{(1-e^2)^2(1-\gamma^2)+\gamma^2}}\int\limits_{-\infty}^{0}\frac{g_p(t,\xi_*)-g_{ap}(t,\xi_*)}{t-s_p}dt$$

$$\qquad(13.28)$$

$$+o(\epsilon_q^{1/2}), \quad (\chi,\xi_*)\in\Omega_e\bigcap\Omega_\epsilon, \quad \omega\ll 1,$$

where $g_{ap}(s_p,\xi_*) = q_0(\gamma_e(\xi_*)+s_p\epsilon_q,\xi_*,\gamma_e(\xi_*),\xi_*) = N_{0g}(\xi_*)\sqrt{-2s_p} + \frac{N_{1g}(\xi_*)}{\sqrt{-2s_p}}+o(1), \quad \epsilon_q\ll 1.$

It is important to realize that the function $N_0(x_\Gamma,y_\Gamma)$ depends only on the behavior of $p_{00}(\chi,\xi)$ from (13.6) and it is independent from the shape of the contact boundary Γ which is close to $x^2+y^2=1$ (see (13.13)). It means that at the border between the inlet zone and the Hertzian region, i.e. as $r_p\to\infty$ while $\chi+\sqrt{1-\xi^2} = O(\epsilon_q)$ the value of $\epsilon_q^{1/2}N_0\sqrt{2r_p} = \sqrt{1-\chi^2-\xi^2}+o(1)$. Substituting in the latter equation $r_p = \frac{\chi+\sqrt{1-\xi^2}-\epsilon_q\alpha_{1p}}{\epsilon_q}$ we obtain the equation

$$N_0\sqrt{2}\sqrt{\chi+\sqrt{1-\xi^2}-\epsilon_q\alpha_{1p}} = \sqrt{1-\chi^2-\xi^2}+o(1), \quad \omega\ll 1. \qquad (13.29)$$

At the boundary between the inlet zone and the Hertzian region, i.e., for $r_p\to\infty$ and $\chi+\sqrt{1-\xi^2} = O(1)$, $\omega\ll 1$, after simple calculations from (13.29) we can conclude that in the inlet zone

$$N_0(\xi) = \sqrt[4]{1-\xi^2}. \qquad (13.30)$$

In a similar fashion in the exit zone we will get that $N_0(\xi)$ also satisfies (13.30). Therefore,

$$N_{0q}(\xi) = N_{0g}(\xi) = \sqrt[4]{1-\xi^2}. \qquad (13.31)$$

On the other hand, the function $N_1(x_\Gamma,y_\Gamma)$ depends on the distance of the contact boundary Γ from the boundary of the Hertzian contact $\chi^2+\xi^2=1$, i.e. functions N_{1q} and N_{1g} are proportional to α_{1p} and β_{1p} (see (13.13)), respectively. In more detail this will be described later (see (13.50)).

The above estimates for the derivatives of p and h are valid for an arbitrary $\xi=\xi_*$ as long as the point (χ,ξ) belongs to the inlet or exit zones outside of the vicinity of points $(\chi,\xi)=(0,-1)$ and $(\chi,\xi)=(0,1)$. For convenience, let us introduce two new dimensionless parameters

$$V_{e\gamma} = V\frac{(1-e^2+e^2\gamma^2)^{3/2}}{(1-e^2)^2(1-\gamma^2)+\gamma^2}, \quad \theta = \frac{D(e)}{1-e^2}\sqrt{\frac{1-e^2+e^2\gamma^2}{(1-e^2)^2(1-\gamma^2)+\gamma^2}}. \qquad (13.32)$$

Then, using the estimates for $H_{00}\frac{\partial h}{\partial \chi}$ from (13.27), (13.28), and (13.22) in the inlet $\Omega_i\bigcap\Omega_\epsilon$ and exit $\Omega_e\bigcap\Omega_\epsilon$ zones from the requirement that the differential and integral terms in the Reynolds equation in (13.5) are of the same order of magnitude for $\omega\ll 1$ we obtain

$$A_p^3 \frac{\partial}{\partial r_p}\left(\frac{h_{qp}^3}{\mu_{qp}}\frac{\partial q_p}{\partial r_p}\right) = \frac{2}{\pi}\theta \int\limits_0^\infty \frac{q_p(t,\xi)-q_{ap}(t,\xi)}{t-r_p}dt, \quad q_p(0,\xi)=0, \qquad (13.33)$$

$$A_p^3 \frac{\partial}{\partial s_p}\left(\frac{h_{gp}^3}{\mu_{gp}}\frac{\partial g_p}{\partial s_p}\right) = \frac{2}{\pi}\theta \int\limits_{-\infty}^0 \frac{g_p(t,\xi)-g_{ap}(t,\xi)}{t-s_p}dt, \qquad (13.34)$$

$$g_p(0,\xi)=0, \quad \frac{\partial g_p(0,\xi)}{\partial s_p}=0,$$

where A_p is a dimensionless coefficient of proportionality in the formula for the film thickness

$$H_{00} = A_p(V_{e\gamma}\epsilon_q^2)^{1/3}, \quad A_p = O(1), \quad \epsilon_q \ll V_{e\gamma}^{2/5}, \quad \omega \ll 1, \qquad (13.35)$$

where $h_{qp}(r_p,\xi)$ and $h_{gp}(s_p,\xi)$ are the gap functions in the inlet $\Omega_i \bigcap \Omega_\epsilon$ and exit $\Omega_e \bigcap \Omega_\epsilon$ zones. The expressions for $h_{qp}(r_p,\xi)$ and $h_{gp}(s_p,\xi)$ depend on the lubrication regime. For starved lubrication regimes the gap distribution $h(\chi,\xi)$ is very close to $h_\infty(\xi)$ in the entire contact region. That corresponds to the case of $\epsilon_q^{3/2} \ll H_{00}$ and

$$h_{qp}(r_p,\xi) = h_\infty(\xi), \quad \epsilon_q \ll \epsilon_f, \qquad (13.36)$$

$$h_{gp}(s_p,\xi) = h_\infty(\xi), \quad \epsilon_q \ll \epsilon_f, \qquad (13.37)$$

$$\epsilon_f = V_{e\gamma}^{2/5}, \qquad (13.38)$$

while for fully flooded lubrication regimes $H_{00} = O(\epsilon_q^{3/2})$, $\omega \ll 1$, and in the inlet and exit zones the gap satisfies the equations (see equations (13.27) and (13.28))

$$A_p\frac{\partial h_{qp}}{\partial r_p} = \frac{2}{\pi}\theta \int\limits_0^\infty \frac{q_p(t,\xi)-q_{ap}(t,\xi)}{t-r_p}dt, \quad \epsilon_q = \epsilon_f, \quad \omega \ll 1, \qquad (13.39)$$

$$A_p\frac{\partial h_{gp}}{\partial s_p} = \frac{2}{\pi}\theta \int\limits_{-\infty}^0 \frac{g_p(t,\xi)-g_{ap}(t,\xi)}{t-s_p}dt, \quad \epsilon_q = \epsilon_f, \quad \omega \ll 1. \qquad (13.40)$$

In (13.33) and (13.34) functions μ_{qp} and μ_{gp} are the main terms of the asymptotic expansions of $\mu(\epsilon_q^{1/2}q_p)$ and $\mu(\epsilon_q^{1/2}g_p)$, respectively.

It is obvious from (13.35) that for fully flooded pre-critical lubrication regimes (characterized by the relationship $H_{00} = O(\epsilon_q^{3/2})$, $\omega \ll 1$) we have

$$H_{00} = A_p V_{e\gamma}^{3/5}, \quad A_p = O(1), \quad \epsilon_q = V_{e\gamma}^{2/5}, \omega \ll 1. \qquad (13.41)$$

Here and in (13.35) the value of the coefficient A_p depends on θ, $\alpha_{1p}(0)$, and on $Q\epsilon_q^{1/2}$ for starved lubrication regimes and on $QV_{e\gamma}^{1/5}$ for fully flooded lubrication regimes, respectively.

It is clear from equations (13.39) and (13.40) that when $r_p \to \infty$ and $s_p \to -\infty$ we get

$$\int_0^\infty [q_p(t,\xi) - q_{ap}(t,\xi)]dt = 0, \qquad (13.42)$$

$$\int_{-\infty}^0 [g_p(t,\xi) - g_{ap}(t,\xi)]dt = 0, \qquad (13.43)$$

because otherwise the functions $h_{qp}(r_p,\xi)$ and $h_{gp}(s_p,\xi)$ would be unbounded as $r_p \to \infty$ and $s_p \to -\infty$, respectively. Equations (13.42) and (13.43) are consistent with the second equation in (13.16).

For fully flooded lubrication regimes the values of functions $h_{qp}(r_p,\xi)$ and $h_{gp}(s_p,\xi)$ are obtained by integrating equations (13.39) and (13.40) with the conditions $h_{qp}(r_p,\xi) \to h_\infty(\xi)$ as $r_p \to \infty$ and $h_{gp}(s_p,\xi) \to h_\infty(\xi)$ as $s_p \to -\infty$ as follows

$$A_p[h_{qp}(r_p,\xi) - h_\infty(\xi)] = \tfrac{2}{\pi}\theta \int_0^\infty [q_p(t,\xi) - q_{ap}(t,\xi)] \ln \tfrac{1}{|t-r_p|} dt, \qquad (13.44)$$

$$A_p[h_{gp}(s_p,\xi) - h_\infty(\xi)] = \tfrac{2}{\pi}\theta \int_{-\infty}^0 [g_p(t,\xi) - g_{ap}(t,\xi)] \ln \tfrac{1}{|t-s_p|} dt. \qquad (13.45)$$

Therefore, in the inlet zone for the given values of θ, $Q\epsilon_q^{1/2}$ (involved in μ_{qp}), $\alpha_{1p}(\xi)$, and ξ we derived the system of equations (13.33), (13.42), and (13.36) for starved lubrication regimes or (13.44) for fully flooded lubrication regimes for functions $q_p(r_p,\xi)$, $h_{qp}(r_p,\xi)$, $h_\infty(\xi)$, and coefficient A_p. In the exit zone for the given values of θ, $Q\epsilon_q^{1/2}$, A_p, $h_\infty(\xi)$, and ξ we need to solve the system of equations (13.34), (13.43), and (13.37) for starved lubrication regimes or (13.45) for fully flooded lubrication regimes for functions $g_p(s_p,\xi)$, $h_{gp}(s_p,\xi)$, and $\beta_{1p}(\xi)$. It is important to keep in mind that constant A_p is the coefficient of proportionality in formula (13.35) for the central film thickness and it is determined for $y = 0$ or $\xi = 0$ (see (13.3)). Also, it means that $h_\infty(0) = 1$ while, generally, $h_\infty(\xi) \neq 1$ for $\xi \neq 0$. The solution procedure should be implemented according to the following sequence: first, the system of asymptotically valid equations in the inlet zone should be solved for $\xi = 0$ which determines the value of constant A_p; after that, the values of function $h_\infty(\xi)$ for $\xi \neq 0$ are determined from the solution of the asymptotically valid system of equations in the inlet zone and these values play the role similar to the one played by constant A_p for $\xi = 0$.

It is important to notice that the asymptotically valid in the inlet and exit zones equations (13.33), (13.34), (13.36)-(13.45) are identical to the corresponding asymptotically valid equations for the case of the lubricant entrainment along the x-axis (i.e., when $\gamma = 1$, see Chapter 12). At the same time, formulas (13.35) and (13.41) for the film thickness H_{00} coincide with the corresponding formulas for H_{00} from Chapter 12 if parameter $V_{e\gamma}$ is replaced by parameter V.

Also, as in case of $\gamma = 1$, the above systems of asymptotic equations in the inlet and exit zones are very similar to the ones derived and analyzed for the cases of line EHL contacts. Namely, by introducing the substitutions

$$\{r, s, \alpha_1, \beta_1\} = \tfrac{1}{r_0}\{r_p, s_p, \alpha_{1p}, \beta_{1p}\}, \ \{q, g\} = \tfrac{1}{N_0 r_0^{1/2}}\{q_p, g_p\},$$

$$\{h_q, h_g\} = \tfrac{1}{h_\infty}\{h_{qp}, h_{gp}\}, \ \{\mu_q(q), \mu_g(g)\} = \{\mu_{qp}(q_p), \mu_{gp}(g_p)\},$$
(13.46)

and the definitions

$$A = A_p[\theta N_0^4(\xi)]^{1/5} h_\infty(\xi), \ H_0(\xi) = H_{00} h_\infty(\xi), \ r_0 = [\theta^2 N_0^3(\xi)]^{-2/5} \quad (13.47)$$

(where θ is determined by (13.32)) we will be able to reduce each of the above systems of valid equations to

$$A^3 \frac{\partial}{\partial r}\left(\frac{h_q^3}{\mu_q}\frac{\partial q}{\partial r}\right) = \frac{2}{\pi}\int\limits_0^\infty \frac{q(t,\xi) - q_a(t,\xi)}{t - r}dt, \ q(0,\xi) = 0,$$

$$\int\limits_0^\infty [q(t,\xi) - q_a(t,\xi)]dt = 0, \ q_a(r,\xi) = \sqrt{2r} + \frac{\alpha_1(\xi)}{\sqrt{2r}}, \quad (13.48)$$

$$h_q(r,\xi) = 1 \ or \ A[h_q(r,\xi) - 1] = \frac{2}{\pi}\int\limits_0^\infty [q(t,\xi) - q_a(t,\xi)]\ln\frac{1}{|t-r|}dt,$$

in the inlet zone and to

$$A^3 \frac{\partial}{\partial s}\left(\frac{h_g^3}{\mu_g}\frac{\partial g}{\partial s}\right) = \frac{2}{\pi}\int\limits_{-\infty}^0 \frac{g(t,\xi) - g_a(t,\xi)}{t - s}dt, \ g(0,\xi) = 0, \ \frac{\partial g(0,\xi)}{\partial s} = 0,$$

$$\int\limits_{-\infty}^0 [g(t,\xi) - g_a(t,\xi)]dt = 0, \ g_a(s,\xi) = \sqrt{-2s} - \frac{\beta_1(\xi)}{\sqrt{-2s}}, \quad (13.49)$$

$$h_g(s,\xi) = 1 \ or \ A[h_g(s,\xi) - 1] = \frac{2}{\pi}\int\limits_{-\infty}^0 [g(t,\xi) - g_a(t,\xi)]\ln\frac{1}{|t-s|}dt,$$

in the exit zone. Here $H_0(\xi)$ is the dimensionless analog of the dimensional film thickness h_0 but at the exit from the contact region as well as in the Hertzian region at the points with the fixed ordinate ξ. In other words, we can expect that there are some variations in the film thickness in the Hertzian region along the ξ-axis. Besides that, we used the fact that

$$\alpha_{1p} = \frac{N_{1q}}{N_{0q}}, \ \beta_{1p} = -\frac{N_{1g}}{N_{0g}}. \quad (13.50)$$

For fully flooded lubrication regimes by integrating equations (13.48) and (13.49) for q and g the gap functions h_q and h_g from (13.48) and (13.49) can be replaced by equations

$$h_q = 1 + A^2\frac{\partial}{\partial r}\left(\frac{h_q^3}{\mu_q}\frac{\partial q}{\partial r}\right), \ h_g = 1 - A^2\frac{\partial}{\partial s}\left(\frac{h_g^3}{\mu_g}\frac{\partial g}{\partial s}\right). \quad (13.51)$$

We have to keep in mind that due to a particular shape of the contact region boundary Γ the inlet coordinate α_1 is generally a function of ξ which makes the exit coordinate β_1 also a certain function of ξ. In addition to that, in the inlet and exit zones the functions of the lubricant viscosity μ_q and μ_g depend on $Q\epsilon_q^{1/2}\theta^{-2/5}(1-\xi^2)^{1/10}q(r,\xi)$ and $Q\epsilon_q^{1/2}\theta^{-2/5}(1-\xi^2)^{1/10}g(s,\xi)$, respectively.

Obviously, in case of α_1 independent of ξ and $\mu_q = \mu_g = 1$ (i.e., in the inlet and exit zones the lubricant viscosity is independent of pressure) we will need to solve systems (13.48) and (13.49) just once and using substitutions (13.46) and (13.47) get the solution of the problem in the inlet and exit zones for all values of ξ such that $(\chi,\xi) \in \Omega_i \bigcap \Omega_\epsilon$ and $(\chi,\xi) \in \Omega_e \bigcap \Omega_\epsilon$.

Systems of equations (13.48) and (13.49) are identical to the corresponding systems of equations derived in the inlet and exit zones of lubricated heavily loaded line contacts (see Chapter 6). Therefore, the whole arsenal of methods developed for the analysis of these equations in Chapters 6 - 10 and the corresponding conclusions made for line contacts are transferable to point contacts. That includes the representation of the systems (13.48) and (13.49) in the equivalent form similar to systems (12.57) and (12.58), validation of the applied asymptotic approach, discussions of the numerical methods suitable for solution of asymptotically valid equations in the inlet and exit zones, solution precision, proper grid sizing, and solution stability/instability issues as well as the regularization approach to obtaining stable numerical solutions for high values of the dimensionless pressure viscosity coefficient Q. Also, these equations are controlled by less number of dimensionless parameters than the equations of the EHL problem in the original (not asymptotic) formulation. A large number of numerical examples can be found in Chapter 6.

The above substitutions and formulas (13.31), (13.35), (13.47), and (13.50) lead to the conclusion that the exit film thickness H_0 besides being a function of α_1, $V_{e\gamma}$, Q, ϵ_q, e and γ is also a function of ξ and

$$H_0(\xi) = A[\theta(1-\xi^2)]^{-1/5}(V_{e\gamma}\epsilon_q^2)^{1/3}, \ A = O(1), \ \omega \ll 1, \qquad (13.52)$$

where A is a constant determined from the solution of the asymptotic system of equations (13.48) in the inlet zone for a line contact (which is dependent on α_1 as well as on $Q\epsilon_q^{1/2}$ and θ from (13.32) through μ_q) while $V_{e\gamma}$ is determined by formulas (13.1) and (13.32). For pre-critical lubrication regimes the value of $Q\epsilon_q^{1/2}$ can be taken as either 0 for $\epsilon_q \ll Q^{-2}$, $\omega \ll 1$, or equal to $Q\epsilon_q^{1/2} = O(1)$, $\omega \ll 1$, if $\epsilon_q = O(Q^{-2})$, $\omega \ll 1$. In a special case of constant viscosity μ or the case of starved lubrication regime when $\epsilon_q \ll Q^{-2}$, $\omega \ll 1$, the situation is simpler because the value of the constant A is dependent only on α_1, i.e., it is independent from $Q\epsilon_q^{1/2}$ and θ. In the latter case the dependence of the exit film thickness H_0 on the contact ellipse eccentricity e is limited to just the one shown in (13.52).

The exit film thickness H_0 and, simultaneously, the film thickness in the Hertzian region slowly increases with $|\xi|$, i.e., farther away from the contact center $(\chi, \xi) = (0,0)$ along the ξ-axis the film thickness is slightly higher than at the center. It is important to keep in mind that formula (13.52) is valid only away from the vicinity of points with coordinates $(\chi, \xi) = (0, -1)$ and $(\chi, \xi) = (0, 1)$.

In a similar fashion for fully flooded heavily loaded lubrication regimes when $\epsilon_q = \epsilon_f = V_{e\gamma}^{2/5}$ we obtain (see (13.52))

$$H_0(\xi) = A[\theta(1 - \xi^2)]^{-1/5} V_{e\gamma}^{3/5}, \ A = O(1), \ \omega \ll 1. \tag{13.53}$$

In the original (x, y) coordinates formula (13.53) is as follows

$$H_0(\xi) = A\{\theta[1 - \frac{(-\sqrt{1-e^2}\sqrt{1-\gamma^2}x+\gamma y)^2}{1-e^2+e^2\gamma^2}]\}^{-1/5} V_{e\gamma}^{3/5},$$

$$A = O(1), \ \omega \ll 1. \tag{13.54}$$

The rest of the analysis of the EHL problem in the inlet and exit zones is done according to the procedures of Chapter 6. Additionally, the uniformly valid approximations of pressure $p(\chi, \xi)$ and gap $h(\chi, \xi)$ as well as the exit film thickness $H_0(\xi)$ in the central part of the contact region can be obtained in the fashion similar to the one described in the end of Section 12.3.

It is remarkable that essentially the lubrication mechanism and the equations describing it in the inlet and exit zones of a point contact with straight and skewed lubricant entrainment are identical. Moreover, for contacts involved in different types of motion the values of constants A in formulas for the exit film thickness H_0 (see formulas (12.55), (12.56), (13.52), and (13.53)) are the same (i.e., independent of the type of motion) as long as the coefficients of q and g in the expressions for the lubricant viscosity μ_q and μ_g in the inlet and exit zones are identical for the corresponding cases of motion.

For pre-critical lubrication regimes with constant viscosity in the inlet zone it is easy to determine the influence of the ellipse eccentricity e and the skewness ratio γ on the film thickness. Let us denote by $H_0(\gamma = 1)$ and $H_0(\gamma < 1)$ the exit film thickness for $y = 0$ and the case of absent skewness in lubricant entrainment $(\gamma = 1)$ and presence of lubricant entrainment skewness $(\gamma < 1)$, respectively. Then based on formula (13.53) for fully flooded pre-critical lubrication regimes we get

$$R_{\gamma e} = \frac{H_0(\gamma<1)}{H_0(\gamma=1)} \mid_{\xi=0} = \frac{(1-e^2+e^2\gamma^2)^{4/5}}{[(1-e^2)^2(1-\gamma^2)+\gamma^2]^{1/2}}. \tag{13.55}$$

For $\gamma = 1$, the expression in the right-hand side of (13.55) is equal to $R_{\gamma e} = 1$. The variations of the ratio $R_{\gamma e}$ for several other values of γ and e are represented in Table 13.1.

At the same time the values of the ratio

$$R_\xi = \frac{H_0(|\xi|>0)}{H_0(\xi=0)} = (1 - \xi^2)^{-1/5}, \tag{13.56}$$

are given in Table 13.2 for several values of variable $| \xi |$.

TABLE 13.1

Variations of the ratio

$R_{\gamma e} = \frac{H_0(\gamma<1)}{H_0(\gamma=1)} \mid_{\xi=0}$ versus the values of γ

and e.

γ/e	0	0.447	0.632	0.775	0.894
0.707	1	1.015	1.014	0.987	0.922
0.5	1	1.028	1.043	1.019	0.908

TABLE 13.2

Variations of the ratio

$R_\xi = \frac{H_0(|\xi|>0)}{H_0(\xi=0)}$ versus the value of

$|\xi|$.

$|\xi|$	0.25	0.5	0.707	0.8
R_ξ	1.013	1.059	1.149	1.227

13.4   Asymptotic Analysis of a Heavily Loaded Lubricated Contact under Over-Critical Lubrication Conditions

Over-critical lubrication regimes occur when lubricant viscosity rapidly grows with pressure which occurs when $Q \gg 1$. In a heavily loaded lubricated contact the maximum characteristic size of the inlet zone ϵ_0 for which the lubrication regime is still a pre-critical one we will call the critical characteristic size of the inlet zone ϵ_q. To determine ϵ_0 we need to consider the definition (13.19) of the pre-critical lubrication regimes. The value of ϵ_0 is such that for any $\epsilon_q \gg \epsilon_0$ the estimate (13.19) is not valid while for $\epsilon_q = \epsilon_0$ this estimate is still valid. If it appears that $\epsilon_0 \ll 1$ for the given problem parameters then it means that the viscosity μ depends on a large parameter Q. Thus, if $\epsilon_0 \ll 1$ then for $\epsilon_q = O(\epsilon_0)$ we have pre-critical regimes and for $\epsilon_q \gg \epsilon_0$ - over-critical regimes. For example, if $\mu = \exp(Qp)$ then $\epsilon_0 = Q^{-2} \ll 1$ for $\omega = Q^{-1} \ll 1$. Obviously, if $\epsilon_0 \gg 1$ or $\epsilon_0 = O(1)$ for $\omega \ll 1$ then over-critical regimes cannot be realized and pre-critical lubrication regimes occur.

Therefore, the formal definition of over-critical lubrication regimes is as follows

$$\mu(\epsilon_q^{1/2}) \gg 1, \; \epsilon_q(\omega) \gg \epsilon_0(\omega), \; \epsilon_0(\omega) \ll 1, \; \omega \ll 1. \tag{13.57}$$

A heavily loaded over-critical lubricated contact has the following structure (see Chapter 6). The inlet zone is represented by two sub-zones: one of a characteristic size $\epsilon_q \gg \epsilon_0$ (inlet ϵ_q-zone) which is adjacent to the the inlet contact boundary Γ_i and a smaller inlet zone of the characteristic size ϵ_0 (inlet ϵ_0-zone) located between the inlet ϵ_q-zone and the Hertzian region where pressure is very close to $p_{00}(\chi, \xi) = \sqrt{1 - \chi^2 - \xi^2}$. On the exit side of the contact the structure of the exit zone is similar to the one of the inlet zone, i.e. a smaller exit ϵ_0-zone is located between the Hertzian region and the larger exit ϵ_q-zone adjacent to the contact exit boundary Γ_e.

Using the same approach as the one used in Chapter 12 we obtain that for starved and fully flooded over-critical lubrication regimes the film thickness is determined by the formulas

$$H_{00} = A_p (V_{e\gamma} Q \epsilon_q^{5/2})^{1/3}, \ \epsilon_q \ll (V_{e\gamma} Q)^{1/2}, \ A_p = O(1), \ Q \gg 1,$$

$$H_{00} = A_f (V_{e\gamma} Q)^{3/4}, \ \epsilon_q = (V_{e\gamma} Q)^{1/2}, \ A_f = O(1), \ Q \gg 1,$$
(13.58)

where the coefficients A_s and A_f depend on θ from (13.32) and $\alpha_{1p}(0)$. In the manner similar to the one used for pre-critical lubrication regimes from formulas (13.58) we obtain formulas for the exit film thickness as follows

$$H_0(\xi) = A_{p1}[\theta(1 - \xi^2)]^{-1/5} (V_{e\gamma} Q \epsilon_q^{5/2})^{1/3},$$

$$\epsilon_q \ll (V_{e\gamma} Q)^{1/2}, \ A_{p1} = O(1), \ Q \gg 1,$$
(13.59)

for starved over-critical lubrication regimes and

$$H_0(\xi) = A_{f1}[\theta(1 - \xi^2)]^{-1/5} (V_{e\gamma} Q)^{3/4},$$

$$\epsilon_q = (V_{e\gamma} Q)^{1/2}, \ A_{f1} = O(1), \ Q \gg 1,$$
(13.60)

for fully flooded over-critical lubrication regimes. In the original (x, y) coordinates formula (13.60) is as follows

$$H_0(\xi) = A_{f1}\{\theta[1 - \frac{(-\sqrt{1-e^2}\sqrt{1-\gamma^2}x + \gamma y)^2}{1 - e^2 + e^2\gamma^2}]\}^{-1/5}(V_{e\gamma}Q)^{3/4},$$

$$A_{f1} = O(1), \ Q \gg 1.$$
(13.61)

TABLE 13.3
Variations of the ratio
$R_{\gamma e} = \frac{H_{00}(\gamma < 1)}{H_{00}(\gamma = 1)} \mid_{\xi = 0}$ versus the values of
γ and e.

γ / e	0	0.447	0.632	0.775	0.894
0.707	1	1.021	1.022	0.989	0.906
0.5	1	1.039	1.061	1.034	0.894

For over-critical lubrication regimes it is easy to determine the influence of the ellipse eccentricity e and the skewness ratio γ on the film thickness. Let us denote by $H_0(\gamma = 1)$ and $H_0(\gamma < 1)$ the exit film thickness for $\xi = 0$ and the case of absent skewness in lubricant entrainment $(\gamma = 1)$ and presence of lubricant entrainment skewness ($\gamma < 1$), respectively. Then based on formula (13.61) for fully flooded over-critical lubrication regimes we get

$$R_{\gamma e} = \frac{H_0(\gamma<1)}{H_0(\gamma=1)} \mid_{\xi=0} = \{ \frac{(1-e^2+e^2\gamma^2)^{41}}{[(1-e^2)^2(1-\gamma^2)+\gamma^2]^{26}} \}^{\frac{1}{40}}. \qquad (13.62)$$

For $\gamma = 1$, the expression in the right-hand side of (13.62) is equal to $R_{\gamma e} = 1$. The variations of the ratio $R_{\gamma e}$ for several other values of γ and e are represented in Table 13.3.

Also, based on formula (13.61) for over-critical lubrication regimes we obtain the same ratio $R_\xi = \frac{H_0(\xi>0)}{H_0(\xi=0)} = (1 - \xi^2)^{-1/5}$ as for the case of pre-critical lubrication regimes (see (13.56)). The values of ratio R_ξ are given in Table 13.2.

Numerical studies of the inlet and exit zones based on the derived asymptotically valid equations can be found in Chapter 6. The numerical results include parametric studies, studies of solution precision and stability, etc.

The values of constants A_p and A_f can be obtained from the solution of the system of asymptotically valid equations in the inlet zone for pre-critical lubrication regimes as it was explained in the end of Chapter 12. Also, the validity of the definition of heavily loaded lubricated contacts (13.8) and (13.9) easily follows from these relationships in $\Omega \setminus \Omega_\epsilon$ (i.e., outside of the inlet and exit zones) due to the fact that the film thickness H_{00} is determined by equations (13.35), (13.41), and (13.58) for starved and fully flooded pre- and over-critical lubrication regimes.

The choice between pre- and over-critical lubrication regimes can be made based on a simple analysis from Section 9.5 while the regularization approaches to numerical solution of isothermal EHL problems are discussed in Section 15.6.

13.5   Closure

An asymptotic approach to isothermal steady heavily loaded point EHL contacts with incompressible Newtonian lubricants and skewed lubricant entrainment is developed. The behavior of the pressure and gap distributions in the inlet and exit zones along "central" streamlines of the lubricant flow is analyzed in detail. Asymptotic analysis of heavily loaded contacts lubricated by a Newtonian fluid recognizes two different regimes of lubrication: pre- and over-critical lubrication regimes. Overall, the difference between these two regimes is in how rapidly the lubricant viscosity grows with pressure and how much lubricant is available at the inlet of the contact. For pre-critical lubrication regimes the asymptotically valid equations for pressure and gap have been derived. It has been shown that by a simple transformation the derived equations can be reduced to equations identical to the corresponding equations for pre-critical lubrication regimes in the inlet and exit zones of line heavily loaded contacts. The latter equations have been validated through the comparison with the direct numerical solutions of the original EHL equations in Chapter 6. For over-critical lubrication regimes another approach has been used which led to formulas for the lubrication film thickness for starved and fully flooded conditions. The asymptotic approaches to EHL problems offer certain distinct advantages in comparison with the direct numerical solutions of EHL problem such as reduction of the number of problem input parameters and the approach to regularization of numerical solutions (see Chapter 6 and Section 8.2). Each of the analytically derived formulas for the central H_{00} and exit H_0 lubrication film thicknesses contains a coefficient of proportionality which can be found by numerically solving the corresponding asymptotically valid equations in the inlet zone or by just curve fitting the experimentally obtained values of the lubrication film thickness.

13.6 Exercises and Problems

1. What is the difference between the asymptotic approaches used for the analysis of heavily loaded point EHL contacts in the cases straight and skew lubricant entrainment and the structures of the lubricated regions?

2. List and discuss all assumptions which in the inlet and exit zones allowed to reduce the equations for the heavily loaded point EHL contact to the equations of the corresponding heavily loaded line EHL contact. Characterize the validity of these assumptions. What is the size of the zones where the proposed asymptotic approach is not valid?

3. What is the difference between pre- and over-critical lubrication regimes? Determine the ratio of the characteristic sizes of the inlet zone for pre- and over-critical lubrication regimes and relate the value of this ratio to the definition of over-critical lubrication regimes.

4. Why for over-critical lubrication regimes there is a necessity to consider not just one inlet and one exit zones but two inlet and two exit zones of different sizes? Give the description of the mechanisms taking place in each of the zones of a lubricated contact under over-critical lubrication regimes.

References

[1] Thorp, N. and Gohar, R. 1972. Oil Film Thickness and Shape for a Ball Sliding in a Grooved Raceway. *ASME J. Lubrication Technology*, July, pp. 199-210.

[2] Thorp, N. and Gohar, R. 1974. Hydrodynamic Friction in Elliptical and Circular Point Contacts. *J. Mech. Engn. Science, Inst. Mech. E.*, Vol. 16, No. 4, pp. 243-249.

[3] Mostofi, A. and Gohar, R. 1982. Oil Film Thickness and Pressure Distribution in Elastohydrodynamic Point Contacts. *J. Mech. Engrs. Science, Insn. Mech. Engrs.*, Vol. 24, No. 4, pp. 173-182.

[4] Chittenden, R.J., Dowson, D., Dunn, J.F., and Taylor, C.M. 1985. A Theoretical Analysis of the Isothermal Elastohydrodynamic Lubrication of Concentrated Contacts. II. Generantrainment along Either Principal Axis of the Hertzian Ellipse or at Some Intermediate Angle. *Proc. Royal Soc. London. Ser. A. Math. and Phys. Sci.*, Vol. 397, No. 1813, pp. 271-294.

[5] Jalali-Vahid, D., Rahnejat., H., Gohar., R., and Jin, Z.M. 2000. Prediction of Oil Film Thickness and Shape in Elliptical Point Contacts under Combined Rolling and Sliding Motion. *Proc. Instn. Mech. Engrs*, Vol. 214, Part J, pp. 427-437.

[6] Lubrecht, A.A. and Venner, C.H. 1999. Elastohydrodynamic lubrication of rough surfaces. *Proc. Instn. Mech. Engrs. J. Eng. Tribology.* V. 213, No. 5, pp. 397-404

[7] Vorovich, I.I., Aleksandrov, V.M., and Babeshko, V.A. 1974. *Non − classical Mixed Problems of Elasticity*. Moscow: Nauka.

[8] Van Dyke, M. 1964. *Perturbation Methods in Fluid Mechanics*. New York: Academic Press.

[9] Kevorkian, J. and Cole, J.D. 1985. *Perturbation Methods in Applied Mathematics. Applied Mathematics Series*, Vol. 34, New York, Springer-Verlag.

14

Asymptotic Analysis of a Lubricated Heavily Loaded Rolling and Spinning Ball in a Grooved Raceway

14.1 Introduction

The chapter is devoted to the application of the early developed asymptotic approach to solution of the steady isothermal EHL problem for heavily loaded point contacts with spinning and rolling elastic ball in an elastic grooved raceway. It is shown that the whole contact region can be subdivided into three subregions: the central one which is far away from the other two regions occupied by the ends of the horse-shoe shaped pressure/gap distribution (HSSPGD). The central region, in turn, can be subdivided into the Hertzian region and two adjacent boundary layers - the inlet and exit zones. Moreover, in the central region in the inlet and exit zones the EHL problem can be reduced to asymptotically valid equations identical to the ones obtained in the inlet and exit zones of heavily loaded line EHL contacts. These equations can be analyzed and numerically solved based on the stable methods using a specific regularization approach which were developed for lubricated line contacts. Cases of pre-critical and over-critical lubrication regimes are considered. Some special cases of the problem are also considered. The byproduct of this asymptotic analysis is an easy analytical derivation of formulas for the lubrication film thickness for pre-critical and over-critical lubrication regimes. The latter allows for simple analysis of the film thickness as a function of spinning angular speed, angle of the entrained lubricant, and other pertinent contact characteristics.

The purpose of this chapter is to establish a relationship between the EHL problems and their solutions for heavily loaded line and point contacts an case of the presence of spinning. It will be done using the asymptotic methods developed in the preceding chapters. It will be shown that along the "central" lubricant flow streamlines in the contact the EHL problem solution behavior for heavily loaded point contacts are very similar if not identical to the ones of the well understood EHL problems for heavily loaded line contacts.

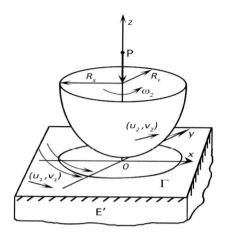

FIGURE 14.1

The general view of a lubricated point contact with spinning.

14.2 Problem Formulation

Let us consider a steady isothermal lubricated contact of an elastic ball rolling and spinning in an elastic parabolic raceway. The ball is spinning in the counterclockwise direction with the angular speed w_2. It means that the solid surface velocities are (u_1, v_1) and $(u_2 - w_2 y, v_2 + w_2 x)$, where (x, y) are the coordinates of a ball surface point introduced the same way as in Chapter 12 (see Fig 14.1). In all other respects the problem formulation is identical to the one described in detail in Chapter 12. Therefore, the equations of the isothermal EHL problem have the form:

$$\frac{\partial}{\partial x}\left(\frac{h^3}{12\mu}\frac{\partial p}{\partial x}\right) + \frac{\partial}{\partial y}\left(\frac{h^3}{12\mu}\frac{\partial p}{\partial y}\right) = \frac{u_1+u_2-w_2 y}{2}\frac{\partial h}{\partial x} + \frac{v_1+v_2+w_2 x}{2}\frac{\partial h}{\partial y},$$

$$p(x, y) \mid_\Gamma = 0, \quad \frac{dp(x,y)}{d\overrightarrow{n}} \mid_{\Gamma_e} = 0,$$

$$h(x, y) = h_0 + \frac{x^2}{2R_x} + \frac{y^2}{2R_y} + \frac{1}{\pi E'}\{\int\int\limits_\Omega \frac{p(x_1,y_1)dx_1 dy_1}{\sqrt{(x-x_1)^2+(y-y_1)^2}} \tag{14.1}$$

$$- \int\int\limits_\Omega \frac{p(x_1,y_1)dx_1 dy_1}{\sqrt{x_1^2+y_1^2}}\}, \quad \int\int\limits_\Omega p(x_1,y_1)dx_1 dy_1 = P,$$

where $p(x, y)$ and $h(x, y)$ are the pressure and gap distributions in the contact, respectively, μ is the lubricant viscosity, $\mu = \mu(p)$, h_0 is the central film thickness at the point with coordinates $(0, 0)$, R_x, R_y, and E' are the effective radii along the x- and y-axes and the effective modulus of elasticity, $\frac{1}{R_x} = \frac{1}{R_{1x}} + \frac{1}{R_{2x}}$, $\frac{1}{R_y} = \frac{1}{R_{1y}} + \frac{1}{R_{2y}}$, and $\frac{1}{E'} = \frac{1-\nu_1^2}{E_1} + \frac{1-\nu_2^2}{E_2}$, Ω and Γ are the contact region and its boundary, respectively, and $\overrightarrow{n}$ is the unit external normal vector to the boundary Γ. The inlet Γ_i and exit Γ_e portions of the contact boundary Γ through which the lubricant enters and leaves the contact, respectively, constitute the whole boundary of Ω, i.e. $\Gamma_i \bigcup \Gamma_e = \Gamma$. The inlet Γ_i and exit Γ_e boundaries are determined by the relationships

$$(x, y) \in \Gamma_i \ if \ \overrightarrow{F_f} \cdot \overrightarrow{n} < 0; \ (x, y) \in \Gamma_e \ if \ \overrightarrow{F_f} \cdot \overrightarrow{n} \geq 0,$$

$$\overrightarrow{F_f} = (u_1 + u_2 - \omega_2 y, v_1 + v_2 + \omega_2 x)\frac{h}{2} - \frac{h^3}{12\mu}(\frac{\partial p}{\partial x}, \frac{\partial p}{\partial y}),$$

$$(14.2)$$

where $\overrightarrow{F_f}$ is the fluid flux through the contact boundary Γ and $\overrightarrow{a} \cdot \overrightarrow{b}$ is the dot product of vectors $\overrightarrow{a}$ and $\overrightarrow{b}$.

It is assumed that the lubricant viscosity μ satisfies the Barus equation (12.4).

The dimensionless variables we will introduce according to (12.5) while the dimensionless parameters will be determined by formulas (13.1) as well as the formula

$$\omega_0 = \frac{\omega_2 b_H}{\sqrt{(u_1+u_2)^2+(v_1+v_2)^2}}, \quad (14.3)$$

where b_H is the Hertzian large semi-axes of the dry elliptical contact (see formulas (12.7)).

Using the dimensionless variables from (12.5) and parameters (13.1) and (14.3) and omitting the primes at the dimensionless variables equations (14.1) and (12.4) assume the form

$$\frac{\partial}{\partial x}(\frac{h^3}{\mu}\frac{\partial p}{\partial x}) + (1 - e^2)\frac{\partial}{\partial y}(\frac{h^3}{\mu}\frac{\partial p}{\partial y})$$

$$= \frac{V}{H_{00}^2}\{(\gamma - \omega_0 y)\frac{\partial h}{\partial x} + \sqrt{1 - e^2}(\sqrt{1 - \gamma^2} + \sqrt{1 - e^2}\omega_0 x)\frac{\partial h}{\partial y}\},$$

$$\mu = \exp(Qp), \ p(x, y) \mid_\Gamma = 0, \ \frac{dp(x,y)}{d\overrightarrow{n}} \mid_{\Gamma_e} = 0,$$

$$(14.4)$$

$$H_{00}[h(x, y) - 1] = x^2 + \frac{\delta}{1-e^2}y^2 + \frac{D(e)}{\pi(1-e^2)^{3/2}}\{\int\int\limits_\Omega \frac{p(x_1,y_1)dx_1 dy_1}{\sqrt{(x-x_1)^2+\frac{(y-y_1)^2}{1-e^2}}}$$

$$- \int\int\limits_\Omega \frac{p(x_1,y_1)dx_1 dy_1}{\sqrt{x_1^2+\frac{y_1^2}{1-e^2}}}\}, \ \int\int\limits_\Omega p(x_1, y_1)dx_1 dy_1 = \frac{2\pi}{3},$$

where e is the dry contact ellipse eccentricity which satisfies the equation (12.8), and parameter γ characterizes the direction of lubricant entrainment.

In particular, $\gamma = 1$ when the lubricant is entrained along the x-axis directed along the small Hertzian semi-axis.

Therefore, for the given inlet boundary Γ_i and constants V, Q, γ, and e we need to find the distributions of pressure $p(x, y)$ and gap $h(x, y)$ as well as the exit boundary Γ_e and the dimensionless central film thickness H_{00}.

The further analysis of the problem will be done in polar variables (ρ, φ) which are introduced as follows

$$x = \frac{1}{\sqrt{1-e^2}}[-\frac{\sqrt{1-\gamma^2}}{\omega_0} + \rho\sin\varphi], \ y = \frac{\gamma}{\omega_0} - \rho\cos\varphi, \ dxdy = \frac{\rho d\rho d\varphi}{\sqrt{1-e^2}}, \quad (14.5)$$

$$\rho = \sqrt{(\sqrt{1-e^2}x + \frac{\sqrt{1-\gamma^2}}{\omega_0})^2 + (y - \frac{\gamma}{\omega_0})^2}, \ \tan\varphi = \frac{\sqrt{1-e^2}x + \frac{\sqrt{1-\gamma^2}}{\omega_0}}{\frac{\gamma}{\omega_0} - y}. \quad (14.6)$$

The reason for such a choice of new independent variables will become clear later when we will be considering the problem solution away from the contact boundaries where with high precision the gap is practically constant along circular arcs $\rho = const$, i.e., $h = h(\rho)$ (see formula (14.13)).

In variables (14.5) equations (14.4) will take the form

$$\frac{H_{00}^2}{V}\{\frac{\partial}{\partial\rho}(\frac{h^3}{\mu}\frac{\partial p}{\partial\rho}) + \frac{1}{\rho}\frac{h^3}{\mu}\frac{\partial p}{\partial\rho} + \frac{1}{\rho^2}\frac{\partial}{\partial\varphi}(\frac{h^3}{\mu}\frac{\partial p}{\partial\varphi})\} = \frac{\omega_0}{\sqrt{1-e^2}}\frac{\partial h}{\partial\varphi}, \ \mu = \exp(Qp),$$

$$p(\rho, \varphi)\ |_\Gamma = 0, \ \frac{dp(\rho,\varphi)}{d\overrightarrow{n}}\ |_{\Gamma_e} = 0, \quad (14.7)$$

$$H_{00}[h(\rho, \varphi) - 1] = x^2(\rho, \varphi) + \frac{\delta}{1-e^2}y^2(\rho, \varphi)$$

$$+ \frac{D(e)}{\pi(1-e^2)}\{\int\int_\Omega \frac{p(\rho_1,\varphi_1)\rho_1 d\rho_1 d\varphi_1}{\sqrt{\rho^2 - 2\rho\rho_1\cos(\varphi-\varphi_1)+\rho_1^2}}$$

$$- \int\int_\Omega \frac{p(\rho_1,\varphi_1)\rho_1 d\rho_1 d\varphi_1}{\sqrt{\rho_1^2 - \frac{2}{\omega_0}\rho_1[\gamma\cos(\varphi_1)+\sqrt{1-\gamma^2}\sin(\varphi_1)]+\frac{1}{\omega_0^2}}}\},$$

$$\int\int_\Omega p(\rho_1, \varphi_1)\rho_1 d\rho_1 d\varphi_1 = \frac{2\pi}{3}\sqrt{1-e^2},$$

where x and y are functions of ρ and φ determined according to (14.5).

14.3   Case of Pure Spinning

Let us consider the case of pure spinning. In this case while introducing dimensionless variables instead of using speed $u_\Sigma = \sqrt{(u_1 + u_2)^2 + (v_1 + v_2)^2}$ which is equal to zero we should use some kind of a characteristic speed $u_\Sigma \neq 0$. Then the dimensionless parameter ω_0 will be equal to $\omega_0 = \frac{\omega_2 b_H}{u_\Sigma}$. For pure spinning conditions $\omega_0 = 0$. Then the Reynolds equation from (14.7) is reduced to

$$\frac{H_{00}^2}{V}\{\frac{\partial}{\partial \rho}(\frac{h^3}{\mu}\frac{\partial p}{\partial \rho}) + \frac{1}{\rho}\frac{h^3}{\mu}\frac{\partial p}{\partial \rho} + \frac{1}{\rho^2}\frac{\partial}{\partial \varphi}(\frac{h^3}{\mu}\frac{\partial p}{\partial \varphi})\} = 0. \tag{14.8}$$

Obviously, equation (14.8) is satisfied when $H_{00} = 0$. That makes the last two equations in (14.7) (or in (14.4)) to be the equations describing a contact problem of elasticity for a dry contact. The solution of such a problem for an indenter of parabolic shape with pressure vanishing at the contact boundary Γ (see the corresponding conditions in (14.7) and (14.4)) is the Hertzian solution $p_{00}(x, y) = \sqrt{1 - x^2 - y^2}$ while the contact region is $\Omega = \{(x, y) : x^2 + y^2 \leq 1\}$. In this case the cavitation boundary condition on pressure $\frac{dp(x,y)}{d\overrightarrow{n}}\mid_{\Gamma_e} = 0$ at the exit boundary Γ_e should be dropped because the lubricant is absent.

Therefore, in case of pure spinning the solution of the problem is the Hertzian solution in a dry contact. It means that for the case of pure spinning no positive lubrication film thickness can be sustained. That is the main reason why the iteration process used in [1] was divergent and the only way to make it convergent was to introduce some lubricant entrainment speed.

14.4   Asymptotic Analysis of a Heavily Loaded Contact under Pre-Critical Lubrication Conditions

Let us consider heavily loaded lubrication conditions with lubricant entrainment and spinning. These conditions can be caused by the presence in the equations of the problem a small or large parameter. From the expressions for V and Q in (12.6) it is clear that heavily loaded conditions can be realized only when V is small and/or Q is large. Therefore, we will assume that system (14.7) contains a small parameter $\omega \ll 1$ which can be equal to $V \ll 1$ and/or $Q^{-1} \ll 1$.

The further analysis of the problem will be done along the lines of the asymptotic procedure presented in the preceding two chapters. In heavily loaded lubricated contacts (i.e., for $\omega \ll 1$) at the points away from the contact boundary Γ the gap distribution $h(\rho, \varphi)$ is very close to being constant along the φ-axis while the pressure distribution $p(\rho, \varphi)$ is close to the pressure distribution $p_0(\rho, \varphi)$ which occurs in dry contacts Ω of a rigid indenter with the radii R_x and R_y and sharp edges with an elastic half-space with the effective elastic modulus E'. Away from the contact boundary Γ this pressure distribution is close to the Hertzian pressure distribution

$$p_{00}(\rho, \varphi) = \sqrt{1 - x^2(\rho, \varphi) - y^2(\rho, \varphi)} \tag{14.9}$$

in a circular contact

$$x^2(\rho, \varphi) + y^2(\rho, \varphi) \leq 1. \tag{14.10}$$

The region where that occurs will be called the Hertzian region.

In heavily loaded lubricated contacts the film thickness H_{00} separating the solids is small as the situation is close to the case of a dry contact where the film thickness is zero. Due to that and the fact that in the Hertzian region, i.e. at the points where the gap $h(\rho, \varphi)$ is almost constant along the circular arcs $\varphi = const$, we have $\partial h / \partial \varphi \ll 1$.

Based on this characterization of a heavily loaded EHL contact and the fact that the film thickness $H_{00} \ll 1$ we will define a contact to be heavily loaded if

$$H_{00} \frac{\partial h(\rho, \varphi)}{\partial \varphi} \ll 1, \ (\rho, \varphi) \in \Omega \setminus \Omega_\epsilon, \tag{14.11}$$

where Ω_ϵ is a narrow ring-like shaped region adjacent to the contact boundary Γ which encompasses the region $\Omega \setminus \Omega_\epsilon$ located in the central (Hertzian) part of the contact. Let the characteristic width of Ω_ϵ be $\epsilon_q = \epsilon_q(\omega) \ll 1$ for $\omega \ll 1$.

Condition (14.11) is equivalent to

$$\frac{H_{00}^3}{V} \left[\frac{\partial}{\partial \rho} \left(\frac{h^3}{\mu} \frac{\partial p}{\partial \rho} \right) + \frac{1}{\rho} \frac{h^3}{\mu} \frac{\partial p}{\partial \rho} + \frac{1}{\rho^2} \frac{\partial}{\partial \varphi} \left(\frac{h^3}{\mu} \frac{\partial p}{\partial \varphi} \right) \right] \ll 1, \ (\rho, \varphi) \in \Omega \setminus \Omega_\epsilon, \tag{14.12}$$

as it follows from condition (14.11) and the Reynolds equation (14.7).

Due to the definitions (14.12) and (14.11) and the fact that in the Hertzian region $\partial h / \partial \varphi \ll 1$ it is clear that in the Hertzian region $\Omega \setminus \Omega_\epsilon$ the gap distribution $h(\rho, \varphi)$ is as follows

$$h(\rho, \varphi) = h_\infty(\rho), \ (\rho, \varphi) \in \Omega \setminus \Omega_\epsilon, \tag{14.13}$$

where $h_\infty(\rho)$ is the exit film thickness for the given ρ, i.e. the film thickness at a point of the exit boundary Γ_e with coordinates (ρ, φ). The latter becomes obvious from the further analysis as the derivatives of p with respect to ρ in the Reynolds equation (14.7) in the exit zone can be neglected (see (14.28)) and the obtained equation can be integrated with respect to φ from φ to $\gamma_e(\rho)$ (which is the φ-coordinate of the exit boundary point) with the boundary condition $\partial p(\rho, \gamma_e(\rho)) / \partial \varphi = 0$ (see (14.19) and (14.28)).

The definitions (14.12) and (14.11) and equations (14.7) indicate that pressure $p_0(\rho, \varphi)$ (generally different from the Hertzian pressure distribution $p_{00}(\rho, \varphi)$ from (14.9) due to the shape of the contact region Ω) satisfies the following equations

$$x^2(\rho, \varphi) + \frac{\delta}{1-e^2} y^2(\rho, \varphi) + \frac{D(e)}{\pi(1-e^2)} \left\{ \int\int_\Omega \frac{p(\rho_1, \varphi_1) \rho_1 d\rho_1 d\varphi_1}{\sqrt{\rho^2 - 2\rho\rho_1 \cos(\varphi - \varphi_1) + \rho_1^2}} \right.$$

$$\left. - \int\int_\Omega \frac{p(\rho_1, \varphi_1) \rho_1 d\rho_1 d\varphi_1}{\sqrt{\rho_1^2 - \frac{2}{\omega_0} \rho_1 [\gamma \cos(\varphi_1) + \sqrt{1-\gamma^2} \sin(\varphi_1)] + \frac{1}{\omega_0^2}}} \right\}, \tag{14.14}$$

$$\int\int_\Omega p(\rho_1, \varphi_1) \rho_1 d\rho_1 d\varphi_1 = \frac{2\pi}{3} \sqrt{1 - e^2}.$$

Physically, function $p_0(\rho, \varphi)$ represents the pressure distribution in a dry fixed contact Ω of a rigid indenter with the radii R_x and R_y and sharp edges with an elastic half-space with the effective elastic modulus E'. In most cases $p_0(\rho, \varphi)$ cannot be determined analytically due to the complexity of the shape of the boundary Γ. Depending on the fixed shape of Γ the solution $p_0(\rho, \varphi)$ of equations (14.14) at some points of Γ may vanish or be singular with the singularity of the type $d^{-1/2}$, where d is the normal distance of a point from the region boundary Γ [2]. Therefore, we can conclude that close to the contact boundary Γ pressure $p_0(\rho, \varphi)$ in the dry contact Ω can be asymptotically represented as follows [2]

$$p_0(\rho, \varphi) = \epsilon_q^{1/2} q_0 + o(\epsilon_q^{1/2}), \quad q_0(\rho, \varphi, \rho_\Gamma, \varphi_\Gamma) = N_0 \sqrt{2d_\epsilon} + \frac{N_1}{\sqrt{2d_\epsilon}},$$

$$N_0 = N_0(\rho_\Gamma, \varphi_\Gamma) = O(1), \quad N_1 = N_1(\rho_\Gamma, \varphi_\Gamma) = O(1),$$

$$d_\epsilon = \frac{\sqrt{[x(\rho,\varphi) - x_\Gamma(\rho,\varphi)]^2 + \frac{[y(\rho,\varphi) - y_\Gamma(\rho,\varphi)]^2}{1 - e^2}}}{\epsilon_q} = O(1), \quad \omega \ll 1,$$

$$(\rho, \varphi) \in \Omega_\epsilon, \quad (\rho_\Gamma, \varphi_\Gamma) \in \Gamma,$$

(14.15)

where the point $(x(\rho, \varphi), y(\rho, \varphi))$ belongs to the line normal to the boundary Γ through the boundary point $(\rho_\Gamma, \varphi_\Gamma)$. The direction of the normal practically coincides with the tangent to a circular arc $r = const$ while the distance between these two points is about ϵ_q.

Now, let us determine the angle φ_H of the intersection of the circular flow streamlines $\rho = const$ with the boundary of the Hertzian region $x^2(\rho, \varphi) + y^2(\rho, \varphi) = 1$. We will obtain the following equation

$$e^2 \rho^2 \cos^2 \varphi_H + \frac{2\gamma(1 - e^2)\rho}{\omega_0} \cos \varphi_H + \frac{2\sqrt{1 - \gamma^2}\rho}{\omega_0} \sin \varphi_H$$

$$+1 - e^2 - \frac{1 - e^2 \gamma^2}{\omega_0^2} - \rho^2 = 0.$$

(14.16)

Obviously, with respect to the solutions of equation (14.16) for given parameters e, γ, ω_0, and variable ρ there are three different situations: (a) the equation has no solutions, (b) the equation has a unique solution when the Hertzian circle is tangent to a lubricant flow streamline ellipse, and (c) the equation has two solutions when the Hertzian circle intersects a lubricant flow streamline ellipse. We are not interested in cases (a) and (b) as they do not represent the "central" flow streamlines we are interested in. We will consider only case (c) when there are two points of intersection. Moreover, we will consider only the "central" lubricant flow streamlines for which the difference between the two solutions of (14.16) is much greater than ϵ_q - the characteristic width of the inlet and exit zones (for more details see below). For $e > 0$ equation (14.16) is generally impossible to solve analytically but that can be easily accomplished numerically. At the same time, this equation can be solved analytically in special cases when $e = 0$ or $\gamma = 1$. In particular, for $e = 0$ (i.e., for truly circular Hertzian contact) we have two solutions

$$\varphi_{H1,2} = \arctan \frac{\sqrt{1-\gamma^2}}{\gamma} \mp \arccos \frac{1+\omega_0^2(\rho^2-1)}{2\omega_0\rho}, \tag{14.17}$$

while for $\gamma = 1$ (i.e., for lubricant entrainment along the x-axis) we have

$$\varphi_{H1,2} = \mp \arccos\{\frac{e^2-1+\sqrt{1-e^2-e^2\omega_0^2(1-e^2-\rho^2)}}{e^2\omega_0\rho}\}. \tag{14.18}$$

Due to the fact that $\gamma > 0$ and $\omega_0 \geq 0$ we will take sign minus in (14.17) and (14.18) to be the angle φ_{H1} on the inlet side and take sign plus for the angle φ_{H2} on the exit side. Further, we will assume that one way or another we can determine the two angles φ_{H1} and φ_{H2}.

Let us make some simple and, at the same time, natural assumptions about the contact region boundary Γ. We will assume that the inlet and exit boundaries can be described by the equations $\varphi = \gamma_i(\rho)$ and $\varphi = \gamma_e(\rho)$ and that the inlet Γ_i and exit Γ_e boundaries of the lubricated contact are close to the boundaries of the dry Hertzian contact, i.e.,

$$\gamma_i(\rho) = \varphi_{H1}(\rho) + \epsilon_q\alpha_{1p}(\rho), \ \alpha_{1p}(\rho) = O(1), \ (\rho, \gamma_i) \in \Gamma_i,$$

$$\gamma_e(\rho) = \varphi_{H2}(\rho) + \epsilon_q\beta_{1p}(\rho), \ \beta_{1p}(\rho) = O(1), \ (\rho, \gamma_e) \in \Gamma_e, \ \omega \ll 1, \tag{14.19}$$

where $\alpha_{1p}(\rho)$ is a given function of ρ while $\beta_{1p}(\rho)$ has to be determined from the problem solution. Moreover, we will assume that $\gamma_i(\rho)$ and $\gamma_e(\rho)$ are differentiable functions and that outside of the ends of HSSPGD zone Ω_{HSSPGD} we have

$$\frac{d\gamma_i(\rho)}{d\rho} = O(1), \ \frac{d\gamma_e(\rho)}{d\rho} = O(1), \ \omega \ll 1. \tag{14.20}$$

These assumptions mean that the variations of the contact region boundary Γ are moderate, i.e. the boundary Γ does not experience abrupt step-type variations.

Relationships (14.19) and (14.20), in turn, lead to the following estimates

$$\frac{dN_0(\rho_\Gamma, \varphi_\Gamma)}{d\overrightarrow{\tau_\Gamma}}, \ \frac{dN_1(\rho_\Gamma, \varphi_\Gamma)}{d\overrightarrow{\tau_\Gamma}} = O(1), \ \omega \ll 1, \tag{14.21}$$

where $\overrightarrow{\tau_\Gamma}$ is a unit vector tangent to the contact boundary Γ at the point with coordinates $(\rho_\Gamma, \varphi_\Gamma) \in \Gamma$.

Using (14.14) the last two equations in (14.7) can be rewritten in the form

$$H_{00}[h(\rho, \varphi) - 1] = \frac{D(e)}{\pi(1-e^2)} \int\int_\Omega [p(\rho_1, \varphi_1) - p_0(\rho_1, \varphi_1)]$$

$$\times \{ \frac{1}{\sqrt{\rho^2 - 2\rho\rho_1\cos(\varphi-\varphi_1)+\rho_1^2}}$$

$$- \frac{1}{\sqrt{\rho_1^2 - \frac{2\rho_1}{\omega_0}[\gamma\cos\varphi_1 + \sqrt{1-\gamma^2}\sin\varphi_1] + \frac{1}{\omega_0^2}}} \}\rho_1 d\rho_1 d\varphi_1, \tag{14.22}$$

$$\int\int_\Omega [p(\rho_1, \varphi_1) - p_0(\rho_1, \varphi_1)]\rho_1 d\rho_1 d\varphi_1 = 0.$$

The proposed below asymptotic approach is valid only for the central region occupied by the "central" flow streamlines which are away from the ends of HSSPGD zone Ω_{HSSPGD}. Also, this approach is valid for $0 \le \gamma \le 1$. In other words, we are proposing an asymptotic analysis of the central part of the lubricated contact which includes the corresponding inlet and exit zones which are occupied by "central" lubricant streamlines.

Now, we are ready for the asymptotic analysis of the EHL problem. It is important to remember that in the region of the contact (inside of the Hertzian contact) which is at a distance from the boundary of the Hertzian contact of much more than $O(\epsilon_q) \ll 1$ the pressure distribution $p(\rho, \varphi)$ is very close to the Hertzian pressure $p_{00}(\rho, \varphi)$. At the same time, in the inlet and exit zones the actual pressure $p(\rho, \varphi)$ is significantly different from the Hertzian pressure $p_{00}(\rho, \varphi)$ as well as from pressure $p_0(\rho, \varphi)$. This situation is typical for application of the method of matched asymptotic expansions [3, 4].

First, we will consider the pre-critical heavily loaded lubrication regimes for which the lubricant viscosity varies with pressure moderately. The very basic fact about the matched asymptotic expansions approach is that the solution of the problem $p(\rho, \varphi)$ in the inlet and exit zones of width $O(\epsilon_q) \ll 1$ is of the order of magnitude of $\epsilon_q^{1/2}$ because otherwise it will not be able to match the external (dry) pressure $p_0(\rho, \varphi)$ outside of the inlet and exit zones (see (14.15)). Therefore, a heavily loaded lubrication regime is pre-critical if in the inlet zone of the contact Ω_i where $p(\rho, \varphi) = O(\epsilon_q^{1/2})$, $\epsilon_q \ll 1$, the following estimate holds

$$\mu(p(\rho, \varphi)) = O(1), \; if \; p(\rho, \varphi) = O(\epsilon_q^{1/2}), \; (\rho, \varphi) \in \Omega_\epsilon, \; \omega \ll 1. \quad (14.23)$$

It is important to realize that according to (14.15) $d_\epsilon = \epsilon_q^{-1}\{(x(\rho, \varphi) - x_\Gamma(\rho, \varphi))^2 + \frac{(y(\rho, \varphi) - y_\Gamma(\rho, \varphi))^2}{1 - e^2}\}^{1/2} = O(1)$, $\omega \ll 1$, as long as $x(\rho, \varphi) - x_\Gamma(\rho, \varphi) = O(\epsilon_q)$ and $y(\rho, \varphi) - y_\Gamma(\rho, \varphi) = O(\epsilon_q)$ for $\omega \ll 1$. Therefore, $p_0(\rho, \varphi) = O(\epsilon_q^{1/2})$ within zone Ω_ϵ. In order for the pressure function $p(\rho, \varphi)$ to match the function $p_0(\rho, \varphi)$ at the inner boundary of the region Ω_ϵ and the outer boundary of the region $\Omega \setminus \Omega_\epsilon$ it is necessary that they are of the same order of magnitude [3, 4]. Therefore, we have

$$p(\rho, \varphi) = O(\epsilon_q^{1/2}), \; (\rho, \varphi) \in \Omega_\epsilon, \; \omega \ll 1. \quad (14.24)$$

Based on (14.23) and (14.24) the definition of the pre-critical regimes can be presented in the form

$$\mu(\epsilon_q^{1/2}) = O(1), \; \omega \ll 1. \quad (14.25)$$

Let us consider the points in the inlet Ω_i and exit Ω_e zones but outside of the ends of HSSPGD zone Ω_{HSSPGD} which belong to the line $\rho = \rho_*$. First, let us consider the inlet zone. We will replace our original coordinate system (ρ, φ) by a new coordinate system (ρ, r_p), where $r_p = \rho_* \frac{\varphi - \gamma_i(\rho_*)}{\epsilon_q}$. Using (14.15)-(14.21), (14.24) and the principle of matching [3, 4] of asymptotic expansions of the problem solutions in the inlet zone $\Omega_i \bigcap \Omega_e$ and the Hertzian region we obtain that

$$p(\rho_*, \varphi) = \epsilon_q^{1/2} q_p(\rho_*, r_p) + o(\epsilon_q^{1/2}), \quad q_p(\rho_*, r_p) = O(1),$$

$$r_p = \rho_* \frac{\varphi - \gamma_i(\rho_*)}{\epsilon_q} = O(1), \quad (\rho_*, \varphi) \in \Omega_i \bigcap \Omega_e, \quad \omega \ll 1, \tag{14.26}$$

where $q_p(\rho_*, r_p)$ is the unknown function which is the main term of the asymptotic expansion of pressure in the inlet zone. In a similar manner we can easily establish that

$$p(\rho_*, \varphi) = \epsilon_q^{1/2} g_p(\rho_*, s_p) + o(\epsilon_q^{1/2}), \quad g_p(\rho_*, s_p) = O(1),$$

$$s_p = \rho_* \frac{\varphi - \gamma_e(\rho_*)}{\epsilon_q} = O(1), \quad (\rho_*, \varphi) \in \Omega_e \bigcap \Omega_e, \quad \omega \ll 1, \tag{14.27}$$

where $g_p(\rho_*, s_p)$ is the unknown function which is the main term of the asymptotic expansion of pressure in the exit zone. Functions $q_p(\rho_*, r_p)$ and $g_p(\rho_*, s_p)$ further for simplicity will be called asymptotic pressure distributions in the inlet and exit zones.

Based on (14.20) and (14.21) in the inlet and exit zones outside of the HSSPGD zone Ω_{HSSPGD} we obtain

$$\frac{\partial}{\partial \varphi} = \rho_* \epsilon_q^{-1} \frac{\partial}{\partial r_p}, \quad r_p = O(1); \quad \frac{\partial}{\partial \varphi} = \rho_* \epsilon_q^{-1} \frac{\partial}{\partial s_p}, \quad s_p = O(1), \quad \omega \ll 1,$$

$$\frac{\partial}{\partial \rho} = O(1), \quad \omega \ll 1. \tag{14.28}$$

Using (14.20) we can determine that the external normal vector $\vec{n} = (n_\rho, n_\varphi)$ has a nonzero ρ-component $n_\rho = O(1)$, $\omega \ll 1$. Therefore, based on (14.27) and (14.28) we can conclude that the condition $\frac{dp(\rho, \varphi)}{d\vec{n}} \mid_{\Gamma_e} = 0$ at the exit boundary Γ_e (see (14.7)) can be represented in the form $\{\epsilon_q^{1/2} \frac{\partial g_p}{\partial \rho} n_\rho + \epsilon_q^{-1/2} \frac{\partial g_p}{\partial s_p} n_\varphi + \ldots\} \mid_{\Gamma_e} = 0$, $\omega \ll 1$. Obviously, for $\omega \ll 1$ the latter condition leads to the following one $\frac{\partial g_p}{\partial s_p} \mid_{s_p=0} = 0$.

Now, let us consider the asymptotic representations for $\frac{\partial h}{\partial \varphi}$ in the inlet and exit zones. From the first equation in (14.22) we have

$$H_{00} \frac{\partial h}{\partial \varphi} = -\frac{D(e)}{\pi(1-e^2)} \rho \int \int\limits_{\Omega} \frac{\rho_1^2 \sin(\varphi - \varphi_1)[p(\rho_1, \varphi_1) - p_0(\rho_1, \varphi_1)] d\rho_1 d\varphi_1}{[\rho^2 - 2\rho\rho_1 \cos(\varphi - \varphi_1) + \rho_1^2]^{3/2}}. \tag{14.29}$$

We will consider the inlet and exit zones (away from the ends of HSSPGD zone Ω_{HSSPGD}) fully extended along the φ-axis (i.e., along the line $\rho = \rho_*$) (see (14.20)) and with characteristic sizes along the ρ-axis of about ϵ_q. Due to (14.15)-(14.21) within these zones

$$p(\rho, \varphi) = p(\rho_*, \varphi) + O(\epsilon_q),$$

$$p(\rho, \varphi) - p_0(\rho, \varphi) = p(\rho_*, \varphi) - p_0(\rho_*, \varphi) + O(\epsilon_q), \tag{14.30}$$

$$(\rho, \varphi) \in (\Omega_i \bigcap \Omega_e) \bigcup (\Omega_e \bigcap \Omega_e), \quad \omega \ll 1.$$

Taking into account definition (14.11) of a heavily loaded EHL contact and related to it assumption that away from the boundary of the Hertzian contact

$$p(\rho, \varphi) - p_0(\rho, \varphi) = O(\epsilon_q^2), \ (\rho, \varphi) \in \Omega \setminus \Omega_\epsilon, \ \omega \ll 1, \tag{14.31}$$

the representations (14.26) and (14.27), as well as the estimate (14.30) and using the local coordinates r_p, s_p, and $\eta = \rho_1 - \rho_*$ (where $\epsilon_q \ll \eta \ll 1$ we can estimate that if the point (ρ, φ) belongs to the inlet or exit zones (away from the ends of the HSSPGD zone Ω_{HSSPGD}) then the contribution of the integral of $\frac{\rho_1^2 \sin(\varphi-\varphi_1)[p(\rho_1,\varphi_1)-p_0(\rho_1,\varphi_1)]}{[\rho_1^2-2\rho\rho_1\cos(\varphi-\varphi_1)+\rho_1^2]^{3/2}}$ (see equation (14.29)) over the entire contact region Ω excluding the considered inlet or exit zone will be of the order of magnitude $O(\epsilon_q^2)$, $\omega \ll 1$, while the contribution of the integral of the above function over the inlet or exit zone will be of the order of $O(\epsilon_q^{1/2})$, $\omega \ll 1$. Moreover, we have $\rho_1^2 d\rho_1 = \rho_*^2 d\eta + O(\rho_*\eta d\eta)$ and $\rho_*^2 - 2\rho_*\rho_1\cos(\varphi-\varphi_1) + \rho_1^2 = (\eta + 2\rho_* \sin^2 \frac{\varphi-\varphi_1}{2})^2 + \rho_*^2 \sin^2(\varphi-\varphi_1)$. Therefore, we can conclude that in the inlet zone

$$H_{00} \frac{\partial h(\rho_*,\varphi)}{\partial \varphi} = -\frac{1}{\pi} \frac{D(e)}{1-e^2} \rho_*^3 \int\limits_\Omega \int [p(\rho_*, \varphi_1) - p_0(\rho_*, \varphi_1)]$$

$$\times \frac{\sin(\varphi-\varphi_1)}{[(\eta+2\rho_* \sin^2 \frac{\varphi-\varphi_1}{2})^2+\rho_*^2 \sin^2(\varphi-\varphi_1)]^{3/2}} d\eta d\varphi_1 + O(\eta) \tag{14.32}$$

$$= -\frac{1}{\pi} \frac{D(e)}{1-e^2} \rho_*^3 \int\limits_{\gamma_i}^{\gamma_e} \sin(\varphi - \varphi_1)[p(\rho_*, \varphi_1) - p_0(\rho_*, \varphi_1)] d\varphi_1$$

$$\times \int\limits_{-\infty}^{\infty} \frac{d\eta}{[(\eta+2\rho_* \sin^2 \frac{\varphi-\varphi_1}{2})^2+\rho_*^2 \sin^2(\varphi-\varphi_1)]^{3/2}} + O(\eta), \ (\rho_*, \varphi) \in \Omega_i \bigcap \Omega_\epsilon.$$

Using the fact that $\int\limits_{-\infty}^{\infty} \frac{d\xi}{(\xi^2+a^2)^{3/2}} = \frac{2}{a^2}$ the last integral in (14.32) is equal to $\frac{2}{\rho_*^2 \sin^2(\varphi-\varphi_1)}$. Therefore, substituting the value of this integral and (14.26) in (14.32) and taking into account that in the inlet zone $\varphi - \varphi_1 = O(\epsilon_q) \ll 1$ we obtain that in the inlet zone

$$H_{00} \frac{\partial h}{\partial \varphi} = \epsilon_q^{1/2} \frac{2}{\pi} \frac{D(e)}{1-e^2} \rho_* \int\limits_0^{\infty} \frac{q_p(\rho_*,t)-q_{ap}(\rho_*,t)}{t-r_p} dt + \dots, \tag{14.33}$$

$$(\rho_*, \varphi) \in \Omega_i \bigcap \Omega_\epsilon,$$

where $q_0(\rho, \varphi, \rho_\Gamma, \varphi_\Gamma)$ is replaced by $q_{ap}(\rho_*, r_p) = q_0(\gamma_i(\rho_*, \rho_*) + r_p\epsilon_q, \rho_*, \gamma_i(\rho_*)) = N_{0q}(\rho_*, \varphi_{H1})\sqrt{2r_p} + \frac{N_{1q}(\rho_*,\varphi_{H1})}{\sqrt{2r_p}}$ due to the fact that the direction of the normal vector $\vec{n}$ to the contact boundary Γ_i in $\Omega_i \bigcap \Omega_\epsilon$ practically coincides with the ρ-axis (see (14.20)). In a similar fashion, in the exit zone $\Omega_e \bigcap \Omega_\epsilon$ we get

$$H_{00} \frac{\partial h}{\partial \varphi} = \epsilon_q^{1/2} \frac{2}{\pi} \frac{D(e)}{1-e^2} \rho_* \int_{-\infty}^{0} \frac{g_p(\rho_*,t) - g_{ap}(\rho_*,t)}{t - s_p} dt + \ldots,$$

(14.34)

$$(\rho_*, \varphi) \in \Omega_e \bigcap \Omega_\epsilon,$$

where $g_{ap}(\rho_*, s_p) = q_0(\rho_*, \gamma_e(\rho_*) + s_p \epsilon_q, \rho_*, \gamma_e(\rho_*)) = N_{0g}(\rho_*, \varphi_{H2}) \sqrt{-2s_p} + \frac{N_{1g}(\rho_*, \varphi_{H2})}{\sqrt{2s_p}} + o(1)$, $\epsilon_q \ll 1$.

It is important to realize that function $N_0(\rho_\Gamma, \varphi_\Gamma)$ depends only on the behavior of function $p_{00}(\rho, \varphi)$ from (14.9) and it is independent from the shape of the contact boundary Γ which is close to $x^2(\rho, \varphi) + y^2(\rho, \varphi) = 1$ (see (14.19)). It means that at the border between the inlet zone and the Hertzian region, i.e., as $r_p \to \infty$ while $\varphi - \gamma_i(\rho) = \varphi - \varphi_{H1}(\rho) - \epsilon_q \alpha_{1p} = O(\eta)$ (where $\epsilon_q \ll \eta \ll$ 1) the value of $\epsilon_q^{1/2} N_0 \sqrt{2r_p} = \sqrt{1 - x^2(\rho, \varphi) - y^2(\rho, \varphi)} + o(1)$. Substituting in the latter equation $r_p = \rho \frac{\varphi - \varphi_{H1}(\rho) - \epsilon_q \alpha_{1p}}{\epsilon_q}$ and using the equations for x and y from (14.5) as well as the equation (14.16) for φ_{H1} we obtain

$$N_0 \sqrt{2\rho(\varphi - \varphi_{H1} - \epsilon_q \alpha_{1p})} = \{ \frac{1}{1-e^2} [(-\frac{\sqrt{1-\gamma^2}}{\omega_0} + \rho \sin \varphi_{H1})^2$$

$$-(-\frac{\sqrt{1-\gamma^2}}{\omega_0} + \rho \sin \varphi)^2] + (\frac{\gamma}{\omega_0} - \rho \cos \varphi_{H1})^2 - (\frac{\gamma}{\omega_0} - \rho \cos \varphi)^2 \}^{1/2}$$

(14.35)

$$+ o(1), \ \eta \ll 1.$$

After using trigonometric identities in (14.35) for $\varphi - \varphi_{H1} = O(\eta)$ (where $\epsilon_q \ll \eta \ll 1$) we obtain

$$N_0(\rho, \varphi_{H1}) \sqrt{2\rho(\varphi - \varphi_{H1})} = \{2\rho(\varphi - \varphi_{H1})[\frac{\sqrt{1-\gamma^2}}{(1-e^2)\omega_0} \cos \varphi_{H1}$$

(14.36)

$$- \frac{\gamma}{\omega_0} \sin \varphi_{H1} - \frac{e^2 \rho}{2(1-e^2)} \sin(2\varphi_{H1})]\}^{1/2} + o(1), \ \eta \ll 1.$$

A similar analysis can be performed on the exit side of the contact. Therefore, equation (14.36) leads to the conclusion that

$$N_0(\rho, \varphi_H) = | \frac{\sqrt{1-\gamma^2}}{(1-e^2)\omega_0} \cos \varphi_H - \frac{\gamma}{\omega_0} \sin \varphi_H - \frac{e^2 \rho}{2(1-e^2)} \sin(2\varphi_H) |^{1/2}, \quad (14.37)$$

where the values of φ_H are determined by the solutions of equation (14.16) (see also (14.17) and (14.18)). Therefore, in the inlet and exit zones we get

$$N_{0q} = | \frac{\sqrt{1-\gamma^2}}{(1-e^2)\omega_0} \cos \varphi_{H1} - \frac{\gamma}{\omega_0} \sin \varphi_{H1} - \frac{e^2 \rho}{2(1-e^2)} \sin(2\varphi_{H1}) |^{1/2},$$

(14.38)

$$N_{0g} = | \frac{\sqrt{1-\gamma^2}}{(1-e^2)\omega_0} \cos \varphi_{H2} - \frac{\gamma}{\omega_0} \sin \varphi_{H2} - \frac{e^2 \rho}{2(1-e^2)} \sin(2\varphi_{H2}) |^{1/2}.$$

Obviously, for $\gamma = 1$ and $e = 0$ from (14.17) (or (14.18)) and (14.37) we obtain that

$$N_0(\rho, \varphi_H) = \frac{1}{\omega_0^{1/2}} \{1 - \frac{[1-\omega_0^2(1-\rho^2)]^2}{4\omega_0^2\rho^2}\}^{1/4}. \tag{14.39}$$

Taking into account the fact that in this case $\omega_0^2\rho^2 = 1 - 2\omega_0 y + \omega_0^2(x^2+y^2)$ and $x^2 + y^2 = O(1)$ for $\omega_0 \to 0$ we obtain $N_0(\rho, \varphi_H) \to \sqrt[4]{1 - y^2} + O(\omega_0^{1/2})$ which coincides with the results in the case of no spin in Chapter 12 (see formula (12.34)). For other cases the expressions for N_0 are more cumbersome.

On the other hand, the function $N_1(\rho_\Gamma, \varphi_\Gamma)$ depends on the distance of the contact boundary Γ from the boundary of the Hertzian contact $x^2 + y^2 = 1$, i.e., functions N_{1q} and N_{1g} are proportional to α_{1p} and β_{1p} (see (14.19)), respectively. In more detail this is described later (see (14.58)).

The above estimates for the derivatives of p and h are valid for an arbitrary $\rho = \rho_*$ as long as the point (ρ, φ) belongs to the inlet or exit zones outside of the ends of HSSPGD zone Ω_{HSSPGD}. For convenience, let introduce two new dimensionless parameters

$$V_e = \frac{V}{\sqrt{1-e^2}}, \ \theta = \frac{D(e)}{1-e^2}. \tag{14.40}$$

Then, using the estimates for $H_{00}\frac{\partial h}{\partial \varphi}$ from (14.33), (14.34), and (14.28) in the inlet $\Omega_i \bigcap \Omega_\epsilon$ and exit $\Omega_e \bigcap \Omega_\epsilon$ zones (i.e., for variables $r_p = \rho\frac{\varphi - \varphi_{H1}(\rho) - \epsilon_q\alpha_{1p}}{\epsilon_q} = O(1)$ and $s_p = \rho_*\frac{\varphi - \gamma_e(\rho_*)}{\epsilon_q} = O(1)$, respectively) from the requirement that the differential and integral terms in the Reynolds equation in (14.7) are of the same order of magnitude for $\omega \ll 1$ we obtain

$$A_p^3\frac{\partial}{\partial r_p}(\frac{h_{qp}^3}{\mu_{qp}}\frac{\partial q_p}{\partial r_p}) = \frac{2}{\pi}\theta\omega_0\rho\int\limits_0^\infty \frac{q_p(\rho,t)-q_{ap}(\rho,t)}{t-r_p}dt, \ q_p(\rho,0) = 0, \tag{14.41}$$

$$A_p^3\frac{\partial}{\partial s_p}(\frac{h_{gp}^3}{\mu_{gp}}\frac{\partial g_p}{\partial s_p}) = \frac{2}{\pi}\theta\omega_0\rho\int\limits_{-\infty}^0 \frac{g_p(\rho,t)-g_{ap}(\rho,t)}{t-s_p}dt, \tag{14.42}$$

$$g_p(\rho,0) = \frac{\partial g_p(\rho,0)}{\partial s_p} = 0,$$

where A_p is a dimensionless coefficient of proportionality in the formula for the film thickness

$$H_{00} = A_p(V_e\epsilon_q^2)^{1/3}, \ A_p = O(1), \ \omega \ll 1, \tag{14.43}$$

where $h_{qp}(\rho, r_p)$ and $h_{gp}(\rho, s_p)$ are the gap functions in the inlet $\Omega_i \bigcap \Omega_\epsilon$ and exit $\Omega_e \bigcap \Omega_\epsilon$ zones. In (14.41) and (14.42) functions μ_{qp} and μ_{gp} are the main terms of the asymptotic expansions of $\mu(\epsilon_q^{1/2}q_p)$ and $\mu(\epsilon_q^{1/2}g_p)$, respectively.

The expressions for $h_{qp}(\rho, r_p)$ and $h_{gp}(\rho, s_p)$ depend on the lubrication regime. For starved lubrication regimes the gap distribution $h(\rho, \varphi)$ is very close to $h_\infty(\rho)$ in the entire contact region. That corresponds to the case of $\epsilon_q^{3/2} \ll H_{00}$ and

$$h_{qp}(\rho, r_p) = h_\infty(\rho), \ \epsilon_q \ll \epsilon_f, \tag{14.44}$$

$$h_{gp}(\rho, s_p) = h_\infty(\rho), \quad \epsilon_q \ll \epsilon_f, \tag{14.45}$$

$$\epsilon_f = V_e^{2/5}, \tag{14.46}$$

while for fully flooded lubrication regimes $H_{00} = O(\epsilon_q^{3/2})$, $\omega \ll 1$, and in the inlet and exit zones the gap satisfies the equations (see (14.33) and (14.34))

$$A_p \frac{\partial h_{qp}}{\partial r_p} = \frac{2}{\pi} \theta \int_0^\infty \frac{q_p(\rho,t) - q_{ap}(\rho,t)}{t - r_p} dt, \quad \epsilon_q = \epsilon_f, \ \omega \ll 1, \tag{14.47}$$

$$A_p \frac{\partial h_{gp}}{\partial s_p} = \frac{2}{\pi} \theta \int_{-\infty}^0 \frac{g_p(\rho,t) - g_{ap}(\rho,t)}{t - s_p} dt, \quad \epsilon_q = \epsilon_f, \ \omega \ll 1. \tag{14.48}$$

Here it is important to realize that arc $\rho = \frac{1}{\omega_0}$ passes through the origin $(x, y) = (0, 0)$. Therefore, $h_\infty(\frac{1}{\omega_0}) = 1$ while, generally, $h_\infty(\rho) \neq 1$ for $\rho \neq \frac{1}{\omega_0}$.

It is obvious from (14.43) that for fully flooded pre-critical lubrication regimes (characterized by the relationship $H_{00} = O(\epsilon_q^{3/2})$, $\omega \ll 1$) we have

$$H_{00} = A_p V_e^{3/5}, \quad \epsilon_q = \epsilon_f = V_e^{2/5}. \tag{14.49}$$

Here and in (14.43) the value of the coefficient A_p depends on θ, $\alpha_{1p}(\frac{1}{\omega_0})$, and on $Q\epsilon_q^{1/2}$ for starved lubrication regimes and on $QV_e^{1/5}$ for fully flooded lubrication regimes, respectively.

It is clear from equations (14.47) and (14.48) that when $r_p \to \infty$ and $s_p \to -\infty$ we get

$$\int_0^\infty [q_p(\rho, t) - q_{ap}(\rho, t)] dt = 0, \tag{14.50}$$

$$\int_{-\infty}^0 [g_p(\rho, t) - g_{ap}(\rho, t)] dt = 0, \tag{14.51}$$

because otherwise the functions $h_{qp}(\rho, r_p)$ and $h_{gp}(\rho, s_p)$ would be unbounded as $r_p \to \infty$ and $s_p \to -\infty$, respectively. Equations (14.50) and (14.51) are consistent with the second equation in (14.22).

For fully flooded lubrication regimes the values of functions $h_{qp}(\rho, r_p,)$ and $h_{gp}(\rho, s_p)$ are obtained by integrating equations (14.47) and (14.48) with the conditions $h_{qp}(\rho, r_p) \to h_\infty(\rho)$ as $r_p \to \infty$ and $h_{gp}(\rho, s_p) \to h_\infty(\rho)$ as $s_p \to -\infty$ as follows

$$A_p[h_{qp}(\rho, r_p) - h_\infty(\rho)] = \frac{2}{\pi} \theta \int_0^\infty [q_p(\rho, t) - q_{ap}(\rho, t)] \ln \frac{1}{|t - r_p|} dt, \tag{14.52}$$

$$A_p[h_{gp}(\rho, s_p) - h_\infty(\rho)] = \frac{2}{\pi} \theta \int_{-\infty}^0 [g_p(\rho, t) - g_{ap}(\rho, t)] \ln \frac{1}{|t - s_p|} dt. \tag{14.53}$$

Here it is appropriate to discuss the applicability of the proposed analysis and some solution properties with respect to the parameter ω_0. For small and moderate values of ω_0 there is a pretty wide range of parameters e and γ for which we can find "central" lubricant flow streamlines and our asymptotic analysis is valid. At the same time, for $\omega_0 \gg 1$ the proposed method is applicable only for $\sqrt{1 - e^2} < \rho < 1$ because the center of the fluid flow streamline ellipses is very close to the origin $(x, y) = (0, 0)$ and the above inequality provides the condition for the existence of the "central" streamlines. For $\omega_0 \to 0$ the value of $\omega_0 \rho$ remains finite, i.e., it does not vanish. Instead, $\omega_0 \rho \to 1$ as $\omega_0 \to 0$ (see formulas (14.6)) which leads to the conclusion that in this limiting case the film thickness H_{00} approaches the one for the case of no spin which was considered in the preceding chapter.

Therefore, in the inlet zone for the given values of ω_0, θ, $Q\epsilon_q^{1/2}$ (which is involved in μ_{qp}), $\alpha_{1p}(\rho)$, and ρ we derived the system of equations (14.41), (14.50), and (14.44) for starved lubrication regimes or (14.52) for fully flooded lubrication regimes for functions $q_p(\rho, r_p)$, $h_{qp}(\rho, r_p)$, $h_\infty(\rho)$, and constant A_p. In the exit zone for the given values of ω_0, θ, $Q\epsilon_q^{1/2}$ (which is involved in μ_{gp}), A_p, $h_\infty(\rho)$, and ρ we need to solve the system of equations (14.42), (14.51), and (14.45) for starved lubrication regimes or (14.53) for fully flooded lubrication regimes for functions $g_p(\rho, s_p)$, $h_{gp}(\rho, s_p)$, and $\beta_{1p}(\rho)$. Obviously, systems of equations (14.41), (14.42), (14.44)-(14.53) are practically identical to the ones derived in the preceding chapter for the case of no spin, $\omega_0 = 0$. At the same time, the formulas (14.43) and (14.49) for the film thickness H_{00} have the structure similar to the corresponding formulas for H_{00} from the preceding two chapters. It is important to keep in mind that constant A_p is the coefficient of proportionality in formula (14.43) for the central film thickness H_{00} and it is determined for $y = 0$ or $\rho = \frac{1}{\omega_0}$ (see (14.6)). Also, it means that $h_\infty(\frac{1}{\omega_0}) = 1$ while, generally, $h_\infty(\rho) \neq 1$ for $\rho \neq \frac{1}{\omega_0}$. The solution procedure should be implemented according to the following sequence: first, the system of asymptotically valid equations in the inlet zone should be solved for $\rho = \frac{1}{\omega_0}$ which determines the value of constant A_p; after that, the values of function $h_\infty(\rho)$ for $\rho \neq \frac{1}{\omega_0}$ are determined from the solution of the asymptotically valid system of equations in the inlet zone and these values play the role similar to the one played by constant A_p for $\rho = \frac{1}{\omega_0}$.

Let us show that these systems of equations can be reduced to systems identical to the ones derived in the inlet and exit zones of heavily loaded line EHL contacts in Chapter 6. Namely, by introducing the substitutions

$$\{r, \alpha_1\} = \tfrac{1}{r_{0q}}\{r_p, \alpha_{1p}\}, \quad \{s, \beta_1\} = \tfrac{1}{r_{0g}}\{s_p, \beta_{1p}\},$$

$$q = \frac{q_p}{N_0 r_{0q}^{1/2}}, \quad g = \frac{g_p}{N_0 r_{0g}^{1/2}}, \tag{14.54}$$

$$\{h_q, h_g\} = \tfrac{1}{h_\infty}\{h_{qp}, h_{gp}\}, \quad \{\mu_q(q), \mu_g(g)\} = \{\mu_{qp}(q_p), \mu_{gp}(g_p)\},$$

and the definitions

$$A_q = A_p \big[\tfrac{\theta N_0^4(\rho,\varphi_{H1})}{\omega_0^3 \rho^3}\big]^{1/5} h_\infty(\rho), \ \ A_g = A_p \big[\tfrac{\theta N_0^4(\rho,\varphi_{H2})}{\omega_0^3 \rho^3}\big]^{1/5} h_\infty(\rho),$$

$$H_0(\rho) = H_{00} h_\infty(\rho), \ \ r_{0q} = \big[\tfrac{\omega_0 \rho}{\theta^2 N_0^3(\rho,\varphi_{H1})}\big]^{2/5}, \ \ r_{0g} = \big[\tfrac{\omega_0 \rho}{\theta^2 N_0^3(\rho,\varphi_{H2})}\big]^{2/5},$$

(14.55)

(where θ is determined by (14.40) and A_q and A_g are constants) we will be able to reduce each of the above systems of asymptotically valid equations to

$$A_q^3 \frac{\partial}{\partial r}\big(\frac{h_q^3}{\mu_q}\frac{\partial q}{\partial r}\big) = \frac{2}{\pi}\int\limits_0^\infty \frac{q(\rho,t)-q_a(\rho,t)}{t-r}dt, \ \ q(\rho,0)=0,$$

$$\int\limits_0^\infty [q(\rho,t)-q_a(\rho,t)]dt = 0,$$

(14.56)

$$q(\rho,r) \to q_a(\rho,r) = \sqrt{2r} + \frac{\alpha_1(\rho)}{\sqrt{2r}}, \ \ r \to \infty,$$

$$h_q(\rho,r) = 1 \ \text{or} \ A_q[h_q(\rho,r)-1] = \frac{2}{\pi}\int\limits_0^\infty [q(\rho,t)-q_a(\rho,t)]\ln\frac{1}{|t-r|}dt,$$

in the inlet zone and to

$$A_g^3 \frac{\partial}{\partial s}\big(\frac{h_g^3}{\mu_g}\frac{\partial g}{\partial s}\big) = \frac{2}{\pi}\int\limits_{-\infty}^0 \frac{g(\rho,t)-g_a(\rho,t)}{t-s}dt, \ \ g(\rho,0)=0, \ \frac{\partial g(\rho,0)}{\partial s}=0,$$

$$\int\limits_{-\infty}^0 [g(\rho,t)-g_a(\rho,t)]dt = 0,$$

(14.57)

$$g(\rho,s) \to g_a(\rho,s) = \sqrt{-2s} - \frac{\beta_1(\rho)}{\sqrt{-2s}}, \ \ s \to -\infty,$$

$$h_g(\rho,s) = 1 \ \text{or} \ A_g[h_g(\rho,s)-1] = \frac{2}{\pi}\int\limits_{-\infty}^0 [g(\rho,t)-g_a(\rho,t)]\ln\frac{1}{|t-s|}dt,$$

in the exit zone. In (14.55) $H_0(\rho)$ is the dimensionless analog of the dimensional film thickness h_0 but at the exit from the contact region as well as in the Hertzian region at the points with the fixed coordinate ρ. In other words, we can expect that there are some variations in the film thickness in the Hertzian region with respect to ρ. Besides that, we used the fact that

$$\alpha_{1p} = \frac{N_{1q}}{N_{0q}}, \ \ \beta_{1p} = -\frac{N_{1g}}{N_{0g}}.$$

(14.58)

For fully flooded lubrication regimes by integrating equations (14.56) and (14.57) for q and g the gap functions h_q and h_g from (14.56) and (14.57) can be replaced by equations

$$h_q = 1 + A_q^2 \frac{\partial}{\partial r}\big(\frac{h_q^3}{\mu_q}\frac{\partial q}{\partial r}\big), \ \ h_g = 1 - A_g^3 \frac{\partial}{\partial s}\big(\frac{h_g^3}{\mu_g}\frac{\partial g}{\partial s}\big).$$

(14.59)

We have to keep in mind that due to a particular shape of the contact region boundary Γ the inlet coordinate α_1 is generally a function of ρ which makes the exit coordinate β_1 also a certain function of ρ. In addition to that, in the inlet and exit zones the functions of the lubricant viscosity μ_q and μ_g depend on $Q\epsilon_q^{1/2}[\frac{\omega_0\rho N_0^2(\rho,\varphi_{H1})}{\theta^2}]^{1/5}q(\rho,r)$ and $Q\epsilon_q^{1/2}[\frac{\omega_0\rho N_0^2(\rho,\varphi_{H2})}{\theta^2}]^{1/5}g(\rho,s)$, respectively.

Obviously, in case of α_1 independent of ρ and $\mu_q = \mu_g = 1$ (i.e., in the inlet and exit zones the lubricant viscosity is independent of pressure) we will need to solve systems (14.56) and (14.57) just once and using substitutions (14.54) and (14.55) we get the solution of the problem in the inlet and exit zones for all values of ρ such that $(\rho,\varphi) \in \Omega_i \bigcap \Omega_\epsilon$ and $(\rho,\varphi) \in \Omega_e \bigcap \Omega_\epsilon$.

Systems of equations (14.56) and (14.57) are identical to the corresponding systems of equations derived in the inlet and exit zones of lubricated heavily loaded line contacts which were derived in Chapter 6. Therefore, the whole arsenal of methods developed for the analysis of these equations in Chapters 6 - 10 and the corresponding conclusions made for line contacts are transferable to point contacts. That includes the equivalent representation of systems (14.56) and (14.57) (similar to systems (12.57) and (12.58)), validation of the applied asymptotic approach, discussions of the numerical methods suitable for solution of asymptotically valid equations in the inlet and exit zones, solution precision and solution stability/instability issues as well as the regularization approach to obtaining stable numerical solutions for high values of the dimensionless pressure viscosity coefficient Q proposed and realized in Chapter 8. Moreover, the uniformly valid approximations of pressure $p(\rho,\varphi)$ and gap $h(\rho,\varphi)$ as well as the exit film thickness $H_0(\rho)$ in the central part of the contact region can be obtained in the fashion similar to the one described in the end of Section 12.3. Also, these equations are controlled by less number of dimensionless parameters. A large number of numerical examples can be found in Chapter 6.

The above substitutions and formulas (14.38), (14.43), (14.55), and (14.58) lead to the conclusion that the exit film thickness H_0 besides being a function of α_1, V_e, Q, ϵ_q, e, γ, and ω_0 is also a function of ρ and

$$H_0(\rho) = A_g[\frac{\omega_0^3\rho^3}{\theta N_0^4(\rho,\varphi_{H2})}]^{1/5}(V_e\epsilon_q^2)^{1/3}, \ A_g = O(1), \ \omega \ll 1, \qquad (14.60)$$

where A_g is a constant determined with the help of the solution of the asymptotic system of equations (14.56) in the inlet zone for a line contact and the simple formula (see formulas (14.55))

$$A_g = A_q[\frac{N_0(\rho,\varphi_{H2})}{N_0(\rho,\varphi_{H1})}]^{4/5}. \qquad (14.61)$$

Coefficient A_g is dependent on α_1 and $Q\epsilon_q^{1/2}[\frac{\omega_0\rho N_0^2(\rho,\varphi_{H1})}{\theta^2}]^{1/5}$ through μ_q) and A_q while V_e and θ are determined by the formulas in (14.40). For precritical lubrication regimes the value of $Q\epsilon_q^{1/2}$ can be taken as either equal to 0 for $\epsilon_q \ll Q^{-2}$, $\omega \ll 1$, or equal to $Q\epsilon_q^{1/2} = O(1)$, $\omega \ll 1$, if $\epsilon_q = O(Q^{-2})$, $\omega \ll 1$. In a special case of constant viscosity μ or the case of starved lubrication regime when $\epsilon_q \ll Q^{-2}$, $\omega \ll 1$, the situation is simpler because the value of the constant A_g is dependent only on α_1, i.e., it is independent from $Q\epsilon_q^{1/2}[\frac{\omega_0\rho N_0^2(\rho,\varphi_{H1})}{\theta^2}]^{1/5}$. In the latter case the dependence of the exit film thickness H_0 on the contact ellipse eccentricity e is limited to just the one shown in (14.60). Moreover, for $e = 0$ or $\gamma = 1$ we have $N_0^4(\rho,\varphi_{H2}) = N_0^4(\rho,\varphi_{H1})$ (see (14.16)-(14.18) and (14.37)) and, therefore, $A_g = A_q$.

The exit film thickness H_0 and, simultaneously, the film thickness in the Hertzian region varies with ρ. It is important to keep in mind that formula (14.60) is valid only outside of the ends of HSSPGD zone Ω_{HSSPGD}.

In a similar fashion for fully flooded heavily loaded lubrication regimes when $H_{00} = O(\epsilon_q^{3/2})$ and, therefore, $\epsilon_q = V_e^{2/5}$ we obtain (see (14.60))

$$H_0(\rho) = A[\frac{\omega_0^3\rho^3}{\theta N_0^4(\rho,\varphi_{H2})}]^{1/5}V_e^{3/5}, \quad A = O(1), \quad \omega \ll 1. \tag{14.62}$$

As an example let us consider the case of constant viscosity $\mu = 1$. Then for $\gamma = 1$ and $e = 0$ in the original (x, y) coordinates formula (14.62) has the form (see formulas (14.6) and (14.39))

$$H_0(x, y) = A\{\frac{\theta}{\omega_0^5\rho^3}[1 - \frac{(1+\omega_0^2(\rho^2-1))^2}{4\omega_0^2\rho^2}]\}^{-1/5}V_e^{3/5},$$
$$\rho^2 = x^2 + (y - \frac{1}{\omega_0})^2, \quad A = O(1), \quad \omega \ll 1. \tag{14.63}$$

Here parameter A depends only on the inlet coordinate α_1 while V_e and θ are determined by (14.40).

Also, it is remarkable that essentially the lubrication mechanism and the equations describing it in the inlet and exit zones of a point contact involved in spinning are the same as in the case of no spinning under the conditions of straight or skewed lubricant entrainment. Moreover, for contacts involved in different types of motion the values of constants A in formulas for the exit film thickness H_0 (see formulas (12.55), (12.56), (13.52), (13.53), (14.60), and (14.62)) are the same (i.e., independent of the type of motion) as long as the coefficients of q and g in the expressions for the lubricant viscosity μ_q and μ_g in the inlet and exit zones are identical for the corresponding cases of motion.

Let us consider a simple example for $\gamma = 1$ and $e = 0$ which would reveal the dependence of the film thickness H_0 on the spinning speed ω_0. Let us denote by $H_0(\omega_0 = 0)$ and $H_0(\omega_0 > 0)$ the exit film thicknesses for the cases of no spin ($\omega_0 = 0$) and presence of spin ($\omega_0 > 0$), respectively. For constant viscosity and a particular value of the inlet coordinate α_1 the values of coefficients A in formulas for the film thickness H_0 with and without spin are equal. Therefore, for pre-critical fully flooded lubrication regimes with constant viscosity from formula (14.63) and the corresponding formula for H_0 from Chapter 12 (which is the limit of the expression in (14.63) as $\omega_0 \to 0$) we obtain

$$R_{\omega\rho} = \frac{H_0(\omega_0>0)}{H_0(\omega_0=0)} = \{\frac{1}{\omega_0^5\rho^3(1-y^2)}[1 - \frac{(1+\omega_0^2(\rho^2-1))^2}{4\omega_0^2\rho^2}]\}^{-1/5},$$
$$\rho^2 = x^2 + (y - \frac{1}{\omega_0})^2. \tag{14.64}$$

For $\omega_0 = 0$ the ratio from (14.64) is $R_{\omega\rho} = 1$. The variations of this ratio for different values of ω_0 and ρ are given in Table 14.1. We have to remember that the difference between the exit film thickness $H_0(\rho)$ and the film thickness on the arc $\rho = const$ is negligibly small. Therefore, the ratio $R_{\omega\rho}$ can be calculated at points $(0, y)$ with different values of y. These local lubrication film thickness variations are caused by local differences in the lubricant entrainment speed. In fact, for $\omega_0 > 0$ spinning promotes growth of film thickness for $y < 0$ where it increases the entrainment speed while for $y > 0$ spinning decreases the entrainment speed and, therefore, decreases the film thickness.

TABLE 14.1
Variations of the ratio $R_{\omega\rho} = \frac{H_0(\omega_0 > 0)}{H_0(\omega_0 = 0)}$ versus the values of ω_0 at different points of the lubricated contact (x, y).

$\omega_0/(x,y)$	(0,0.5)	(0,0)	(0,-0.5)	(0,-0.75)	(0,-0.85))
0	1	1	1	1	1
0.5	0.805	1.013	1.205	1.295	1.331
1	0.631	1.059	1.421	1.585	1.648
2	-	-	1.913	2.163	2.262

14.5 Asymptotic Analysis of a Heavily Loaded Contact under Over-Critical Lubrication Conditions

Over-critical lubrication regimes occur when lubricant viscosity rapidly grows with pressure which occurs when $Q \gg 1$. In a heavily loaded lubricated contact the maximum characteristic size of the inlet zone ϵ_0 for which the lubrication regime is still a pre-critical one we will call the critical characteristic size of the inlet zone ϵ_q. To determine ϵ_0 we need to consider the definition (14.25) of the pre-critical lubrication regimes. The value of ϵ_0 is such that for any $\epsilon_q \gg \epsilon_0$ the estimate (14.25) is not valid while for $\epsilon_q = \epsilon_0$ this estimate is still valid. If it appears that $\epsilon_0 \ll 1$ for the given problem parameters then it means that the viscosity μ depends on a large parameter Q. Thus, if $\epsilon_0 \ll 1$ then for $\epsilon_q = O(\epsilon_0)$ we have pre-critical regimes and for $\epsilon_q \gg \epsilon_0$ - over-critical regimes. For example, if $\mu = \exp(Qp)$ then $\epsilon_0 = Q^{-2} \ll 1$ for $\omega = Q^{-1} \ll 1$. Obviously, if $\epsilon_0 \gg 1$ or $\epsilon_0 = O(1)$ for $\omega \ll 1$ then over-critical regimes cannot be realized and pre-critical lubrication regimes occur.

Therefore, the formal definition of over-critical lubrication regimes is as follows

$$\mu(\epsilon_q^{1/2}) \gg 1, \quad \epsilon_q(\omega) \gg \epsilon_0(\omega), \quad \epsilon_0(\omega) \ll 1, \quad \omega \ll 1. \tag{14.65}$$

A heavily loaded over-critical lubricated contact has the following structure described in Chapter 6. The inlet zone is represented by two sub-zones: one of a characteristic size $\epsilon_q \gg \epsilon_0$ (inlet ϵ_q-zone) which is adjacent to the inlet contact boundary Γ_i and a smaller inlet zone of the characteristic size ϵ_0 (inlet ϵ_0-zone) located between the inlet ϵ_q-zone and the Hertzian region where pressure is very close to $p_{00}(x, y) = \sqrt{1 - x^2 - y^2}$. On the exit side of the contact the structure of the exit zone is similar to the one of the inlet zone, i.e. a smaller exit ϵ_0-zone is located between the Hertzian region and the larger exit ϵ_q-zone adjacent to the contact exit boundary Γ_e.

Using the same approach as the one used in Chapter 6 we obtain that for starved and fully flooded over-critical lubrication regimes the film thickness is determined by the formulas

$$H_{00} = A_s(V_e Q \epsilon_q^{5/2})^{1/3}, \ \epsilon_q \ll (V_e Q)^{1/2}, \ A_s = O(1), \ Q \gg 1,$$
$$H_{00} = A_f(V_e Q)^{3/4}, \ \epsilon_q = (V_e Q)^{1/2}, \ A_f = O(1), \ Q \gg 1,$$
(14.66)

where the coefficients A_s and A_f depend on θ and $\alpha_{1p}(\frac{1}{\omega_0})$. In the manner similar to the one used for pre-critical lubrication regimes from formulas (14.66) we obtain formulas for the exit film thickness as follows

$$H_0(\rho) = A_{s0}[\frac{\omega_0^3 \rho^3}{\theta N_0^4(\rho, \varphi_{H2})}]^{1/5}(V_e Q \epsilon_q^{5/2})^{1/3}, \ \epsilon_q \ll (V_e Q)^{1/2},$$
$$A_{s0} = O(1), \ Q \gg 1,$$
(14.67)

for starved over-critical lubrication regimes and

$$H_0(\rho) = A_{f0}[\frac{\omega_0^3 \rho^3}{\theta N_0^4(\rho, \varphi_{H2})}]^{1/5}(V_e Q)^{3/4}, \ \epsilon_q = (V_e Q)^{1/2},$$
$$A_{f0} = O(1), \ Q \gg 1,$$
(14.68)

for fully flooded over-critical lubrication regimes. Here V_e and θ are determined by (14.40).

In a particular case of $\gamma = 1$ and $e = 0$ in the original (x, y) coordinates formula (14.68) is as follows

$$H_0(x, y) = A_{f0}\{\frac{\theta}{\omega_0^5 \rho^3}[1 - \frac{(1+\omega_0^2(\rho^2-1))^2}{4\omega_0^2 \rho^2}]\}^{-1/5}(V_e Q)^{3/4},$$
$$\rho^2 = x^2 + (y - \frac{1}{\omega_0})^2, \ A_{f0} = O(1), \ Q \gg 1,$$
(14.69)

where the value of constant A_{f0} depends on parameters θ, γ and variable ρ through the dependence of the viscosity μ_q in the inlet zone on pressure p.

The choice between pre- and over-critical lubrication regimes can be made based on a simple analysis from Section 9.5 while the regularization approaches to numerical solution of isothermal EHL problems are discussed in Section 15.6.

14.6   Closure

An asymptotic approach to isothermal steady heavily loaded point EHL contacts with incompressible Newtonian lubricants with skewed lubricant entrainment and spinning ball is developed. The behavior of the pressure and gap distributions in the inlet and exit zones along the "central" streamlines of the lubricant flow (represented by arcs of certain ellipses) are analyzed in detail. Asymptotic analysis of heavily loaded contacts lubricated by a Newtonian fluid recognizes two different regimes of lubrication: pre- and over-critical lubrication regimes. Overall, the difference between these two regimes is in how rapidly the lubricant viscosity grows with pressure and how much lubricant is available at the inlet of the contact. For pre-critical lubrication regimes the asymptotically valid equations for pressure and gap have been derived. It has been shown that by a simple transformation these equations can be reduced to equations identical to the corresponding equations for pre-critical lubrication regimes in the inlet and exit zones of line heavily loaded contacts. The latter equations have been validated through the comparison with the direct numerical solutions of the original EHL equations. For over-critical lubrication regimes another approach was used which led to formulas for the lubrication film thickness for starved and fully flooded conditions. The asymptotic approaches to EHL problems offer certain distinct advantages in comparison with the direct numerical solutions of EHL problem among which are a reduction of the number of problem input parameters and the approach to regularization of numerical solutions. Each of the analytically derived formulas for the central H_{00} and exit H_0 lubrication film thicknesses contains a coefficient of proportionality which can be found by numerically solving the corresponding asymptotically valid equations in the inlet zone or by just curve fitting the experimentally obtained values of the lubrication film thickness.

14.7 Exercises and Problems

1. Explain why it is beneficial to use the new independent variables (ρ, φ) determined by equations (14.5) and (14.6). (Hint: Consider the approximate solution of the EHL problem in the Hertzian region.)

2. Explain the concept and the details of how the expression for function N_0 from (14.15) is obtained.

3. For $\gamma = 1$ and $e > 0$ determine the limit of the function $N_0(\rho)$ for $\omega_0 \to 0$ and compare it with the expression for $N_0(y)$ from (12.34).

4. Consider the distribution of the lubricant entrainment speed over the contact region in the presence of spin. Explain what is the physical mechanism for the dependence of the exit lubrication film thickness $H_0(x, y)$ on the spinning parameter ω_0 presented in Table 14.1.

References

[1] Mostofi, A. and Gohar, R. 1982. Oil Film Thickness and Pressure Distribution in Elastohydrodynamic Point Contacts. *J. Mech. Engrs. Science, Insn. Mech. Engrs.*, Vol. 24, No. 4, pp. 173-182.

[2] Vorovich, I.I., Aleksandrov, V.M., and Babeshko, V.A. 1974. *Non − classical Mixed Problems of Elasticity.* Moscow: Nauka.

[3] Van Dyke, M. 1964. *Perturbation Methods in Fluid Mechanics.* New York: Academic Press.

[4] Kevorkian, J. and Cole, J.D. 1985. *Perturbation Methods in Applied Mathematics. Applied Mathematics Series*, Vol. 34, New York, Springer-Verlag.

15

Thermal EHL Problems for Heavily Loaded Point Contacts with Newtonian Lubricants

15.1 Introduction

In this chapter, using the approaches to thermal EHL problems for line contacts developed in Chapter 7, a similar approach is developed in application to point contacts. The lubricant rheology is considered to be Newtonian. The EHL problems are assumed to be under the condition of high slide-to-roll ratio. The three types of problems considered in Chapters 12, 13, and 14 in isothermal formulation here are analyzed here under thermal conditions. Along lubricant "central" flow streamlines the point EHL problems are reduced to the corresponding EHL problems for line contacts very similar to the ones for isothermal and thermal contacts. Analytical formulas for the lubrication film thickness are derived.

15.2 Problem Formulation

Let us consider a steady thermal lubricated contact with skewed lubricant entrainment of an elastic ball rolling and spinning in an elastic parabolic raceway. The ball is spinning in the counterclockwise direction with the angular speed ω_2. It means that the solid surface velocities are (u_1, v_1) and $(u_2 - \omega_2 y, v_2 + \omega_2 x)$, where (x, y) are the coordinates of a ball surface point introduced the same way as in Chapter 12. In this coordinate system the shapes of the solid surfaces can be approximated by elliptical paraboloids: $z = z_2(x, y) = \frac{x^2}{2R_{2x}} + \frac{y^2}{2R_{2y}}$ and $z = z_1(x, y) = -\frac{x^2}{2R_{1x}} - \frac{y^2}{2R_{1y}}$, where R_{1x}, R_{2x}, R_{1y}, and R_{2y} are the surface curvature radii of the solid surfaces along the x- and y-axes, respectively. The solids' elastic moduli and Poisson's ratios are E_1, E_2 and ν_1, ν_2, respectively. The two solids are loaded by a normal force P acting along the z-axis. The solids are separated by a layer of an incompressible viscous lubricant with Newtonian rheology. Using the traditional assumptions that the solids' motions are steady and relatively slow, the size of the contact region is significantly smaller than the solid radii, and the lubri-cation film thickness is much smaller than the contact size [1] we obtain the continuity equations (4.4) and the momentum equations (4.11) (see Chapter 4). Integrating the latter ones we obtain

$$p_{xz} = f_x + z\frac{\partial p}{\partial x}, \ p_{yz} = f_y + z\frac{\partial p}{\partial y}, \tag{15.1}$$

where $f_x(x,y)$ and $f_y(x,y)$ are the unknown x- and y-components of the lubricant sliding frictional stress which are determined from the second pair of boundary conditions on u and v at $z = \frac{h}{2}$ and the tensor components p_{xz} and p_{yz} are determined by the Newtonian rheology of the lubricant as follows

$$p_{xz} = \mu\frac{\partial u}{\partial z}, \ p_{yz} = \mu\frac{\partial v}{\partial z}. \tag{15.2}$$

Integrating equations (16.2) and using the expressions (16.3) together with the no slippage conditions (12.1) on u and v we determine

$$u = u_1 + \int\limits_{-h/2}^{z} \frac{1}{\mu}(f_x + s\frac{\partial p}{\partial x})ds, \ v = v_1 + \int\limits_{-h/2}^{z} \frac{1}{\mu}(f_y + s\frac{\partial p}{\partial y})ds. \tag{15.3}$$

By satisfying the boundary conditions on u and v at $z = \frac{h}{2}$ we obtain expres-sions for stresses f_x and f_y in the form

$$f_x = \{u_2 - u_1 - \omega_2 y - \frac{\partial p}{\partial x}\int\limits_{-h/2}^{h/2} \frac{sds}{\mu}\}/\int\limits_{-h/2}^{h/2} \frac{ds}{\mu},$$

$$\tag{15.4}$$

$$f_y = \{v_2 - v_1 + \omega_2 x - \frac{\partial p}{\partial y}\int\limits_{-h/2}^{h/2} \frac{sds}{\mu}\}/\int\limits_{-h/2}^{h/2} \frac{ds}{\mu}.$$

Integrating the continuity equation (4.4) with respect to z from $-\frac{h}{2}$ to $\frac{h}{2}$, changing the order of differentiation and integration and using the last pair of boundary conditions from the no slippage conditions (12.1) as well as the integrals of u and v from (15.3) with respect to z from $-\frac{h}{2}$ to $\frac{h}{2}$ we obtain the generalized Reynolds equation

$$\frac{\partial}{\partial x}\{[\int\limits_{-h/2}^{h/2} \frac{s^2 ds}{\mu} - (\int\limits_{-h/2}^{h/2} \frac{sds}{\mu})^2/\int\limits_{-h/2}^{h/2} \frac{ds}{\mu}]\frac{\partial p}{\partial x}\}$$

$$+\frac{\partial}{\partial y}\{[\int\limits_{-h/2}^{h/2} \frac{s^2 ds}{\mu} - (\int\limits_{-h/2}^{h/2} \frac{sds}{\mu})^2/\int\limits_{-h/2}^{h/2} \frac{ds}{\mu}]\frac{\partial p}{\partial y}\}$$

$$\tag{15.5}$$

$$\frac{\partial}{\partial x}\{\frac{u_1+u_2-\omega_2 y}{2}h - (u_2 - u_1 - \omega_2 y)\int\limits_{-h/2}^{h/2} \frac{sds}{\mu}/\int\limits_{-h/2}^{h/2} \frac{ds}{\mu}\}$$

$$+\frac{\partial}{\partial y}\{\frac{v_1+v_2+\omega_2 x}{2}h - (v_2 - v_1 + \omega_2 x)\int\limits_{-h/2}^{h/2} \frac{sds}{\mu}/\int\limits_{-h/2}^{h/2} \frac{ds}{\mu}\}.$$

To simplify the Reynolds equation the repeated integrals were reduced to single integrals.

Substituting the expressions for tensors p_{xz} and p_{yz} from (15.1) and stresses f_x and f_y from (15.4) into the energy equation (4.12) and imposing the boundary conditions on lubricant temperature $T(x, y, z)$ we obtain

$$\frac{\partial}{\partial z}(\lambda \frac{\partial T}{\partial z}) = -\frac{1}{\mu}\{[u_2 - u_1 - \omega_2 y + \frac{\partial p}{\partial x}(z\int\limits_{-h/2}^{h/2}\frac{ds}{\mu} - \int\limits_{-h/2}^{h/2}\frac{sds}{\mu})]^2$$

$$+[v_2 - v_1 + \omega_2 x + \frac{\partial p}{\partial y}(z\int\limits_{-h/2}^{h/2}\frac{ds}{\mu} - \int\limits_{-h/2}^{h/2}\frac{sds}{\mu})]^2\}/(\int\limits_{-h/2}^{h/2}\frac{ds}{\mu})^2, \tag{15.6}$$

$$T(x, y, -\tfrac{h}{2}) = T_{w1}(x, y), \ T(x, y, \tfrac{h}{2}) = T_{w2}(x, y),$$

where $T_{w1}(x, y)$ and $T_{w2}(x, y)$ are the surface temperatures of the lower and upper solids.

The rest of the problem formulation is done the same way as in Chapter 12. As a result of this derivation the thermal EHL problem for Newtonian lubricants with viscosity μ determined as follows

$$\mu(p, T) = \mu^0(p)\exp[-\alpha_T(T - T_a)], \ \mu^0(p) = \mu_a\exp(\alpha p), \tag{15.7}$$

is reduced to the equations

$$\frac{\partial}{\partial x}\{[\int\limits_{-h/2}^{h/2}\frac{s^2 ds}{\mu} - (\int\limits_{-h/2}^{h/2}\frac{sds}{\mu})^2/\int\limits_{-h/2}^{h/2}\frac{ds}{\mu}]\frac{\partial p}{\partial x}\}$$

$$+\frac{\partial}{\partial y}\{[\int\limits_{-h/2}^{h/2}\frac{s^2 ds}{\mu} - (\int\limits_{-h/2}^{h/2}\frac{sds}{\mu})^2/\int\limits_{-h/2}^{h/2}\frac{ds}{\mu}]\frac{\partial p}{\partial y}\}$$

$$=\frac{\partial}{\partial x}\{\frac{u_1 + u_2 - \omega_2 y}{2}h - (u_2 - u_1 - \omega_2 y)\int\limits_{-h/2}^{h/2}\frac{sds}{\mu}/\int\limits_{-h/2}^{h/2}\frac{ds}{\mu}\}$$

$$+\frac{\partial}{\partial y}\{\frac{v_1 + v_2 + \omega_2 x}{2}h - (v_2 - v_1 + \omega_2 x)\int\limits_{-h/2}^{h/2}\frac{sds}{\mu}/\int\limits_{-h/2}^{h/2}\frac{ds}{\mu}\}, \tag{15.8}$$

$$p(x, y)\ |_\Gamma = 0, \ \frac{dp(x,y)}{d\overrightarrow{n}}\ |_{\Gamma_e} = 0,$$

$$f_x = \{u_2 - u_1 - \omega_2 y - \frac{\partial p}{\partial x}\int\limits_{-h/2}^{h/2}\frac{sds}{\mu}\}/\int\limits_{-h/2}^{h/2}\frac{ds}{\mu},$$

$$f_y = \{v_2 - v_1 + \omega_2 x - \frac{\partial p}{\partial y}\int\limits_{-h/2}^{h/2}\frac{sds}{\mu}\}/\int\limits_{-h/2}^{h/2}\frac{ds}{\mu},$$

$$h(x,y) = h_0 + \frac{x^2}{2R_x} + \frac{y^2}{2R_y} + \frac{1}{\pi E'}\{\iint\limits_{\Omega} \frac{p(x_1,y_1)dx_1dy_1}{\sqrt{(x-x_1)^2+(y-y_1)^2}}$$

$$- \iint\limits_{\Omega} \frac{p(x_1,y_1)dx_1dy_1}{\sqrt{x_1^2+y_1^2}}\}, \quad \iint\limits_{\Omega} p(x_1,y_1)dx_1dy_1 = P,$$

$$\frac{\partial}{\partial z}(\lambda\frac{\partial T}{\partial z}) = -\frac{1}{\mu}\{[u_2 - u_1 - \omega_2 y + \frac{\partial p}{\partial x}(z\int\limits_{-h/2}^{h/2}\frac{ds}{\mu} - \int\limits_{-h/2}^{h/2}\frac{sds}{\mu})]^2$$

$$+[v_2 - v_1 + \omega_2 x + \frac{\partial p}{\partial y}(z\int\limits_{-h/2}^{h/2}\frac{ds}{\mu} - \int\limits_{-h/2}^{h/2}\frac{sds}{\mu})]^2\}/(\int\limits_{-h/2}^{h/2}\frac{ds}{\mu})^2,$$

$$T(x,y,-\tfrac{h}{2}) = T_{w1}(x,y), \quad T(x,y,\tfrac{h}{2}) = T_{w2}(x,y),$$

where μ_a, α, α_T are the ambient lubricant viscosity, the pressure and temperature coefficients of viscosity, respectively, h_0 is the central film thickness at the point with coordinates $(0,0)$, R_x, R_y, and E' are the effective radii along the x- and y-axes and the effective modulus of elasticity, $\frac{1}{R_x} = \frac{1}{R_{1x}} + \frac{1}{R_{2x}}$, $\frac{1}{R_y} = \frac{1}{R_{1y}} + \frac{1}{R_{2y}}$, and $\frac{1}{E'} = \frac{1-\nu_1^2}{E_1} + \frac{1-\nu_2^2}{E_2}$, respectively, Ω and Γ are the contact region and its boundary, respectively, and $\overrightarrow{n}$ is the unit external normal vector to the boundary Γ. The inlet Γ_i and exit Γ_e portions of the contact boundary Γ through which the lubricant enters and leaves the contact, respectively, constitute the whole boundary of Ω, i.e. $\Gamma_i \bigcup \Gamma_e = \Gamma$. The inlet Γ_i and exit Γ_e boundaries are determined by the relationships

$$(x,y) \in \Gamma_i \ if \ \overrightarrow{F_f} \cdot \overrightarrow{n} < 0; \ (x,y) \in \Gamma_e \ if \ \overrightarrow{F_f} \cdot \overrightarrow{n} \geq 0,$$

$$\overrightarrow{F_f} = (u_1 + u_2 - \omega_2 y, v_1 + v_2 + \omega_2 x)\frac{h}{2}$$

$$-(u_2 - u_1 - \omega_2 y, v_2 - v_1 + \omega_2 x)\int\limits_{-h/2}^{h/2}\frac{sds}{\mu}/\int\limits_{-h/2}^{h/2}\frac{ds}{\mu} \qquad (15.9)$$

$$-(\frac{\partial p}{\partial x}, \frac{\partial p}{\partial y})[\int\limits_{-h/2}^{h/2}\frac{s^2ds}{\mu} - (\int\limits_{-h/2}^{h/2}\frac{sds}{\mu})^2/\int\limits_{-h/2}^{h/2}\frac{ds}{\mu}],$$

where $\overrightarrow{F_f}$ is the fluid flux through the contact boundary Γ and $\overrightarrow{a} \cdot \overrightarrow{b}$ is the dot product of vectors $\overrightarrow{a}$ and $\overrightarrow{b}$.

Let us introduce the dimensionless variables according to (12.5) and also

$$\{f_x', f_y'\} = \frac{2R_x}{a_H p_H}\{f_x, f_y\}, \quad \{T', T_{w1}', T_{w2}'\} = \frac{1}{T_a}\{T, T_{w1}, T_{w2}\} - 1, \qquad (15.10)$$

as well as use the dimensionless parameters (13.1), (14.3), and

$$s_x = \frac{2(u_2-u_1)}{\sqrt{(u_1+u_2)^2+(v_1+v_2)^2}}, \quad s_y = \frac{2(v_2-v_1)}{\sqrt{(u_1+u_2)^2+(v_1+v_2)^2}},$$

$$(15.11)$$

$$\kappa = \frac{\mu_a[(u_1+u_2)^2+(v_1+v_2)^2]}{\lambda_a T_a}, \quad \delta_T = \alpha_T T_a,$$

where a_H, b_H, and p_H are the Hertzian semi-axes (small and large) of the dry elliptical contact and the Hertzian maximum pressure, respectively, which are determined by equations (12.7) and (12.8). Using these dimensionless variables and parameters (primes are dropped for convenience) the equations of the problem are reduced to the following form

$$\frac{\partial}{\partial x}\{[\int\limits_{-h/2}^{h/2} e^{\delta_T T} s^2 ds - (\int\limits_{-h/2}^{h/2} e^{\delta_T T} s ds)^2 / \int\limits_{-h/2}^{h/2} e^{\delta_T T} ds] \frac{1}{\mu^0} \frac{\partial p}{\partial x}\}$$

$$+\frac{\partial}{\partial y}\{[\int\limits_{-h/2}^{h/2} e^{\delta_T T} s^2 ds - (\int\limits_{-h/2}^{h/2} e^{\delta_T T} s ds)^2 / \int\limits_{-h/2}^{h/2} e^{\delta_T T} ds] \frac{1}{\mu^0} \frac{\partial p}{\partial y}\}$$

$$= \frac{V}{6H_{00}^2} \frac{\partial}{\partial x}\{\frac{\gamma - \omega_0 y}{2} h - (\frac{s_x}{2} - \omega_0 y) \int\limits_{-h/2}^{h/2} e^{\delta_T T} s ds / \int\limits_{-h/2}^{h/2} e^{\delta_T T} ds\}$$

$$+ \frac{V}{6H_{00}^2} \sqrt{1-e^2} \frac{\partial}{\partial y}\{\frac{\sqrt{1-\gamma^2}-\sqrt{1-e^2}\omega_0 x}{2} h - (\frac{s_y}{2} + \sqrt{1-e^2}\omega_0 x)$$

$$\times \int\limits_{-h/2}^{h/2} e^{\delta_T T} s ds / \int\limits_{-h/2}^{h/2} e^{\delta_T T} ds\}, \quad \mu^0 = \exp(Qp),$$

$$p(x,y)\mid_{\Gamma} = 0, \quad \frac{dp(x,y)}{d\overrightarrow{n}}\mid_{\Gamma_e} = 0,$$

$$f_x = \frac{V}{6H_{00}}\{\frac{s_x}{2} - \omega_0 y - \frac{6H_{00}^2}{V} \frac{\partial p}{\partial x} \int\limits_{-h/2}^{h/2} e^{\delta_T T} s ds\} / \int\limits_{-h/2}^{h/2} e^{\delta_T T} ds,$$

$$f_y = \frac{V}{6H_{00}}\{\frac{s_y}{2} + \sqrt{1-e^2}\omega_0 x$$

$$-\sqrt{1-e^2}\frac{6H_{00}^2}{V} \frac{\partial p}{\partial y} \int\limits_{-h/2}^{h/2} e^{\delta_T T} s ds\} / \int\limits_{-h/2}^{h/2} e^{\delta_T T} ds,$$

$$H_{00}[h(x,y)-1] = x^2 + \frac{\delta}{1-e^2}y^2 + \frac{D(e)}{\pi(1-e^2)^{3/2}}\{\int\int\limits_{\Omega} \frac{p(x_1,y_1)dx_1 dy_1}{\sqrt{(x-x_1)^2+\frac{(y-y_1)^2}{1-e^2}}}$$

$$-\int\int\limits_{\Omega} \frac{p(x_1,y_1)dx_1 dy_1}{\sqrt{x_1^2+\frac{y_1^2}{1-e^2}}}\}, \quad \int\int\limits_{\Omega} p(x_1,y_1)dx_1 dy_1 = \frac{2\pi}{3}$$

$$\frac{\partial}{\partial z}(\lambda \frac{\partial T}{\partial z}) = -\kappa \mu^0 e^{\delta_T T}\{[\frac{s_x}{2} - \omega_0 y + \frac{6H_{00}^2}{V}\frac{1}{\mu^0}\frac{\partial p}{\partial x}(z \int\limits_{-h/2}^{h/2} e^{\delta_T T} ds$$

$$-\int\limits_{-h/2}^{h/2} e^{\delta_T T} s ds)]^2 + [\frac{s_y}{2} + \sqrt{1-e^2}\omega_0 x + \sqrt{1-e^2}\frac{6H_{00}^2}{V}\frac{1}{\mu^0}\frac{\partial p}{\partial y}$$

(15.12)

$$\times (z \int_{-h/2}^{h/2} e^{\delta_T T} ds - \int_{-h/2}^{h/2} e^{\delta_T T} s ds)]^2 \} / (\int_{-h/2}^{h/2} e^{\delta_T T} ds)^2,$$

$$T(x, y, -\tfrac{h}{2}) = T_{w1}(x, y), \ T(x, y, \tfrac{h}{2}) = T_{w2}(x, y).$$

Therefore, for the given values of parameters V, Q, δ_T, e, γ, ω_0, s_x, s_y, function F, and the inlet boundary Γ_i we need to determine functions $p(x, y)$, $h(x, y)$, $f_x(x, y)$, $f_y(x, y)$, constant H_{00}, and the exit boundary Γ_e.

It is important to remember that the EHL problem equations for heavily loaded contacts contain a small parameter ω, which may be equal to $\omega = V \ll 1$ and/or $\omega = Q^{-1} \ll 1$.

15.3 Asymptotic Solution for the Lubricant Temperature T. Case of No Spinning EHL Contact

Let us assume that there is no spinning, i.e., $\omega_0 = 0$. The case with spinning will be considered separately.

The idea of the further problem analysis is to determine the lubricant temperature T and then using it to reduce the Reynolds equation in (15.12) to an analog of the Reynolds equation for the corresponding isothermal EHL problem. Therefore, first we need to asymptotically solve the equation for T from (15.12). It can be done relatively easy for the case when the amount of heat produced in the lubrication layer by the sliding frictional stress in the dominant direction compared to the amount of heat produced by the rolling stress in this direction is much higher. The dominant direction is determined by the problem solution behavior in the Hertzian region - the region away from the contact boundaries. In certain cases it is determined by the first terms in the derivatives with respect to x and y in the right-hand side of the Reynolds equation in (15.12) and this direction is just a straight line. In other words, in these cases, in the Hertzian region $\frac{\partial}{\partial x} \frac{\gamma h}{2} + \frac{\partial}{\partial y} \frac{\sqrt{1-\gamma^2} h}{2} \approx 0$ which determines the dominant direction while the other terms from the right-hand side of the Reynolds equation, as it will be shown later, contribute only to the terms proportional to $\frac{\partial p}{\partial x}$ and $\frac{\partial p}{\partial y}$ from the left-hand side of this equation. The specific conditions for which this is true will be stated later. Therefore in these cases we can utilize the rotation of the coordinate system about the z-axis defined in (13.3) and (13.4)

$$\chi = \frac{\gamma x + \sqrt{1-e^2}\sqrt{1-\gamma^2} y}{\sqrt{1-e^2+e^2\gamma^2}}, \ \xi = \frac{-\sqrt{1-e^2}\sqrt{1-\gamma^2} x + \gamma y}{\sqrt{1-e^2+e^2\gamma^2}}, \ d\chi d\xi = dx dy, \qquad (15.13)$$

$$x = \frac{\gamma \chi - \sqrt{1-e^2}\sqrt{1-\gamma^2}\xi}{\sqrt{1-e^2+e^2\gamma^2}}, \ y = \frac{\sqrt{1-e^2}\sqrt{1-\gamma^2}\chi + \gamma \xi}{\sqrt{1-e^2+e^2\gamma^2}}. \qquad (15.14)$$

In variables (15.13) equations (15.12) will take the form

$$[(1-e^2)^2(1-\gamma^2)+\gamma^2]\frac{\partial}{\partial\chi}(\frac{R}{\mu^0}\frac{\partial p}{\partial\chi})-e^2\sqrt{1-e^2}\gamma\sqrt{1-\gamma^2}[\frac{\partial}{\partial\chi}(\frac{R}{\mu^0}\frac{\partial p}{\partial\xi})$$

$$+\frac{\partial}{\partial\xi}(\frac{R}{\mu^0}\frac{\partial p}{\partial\chi})]+(1-e^2)\frac{\partial}{\partial\xi}(\frac{R}{\mu^0}\frac{\partial p}{\partial\xi})$$

$$=\frac{V\sqrt{1-e^2+e^2\gamma^2}}{12H_{00}^2}\{(1-e^2+e^2\gamma^2)\frac{\partial h}{\partial\chi}$$

$$-[\gamma s_x+(1-e^2)\sqrt{1-\gamma^2}s_y]\frac{\partial S}{\partial\chi}+\sqrt{1-e^2}[\sqrt{1-\gamma^2}s_x-\gamma s_y]\frac{\partial S}{\partial\xi}\},$$

$$R=\int\limits_{-h/2}^{h/2}e^{\delta_T T}s^2ds-(\int\limits_{-h/2}^{h/2}e^{\delta_T T}sds)^2/\int\limits_{-h/2}^{h/2}e^{\delta_T T}ds,$$

$$S=\int\limits_{-h/2}^{h/2}e^{\delta_T T}sds/\int\limits_{-h/2}^{h/2}e^{\delta_T T}ds,\ \mu^0=\exp(Qp),$$

$$p(\chi,\xi)\mid_\Gamma=0,\ \frac{dp(\chi,\xi)}{d\overrightarrow{n}}\mid_{\Gamma_e}=0,$$

$$H_{00}[h(\chi,\xi)-1]=x^2(\chi,\xi)+\frac{\delta}{1-e^2}y^2(\chi,\xi)$$

$$+\frac{D(e)}{\pi(1-e^2)^{3/2}}\{\int\int\limits_\Omega\frac{p(\chi_1,\xi_1)d\chi_1 d\xi_1}{\sqrt{(\chi-\chi_1)^2+\frac{(\xi-\xi_1)^2}{1-e^2}}}-\int\int\limits_\Omega\frac{p(\chi_1,\xi_1)d\chi_1 d\xi_1}{\sqrt{\chi_1^2+\frac{\xi_1^2}{1-e^2}}}\},$$

$$\int\int\limits_\Omega p(\chi_1,\xi_1)d\chi_1 d\xi_1=\frac{2\pi}{3},$$

$$\frac{\partial}{\partial z}(\lambda\frac{\partial T}{\partial z})=-\kappa\mu^0e^{\delta_T T}\{[\frac{s_x}{2}+\frac{6H_{00}^2}{V\sqrt{1-e^2+e^2\gamma^2}}\frac{1}{\mu^0}(\gamma\frac{\partial p}{\partial\chi}$$

$$-\sqrt{1-e^2}\sqrt{1-\gamma^2}\frac{\partial p}{\partial\xi})(z\int\limits_{-h/2}^{h/2}e^{\delta_T T}ds-\int\limits_{-h/2}^{h/2}e^{\delta_T T}sds)]^2$$

$$+[\frac{s_y}{2}+\frac{6\sqrt{1-e^2}H_{00}^2}{V\sqrt{1-e^2+e^2\gamma^2}}\frac{1}{\mu^0}(\sqrt{1-e^2}\sqrt{1-\gamma^2}\frac{\partial p}{\partial\chi}+\gamma\frac{\partial p}{\partial\xi})$$

$$\times(z\int\limits_{-h/2}^{h/2}e^{\delta_T T}ds-\int\limits_{-h/2}^{h/2}e^{\delta_T T}sds)]^2\}/(\int\limits_{-h/2}^{h/2}e^{\delta_T T}ds)^2,$$

$$T(x,y,-\frac{h}{2})=T_{w1}(x,y),\ T(x,y,\frac{h}{2})=T_{w2}(x,y),$$

(15.15)

where x, y and x_1, y_1 are functions of χ, ξ and χ_1, ξ_1 determined according to (13.4), respectively, while $D(e)$ and e are determined by equations (12.7) and (12.8).

Now, our goal is to make the system of equations (13.5) more manageable for analysis. To do that we will assume that the heat produced by the rolling friction along the χ-axis is much smaller than the one produced by the sliding friction in this direction, i.e., we will introduce a ratio

$$\nu = \frac{12H_{00}^2}{Vs_*\sqrt{1-e^2+e^2\gamma^2}} \frac{1}{\mu^0} \frac{\partial p}{\partial \chi}, \qquad (15.16)$$

which we will consider to be small in the inlet zone and, therefore, everywhere else in a heavily loaded EHL contact. In (15.16) and for further derivation we introduced

$$s_* = \sqrt{s_x^2 + s_y^2}, \quad s_{*x} = \frac{s_x}{s_*}, \quad s_{*y} = \frac{s_y}{s_*}, \quad \kappa_{N*} = \frac{\kappa s_*^2}{4}. \qquad (15.17)$$

Also, we will assume that in the inlet and exit zones of heavily loaded contacts

$$\frac{\partial p}{\partial y} = O(\frac{\partial p}{\partial x}). \qquad (15.18)$$

Let us try to find the solution of the equation for T in the form of an asymptotic expansion in powers of $\nu \ll 1$

$$T(\chi,\xi,z) = T_0(\chi,\xi,z) + \nu T_1(\chi,\xi,z) + \dots$$

$$T_0(\chi,\xi,z), \; T_1(\chi,\xi,z) = O(1), \; \nu \ll 1, \qquad (15.19)$$

where $T_0(\chi,\xi,z)$ and $T_1(\chi,\xi,z)$ are unknown and have to be found from the solution of the problem.

Let us for simplicity assume that $T_{w1}(\chi,\xi,z) = T_{w2}(\chi,\xi,z) = T_{w0}(\chi,\xi,z)$ which is independent of ν. The general case of such a problem for line EHL contacts is considered in Section 10.3. If the latter condition is not true then the dominant direction in the Hertzian region is not a straight line and the terms from the right-hand side of the Reynolds equation contribute not only to the left-hand side of the Reynolds equation from (15.15) but also contribute to the terms in the right-hand side of the equation in a nonlinear manner. That makes it practically impossible to determine the flow dominant direction in the Hertzian region analytically. On the other hand, it is reasonable to assume that even when the surface temperatures T_{w1} and T_{w2} are not equal their difference is of the order of magnitude of $\nu \ll 1$ which creates the conditions when the terms from the right-hand side of the Reynolds equation contribute only to the left-hand side of the Reynolds equation from (15.15) like in the case of equality of the surface temperatures.

Also, we will assume that λ may be a constant ($\lambda = 1$) or it can be a function of pressure p but it is independent of temperature T. Then temporarily assuming that $p(\chi,\xi)$ and $h(\chi,\xi)$ are known and substituting the expansion (15.19) into the equation for T from (15.15), taking into account (15.18), expanding the equation in powers of ν, and equating the corresponding terms we obtain the following systems of equations

$$\lambda \frac{\partial^2 T_0}{\partial z^2} = -\frac{\kappa_{N*} \mu^0 e^{\delta_T T_0}}{(\int\limits_{-h/2}^{h/2} e^{\delta_T T_0} ds)^2}, \quad T_0(x, y, -\tfrac{h}{2}) = T_0(x, y, \tfrac{h}{2}) = T_{w0}(x, y),$$
$$\tag{15.20}$$

$$\lambda \frac{\partial^2 T_1}{\partial z^2} = -\kappa_{N*} \mu^0 e^{\delta_T T_0} \{\delta_T T_1 + 2[s_{*x}(\gamma - \sqrt{1-e^2}\sqrt{1-\gamma^2} \frac{\partial p}{\partial \xi} / \frac{\partial p}{\partial \chi})$$

$$+ s_{*y}\sqrt{1-e^2}(\sqrt{1-e^2}\sqrt{1-\gamma^2} + \gamma \frac{\partial p}{\partial \xi} / \frac{\partial p}{\partial \chi})][z \int\limits_{-h/2}^{h/2} e^{\delta_T T_0} ds$$

$$\tag{15.21}$$

$$- \int\limits_{-h/2}^{h/2} e^{\delta_T T_0} s\, ds)] - 2\delta_T \int\limits_{-h/2}^{h/2} e^{\delta_T T_0} T_1 ds / \int\limits_{-h/2}^{h/2} e^{\delta_T T_0} ds\}$$

$$/(\int\limits_{-h/2}^{h/2} e^{\delta_T T_0} ds)^2, \quad T_1(x, y, -\tfrac{h}{2}) = T_1(x, y, \tfrac{h}{2}) = 0, \ldots$$

The solution of problem (15.20) with different parameters is obtained in (7.23) and here it can be represented in the form

$$T_0 = T_{w0} + \tfrac{2}{\delta_T} \ln\{\cosh(\tfrac{Bh}{2}) / \cosh(Bz)\},$$

$$\tag{15.22}$$

$$B = \tfrac{2}{h} \ln\{\eta + \sqrt{\eta^2 + 1}\}, \quad \eta = \tfrac{1}{2}\sqrt{\frac{\kappa_{N*}\delta_T \mu^0}{2\lambda} e^{-\delta_T T_{w0}}}.$$

Function T_0 is even with respect to variable z. Therefore,

$$\int\limits_{-h/2}^{h/2} e^{\delta_T T_0} ds = e^{\delta_T T_{w0}} \frac{\sinh(Bh)}{B}, \quad \int\limits_{-h/2}^{h/2} s e^{\delta_T T_0} ds = 0. \tag{15.23}$$

That allows to simplify equation for T_1 from (15.21). After that it is easy to show that there exists an odd in z solution of the boundary problem (15.21). Using the linearity of the problem and (15.23) we can find the solution of problem (15.21) in the form

$$T_1 = \frac{h e^{\delta_T T_{w0}} \sinh(Bh)}{\delta_T B} [s_{*x}(\gamma - \sqrt{1-e^2}\sqrt{1-\gamma^2} \frac{\partial p}{\partial \xi} / \frac{\partial p}{\partial \chi})$$

$$\tag{15.24}$$

$$+ s_{*y}\sqrt{1-e^2}(\sqrt{1-e^2}\sqrt{1-\gamma^2} + \gamma \frac{\partial p}{\partial \xi} / \frac{\partial p}{\partial \chi})]\{\frac{\tanh(Bz)}{\tanh \frac{Bh}{2}} - \frac{2z}{h}\}.$$

For the case of line EHL contacts, i.e., for $\gamma = 1$, $s_{*x} = \pm 1$, $s_{*y} = 0$, the expression for νT_1 obtained in this section is identical with the corresponding expression obtained for a line contact in Section 7.3.

Further, it will become clear that it is always sufficient to find a two-term approximation for T to properly approximate the Reynolds equation. Now, we can substitute the expansion (15.20), (15.22), and (15.24) into the Reynolds equation from (15.15), expand the necessary functions in power series in $\nu \ll 1$ and retain just the first nonzero terms of functions R and S in (15.15). That will lead us to an approximate Reynolds equation. However, it can be simplified further in the inlet and exit zones where the derivatives along the flow dominant direction (the χ-axis) are much larger than the corresponding derivatives along the perpendicular to it direction of the ξ-axis, i.e., $\frac{\partial}{\partial\chi} \gg \frac{\partial}{\partial\xi}$. Therefore, retaining in the inlet and exit zones just the main terms in the Reynolds equation in these zones we get the following approximation of the original thermal EHL problem

$$\frac{H_{00}^2}{V}\frac{\partial}{\partial\chi}\{\frac{h^3}{\mu^0}R_{pT}(\chi)\frac{\partial p}{\partial\chi}\} = (1-e^2+e^2\gamma^2)^{3/2}\frac{\partial h}{\partial\chi}, \quad \mu^0=\exp(Qp),$$

$$p(\chi,\xi)\mid_\Gamma=0, \quad \frac{dp(\chi,\xi)}{d\vec{n}}\mid_{\Gamma_e}=0,$$

$$H_{00}[h(\chi,\xi)-1] = x^2(\chi,\xi)+\frac{\delta}{1-e^2}y^2(\chi,\xi)$$

(15.25)

$$+\frac{D(e)}{\pi(1-e^2)^{3/2}}\{\int\int_\Omega \frac{p(\chi_1,\xi_1)d\chi_1 d\xi_1}{\sqrt{(x-\chi_1)^2+\frac{(\xi-\xi_1)^2}{1-e^2}}} - \int\int_\Omega \frac{p(\chi_1,\xi_1)d\chi_1 d\xi_1}{\sqrt{\chi_1^2+\frac{\xi_1^2}{1-e^2}}}\},$$

$$\int\int_\Omega p(\chi_1,\xi_1)d\chi_1 d\xi_1 = \frac{2\pi}{3},$$

where function $R_{pT}(\chi)$ is determined by the expression

$$R_{pT}(\chi) = e^{\delta_T T_{w0}}\frac{3(1+\eta^2)}{\ln^2(\eta+\sqrt{\eta^2+1})}\{[\gamma^2+(1-e^2)^2(1-\gamma^2)]$$

$$\times[\frac{\eta\ln(\eta+\sqrt{\eta^2+1})}{\sqrt{\eta^2+1}} - \ln(1+\eta^2) + 2\frac{\int_0^{\ln(\eta+\sqrt{\eta^2+1})}\ln\cosh t dt}{\ln(\eta+\sqrt{\eta^2+1})}]$$

$$+[\gamma s_{*x}+(1-e^2)\sqrt{1-\gamma^2}s_{*y}]^2[1-(\frac{\eta}{\sqrt{\eta^2+1}}+\frac{\sqrt{\eta^2+1}}{\eta})$$

(15.26)

$$\times\ln(\eta+\sqrt{\eta^2+1})+2\ln(1+\eta^2)-4\frac{\int_0^{\ln(\eta+\sqrt{\eta^2+1})}\ln\cosh t dt}{\ln(\eta+\sqrt{\eta^2+1})}]\},$$

$$\eta = \frac{1}{2}\sqrt{\frac{\kappa N_*\delta_T\mu^0}{2\lambda}}e^{-\delta_T T_{w0}},$$

which is obviously much simpler than the original problem (15.15). In (15.25) we have

$$\int_0^z \ln(\cosh(t))dt = \frac{z^2}{2} - z\ln 2 + \frac{\pi^2}{24} - \frac{1}{2}\sum_{k=1}^\infty \frac{(-1)^{k+1}}{k^2}e^{-2kz},$$

(15.27)

$$\int_0^z \ln(\cosh(t))dt = \frac{z^3}{6} + O(z^4), \quad z \ll 1.$$

Here the meaning of the function $R_{pT}(\chi)$ is similar to the one for function $R_T(x)$ which was obtained for a line thermal EHL problem in Section 7.3 (see formula (7.27)).

For the case of a line contact in (15.26) we have to take $e = 1$, $\gamma = 1$, $s_{*x} = \pm 1$, and $s_{*y} = 0$ which will make $R_{pT}(\chi) = R_T(\chi)$ from (7.27).

It can be shown that if $T_{w0} = 0$ and $\eta \to 0$ then $R_{pT}(\chi) \to \gamma^2 + (1 - e^2)^2(1 - \gamma^2)$, which corresponds to the case of isothermal regime for Newtonian lubricant. On the other hand, for $\eta \to \infty$ (which represents the conditions of high heat generation in the lubrication layer), we have

$$R_{pT}(\chi) \to R_\infty(\chi),$$

$$R_\infty(\chi) = \frac{3(1+\eta^2)}{\ln^2(\eta+\sqrt{\eta^2+1})}[\gamma s_{*x} + (1 - e^2)\sqrt{1 - \gamma^2}s_{*y}]^2, \qquad (15.28)$$

$$\eta = \frac{1}{2}\sqrt{\frac{\kappa_{N*}\delta_T\mu^0}{2\lambda}}e^{-\delta_T T_{w0}}.$$

Obviously, there is a special case when

$$[\gamma s_{*x} + (1 - e^2)\sqrt{1 - \gamma^2}s_{*y}]^2 = \gamma^2 + (1 - e^2)^2(1 - \gamma^2), \qquad (15.29)$$

for which function $R_{pT}(\chi)$ takes the form

$$R_{pT}(\chi) = [\gamma^2 + (1 - e^2)^2(1 - \gamma^2)]R_T(\chi),$$

$$R_T(\chi) = e^{\delta_T T_{w0}}\frac{3(1+\eta^2)}{\ln^2(\eta+\sqrt{\eta^2+1})}\{1 - \frac{\sqrt{\eta^2+1}}{\eta}\ln(\eta + \sqrt{\eta^2 + 1})$$

$$\qquad (15.30)$$

$$+ \ln(1 + \eta^2) - 2\frac{\int_0^{\ln(\eta+\sqrt{\eta^2+1})}\ln\cosh t\,dt}{\ln(\eta+\sqrt{\eta^2+1})}\}.$$

Here function $R_T(\chi)$ coincides with the one determined in (7.27) for thermal line EHL contacts while the Reynolds equation differs from the corresponding Reynolds equation obtained in the inlet and exit zones of isothermal point EHL contacts with skewed lubricant entrainment by just the presence of function $R_T(\chi)$ (compare to equation (13.5)) while all other problem equations are identical. Therefore, it can be easily shown that all results obtained in Chapter 13 are valid with a simple substitution of the term $\frac{\partial}{\partial\chi}\{\frac{h^3}{\mu^0}R_T(\chi)\frac{\partial p}{\partial\chi}\}$ for $\frac{\partial}{\partial\chi}\{\frac{h^3}{\mu^0}\frac{\partial p}{\partial\chi}\}$ in the Reynolds equation and similar substitutions in the asymptotic Reynolds equations in the inlet and exit zones in Chapter 13. In particular, formulas for the film thickness H_{00} and H_0 from Chapter 13 obtained for pre- and over-critical lubrication regimes remain structurally the same with the only difference in the values of the coefficients of proportionality A, A_s, A_f. Specifically, in addition to the parameters these coefficients depend on in Chapter 13 here they also depend on T_{w0} and η (see (15.25)), i.e., on the parameters responsible for heat generation in the lubrication layer.

Another special case is when the components of the slide-to-roll ratios are such that

$$\gamma s_{*x} + (1 - e^2)\sqrt{1 - \gamma^2}s_{*y} = 0. \tag{15.31}$$

In this case function $R_{pT}(\chi)$ takes the form

$$R_{pT}(\chi) = [\gamma^2 + (1 - e^2)^2(1 - \gamma^2)]e^{\delta_T T_{w0}} \frac{3(1+\eta^2)}{\ln^2(\eta+\sqrt{\eta^2+1})}$$

$$\times \left\{ \frac{\eta \ln(\eta+\sqrt{\eta^2+1})}{\sqrt{\eta^2+1}} - \ln(1 + \eta^2) + 2\frac{\int_0^{\ln(\eta+\sqrt{\eta^2+1})} \ln\cosh t\, dt}{\ln(\eta+\sqrt{\eta^2+1})} \right\}. \tag{15.32}$$

In the general case different from (15.29) the asymptotic analysis of the problem (15.25) follows precisely the analysis of Chapter 13. Again, all the results here are different from the ones obtained in Chapter 13 only by the substitution of $\frac{\partial}{\partial\chi}\{\frac{h^3}{\mu^0}R_{pT}(\chi)\frac{\partial p}{\partial\chi}\}$ for $\frac{\partial}{\partial\chi}\{\frac{h^3}{\mu^0}\frac{\partial p}{\partial\chi}\}$ in the Reynolds equation and similar substitutions in the asymptotic Reynolds equations in the inlet and exit zones. Therefore, formulas for the film thickness H_{00} and H_0 from Chapter 13 obtained for pre- and over-critical lubrication regimes remain structurally the same with the only difference in the values of the coefficients of proportionality A, A_s, A_f. Specifically, in addition to the parameters these coefficients depend on in Chapter 13, here they also depend on T_{w0}, η, s_{*x}, and s_{*y} (see (15.25)), i.e., on the parameters responsible for heat generation in the lubrication film.

Moreover, the above results are valid if $\epsilon_q \ll \epsilon_\nu$, where ϵ_ν is the solution of the equation $\nu(x) = O(1)$ in the inlet zone for pre-critical lubrication regimes and in the ϵ_0-inlet zone for over-critical lubrication regimes. Also, we can determine ϵ_f as the size of the inlet (or ϵ_q-inlet) zone for which we have a fully flooded lubrication regime. Obviously, ϵ_f is determined by the solution of the equation $H_{00} = O(\epsilon_q^{3/2})$. Therefore, if $\epsilon_\nu \ll \epsilon_f$ then both starved and fully flooded lubrication regimes can be realized while if $\epsilon_f \ll \epsilon_\nu$ or $\epsilon_f = O(\epsilon_\nu)$ then only starved lubrication regimes can take place. The details of these limitations are left as an exercise for the reader.

The point EHL problem formulation for the case of straight lubricant entrainment is obtained from equations (15.15) for $\gamma = 1$. For this case all the results of the analysis follow directly from the results obtained for the case of skewed lubricant entrainment in Chapter 13 by substituting in them $\gamma = 1$.

The behavior of the numerical solutions of the point EHL problem along the lubricant "central" flow streamlines is very similar to the one studies in Chapter 7.

15.4 Approximations of Sliding Frictional Stresses. No Spinning EHL Contact Case

For the cases when $\nu \ll 1$ in the entire contact and $\omega_0 = 0$ the single-term approximations for the components of the sliding frictional stress follows from (15.12), (15.22), and (15.23). These expressions take the form

$$f_x = \frac{V s_x}{12 H_{00}} e^{-\delta_T T_{w0}} \frac{B}{\sinh(Bh)}, \quad f_y = \frac{V s_y}{12 H_{00}} e^{-\delta_T T_{w0}} \frac{B}{\sinh(Bh)},$$

$$B = \frac{2}{h} \ln\{\eta + \sqrt{\eta^2 + 1}\}, \quad \eta = \frac{1}{2}\sqrt{\frac{\delta_T \kappa_{N*} \mu^0}{2\lambda}} e^{-\delta_T T_{w0}}.$$

$$(15.33)$$

Depending on the parameters of the problem and limitations of the method mentioned in the preceding section formulas (15.33) may be valid for starved and fully flooded lubrication regimes. Therefore it makes sense to analyze the dependence of f_x and f_y on the slide-to-roll ratios s_x and s_y, respectively. Obviously, keeping in mind that κ_* is proportional to $s_*^2 = s_x^2 + s_y^2$ (see (15.17)) for small s_x and s_y the values of f_x and f_y are proportional to s_x and s_y, respectively. For large values of s_* values of η and, therefore, B (see (15.33)) are large which makes $B/\sinh(Bh)$ exponentially small. Therefore, for large values of s_x and s_y the values of f_x and f_y vanish exponentially fast due to very fast increase of heat generation in the lubrication layer and rapid drop in lubricant viscosity. Now, we can conclude that as the sliding frictional stresses first grow almost linearly with the slide-to-roll ratios, then they reach their maximum values, and after that for higher slide-to-roll ratios they decrease precipitously.

The analysis of contact fatigue performed in [2] showed that contact fatigue life depends significantly on the contact frictional force. It means that for subsurface initiated contact fatigue the details of the distribution of the frictional contact stress are insignificant as long as the frictional force is determined sufficiently precisely. Due to the fact that the provided above asymptotic analysis of the heavily loaded point EHL contacts is valid along the "central" flow streamlines and fails to produce an accurate solution approximations only in the zones the areas of which are of the order of magnitude of $O(\epsilon_q^2) \ll 1$, $\omega \ll 1$ (for example, for the case of straight lubricant entrainment these small zones are in the vicinity of the points with coordinates $(0, -1)$ and $(0, 1)$). The latter zones produce a negligibly small contribution $O(\epsilon_q^2) \ll 1$, $\omega \ll 1$, to the dominating sliding frictional force. Therefore, the above asymptotic analysis provides sufficient information for an accurate approximation of frictional forces and, thus, for contact fatigue life.

15.5 Asymptotic Solution for the Lubricant Temperature T. Case of EHL Contact with Spinning

Let us consider the case of spinning, i.e., $\omega_0 > 0$. Using the parameters determined in (15.17) from the last equation in (15.12) for the lubricant temperature T we obtain

$$\frac{\partial}{\partial z}(\lambda \frac{\partial T}{\partial z}) = -\kappa_{N*} \mu^0 e^{\delta_T T} \{ [s_{*x} - 2\frac{\omega_0 y}{s_*} + \frac{12H_{00}^2}{Vs_*} \frac{1}{\mu^0} \frac{\partial p}{\partial x}(z \int_{-h/2}^{h/2} e^{\delta_T T} ds$$

$$- \int_{-h/2}^{h/2} e^{\delta_T T} sds)]^2 + [s_{*y} + 2\sqrt{1-e^2}\frac{\omega_0 x}{s_*} + \sqrt{1-e^2}\frac{12H_{00}^2}{Vs_*} \frac{1}{\mu^0} \frac{\partial p}{\partial y}$$

$$\tag{15.34}$$

$$\times (z \int_{-h/2}^{h/2} e^{\delta_T T} ds - \int_{-h/2}^{h/2} e^{\delta_T T} sds)]^2 \} / (\int_{-h/2}^{h/2} e^{\delta_T T} ds)^2,$$

$$T(x, y, -\tfrac{h}{2}) = T_{w1}(x, y), \ T(x, y, \tfrac{h}{2}) = T_{w2}(x, y).$$

For simplicity we will consider just the case of equal surface temperatures

$$T(x, y, -\tfrac{h}{2}) = T(x, y, \tfrac{h}{2}) = T_{w0}(x, y). \tag{15.35}$$

The idea of the problem solution is exactly the same as in the previous cases. First, we need to asymptotically solve the equation for the lubricant temperature T for the case of sliding frictional stresses much higher than the rolling frictional stresses. This solution for T will allow us to simplify the Reynolds equation and to proceed with our further asymptotic analysis. To be precise, let us introduce the function

$$\nu = \frac{12H_{00}^2}{Vs_*} \frac{1}{\mu^0} \frac{\partial p}{\partial x}. \tag{15.36}$$

This function $\nu(x, y)$ represents the ratio of the rolling to sliding frictional stresses. We will assume that in the inlet and exit zones $\nu \ll 1$. That assumption makes $\nu \ll 1$ in the entire contact region. In addition to that we will assume that (see (15.18))

$$\frac{\partial p}{\partial y} = O(\frac{\partial p}{\partial x}). \tag{15.37}$$

After that the solution for T can be searched in the form of a regular asymptotic expansion in $\nu \ll 1$ (see (15.19))

$$T(x, y, z) = T_0(x, y, z) + \nu T_1(x, y, z) + \dots$$

$$\tag{15.38}$$

$$T_0(x, y, z), \ T_1(x, y, z) = O(1), \ \nu \ll 1,$$

where $T_0(x, y, z)$ and $T_1(x, y, z)$ are unknown and have to be found from the solution of the problem.

We will assume that λ may be a constant ($\lambda = 1$) or it can be a function of pressure p but it is independent of temperature T. Then temporarily assuming that $p(x, y)$ and $h(x, y)$ are known and substituting the expansion (15.38) into the equation for T from (15.34), taking into account (15.35), expanding the equation in powers of ν, and equating the corresponding terms we obtain the following systems of equations

$$\lambda \frac{\partial^2 T_0}{\partial z^2} = -\frac{\kappa_{N*} \mu^0 e^{\delta_T T_0}(s_{x1}^2 + s_{y1}^2)}{(\int\limits_{-h/2}^{h/2} e^{\delta_T T_0} ds)^2},$$
(15.39)

$$T_0(x, y, -\tfrac{h}{2}) = T_0(x, y, \tfrac{h}{2}) = T_{w0}(x, y),$$

$$\lambda \frac{\partial^2 T_1}{\partial z^2} = -\kappa_{N*} \mu^0 e^{\delta_T T_0} \{(s_{x1}^2 + s_{y1}^2) \delta_T T_1 + [2s_{x1}$$

$$+2s_{y1} \sqrt{1 - e^2} \frac{\partial p}{\partial y} / \frac{\partial p}{\partial x}]z \int\limits_{-h/2}^{h/2} e^{\delta_T T_0} ds \} / (\int\limits_{-h/2}^{h/2} e^{\delta_T T_0} ds)^2,$$
(15.40)

$$T_1(x, y, -\tfrac{h}{2}) = T_1(x, y, \tfrac{h}{2}) = 0,$$

where it is already taken into account the fact that $T_1(x, y, z)$ is an odd function with respect to z (i.e., $T_{(x, y, -z)} = -T_1(x, y, z)$) and

$$s_{x1} = s_{*x} - 2\frac{\omega_0 y}{s_*}, \quad s_{y1} = s_{*y} + 2\sqrt{1 - e^2}\frac{\omega_0 x}{s_*}.$$
(15.41)

The solution of problem (15.39) with different parameters is obtained in (7.23) and here it can be represented in the form

$$T_0 = T_{w0} + \tfrac{2}{\delta_T} \ln\{\cosh(\tfrac{Bh}{2}) / \cosh(Bz)\},$$

$$B = \tfrac{2}{h} \ln\{\eta + \sqrt{\eta^2 + 1}\}, \quad \eta = \tfrac{1}{2}\sqrt{\frac{\delta_T \kappa_{N*}(s_{x1}^2 + s_{y1}^2)\mu^0}{2\lambda}} e^{-\delta_T T_{w0}}.$$
(15.42)

Function T_0 is even with respect to variable z. Therefore,

$$\int\limits_{-h/2}^{h/2} e^{\delta_T T_0} ds = e^{\delta_T T_{w0}} \frac{\sinh(Bh)}{B}.$$
(15.43)

Using the linearity of problem (15.40) and (15.23) the solution for T_1 can be found in the form

$$T_1 = \frac{he^{\delta_T T_{w0}} \sinh(Bh)}{\delta_T B(s_{x1}^2 + s_{y1}^2)}[s_{x1} + s_{y1}\sqrt{1 - e^2}\frac{\partial p}{\partial y} / \frac{\partial p}{\partial x}]\{\frac{\tanh(Bz)}{\tanh\frac{Bh}{2}} - \frac{2z}{h}\}.$$
(15.44)

The solution for the general case of unequal surface temperatures T_{w1} and T_{w2} is considered in Section 10.3.

For the case of line EHL contacts, i.e., for $e = 1$, $\omega_{*0} = 0$, $s_{*x} = \pm 1$ and $s_{*y} = 0$, the expression for νT_1 obtained in this section is identical with the corresponding expression obtained for a line contact in Section 7.3.

Now, we can determine the approximate Reynolds equation if we substitute the expressions for T, T_0, and T_1 from (15.38), (15.41), (15.41), and (15.44) in the terms of the Reynolds equation (15.12) and retain only the terms proportional to $\nu^0 = 1$ and ν. This substitution produces the following approximate Reynolds equation

$$\frac{\partial}{\partial x}\{W_1[W_2 + \frac{s_{x1}^2 W_3}{4}]\frac{\partial p}{\partial x}\} + (1 - e^2)\frac{\partial}{\partial y}\{W_1[W_2 + \frac{s_{y1}^2 W_3}{4}]\frac{\partial p}{\partial y}\}$$

$$+\frac{\sqrt{1-e^2}}{4}\{\frac{\partial}{\partial x}[W_1 W_3 s_{x1} s_{y1}\frac{\partial p}{\partial y}] + \frac{\partial}{\partial y}[W_1 W_3 s_{x1} s_{y1}\frac{\partial p}{\partial x}]\} + \ldots \qquad (15.45)$$

$$= \frac{V}{H_{00}^2}\{(\gamma - \omega_0 y)\frac{\partial h}{\partial x} + \sqrt{1 - e^2}(\sqrt{1 - \gamma^2} + \sqrt{1 - e^2}\omega_0 x)\frac{\partial h}{\partial y}\},$$

where functions W_1, W_2, and W_3 are determined by the relationships

$$W_1 = 12\frac{e^{\delta_T T_{w0}}}{\mu^0}\frac{h \cosh^2 \frac{Bh}{2}}{B^2}, \quad W_2 = \frac{Bh}{2}\tanh\frac{Bh}{2} - 2\ln\cosh\frac{Bh}{2}$$

$$+\frac{4}{Bh}\int_0^{Bh/2}\ln\cosh t\, dt, \quad W_3 = 1 - \frac{Bh}{\sinh(Bh)} - 2W_2. \qquad (15.46)$$

In equation (15.45) we properly retained the derivatives of functions with respect to x and y. That will allow us to consider the approximate Reynolds equation in a more convenient coordinate system. In particular, as in the case of the isothermal EHL problem for heavily loaded contact with spinning (see Chapter 14) in this case it is convenient to consider the problem in polar coordinates (ρ, φ) which are introduced as follows (see (14.5) and (14.6))

$$x = \frac{1}{\sqrt{1-e^2}}[-\frac{\sqrt{1-\gamma^2}}{\omega_0} + \rho\sin\varphi], \quad y = \frac{\gamma}{\omega_0} - \rho\cos\varphi, \quad dxdy = \frac{\rho d\rho d\varphi}{\sqrt{1-e^2}}, \qquad (15.47)$$

$$\rho = \sqrt{(\sqrt{1 - e^2}x + \frac{\sqrt{1-\gamma^2}}{\omega_0})^2 + (y - \frac{\gamma}{\omega_0})^2}, \quad \tan\varphi = \frac{\sqrt{1-e^2}x+\frac{\sqrt{1-\gamma^2}}{\omega_0}}{\frac{\gamma}{\omega_0}-y}. \qquad (15.48)$$

The reason for such a choice of new independent variables is the fact that in the Hertzian region (i.e., away from the inlet and exit contact boundaries) the lubricant moves along circular arcs $\rho = const$ and the gap h along these arcs is practically constant, i.e., $h = h(\rho)$.

In variables (14.66) equation (15.45), the boundary conditions on pressure p from (15.12), and the last two equations from (15.12) will take the form

$$(1 - e^2)[\sin\varphi\frac{\partial}{\partial\rho} + \frac{\cos\varphi}{\rho}\frac{\partial}{\partial\varphi}]\{W_1(W_2 + \frac{s_{x1}^2}{4}W_3)[\sin\varphi\frac{\partial p}{\partial\rho}$$

$$+\frac{\cos\varphi}{\rho}\frac{\partial p}{\partial\varphi}]\} + \frac{1-e^2}{4}[\sin\varphi\frac{\partial}{\partial\rho} + \frac{\cos\varphi}{\rho}\frac{\partial}{\partial\varphi}]\{W_1 W_3 s_{x1} s_{y1}[-\cos\varphi\frac{\partial p}{\partial\rho} \qquad (15.49)$$

$$+ \tfrac{\sin\varphi}{\rho}\tfrac{\partial p}{\partial \varphi}]\} + (1-e^2)[-\cos\varphi\tfrac{\partial}{\partial \rho} + \tfrac{\sin\varphi}{\rho}\tfrac{\partial}{\partial \varphi}]\{W_1(W_2 + \tfrac{s_{y1}^2}{4}W_3)$$

$$\times[-\cos\varphi\tfrac{\partial p}{\partial \rho} + \tfrac{\sin\varphi}{\rho}\tfrac{\partial p}{\partial \varphi}]\} + \tfrac{1-e^2}{4}[-\cos\varphi\tfrac{\partial}{\partial \rho} + \tfrac{\sin\varphi}{\rho}\tfrac{\partial}{\partial \varphi}]\{W_1 W_3 s_{x1} s_{y1}$$

$$\times[\sin\varphi\tfrac{\partial p}{\partial \rho} + \tfrac{\cos\varphi}{\rho}\tfrac{\partial p}{\partial \varphi}]\} = \tfrac{V\omega_0(1-e^2)^{3/2}}{H_{00}^2}\tfrac{\partial h}{\partial \varphi}, \quad \mu = \exp(Qp),$$

$$p(\rho,\varphi)\mid_\Gamma = 0, \quad \tfrac{dp(\rho,\varphi)}{d\overrightarrow{n}}\mid_{\Gamma_e} = 0,$$

$$H_{00}[h(\rho,\varphi) - 1] = x^2(\rho,\varphi) + \tfrac{\delta}{1-e^2}y^2(\rho,\varphi)$$

$$+ \tfrac{D(e)}{\pi(1-e^2)}\{\int\int_\Omega \tfrac{p(\rho_1,\varphi_1)\rho_1 d\rho_1 d\varphi_1}{\sqrt{\rho^2 - 2\rho\rho_1\cos(\varphi - \varphi_1) + \rho_1^2}}$$

$$- \int\int_\Omega \tfrac{p(\rho_1,\varphi_1)\rho_1 d\rho_1 d\varphi_1}{\sqrt{\rho_1^2 - \tfrac{2}{\omega_0}\rho_1[\gamma\cos(\varphi_1) + \sqrt{1-\gamma^2}\sin(\varphi_1)] + \tfrac{1}{\omega_0^2}}}\}, \tag{15.50}$$

$$\int\int_\Omega p(\rho_1,\varphi_1)\rho_1 d\rho_1 d\varphi_1 = \tfrac{2\pi}{3}\sqrt{1-e^2},$$

where x and y are functions of ρ and φ determined according to (15.47).

Now, we are ready to further engage in the asymptotic analysis of the problem. As it was done earlier for the isothermal EHL problem with spinning away from the inlet Ω_i and exit Ω_e zones along the "central" flow streamlines we have the condition (14.11). That leads to the fact that in the Hertzian region of a heavily loaded point EHL contact along these "central" streamlines with high precision $h(\rho,\varphi) = h_\infty(\rho)$, where $h_\infty(\rho)$ is the exit film thickness at the point $(\rho,\varphi) \in \Gamma_e$ practically independent from φ. After that we can conduct the asymptotic analysis which essentially is identical to the one performed in Chapter 14 and it leads to results very similar to the ones obtained in Chapter 14. In particular, keeping in mind that in the inlet Ω_i and exit Ω_e zones derivatives with respect to φ are much larger than the corresponding derivatives with respect to ρ (i.e., $\tfrac{\partial}{\partial \varphi} \gg \tfrac{\partial}{\partial \rho}$) in these zones the main terms of the Reynolds equation (15.49) are as follows

$$\tfrac{H_{00}^2}{V}\tfrac{1}{\rho^2}\tfrac{\partial}{\partial \varphi}\{R_s(\rho,\varphi)\tfrac{\partial p}{\partial \varphi}\} = \tfrac{\omega_0}{\sqrt{1-e^2}}\tfrac{\partial h}{\partial \varphi},$$

$$R_s(\rho,\varphi) = W_1[W_2 + \tfrac{W_3}{4}(s_{x1}^2\cos^2\varphi + s_{y1}^2\sin^2\varphi)], \tag{15.51}$$

$$s_{x1} = s_{*x} - \tfrac{2(\gamma - \omega_0\rho\cos\varphi)}{s_*}, \quad s_{y1} = s_{*y} - \tfrac{2(\sqrt{1-\gamma^2} - \omega_0\rho\sin\varphi)}{s_*}, \tag{15.52}$$

where functions W_1, W_2, and W_3 are determined by (15.46).

In a special case of $s_x = 2\gamma$ and $s_y = 2\sqrt{1-\gamma^2}$ the expression for function R_s assumes a simpler form

$$R_s(\rho,\varphi) = W_1[W_2 + \tfrac{W_3\omega_0^2\rho^2}{s_*^2}(\cos^4\varphi + \sin^4\varphi)]. \tag{15.53}$$

It can be shown that $W_2(p,h) \to \frac{B^2 h^2}{12}$ and $W_3(p,h) = O(B^4 h^4)$ as $Bh \to 0$. Therefore, for $\delta_T = 0$ and $Bh \to 0$ we have $R_s(\rho, \varphi) \to \frac{h^3}{\mu^0}$ which corresponds to the isothermal EHL problem for lubricant with Newtonian rheology.

In the general case, the rest of the asymptotic analysis is done exactly as in Chapter 14. All the results following from this asymptotic analysis are different from the ones obtained in Chapter 14 only by the substitution in the Reynolds equation of $\frac{\partial}{\partial r_p}\{R_s(\rho, r_p)\frac{\partial q_p}{\partial r_p}\}$ for $\frac{\partial}{\partial r_p}\{\frac{h_{qp}^3}{\mu_q^0}\frac{\partial q_p}{\partial r_p}\}$ in the inlet zone and $\frac{\partial}{\partial s_p}\{R_s(\rho, s_p)\frac{\partial g_p}{\partial s_p}\}$ for $\frac{\partial}{\partial s_p}\{\frac{h_{gp}^3}{\mu_g^0}\frac{\partial g_p}{\partial s_p}\}$ in the exit zone, where $R_s(\rho, r_p)$ and $R_s(\rho, s_p)$ are the expressions for $R_s(\rho, \varphi)$ in the inlet zone (i.e., when $\varphi = \gamma_i(\rho) + \epsilon_q \frac{r_p}{\rho}$) and exit zone (i.e., when $\varphi = \gamma_e(\rho) + \epsilon_q \frac{s_p}{\rho}$). Therefore, formulas for the film thickness H_{00} and H_0 from Chapter 14 obtained for pre- and over-critical lubrication regimes remain structurally the same with the only difference in the values of the coefficients of proportionality A, A_s, A_f, etc. which reflect the thermal nature of the problem. Specifically, in addition to the parameters these coefficients depend on in Chapter 14, here they also depend on T_{w0}, η, s_{*x}, s_{*y}, and ω_0, i.e., on the parameters responsible for heat generation in the lubrication film.

Moreover, the above results are valid if $\epsilon_q \ll \epsilon_\nu$, where ϵ_ν is the solution of the equation $\nu(x) = O(1)$ in the inlet zone for pre-critical lubrication regimes and in the ϵ_0-inlet zone for over-critical lubrication regimes. Also, we can determine ϵ_f as the size of the inlet (or ϵ_q-inlet) zone for which we have a fully flooded lubrication regime. Obviously, ϵ_f is determined by the solution of the equation $H_{00} = O(\epsilon_q^{3/2})$. Therefore, if $\epsilon_\nu \ll \epsilon_f$ then both starved and fully flooded lubrication regimes can be realized while if $\epsilon_f \ll \epsilon_\nu$ or $\epsilon_f = O(\epsilon_\nu)$ then only starved lubrication regimes can take place. The details of these limitations are left as an exercise for the reader.

Again, here the behavior of numerical solutions of the point EHL problem along the lubricant "central" flow streamlines is very similar to the one considered for the thermal line EHL contacts in Chapter 7.

15.6 Regularization Approaches to Solution of Isothermal Point EHL Problems for Newtonian Lubricants

Numerical solution of isothermal EHL problems for heavily loaded line and point contacts tend to be unstable. The only possible remedy for such a situation is regularization of the numerical solution of such EHL problems. As it has been shown for heavily loaded line EHL contacts such a regularization can be successfully achieved by introducing in the Reynolds equation a relatively small heat generation. Therefore, a similar approach can be used in the case of an isothermal EHL problem for a heavily loaded point contact which would require a certain modification of just the Reynolds equation while the rest of the EHL problem equations would remain the same.

Following the regularization idea realized for line EHL contacts we will assume that $T_{w0} = 0$ and we will consider the cases with and without spinning separately.

In the exit zone the systems of asymptotically valid equations for the isothermal EHL problems can be regularized by replacing the corresponding function M_0 in these equations by its thermal analog while assuming that $\eta = \beta\sqrt{\mu^0}$, where μ^0 coincides with the isothermal lubricant viscosity while β is a sufficiently small positive constant which would provide stability to the problem numerical solution. The value of β is determined empirically.

Suppose we have a heavily loaded lubricated point contact with straight of skewed lubricant entrainment but without spinning. Then the Reynolds equation for the isothermal problem can be replaced by the regularized Reynolds equation which follows from equation (15.15), where T is taken in the form of the asymptotic expansion (15.19), (15.22), and (15.24). Specifically, in this case the Reynolds equation can be taken in the form of equation (15.15) where the expressions for R and S should be taken in the form (see equations (15.42))

$$R = \frac{h^3}{\mu^0} \frac{1+\eta^2}{4\ln^3(\eta+\sqrt{\eta^2+1})} \{ \frac{\eta}{\sqrt{\eta^2+1}} \ln^2(\eta+\sqrt{\eta^2+1})$$

$$- \ln(\eta^2+1)\ln(\eta+\sqrt{\eta^2+1}) + 2 \int\limits_0^{\ln(\eta+\sqrt{\eta^2+1})} \ln\cosh y\, dy \},$$

$$S = \frac{H_{00}^2}{V s_* \sqrt{1-e^2+e^2\gamma^2}} \frac{h^3}{\mu^0} \frac{3(1+\eta^2)}{\ln^2(\eta+\sqrt{\eta^2+1})} [s_{*x}(\gamma\frac{\partial p}{\partial \chi} - \sqrt{1-e^2}\sqrt{1-\gamma^2}\frac{\partial p}{\partial \xi})$$

$$+ s_{*y}\sqrt{1-e^2}(\sqrt{1-e^2}\sqrt{1-\gamma^2}\frac{\partial p}{\partial \chi} + \gamma\frac{\partial p}{\partial \xi})]\{1 - (\frac{\eta}{\sqrt{\eta^2+1}} + \frac{\sqrt{\eta^2+1}}{\eta})$$

$$\times \ln(\eta+\sqrt{\eta^2+1}) + 2\ln(\eta^2+1) - 4\frac{\int\limits_0^{\ln(\eta+\sqrt{\eta^2+1})} \ln\cosh y\, dy}{\ln(\eta+\sqrt{\eta^2+1})} \},$$

$$\eta = \beta\sqrt{\mu^0},$$

$$(15.54)$$

where β is a sufficiently small positive number which provides stability to the solution of the EHL problem for heavily loaded contact while μ^0 coincides with the lubricant viscosity μ in the isothermal case. It can be shown that the Reynolds equation (15.15) with functions R and S from (15.54) approaches the Reynolds equation (13.5) for the isothermal EHL problem as $\beta \to 0$. The rest of the regularized EHL problem equations remain the same as for the isothermal EHL problem.

Therefore, equations (15.15), (15.54), and the last three equations in (13.2) represent the regularized EHL problem for an isothermal heavily loaded point with Newtonian lubricant. By choosing the appropriate value of the positive constant β the solution of this EHL problem will be stable.

Now, let us consider a heavily loaded lubricated point contact with straight or skewed lubricant entrainment and spinning. Then the Reynolds equation for isothermal EHL problem from (14.7) can be regularized by replacing it with the Reynolds equation (15.49) in which functions W_1, W_2, and W_3 follow from (15.46) and can be represented in the form

$$W_1 = e^{\delta_T T_{W0}} \frac{h^3}{\mu^0} \frac{3(1+\eta^2)}{\ln^2(\eta+\sqrt{\eta^2+1})}, \quad W_2 = \frac{\eta}{\sqrt{\eta^2+1}} \ln(\eta + \sqrt{\eta^2 + 1})$$

$$- \ln(\eta^2 + 1) + 2 \frac{\int_0^{\ln(\eta+\sqrt{\eta^2+1})} \ln \cosh y \, dy}{\ln(\eta+\sqrt{\eta^2+1})}, \tag{15.55}$$

$$W_3 = 1 - \frac{\ln(\eta+\sqrt{\eta^2+1})}{\eta\sqrt{\eta^2+1}} - 2W_2, \quad \eta = \beta\sqrt{\mu^0},$$

where β is a sufficiently small positive number which provides stability to the solution of the EHL problem for heavily loaded contact coincides with the lubricant viscosity μ in the isothermal case. It can be shown that the Reynolds equation from (15.49) approaches the Reynolds equation (14.7) for the isothermal EHL problem as $\beta \to 0$. The rest of the regularized EHL problem equations remain the same as for the isothermal EHL problem.

15.7   Closure

An asymptotic approach to thermal steady heavily loaded point EHL contacts with incompressible Newtonian lubricants with skewed lubricant entrainment and spinning ball is developed. The asymptotic solution for lubricant temperature T is obtained analytically. This asymptotic solution for T is used to simplify and reduce the original Reynolds equation to the form similar to the one for an isothermal case. Based on this reduced Reynolds equation the behavior of the pressure and gap distributions in the inlet and exit zones along the "central" streamlines of the lubricant flow are analyzed in detail. It has been shown that in a certain sense the behavior of the lubrication parameters in a thermal point EHL contact is similar to the one in an isothermal point EHL contact which in turn is similar to the behavior of these parameters in a corresponding line EHL contact. Asymptotic analysis of heavily loaded contacts lubricated by a Newtonian fluid recognizes two different regimes of lubrication: pre- and over-critical lubrication regimes. Overall, the difference between these two regimes is in how rapidly the lubricant viscosity grows with pressure and how much lubricant is available at the inlet of the contact. For pre-critical lubrication regimes the asymptotically valid equations for pressure and gap can be derived by following the procedure described in Chapter 14. For over-critical lubrication regimes another approach can be used which (see Chapter 14) which leads to formulas for the lubrication film thickness for starved and fully flooded conditions. The asymptotic approaches to EHL problems offer certain distinct advantages in comparison with the direct numerical solutions of EHL problem among which are the significant simplification of the original thermal EHL problem, a reduction of the number of problem input parameters, and the approach to regularization of numerical solutions. Each of the analytically derived formulas for the central H_{00} lubrication film thickness contains a coefficient of proportionality which can be found by numerically solving the corresponding asymptotically valid equations in the inlet zone or by just curve fitting the experimentally obtained values of the lubrication film thickness.

15.8 Exercises and Problems

1. (a) Derive the expressions for ϵ_f and ϵ_ν for the cases of pre- and over-critical lubrication regimes.

(b) Determine the limitations of the presented approach for the cases of pre- and over-critical lubrication regimes when both starved and fully flooded lubrication regimes can be realized. (Hint: consider the relationship $\epsilon_\nu \ll \epsilon_f$.)

(c) Determine the limitations of the presented approach for the cases of pre- and over-critical lubrication regimes when only starved lubrication regimes can be realized. (Hint: consider the relationship $\epsilon_f \ll \epsilon_\nu$ or $\epsilon_f = O(\epsilon_\nu)$.)

2. Explain why in case when $\epsilon_f = O(\epsilon_\nu)$ and the above analysis is valid only for starved lubrication regimes nonetheless for the case of fully flooded lubrication regimes the structure of the formula for the film thickness H_{00} remains the same.

3. Fix all parameters involved in formulas (15.33) for f_x and f_y except for s_x and s_y. Graph and analyze the functions of f_x and f_y versus s_x and s_y. Determine at which slide-to-roll ratios s_x and s_y frictional sliding stresses f_x and f_y reach their maximum values.

References

[1] Hamrock, B.J. 1994. *Fundamentals of Fluid Film Lubrication*. New York: McGraw-Hill.

[2] Kudish, I.I. and Covitch, M.J. 2010. *Modeling and Analytical Methods in Tribology*. Chapman & Hall/CRC.

16

Isothermal EHL Problems for Heavily Loaded Point Contacts. Non-Newtonian Lubricants

---

16.1 Introduction

The chapter is devoted to the extension of the techniques developed in Chapter 9 for line heavily loaded EHL contacts and in Chapters 12, 13, and 14 for point heavily loaded EHL contacts on the case of isothermal point EHL problems for contacts with non-Newtonian lubricants. The EHL problems of Chapters 12, 13, and 14 are considered under isothermal conditions when lubricants are non-Newtonian. The problems are analyzed for the case of high slide-to-roll ratio. The point EHL problem equations are reduced to equations of problems similar to the line EHL problems for the corresponding non-Newtonian lubricant. In the inlet and exit zones along the lubricant "central" flow streamlines the point EHL problems are reduced to asymptotically valid equations similar to the ones obtained in Chapter 9 for line EHL contacts. Analytical formulas for the lubrication film thickness are derived.

16.2 Problem Formulation

Let us consider a steady isothermal point EHL contact with non-Newtonian lubricant. We will assume that the solids in contact are involved in rolling, sliding, and spinning motion. Specifically, we will assume that the ball is spinning in the counterclockwise direction with the angular speed ω_2 while the translational velocities of the solids are (u_1, v_1) and (u_2, v_2). Therefore, the total solid surface velocities are (u_1, v_1) and $(u_2 - \omega_2 y, v_2 + \omega_2 x)$, where $(x, y))$ are the coordinates of a ball surface point. The coordinate system is introduced in such a way that the x- and y-axes belong to the contact middle plane while the z-axis is perpendicular to this plane and directed upward passing through the ball center (see Chapter 12). In this coordinate system the shapes of the solid surfaces can be approximated by elliptical paraboloids: $z = z_2(x, y) = \frac{x^2}{2R_{2x}} + \frac{y^2}{2R_{2y}}$ and $z = z_1(x, y) = -\frac{x^2}{2R_{1x}} - \frac{y^2}{2R_{1y}}$, where R_{1x}, R_{2x}, R_{1y}, and R_{2y} are the surface curvature radii of the solid surfaces along the x- and y-axes, respectively. The solids' elastic moduli and Poisson's ratios are E_1, E_2 and ν_1, ν_2, respectively. The two solids are loaded by a normal force P acting along the z-axis. The solids are separated by a layer of an incompressible viscous lubricant with non-Newtonian rheology. Using the traditional assumptions that the solids' motions are steady and relatively slow, the size of the contact region is significantly smaller than the solid radii, and the lubrication film thickness is much smaller than the contact size [1] we obtain the continuity equations (4.4) and the momentum equations (4.11) (see Chapter 4)

$$-\frac{\partial p}{\partial x} + \frac{\partial p_{xz}}{\partial z} = 0, \quad -\frac{\partial p}{\partial y} + \frac{\partial p_{yz}}{\partial z} = 0, \quad \frac{\partial p}{\partial z} = 0, \tag{16.1}$$

where the tensor components p_{xz} and p_{yz} are determined by the lubricant rheology as follows

$$\mu \frac{\partial u}{\partial z} = F(p_{xz}), \quad \mu \frac{\partial v}{\partial z} = F(p_{yz}), \tag{16.2}$$

where function F describes the lubricant rheology and it is an odd monotonically increasing function such that $F(0) = 0$. Obviously, this function F has an inverse Φ, $\Phi(F(x)) = x$.

Integrating equations (16.1) we obtain

$$p_{xz} = f_x + z\frac{\partial p}{\partial x}, \quad p_{yz} = f_y + z\frac{\partial p}{\partial y}, \tag{16.3}$$

where $f_x(x,y)$ and $f_y(x,y)$ are the unknown x- and y-components of the lubricant sliding frictional stress which are determined from the second pair of boundary conditions on u and v at $z = \frac{h}{2}$. Integrating equations (16.2) and using the expressions (16.3) together with the no slippage conditions (12.1) on u and v we determine

$$u = u_1 + \frac{1}{\mu} \int_{-h/2}^{z} F(f_x + s\frac{\partial p}{\partial x})ds, \quad v = v_1 + \frac{1}{\mu} \int_{-h/2}^{z} F(f_y + s\frac{\partial p}{\partial y})ds. \tag{16.4}$$

Integrating the continuity equation (4.4) with respect to z from $-\frac{h}{2}$ to $\frac{h}{2}$, changing the order of differentiation and integration and using the last pair of the boundary no slippage conditions which is due to ball spinning instead of having the form of (12.1) actually are

$$u(x,y,\tfrac{h}{2}) = u_2 - \omega_2 y, \quad v(x,y,\tfrac{h}{2}) = v_2 + \omega_2 x, \tag{16.5}$$

as well as using the integrals of u and v from (16.4) with respect to z from $-\frac{h}{2}$ to $\frac{h}{2}$ we obtain the generalized Reynolds equation which describes variations in pressure $p(x,y)$ as a function of variations in gap $h(x,y)$ between the contact surfaces. To simplify the form of the Reynolds equation repeated integrals are reduced to single integrals. The rest of the problem formulation is done the same way as in Chapter 12. As a result of this derivation the EHL problem for the case of non-Newtonian lubricants is reduced to the equations

$$\frac{\partial}{\partial x}\{\frac{1}{\mu}\int\limits_{-h/2}^{h/2} zF(f_x + z\frac{\partial p}{\partial x})dz\} + \frac{\partial}{\partial y}\{\frac{1}{\mu}\int\limits_{-h/2}^{h/2} zF(f_y + z\frac{\partial p}{\partial y})dz\}$$

$$= \frac{u_1+u_2-\omega_2 y}{2}\frac{\partial h}{\partial x} + \frac{v_1+v_2+\omega_2 x}{2}\frac{\partial h}{\partial y}, \; \mu = \mu_a e^{\alpha p},$$

$$p(x,y)\mid_\Gamma = 0, \; \frac{dp(x,y)}{d\overrightarrow{n}}\mid_{\Gamma_e} = 0,$$

$$\int\limits_{-h/2}^{h/2} F(f_x + z\frac{\partial p}{\partial x})dz = \mu(u_2 - u_1 - \omega_2 y), \quad (16.6)$$

$$\int\limits_{-h/2}^{h/2} F(f_y + z\frac{\partial p}{\partial y})dz = \mu(v_2 - v_1 + \omega_2 x),$$

$$h(x,y) = h_0 + \frac{x^2}{2R_x} + \frac{y^2}{2R_y} + \frac{1}{\pi E'}\{\int\int\limits_\Omega \frac{p(x_1,y_1)dx_1 dy_1}{\sqrt{(x-x_1)^2+(y-y_1)^2}}$$

$$- \int\int\limits_\Omega \frac{p(x_1,y_1)dx_1 dy_1}{\sqrt{x_1^2+y_1^2}}\}, \; \int\int\limits_\Omega p(x_1,y_1)dx_1 dy_1 = P,$$

where μ is the Barus lubricant viscosity, μ_a and α are the ambient lubricant viscosity and the pressure coefficient of viscosity, respectively, h_0 is the central film thickness at the point with coordinates $(0,0)$, R_x, R_y, and E' are the effective radii along the x- and y-axes and the effective modulus of elasticity, $\frac{1}{R_x} = \frac{1}{R_{1x}} + \frac{1}{R_{2x}}$, $\frac{1}{R_y} = \frac{1}{R_{1y}} + \frac{1}{R_{2y}}$, and $\frac{1}{E'} = \frac{1-\nu_1^2}{E_1} + \frac{1-\nu_2^2}{E_2}$, respectively, Ω and Γ are the contact region and its boundary, respectively, and $\overrightarrow{n}$ is the unit external normal vector to the boundary Γ. The inlet Γ_i and exit Γ_e portions of the contact boundary Γ through which the lubricant enters and leaves the contact, respectively, constitute the whole boundary of Ω, i.e. $\Gamma_i \bigcup \Gamma_e = \Gamma$. The inlet Γ_i and exit Γ_e boundaries are determined by the relationships

$$(x,y) \in \Gamma_i \; if \; \overrightarrow{F_f} \cdot \overrightarrow{n} < 0; \; (x,y) \in \Gamma_e \; if \; \overrightarrow{F_f} \cdot \overrightarrow{n} \geq 0,$$

$$\overrightarrow{F_f} = (u_1 + u_2 - \omega_2 y, v_1 + v_2 + \omega_2 x)\frac{h}{2}$$

$$-(\frac{1}{\mu}\int\limits_{-h/2}^{h/2} zF(f_x + z\frac{\partial p}{\partial x})dz, \frac{1}{\mu}\int\limits_{-h/2}^{h/2} zF(f_y + z\frac{\partial p}{\partial y})dz), \quad (16.7)$$

where $\overrightarrow{F_f}$ is the fluid flux through the contact boundary Γ and $\overrightarrow{a} \cdot \overrightarrow{b}$ is the dot product of vectors $\overrightarrow{a}$ and $\overrightarrow{b}$.

Let us introduce the dimensionless variables according to (12.5) and also

$$F' = \frac{h_0}{\mu_a \sqrt{(u_1+u_2)^2+(v_1+v_2)^2}}F, \; f_x' = \frac{2R_x}{a_H p_H}f_x, \; f_y' = \frac{2R_x}{a_H p_H}f_y, \quad (16.8)$$

as well as use the dimensionless parameters from (13.1) and (15.11), where a_H, b_H, and p_H are the Hertzian semi-axes (small and large) of the dry elliptical contact and the Hertzian maximum pressure, respectively, which are determined by equations (12.7) and (12.8). Therefore, in dimensionless variables (12.5) and parameters (14.3), (13.1) (primes are dropped for convenience) the equations of the problem are reduced to the following form

$$\frac{\partial}{\partial x}\{\frac{1}{\mu}\int_{-h/2}^{h/2} zF(f_x + H_{00}z\frac{\partial p}{\partial x})dz\}$$

$$+\sqrt{1-e^2}\frac{\partial}{\partial y}\{\frac{1}{\mu}\int_{-h/2}^{h/2} zF(f_y + \sqrt{1-e^2}H_{00}z\frac{\partial p}{\partial y})dz\}$$

$$= \frac{\gamma - \omega_0 y}{2}\frac{\partial h}{\partial x} + \sqrt{1-e^2}\frac{\sqrt{1-\gamma^2}+\sqrt{1-e^2}\omega_0 x}{2}\frac{\partial h}{\partial y}, \quad \mu = \exp(Qp),$$

$$p(x,y)\mid_{\Gamma} = 0, \quad \frac{dp(x,y)}{d\overrightarrow{n}}\mid_{\Gamma_e} = 0,$$

$$\int_{-h/2}^{h/2} F(f_x + H_{00}z\frac{\partial p}{\partial x})dz = \mu(\frac{s_x}{2} - \omega_0 y),$$

(16.9)

$$\int_{-h/2}^{h/2} F(f_y + \sqrt{1-e^2}H_{00}z\frac{\partial p}{\partial y})dz = \mu(\frac{s_y}{2} + \sqrt{1-e^2}\omega_0 x),$$

$$H_{00}[h(x,y) - 1] = x^2 + \frac{\delta}{1-e^2}y^2 + \frac{D(e)}{\pi(1-e^2)^{3/2}}\{\int\int_\Omega \frac{p(x_1,y_1)dx_1 dy_1}{\sqrt{(x-x_1)^2+\frac{(y-y_1)^2}{1-e^2}}}$$

$$-\int\int_\Omega \frac{p(x_1,y_1)dx_1 dy_1}{\sqrt{x_1^2+\frac{y_1^2}{1-e^2}}}\}, \quad \int\int_\Omega p(x_1,y_1)dx_1 dy_1 = \frac{2\pi}{3}.$$

Therefore, for the given values of parameters V, Q, δ, e, γ, ω_0, s_x, s_y, function F, and the inlet boundary Γ_i we need to determine functions $p(x,y)$, $h(x,y)$, $f_x(x,y)$, $f_y(x,y)$, constant H_{00}, and exit boundary Γ_e.

It is important to remember that the EHL problem equations for heavily loaded contacts contain a small parameter ω, which may be equal to $\omega = V \ll 1$ and/or $\omega = Q^{-1} \ll 1$.

16.3 Case of Pure Rolling

Let us consider the case of pure rolling when $s_x = s_y = \omega_0 = 0$. Then due to the fact that F is an odd function the solutions of the equations for f_x and f_y from (16.9) are $f_x = f_y = 0$. Therefore, the Reynolds equation from (16.9) is reduced to the form

$$\frac{\partial}{\partial x}\{\frac{1}{\mu} \int\limits_{-h/2}^{h/2} z F(H_{00} z \frac{\partial p}{\partial x}) dz\}$$

$$+\sqrt{1-e^2}\frac{\partial}{\partial y}\{\frac{1}{\mu} \int\limits_{-h/2}^{h/2} z F(\sqrt{1-e^2} H_{00} z \frac{\partial p}{\partial y}) dz\} \qquad (16.10)$$

$$= \frac{\gamma}{2}\frac{\partial h}{\partial x} + \frac{\sqrt{1-e^2}\sqrt{1-\gamma^2}}{2}\frac{\partial h}{\partial y}.$$

Here the approach to the asymptotic analysis of the problem coincides with the one used in Chapters 12 and 13 for the case of lubricants with Newtonian rheology. The analysis of the solution in the Hertzian region, i.e., away from the inlet and exit boundaries is identical with the one in the above mentioned chapters. Here we will consider the case of skewed lubricant entrainment which includes the case of straight lubricant entrainment as just a special case. As in Chapter 13 it is clear that in a heavily loaded contact away from its boundaries we will have an equation (see the definitions of heavily loaded contact (13.8) and (13.9))

$$\frac{\gamma}{2}\frac{\partial h}{\partial x} + \frac{\sqrt{1-e^2}\sqrt{1-\gamma^2}}{2}\frac{\partial h}{\partial y} = 0. \qquad (16.11)$$

The easiest way to consider solution of this equation and to proceed with the entire asymptotic analysis is to rewrite the problem equations in new independent variables (χ, ξ) which are related to variables (x, y) in a linear fashion (see (13.3) and (13.4))

$$\chi = \frac{\gamma x + \sqrt{1-e^2}\sqrt{1-\gamma^2}y}{\sqrt{1-e^2+e^2\gamma^2}}, \quad \xi = \frac{-\sqrt{1-e^2}\sqrt{1-\gamma^2}x + \gamma y}{\sqrt{1-e^2+e^2\gamma^2}}, \quad d\chi d\xi = dx dy, \qquad (16.12)$$

where

$$x = \frac{\gamma\chi - \sqrt{1-e^2}\sqrt{1-\gamma^2}\xi}{\sqrt{1-e^2+e^2\gamma^2}}, \quad y = \frac{\sqrt{1-e^2}\sqrt{1-\gamma^2}\chi + \gamma\xi}{\sqrt{1-e^2+e^2\gamma^2}}. \qquad (16.13)$$

In these new variables, away from the contact boundaries where with high precision the gap is practically constant along straight lines $\xi(x, y) = const$, i.e., $h(\chi, \xi) = h(\xi)$ (see equation (16.11) which is equivalent to equation $\frac{\partial h}{\partial \chi} = 0$).

In variables (16.12) the Reynolds equation (16.10) and other equations of the EHL problem (see Chapter 13) takes the form

$$\frac{1}{\sqrt{1-e^2+e^2\gamma^2}}[\gamma\frac{\partial}{\partial\chi} - \sqrt{1-e^2}\sqrt{1-\gamma^2}\frac{\partial}{\partial\xi}]$$

$$\times\{\frac{1}{\mu}\int_{-h/2}^{h/2} zF(H_{00}z\frac{1}{\sqrt{1-e^2+e^2\gamma^2}}[\gamma\frac{\partial p}{\partial\chi} - \sqrt{1-e^2}\sqrt{1-\gamma^2}\frac{\partial p}{\partial\xi}])dz\}$$

$$+\frac{\sqrt{1-e^2}}{\sqrt{1-e^2+e^2\gamma^2}}[\sqrt{1-e^2}\sqrt{1-\gamma^2}\frac{\partial}{\partial\chi} + \gamma\frac{\partial}{\partial\xi}]$$

$$\times\{\frac{1}{\mu}\int_{-h/2}^{h/2} zF(\frac{\sqrt{1-e^2}}{\sqrt{1-e^2+e^2\gamma^2}}H_{00}z[\sqrt{1-e^2}\sqrt{1-\gamma^2}\frac{\partial p}{\partial\chi} + \gamma\frac{\partial p}{\partial\xi}])dz\}$$

$$= \frac{\sqrt{1-e^2+e^2\gamma^2}}{2}\frac{\partial h}{\partial\chi}, \quad \mu = \exp(Qp), \quad p(\chi,\xi)\mid_\Gamma = 0, \quad \frac{dp(\chi,\xi)}{d\vec{n}}\mid_{\Gamma_e} = 0,$$

$$H_{00}[h(\chi,\xi) - 1] = x^2(\chi,\xi) + \frac{\delta}{1-e^2}y^2(\chi,\xi)$$

$$+\frac{D(e)}{\pi(1-e^2)^{3/2}}\{\int\int_\Omega \frac{p(\chi_1,\xi_1)d\chi_1 d\xi_1}{\sqrt{(\chi-\chi_1)^2+\frac{(\xi-\xi_1)^2}{1-e^2}}} - \int\int_\Omega \frac{p(\chi_1,\xi_1)d\chi_1 d\xi_1}{\sqrt{\chi_1^2+\frac{\xi_1^2}{1-e^2}}}\},$$

$$\int\int_\Omega p(\chi_1,\xi_1)d\chi_1 d\xi_1 = \frac{2\pi}{3},$$

(16.14)

where x, y and x_1, y_1 are functions of χ, ξ and χ_1, ξ_1 determined according to (13.4), respectively.

It is clear that for the case of lubricant entrainment along the x-axis we have $\gamma = 1$ and the EHL problem with skewed lubricant entrainment described by equations (13.2) and (13.5) gets reduced to the problem for straight lubricant entrainment (see similar EHL problem for Newtonian fluid in Chapter 12).

Taking into account that in the inlet Ω_i and exit Ω_e zones (see Chapter 13) $\frac{\partial}{\partial\chi} \gg \frac{\partial}{\partial\xi}$ from the first equation in (16.15) in the inlet and exit zones we get a reduced Reynolds equation in the form

$$\frac{\gamma}{\sqrt{1-e^2+e^2\gamma^2}}\frac{\partial}{\partial\chi}\{\frac{1}{\mu}\int_{-h/2}^{h/2} zF(\frac{\gamma}{\sqrt{1-e^2+e^2\gamma^2}}H_{00}z\frac{\partial p}{\partial\chi})dz\}$$

$$+\frac{(1-e^2)\sqrt{1-\gamma^2}}{\sqrt{1-e^2+e^2\gamma^2}}\frac{\partial}{\partial\chi}\{\frac{1}{\mu}\int_{-h/2}^{h/2} zF(\frac{(1-e^2)\sqrt{1-\gamma^2}}{\sqrt{1-e^2+e^2\gamma^2}}H_{00}z\frac{\partial p}{\partial\chi})dz\} + \dots$$

(16.15)

$$= \frac{\sqrt{1-e^2+e^2\gamma^2}}{2}\frac{\partial h}{\partial\chi}.$$

Now, we are ready for the further asymptotic analysis.

16.3.1 Pre-Critical Lubrication Regimes

Let us consider the pre-critical lubrication regimes described by the relationship (13.17).

Due to the fact that the equation for gap h between the solids and the equilibrium equation (see the last two equations in (16.9)) remain the same as in the case of lubricant with Newtonian rheology (see Chapter 12) the solution analysis in the Hertzian region is identical to the one conducted in Section 13.3. This analysis leads to the conclusion that the inlet Γ_i and exit Γ_e contact boundaries are close to the boundaries of the dry circular contact and can be described by the expressions $\chi = \gamma_i(\xi)$ and $\chi = \gamma_e(\xi)$, respectively, (see (13.13))

$$\gamma_i(\xi) = -\sqrt{1 - \xi^2} + \epsilon_q \alpha_{1p}(\xi), \ \alpha_{1p}(\xi) = O(1), \ (\chi, \xi) \in \Gamma_i,$$

$$\gamma_e(\xi) = \sqrt{1 - \xi^2} + \epsilon_q \beta_{1p}(\xi), \ \beta_{1p}(\xi) = O(1), \ (\chi, \xi) \in \Gamma_e, \ \omega \ll 1, \tag{16.16}$$

where $\alpha_{1p}(\xi)$ is a given function of ξ while $\beta_{1p}(\xi)$ has to be determined from the problem solution. Concerning these boundaries we will make the assumption that they vary moderately, i.e., we will assume that outside of the ends of HSSPGD zone Ω_{HSSPGD} assumptions (13.14) and (13.15) are true.

The analysis of the Hertzian region and the requirement of asymptotic matching of the inlet and exit solutions with the problem solution in the Hertzian region indicate that the solution for pressure p away from the the ϵ_q sized vicinity of points $(\chi, \xi) = (0, -1)$ and $(\chi, \xi) = (0, 1)$ should be searched in the inlet zone the form

$$p(\chi, \xi) = \epsilon_q^{1/2} q_p(r_p, \xi) + o(\epsilon_q^{1/2}), \ q_p(r_p, \xi) = O(1),$$

$$r_p = \frac{\chi - \gamma_i(\xi)}{\epsilon_q} = O(1), \ (\chi, \xi) \in \Omega_i \bigcap \Omega_\epsilon, \ \omega \ll 1, \tag{16.17}$$

where $q_p(r_p, \xi)$ is the unknown function which is the main term of the asymptotic expansion of pressure in the inlet zone while in the exit zone it should be searched in the form

$$p(\chi, \xi) = \epsilon_q^{1/2} g_p(s_p, \xi) + o(\epsilon_q^{1/2}), \ g_p(s_p, \xi) = O(1),$$

$$s_p = \frac{\chi - \gamma_e(\xi)}{\epsilon_q} = O(1), \ (\chi, \xi) \in \Omega_e \bigcap \Omega_\epsilon, \ \omega \ll 1, \tag{16.18}$$

where $g_p(s_p, \xi)$ is the unknown function which is the main term of the asymptotic expansion of pressure in the exit zone. Further, for simplicity functions $q_p(r_p, \xi_*)$ and $g_p(s_p, \xi_*)$ will be called asymptotic pressure distributions in the inlet and exit zones. Moreover, these functions q_p and g_p satisfy the matching conditions (see (13.30), (13.12) and (13.50))

$$q_p(r_p, \xi) \to q_{ap}(r_p, \xi) = N_0 \sqrt{2r_p} + \frac{N_1}{\sqrt{2r_p}}, \ r_p \to \infty;$$

$$g_p(s_p, \xi) \to g_{ap}(s_p, \xi) = N_0 \sqrt{-2s_p} + \frac{N_1}{\sqrt{-2s_p}}, \ s_p \to -\infty, \tag{16.19}$$

where

$$N_0(\xi) = \sqrt[4]{1-\xi^2}, \ \alpha_{1p} = \tfrac{N_1}{N_0}, \ \beta_{1p} = -\tfrac{N_1}{N_0}. \tag{16.20}$$

Estimation of the term $H_{00}\frac{\partial h}{\partial \chi}$ away from points $(\chi, \xi) = (0, -1)$ and $(\chi, \xi) = (0, 1)$ in the inlet zone gives (see (13.27))

$$H_{00}\frac{\partial h}{\partial \chi} = \epsilon_q^{1/2} \tfrac{2}{\pi}\theta \int\limits_0^\infty \frac{q_p(t,\xi) - q_{ap}(t,\xi)}{t - r_p} dt + o(\epsilon_q^{1/2}), \ (\chi,\xi) \in \Omega_i \bigcap \Omega_\epsilon, \tag{16.21}$$

$$\theta = \frac{D(e)}{1-e^2}\sqrt{\frac{1 - e^2 + e^2\gamma^2}{(1-e^2)^2(1-\gamma^2) + \gamma^2}}, \tag{16.22}$$

while in the exit zone it gives (see (13.28))

$$H_{00}\frac{\partial h}{\partial \chi} = \epsilon_q^{1/2} \tfrac{2}{\pi}\theta \int\limits_{-\infty}^0 \frac{q_p(t,\xi) - q_{ap}(t,\xi)}{t - s_p} dt + o(\epsilon_q^{1/2}), \ (\chi,\xi) \in \Omega_e \bigcap \Omega_\epsilon. \tag{16.23}$$

To continue the asymptotic analysis we have to know some characteristics of the behavior of the rheology function F. Therefore, we will assume that (see (9.35))

$$F(H_{00}\epsilon_q^{-1/2}w(t)) = V^{-k}(\omega^{-l}\epsilon_q^{-1/2}H_{00}^{n+1})^{1/m}F_0(w(t)) + \ldots,$$

$$\tag{16.24}$$

$$F_0(w(t)) = O(1) \ for \ w(t) = O(1) \ and \ H_{00}\epsilon_q^{-1/2} \ll 1,$$

where k, l, m, and n are constants, $m > 0$, $w(t)$ is some function of order of unity, and $F_0(w(t))$ is a certain function of order of unity which is determined by function F.

Now, by making the necessary estimates in the inlet zone Ω_i and following the analysis of Chapter 13, we will arrive at the formula for the film thickness

$$H_{00} = A_p\Big\{\frac{(1 - e^2 + e^2\gamma^2)^{\frac{2m+1}{2}}}{[\gamma^{\frac{m+1}{m}} + ((1-e^2)\sqrt{1-\gamma^2})^{\frac{m+1}{m}}]^m} V^{km}\omega^l\epsilon_q^{\frac{3m+1}{2}}\Big\}^{\frac{1}{m+n+1}} + \ldots,$$

$$\tag{16.25}$$

$$A_p(\alpha_{1p}) = O(1), \ \omega \ll 1,$$

where $A_p(\alpha_{1p})$ is an unknown nonnegative constant independent from ω and ϵ_q, which is determined by the solution of the problem in the inlet zone. Furthermore, this analysis (see Section 13.3) in the inlet and exit zones lead to the following asymptotically valid equations:

$$\frac{dM_0(q_p, h_{qp}, \mu_{qp}, r_p)}{dr_p} = \tfrac{2}{\pi}\theta \int\limits_0^\infty \frac{q_p(t,\xi) - q_{ap}(t,\xi)}{t - r_p} dt,$$

$$q_p(0,\xi) = 0, \ q_p(r_p,\xi) \to q_{ap}(r_p,\xi) \ as \ r_p \to \infty, \tag{16.26}$$

$$\int\limits_0^\infty [q_p(t,\xi) - q_{ap}(t,\xi)]dt = 0,$$

$$\frac{dM_0(g_p, h_{gp}, \mu_{gp}, s_p)}{ds_p} = \frac{2}{\pi}\theta \int_{-\infty}^{0} \frac{g_p(t,\xi) - g_{ap}(t,\xi)}{t - s_p} dt,$$

$$g_p(0,\xi) = \frac{\partial g_p(0,\xi)}{\partial s_p} = 0, \ g_p(s_p) \to g_{ap}(s_p) \ as \ s_p \to -\infty, \tag{16.27}$$

$$\int_{-\infty}^{0} [g_p(t,\xi) - g_{ap}(t,\xi)]dt = 0,$$

where

$$M_0(p, h, \mu, x) = \frac{A_p^{\frac{m+n+1}{m}}}{\mu} \int_{-h/2}^{h/2} z F_0(z\frac{\partial p}{\partial x})dz, \tag{16.28}$$

$$q_{ap}(r_p, \xi) = N_0(\xi)[\sqrt{2r_p} + \frac{\alpha_{1p}}{\sqrt{2r_p}}], \tag{16.29}$$

$$g_{ap}(s_p, \xi) = N_0(\xi)[\sqrt{-2s_p} - \frac{\beta_{1p}}{\sqrt{-2s_p}}], \tag{16.30}$$

and functions μ_{qp} and μ_{gp} are the main terms of the asymptotic expansions of $\mu(\epsilon_q^{1/2} q_p)$ and $\mu(\epsilon_q^{1/2} g_p)$, respectively. Furthermore, $h_{qp}(r_p, \xi)$ and $h_{gp}(s_p, \xi)$ are the gap functions in the inlet $\Omega_i \bigcap \Omega_e$ and exit $\Omega_e \bigcap \Omega_e$ zones. The expressions for $h_{qp}(r_p, \xi)$ and $h_{gp}(s_p, \xi)$ depend on the lubrication regime. For starved lubrication regimes the gap distribution $h(\chi, \xi)$ is very close to $h_\infty(\xi)$ in the entire contact region. That corresponds to the case of $\epsilon_q^{3/2} \ll H_{00}$ and

$$h_{qp}(r_p, \xi) = h_\infty(\xi), \ \epsilon_q \ll \epsilon_f, \tag{16.31}$$

$$h_{gp}(s_p, \xi) = h_\infty(\xi), \ \epsilon_q \ll \epsilon_f, \tag{16.32}$$

where

$$\epsilon_f = \{\frac{(1-e^2+e^2\gamma^2)^{\frac{2m+1}{2}}}{[\gamma^{\frac{m+1}{m}} + ((1-e^2)\sqrt{1-\gamma^2})^{\frac{m+1}{m}}]^m}\}^{\frac{2}{3n+2}}(V^{km}\omega^l)^{\frac{2}{3n+2}}. \tag{16.33}$$

At the same time, for fully flooded lubrication regimes $H_{00} = O(\epsilon_q^{3/2})$, $\omega \ll 1$, and in the inlet and exit zones the gap satisfies the equations (see equations (13.27) and (13.28))

$$A_p \frac{\partial h_{qp}}{\partial r_p} = \frac{2}{\pi}\theta \int_{0}^{\infty} \frac{q_p(t,\xi) - q_{ap}(t,\xi)}{t - r_p} dt, \ \epsilon_q = \epsilon_f, \ \omega \ll 1, \tag{16.34}$$

$$A_p \frac{\partial h_{gp}}{\partial s_p} = \frac{2}{\pi}\theta \int_{-\infty}^{0} \frac{g_p(t,\xi) - g_{ap}(t,\xi)}{t - s_p} dt, \ \epsilon_q = \epsilon_f, \ \omega \ll 1. \tag{16.35}$$

For the fully flooded lubrication regimes (see formula (16.25) for $\epsilon_q = \epsilon_f$) the formula for the film thickness H_{00} takes the form

$$H_{00} = A_f\{\frac{(1-e^2+e^2\gamma^2)^{\frac{2m+1}{2}}}{[\gamma^{\frac{m+1}{m}} + ((1-e^2)\sqrt{1-\gamma^2})^{\frac{m+1}{m}}]^m}V^{km}\omega^l\}^{\frac{3}{3n+2}} \dots,$$

$$\tag{16.36}$$

$$A_f(\alpha_{1p}) = O(1), \ \omega \ll 1,$$

where constant A_f is determined from the solution of the system (16.26), (16.28), (16.29), and (16.34). Here we need to notice that equations (16.34) and (16.35) can be integrated with respect to r_p and s_p by taking into account the conditions

$$h_{qp}(r_p, \xi) \to h_\infty(\xi), \ r_p \to \infty; \ h_{gp}(s_p, \xi) \to h_\infty(\xi), \ s_p \to -\infty, \qquad (16.37)$$

respectively. This integration produces the following equations for h_{qp} and h_{gp} (see (13.44) and (13.45)):

$$A_p[h_{qp}(r_p, \xi) - h_\infty(\xi)] = \tfrac{2}{\pi}\theta \int_0^\infty [q_p(t, \xi) - q_{ap}(t, \xi)] \ln \tfrac{1}{|t - r_p|} dt, \qquad (16.38)$$

$$A_p[h_{gp}(s_p, \xi) - h_\infty(\xi)] = \tfrac{2}{\pi}\theta \int_{-\infty}^0 [g_p(t, \xi) - g_{ap}(t, \xi)] \ln \tfrac{1}{|t - s_p|} dt. \qquad (16.39)$$

For fully flooded lubrication regimes by integrating equations (16.26) and (16.27) for q and g the gap functions h_q and h_g from (16.38) and (16.39) can be replaced by equations

$$h_{qp} = h_\infty(\xi) + \tfrac{1}{A_p} M_0(q_p, h_{qp}, \mu_{qp}, r_p),$$

$$h_{gp} = h_\infty(\xi) - \tfrac{1}{A_p} M_0(g_p, h_{gp}, \mu_{gp}, s_p). \qquad (16.40)$$

Therefore, in the inlet zone for the given values of e, γ, θ, $Q\epsilon_q^{1/2}$ (involved in μ_{qp}), $\alpha_{1p}(\xi)$, and ξ we derived the system of equations (16.26), (16.28), (16.29), and (16.31) for starved lubrication regimes or (16.38) for fully flooded lubrication regimes for functions $q_p(r_p, \xi)$, $h_{qp}(r_p, \xi)$, $h_\infty(\xi)$, and coefficient A_p. In the exit zone for the given values of e, γ, θ, $Q\epsilon_q^{1/2}$, A_p, $h_\infty(\xi)$, and ξ we need to solve the system of equations (16.27), (16.28), (16.30), and (16.32) for starved lubrication regimes or (16.39) for fully flooded lubrication regimes for functions $g_p(s_p, \xi)$, $h_{gp}(s_p, \xi)$, and $\beta_{1p}(\xi)$. It is important to keep in mind that constant A_p is the coefficient of proportionality in formula (16.25) for the central film thickness H_{00} and it is determined for $\xi = 0$ when $h_\infty(\xi) =$. Also, it means that $h_\infty(0) = 1$ while, generally, $h_\infty(\xi) \neq 1$ for $\xi \neq 0$. The solution procedure should be implemented according to the following sequence: first, the system of asymptotically valid equations in the inlet zone should be solved for $\xi = 0$ which determines the value of constant A_p; after that, the values of function $h_\infty(\xi)$ for $\xi \neq 0$ are determined from the solution of the asymptotically valid system of equations in the inlet zone and these values play the role similar to the one played by constant A_p for $\xi = 0$.

The above systems of asymptotically valid equations derived in the inlet $\Omega_i \bigcap \Omega_\epsilon$ and exit $\Omega_e \bigcap \Omega_\epsilon$ zones in heavily loaded point EHL contacts allow for the direct usage of the entire apparatus of numerical analysis of these systems as well as of the original equations developed in Chapters 6 - 10 for heavily loaded line EHL contacts. These methods include usage of the equivalent form of these equations, design of stable numerical schemes based on properly regularized Reynolds equation, proper choice of the grid sizes for the given EHL problem parameters, etc.

In the case of a lubricant with Newtonian rheology, i.e., when $F(x) = \frac{12H_{00}}{V}x$ and $l = 0$, $k = m = n = 1$ or $k = 0$, $l = m = n = 1$ and $\omega = V$, we have $F_0(w(t)) = 12\frac{\partial w}{\partial t}$ and formula (16.25) for the lubrication film thickness H_{00} coincides with the formula for H_{00} obtained for the case of Newtonian lubricant (see formulas (13.32) and (13.35)). The same is true about the systems of asymptotically valid equations in the inlet and exit zones.

The above systems of asymptotic equations in the inlet and exit zones are very similar to the ones derived and analyzed for the cases of line EHL contacts. Namely, by introducing the substitutions

$$\{r, s, \alpha_1, \beta_1\} = \frac{1}{r_0}\{r_p, s_p, \alpha_{1p}, \beta_{1p}\}, \quad \{q, g\} = \frac{1}{N_0 r_0^{1/2}}\{q_p, g_p\},$$

$$\{h_q, h_g\} = \frac{1}{h_\infty}\{h_{qp}, h_{gp}\}, \quad \{\mu_q(q), \mu_g(g)\} = \{\mu_{qp}(q_p), \mu_{gp}(g_p)\}, \tag{16.41}$$

and the definition

$$H_0(\xi) = H_{00}h_\infty(\xi), \tag{16.42}$$

and choosing the proper definition for r_0 (see, for example, (13.47) for the case of Newtonian lubricant) we will be able to reduce each of the above systems to systems of equations similar to the ones derived for the case of line EHL contact in which h_∞ is gone. It will also allow us to obtain a formula for $H_0(\xi)$ which represents the dimensionless analog of the dimensional film thickness h_0 but at the exit from the contact region.

16.3.2 Asymptotic Analysis of a Heavily Loaded Lubricated Contact under Over-Critical Lubrication Conditions

Over-critical lubrication regimes occur when lubricant viscosity rapidly grows with pressure which occurs when $Q \gg 1$. In a heavily loaded lubricated contact the maximum characteristic size of the inlet zone ϵ_0 for which lubrication regime remains pre-critical can be determine according to the procedure described in Section 13.4. Therefore, the formal definition of over-critical lubrication regimes is as follows

$$\mu(\epsilon_q^{1/2}) \gg 1, \quad \epsilon_q(\omega) \gg \epsilon_0(\omega), \quad \epsilon_0(\omega) \ll 1, \quad \omega \ll 1. \tag{16.43}$$

A heavily loaded over-critical lubricated contact has the following structure (see Chapter 6). The inlet zone is represented by two sub-zones: one of a characteristic size $\epsilon_q \gg \epsilon_0$ (inlet ϵ_q-zone) which is adjacent to the the the inlet contact boundary Γ_i and a smaller inlet zone of the characteristic size ϵ_0 (inlet ϵ_0-zone) located between the inlet ϵ_q-zone and the Hertzian region where pressure is very close to the Hertzian pressure $p_{00}(\chi, \xi) = \sqrt{1 - \chi^2 - \xi^2}$. On the exit side of the contact the structure of the exit zone is similar to the one of the inlet zone, i.e. a smaller exit ϵ_0-zone is located between the Hertzian region and the larger exit ϵ_q-zone adjacent to the contact exit boundary Γ_e.

Most of the analysis of Section 6.6 deals with the properties of the solutions of the integral equations related to the equation for gap h. This analysis remains exactly the same for the problem at hand. Also, the reduced Reynolds equation is still represented by equation (16.15). However, some changes have to be made in the analysis in the inlet and exit ϵ_0- and ϵ_q-zones. To proceed further along the lines of the analysis of over-critical lubrication regimes outlined in Section 6.6 and to obtain formulas for the lubrication film thickness H_{00}, we need to make two assumptions about the behavior of the rheology function F in the ϵ_q-inlet zone. This assumption can be derived from the assumption (16.24) by replacing in it $\epsilon_q^{-1/2}$ by $\epsilon_0^{1/2}\epsilon_q^{-1}$. As a result of that we get

$$F(H_0\epsilon_0^{1/2}\epsilon_q^{-1}y(t)) = V^{-k}(\omega^{-l}\epsilon_0^{1/2}\epsilon_q^{-1}H_0^{n+1})^{1/m}F_0(y(t)) + \ldots,$$

$$F_0(y(t)) = O(1) \ for \ y(t) = O(1) \ and \ H_0\epsilon_0^{1/2}\epsilon_q^{-1} \ll 1,$$

(16.44)

where k, l, m, and n are certain constants. The solution of the problem in the ϵ_q-inlet zone will be searched in the form of asymptotic representation (see Section 6.6).

Using the same approach as the one used in Sections 6.6.3, 6.6.4, and Section 12.4 we obtain that for starved and fully flooded over-critical lubrication regimes in the ϵ_q-inlet zone the pressure $p = O(\epsilon_0^{1/2})$ (here $\epsilon_0 = Q^{-2} \ll 1$) and the gap $h = O(\epsilon_q^{3/2})$. Therefore, by balancing the order of the main terms of the reduced Reynolds equation in the ϵ_q-inlet zone we obtain the following formulas for the film thickness H_{00} for starved lubrication regimes

$$H_{00} = A_p\{\frac{(1-e^2+e^2\gamma^2)^{\frac{2m+1}{2}}}{[\gamma^{\frac{m+1}{m}}+((1-e^2)\sqrt{1-\gamma^2})^{\frac{m+1}{m}}]^m}V^{km}\omega^l Q\epsilon_q^{\frac{3m+2}{2}}\}^{\frac{1}{n+m+1}} + \ldots,$$

(16.45)

$$A_p = O(1), \ Q \gg 1,$$

while for fully flooded lubrication regimes we get

$$H_{00} = A_f\{\frac{(1-e^2+e^2\gamma^2)^{\frac{2m+1}{2}}}{[\gamma^{\frac{m+1}{m}}+((1-e^2)\sqrt{1-\gamma^2})^{\frac{m+1}{m}}]^m}V^{km}\omega^l Q\}^{\frac{3}{3n+1}} + \ldots,$$

(16.46)

$$A_f = O(1), \ Q \gg 1,$$

where coefficients A_p and A_f depend on θ, and $\alpha_{1p}(0)$. Here the expression (16.46) for the film thickness H_{00} for fully flooded lubrication regimes is obtained from formula (16.45) and the relationship $H_{00} = O(\epsilon_q^{3/2})$.

It is easy to see that for a lubricant with Newtonian rheology, i.e., when $F(x) = \frac{12H_{00}}{V}x$ and $l = 0$, $k = m = n = 1$ or $k = 0$, $l = m = n = 1$ and

$\omega = V$, we have formulas (16.45) and (16.46) for the lubrication film thickness H_{00} coincides with the formula for H_{00} obtained for the case of Newtonian lubricant (see formulas in (13.58)).

Numerical studies of the inlet and exit zones for pre- and over-critical lubrication regimes based on the derived asymptotically valid equations can be found in Chapter 6. The numerical results include parametric studies, studies of solution precision and stability, etc.

The choice between the pre- and over-critical lubrication regimes can be made in the fashion similar to the one described in Section 6.7.

16.4 Case of High Slide-to-Roll Ratio. Approximation of the Reynolds Equation

The idea of the further analysis is based on asymptotic solution of the equations for f_x and f_y from (16.9) and substituting these solutions in the Reynolds equation in (16.9) to simplify it. It can be easily done in the case of high slide-to-roll ratio, i.e., when the rolling frictional stress $\frac{H_{00}h}{2}\frac{\partial p}{\partial x}$ is much smaller than the sliding frictional stress f_x. Therefore, in the inlet zone of a heavily loaded contact we can introduce a small function (compare to the ones in (7.17) and (9.94))

$$\nu = \frac{H_{00}h}{2f_x}\frac{\partial p}{\partial x} \ll 1, \ \omega \ll 1. \tag{16.47}$$

It can be shown that if the estimate (16.47) is valid in the inlet zone of a lubricated contact then it is valid everywhere in the contact.

Also, we will assume that the rheological function F is continuously differentiable and in the inlet zone $\frac{\partial p}{\partial y} = O(\frac{\partial p}{\partial y})$, $\omega \ll 1$. The latter condition guarantees that for any dominant direction of the lubricant flow (not necessarily the direction along the x-axis) the resulting approximate Reynolds equation will be valid as long as we retain terms proportional to both first derivatives of pressure.

Then, we can search for functions f_x and f_y in the form of asymptotic expansions in powers of ν

$$f_x = f_{x0} + \nu f_{x1} + \dots, \ f_y = f_{y0} + \nu f_{y1} + \dots,$$
$$f_{x0}, \ f_{x1}, \ f_{y0}, \ f_{y1} = O(1), \ \omega \ll 1. \tag{16.48}$$

Substituting expansions (16.48) into equations for functions f_x and f_y in (16.9), expanding functions F in Taylor series, and equating coefficients of these series with the same powers of ν we obtain

$$hF(f_{x0}) = \mu(\tfrac{s_x}{2} - \omega_0 y), \ f_{x1} = 0, \ldots,$$

$$hF(f_{y0}) = \mu(\tfrac{s_y}{2} + \sqrt{1-e^2}\omega_0 x), \ f_{y1} = 0, \ldots \tag{16.49}$$

Applying function Φ (which is the inverse of function F) from (16.49) we determine

$$f_{x0} = \Phi(\tfrac{\mu}{h}(\tfrac{s_x}{2} - \omega_0 y)), \ f_{x1} = 0, \ldots,$$

$$f_{y0} = \Phi(\tfrac{\mu}{h}(\tfrac{s_y}{2} + \sqrt{1-e^2}\omega_0 x)), \ f_{y1} = 0, \ldots \tag{16.50}$$

Now, substituting expansions (16.48) and (16.50) into the Reynolds equation in (16.9), expanding functions F in the Taylor series, retaining (16.9) only the terms proportional to ν, and using the expression for ν from (16.47) we obtain the asymptotically valid approximate generalized Reynolds equation in the form

$$\frac{\partial}{\partial x}\{\tfrac{H_{00}h^3}{6\mu}F'(f_{x0})\tfrac{\partial p}{\partial x}\} + (1-e^2)\frac{\partial}{\partial y}\{\tfrac{H_{00}h^3}{6\mu}F'(f_{y0})\tfrac{\partial p}{\partial y}\} + \ldots$$

$$= (\gamma - \omega_0 y)\tfrac{\partial h}{\partial x} + \sqrt{1-e^2}(\sqrt{1-\gamma^2} + \sqrt{1-e^2}\omega_0 x)\tfrac{\partial h}{\partial y}. \tag{16.51}$$

Obviously, for Newtonian lubricants the rheological functions $F(x) = \frac{6H_{00}}{V}x$ and $\Phi(x) = \frac{V}{6H_{00}}x$ (keep in mind that here the dimensionless variables for functions F and Φ are introduced slightly differently from (9.2)) the approximate generalized Reynolds equation (16.51) becomes identical to the exact (not approximate) Reynolds equation (14.4) obtained for isothermal contact with Newtonian lubricant.

Now, it is clear that the complete system to which problem (16.9) is reduced has the form

$$\frac{\partial}{\partial x}\{\tfrac{H_{00}h^3}{6\mu}F'(f_{x0})\tfrac{\partial p}{\partial x}\} + (1-e^2)\frac{\partial}{\partial y}\{\tfrac{H_{00}h^3}{6\mu}F'(f_{y0})\tfrac{\partial p}{\partial y}\} + \ldots$$

$$= (\gamma - \omega_0 y)\tfrac{\partial h}{\partial x} + \sqrt{1-e^2}(\sqrt{1-\gamma^2} + \sqrt{1-e^2}\omega_0 x)\tfrac{\partial h}{\partial y},$$

$$\mu = \exp(Qp), \ p(x,y)\mid_\Gamma = 0, \ \tfrac{dp(x,y)}{d\overrightarrow{n}}\mid_{\Gamma_e} = 0,$$

$$H_{00}[h(x,y) - 1] = x^2 + \tfrac{\delta}{1-e^2}y^2 + \tfrac{D(e)}{\pi(1-e^2)^{3/2}}\{\int\int\limits_\Omega \tfrac{p(x_1,y_1)dx_1dy_1}{\sqrt{(x-x_1)^2 + \frac{(y-y_1)^2}{1-e^2}}} \tag{16.52}$$

$$- \int\int\limits_\Omega \tfrac{p(x_1,y_1)dx_1dy_1}{\sqrt{x_1^2 + \frac{y_1^2}{1-e^2}}}\}, \ \int\int\limits_\Omega p(x_1,y_1)dx_1dy_1 = \tfrac{2\pi}{3}.$$

Obviously, this system of equations (16.52) and the Reynolds equation itself (16.51) obtained for non-Newtonian lubricants resemble very much the problems for the Reynolds equations (12.9), (13.2), and (14.4) for the cases of straight and skewed lubricant entrainment as well as for the case of skewed lubricant entrainment with spinning considered for Newtonian lubricant rheology earlier. Therefore, to study the EHL problem in this case we will be able to use exactly the same methodology as the earlier developed one and, in a certain sense, it is not hard to predict the results of such an analysis.

To continue with our asymptotic analysis we need to realize that

$$F'(f_{x0}) = [\Phi'(\tfrac{\mu}{h}(\tfrac{s_x}{2} - \omega_0 y))]^{-1},$$

$$F'(f_{y0}) = [\Phi'(\tfrac{\mu}{h}(\tfrac{s_y}{2} + \sqrt{1 - e^2}\omega_0 x))]^{-1}, \qquad (16.53)$$

and to make certain assumptions concerning the behavior of the rheology function Φ in the inlet and exit zones. However, to avoid unnecessary algebraic calculations instead of making a general assumption and specify it for each case of lubricant motion we will do it separately for each case. The latter estimates will allow to calculate the coefficients of the Reynolds equation and provide an opportunity to study the problem in the inlet and exit zones of heavily loaded EHL contacts.

16.4.1 Case of Straight Lubricant Entrainment

In the case of straight lubricant entrainment without spinning we have $\gamma = 1$ and $\omega_0 = 0$. For pre-critical heavily loaded lubrication regimes the whole analysis of the inlet and exit zones presented in Chapter 12 can be literally repeated here with just a few modifications due to the specific expressions for the rheological functions F and Φ. First of all, the rheology of real lubricants is non-Newtonian usually only for large frictional stresses while for small stresses it is Newtonian. Therefore, we will assume that

$$F'(0) > \psi_F, \quad \Phi'(0) > \psi_\Phi, \qquad (16.54)$$

where ψ_F and ψ_Φ are certain positive constants which may depend on some input parameters of the EHL problem and on H_{00}. In addition to that let us assume that in the inlet and exit zones of a heavily loaded contact we have

$$\Phi(\tfrac{\mu s_x}{2h}) = \tfrac{V^k \omega^l}{H_{00}^m}\Phi_0(\tfrac{\mu s_x}{2h}), \quad \Phi_0(\tfrac{\mu s_x}{2h}) = O(1), \quad \omega \ll 1, \qquad (16.55)$$

where function Φ_0 depends on the inner/local variable in the inlet or exit zone while k, l, and m are certain constants. In this case from (16.55) we obtain the behavior of function Φ' in the inlet and exit zones in the form

$$\Phi'(\tfrac{\mu s_x}{2h}) = \tfrac{V^k \omega^l}{H_{00}^m}\Phi_1(\tfrac{\mu s_x}{2h}), \quad \Phi_1(\tfrac{\mu s_x}{2h}) = O(1), \quad \omega \ll 1, \qquad (16.56)$$

where function Φ_1 depends on the inner/local variable in the inlet or exit zone and it is independent of V, ω, and H_{00}.

In this case, as it follows from the Reynolds equation in (16.51) for $\gamma = 1$ and $\omega_0 = 0$ in the Hertzian region, the dominant direction of the lubricant flow is along the x-axis which means that in the inlet and exit zones of a heavily loaded point EHL contact $\frac{\partial}{\partial y} \ll \frac{\partial}{\partial x}$. Therefore, due to estimates (16.56) for pre-critical lubrication regimes in the inlet and exit zones we will be able to derive the asymptotically valid systems of equations (12.36) and (12.37) valid for line contacts (see Section 9.4) in which function M_0 from (12.38) should be replaced by

$$M_0(A, p, h, \mu, x, y) = A^{m+2} \frac{h^3(x,y)}{6\mu} \Big/ \Phi_1\Big(\frac{\mu s_x}{2h}\Big)\frac{\partial p}{\partial x}, \qquad (16.57)$$

while the formula for H_{00} will take the form

$$H_{00} = A_p(V^k \omega^l \epsilon_q^2)^{\frac{1}{m+2}}, \quad A_p = O(1), \quad \omega \ll 1, \qquad (16.58)$$

were A_p is a constant obtained from the solution of equations (12.36). This analysis is valid as long as $\nu(x, y) \ll 1$ in the inlet zone which is based on (16.47) and (16.58) leads to the condition

$$\epsilon_q \ll \epsilon_\nu = \Phi_0^{\frac{2(m+2)}{3m+2}}(s_x)(V^k \omega^l)^{\frac{2}{3m+2}}, \quad \omega \ll 1. \qquad (16.59)$$

Here it is taken into account that in the inlet region $\mu = O(1)$ and $h = O(1)$, $\omega \ll 1$. As it follows from the formula (16.58) for the film thickness H_{00} if

$$\epsilon_q \ll \epsilon_f = (V^k \omega^l)^{\frac{2}{3m+2}}, \quad \omega \ll 1, \qquad (16.60)$$

then a starved lubrication regime is realized as it follows from the definition of the starved lubrication regimes described by the estimate $\epsilon_q^{3/2} \ll H_{00}$ for $\omega \ll 1$.

Now, let us determine the conditions when the above asymptotic method is applicable to only starved lubrication regimes and when it is applicable to both starved and fully flooded lubrication regimes. To do that we need to compare the values of ϵ_ν and ϵ_f from (16.59) and (16.60), respectively. If $\epsilon_\nu \ll \epsilon_f$, i.e., if

$$\Phi_0(s_x) \ll 1 \qquad (16.61)$$

then the applied method is valid only for starved lubrication regimes. On the other hand, if $\epsilon_\nu = O(\epsilon_f)$ or $\epsilon_\nu \gg \epsilon_f$, i.e., if

$$\Phi_0(s_x) = O(1) \ or \ \Phi_0(s_x) \gg 1 \qquad (16.62)$$

then the analysis is valid for both starved and fully flooded lubrication regimes. For starved lubrication regimes for the film thickness we have formula (16.58) as long as $\epsilon_q \ll \epsilon_f$ while for fully flooded lubrication regimes defined by the estimate $\epsilon_q^{3/2} = O(H_{00})$ for $\omega \ll 1$ we have

$$H_{00} = A_f(V^k \omega^l)^{\frac{3}{3m+2}}, \quad A_f = O(1), \quad \epsilon_q = (V^k \omega^l)^{\frac{2}{3m+2}}, \quad \omega \ll 1. \qquad (16.63)$$

Obviously, by solving equations (12.36) and (12.37) with respect to functions q_p and g_p involved in the integrals in the right-hand sides of these equations they can be transformed into equivalent equations similar to (12.57) and

(12.58) which are more convenient for numerical solution and can be regularized and numerically solved in accordance with the approaches described in Sections 6.9 and 17.9. Moreover, if function $\Phi_1(\frac{\mu s_x}{2h})$ is a homogeneous function of its argument (which is often the case in one approximation of the rheology function Φ or another) then we can make a substitution similar to (12.50) and (12.51) and make systems of equations (12.36) and (12.37) identical with the ones derived for the case of a corresponding line contact.

If $\mu = e^{Qp}$ then for over-critical lubrication regimes described by the condition (16.43) (i.e., $\epsilon_0 = Q^{-2}$) using an approach similar to the one used in Section 12.4 we obtain the following formula for the film thickness

$$H_{00} = A_p (V^k \omega^l Q \epsilon_q^{5/2})^{\frac{1}{m+2}}, \ A_p = O(1), \ \omega \ll 1, \tag{16.64}$$

where A_p is a constant independent of V, Q, and ω. In the inlet ϵ_q-zone the condition $\nu(x, y) \ll 1$ is satisfied if

$$\epsilon_q \ll \epsilon_\nu = \Phi_0^{\frac{2(m+2)}{3m+1}} (s_x)(V^k \omega^l Q)^{\frac{2}{3m+1}}, \ \omega \ll 1. \tag{16.65}$$

Under starved lubrication conditions the film thickness is determined by (16.64) while

$$\epsilon_q \ll \epsilon_f = (V^k \omega^l Q)^{\frac{2}{3m+1}}, \ \omega \ll 1. \tag{16.66}$$

Let us determine the conditions when the above asymptotic method is applicable to only starved lubrication regimes and when it is applicable to both starved and fully flooded lubrication regimes. To do that we need to compare the values of ϵ_ν and ϵ_f from (16.65) and (16.80), respectively. If $\epsilon_\nu \ll \epsilon_f$, i.e., if condition (16.61) is satisfied then the applied method is valid only for starved lubrication regimes. On the other hand, if $\epsilon_\nu = O(\epsilon_f)$ or $\epsilon_\nu \gg \epsilon_f$, i.e., if conditions (16.62) are satisfied then the analysis is valid for both starved and fully flooded lubrication regimes. For starved lubrication regimes for the film thickness we have formula (16.64) as long as $\epsilon_q \ll \epsilon_f$ while for fully flooded lubrication regimes defined by the estimate $\epsilon_q^{3/2} = O(H_{00})$ for $\omega \ll 1$ we have

$$H_{00} = A_f (V^k \omega^l Q)^{\frac{3}{3m+1}}, \ A_f = O(1), \ \epsilon_q = (V^k \omega^l Q)^{\frac{2}{3m+1}}, \ \omega \ll 1. \tag{16.67}$$

For the case of a lubricant with Newtonian rheology for pre- and over-critical lubrication regimes the formulas for the film thickness H_{00} derived in this section coincide with the corresponding formulas for the film thickness H_{00} obtained in Chapter 12.

Several specific examples for particular non-Newtonian lubricant rheology are given in Section 9.4.

16.4.2 Case of Skewed Lubricant Entrainment without Spinning

The case of skewed lubricant entrainment without spinning is characterized by $\omega_0 = 0$. To simplify the analysis of this problem we can make the rotation (13.3) and (13.4) (see also (15.13) and (15.14)) of the coordinate system

(x, y, z) about the z-axis in the approximate Reynolds equation in (16.9). In variables (15.13) the Reynolds equation (16.51) will take the form

$$\frac{\partial}{\partial \chi} \{ \frac{H_{00} h^3}{6\mu} [\gamma^2 F'(f_{x0}) + (1 - e^2)^2 (1 - \gamma^2) F'(f_{y0})] \frac{\partial p}{\partial \chi} \}$$

$$+\gamma \sqrt{1 - e^2} \sqrt{1 - \gamma^2} \{ \frac{\partial}{\partial \chi} [\frac{H_{00} h^3}{6\mu} [(1 - e^2) F'(f_{y0}) - F'(f_{x0})] \frac{\partial p}{\partial \xi}]$$

$$+\frac{\partial}{\partial \xi} [\frac{H_{00} h^3}{6\mu} [(1 - e^2) F'(f_{y0}) - F'(f_{x0})] \frac{\partial p}{\partial \chi}] \} \qquad (16.68)$$

$$+(1 - e^2) \frac{\partial}{\partial \xi} \{ \frac{H_{00} h^3}{6\mu} [(1 - \gamma^2) F'(f_{x0}) + \gamma^2 F'(f_{y0})] \frac{\partial p}{\partial \xi} \}$$

$$= (1 - e^2 + e^2 \gamma^2)^{3/2} \frac{\partial h}{\partial \chi}.$$

Taking into account the fact that the dominant lubricant flow direction is along the χ-axis and that in the inlet and exit zones of the contact $\frac{\partial}{\partial \xi} \ll \frac{\partial}{\partial \chi}$ in these zones we obtain the following system of equations

$$\frac{\partial}{\partial \chi} \{ \frac{H_{00} h^3}{6\mu} [\gamma^2 F'(f_{x0}) + (1 - e^2)^2 (1 - \gamma^2) F'(f_{y0})] \frac{\partial p}{\partial \chi} \}$$

$$= (1 - e^2 + e^2 \gamma^2)^{3/2} \frac{\partial h}{\partial \chi}, \quad \mu = \exp(Qp),$$

$$p(\chi, \xi) |_{\Gamma} = 0, \quad \frac{dp(\chi, \xi)}{d\vec{n}} |_{\Gamma_e} = 0, \qquad (16.69)$$

$$H_{00}[h(\chi, \xi) - 1] = x^2(\chi, \xi) + \frac{\delta}{1 - e^2} y^2(\chi, \xi)$$

$$+\frac{D(e)}{\pi (1 - e^2)^{3/2}} \{ \int \int_{\Omega} \frac{p(\chi_1, \xi_1) d\chi_1 d\xi_1}{\sqrt{(\chi - \chi_1)^2 + \frac{(\xi - \xi_1)^2}{1 - e^2}}} - \int \int_{\Omega} \frac{p(\chi_1, \xi_1) d\chi_1 d\xi_1}{\sqrt{\chi_1^2 + \frac{\xi_1^2}{1 - e^2}}} \},$$

$$\int \int_{\Omega} p(\chi_1, \xi_1) d\chi_1 d\xi_1 = \frac{2\pi}{3},$$

where x, y and x_1, y_1 are functions of χ, ξ and χ_1, ξ_1 determined according to (13.4), respectively, while $D(e)$ and e are determined by equations (12.7) and (12.8).

Obviously, for straight entrained lubricant $\gamma = 1$ and problem (16.83) is reduced to the EHL problem from the preceding section.

We will assume that

$$s_y = O(s_x), \quad \omega \ll 1. \qquad (16.70)$$

To analyze the problem further we will assume that in the inlet and exit zones of a heavily loaded EHL contact the rheology functions F and Φ satisfy (16.54) and Φ is represented by (16.55). As a result of that in the inlet and exit zones we have (16.56) where function Φ_1 depends on the inner/local variable in the inlet or exit zone and it is independent of V, ω, and H_{00}. After that we have to take into account formulas (16.53).

In this case, as it follows from the Reynolds equation (16.83) for $\omega_0 = 0$ the dominant direction of the lubricant flow is along the χ-axis which means that in the inlet and exit zones of a heavily loaded point EHL contact $\frac{\partial}{\partial \xi} \ll \frac{\partial}{\partial \chi}$. Therefore, due to estimates (16.56) for pre-critical lubrication regimes (defined by (14.25)) in the inlet and exit zones we will be able to derive the asymptotically valid systems of equations (12.36) and (12.37) valid for line contacts (see Section 9.4) in which function M_0 from (12.38) should be replaced by

$$M_0(A, p, h, \mu, x, y) = A^{m+2} \frac{h^3(x,y)}{6\mu} \left\{ \frac{\gamma^2}{\Phi_1(\frac{\mu s_x}{2h})} + \frac{(1-e^2)^2(1-\gamma^2)}{\Phi_1(\frac{\mu s_y}{2h})} \right\} \frac{\partial p}{\partial x} \quad (16.71)$$

and θ is determined by (13.32) while the formula for H_{00} take the form of

$$H_{00} = A_p((1 - e^2 + e^2\gamma^2)^{3/2} V^k \omega^l \epsilon_q^2)^{\frac{1}{m+2}}, \quad A_p = O(1), \quad \omega \ll 1, \quad (16.72)$$

were A_p is a constant obtained from the solution of equations (12.36) which depends on e, γ, Q, s_x, and s_y. This analysis is valid as long as $\nu(x,y) \ll 1$ in the inlet zone which is based on (16.47) and (16.85) leads to the condition

$$\epsilon_q \ll \epsilon_\nu = \Phi_0^{\frac{2(m+2)}{3m+2}} (s_x)(1 - e^2 + e^2\gamma^2)^{-\frac{3(m+1)}{3m+2}} (V^k \omega^l)^{\frac{2}{3m+2}}, \quad \omega \ll 1. \quad (16.73)$$

As it follows from the formula (16.85) for the film thickness H_{00} if

$$\epsilon_q \ll \epsilon_f = (1 - e^2 + e^2\gamma^2)^{\frac{3}{3m+2}} (V^k \omega^l)^{\frac{2}{3m+2}}, \quad \omega \ll 1, \quad (16.74)$$

then the above analysis describes only starved lubrication regimes defined by the estimate $\epsilon_q^{3/2} \ll H_{00}$ for $\omega \ll 1$ and the film thickness is described by formula (16.85). To determine the actual lubrication regime that is realized we need to compare the values of ϵ_{nu} and ϵ_f. If $\epsilon_{nu} \ll \epsilon_f$ then

$$\Phi_0(s_x) \ll (1 - e^2 + e^2\gamma^2)^{3/2} \quad (16.75)$$

and the applied method is valid only for starved lubrication regimes. On the other hand, if $\epsilon_\nu = O(\epsilon_f)$ or $\epsilon_\nu \gg \epsilon_f$, i.e., if

$$\Phi_0(s_x) = O((1 - e^2 + e^2\gamma^2)^{3/2}) \text{ or } \Phi_0(s_x) \gg (1 - e^2 + e^2\gamma^2)^{3/2} \quad (16.76)$$

then the analysis is valid for both starved and fully flooded lubrication regimes. For starved lubrication regimes for the film thickness we have formula (16.85) as long as $\epsilon_q \ll \epsilon_f$ while for fully flooded lubrication regimes defined by the estimate $\epsilon_q^{3/2} = O(H_{00})$ for $\omega \ll 1$ we have

$$H_{00} = A_f((1 - e^2 + e^2\gamma^2)^{3/2} V^k \omega^l)^{\frac{3}{3m+2}}, \quad A_f = O(1),$$

$$\epsilon_q = (1 - e^2 + e^2\gamma^2)^{\frac{3}{3m+2}} (V^k \omega^l)^{\frac{2}{3m+2}}, \quad \omega \ll 1. \quad (16.77)$$

Obviously, by solving equations (12.36) and (12.37) with respect to functions q_p and g_p involved in the integrals in the right-hand sides of these equations they can be transformed into equivalent ones similar to equations (12.57) and (12.58) which are more convenient for numerical solution and can be regularized and numerically solved by the methods described in Sections 6.9 and 17.9. Moreover, substitution (16.41) and (16.42) and proper definition for r_0 (see, for example, (13.47) for the case of Newtonian lubricant) will reduce systems (17.27)-(17.29) to the form similar to the ones derived and analyzed in Chapter 9. For certain rheology functions F and Φ this substitution can reduce the latter systems to systems identical to the ones from Chapter 9. Also, this substitution will allow to obtain a formula for $H_0(\xi)$ which represents the dimensionless analog of the dimensional film thickness h_0 but at the exit from the contact region.

In a special case of a lubricant with Newtonian rheology formulas (16.84) and (16.85) for M_0 and H_{00} can be rewritten in the form identical to the corresponding formulas in Chapter 12.

If $\mu = e^{Qp}$ then for over-critical lubrication regimes described by the condition (16.43) (i.e., $\epsilon_0 = Q^{-2}$) using an approach similar to the one used in Section 12.4 we obtain the following formula for the film thickness

$$H_{00} = A_p(1 - e^2 + e^2\gamma^2)^{\frac{3}{2(m+2)}} (V^k\omega^l Q \epsilon_q^{5/2})^{\frac{1}{m+2}},$$

$$A_p = O(1), \ \omega \ll 1,$$

(16.78)

where A_p is a constant independent of V, Q, and ω. In the inlet ϵ_q-zone the condition $\nu(x,y) \ll 1$ is satisfied if

$$\epsilon_q \ll \epsilon_\nu = \Phi_0^{\frac{2(m+2)}{3m+1}} (s_x)(1 - e^2 + e^2\gamma^2)^{-\frac{3(m+1)}{3m+1}} (V^k\omega^l Q)^{\frac{2}{3m+1}},$$

$$\omega \ll 1.$$

(16.79)

Under starved lubrication conditions the film thickness is determined by (16.91) while

$$\epsilon_q \ll \epsilon_f = (1 - e^2 + e^2\gamma^2)^{\frac{3}{3m+1}} (V^k\omega^l Q)^{\frac{2}{3m+1}}, \ \omega \ll 1. \tag{16.80}$$

Therefore, the above asymptotic analysis is valid only for starved lubrication conditions if $\epsilon_\nu \ll \epsilon_f$ which is equivalent to condition (16.88). Otherwise, if $\epsilon_\nu = O(\epsilon_f)$ or $\epsilon_\nu \gg \epsilon_f$, i.e., if the conditions (16.89) are satisfied, then regimes of fully flooded lubrication are realized and the film thickness is determined by the formula

$$H_{00} = A_f(1 - e^2 + e^2\gamma^2)^{\frac{9}{2(3m+1)}} (V^k\omega^l Q)^{\frac{3}{3m+1}},$$

$$\epsilon_q = (1 - e^2 + e^2\gamma^2)^{-\frac{3}{3m+1}} (V^k\omega^l Q)^{\frac{2}{3m+1}}, \ \omega \ll 1,$$

(16.81)

where A_f is a constant independent of V, Q, and ω.

In most cases the value of $1 - e^2 + e^2\gamma^2$ is of the order of 1. Therefore, according to (16.88) and (16.89) in these cases to determine the applicability of the developed method it is sufficient to compare $\Phi_0(s_x)$ with 1.

Several specific examples for particular non-Newtonian lubricant rheology are given in Section 9.4.

16.4.3   Case of Skewed Lubricant Entrainment with Spinning

To consider the case of spinning we need to rewrite our simplified system of equations (16.52) in polar coordinates (see (14.5) and (14.6))

$$x = \frac{1}{\sqrt{1-e^2}}[-\frac{\sqrt{1-\gamma^2}}{\omega_0} + \rho\sin\varphi], \ y = \frac{\gamma}{\omega_0} - \rho\cos\varphi, \ dxdy = \frac{\rho d\rho d\varphi}{\sqrt{1-e^2}}, \quad (16.82)$$

$$\rho = \sqrt{(\sqrt{1-e^2}x + \frac{\sqrt{1-\gamma^2}}{\omega_0})^2 + (y - \frac{\gamma}{\omega_0})^2}, \ \tan\varphi = \frac{\sqrt{1-e^2}x + \frac{\sqrt{1-\gamma^2}}{\omega_0}}{\frac{\gamma}{\omega_0} - y}. \quad (16.83)$$

The reason for such a choice of new independent variables is the fact that in the Hertzian region (i.e., away from the inlet and exit contact boundaries) the lubricant moves along circular arcs $\rho = const$ and the gap h along these arcs is practically constant, i.e., $h = h(\rho)$.

In these variables equations (16.52) assume the following form

$$(\sin\varphi\frac{\partial}{\partial\rho} + \frac{\cos\varphi}{\rho}\frac{\partial}{\partial\varphi})\{\frac{H_{00}h^3}{6\mu}F'(f_{x0})[\sin\varphi\frac{\partial p}{\partial\rho} + \frac{\cos\varphi}{\rho}\frac{\partial p}{\partial\varphi}]\}$$

$$+(-\cos\varphi\frac{\partial}{\partial\rho} + \frac{\sin\varphi}{\rho}\frac{\partial}{\partial\varphi})\{\frac{H_{00}h^3}{6\mu}F'(f_{y0})[-\cos\varphi\frac{\partial p}{\partial\rho} + \frac{\sin\varphi}{\rho}\frac{\partial p}{\partial\varphi}]\} + \cdots$$

$$= \frac{\omega_0}{\sqrt{1-e^2}}\frac{\partial h}{\partial\varphi}, \ \mu = \exp(Qp), \ p(\rho,\varphi)\mid_\Gamma = 0, \ \frac{dp(\rho,\varphi)}{d\vec{n}}\mid_{\Gamma_e} = 0,$$

$$H_{00}[h(\rho,\varphi) - 1] = x^2(\rho,\varphi) + \frac{\delta}{1-e^2}y^2(\rho,\varphi)$$

$$+\frac{D(e)}{\pi(1-e^2)}\{\int\int_\Omega \frac{p(\rho_1,\varphi_1)\rho_1 d\rho_1 d\varphi_1}{\sqrt{\rho^2 - 2\rho\rho_1\cos(\varphi-\varphi_1) + \rho_1^2}}$$

$$-\int\int_\Omega \frac{p(\rho_1,\varphi_1)\rho_1 d\rho_1 d\varphi_1}{\sqrt{\rho_1^2 - \frac{2}{\omega_0}\rho_1[\gamma\cos(\varphi_1) + \sqrt{1-\gamma^2}\sin(\varphi_1)] + \frac{1}{\omega_0^2}}}\},$$

$$\int\int_\Omega p(\rho_1,\varphi_1)\rho_1 d\rho_1 d\varphi_1 = \frac{2\pi}{3}\sqrt{1-e^2},$$

(16.84)

where x and y are functions of ρ and φ determined according to (16.82).

First, let us consider pre-critical lubrication regimes defined by (14.25). In (16.84) the equation for h and the balance equation (the last equation in (16.84)) are identical with those in (14.7). Therefore, the analysis of the Hertzian region of such a heavily loaded contact as well as the form in which the solutions for pressure p should be searched in the inlet Ω_i and exit Ω_e zones along the "central" lubricant flow streamlines are identical with those from Chapter 14. The latter is the result of the requirement that in intermediate zones between the inlet and exit zones on one hand and the Hertzian region on the other the problem solutions in the inlet and exit zones must match the problem solution in the Hertzian region. The form in which the solutions are searched indicates that in the inlet Ω_i and exit Ω_e zones the derivatives with respect to φ are much larger than the corresponding derivatives with respect to ρ, i.e., $\frac{\partial}{\partial \varphi} \gg \frac{\partial}{\partial \rho}$. Using this fact in the inlet Ω_i and exit Ω_e zones the Reynolds equation from (16.84) can be reduced to the following one:

$$\frac{1}{\rho^2}\{\cos\varphi\frac{\partial}{\partial\varphi}[\frac{H_{00}h^3}{6\mu}F'(f_{x0})\cos\varphi\frac{\partial p}{\partial\varphi}]$$
$$+\sin\varphi\frac{\partial}{\partial\varphi}[\frac{H_{00}h^3}{6\mu}F'(f_{y0})\sin\varphi\frac{\partial p}{\partial\varphi}]\}+\ldots=\frac{\omega_0}{\sqrt{1-e^2}}\frac{\partial h}{\partial\varphi}, \tag{16.85}$$

where functions f_{x0} and f_{y0} are determined by (16.50) which in the polar coordinates can be expressed as follows

$$f_{x0} = \Phi(\frac{\mu}{h}(\frac{s_x}{2}-\gamma+\omega_0\rho\cos\varphi)),$$
$$f_{y0} = \Phi(\frac{\mu}{h}(\frac{s_y}{2}-\sqrt{1-\gamma^2}+\omega_0\rho\sin\varphi)). \tag{16.86}$$

Based on the analysis of the problem solution in the Hertzian region (see Chapter 14) it is clear that in the inlet $\Omega_i \bigcap \Omega_\epsilon$ and exit $\Omega_e \bigcap \Omega_\epsilon$ zones the pressure distribution p should be searched in the form (see (14.26) and (14.27))

$$p(\rho,\varphi) = \epsilon_q^{1/2}q_p(\rho,r_p)+o(\epsilon_q^{1/2}), \ \ q_p(\rho,r_p)=O(1),$$
$$r_p = \rho\frac{\varphi-\gamma_i(\rho)}{\epsilon_q} = O(1), \ (\rho,\varphi)\in\Omega_i\bigcap\Omega_\epsilon, \ \omega\ll 1, \tag{16.87}$$

$$p(\rho,\varphi) = \epsilon_q^{1/2}g_p(\rho,s_p)+o(\epsilon_q^{1/2}), \ \ g_p(\rho,s_p)=O(1),$$
$$s_p = \rho\frac{\varphi-\gamma_e(\rho)}{\epsilon_q} = O(1), \ (\rho,\varphi)\in\Omega_e\bigcap\Omega_\epsilon, \ \omega\ll 1, \tag{16.88}$$

where $q_p(\rho,r_p)$ and $g_p(\rho,s_p)$ are the unknown functions which are the main terms of the asymptotic expansions of pressure in the inlet and exit zones, respectively, while $\varphi = \gamma_i(\rho)$ and $\varphi = \gamma_e(\rho)$ are the equations expressing the inlet and exit boundaries (see (14.19)). Functions $q_p(\rho,r_p)$ and $g_p(\rho,s_p)$ further for simplicity will be called asymptotic pressure distributions in the inlet and exit zones.

The asymptotic representations for $\frac{\partial h}{\partial \varphi}$ in the inlet and exit zones have the form (see (14.33) and (14.34))

$$H_{00}\frac{\partial h}{\partial\varphi} = \epsilon_q^{1/2}\frac{2}{\pi}\frac{D(e)}{1-e^2}\rho\int\limits_0^\infty\frac{q_p(\rho,t)-q_{ap}(\rho,t)}{t-r_p}dt+\ldots, \ (\rho,\varphi)\in\Omega_i\bigcap\Omega_\epsilon, \tag{16.89}$$

$$H_{00}\frac{\partial h}{\partial \varphi} = \epsilon_q^{1/2}\frac{2}{\pi}\frac{D(e)}{1-e^2}\rho \int\limits_{-\infty}^{0} \frac{g_p(\rho,t)-g_{ap}(\rho,t)}{t-s_p}dt + \ldots, \quad (\rho,\varphi)\in \Omega_e\bigcap\Omega_\epsilon, \quad (16.90)$$

where $q_0(\rho,\varphi,\rho_\Gamma,\varphi_\Gamma)$ is replaced by $q_{ap}(\rho,r_p) = q_0(\rho,\gamma_i(\rho)+r_p\epsilon_q,\rho,\gamma_i(\rho)) = N_0(\rho,\varphi_{H1})\sqrt{2r_p}+\frac{N_1(\rho)}{\sqrt{2r_p}}+o(1)$ and $g_{ap}(\rho,s_p) = q_0(\rho,\gamma_e(\rho)+s_p\epsilon_q,\rho,\gamma_e(\rho)) = N_0(\rho,\varphi_{H2})\sqrt{-2s_p}+\frac{N_1(\rho)}{\sqrt{-2s_p}}+o(1)$, $\epsilon_q \ll 1$ due to the fact that the direction of the normal vector $\overrightarrow{n}$ to the contact boundary Γ_i in $\Omega_i\bigcap\Omega_\epsilon$ and Γ_e in $\Omega_e\bigcap\Omega_\epsilon$ practically coincides with the ρ-axis (see (14.20)). Here function N_0 can be determined from matching considerations and it has the form (see (14.37))

$$N_0(\rho,\varphi_H) = \mid \frac{\sqrt{1-\gamma^2}}{\omega_0(1-e^2)}\cos\varphi_H - \frac{\gamma}{\omega_0}\sin\varphi_H - \frac{e^2\rho}{2(1-e^2)}\sin(2\varphi_H)\mid^{1/2}, \quad (16.91)$$

where the values of two angles φ_{H1} and φ_{H2} are determined by the solutions of equation (14.16) (see also (14.17) and (14.18)). Also, we can use the fact that (see (14.58))

$$\alpha_{1p} = \frac{N_{1q}}{N_{0q}}, \quad \beta_{1p} = -\frac{N_{1g}}{N_{0g}}, \quad (16.92)$$

where N_{0q}, N_{1q} and N_{0g}, N_{1g} are calculated for $\varphi_H = \varphi_{H1}$ and $\varphi_H = \varphi_{H2}$, respectively (i.e., in the inlet and exit zones).

Using the estimates (16.87) for p and (16.89) for $H_{00}\frac{\partial h}{\partial \varphi}$ as well as the assumption (16.56) concerning the rheology function Φ in the inlet zone from the Reynolds equation (16.85) by comparing the orders of the Reynolds equation terms we derive the formula for the film thickness H_{00} as follows

$$H_{00} = A_p\left[\frac{V^k\omega^l\epsilon_q^2}{\sqrt{1-e^2}}\right]^{\frac{1}{m+2}}, \quad A_p = O(1), \quad \omega \ll 1, \quad (16.93)$$

where A_p is a constant which depends on θ, $Q\epsilon_q^{1/2}$, and $\alpha_{1p}(\frac{1}{\omega_0})$. Moreover, in the inlet $\Omega_i\bigcap\Omega_\epsilon$ and the exit $\Omega_e\bigcap\Omega_\epsilon$ zones based on the asymptotic procedure described in detail in Chapter 14 we obtain the asymptotically valid equations

$$\frac{\partial}{\partial r_p}M_0(A_p,q_p,h_{qp},\mu_{qp},\rho,r_p,\varphi_{H1}) = \frac{2}{\pi}\omega_0\rho\theta\int\limits_{0}^{\infty}\frac{q_p(\rho,t)-q_{ap}(\rho,t)}{t-r_p}dt,$$

$$q_p(\rho,0) = 0, \quad (16.94)$$

$$q_p(\rho,r_p) \to q_{ap}(\rho,r_p) = N_0(\rho,\varphi_{H1})[\sqrt{2r_p}+\frac{\alpha_{1p}}{\sqrt{2r_p}}], \quad r_p \to \infty,$$

$$\frac{\partial}{\partial s_p}M_0(A_p,g_p,h_{gp},\mu_{gp},\rho,s_p,\varphi_{H2}) = \frac{2}{\pi}\omega_0\rho\theta\int\limits_{-\infty}^{0}\frac{g_p(\rho,t)-g_{ap}(\rho,t)}{t-s_p}dt,$$

$$g_p(\rho,0) = \frac{\partial g_p(\rho,0)}{\partial s_p} = 0, \quad (16.95)$$

$$g_p(\rho,s_p) \to g_{ap}(\rho,s_p) = N_0(\rho,\varphi_{H2})[\sqrt{-2s_p}-\frac{\beta_{1p}}{\sqrt{-2s_p}}], \quad s_p \to -\infty,$$

$$M_0(A, p, h, \mu, x, y, \varphi_H) = \frac{A^{m+2}}{6} \frac{h^3}{\mu} \Big\{ \frac{\cos^2 \varphi_H}{\Phi_1(\frac{\mu}{h}(\frac{s_x}{2} - \gamma + \omega_0 x \cos \varphi_H))}$$

$$+ \frac{\sin^2 \varphi_H}{\Phi_1(\frac{\mu}{h}(\frac{s_y}{2} - \sqrt{1 - \gamma^2} + \omega_0 x \sin \varphi_H))} \Big\} \frac{\partial p}{\partial y}, \quad \theta = \frac{D(e)}{1 - e^2}, \tag{16.96}$$

where $h_{qp}(\rho, r_p)$ and $h_{gp}(\rho, s_p)$ are the gap functions in the inlet $\Omega_i \bigcap \Omega_\epsilon$ and exit $\Omega_e \bigcap \Omega_\epsilon$ zones, functions μ_{qp} and μ_{gp} are the main terms of the asymptotic expansions of $\mu(\epsilon_q^{1/2} q_p)$ and $\mu(\epsilon_q^{1/2} g_p)$, respectively. The expressions for $h_{qp}(\rho, r_p)$ and $h_{gp}(\rho, s_p)$ depend on the lubrication regime. For starved lubrication regimes the gap distribution $h(\rho, \varphi)$ is very close to $h_\infty(\rho)$ in the entire contact region. That corresponds to the case of $\epsilon_q^{3/2} \ll H_{00}$ and

$$h_{qp}(\rho, r_p) = h_\infty(\rho), \quad \epsilon_q \ll \epsilon_f, \tag{16.97}$$

$$h_{gp}(\rho, s_p) = h_\infty(\rho), \quad \epsilon_q \ll \epsilon_f, \tag{16.98}$$

$$\epsilon_f = \Big[\frac{V^k \omega^l}{\sqrt{1 - e^2}} \Big]^{\frac{2}{3m+2}}, \tag{16.99}$$

while for fully flooded lubrication regimes $H_{00} = O(\epsilon_q^{3/2})$, $\epsilon_q = \epsilon_f$, $\omega \ll 1$, and in the inlet and exit zones the gap satisfies the equations (see equations (14.33) and (14.34))

$$A_p \frac{\partial h_{qp}}{\partial r_p} = \frac{2}{\pi} \theta \int_0^\infty \frac{q_p(\rho, t) - q_{ap}(\rho, t)}{t - r_p} dt, \quad \epsilon_q = \epsilon_f, \ \omega \ll 1, \tag{16.100}$$

$$A_p \frac{\partial h_{gp}}{\partial s_p} = \frac{2}{\pi} \theta \int_{-\infty}^0 \frac{g_p(\rho, t) - g_{ap}(\rho, t)}{t - s_p} dt, \quad \epsilon_q = \epsilon_f, \ \omega \ll 1. \tag{16.101}$$

In (16.97) and (16.99) ϵ_f is the characteristic width of the inlet Ω_i and exit Ω_e zones in fully flooded lubrication regimes.

For fully flooded pre-critical lubrication regimes (characterized by the relationship $H_{00} = O(\epsilon_f^{3/2})$, $\omega \ll 1$) we have

$$H_{00} = A_f \Big[\frac{V^k \omega^l}{\sqrt{1 - e^2}} \Big]^{\frac{3}{3m+2}}, \quad A_f = O(1), \ \epsilon_q = \epsilon_f, \ \omega \ll 1. \tag{16.102}$$

Here and the value of the coefficient A_f depends on θ, $\alpha_{1p}(\frac{1}{\omega_0})$, and $Q\epsilon_f^{1/2}$.

It is clear from equations (16.99) and (16.101) that when $r_p \to \infty$ and $s_p \to -\infty$ we get

$$\int_0^\infty [q_p(\rho, t) - q_{ap}(\rho, t)] dt = 0, \tag{16.103}$$

$$\int_{-\infty}^0 [g_p(\rho, t) - g_{ap}(\rho, t)] dt = 0, \tag{16.104}$$

because otherwise the functions $h_{qp}(\rho, r_p)$ and $h_{gp}(\rho, s_p)$ would be unbounded as $r_p \to \infty$ and $s_p \to -\infty$, respectively. Equations (16.103) and (16.104) are consistent with the last equation in (16.84).

For fully flooded lubrication regimes the values of functions $h_{qp}(\rho, r_p,)$ and $h_{gp}(\rho, s_p)$ are obtained by integrating equations (16.100) and (16.101) with the conditions $h_{qp}(\rho, r_p) \to h_{\infty}(\rho)$ as $r_p \to \infty$ and $h_{gp}(\rho, s_p) \to h_{\infty}(\rho)$ as $s_p \to -\infty$ as follows

$$A_p[h_{qp}(\rho, r_p) - h_{\infty}(\rho)] = \frac{2}{\pi}\theta \int_0^{\infty} [q_p(\rho, t) - q_{ap}(\rho, t)] \ln \frac{1}{|t - r_p|} dt, \qquad (16.105)$$

$$A_p[h_{gp}(\rho, s_p) - h_{\infty}(\rho)] = \frac{2}{\pi}\theta \int_{-\infty}^0 [g_p(\rho, t) - g_{ap}(\rho, t)] \ln \frac{1}{|t - s_p|} dt. \qquad (16.106)$$

For fully flooded lubrication regimes by integrating equations (16.94) and (16.95) for q_p and g_p the gap functions h_{qp} and h_{gp} from (16.105) and (16.106) can be replaced by equations

$$h_{qp} = h_{\infty}(\rho) + \frac{1}{A_p} M_0(A_p, q_p, h_{qp}, \mu_{qp}, \rho, r_p, \varphi_{H1}),$$

$$h_{gp} = h_{\infty}(\rho) - \frac{1}{A_p} M_0(A_p, g_p, h_{gp}, \mu_{gp}, \rho, s_p, \varphi_{H2}),$$

$$(16.107)$$

where $h_{\infty}(\frac{1}{\omega_0}) = 1$ while $h_{\infty}(\rho)$ for $\rho \neq \frac{1}{\omega_0}$ is determined from solution of system (16.94), (16.96), (16.103), and (16.97) or (16.107) for starved or fully flooded lubrication regimes, respectively, for earlier determined value of coefficient A_p which is a part of the solution of the same system for $\rho = \frac{1}{\omega_0}$.

Here it is appropriate to discuss the applicability of the proposed analysis and some solution properties with respect to the parameter ω_0. For small and moderate values of ω_0 there is a pretty wide range of parameters e and γ for which we can find "central" lubricant flow streamlines and our asymptotic analysis is valid. At the same time, for $\omega_0 \gg 1$ the proposed method is applicable only for $\sqrt{1 - e^2} < \rho < 1$ because the center of the fluid flow streamline ellipses is very close to the origin $(x, y) = (0, 0)$ and the above inequality provides the condition for the existence of the "central" streamlines. For $\omega_0 \to 0$ the value of $\omega_0 \rho$ remains finite, i.e., it does not vanish. Instead, $\omega_0 \rho \to 1$ as $\omega_0 \to 0$ (see formulas (16.83)) which leads to the conclusion that in this limiting case the film thickness H_{00} approaches the one for the case of no spin which was considered in the preceding section.

Therefore, in the inlet zone for the given values of θ, $Q\epsilon_q^{1/2}$ (involved in μ_{qp}), $\alpha_{1p}(\rho)$, and ρ we derived the system of equations (16.94), (16.96), (16.97), and (16.103) for starved lubrication regimes or (16.105) for fully flooded lubrication regimes for functions $q_p(\rho, r_p)$, $h_{qp}(\rho, r_p)$, $h_{\infty}(\rho)$, and constant A_p. In the exit zone for the given values of θ, $Q\epsilon_q^{1/2}$, A_p, $h_{\infty}(\rho)$, and ρ we need to solve the system of equations (16.95), (16.96), (16.99), and (16.104) for starved lubrication regimes or (16.106) for fully flooded lubrication regimes for functions $g_p(\rho, s_p)$, $h_{gp}(\rho, s_p)$, and $\beta_{1p}(\rho)$. Obviously, these systems of equations are practically identical to the ones derived in the preceding section for the case of no spin, $\omega_0 = 0$. At the same time, the formulas (16.93) and (16.102) for the film thickness H_{00} have the structure in a certain sense similar to the corresponding formulas for H_{00} from the preceding section. It is important to keep in mind that constant A_p is the coefficient of proportionality in formula (16.93) for the central film thickness and it is determined for $y = 0$ or $\rho = \frac{1}{\omega_0}$ (see (16.83)). Also, it means that $h_\infty(\frac{1}{\omega_0}) = 1$ while, generally, $h_\infty(\rho) \neq 1$ for $\rho \neq \frac{1}{\omega_0}$. The solution procedure should be implemented according to the following sequence: first, the system of asymptotically valid equations in the inlet zone should be solved for $\rho = \frac{1}{\omega_0}$ which determines the value of constant A_p; after that, the values of function $h_\infty(\rho)$ for $\rho \neq \frac{1}{\omega_0}$ are determined from the solution of the asymptotically valid system of equations in the inlet zone and these values play the role similar to the one played by constant A_p for $\rho = \frac{1}{\omega_0}$.

Substitution (16.41) and (16.42) for $\xi = \rho$ and proper definition for r_0 (see, for example, (13.47) for the case of Newtonian lubricant) will reduce systems (17.27)-(17.29) to the form similar to the ones derived and analyzed in Chapter 9. For certain rheology functions F and Φ this substitution can reduce the latter systems to systems identical to the ones from Chapter 9. Moreover, this substitution will allow to obtain a formula for $H_0(y)$ which represents the dimensionless analog of the dimensional film thickness h_0 but at the exit from the contact region.

The above systems of asymptotically valid equations derived in the inlet $\Omega_i \bigcap \Omega_\epsilon$ and exit $\Omega_e \bigcap \Omega_\epsilon$ zones in heavily loaded point EHL contacts allow for the direct usage of the entire apparatus of numerical analysis of these systems as well as of the original equations developed in Chapters 6 - 10 for heavily loaded line EHL contacts. These methods include usage of systems of equations represented in equivalent forms, their regularization and design of stable numerical schemes, proper choice of the grid sizes for the given EHL problem parameters, etc.

Now, let us consider over-critical lubrication regimes defined by (16.43). In this case the structure of the lubricated region as well as the whole analysis are identical to the ones described in Sections 12.4 and 12.4. Applying this approach to the Reynolds equation (16.85), assumption (16.56) concerning the behavior of the rheology function Φ in the ϵ_q-inlet zone as well as using in this zone the estimates $p = O(\epsilon_0^{1/2})$ (where $\epsilon_0 = Q^{-1}$) and $H_{00}\frac{\partial h}{\partial \varphi} = O(\epsilon_q^{1/2})$ for starved lubrication regimes for the film thickness H_{00} we obtain the following formula

$$H_{00} = A_p\left[\frac{V^k\omega^l Q\epsilon_q^{5/2}}{\sqrt{1-e^2}}\right]^{\frac{1}{m+2}}, \quad A_p = O(1), \quad \epsilon_q \ll \epsilon_{fo}, \quad \omega \ll 1, \qquad (16.108)$$

$$\epsilon_{fo} = \left[\frac{V^k\omega^l Q}{\sqrt{1-e^2}}\right]^{\frac{2}{3m+1}}, \qquad (16.109)$$

where the value of constant A_p depends on e, s_x, s_y, and $\alpha_{1p}(\frac{1}{\omega_0})$ while ϵ_{fo} is the characteristic width of inlet zone in fully flooded lubrication regime. For fully flooded over-critical lubrication regime the formula for the film thickness

H_{00} follows from formula (16.108) and estimate $H_{00} = O(\epsilon_q^{3/2})$ and it takes the form

$$H_{00} = A_f \left[\frac{V^k \omega^l Q}{\sqrt{1-e^2}} \right]^{\frac{3}{3m+1}}, \quad A_f = O(1), \quad \epsilon_q = \epsilon_{fo}, \quad \omega \ll 1, \tag{16.110}$$

where constant A_f also depends on e, s_x, s_y, and $\alpha_{1p}(\frac{1}{\omega_0})$.

The values of constants A_p and A_f can be obtained from the solution of the system of asymptotically valid equations in the inlet zone for pre-critical lubrication regimes. For example, for fully flooded over-critical lubrication regime the value of A_f can be obtained the following way. For pre-critical fully flooded lubrication regimes $\gamma_i = \varphi_{H1} + \alpha_{1p}\epsilon_f$ (see expression for ϵ_f in (16.99)) while simultaneously for over-critical fully flooded lubrication regimes $\gamma_i = \varphi_{H1} + \alpha_{1po}\epsilon_{fo}$ (see expression for ϵ_{fo} in (16.109)). Therefore, we have the corresponding inlet coordinate α_{1p} for pre-critical lubrication regimes $\alpha_{1p} = \alpha_{1po}[(\frac{V^k \omega^l}{\sqrt{1-e^2}})^{\frac{1}{3m+2}}Q]^{\frac{2}{3m+1}}$. For this inlet coordinate the value of constant $A_f(\alpha_{1p})$ from the solution in the inlet zone of asymptotically valid equations for pre-critical lubrication regimes will produce the value of $A_f(\alpha_{1po})$ in formula (16.110) for over-critical lubrication regimes if the value of $A_f(\alpha_{1p})$ would be substituted in formula (16.102) for H_{00} for pre-critical lubrication regimes and compared with formula (16.110) for H_{00} for over-critical lubrication regimes. To obtain the value of constant $A_p(\alpha_{1po})$ in formula (16.108) we need to solve in the inlet zone the asymptotically valid equations for pre-critical lubrication regimes with the inlet coordinate $\alpha_{1p} = \alpha_{1p0}\frac{\epsilon_{qo}}{\epsilon_q}$ to obtain the value of $A_p(\alpha_{1p})$ where for pre-critical regime calculations $\gamma_i = \varphi_{H1} + \alpha_{1p}\epsilon_q$ while for over-critical regime calculations $\gamma_i = \varphi_{H1} + \alpha_{1po}\epsilon_{qo}$ and $\epsilon_q = O(\epsilon_0) \ll \epsilon_{qo}$. After that, the comparison of formulas (16.93) and (16.108) will produce the value of $A_p(\alpha_{1po})$ in equation (16.108) for the film thickness H_{00} for over-critical starved lubrication regimes. For more details see Section 6.10.

Finally, we need to check the applicability of the above asymptotic analysis. Our analysis involved only one essential assumption (if we disregard the assumption made about the behavior of the rheology function Φ) represented by the estimate (16.47). Therefore, for both pre- and over-critical lubrication regimes we need to determine the maximum characteristic sizes ϵ_ν and $\epsilon_{\nu o}$ of the inlet zone from solution of equation $\nu = 1$ in these zones for pre- and over-critical lubrication regimes, respectively. After that, the conclusion about the applicability of the proposed analysis for starved and fully flooded lubrication regimes can be done based on the validity of the estimates $\epsilon_f \gg \epsilon_\nu$ and $\epsilon_f \ll \epsilon_\nu$ for pre-critical lubrication regimes and based on the validity of $\epsilon_{fo} \gg \epsilon_{\nu o}$ and $\epsilon_{fo} \ll \epsilon_{\nu o}$ for over-critical lubrication regimes, respectively. The specific calculations are left for the reader to perform as an exercise.

For the case of a lubricant with Newtonian rheology for pre- and over-critical lubrication regimes the formulas for the film thickness H_{00} derived in this section coincide with the corresponding formulas for the film thickness H_{00} obtained in Chapter 14.

The choice between pre- and over-critical lubrication regimes can be made based on a simple analysis from Section 9.5.

The regularization of the numerical solutions of the reduced original and asymptotic equations can be done by introducing in the reduced or asymptotically valid analog of the Reynolds equation a small heat dissipation based on the formulas derived in the next chapter in the way similar to the one used in Chapter 8.

16.5 Closure

An asymptotic approach to isothermal steady heavily loaded point EHL contacts with general incompressible non-Newtonian lubricants with skewed lubricant entrainment and spinning ball is developed. The cases of pure rolling and high slide-to-roll ration are considered. The original Reynolds equation for such EHL problems is reduced to an analog of the corresponding Reynolds equation for a Newtonian lubricant. For the case of high slide-to-roll ratio using the problem formulation with the reduced Reynolds equation the further asymptotic analysis has been carried out and the behavior of the pressure and gap distributions in the inlet and exit zones along the "central" streamlines of the lubricant flow have been analyzed in detail. Asymptotic analysis of heavily loaded contacts lubricated by a non-Newtonian fluid recognizes two different regimes of lubrication: pre- and over-critical lubrication regimes. Overall, the difference between these two regimes is in how rapidly the lubricant viscosity grows with pressure and how much lubricant is available at the inlet of the contact. For pre-critical lubrication regimes the asymptotically valid equations for pressure and gap have been derived. These equations are similar to the corresponding equations for pre-critical lubrication regimes in the inlet and exit zones of line heavily loaded contacts. The latter equations have been validated through the comparison with the direct numerical solutions of the original EHL equations in Chapter 6. For over-critical lubrication regimes another approach has been used which led to formulas for the lubrication film thickness for starved and fully flooded conditions. The asymptotic approaches to EHL problems offer certain distinct advantages in comparison with the direct numerical solutions of EHL problem such a reduction of the number of problem input parameters and the approach to regularization of numerical solutions (see Chapter 6 and Section 8.2). Each of the analytically derived formulas for the central H_{00} lubrication film thickness contains a number of parameters characterizing lubricant rheology as well as a coefficient of proportionality which can be found by numerically solving the corresponding asymptotically valid equations in the inlet zone or by just curve fitting the experimentally obtained values of the lubrication film thickness.

16.6   Exercises and Problems

1. What are the differences and similarities between the asymptotic approaches used for the analysis of heavily loaded point EHL contacts in the cases straight and skewed lubricant entrainment as well as in the case of spinning motion? For each case describe in detail the structure of the lubricated region.

2. List and discuss all assumptions which in the inlet and exit zones allowed to reduce the equations for the heavily loaded point EHL contact to the equations of the corresponding heavily loaded line EHL contact. Characterize the validity of these assumptions. What is the size of the zones where the proposed asymptotic approach is not valid?

3. Perform the detailed analysis of the applicability of the employed asymptotic method for pre- and over-critical lubrication regimes with spinning.

References

[1] Hamrock, B.J. 1994. *Fundamentals of Fluid Film Lubrication*. New York: McGraw-Hill.

17

Thermal EHL Problems for Heavily Loaded Point Contacts. Non-Newtonian Lubricants

17.1 Introduction

In this chapter the techniques developed in Chapter 10 for heavily loaded line EHL contacts and in Chapters 12 - 14, and 16 for point heavily loaded EHL contacts are extended on the case of thermal point EHL problems for contacts with non-Newtonian lubricants. The problems for point EHL contacts of Chapters 13 and 14 are considered under thermal conditions when lubricants are non-Newtonian. The problems are analyzed for the case of high slide-to-roll ratio. The point EHL problem equations are reduced to equations of problems similar to the thermal line EHL problems for the corresponding non-Newtonian lubricants. In the inlet and exit zones of heavily loaded point EHL contacts along the lubricant "central" flow streamlines the problems are reduced to asymptotically valid equations similar to the ones obtained in Chapter 10. Analytical formulas for the lubrication film thickness are derived. A specific regularization technique for stabilizing numerical solutions of point EHL problems for heavily loaded contacts is proposed.

17.2 Problem Formulation

Let us consider a steady thermal lubricated contact with skewed lubricant entrainment of an elastic ball rolling and spinning in an elastic grooved raceway. The lubricant is considered to be a non-Newtonian fluid. In other respects the problem formulation is exactly the same as described in Chapter 16. Let us assume that the lubricant viscosity μ depends on lubricant temperature $T(x, y, z)$ according to the relationship

$$\mu = \mu^0(p) \exp[-\alpha_T(T - T_a)], \ \mu^0(p) = \mu_a e^{\alpha p}, \tag{17.1}$$

where μ_a is the lubricant viscosity at pressure $p = 0$ and ambient temperature T_a, α and α_T are the pressure and temperature viscosity coefficients, respectively. Repeating the Chapter 16 derivation of the problem equations and taking into account the fact that in this case μ depends on z we obtain

$$\frac{\partial}{\partial x}\{\int_{-h/2}^{h/2} \frac{z}{\mu}F(f_x + z\frac{\partial p}{\partial x})dz\} + \frac{\partial}{\partial y}\{\int_{-h/2}^{h/2} \frac{z}{\mu}F(f_y + z\frac{\partial p}{\partial y})dz\}$$

$$= \frac{u_1+u_2-\omega_2 y}{2}\frac{\partial h}{\partial x} + \frac{v_1+v_2+\omega_2 x}{2}\frac{\partial h}{\partial y}, \quad \mu = \mu^0(p)\exp\left[-\alpha_T(T-T_a)\right],$$

$$p(x,y)\,|_{\Gamma} = 0, \quad \frac{dp(x,y)}{d\overrightarrow{n}}\,|_{\Gamma_e} = 0,$$

$$\int_{-h/2}^{h/2} \frac{1}{\mu}F(f_x + z\frac{\partial p}{\partial x})dz = u_2 - u_1,$$

$$\int_{-h/2}^{h/2} \frac{1}{\mu}F(f_y + z\frac{\partial p}{\partial y})dz = v_2 - v_1, \tag{17.2}$$

$$h(x,y) = h_0 + \frac{x^2}{2R_x} + \frac{y^2}{2R_y} + \frac{1}{\pi E'}\{\int\int_{\Omega} \frac{p(x_1,y_1)dx_1 dy_1}{\sqrt{(x-x_1)^2+(y-y_1)^2}}$$

$$- \int\int_{\Omega} \frac{p(x_1,y_1)dx_1 dy_1}{\sqrt{x_1^2+y_1^2}}\}, \quad \int\int_{\Omega} p(x_1,y_1)dx_1 dy_1 = P,$$

$$\frac{\partial}{\partial z}[\lambda\frac{\partial T}{\partial z}] = -\frac{1}{\mu}[(f_x + z\frac{\partial p}{\partial x})F(f_x + z\frac{\partial p}{\partial x}) + (f_y + z\frac{\partial p}{\partial y})F(f_y + z\frac{\partial p}{\partial y})],$$

$$T(x,y,-\frac{h}{2}) = T_{w1}(x,y), \quad T(x,y,\frac{h}{2}) = T_{w2}(x,y),$$

where for the derivation of the equation for the temperature T we used the energy equation (4.12) and the formulas (16.2) and (16.3).

The inlet Γ_i and exit Γ_e portions of the contact boundary Γ through which the lubricant enters and leaves the contact, respectively, constitute the whole boundary of Ω, i.e. $\Gamma_i \bigcup \Gamma_e = \Gamma$. The inlet Γ_i and exit Γ_e boundaries are determined by the relationships

$$(x,y) \in \Gamma_i \text{ if } \overrightarrow{F_f} \cdot \overrightarrow{n} < 0; \quad (x,y) \in \Gamma_e \text{ if } \overrightarrow{F_f} \cdot \overrightarrow{n} \geq 0,$$

$$\overrightarrow{F_f} = (u_1 + u_2 - \omega_2 y, v_1 + v_2 + \omega_2 x)\frac{h}{2}$$

$$-(\int_{-h/2}^{h/2} \frac{z}{\mu}F(f_x + z\frac{\partial p}{\partial x})dz, \int_{-h/2}^{h/2} \frac{z}{\mu}F(f_y + z\frac{\partial p}{\partial y})dz), \tag{17.3}$$

where $\overrightarrow{F_f}$ is the fluid flux through the contact boundary Γ and $\overrightarrow{a} \cdot \overrightarrow{b}$ is the dot product of vectors $\overrightarrow{a}$ and $\overrightarrow{b}$.

In equations (17.2) and (17.3) we have function F describes the lubricant rheology and it is an odd monotonically increasing function such that $F(0) =$

0. Obviously, this function F has an inverse Φ, $\Phi(F(x)) = x$, h_0 is the central film thickness at the point with coordinates $(0, 0)$, R_x, R_y, and E' are the effective radii along the x- and y-axes and the effective modulus of elasticity, $\frac{1}{R_x} = \frac{1}{R_{1x}} + \frac{1}{R_{2x}}$, $\frac{1}{R_y} = \frac{1}{R_{1y}} + \frac{1}{R_{2y}}$, and $\frac{1}{E'} = \frac{1-\nu_1^2}{E_1} + \frac{1-\nu_2^2}{E_2}$, respectively, Ω and Γ are the contact region and its boundary, respectively, and $\vec{n}$ is the unit external normal vector to the boundary Γ.

Let us introduce the dimensionless variables according to (12.5), (16.8), and

$$T' = \frac{T}{T_a} - 1, \quad T'_{w1} = \frac{T_{w1}}{T_a} - 1, \quad T'_{w2} = \frac{T_{w2}}{T_a} - 1, \tag{17.4}$$

as well as use the dimensionless parameters (13.1) and

$$s_x = \frac{2(u_2 - u_1)}{\sqrt{(u_1+u_2)^2 + (v_1+v_2)^2}}, \quad s_y = \frac{2(v_2 - v_1)}{\sqrt{(u_1+u_2)^2 + (v_1+v_2)^2}}, \quad \delta_T = \alpha_T T_a,$$

$$\kappa = \frac{a_H^3 p_H \sqrt{(u_1+u_2)^2 + (v_1+v_2)^2}}{4\lambda_a T_a R_x^2}, \tag{17.5}$$

where a_H and p_H are the Hertzian small semi-axes and the maximum pressure in the dry elliptical Hertzian contact, respectively, which are determined by equations (12.7) and (12.8), and λ_a is the coefficient of lubricant heat conductivity at pressure $p = 0$ and ambient temperature T_a. Therefore, in these dimensionless variables and parameters (primes are dropped for convenience) the equations of the problem are reduced to the following form

$$\frac{\partial}{\partial x} \{ \int\limits_{-h/2}^{h/2} \frac{z}{\mu} F(f_x + H_{00} z \frac{\partial p}{\partial x}) dz \}$$

$$+ \sqrt{1 - e^2} \frac{\partial}{\partial y} \{ \int\limits_{-h/2}^{h/2} \frac{z}{\mu} F(f_y + \sqrt{1 - e^2} H_{00} z \frac{\partial p}{\partial y}) dz \}$$

$$= \frac{\gamma - \omega_0 y}{2} \frac{\partial h}{\partial x} + \sqrt{1 - e^2} \frac{\sqrt{1-\gamma^2} + \sqrt{1-e^2} \omega_0 x}{2} \frac{\partial h}{\partial y}, \quad \mu = \mu^0 e^{-\delta_T T}, \quad \mu^0 = e^{Qp},$$

$$p(x, y) \mid_\Gamma = 0, \quad \frac{dp(x,y)}{d\vec{n}} \mid_{\Gamma_e} = 0, \tag{17.6}$$

$$\int\limits_{-h/2}^{h/2} F(f_x + H_{00} z \frac{\partial p}{\partial x}) dz = \frac{\mu s_x}{2},$$

$$\int\limits_{-h/2}^{h/2} F(f_y + \sqrt{1 - e^2} H_{00} z \frac{\partial p}{\partial y}) dz = \frac{\mu s_y}{2},$$

$$H_{00}[h(x, y) - 1] = x^2 + \frac{\delta}{1 - e^2} y^2 + \frac{D(e)}{\pi(1-e^2)^{3/2}} \{ \int\int\limits_\Omega \frac{p(x_1, y_1) dx_1 dy_1}{\sqrt{(x-x_1)^2 + \frac{(y-y_1)^2}{1-e^2}}}$$

$$-\int\int_\Omega \frac{p(x_1,y_1)dx_1dy_1}{\sqrt{x_1^2+\frac{y_1^2}{1-e^2}}}\}, \quad \int\int_\Omega p(x_1,y_1)dx_1dy_1 = \frac{2\pi}{3},$$

$$\frac{\partial}{\partial z}[\lambda\frac{\partial T}{\partial z}] = -\frac{\kappa H_{00}}{\mu}[(f_x + H_{00}z\frac{\partial p}{\partial x})F(f_x + H_{00}z\frac{\partial p}{\partial x})$$

$$+(f_y + \sqrt{1-e^2}H_{00}z\frac{\partial p}{\partial y})F(f_y + \sqrt{1-e^2}H_{00}z\frac{\partial p}{\partial y})], \qquad (17.7)$$

$$T(x,y,-\tfrac{h}{2}) = T_{w1}(x,y), \quad T(x,y,\tfrac{h}{2}) = T_{w2}(x,y),$$

Therefore, for the given values of parameters V, Q, δ_T, δ, e, γ, ω_0, s_x, s_y, κ and functions F, λ, T_{w1}, T_{w2} and the inlet boundary Γ_i we need to determine functions $p(x,y)$, $h(x,y)$, $f_x(x,y)$, $f_y(x,y)$, $T(x,y,z)$, constant H_{00}, and exit boundary Γ_e.

To simplify further analysis we will assume that λ is independent of z. In other words, λ is either a constant ($\lambda = 1$) or it is a function of pressure p, i.e., $\lambda = \lambda(p)$.

The idea of the analysis of the problem is to reduce it to an analog of the isothermal EHL problem and, then, to analyze it asymptotically and numerically. To do that we need to obtain the expressions for the sliding frictional stresses f_x and f_y which would allow us to determine the asymptotic approximation for the lubricant temperature $T(x,y,z)$.

17.3 Solutions for f_x and f_y. High Slide-to-Roll Ratio

First, let us solve equations for f_x and f_y. It can be easily done in the case of high slide-to-roll ratio. Therefore, let us assume that in the inlet (and, thus, in the exit) zone

$$\nu(x,y) = \frac{H_{00}h}{2f_x}\frac{\partial p}{\partial x} \ll 1. \qquad (17.8)$$

The solutions for f_x and f_y we will search in the form of a perturbation series

$$f_x = f_{x0} + \nu f_{x1} + \dots, \quad f_y = f_{y0} + \nu f_{y1} + \dots,$$

$$f_{x0}, \ f_{x1}, \ f_{y0}, \ f_{y1} = O(1), \ \nu \ll 1, \qquad (17.9)$$

where functions f_{x0}, f_{x1}, f_{y0}, and f_{y1} depend on x and y and have to be determined from the solution of the third and fourth integral conditions in (17.6) while the lubricant temperature T from (17.8) we will search in the form

$$T = T_0 + \nu T_1 + \dots, \quad T_0, \ T_1 = O(1), \ \nu \ll 1, \qquad (17.10)$$

where functions T_0 and T_1 depend on x, y, z and have to be determined from the solution of the equations for T in (17.8). While determining functions f_x

and f_y we will temporarily assume that all other parameters of the lubricated contact are known, i.e., we will assume that p, h, T_0, T_1, and Γ_e are known.

Substituting expressions (17.9) and (17.10) in the mentioned integral conditions we obtain

$$f_{x0} = \Phi\{\frac{\mu^0(\frac{s_x}{2} - \omega_0 y)}{\int_{-h/2}^{h/2} e^{\delta_T T_0} dz}\}, \quad f_{y0} = \Phi\{\frac{\mu^0(\frac{s_y}{2} + \sqrt{1-e^2}\omega_0 x)}{\int_{-h/2}^{h/2} e^{\delta_T T_0} dz}\},$$

$$F(f_{x0})\delta_T \int_{-h/2}^{h/2} T_1 e^{\delta_T T_0} dz + F'(f_{x0})\{f_{x1} \int_{-h/2}^{h/2} e^{\delta_T T_0} dz$$

$$+ \frac{2f_{x0}}{h} \int_{-h/2}^{h/2} z e^{\delta_T T_0} dz\} = 0, \quad \dots, \tag{17.11}$$

$$F(f_{y0})\delta_T \int_{-h/2}^{h/2} T_1 e^{\delta_T T_0} dz + F'(f_{y0})\{f_{y1} \int_{-h/2}^{h/2} e^{\delta_T T_0} dz$$

$$+ \sqrt{1-e^2}\frac{2f_{x0}}{h} \int_{-h/2}^{h/2} z e^{\delta_T T_0} dz \frac{\partial p}{\partial y} / \frac{\partial p}{\partial x}\} = 0, \quad \dots,$$

where Φ is the function inverse to function F, i.e., $\Phi(F(x)) = x$.

To simplify further the solution process we will consider a special case of equal surface temperatures, i.e., the case when

$$T_{w1}(x, y) = T_{w2}(x, y) = T_w(x, y) \tag{17.12}$$

(the general case can be considered in accordance with Section 10.3). It can be shown (see the next section) that in this case $T_0(x, y, -z) = T_0(x, y, z)$ and $T_1(x, y, -z) = -T_1(x, y, z)$. Using these properties of functions T_0 and T_1 from (17.11) we obtain that

$$f_{x1} = f_{y1} = 0, \quad \dots \tag{17.13}$$

Therefore, in the above case in the inlet and exit zones from (17.9), (17.11), and (17.13) we have

$$f_x = \Phi\{\mu^0(\frac{s_x}{2} - \omega_0 y) / \int_{-h/2}^{h/2} e^{\delta_T T_0} dz\} + O(\nu^2),$$

$$\tag{17.14}$$

$$f_y = \Phi\{\mu^0(\frac{s_y}{2} + \sqrt{1-e^2}\omega_0 x) / \int_{-h/2}^{h/2} e^{\delta_T T_0} dz\} + O(\nu^2), \quad \nu \ll 1.$$

In the next section we will determine the expression for function T_0. Using this expression we can rewrite the formulas for f_x and f_y in the form (see

(17.20))

$$f_x = \Phi\{\tfrac{\mu^0(s_x - 2\omega_0 y)}{2h} e^{-\delta_T T_w} \tfrac{Bh}{\sinh Bh}\} + O(\nu^2),$$

$$f_y = \Phi\{\tfrac{\mu^0(s_y + 2\sqrt{1 - e^2}\omega_0 x)}{2h} e^{-\delta_T T_w} \tfrac{Bh}{\sinh Bh}\} + O(\nu^2), \ \nu \ll 1.$$

(17.15)

17.4 Solution for Lubricant Temperature T. High Slide-to-Roll Ratio

The idea of the further analysis is based on asymptotic solution of the problem (17.8) for lubricant temperature T and then using that solution to simplify the Reynolds equation in (17.6). For convenience let us introduce two notations

$$G = f_{x0}F(f_{x0}) + f_{y0}F(f_{y0}), \ H = f_{x0}[F(f_{x0}) + f_{x0}F'(f_{x0})]$$

$$+\sqrt{1 - e^2}f_{x0}[F(f_{y0}) + f_{y0}F'(f_{y0})]\tfrac{\partial p}{\partial y}/\tfrac{\partial p}{\partial x}.$$

(17.16)

Substituting (17.10) in (17.8) and (17.12), expanding the equation for T in (17.8) in powers of $\nu \ll 1$, and equating terms with the same powers of ν we obtain

$$\tfrac{\partial^2 T_0}{\partial z^2} = -\tfrac{\kappa H_{00} G}{\lambda \mu^0} e^{\delta_T T_0}, \ T_0(x, y, \pm\tfrac{h}{2}) = T_w(x, y),$$

$$\tfrac{\partial^2 T_1}{\partial z^2} = -\tfrac{\kappa H_{00}}{\lambda \mu^0} e^{\delta_T T_0}[G\delta_T T_1 + \tfrac{2z}{h} H], \ T_1(x, y, \pm\tfrac{h}{2}) = 0, \ \dots$$

(17.17)

Solutions of the boundary value problems (17.17) can be obtained in the form (see Section 7.3)

$$T_0 = T_w + \tfrac{2}{\delta_T} \ln \tfrac{\cosh \frac{Bh}{2}}{\cosh Bz},$$

$$\tfrac{G}{B^2 h^2} \cosh^2 \tfrac{Bh}{2} = \tfrac{2\lambda \mu^0 e^{-\delta_T T_w}}{\kappa \delta_T H_{00} h^2},$$

(17.18)

$$T_1 = \tfrac{H}{\delta_T G}\{\tfrac{\tanh Bz}{\tanh \frac{Bh}{2}} - \tfrac{2z}{h}\},$$

(17.19)

where

$$\int\limits_{-h/2}^{h/2} e^{\delta_T T_0} dz = I_0(B, p, h) = e^{\delta_T T_w}\tfrac{\sinh Bh}{B},$$

$$\int\limits_{-h/2}^{h/2} z^2 e^{\delta_T T_0} dz = I_z(B, p, h) = \tfrac{2}{B^3} e^{\delta_T T_w} \cosh^2 \tfrac{Bh}{2}\{\tfrac{B^2 h^2}{4} \tanh \tfrac{Bh}{2}$$

(17.20)

$$-Bh \ln \cosh \tfrac{Bh}{2} + 2\int\limits_0^{Bh/2} \ln \cosh t \, dt\},$$

$$\int_{-h/2}^{h/2} z T_1 e^{\delta_T T_0} dz = I_{zT}(B, p, h, x, y) = \frac{4H}{B^3 h \delta_T G} e^{\delta_T T_w} \cosh^2 \frac{Bh}{2}$$

$$\times \left\{ \frac{Bh}{4 \tanh \frac{Bh}{2}} \left[-\frac{Bh}{2} - \frac{Bh}{2} \tanh^2 \frac{Bh}{2} + \tanh \frac{Bh}{2} \right] + Bh \ln \cosh \frac{Bh}{2} \right.$$

$$\left. -2 \int_0^{Bh/2} \ln \cosh t \, dt \right\},$$

where the dependence of I_0, I_z and I_{zT} on p and h is due to the dependence of B on p and h while the dependence of I_{zT} on x and y is due to the fact that I_{zT} depends on H (see (17.16)) which, in turn, depends on $\frac{\partial p}{\partial x}$ and $\frac{\partial p}{\partial y}$.

17.5 Approximate Reynolds Equation. High Slide-to-Roll Ratio

Now, we are ready to approximate the Reynolds equation from (17.6). Substituting the expressions from (17.10), (17.14), (17.18), and (17.19), expanding all expressions in series in powers of $\nu \ll 1$, and retaining only terms proportional to ν we obtain the approximate Reynolds equation in the form

$$\frac{\partial}{\partial x} \left\{ \frac{H_{00}}{\mu^0} \left[\frac{h \delta_T F(f_{x0})}{2 f_{x0}} I_{zT}(B, p, h, x, y) + F'(f_{x0}) I_z(B, p, h) \right] \frac{\partial p}{\partial x} \right\}$$

$$+ \sqrt{1 - e^2} \frac{\partial}{\partial y} \left\{ \frac{H_{00}}{\mu^0} \left[\frac{h \delta_T F(f_{y0})}{2 f_{x0}} I_{zT}(B, p, h, x, y) \frac{\partial p}{\partial x} \right. \right.$$

$$\left. \left. + \sqrt{1 - e^2} F'(f_{y0}) I_z(B, p, h) \frac{\partial p}{\partial y} \right] \right\} + \dots \tag{17.21}$$

$$= \frac{\gamma - \omega_0 y}{2} \frac{\partial h}{\partial x} + \sqrt{1 - e^2} \frac{\sqrt{1 - \gamma^2} + \sqrt{1 - e^2} \omega_0 x}{2} \frac{\partial h}{\partial y}.$$

It is easy to verify that for $\delta_T = 0$ (i.e., for the case an isothermal lubrication regime) the approximate Reynolds equation from (17.21) is reduced to the approximate Reynolds equation (16.51) for the case of isothermal lubrication by a non-Newtonian fluid.

It is important to remember that if $\nu \ll 1$ in the inlet zone then $\nu \ll 1$ in the entire contact region and the approximate Reynolds equation is asymptotically valid in the entire contact region. Therefore, our original thermal EHL problem for a point heavily loaded contact is reduced to solution of the following analog of the isothermal EHL problem for heavily loaded contact (see equations (17.6), (17.8), and (17.21))

$$\frac{\partial}{\partial x}\{\frac{H_{00}}{\mu^0}[\frac{h\delta_T F(f_{x0})}{2f_{x0}}I_{zT}(B,p,h,x,y)+F'(f_{x0})I_z(B,p,h)]\frac{\partial p}{\partial x}\}$$

$$+\sqrt{1-e^2}\frac{\partial}{\partial y}\{\frac{H_{00}}{\mu^0}[\frac{h\delta_T F(f_{y0})}{2f_{x0}}I_{zT}(B,p,h,x,y)\frac{\partial p}{\partial x}$$

$$+\sqrt{1-e^2}F'(f_{y0})I_z(B,p,h)\frac{\partial p}{\partial y}]\}+\dots \qquad (17.22)$$

$$=\frac{\gamma-\omega_0 y}{2}\frac{\partial h}{\partial x}+\sqrt{1-e^2}\frac{\sqrt{1-\gamma^2}+\sqrt{1-e^2}\omega_0 x}{2}\frac{\partial h}{\partial y},$$

$$p(x,y)\mid_\Gamma=0,\ \frac{dp(x,y)}{d\overrightarrow{n}}\mid_{\Gamma_e}=0,$$

$$H_{00}[h(x,y)-1]=x^2+\frac{\delta}{1-e^2}y^2+\frac{D(e)}{\pi(1-e^2)^{3/2}}\{\int\int_\Omega\frac{p(x_1,y_1)dx_1 dy_1}{\sqrt{(x-x_1)^2+\frac{(y-y_1)^2}{1-e^2}}}$$

$$-\int\int_\Omega\frac{p(x_1,y_1)dx_1 dy_1}{\sqrt{x_1^2+\frac{y_1^2}{1-e^2}}}\},\ \int\int_\Omega p(x_1,y_1)dx_1 dy_1=\frac{2\pi}{3}.$$

Now, we are ready to consider different types of motion the lubricated contact can be involved in.

17.6 Straight Lubricant Entrainment. High Slide-to-Roll Ratio

Let us consider the case of lubricant straight entrainment, i.e., $\gamma=1$, $s_y=\omega_0=0$. Then according to (17.14) we have $f_{y0}=0$ and $F(f_{y0})=F'(f_{y0})=0$. The latter simplifies the expressions for G and H from (17.16) and reduces them to

$$G=f_{x0}F(f_{x0}),\ H=f_{x0}[F(f_{x0})+f_{x0}F'(f_{x0})], \qquad (17.23)$$

while the Reynolds equation from (17.22) is reduced to

$$\frac{\partial}{\partial x}\{\frac{H_{00}}{\mu^0}[\frac{h\delta_T F(f_{x0})}{2f_{x0}}I_{zT}(B,p,h,x,y)+F'(f_{x0})I_z(B,p,h)]\frac{\partial p}{\partial x}\}$$

$$+(1-e^2)\frac{\partial}{\partial y}\{\frac{H_{00}}{\mu^0}F'(f_{y0})I_z(B,p,h)\frac{\partial p}{\partial y}\}+\dots=\frac{1}{2}\frac{\partial h}{\partial x}. \qquad (17.24)$$

The further asymptotic analysis of pre-critical lubrication regimes can be done by precisely following the methodology presented in Chapters 12 and 16. Namely, the analysis of the gap h behavior in the inlet and exit zones away from the ends of HSSPGD zone is identical to the one done in Chapters 12 and 16. The asymptotic analysis of the Reynolds equation (17.24) requires assumptions about the behavior in the inlet and exit zones of functions F and F' which can be expressed in terms of functions Φ and Φ'. Specifically, we have (see (17.11))

$$\frac{F(f_{x0})}{f_{x0}} = \frac{\mu^0 s_x}{2h} e^{-\delta_T T_w} \frac{Bh}{\sinh Bh} / \Phi\left(\frac{\mu^0 s_x}{2h} e^{-\delta_T T_w} \frac{Bh}{\sinh Bh}\right),$$

$$F'(f_{x0}) = \left[\Phi'\left(\frac{\mu^0 s_x}{2h} e^{-\delta_T T_w} \frac{Bh}{\sinh Bh}\right)\right]^{-1}, \qquad (17.25)$$

$$F'(f_{y0}) = \left[\Phi'\left(\frac{\mu^0 s_y}{2h} e^{-\delta_T T_w} \frac{Bh}{\sinh Bh}\right)\right]^{-1}.$$

In the inlet and exit zones we will introduce local independent variables $r_p = \frac{x - \gamma_i(y)}{\epsilon_q}$ and $s_p = \frac{x - \gamma_e(y)}{\epsilon_q}$, respectively, where $x = \gamma_i(y)$ and $x = \gamma_e(y)$ are the inlet and exit contact boundaries, respectively, while ϵ_q is the characteristic width of the inlet and exit zones (see Chapter 12). Then pressure $p(x,y)$ and gap $h(x,y)$ will be searched in the form of $p = \epsilon_q^{1/2} q_p(r_p, y) + o(1)$ and $h = h_{qp}(r_p, y) + o(1)$, $\omega \ll 1$, (ω is the small parameter of the problem), in the inlet zone while in the exit zone these functions will be searched in the form $p = \epsilon_q^{1/2} g_p(s_p, y) + o(1)$ and $h = h_{gp}(s_p, y) + o(1)$, $\omega \ll 1$, where q_p, g_p, h_{qp}, and h_{gp} are new unknowns (see Chapter 12). If the assumptions concerning the behavior of functions Φ and Φ' are taken in the form of (16.55) and (16.56), assuming that in the inlet and exit zones $f_{x0} = O(1)$, $\omega \ll 1$, and taking into account the fact that in the inlet and exit zones $\frac{\partial}{\partial x} \gg \frac{\partial}{\partial y}$ by balancing the terms of the Reynolds equation (17.24) along the lubrication "central" flow streamlines for the central film thickness H_{00} we obtain (compare to formula (16.58) as well as to formula (12.44) for the case of Newtonian lubricant when $k = 1$ and $l = 0$)

$$H_{00} = A_p (V^k \omega^l \epsilon_q^2)^{\frac{1}{m+2}}, \quad A_p = O(1), \quad \omega \ll 1, \qquad (17.26)$$

where ϵ_q is the characteristic width of the inlet zone (see Chapters 12 and 16) and constant A_p depends not only on the lubricant rheology and the position of the inlet coordinate but also on parameters δ_T, s_x, T_w, and B (B satisfies equation (17.18)).

Again, following the asymptotic analysis of the preceding chapters along the lubricant "central" flow streamlines in the inlet zone we obtain the following asymptotically valid equations (see (12.36) and (12.37))

$$\frac{\partial}{\partial r_p} M_0(A_p, q_p, h_{qp}, \mu_{qp}^0, B_{qp}, r_p, y) = \frac{2}{\pi} \theta \int_0^\infty \frac{q_p(t,y) - q_{ap}(t,y)}{t - r_p} dt,$$

$$q_p(0, y) = 0, \qquad (17.27)$$

$$q_p(r_p, y) \to q_{ap}(r_p, y) = N_0(y)[\sqrt{2r_p} + \frac{\alpha_{1p}}{\sqrt{2r_p}}], \quad r_p \to \infty,$$

while in the exit zone the asymptotically valid equations assume the form

$$\frac{\partial}{\partial s_p} M_0(A_p, g_p, h_{gp}, \mu_{gp}^0, B_{gp}, s_p, y) = \frac{2}{\pi}\theta \int_{-\infty}^{0} \frac{g_p(t,y) - g_{ap}(t,y)}{t - s_p} dt,$$

$$g_p(0, y) = 0, \quad \frac{\partial g_p(0,y)}{\partial s_p} = 0, \tag{17.28}$$

$$g_p(s_p, y) \to g_{ap}(s_p, y) = N_0(y)[\sqrt{-2s_p} - \frac{\beta_{1p}}{\sqrt{-2s_p}}], \quad s_p \to -\infty,$$

where

$$M_0(A, p, h, \mu, B, x, y) = \frac{A^{m+2}}{\mu^0}\{\frac{\mu^0 \overline{s_x}\delta_T}{2} I_{zT}(B, p, h, x, y)/\Phi_0(\frac{\mu^0 \overline{s_x}}{2h})$$

$$\tag{17.29}$$

$$+ 2I_z(B, p, h)/\Phi_1(\frac{\mu^0 \overline{s_x}}{2h})\}\frac{\partial p}{\partial x}, \quad \overline{s_x} = s_x e^{-\delta_T T_w} \frac{Bh}{\sinh Bh},$$

and θ is determined by (13.32) while B_{qp} and B_{gp} are the values of function B from (17.18) determined for $p = \epsilon_q^{1/2} q_p + \ldots$, $h = h_{qp} + \ldots$ and $p = \epsilon_q^{1/2} g_p + \ldots$, $h = h_{gp} + \ldots$, respectfully. For details on the method and notations used please see the preceding chapters considering the case of straight lubricant entrainment in a heavily loaded point EHL contact.

The expressions for the gap functions h_{qp} and h_{gp} for starved and fully flooded lubrication regimes are given in Chapter 12. In particular, for starved lubrication regimes we have

$$h_{qp} = h_\infty(y), \quad h_{gp} = h_\infty(y), \tag{17.30}$$

while for fully flooded lubrication regimes they can be replaced by equations

$$h_{qp} = h_\infty(y) + \frac{1}{A_p} M_0(A_p, q_p, h_{qp}, \mu_{qp}^0, B_{qp}, r_p, y),$$

$$\tag{17.31}$$

$$h_{gp} = h_\infty(y) - \frac{1}{A_p} M_0(A_p, g_p, h_{gp}, \mu_{gp}^0, B_{gp}, s_p, y),$$

where $h_\infty(y)$ is the gap between the contact solids in the Hertzian region (for example, at $x = 0$). Here $h_\infty(0) = 1$ while $h_\infty(y)$ for $y \neq 0$ is determined from the solution of system (17.27) and (17.30) for starved or (17.31) for fully flooded lubrication conditions and for the known coefficient A_p the value of which is determined from the same system of equations for $y = 0$ when $h_\infty(0) = 1$.

Substitution (16.41) and (16.42) for $\xi = y$ and proper definition for r_0 (see, for example, (13.47) for the case of Newtonian lubricant) will reduce systems (17.27)-(17.29) to the form similar to the ones derived and analyzed in Chapter 9. For certain rheology functions F and Φ this substitution can reduce the latter systems to systems identical to the ones from Chapter 9. Moreover, this substitution will allow to obtain a formula for $H_0(y)$ which represents the dimensionless analog of the dimensional film thickness h_0 but at the exit from the contact region.

For fully flooded pre-critical lubrication regimes characterized by the estimate $H_{00} = O(\epsilon_q^{3/2})$, $\omega \ll 1$, using formula (17.26) for the lubrication film thickness we obtain the formula

$$H_{00} = A_f (V^k \omega^l)^{\frac{3}{3m+2}}, \quad A_f = O(1), \quad \epsilon_q = \epsilon_f = (V^k \omega^l)^{\frac{2}{3m+2}}, \quad \omega \ll 1, \quad (17.32)$$

where constant A_f is independent from V and ω but depends on parameters $\alpha_{1p}(0)$, θ (see Chapter 16), as well as the lubricant rheology function Φ and constant A_f is obtained from the solution of equations (17.27) and (17.29).

However, it is important to remember that the above approach is applicable only as long as $\epsilon_q \ll \epsilon_\nu$, where ϵ_ν satisfies the estimate $\nu = O(1)$, $\omega \ll 1$. Due to the fact that for some problem input parameters we may have $\epsilon_f \ll \epsilon_\nu$ while for other parameters $\epsilon_f = O(\epsilon_\nu)$ or $\epsilon_f \gg \epsilon_\nu$ only starved or both starved and fully flooded lubrication regimes can be realized. Namely, if $\epsilon_f \ll \epsilon_\nu$ then both starved and fully flooded lubrication regimes are realized and the entire asymptotic analysis is valid. On the other hand, if $\epsilon_f = O(\epsilon_\nu)$ then formula (17.32) for the central film thickness H_{00} in a fully flooded lubrication regime is still valid while equations (17.27)-(17.29) are not. Finally, if $\epsilon_f \gg \epsilon_\nu$ then only starved lubrication regimes can be realized and formulas (17.27)-(17.29) are valid. The specific conditions which satisfy one of the above three conditions can be determined in the fashion similar to the one used in Section 16.4.1 and left as an exercise for the reader.

Obviously, under thermal lubrication conditions the structure of formula (17.26) for the lubrication film thickness is identical to (16.58) which is valid for the case of isothermal lubrication conditions. The only difference between these formulas is the dependence of the proportionality coefficient A_p on additional problem parameters in the case of TEHL problem. Similarly, for the case of TEHL problem along the lubricant "central" flow streamlines the asymptotically valid in the inlet and exit zones equations (17.27) and (17.28) resemble the corresponding equations (12.36) and (12.37) derived for the isothermal EHL problem (see Section 9.4) in which function M_0 should be taken from (16.57).

In the inlet and exit zones along the "central" flow streamlines the above asymptotically valid systems of equations (17.27)-(17.30) and/or (17.31) can be transformed into equivalent ones explicitly resolved with respect to functions q_p (q) and g_p (g) involved in the integrals for h_{qp} (h_q) and h_{gp} (h_g) (see Section 6.3). The numerical solution of the equivalent systems of equations can be done according to the procedure described in Sections 6.9 and 8.2 while the numerical solutions themselves are very similar to the one obtained in Section 7.4 and Section 9.3.2 for line contacts. The numerical solution of the EHL problems in the original (non-asymptotic) formulation based on reduced Reynolds equation can be done in a way similar to Chapter 8. The choice between pre- and over-critical lubrication regimes can be made based on a simple analysis from Section 9.5.

For $\mu^0 = e^{Qp}$ and over-critical lubrication regimes (i.e., for $\epsilon_q \gg Q^{-2}$) using the same approach as in Section 12.4 and Section 16.4.1 for the central film thickness we obtain the formula (see (16.64))

$$H_{00} = A_p(V^k\omega^l Q\epsilon_q^{5/2})^{\frac{1}{m+2}},$$

$$A_p = O(1), \quad \epsilon_q \ll (V^k\omega^l Q)^{\frac{2}{3m+1}}, \quad \omega \ll 1, \tag{17.33}$$

where A_p is a constant independent of V, Q, and ω. For fully flooded over-critical lubrication regimes (when $H_{00} = O(\epsilon_q^{3/2})$, $\omega \ll 1$) from (17.33) we obtain (see (16.83))

$$H_{00} = A_f(V^k\omega^l Q)^{\frac{3}{3m+1}}, \quad A_f = O(1), \quad \epsilon_q = (V^k\omega^l Q)^{\frac{2}{3m+1}}, \quad \omega \ll 1. \tag{17.34}$$

The analysis of the possibility for these over-critical regimes to be realized can be done the same way it was done in Section 16.4.1 by comparing the sizes of the inlet zone ϵ_q for which over-critical regimes are realized and for which $\nu(x,y) \ll 1$ in the inlet zone and it is left as an exercise for the reader.

17.7   Skewed Lubricant Entrainment without Spinning. High Slide-to-Roll Ratio

Let us consider the case of skewed lubricant entrainment without spinning, i.e., $\omega_0 = 0$. Then we can introduce a new coordinate system according to (16.12) and (16.13)

$$\chi = \frac{\gamma x + \sqrt{1-e^2}\sqrt{1-\gamma^2}y}{\sqrt{1-e^2+e^2\gamma^2}}, \quad \xi = \frac{-\sqrt{1-e^2}\sqrt{1-\gamma^2}x+\gamma y}{\sqrt{1-e^2+e^2\gamma^2}}, \quad d\chi d\xi = dxdy, \tag{17.35}$$

where

$$x = \frac{\gamma\chi - \sqrt{1-e^2}\sqrt{1-\gamma^2}\xi}{\sqrt{1-e^2+e^2\gamma^2}}, \quad y = \frac{\sqrt{1-e^2}\sqrt{1-\gamma^2}\chi+\gamma\xi}{\sqrt{1-e^2+e^2\gamma^2}}. \tag{17.36}$$

In variables (17.35) the Reynolds equation and other equations of the EHL problem (17.22) can be rewritten in the form

$$[\gamma\frac{\partial}{\partial\chi} - \sqrt{1-e^2}\sqrt{1-\gamma^2}\frac{\partial}{\partial\xi}]\{\frac{H_{00}}{\mu^0}[\frac{h\delta_T F(f_{x0})}{2f_{x0}}I_{zT}(B,h,p,\chi,\xi)$$

$$+F'(f_{x0})I_z(B,h)][\gamma\frac{\partial p}{\partial\chi} - \sqrt{1-e^2}\sqrt{1-\gamma^2}\frac{\partial p}{\partial\xi}]\}$$

$$+\sqrt{1-e^2}[\sqrt{1-e^2}\sqrt{1-\gamma^2}\frac{\partial}{\partial\chi} + \gamma\frac{\partial}{\partial\xi}] \tag{17.37}$$

$$\{\frac{H_{00}}{\mu^0}[\frac{h\delta_T F(f_{y0})}{2f_{x0}}I_{zT}(B,h,p,\chi,\xi) + \sqrt{1-e^2}F'(f_{y0})I_z(B,h)]$$

$$\times[\sqrt{1-e^2}\sqrt{1-\gamma^2}\frac{\partial}{\partial\chi} + \gamma\frac{\partial}{\partial\xi}]\} + \ldots = \frac{(1-e^2+e^2\gamma^2)^{3/2}}{2}\frac{\partial h}{\partial\chi},$$

$$\mu = \exp(Qp), \ p(\chi,\xi)\mid_\Gamma = 0, \ \tfrac{dp(\chi,\xi)}{d\overrightarrow{n}}\mid_{\Gamma_e} = 0,$$

$$H_{00}[h(\chi,\xi) - 1] = x^2(\chi,\xi) + \tfrac{\delta}{1-e^2}y^2(\chi,\xi)$$

$$+\tfrac{D(e)}{\pi(1-e^2)^{3/2}}\Big\{\int\!\!\int_\Omega \frac{p(\chi_1,\xi_1)d\chi_1 d\xi_1}{\sqrt{(\chi-\chi_1)^2+\frac{(\xi-\xi_1)^2}{1-e^2}}} - \int\!\!\int_\Omega \frac{p(\chi_1,\xi_1)d\chi_1 d\xi_1}{\sqrt{\chi_1^2+\frac{\xi_1^2}{1-e^2}}}\Big\},$$

$$\int\!\!\int_\Omega p(\chi_1,\xi_1)d\chi_1 d\xi_1 = \tfrac{2\pi}{3},$$

where x, y and x_1, y_1 are functions of χ, ξ and χ_1, ξ_1 determined according to (17.36), respectively.

Taking into account the fact that in the inlet and exit zones $\frac{\partial}{\partial \chi} \gg \frac{\partial}{\partial \xi}$ from (17.16) we obtain

$$G = f_{x0}F(f_{x0}) + f_{y0}F(f_{y0}), \ H = f_{x0}[F(f_{x0}) + f_{x0}F'(f_{x0})]$$

$$+\frac{(1-e^2)\sqrt{1-\gamma^2}}{\gamma}f_{x0}[F(f_{y0}) + f_{y0}F'(f_{y0})], \tag{17.38}$$

and equation (17.37) in these zones can be reduced to a simpler equation

$$\frac{\partial}{\partial \chi}\langle \frac{H_{00}}{\mu^0}\{\frac{h\delta_T}{2f_{x0}}[\gamma^2 F(f_{x0}) + (1-e^2)^{3/2}(1-\gamma^2)F(f_{y0})]$$

$$\times I_{zT}(B,h,p,\chi,\xi) + [\gamma^2 F'(f_{x0}) + (1-e^2)^2(1-\gamma^2)F'(f_{y0})] \tag{17.39}$$

$$\times I_z(B,h)\}\frac{\partial p}{\partial \chi}\rangle + \ldots = \frac{(1-e^2+e^2\gamma^2)^{3/2}}{2}\frac{\partial h}{\partial \chi},$$

where functions G and H are determined by equations (17.38). Obviously, the latter equation coincides with the approximate Reynolds equation (16.83) valid for the case of isothermal conditions when $\delta_T = 0$.

First, let us consider pre-critical lubrication regimes. We will use the asymptotic methodology developed in Chapters 13, 15, and 16. In the inlet and exit zones we will introduce local independent variables $r_p = \frac{x-\gamma_i(\xi)}{\epsilon_q}$ and $s_p = \frac{x-\gamma_e(\xi)}{\epsilon_q}$, respectively, where $\chi = \gamma_i(\xi)$ and $\chi = \gamma_e(\xi)$ are the inlet and exit contact boundaries, respectively, while ϵ_q is the characteristic width of the inlet and exit zones (see Chapter 13). Then pressure $p(\chi,\xi)$ and gap $h(\chi,\xi)$ will be searched in the form of $p = \epsilon_q^{1/2}q_p(r_p,\xi) + o(1)$ and $h = h_{qp}(r_p,\xi) + o(1)$, $\omega \ll 1$, in the inlet zone while in the exit zone these functions will be searched in the form $p = \epsilon_q^{1/2}g_p(s_p,\xi) + o(1)$ and $h = h_{gp}(s_p,\xi) + o(1)$, $\omega \ll 1$, where q_p, g_p, h_{qp}, and h_{gp} are new unknowns (see Chapter 13). Using formulas (17.20) as well as assumptions (16.55) and (16.56) concerning the behavior of functions Φ and Φ' after estimating the terms in the left- and right-hand sides of equation (17.39) and require them to be of the same order of magnitude as $\omega \to 0$ we obtain an estimate for the lubrication film thickness as follows

$$H_{00} = A_p[(1 - e^2 + e^2\gamma^2)^{3/2}V^k\omega^l\epsilon_q^2]^{\frac{1}{m+2}}, \quad A_p = O(1),$$

$$\epsilon_q \ll \epsilon_f = [(1 - e^2 + e^2\gamma^2)^{3/2}V^k\omega^l]^{\frac{2}{3m+2}}), \quad \omega \ll 1,$$

(17.40)

while the asymptotically valid in the inlet zone equations assume the form

$$\frac{\partial}{\partial r_p}M_0(A_p, q_p, h_{qp}, \mu_{qp}^0, B_{qp}, r_p, \xi) - \frac{2}{\pi}\theta\int_0^\infty \frac{q_p(t,\xi) - q_{ap}(t,\xi)}{t - r_p}dt,$$

$$q_p(0, \xi) = 0,$$

(17.41)

$$q_p(r_p, \xi) \to q_{ap}(r_p, \xi) = N_0(\xi)[\sqrt{2r_p} + \frac{\alpha_{1p}}{\sqrt{2r_p}}], \quad r_p \to \infty,$$

while in the exit zone the asymptotically valid equations assume the form

$$\frac{\partial}{\partial s_p}M_0(A_p, g_p, h_{gp}, \mu_{gp}^0, B_{gp}, s_p, \xi) = \frac{2}{\pi}\theta\int_{-\infty}^0 \frac{g_p(t,\xi) - g_{ap}(t,\xi)}{t - s_p}dt,$$

$$g_p(0, \xi) = 0, \quad \frac{\partial g_p(0,\xi)}{\partial s_p} = 0,$$

(17.42)

$$g_p(s_p, \xi) \to g_{ap}(s_p, \xi) = N_0(\xi)[\sqrt{-2s_p} - \frac{\beta_{1p}}{\sqrt{-2s_p}}], \quad s_p \to -\infty,$$

where

$$M_0(A, p, h, \mu, B, x, y) = \frac{A^{m+2}}{\mu^0}\{\frac{\mu^0\delta_T}{2}[\gamma^2\overline{s_x}$$

$$+(1 - e^2)^{3/2}(1 - \gamma^2)\overline{s_y}]I_zT(B, p, h, x, y)/\Phi_0(\frac{\mu^0\overline{s_x}}{2h})$$

$$+2[\gamma^2/\Phi_1(\frac{\mu^0\overline{s_x}}{2h}) + (1 - e^2)^2(1 - \gamma^2)/\Phi_1(\frac{\mu^0\overline{s_y}}{2h})]I_z(B, p, h)\}\frac{\partial p}{\partial x},$$

(17.43)

$$\overline{s_x} = s_x e^{-\delta_T T_w}\frac{Bh}{\sinh Bh}, \quad \overline{s_y} = s_y e^{-\delta_T T_w}\frac{Bh}{\sinh Bh},$$

For starved lubrication regimes we have

$$h_{qp} = h_\infty(\xi), \quad h_{gp} = h_\infty(\xi),$$

(17.44)

For fully flooded lubrication regimes by integrating equations (17.42) and (17.43) for q_p and g_p the equations for the gap functions h_{qp} and h_{gp} can be replaced by equations

$$h_{qp} = h_\infty(\xi) + \frac{1}{A_p}M_0(A_p, q_p, h_{qp}, \mu_{qp}^0, B_{qp}, r_p, \xi),$$

$$h_{gp} = h_\infty(\xi) - \frac{1}{A_p}M_0(A_p, g_p, h_{gp}, \mu_{gp}^0, B_{gp}, s_p, \xi),$$

(17.45)

where $h_\infty(y)$ is the gap between the contact solids in the Hertzian region (for example, at $x = 0$). Here $h_\infty(0) = 1$ while $h_\infty(y)$ for $y \neq 0$ is determined from the solution of system (17.41), (17.43) and (17.44) for starved or (17.45) for fully flooded lubrication conditions and for the known coefficient A_p the value of which is determined from the same system of equations for $y = 0$ when $h_\infty(0) = 1$.

For fully flooded lubrication regimes $H_{00} = O(\epsilon_q^{3/2})$ which takes place for $\epsilon_q = \epsilon_f$ and the film thickness

$$H_{00} = A_f[(1 - e^2 + e^2\gamma^2)^{3/2}V^k\omega^l]^{\frac{3}{3m+2}}, \ A_f = O(1), \ \epsilon_q = \epsilon_f. \qquad (17.46)$$

Here formulas for the film thickness (17.40) for starved lubrication regimes and (17.46) for fully flooded lubrication regimes are structurally identical to the corresponding formulas (16.85) and (16.90), respectively. The difference between these two formulas is in the values of the coefficients of proportionality A_p and A_f and their dependence on the problem input parameters.

The rest of the analysis can be done along the lines of the studies conducted in Chapters 15 and 16. In particular, substitution (16.41) and (16.42) and proper definition for parameter r_0 (see, for example, (13.47) for the case of Newtonian lubricant) will reduce systems (17.41)-(17.45) to the form similar to the ones derived and analyzed in Chapter 9. For certain rheology functions F and Φ this substitution can reduce the latter systems to systems identical to the ones from Chapter 9. Moreover, this substitution will allow to obtain a formula for $H_0(y)$ which represents the dimensionless analog of the dimensional film thickness h_0 but at the exit from the contact region.

In the inlet and exit zones along the "central" flow streamlines the above asymptotically valid equations can be transformed into equivalent ones resolved with respect to functions q_p (q) and g_p (g) involved in the integrals for h_{qp} (h_q) and h_{gp} (h_g) (see Section 6.3) which are more convenient for numerical analysis and then solved numerically according to the procedures described in Section 6.9. The numerical solution of the EHL problems in the original formulation based on reduced (non-asymptotic) Reynolds equation can be done in a way similar to Chapter 8.

For the lubricant isothermal viscosity $\mu^0 = e^{Qp}$ and over-critical lubrication regimes (i.e., for $\epsilon_q \gg Q^{-2}$) using the same approach as in Sections 12.4 and 16.4.1 for the central film thickness we obtain the formula

$$H_{00} = A_p((1 - e^2 + e^2\gamma^2)^{3/2}V^k\omega^l Q\epsilon_q^{5/2})^{\frac{1}{m+2}},$$

$$(17.47)$$

$$A_p = O(1), \ \epsilon_q \ll \epsilon_f = ((1 - e^2 + e^2\gamma^2)^{3/2}V^k\omega^l Q)^{\frac{2}{3m+1}}, \ \omega \ll 1,$$

where A_p is a constant independent of V, Q, and ω. For fully flooded over-critical lubrication regimes (when $H_{00} = O(\epsilon_q^{3/2})$, $\omega \ll 1$) from (17.47) we obtain

$$H_{00} = A_f((1 - e^2 + e^2\gamma^2)^{3/2} V^k \omega^l Q)^{\frac{3}{3m+1}},$$

(17.48)

$$A_f = O(1), \quad \epsilon_q = \epsilon_f, \quad \omega \ll 1.$$

The analysis of the possibility for these over-critical regimes to be realized can be done the same way it was done in Section 16.4.1 by comparing the sizes of the inlet zone ϵ_q for which over-critical regimes are realized and for which $\nu(x, y) \ll 1$ in the inlet zone and it is left as an exercise for the reader.

The choice between pre- and over-critical lubrication regimes can be done the same way it was done in Section 6.7. The uniformly valid solution approximations can be determined in accordance with Section 6.4.

17.8   Skewed Lubricant Entrainment with Spinning. High Slide-to-Roll Ratio

To consider the case of spinning we need to rewrite our simplified system of equations (17.22) in polar coordinates (see (14.5) and (14.6))

$$x = \frac{1}{\sqrt{1-e^2}}[-\frac{\sqrt{1-\gamma^2}}{\omega_0} + \rho\sin\varphi], \quad y = \frac{\gamma}{\omega_0} - \rho\cos\varphi, \quad dxdy = \frac{\rho d\rho d\varphi}{\sqrt{1-e^2}}, \quad (17.49)$$

$$\rho = \sqrt{(\sqrt{1-e^2}x + \frac{\sqrt{1-\gamma^2}}{\omega_0})^2 + (y - \frac{\gamma}{\omega_0})^2}, \quad \tan\varphi = \frac{\sqrt{1-e^2}x + \frac{\sqrt{1-\gamma^2}}{\omega_0}}{\frac{\gamma}{\omega_0} - y}. \quad (17.50)$$

The reason for such a choice of new independent variables is the fact that in the Hertzian region (i.e., away from the inlet and exit contact boundaries) the lubricant moves along circular arcs $\rho = const$ which represent the lubricant "central" flow streamlines and the gap h along these arcs is practically constant, i.e., $h = h(\rho)$.

In these variables equations (17.22) assume the following form

$$(\sin\varphi\frac{\partial}{\partial\rho} + \frac{\cos\varphi}{\rho}\frac{\partial}{\partial\varphi})\{\frac{H_{00}}{\mu^0}[\frac{h\delta_T F(f_{x0})}{2f_{x0}}I_{zT}(B, h, p, \rho, \varphi)$$

$$+F'(f_{x0})I_z(B, h, p)][\sin\varphi\frac{\partial p}{\partial\rho} + \frac{\cos\varphi}{\rho}\frac{\partial p}{\partial\varphi}]\}$$

$$+(-\cos\varphi\frac{\partial}{\partial\rho} + \frac{\sin\varphi}{\rho}\frac{\partial}{\partial\varphi})\{\frac{H_{00}}{\mu^0}[\frac{h\delta_T F(f_{y0})}{2\sqrt{1-e^2}f_{x0}}I_{zT}(B, h, p, \rho, \varphi) \qquad (17.51)$$

$$+F'(f_{y0})I_{zT}(B, h, p)][-\cos\varphi\frac{\partial p}{\partial\rho} + \frac{\sin\varphi}{\rho}\frac{\partial p}{\partial\varphi}]\} + \ldots = \frac{\omega_0}{2\sqrt{1-e^2}}\frac{\partial h}{\partial\varphi},$$

$$\mu = \exp(Qp), \quad p(\rho, \varphi)\mid_\Gamma = 0, \quad \frac{dp(\rho,\varphi)}{d\overrightarrow{n}}\mid_{\Gamma_e} = 0,$$

$$H_{00}[h(\rho, \varphi) - 1] = x^2(\rho, \varphi) + \frac{\delta}{1-e^2}y^2(\rho, \varphi)$$

$$+\frac{D(e)}{\pi(1-e^2)}\{\int\int\limits_\Omega \frac{p(\rho_1,\varphi_1)\rho_1 d\rho_1 d\varphi_1}{\sqrt{\rho^2-2\rho\rho_1\cos(\varphi-\varphi_1)+\rho_1^2}}$$

$$-\int\int\limits_\Omega \frac{p(\rho_1,\varphi_1)\rho_1 d\rho_1 d\varphi_1}{\sqrt{\rho_1^2-\frac{2}{\omega_0}\rho_1[\gamma\cos(\varphi_1)+\sqrt{1-\gamma^2}\sin(\varphi_1)]+\frac{1}{\omega_0^2}}}\},$$

$$\int\int\limits_\Omega p(\rho_1,\varphi_1)\rho_1 d\rho_1 d\varphi_1 = \frac{2\pi}{3}\sqrt{1-e^2},$$

where x and y are functions of ρ and φ determined according to (17.50).

In the inlet and exit zones of such a lubricated contact $\frac{\partial}{\partial\varphi} \gg \frac{\partial}{\partial\rho}$. Therefore, in these zones the Reynolds equation from (17.51) can be simplified and rewritten in the form

$$\frac{\cos\varphi}{\rho^2}\frac{\partial}{\partial\varphi}\{\frac{H_{00}}{\mu^0}[\frac{h\delta_T F(f_{x0})}{2f_{x0}}I_{zT}(B,h,p,\rho,\varphi)+F'(f_{x0})I_z(B,h)]\cos\varphi\frac{\partial p}{\partial\varphi}\}$$

$$+\frac{\sin\varphi}{\rho^2}\frac{\partial}{\partial\varphi}\{\frac{H_{00}}{\mu^0}[\frac{h\delta_T F(f_{y0})}{2\sqrt{1-e^2}f_{x0}}I_{zT}(B,h,p,\rho,\varphi) \qquad (17.52)$$

$$+F'(f_{y0})I_z(B,h)]\sin\varphi\frac{\partial p}{\partial\varphi}\}+\ldots = \frac{\omega_0}{2\sqrt{1-e^2}}\frac{\partial h}{\partial\varphi}.$$

Obviously, the latter equation coincides with the approximate Reynolds equation (16.85) valid for the case of an isothermal EHL problem which is realized for $\delta_T = 0$.

Here it is important to realize that the value of H from (17.16) depends on $\frac{\partial p}{\partial y}/\frac{\partial p}{\partial x}$. In the case of skewed lubricant entrainment with spinning in the inlet and exit zones $\frac{\partial p}{\partial\varphi} \gg \frac{\partial p}{\partial\rho}$ which means that in these zones

$$H = f_{x0}[F(f_{x0})+f_{x0}F'(f_{x0})]$$

$$+\sqrt{1-e^2}f_{x0}[F(f_{y0})+f_{y0}F'(f_{y0})]\tan\varphi+\ldots, \qquad (17.53)$$

while G is still determined by (17.16).

Moreover, using (17.11) and (17.20) we obtain

$$\frac{F(f_{x0})}{f_{x0}} = \frac{\mu^0\overline{s_x}(\rho,\varphi)}{2h}/\Phi(\frac{\mu^0\overline{s_x}(\rho,\varphi)}{2h}), \quad \frac{F(f_{y0})}{f_{y0}} = \frac{\mu^0\overline{s_y}(\rho,\varphi)}{2h}/\Phi(\frac{\mu^0\overline{s_x}(\rho,\varphi)}{2h}),$$

$$F'(f_{x0}) = 1/\Phi'(\frac{\mu^0\overline{s_x}(\rho,\varphi)}{2h}), \quad F'(f_{y0}) = 1/\Phi'(\frac{\mu^0\overline{s_y}(\rho,\varphi)}{2h}),$$

$$\overline{s_x}(\rho,\varphi) = (s_x-2\gamma+2\omega_0\rho\cos\varphi)e^{-\delta_T T_w}\frac{Bh}{\sinh Bh}, \qquad (17.54)$$

$$\overline{s_y}(\rho,\varphi) = (s_y-2\sqrt{1-\gamma^2}+2\omega_0\rho\sin\varphi)e^{-\delta_T T_w}\frac{Bh}{\sinh Bh}.$$

Let us consider pre-critical lubrication regimes. Then, using representation (16.87) for p, estimate (16.89) for $\frac{\partial h}{\partial\varphi}$, formulas (17.20) as well as assumptions

(16.55) and (16.56) concerning the behavior of functions Φ and Φ' and equating the orders of terms in the Reynolds equation (17.52) in the inlet zone we obtain

$$H_{00} = A_p \left[\frac{V^k \omega^l \epsilon_q^2}{\sqrt{1-e^2}} \right]^{\frac{1}{m+2}}, \ A_p = O(1),$$

$$\epsilon_q \ll \epsilon_f = \left[\frac{V^k \omega^l \epsilon_q^2}{\sqrt{1-e^2}} \right]^{\frac{2}{3m+2}}, \ \omega \ll 1, \tag{17.55}$$

where coefficient of proportionality A_p depends on the parameter of μ^0 (for example, for $\mu^0 = e^{Qp}$ the parameter is $\epsilon_q^{1/2}Q$), e, γ, and ω_0 as well as the rheology of the lubricating fluid. At the same time, constant A_p is independent from V, ω, and ϵ_q.

For fully flooded lubrication regimes $H_{00} = O(\epsilon_q^{3/2})$, $\omega\epsilon \ll 1$, for the central lubrication film thickness we obtain

$$H_{00} = A_f \left[\frac{V^k \omega^l}{\sqrt{1-e^2}} \right]^{\frac{3}{3m+2}}, \ A_f = O(1), \ \epsilon_q = \epsilon_f, \ \omega \ll 1, \tag{17.56}$$

where coefficient of proportionality A_f depends on the parameter of μ^0 (for example, for $\mu^0 = e^{Qp}$ the parameter is $Q\left[\frac{V^k \omega^l}{\sqrt{1-e^2}} \right]^{\frac{1}{3m+2}}$, e, γ, and ω_0 as well as the rheology of the lubricating fluid. At the same time, constant A_p is independent from V, ω, and ϵ_q.

Formulas (17.55) and (17.56) derived for starved and fully flooded lubrication regimes in a thermal EHL contact are identical in structure with formulas (16.93) and (16.102), respectively, obtained for an isothermal EHL contact and they differ from each other only by the values of coefficients A_p and A_f involved in these formulas.

Under the same assumptions a more detailed analysis in the inlet and exit zones leads to the following asymptotically valid equations

$$\frac{\partial}{\partial r_p} M_0(A_p, q_p, h_{qp}, \mu^0_{qp}, B_{qp}, \rho, r_p, \varphi_{H1}) = \frac{2}{\pi}\omega_0\rho\theta \int\limits_0^\infty \frac{q_p(\rho,t)-q_{ap}(\rho,t)}{t-r_p}dt,$$

$$q_p(\rho, 0) = 0, \tag{17.57}$$

$$q_p(\rho, r_p) \to q_{ap}(\rho, r_p) = N_0(\rho, \varphi_{H1})[\sqrt{2r_p} + \frac{\alpha_{1p}}{\sqrt{2r_p}}], \ r_p \to \infty,$$

in the inlet zone along the "central" lubricant flow streamlines and

$$\frac{\partial}{\partial s_p} M_0(A_p, g_p, h_{gp}, \mu^0_{gp}, B_{gp}, \rho, s_p, \varphi_{H2})$$

$$= \frac{2}{\pi}\omega_0\rho\theta \int\limits_{-\infty}^0 \frac{g_p(\rho,t)-g_{ap}(\rho,t)}{t-s_p}dt, \ g_p(\rho,0) = \frac{\partial g_p(\rho,0)}{\partial s_p} = 0, \tag{17.58}$$

$$g_p(\rho, s_p) \to g_{ap}(\rho, s_p) = N_0(\rho, \varphi_{H2})[\sqrt{-2s_p} - \frac{\beta_{1p}}{\sqrt{-2s_p}}], \ s_p \to -\infty,$$

in the exit zone along the "central" lubricant flow streamlines and

$$M_0(A, p, h, \mu^0, B, x, y, \varphi_H) =$$

$$\frac{A^{m+2}}{\mu^0 \rho^2} \{\cos^2 \varphi_H [\frac{\delta_T \mu^0 \overline{s_x}(x, \varphi_H)}{2} I_{zT}(B, p, h, x, \varphi_H)/\Phi_0(\frac{\mu^0 \overline{s_x}(x, \varphi_H)}{2h})$$

$$+2I_z(B, p, h)/\Phi_1(\frac{\mu^0 \overline{s_x}(x, \varphi_H)}{2h})] \tag{17.59}$$

$$+\sin^2 \varphi_H [\frac{\delta_T \mu^0 \overline{s_y}(x, \varphi_H)}{2\sqrt{1-e^2}} I_{zT}(B, p, h, x, \varphi_H)/\Phi_0(\frac{\mu^0 \overline{s_x}(x, \varphi_H)}{2h})$$

$$+2I_z(B, p, h)/\Phi_1(\frac{\mu^0 \overline{s_y}(x, \varphi_H)}{2h})]\} \frac{\partial p}{\partial y}, \quad \theta = \frac{D(e)}{1-e^2}$$

where $N_0(\rho, \varphi_H)$ is determined by formula (14.37), $h_{qp}(\rho, r_p)$ and $h_{gp}(\rho, s_p)$ are the gap functions in the inlet $\Omega_i \bigcap \Omega_\epsilon$ and exit $\Omega_e \bigcap \Omega_\epsilon$ zones, functions μ_{qp} and μ_{gp} are the main terms of the asymptotic expansions of $\mu(\epsilon_q^{1/2} q_p)$ and $\mu(\epsilon_q^{1/2} g_p)$, respectively. The expressions for $h_{qp}(\rho, r_p)$ and $h_{gp}(\rho, s_p)$ depend on the lubrication regime. For starved lubrication regimes the gap distribution $h(\rho, \varphi)$ is very close to $h_\infty(\rho)$ in the entire contact region. That corresponds to the case of $\epsilon_q^{3/2} \ll H_{00}$ and

$$h_{qp}(\rho, r_p) = h_\infty(\rho), \quad \epsilon_q \ll \epsilon_f, \tag{17.60}$$

$$h_{gp}(\rho, s_p) = h_\infty(\rho), \quad \epsilon_q \ll \epsilon_f, \tag{17.61}$$

$$\epsilon_f = [\frac{V^k \omega^l}{\sqrt{1-e^2}}]^{\frac{2}{3m+2}}, \tag{17.62}$$

while for fully flooded lubrication regimes $H_{00} = O(\epsilon_q^{3/2})$, $\epsilon_q = \epsilon_f$, $\omega \ll 1$, and in the inlet and exit zones the gap satisfies the equations (see equations (14.33) and (14.34))

$$A_p \frac{\partial h_{qp}}{\partial r_p} = \frac{2}{\pi} \theta \rho \int_0^\infty \frac{q_p(\rho, t) - q_{ap}(\rho, t)}{t - r_p} dt, \quad \epsilon_q = \epsilon_f, \quad \omega \ll 1, \tag{17.63}$$

$$A_p \frac{\partial h_{gp}}{\partial s_p} = \frac{2}{\pi} \theta \rho \int_{-\infty}^0 \frac{g_p(\rho, t) - g_{ap}(\rho, t)}{t - s_p} dt, \quad \epsilon_q = \epsilon_f, \quad \omega \ll 1. \tag{17.64}$$

The latter equations can be replaced by equations

$$h_{qp} = h_\infty(\rho) + \frac{1}{A_p} M_0(A_p, q_p, h_{qp}, \mu_{qp}^0, B_{qp}, \rho, r_p, \varphi_{H1}),$$

$$h_{gp} = h_\infty(\rho) - \frac{1}{A_p} M_0(A_p, g_p, h_{gp}, \mu_{gp}^0, B_{gp}, \rho, s_p, \varphi_{H2}). \tag{17.65}$$

In (17.60) and (17.61) ϵ_f is the characteristic width of the inlet Ω_i and exit Ω_e zones in fully flooded lubrication regimes.

For the lubricant isothermal viscosity $\mu^0 = e^{Qp}$ and over-critical lubrication regimes (i.e., for $\epsilon_q \gg Q^{-2}$) using the same approach as the approach used before for over-critical lubrication regimes for the central film thickness we obtain the formula

$$H_{00} = A_p \Big[\frac{V^k \omega^l Q \epsilon_q^{5/2}}{\sqrt{1-e^2}} \Big]^{\frac{1}{m+2}},$$

$$A_p = O(1), \ \epsilon_q \ll \epsilon_f = \Big[\frac{V^k \omega^l Q}{\sqrt{1-e^2}} \Big]^{\frac{2}{3m+1}}, \ \omega \ll 1,$$

(17.66)

where A_p is a constant independent of V, Q, and ω. For fully flooded over-critical lubrication regimes (when $H_{00} = O(\epsilon_q^{3/2})$, $\omega \ll 1$) from (17.66) we obtain

$$H_{00} = A_f \Big[\frac{V^k \omega^l Q}{\sqrt{1-e^2}} \Big]^{\frac{3}{3m+1}}, \ A_f = O(1), \ \epsilon_q = \epsilon_f, \ \omega \ll 1. \tag{17.67}$$

The analysis of the possibility for these over-critical regimes to be realized can be done the same way it was done in Section 16.4.1 by comparing the sizes of the inlet zone ϵ_q for which over-critical regimes are realized and for which $\nu(x,y) \ll 1$ in the inlet zone and it is left as an exercise for the reader.

The choice between pre- and over-critical lubrication regimes can be done the same way it was done in Section 6.7. Substitution (16.41) and (16.42) for proper definitions of angles φ_H and parameter r_0 (see, for example, (13.47) for the case of Newtonian lubricant) will reduce systems (17.57)-(17.61) and/or (17.65) to the form similar to the ones derived and analyzed in Chapter 9. The uniformly valid solution approximations can be determined in accordance with Section 6.4. Systems of asymptotic equations (17.57)-(17.61) and/or (17.65) can be rewritten in the equivalent form explicitly resolved for $q_p(r_p)$ and $g_p(s_p)$ (see Section 6.3) which is more convenient for numerical analysis. The numerical solution of asymptotically valid equations can be done in accordance with Section 6.9 while the numerical solution of EHL problems based on reduced Reynolds equation can be done in a way similar to Chapter 8.

17.9 Regularization Approach

In the inlet and exit zones the regularization and numerical solution of asymptotically valid systems of equations can be done by replacing the isothermal asymptotic versions of the Reynolds equations corresponding to a particular type of motion in the contact (straight or skewed lubricant entrainment with or without spinning) by their thermal versions with a sufficient small fictitious heat generation. Practically, this replacement is done by replacing in the isothermal asymptotic versions of the Reynolds equations functions M_0 by their corresponding thermal analogs and by taking $T_W = 0$. The degree of the regularization is controlled by the value of Bh obtained as a solution of equation (see (17.18))

$$\frac{G}{B^2 h^2} \cosh^2 \frac{Bh}{2} = \beta \frac{\mu^0}{H_{00} h^2}, \qquad (17.68)$$

which, in turn, is controlled by a sufficiently small positive constant β which makes the numerical solution of the asymptotic analog of the isothermal EHL problem in the exit zone stable.

For lubricants with non-Newtonian rheology the regularization of the numerical solution of isothermal EHL problems for heavily loaded point contacts in the original (non-asymptotic) formulation can be done in the way similar to how it was done for EHL problems in case of lubricants with Newtonian rheology. It means that an isothermal EHL problem can be replaced by a similar thermal one with a small fictitious heat generation which will be sufficient to make the numerical solution of the regularized problem stable. Essentially, that requires the replacement of the corresponding Reynolds equation for the isothermal case by a Reynolds equation derived for the case of this fictitious small heat generation. The derivation of such a Reynolds equation for a particular case of contact surface motion would require the repetition of the analysis done for the case of the corresponding thermal problems. In particular, assuming that the ratio ν of the rolling and sliding frictional stresses is a small one would have to get the expressions for the sliding frictional stress components (see (17.11) and (17.13)) and two-term approximation for the lubricant temperature T assuming that the surface temperatures $T_{W1} = T_{W2} = T_W = 0$ (see (17.18) and (17.19)), where the value of Bh is determined by the solution of equation (17.68). After that the isothermal Reynolds equation (in case of pure rolling or sliding and rolling) should be replaced by the corresponding thermal one in which the temperature T should be taken as the above derived two-term approximate expression and the expression for $e^{\delta_T T}$ should be replaced by $e^{\delta_T T_0}[1 + \nu \delta_T T_1]$, where T_0 and T_1 are the consequent terms in the expansion of $T = T_0 + \nu T_1$ from (17.18) and (17.19). After this replacement is done it is irrelevant whether the input problem parameters characterize the case of high slide-to-roll ratio or not. This regularization would work for any value of the slide-to-roll ratio. At the same time, all physical constants characterizing the specifics of the heat generation and transfer such as lubricant heat conductivity, viscosity temperature coefficient, etc., should be clamped together into possibly one constant (such as β in (17.68)) as it was done before which would control the degree of solution regularization. The value of this constant(s) should be chosen empirically to provide stability to the numerical solution of the regularized isothermal EHL problem.

17.10 Closure

An asymptotic approach to thermal steady heavily loaded point EHL contacts with general incompressible non-Newtonian lubricants with straight as well as skewed lubricant entrainment and spinning ball is developed. The analytical expression for the lubricant temperature T has been obtained asymptotically using regular perturbation expansions. By using this asymptotic solution for T the original Reynolds equation has been simplified and reduced to the form similar to the Reynolds equation for Newtonian lubricants in the case of isothermal conditions. Based on the problem formulation with this reduced Reynolds equation the behavior of the pressure and gap distributions in the inlet and exit zones along the "central" streamlines of the lubricant flow has been analyzed in detail. Asymptotic analysis of heavily loaded contacts lubricated by a non-Newtonian fluid recognizes two different regimes of lubrication: pre- and over-critical lubrication regimes. Overall, the difference between these two regimes is in how rapidly the lubricant viscosity grows with pressure and how much lubricant is available at the inlet of the contact. For pre-critical lubrication regimes the asymptotically valid equations for pressure and gap have been derived. These equations can be reduced to equations very similar to the corresponding equations for pre-critical lubrication regimes in the inlet and exit zones of heavily loaded line contacts. The latter asymptotic equations have been validated through the comparison with the direct numerical solutions of the original EHL equations. For over-critical lubrication regimes another approach has been used which led to formulas for the lubrication film thickness for starved and fully flooded conditions. The asymptotic approaches to EHL problems offer certain distinct advantages in comparison with the direct numerical solutions of EHL problem such a reduction of the number of problem input parameters and the approach to regularization of numerical solutions.

Each of the analytically derived formulas for the central H_{00} lubrication film thickness contains a number of parameters characterizing lubricant rheology as well as a coefficient of proportionality A_p or A_f which can be found by numerically solving the corresponding asymptotically valid equations in the inlet zone or by just curve fitting the experimentally obtained values of the lubrication film thickness. It has been shown that for thermal EHL problems formulas for the lubrication film thickness have the structure identical to the ones for the corresponding isothermal EHL problem. The difference between these formulas is in the dependence of the proportionality coefficients A_p and A_f in these formulas on problem parameters.

17.11 Exercises and Problems

1. Try to provide arguments as of why relatively small to moderate deviations of the dependence of lubricant viscosity μ on temperature T from the exponential (17.1) would make a small impact on lubrication film thickness.

2. Provide a detailed order analysis of the inlet zone of a thermal heavily loaded point EHL contact and derive lubrication film thickness formulas

(a) (17.26) for the case of straight lubricant entrainment,

(b) (17.55) for the case of skew lubricant entrainment without spinning, and

(c) (17.40) for the case of skew lubricant entrainment with spinning. Describe the differences in the geometry of lubricant flow in all three cases.

3. Provide a detailed asymptotic analysis of the inlet and exit zones of a thermal heavily loaded point EHL contact and derive the asymptotically valid in these zones equations

(a) (17.27)-(17.31) for the case of straight lubricant entrainment,

(b) (17.41)-(17.45) for the case of skew lubricant entrainment without spinning, and

(c) (17.57)-(17.61) and/or (17.65) for the case of skew lubricant entrainment with spinning.

4. Determine the possibility to apply the employed above asymptotic method to studying fully flooded pre- and over-critical lubrication regimes by comparing the values of ϵ_f and ϵ_ν.

5. Determine a proper size of a numerical grid along the lubricant entrainment direction. (Hint: Relate the grid size to the characteristic width of the inlet zone along the entrainment direction. Consult Section 6.11.)

6. For each particular type of motion (straight and skewed lubricant entrainment with and without spinning) propose several different modifications of the reduced and asymptotically valid Reynolds equations which would provide regularization of the numerical solutions for heavily loaded isothermal EHL problems. (Hint: Use the ideas presented in Chapter 8.)

References

[1] Hamrock, B.J. 1994. *Fundamentals of Fluid Film Lubrication*. New York: McGraw-Hill.

Part VIII

Some Other Topics in Elastohydrodynamic Lubrication

In this part of the monograph we will focus on some EHL problems which are significantly different from the problems considered in the preceding chapters. In particular, we will consider EHL problems for soft elastic solids which explain the experimentally observed "dimple" phenomena. In addition, we will discuss modeling of grease lubricated contacts as well as some non-steady EHL problems and conditions of starvation in point EHL contacts.

18

Analysis of EHL Contacts for Soft Solids

18.1 Introduction

The purpose of this chapter is to provide an opportunity of a new prospective on classic EHL problems. We propose and analyze a new formulation of EHL problems that explains and predicts the abnormal phenomena observed in experiments with hard and soft lubricated solids by M. Kaneta and associates [1, 2]. For the first time, they have described the formation of a dimple in a heavily loaded lubricated contact of a relatively soft elastic solid with a hard one. The occurrence of a dimple cannot be predicted by the classic EHL theory. Kaneta et al. [1, 2] believe that the abnormal behavior of lubricated elastic surfaces can be explained by invoking solidification of oil in contact. Another interpretation of the abnormal phenomena based on a non-steady approach to the EHL problem is offered by Cermk [3]. However, the results of this approach contradict the steady nature of the experimental data presented by Kaneta et al. [1, 2]. Recently a number of numerical studies on this subject were published by Kaneta and Young [4, 5]. The main idea of the latter studies is that the dimple phenomenon is caused by the heat generation in the lubrication layer and its dissipation in the solids. However, the formulation of the problem in which it is studied in [4, 5] has a serious defect. Namely, for sufficiently high slide-to-roll ratios, the lubricant flows in opposite directions near two contact surfaces. Therefore, at the separatrix between the two oppositely directed flows, a condition for continuity of heat flux must be imposed. That condition was not used by Kaneta and Young in their studies. In spite of the fact that heat generation is important under high sliding conditions the lack of the above–mentioned boundary condition invalidates their analysis. Moreover, in spite of the importance of lubricant heat generation and its dissipation in solids the Kaneta and Young [4, 5] solution does not and cannot predict all the features (a)-(d) of such a lubricated contact observed in experiments (see Section 18.6).

The new formulation of the problem [6] is based on the traditional for the EHL theory approach to steady lubricated contacts as contacts of two elastic solids separated by a fluid film without oil solidification phenomenon. A new formulation of a steady problem for lubricated rollers made of elastic materials with low Young's modulus is considered. The distinct feature of

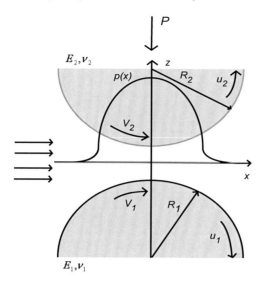

FIGURE 18.1

The general view of a lubricated contact.

this formulation that differs it from other ones is that the linear velocities of the surfaces take into account the tangential displacements of elastic surfaces. Therefore, in Reynolds equation the surface linear velocities are presented by functions of the location in the contact region instead of constants. This formulation explains a deep depression (dimple) produced in contact surfaces of soft materials. These dimples occur in place of flat surfaces which suppose to occur according to the classic EHL theory. A qualitative and numerical analysis of the new EHL problem formulation is presented. The dependence of dimple sizes on problem parameters is considered.

18.2   Formulation of an EHL Problem for Soft Solids

Let us consider a pretty much standard formulation of a plane EHL problem. Suppose two elastic circular cylinders, slowly rolling one over another, with parallel axes and radii, R_1 and R_2. Cylinder 1 is acted upon by force P along the z-axis through the cylinders centers in a Cartesian coordinate system with the x-axis along the lubrication layer. The linear velocities of the surface points, located far away from the contact zone, are u_1 for cylinder 1 and u_2 for cylinder 2.

The cylinders are fully separated by the lubrication layer $h(x)$ (see Fig.

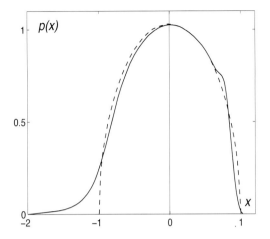

FIGURE 18.2
Distributions of pressure $p(x)$ (solid curve) in a lubricated contact ($V = 0.1$, $Q = 4$) and the Hertzian pressure (dotted curve) in a dry contact (after Kudish [6]). Reprinted with permission from the ASME.

18.1), which is formed by a Newtonian incompressible fluid with viscosity μ, which depends on lubricant pressure p. The lubrication process is isothermal. Because of slow motion of the contact surfaces in lubrication layer, the inertial forces are much smaller than viscous forces. Let us assume that the lubrication layer is relatively small with regard to the characteristic size of the contact, which, in turn, is much smaller than the cylinders' radii.

Under the above assumptions in the dimensional variables, the contact with Newtonian lubricant (which according to (5.1) corresponds to the rheological function $F(x) = x$) is described by equations equations very similar to equations (5.6)-(5.10)

$$\frac{d}{dx}\left[\frac{h^3}{12\mu}\frac{dp}{dx} - \frac{v_1+v_2}{2}h\right] = 0, \tag{18.1}$$

$$p(x_i) = p(x_e) = \frac{dp(x_e)}{dx} = 0, \tag{18.2}$$

$$h = h_e + \frac{x^2-x_e^2}{2R'} + \frac{2}{\pi E'}\int\limits_{x_i}^{x_e} p(t)\ln\mid\frac{x_e-t}{x-t}\mid dt, \tag{18.3}$$

$$\int\limits_{x_i}^{x_e} p(t)dt = P, \tag{18.4}$$

where v_1 and v_2 are linear velocities of contact surfaces within the contact region, which may depend on x, x_i and x_e are the inlet and exit coordinates

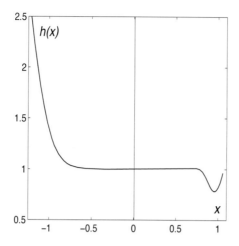

FIGURE 18.3
Gap distributions $h(x)$ in a lubricated contact ($V = 0.1$, $Q = 4$) (after Kudish [6]). Reprinted with permission from the ASME.

of the contact, h_e is the exit film thickness, R' and E' are the effective radius and elastic modulus of the solids (see Section 6.2.1).

18.3 Qualitative Analysis of the EHL Problem for Soft Solids

The Reynolds equation (18.1) can be rewritten in the form

$$\frac{dQ_x(x)}{dx} = 0, \quad Q_x(x) = U(x)h(x), \quad U(x) = \frac{v_1+v_2}{2} - \frac{h^2}{12\mu}\frac{dp}{dx}, \tag{18.5}$$

where Q_x is the x-component of the lubricant flux and $U(x)$ is the average of the lubricant flow velocity. Integration of the above equation gives

$$U(x) = \frac{U_0}{h(x)}, \quad U_0 = U(x_i) = constant. \tag{18.6}$$

For a heavily loaded contact in the central part of the contact region, Hertzian region (far from the inlet x_i and exit x_e points), the second term of the equation for $U(x)$ is negligibly small in comparison to the first one (see the second estimate in (6.30)). Therefore,

$$U(x) \approx \frac{v_1+v_2}{2}, \tag{18.7}$$

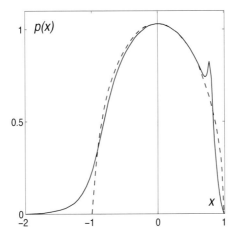

FIGURE 18.4
Distributions of pressure $p(x)$ (solid curve) in a lubricated contact ($V = 0.1$, $Q = 6.25$) and the Hertzian pressure (dotted curve) in a dry contact (after Kudish [6]). Reprinted with permission from the ASME.

and in the central part of the contact region equation (18.6) gives

$$\frac{v_1 + v_2}{2} \approx \frac{U_0}{h(x)}. \tag{18.8}$$

Thus, the conclusion that the linear velocities of contact surfaces, v_1 and v_2, are dependent upon x follows from Kaneta's experimental fact that gap h is a function of x. This result is true only for the Hertzian zone in a heavily loaded lubricated contact.

If the shear moduli of elasticity G_k ($k = 1, 2$) of the solid materials in contact are large enough then the tangential displacements of contact surfaces are negligibly small. In this case the surface linear velocities, v_1 and v_2, are practically independent from x and can be accurately approximated by the "rigid" linear velocities, u_1 and u_2. Therefore, according to the classic formulation of the EHL problem the gap is flat in the Hertzian zone of the contact region, i.e., it is independent from x.

The experimental data presented by Kaneta et al. [2] shows that in certain cases of contacts between soft and hard elastic materials gap $h(x)$ significantly varies within the central part of the contact region. The above variations of $h(x)$ represent an abnormal shape of the contact surfaces in a heavily loaded EHL contact. As it follows from the latter equation the average surface velocity, $(v_1 + v_2)/2$, varies significantly with x. This phenomenon is caused by relatively large tangential displacements of soft elastic surfaces.

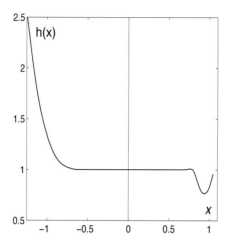

FIGURE 18.5

Distribution of gap $h(x)$ in a lubricated contact ($V = 0.1$, $Q = 6.25$) (after
Kudish [6]). Reprinted with permission from the ASME.

The conclusions made for incompressible lubricants are also valid for compressible lubricants and lubricants with non-Newtonian rheology. This is based
on the fact that the lubricant compressibility and rheology affect the gap between the contact surfaces in the inlet and exit zones of a heavily loaded
contact and practically do not affect the gap in the Hertzian region of the
contact. The same conclusions are valid for non-isothermal EHL problems as
well.

The described variations in surface velocities, v_1 and v_2, reflect tangential
displacements of contact surfaces. Therefore, in order to describe significant
variations in surface velocities, the tangential displacements of relatively soft
elastic surfaces have to be taken into consideration.

18.4   Surface Velocities for Soft Solids

Let us take into account the elastic displacements of contact surfaces, U_k ($k =
1, 2$), in the tangential to the surfaces direction. Then under steady conditions
the equations for surface linear velocities can be written as follows (see Galin

[7]):

$$v_k = u_k(1 + \frac{dU_k}{dx}), \quad \frac{\pi E_k}{2(1-\nu_k^2)}\frac{dU_k}{dx} = -\frac{1-2\nu_k}{2-2\nu_k}\pi p + \int_{x_i}^{x_e} \frac{\tau_k dt}{t-x}, \quad (18.9)$$

where τ_k $(k = 1, 2)$ are the tangential stresses created in a lubricant and acting upon the adjacent material surfaces

$$\tau_1 = \frac{\mu(v_2 - v_1)}{h} - \frac{h}{2}\frac{dp}{dx}, \quad \tau_2 = -\frac{\mu(v_2 - v_1)}{h} - \frac{h}{2}\frac{dp}{dx}. \quad (18.10)$$

Using equations (18.9) and (18.10) the following linear singular integral equation can be derived for the slip velocity $s = v_2 - v_1$:

$$s = [\frac{(1-2\nu_1)u_1}{G_1} - \frac{(1-2\nu_2)u_2}{G_2}]p - \frac{2}{\pi}[\frac{(1-\nu_1)u_1}{G_1} + \frac{(1-\nu_2)u_2}{G_2}]\int_{x_i}^{x_e} \frac{\mu s}{h}\frac{dt}{t-x}$$

$$+\frac{1}{\pi}[\frac{(1-\nu_1)u_1}{G_1} - \frac{(1-\nu_2)u_2}{G_2}]\int_{x_i}^{x_e} h\frac{dp}{dt}\frac{dt}{t-x} + u_2 - u_1. \quad (18.11)$$

After equation (18.11) is solved for s from equations (18.9) and (18.10) one can determine

$$\frac{v_1+v_2}{2} = \frac{u_1+u_2}{2} - \frac{1}{2}[\frac{(1-2\nu_1)u_1}{G_1} + \frac{(1-2\nu_2)u_2}{G_2}]p$$

$$+\frac{1}{\pi}[\frac{(1-\nu_1)u_1}{G_1} - \frac{(1-\nu_2)u_2}{G_2}]\int_{x_i}^{x_e} \frac{\mu s}{h}\frac{dt}{t-x} \quad (18.12)$$

$$-\frac{1}{2\pi}[\frac{(1-\nu_1)u_1}{G_1} + \frac{(1-\nu_2)u_2}{G_2}]\int_{x_i}^{x_e} h\frac{dp}{dt}\frac{dt}{t-x}.$$

Obviously, if $u_1 = u_2$, $G_1 = G_2$, and $v_1 = v_2$ the solution of equation (18.11) is $s(x) = 0$ and the expression (18.12) for the average surface velocity can be simplified. Furthermore, if the shear moduli G_1 and G_2 of contact materials are much larger than the maximum Hertzian pressure p_H then $v_k = u_k + O(p_H/G_1, p_H/G_2)$ and the classic EHL problem formulation is valid. However, if p_H/G_1 and/or p_H/G_2 are not very small (the shear moduli G_1 and G_2 are comparable with p_H), then the actual average of surface velocities $(v_1+v_2)/2$ may significantly differ from the "rigid" one $(u_1+u_2)/2$. Therefore, the new problem formulation described by equations (18.1)-(18.4), (18.11), and (18.12) must be used.

Similarly, a spacial EHL problem, which takes into account tangential deformations of elastic contact surfaces, can be formulated.

18.5   Dimensionless Variables and Numerical Method

By using the dimensionless variables and parameters

$$\{s', v'_{1,2}\} = \{s, v_{1,2}\}\tfrac{2}{u_1+u_2}, \ \ s_0 = 2\tfrac{u_2-u_1}{u_2+u_1}, \ \ m_{1,2} = \tfrac{p_H}{G_{1,2}},$$

$$\alpha_{1,2} = (1-\nu_{1,2})m_{1,2}, \ \ \beta_{1,2} = (1-2\nu_{1,2})m_{1,2}, \ \ A_0 = \alpha_1 + \alpha_2,$$

$$\{A_1, B_1\} = \{\alpha_1, \beta_1\}(1-\tfrac{s_0}{2}), \ \ \{A_2, B_2\} = \{\alpha_2, \beta_2\}(1+\tfrac{s_0}{2}),$$

$$A = A_0(A_1 + A_2), \ \ B = B_1 - B_2,$$

$$C = A_0(A_1 - A_2), \ \ D = B_1 + B_2,$$

$$\text{(18.13)}$$

in addition to dimensionless variables (6.3) and (6.4), the problem can be reduced to the system of integro-differential equations (further the primes at dimensionless variables are omitted)

$$\frac{H_0^2}{V}\frac{h^3}{\mu}\frac{dp}{dx} = \tfrac{1}{2}[(v_1 + v_2)h - v_1(c) - v_2(c)], \ \ p(a) = p(c) = 0, \qquad (18.14)$$

$$\frac{v_1+v_2}{2} = 1 - \frac{D}{2}p + \frac{CV}{12\pi H_0}\int\limits_a^c \frac{\mu s}{h}\frac{dt}{t-x} - \frac{AH_0}{2\pi}\int\limits_a^c \frac{h\,dp}{t-x}, \qquad (18.15)$$

$$s + \frac{AV}{2\pi H_0}\int\limits_a^c \frac{\mu s}{h}\frac{dt}{t-x} = s_0 + Bp + \frac{CH_0}{\pi}\int\limits_a^c \frac{h\,dp}{t-x}, \qquad (18.16)$$

$$H_0(h - 1) = x^2 - c^2 + \tfrac{2}{\pi}\int\limits_a^c p(t)\ln\mid\tfrac{c-t}{x-t}\mid dt, \qquad (18.17)$$

$$\int\limits_a^c p(t)dt = \tfrac{\pi}{2}. \qquad (18.18)$$

In equation (18.17) for gap $h(x)$ we still neglected the influence of the surface tangential displacements.

The problem is completely described by equations (18.14)-(18.18). In addition to these equations the expression for the lubricant viscosity $\mu(p)$ as a function of pressure p and the values of constants a, V, A, B, C, and D must be given. The solution of the problem comprises pressure $p(x)$, gap $h(x)$, and slip velocity $s(x)$ distributions as well as coordinate of the exit boundary c and and film thickness H_0.

The described problem can be solved by using iterations. Let us give a general description of the calculation process for one iteration. Suppose the initial approximations for $p(x)$, $h(x)$, H_0, and c are known. First, we introduce a substitution

$$x = \frac{c+a}{2} + \frac{c-a}{2}y,$$

which maps interval $[a, c]$ onto $[-1, 1]$ to replace the problem with the unknown boundary $x = c$ by a problem with known boundaries. That allows us to introduce two partitions: $\{y_i\}$, $i = 1, \dots, I$, $y_1 = -1$, $y_I = 1$ and $y_{i+1/2} = (y_i + y_{i+1})/2$, $i = 1, \dots, I - 1$. The solution of equation (18.16), $s_k = s(y_k)$, is determined at nodes $\{y_i\}$, $i = 1, \dots, I$, from the system of I linear algebraic equations

$$s_k + \frac{AV}{24\pi H_0} \sum_{i=1}^{I-1} \frac{\mu_{i+1/2}(s_i + s_{i+1})}{h_{i+1/2}} \frac{y_{i+1} - y_i}{y_{i+1/2} - y_k} = Bp_k$$

(18.19)

$$+ \frac{CH_0}{\pi} \sum_{i=1}^{I-1} \frac{h_{i+1/2}(p_{i+1} - p_i)}{y_{i+1/2} - y_k}, \quad k = 1, \dots, I.$$

In equation (18.19) we used a quadrature formula similar to the ones used in [8]. Having the values of s_k and using the proper quadrature formulas for calculation of singular integrals [8] the average velocity $(v_1(y) + v_2(y))/2$ is determined from equation (18.15) at nodes $\{y_{i+1/2}\}$, $i = 1, \dots, I - 1$, and for $y_I = 1$ using a similar quadrature formula for calculation of singular integrals. After that equations (18.14), (18.17), and (18.18) are solved for $p(y), h(y), H_0$, and c by one of the well-known methods, for instance, the method presented in Section 6.2. The iterations stop when consecutive iterates satisfy the desired accuracy.

18.6 Numerical Results and Discussion

According to Kaneta et al. [2], certain specific features (abnormal phenomena) of a lubricated contact were observed experimentally and were linked to the existence of a dimple in EHL contacts:

(a) in some cases when a soft material is involved in a contact a dimple was produced while in other cases when both surfaces are made from a hard material no dimple was observed,

(b) no noticeable dimple was observed when a surface made from a hard material was moving and its counterpart made from a soft material was stationary,

(c) a dimple was observed in the contact region between hard and soft materials for slide-to-roll ratio $s_0 = 2$ and it was not observed for $s_0 = -2$;

(d) a dimple got more pronounced with an increase in slide-to-roll ratio s_0 and its position moved toward the inlet point while the minimum film thickness decreased only slightly.

For further analysis the lubricant viscosity, μ, will be used according to the exponential law, $\mu = \exp(Qp)$, where Q is the dimensionless pressure coefficient of viscosity. Moreover, it will be assumed that the inlet coordinate

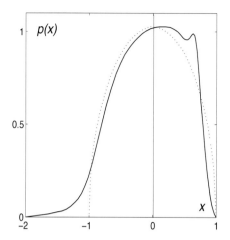

FIGURE 18.6
Distributions of pressure $p(x)$ (solid curve) in a lubricated contact ($\nu_1 = 0.3$, $\nu_2 = 0.1$, $m_1 = 0.003$, $m_2 = 0.25$, $A = 0.1583$, $B = -0.5646$, $C = -0.1591$, $D = 0.5629$, $s_0 = 3.125$, $V = 0.1$, $Q = 4$) and the Hertzian pressure (dotted curve) in a dry contact (after Kudish [6]). Reprinted with permission from the ASME.

$a = -2$, the total number of computational nodes $I = 100$, and the relative precision of numerical solutions is 10^{-4}.

Let us consider each of the aforementioned features of the phenomenon and then find out whether the solution of an EHL problem in the new formulation (18.14)-(18.18) possesses these features. It is necessary to keep in mind that in this section the notions of soft and hard materials involve not only their elastic properties but also the applied loading (see formulas for parameters $m_{1,2}$, A, B, C, and D in (18.13)). The materials are called hard if $m_{1,2}$, A, B, C, and D are very small and soft otherwise.

Feature (a). Let us consider a lubricated contact of two hard materials. Then, according to (18.13) $m_{1,2} \ll 1$ and constants A, B, C, and D in equations (18.15) and (18.16) are small. Therefore, s and $(v_1 + v_2)/2$ are approximately equal to the "rigid" slip s_0 and 1, respectively. As a result, the EHL problem is identical to the classic one. In Sections 6.3 and 6.6 based on the classic formulation it has been shown that the film thickness is flat in the Hertzian zone of a heavily loaded contact. Two examples of such classic solutions for hard surfaces (with constants $A = B = C = D = 0$), and parameters $V = 0.1$, $Q = 4$, and $Q = 6.25$, respectively, are represented in Figs. 18.2, 18.3 and 18.4, 18.5. These solutions are independent from the slide-to-roll ratio s_0. According to the classic problem formulation gap distribution

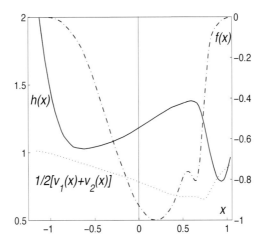

FIGURE 18.7

Distributions of gap $h(x)$ (solid curve), average velocity $\frac{v_1(x)+v_2(x)}{2}$ (dotted curve), and sliding frictional stress $f(x)$ (dash-dotted curve) in a lubricated contact ($\nu_1 = 0.3$, $\nu_2 = 0.1$, $m_1 = 0.003$, $m_2 = 0.25$, $A = 0.1583$, $B = -0.5646$, $C = -0.1591$, $D = 0.5629$, $s_0 = 3.125$, $V = 0.1$, $Q = 4$) (after Kudish [6]). Reprinted with permission from the ASME.

$h(x)$ is virtually flat in the central part of the contact as shown in Fig. 18.3 and 18.5.

A similar behavior of a solution for equations (18.14)-(18.18) is observed for hard materials ($\nu_{1,2} = 0.3$, $m_{1,2} = 0.003$, $A = 0.00003$, $B = -0.0037$, $C = -0.00004$, $D = 0.00312$, $s_0 = 2.375$, $V = 0.1$, and $Q = 4$). In particular, the average velocity $(v_1 + v_2)/2$ differs from 1 by less than 0.002, and in the Hertzian region the distribution of gap $h(x)$ differs from 1 by less than 0.008. A dimple is not observed in these solutions of classic and modified problems. On the contrary, for surfaces made from soft and hard materials (see Figs. 18.6 and 18.7)* with $\nu_1 = 0.3$, $\nu_2 = 0.1$, $m_1 = 0.003$, $m_2 = 0.25$, $A = 0.15832$, $B = -0.56463$, $C = -0.15908$, $D = 0.56287$, $s_0 = 3.125$, $V = 0.1$, and $Q = 4$ there is a pronounced dimple. The surface average velocity, $(v_1 + v_2)/2$, differs from 1 significantly in the central part of the contact. In Figure 18.7 and in the other graphs that follow the frictional stress which is usually much greater than the rolling frictional stress in the Hertzian region of the contact is represented by just its sliding part. On the other hand, the numerical results

---

*In Figs. 18.7, 18.9, 18.11-18.13 the left scale is for values of gap $h(x)$ and average velocity $\frac{v_1(x)+v_2(x)}{2}$ while the right scale is for sliding frictional stress $f(x) = \frac{V}{12H_0}\frac{\mu s}{h}$.

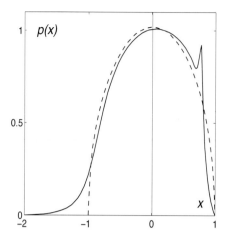

FIGURE 18.8

Distributions of pressure $p(x)$ (solid curve) in a lubricated contact ($\nu_{1,2} = 0.3$, $m_1 = 0.003$, $m_2 = 0.15$, $A = -C = 0.3801$, $B = -D = -0.156$, $V = 0.1$, $Q = 6.25$, $s_0 = 2$) and the Hertzian pressure (dotted curve) in a dry contact (after Kudish [6]). Reprinted with permission from the ASME.

show a significant reduction in the calculated frictional stress $f(x)$ due to the presence of tangential surface displacements which makes its value more reasonable compared to the case when tangential surface displacements are not taken into account. These frictional stress reductions are caused by significant changes in local speeds of solid surface points.

Therefore, the EHL model described by equations (18.14)-(18.18) represents the experimentally observed by Kaneta et al. [2] **Feature (a)** adequately.

Feature (b). Let us consider a stationary surface made from a soft material, i.e., $u_1 = 0$. Then from equation (18.9) follows that $v_1 = 0$. In equations (18.9) for the other surface made from a hard material the actual surface velocity v_2 is practically equal to the "rigid" one u_2. In such cases the values of dimensionless parameters A, B, C, and D in (18.15) are small and $(v_1 + v_2)/2 \approx 1$. That brings us back to the classic EHL problem. Therefore, similarly to the classic formulation of the EHL problem the gap is flat in the Hertzian region of a heavily loaded contact and a dimple is not observed. This finding is in agreement with the experimental data by Kaneta et al. [2].

Feature (c). In the case of a contact of soft and hard materials represented in Figs. 18.6 and 18.7 which are obtained for $\nu_{1,2} = 0.3$, $m_1 = 0.003$, $m_2 = 0.15$, $A = -C = 0.3801$, $B = -D = -0.156$, $V = 0.1$, and $Q = 6.25$ for the slide-to-roll ratio $s_0 = 2$ a pronounced dimple is produced in the contact region. However, in Figs. 18.8 and 18.9 for the same materials and loading

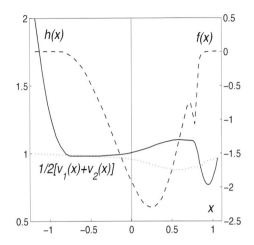

FIGURE 18.9

Distributions of gap $h(x)$ (solid curve), average velocity $\frac{v_1(x)+v_2(x)}{2}$ (dotted curve), and sliding frictional stress $f(x)$ (dash-dotted curve) in a lubricated contact ($\nu_{1,2} = 0.3$, $m_1 = 0.003$, $m_2 = 0.15$, $A = -C = 0.3801$, $B = -D = -0.156$, $V = 0.1$, $Q = 6.25$, $s_0 = 2$) (after Kudish [6]). Reprinted with permission from the ASME.

parameters as in Figs. 18.8 and 18.9 and the slide-to-roll ratio $s_0 = -2$ ($\nu_{1,2} = 0.3$, $m_1 = 0.003$, $m_2 = 0.15$, $A = C = 0.00076$, $B = D = 0.00312$, $V = 0.1$, and $Q = 6.25$) no noticeable dimple is observed. This finding is in agreement with the experimental data by Kaneta et al. [2].

Feature (d). As it follows from Figs. 18.12 and 18.13 a dimple gets more pronounced with increase of the slide-to-roll ratio s_0 and its center moves toward the inlet point. The graphs represented in Figs. 18.12 and 18.13 are obtained for $\nu_1 = 0.3$, $\nu_2 = 0.1$, $m_1 = 0.003$, $m_2 = 0.25$, $V = 0.1$, $Q = 4$, $s_0 = 0$ (i.e., $A = 0.06262$, $B = -0.21844$, $C = -0.06125$, $D = 0.22156$) and $s_0 = 2.375$ (i.e., $A = 0.13535$, $B = -0.48154$, $C = -0.1356$, $D = 0.48096$), respectively. In these graphs the minimal film thickness decreases slightly as s_0 increases. Therefore, the EHL model described by equations (18.14)-(18.18) represents **Feature (d)** experimentally observed by Kaneta et al. [2] adequately.

To give a better idea of how the depth of a dimple changes with the value of the slide-to-roll ratio s_0 some numerical data for two series of problem parameters (Series 1: $\nu_{1,2} = 0.3$, $m_1 = 0.003$, $m_2 = 0.25$, $V = 0.1$, $Q = 5$ and Series 2: $\nu_1 = 0.3$, $\nu_2 = 0.1$, $m_1 = 0.003$, $m_2 = 0.25$, $V = 0.1$, $Q = 4$) is presented in Tables 18.1 and 18.2, respectively. The depth of a dimple, d, is determined as the difference between the maximum gap in the Hertzian zone

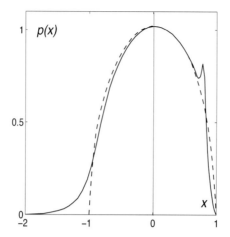

FIGURE 18.10

Distributions of pressure $p(x)$ (solid curve) in a lubricated contact ($\nu_{1,2} =$ 0.3, $m_1 = 0.003$, $m_2 = 0.15$, $A = C = 0.00076$, $B = D = 0.00312$, $V =$ 0.1, $Q = 6.25$, $s_0 = -2$) and the Hertzian pressure (dotted curve) in a dry contact (after Kudish [6]). Reprinted with permission from the ASME.

of the contact and the minimum gap in the inlet zone of the contact. The data are presented in absolute values and as a ratio to the minimum value of the gap in the inlet zone of the contact.

The new formulation of the EHL problem presented in this chapter gives an adequate generalization of the classic EHL problem. The new problem formulation is validated by an analytical and numerical analysis as well as by the experimental data obtained by Kaneta et al. [2]. The quantitative differences between the data described in this section and the data obtained by Kaneta et al. [2] are caused by a variety of reasons such as different contact geometry, different contact operating parameters, different elasticity parameters, oil compressibility, thermal deformations, and onset of plastic deformations. The onset of material plastic deformations can be realized under certain conditions, in particular, when the material is soft and frictional stresses are sufficiently high. The latter may happen when the pressure coefficient of viscosity is large. Different from Kaneta et al. [2] operating parameters are chosen due to numerical stability considerations.

Moreover, it is important to mention that the occurrence of the abnormal phenomena (dimples) does not seem to be linked to oil solidification. In order to illustrate this fact let us perform a simple analysis. It is widely accepted that oil solidification is observed under high pressures regardless of contact kinematics. Such high pressures can be created in a lubricated contact region

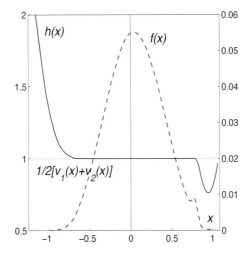

FIGURE 18.11

Distributions of gap $h(x)$ (solid curve), average velocity $\frac{v_1(x)+v_2(x)}{2}$ (dotted curve), and sliding frictional stress $f(x)$ (dash-dotted curve) in a lubricated contact ($\nu_{1,2} = 0.3$, $m_1 = 0.003$, $m_2 = 0.15$, $A = C = 0.00076$, $B = D = 0.00312$, $V = 0.1$, $Q = 6.25$, $s_0 = -2$) (after Kudish [6]). Reprinted with permission from the ASME.

TABLE 18.1
The dependence of dimple depth d on
slide-to-roll ratio s_0 for Series 1 (after Kudish
[6]). Reprinted with permission from the
ASME.

s_0	-2	0	2	2.375
d	no dimple	no dimple	0.1753	0.2088
%	0	0	18.1%	21.9%

between a moving steel ball (with high elasticity modulus) and a stationary glass (with low elasticity modulus). Therefore, under such conditions oil solidification is possible. However, in the aforementioned case of high pressures a noticeable dimple is not produced. This has been described by the analysis of **Feature (b)** without taking into account the phenomenon of oil solidification.

In general, in heavily loaded contacts an EHL problem solution for the new formulation (18.14)-(18.18) has properties similar to those of the classic EHL problem solution except for the dimple phenomenon. In particular, the similarities are noted for certain problem parameters when a spike of pressure is observed in the exit zone of the contact. Using asymptotic approach of the preceding sections, it can be shown that in fully flooded heavily loaded

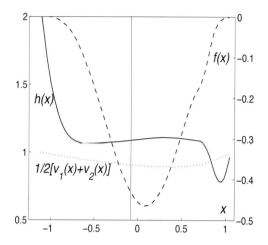

FIGURE 18.12

Distributions of gap $h(x)$ (solid curve), average velocity $\frac{v_1(x)+v_2(x)}{2}$ (dotted curve), and sliding frictional stress $f(x)$ (dash-dotted curve) in a lubricated contact ($\nu_1 = 0.3, \nu_2 = 0.1$, $m_1 = 0.003$, $m_2 = 0.25$, $A = 0.06262$, $B = -0.21844$, $C = -0.06125$, $D = 0.22156$, $V = 0.1$, $Q = 4$, $s_0 = 0$) (after Kudish [6]). Reprinted with permission from the ASME.

TABLE 18.2
The dependence of dimple depth d on slide-to-roll ratio s_0 for Series 2 (after Kudish [6]). Reprinted with permission from the ASME.

s_0	-2	0	$.3752$	3.125
d	no dimple	0.0404	0.2364	0.3431
$\%$	0	3.7%	22.5%	33.3%

lubrication regimes (and in most other cases too), which are described by equations (18.14)-(18.18), the pressure distribution in the Hertzian zone of the contact region is close to the Hertzian one. Moreover, it can be shown that for these regimes lubrication film thickness H_0 is determined by elastic and lubrication processes in the inlet zone of the contact region. This fact explains why H_0 is practically independent from the slide-to-roll ratio s_0. For example, $H_0 = 0.18004$ for $s_0 = 0$ and $H_0 = 0.19123$ for $s_0 = 2.375$, while parameters $\nu_{1,2}$, $m_{1,2}$, V, and Q are the same as in Figs. 18.12 and 18.13, $H_0 = 0.22013$ for $s_0 = -2$ and $H_0 = 0.23224$ for $s_0 = 2$ while parameters $\nu_{1,2}$, $m_{1,2}$, V, and Q are the same as in Figs. 18.11 and 18.9. Therefore, for the aforementioned cases the solutions of the modified and classic EHL problems

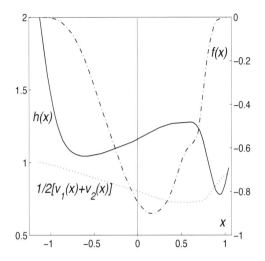

FIGURE 18.13

Distributions of gap $h(x)$ (solid curve), average velocity $\frac{v_1(x)+v_2(x)}{2}$ (dotted curve), and sliding frictional stress $f(x)$ (dash-dotted curve) in a lubricated contact ($\nu_1 = 0.3$, $\nu_2 = 0.1$, $m_1 = 0.003$, $m_2 = 0.25$, $A = 0.13542$, $B = -0.48166$, $C = -0.13569$, $D = 0.48084$, $V = 0.1$, $Q = 4$, $s_0 = 2.375$) (after Kudish [6]). Reprinted with permission from the ASME.

are practically identical in the inlet zone. As a result, the film thickness H_0 increases with Q and V according to the formulas obtained in the preceding sections. The difference between the solutions of the classic and new EHL problems mainly occurs (if at all) in the Hertzian region and it depends solely on the values of parameters A, B, C, D, and s_0.

The material of the section presents a new adequate formulation of EHL problems for soft elastic materials. The properties of solutions of the new EHL problem (in particular, the existence of a dimple) are consistent with those observed in experiments. A qualitative analysis of a lubricated contact for soft elastic materials has shown that if the gap between lubricated contact surfaces varies substantially in the Hertzian region of a heavily loaded contact then the surface linear velocities must vary significantly as well. Therefore, the tangential displacements of elastic surfaces have been taken into proper account in a new formulation of the EHL problem, which has been analyzed both numerically and analytically. Furthermore, the section shows that oil solidification does not seem to relate to dimple existence.

18.7   Closure

A model of EHL soft elastic solids is presented. It is shown that for solids with low Young's modulus it is necessary to take into account not only the normal but also the tangential displacements of solid surfaces. A complete model of such lubricated contact interactions is developed and analyzed numerically. All four distinct features of lubricated surface behavior observed experimentally by Kaneta were replicated numerically. The numerical results showed a significant reduction in the calculated frictional stress due to the presence of tangential surface displacements which makes its value more reasonable compared to the case when tangential surface displacements are not taken into account. These frictional stress reductions are caused by a significant changes in local speeds of solid surface points.

18.8   Exercises and Problems

1. Provide the reason as to why surface tangential displacements and modified surface linear velocities must be included in the formulation of the EHL problem for soft solids to explain the Kaneta "dimple" phenomenon.

2. Explain why the thermal effects alone without taking into account the tangential surface displacements cannot explain different lubrication features occurring in lubricants and observed in experiments.

References

[1] Kaneta, M., Nishikawa, H., Kameishi, K., Sakai, T., and Ohno, N. 1992. Effects of Elastic Moduli of Contact Surfaces in Elastohydrodynamic Lubrication. *ASME J. Tribology* 114:75-80.

[2] Kaneta, M., Nishikawa, H., Kanada, T., and Matsuda, K. 1996. Abnormal Phenomena Appearing in EHL Contacts. *ASME J. Tribology.* 118:886-892.

[3] Cermak, J. 1998. Abnormal Phenomena Appearing in EHL Contacts. Discussion on a previously published paper. *ASME J. Tribology* 120:143-144.

[4] Wang, J., Yang, P., Kaneta, M., and Nishikawa, H. 2003. On the Surface Dimple Phenomena in Elliptical TEHL Contacts with Arbitrary Entrainment. *ASME J. Tribology* 125, No. 1:102-109.

[5] Kaneta, M., Shigeta, T., and Young, P. 2006. Film Pressure Distributions in Point Contacts Predicted by Thermal EHL Analysis. *Tribology Intern.* 39:812-819.

[6] Kudish, I.I. 2000. Formulation and Analysis of EHL Problems for Soft Materials. *ASME J. Tribology* 122, No. 3:705-710.

[7] Galin, L.A. 1980. *Contact Problems in the Theory of Elasticity and Viscoelasticity*. Moscow: Nauka.

[8] Belotserkovsky, S.M. and Lifanov, I.K. 2000. *Method of Discrete Vortices*. Boca Raton: CRC Press.

19

Non-Newtonian Lubricants and Scale Effects

19.1 Introduction

The estimation or prediction of EHL film thickness requires knowledge of the lubricant properties. Today, in many instances, the properties have been obtained from a measurement of the central film thickness in an optical EHL point contact simulator and the assumption of a classical Newtonian film thickness formula. This technique has the practical advantage of using an effective pressure viscosity coefficient that compensates for shear-thinning. It is shown below that the practice of extrapolating from a laboratory scale measurement of film thickness to the film thickness of an operating contact within a real machine may substantially overestimate the film thickness in the real machine if the machine scale is smaller and the lubricant is shear-thinning within the inlet zone.

The film thickness in EHL concentrated contacts has implications for friction and for the wear and fatigue life of the rollers. The estimation or prediction of EHL film thickness requires knowledge of the lubricant properties. If the lubricant is Newtonian within the pressure-boosting inlet zone, the film thickness may be calculated from a classical film thickness equation such as that offered by Dowson and Higginson [1]. The required liquid properties are the ambient low-shear viscosity μ_0 and the pressure viscosity coefficient α.

The pressure viscosity coefficients that are reported in handbooks and journals come from two sources. Originally, the pressure viscosity coefficient was calculated from measurements of viscosity as a function of pressure at fixed temperature using various definitions of the pressure viscosity coefficient [2]. These are viscosity-derived pressure viscosity coefficients α. Today, in many instances, the reported coefficient has been obtained from a measurement of the central film thickness h_c by an optical EHL point contact simulator and the assumption of a classical Newtonian film thickness formula. Typically, the Newtonian Hamrock and Dowson formula [3] is solved for the value of α which will give the measured h_c. These are film-derived effective pressure viscosity coefficients α_e.

Not surprisingly, it is often found that $\alpha_e < \alpha$ [4] because of the Newtonian assumption implicit in the α_e calculation [5]. The practical advantage of using an effective coefficient that compensates for shear-thinning is obvious. The film-thickness under the same conditions may then be estimated easily using the classical Newtonian formula. The disadvantage stems from the fact that the shear-thinning behavior will change the response of the film thickness to variations in rolling velocity [6], sliding velocity [6], and perhaps to geometry and other material properties.

The material of this chapter sounds a warning regarding the extrapolation of measurements of film thickness using Newtonian formulas in order to estimate the film thickness between machine elements of a different scale (and perhaps elastic modulus) from the original measurement. Significant over-estimations may result at smaller scales. In a departure from convention, the maximum Hertzian pressure p_H will be used to quantify the contact loading rather than the normal force P since the pressure of an actual machine element contact would be more closely simulated in an experimental measurement than the normal force and, of course, the rheology is dependent on pressure, not normal force. The generalized Newtonian constitutive equation utilized for shear-thinning will be the single-Newtonian Carreau-Yasuda form, which accurately describes the shear dependence of the viscosity that is measured for base oils in viscosimeters [2, 5] - [8].

We will consider the results of an analytical (asymptotic) treatment of the problem, which covers two limiting cases of relatively small and large shear stresses [9]. Then we proceed to a numerical solution, which, in turn, is well suited to covering the intermediate case of moderate shear stresses. A line contact is assumed for simplicity. Extension to the point contact that is usually used in EHL measurements can be accomplished and should not substantially change the conclusions.

19.2   Results of Perturbation Analysis

Let us consider some results of the perturbation analysis for heavily loaded contacts lubricated with Newtonian and non-Newtonian lubricants. These results were obtained in the preceding sections.

Let us consider a fluid with single-Newtonian Carreau rheology, which, in dimensionless variables in a narrow gap between two heavily loaded elastic solids, is described by equations

$$\tau = \Phi(\mu \tfrac{\partial u}{\partial z}), \quad \Phi(t) = \tfrac{V}{12H_0} t \{1 + (\tfrac{V}{12H_0 G_0} \mid t \mid)^m\}^{\frac{n-1}{m}}, \qquad (19.1)$$

where the dimensionless parameters are introduced based on the Hertzian contact half-width a_H and Hertzian maximum pressure p_H as well as on the ambient low-shear viscosity μ_a, and the central film thickness h_c. The dimensionless parameters in equation (19.1) are defined as follows

$$V = \tfrac{24\mu_a (u_1 + u_2) R'^2}{a_H^3 p_H}, \quad H_0 = \tfrac{2R' h_c}{a_H^2}, \quad G_0 = \tfrac{\pi R' G}{P}. \qquad (19.2)$$

In equation (19.1), parameters m and n are certain positive constants. It is obvious that the rheological equation (19.1) provides for two limiting cases of relatively small and large shear stresses. The case of relatively small shear stresses corresponds to the Newtonian behavior of the lubricant and is described by the relations

$$\tau = \frac{V}{12H_0}\mu\frac{\partial u}{\partial z} \ if \ \frac{V}{12H_0G} \mid \frac{\partial u}{\partial z} \mid \ll 1, \tag{19.3}$$

while the case of relatively large shear stresses leads to the Ostwald-de Waele (power law) rheology as follows

$$\tau = \Phi(\mu\frac{\partial u}{\partial z}), \ \Phi(t) = \frac{V}{12H_0}t(\frac{V}{12H_0G_0} \mid t \mid)^{n-1}, \tag{19.4}$$

which coincides with (9.108) in which V_n is replaced by $VG_0^{\frac{1-n}{n}}$. In case of $n = 1$ equation (19.4) coincides with equation (19.3) describing the Newtonian rheology.

In practice, in the inlet zone of a heavily loaded lubricated contact there is a continuous transition from small to large shear stresses. Therefore, in real situations we can expect that the results (such as film thickness) are somewhere between the results obtained for the two limiting cases. In this subsection we will concentrate on the results of the analytical analysis of the two limiting cases. The value of this analytical analysis is based on the fact that we will be able to derive some conclusions from asymptotically valid (for heavy loading) analytical formulas for the film thickness for the two limiting cases of small and large shear stresses. The case of moderate shear stresses is hardly possible to analyze analytically and it will be done numerically and compared to the analytical results.

Let us consider the case of pure rolling for a heavily loaded contact lubricated by iso-viscous fluids or fluids with moderate pressure dependence of viscosity, i.e., the case of pre-critical lubrication regimes. We will assume that $\mu = \exp(Qp)$, where $Q = \alpha p_H$ is the dimensionless pressure viscosity coefficient. Using the above developed perturbation methods for a Newtonian lubricant under fully flooded conditions, we obtain the formula for the film thickness (see equation (6.63) and (9.55))

$$H_0 = A_f V^{3/5}, \ A_f = O(1), \tag{19.5}$$

where A_f is a constant independent from V. This formula is valid if the following condition are satisfied (see (9.91)

$$Q = O(V^{-1/5}), \ V \ll 1. \tag{19.6}$$

For example, if $V = 0.01$, then the dimensionless pressure viscosity coefficient Q should not be greater than $5 - 10$ because $V^{-1/5} = 2.51$. Condition (19.6) defines lubricants with viscosity moderately dependent on pressure, i.e., precritical lubrication regimes. Moreover, if $Q \ll V^{-1/5}$, then constant A_f in formula (19.5) for film thickness H_0 is also independent from Q while for $Q = O(V^{-1/5})$ constant A_f is a function of Q.

Let us consider a case of a power law fluid under the conditions of pure rolling. For heavily loaded contacts under fully flooded conditions, the formula for the film thickness has the form (see equation (9.57))

$$H_0 = A_f(V^n G_0^{1-n})^{\frac{3}{3n+2}}, \; A_f = O(1), \; V \ll 1, \tag{19.7}$$

where A_f is a dimensionless constant dependent on n but independent of V and G_0. In this case the condition (see (9.91)

$$Q = O((V^n G_0^{1-n})^{-\frac{1}{3n+2}}), \; V \ll 1, \tag{19.8}$$

defines lubricants with viscosity moderately dependent on pressure. In case of $Q \ll (V^n G_0^{1-n})^{-\frac{1}{3n+2}}$ constant A_f is also independent from Q. For small slide-to-roll ratios s_0 formula (19.8) remains in force.

Now, let us consider the pure rolling lubrication regimes with high pressure dependence of viscosity, i.e., over-critical lubrication regimes. Using the developed perturbation methods for power law rheology in a contact under fully flooded lubrication conditions, the formula for the film thickness has the form (see equation (9.93))

$$H_0 = A_f(V^n G_0^{1-n} Q)^{\frac{3}{3n+1}}, \; A_f = O(1), \tag{19.9}$$

where A_f is a dimensionless constant dependent on n but independent of V, G_0, and Q. In this case the condition

$$Q \gg (V^n G_0^{1-n})^{-\frac{1}{3n+2}}, \; V \ll 1, \tag{19.10}$$

defines lubricants with viscosity strongly dependent on pressure. For the Newtonian rheology $n = 1$ and the film thickness formula from equation (19.9) assumes the form

$$H_0 = A_f(VQ)^{3/4}, \; A_f = O(1), \tag{19.11}$$

where A_f is a dimensionless constant independent of V and Q. From condition (19.11) for $n = 1$ and, for example, for $V = 0.01$ the dimensionless pressure viscosity coefficient Q should be greater than 10 because $V^{-1/5} = 2.51$. Structurally, formula (19.11) coincides with the Ertel-Grubin formula (see [10, 11]).

It is worth mentioning that for Newtonian lubricants formulas for the film thickness (19.5) as well as formula (19.11) are valid for any slide-to-roll ratio s_0. Also, it is important to realize that introduction of lubricant compressibility does not change the structural formulas for the film thickness derived above.

In case of high slide-to-roll ratio $s_0 \gg 1$ for the limiting case of a lubricant with power rheology the structural formula for the film thickness in a fully flooded lubrication regime for iso-viscous fluids or fluids with viscosity moderately dependent on pressure (pre-critical lubrication regimes) is as follows (see equation (9.129))

$$H_0 = A_f(V^n G_0^{1-n} \mid s_0 \mid^{n-1})^{\frac{3}{3n+2}}, \; A_f = O(1), \; V \ll 1, \tag{19.12}$$

while for fluids with viscosity strongly dependent on pressure (over-critical lubrication regimes) it is as follows (see equation (9.139))

$$H_0 = A_f(V_n^n Q \mid s_0 \mid^{n-1})^{\frac{3}{3n+1}}, \ A_f = O(1), \ V \ll 1, \tag{19.13}$$

where constants A_f in (19.12) and (19.13) are different dimensionless constants independent of V, G_0, and s_0 but dependent on n. In the latter case A_f is also independent from Q.

A similar asymptotic analysis of lubricated point contacts leads to exactly the same formulas for the film thickness. The only difference between the formulas for line and point contacts is due to the difference in the relationships for the Hertzian half-width for line contact and radius of point contact a_H respectively, and the maximum Hertzian pressures p_H for line and point contacts.

Notice that due to different lubrication regimes all of the above formulas for the film thickness for the power law fluids and Newtonian fluids are different. However, what they do have in common is for all considered cases of non-Newtonian lubricant behavior the film thickness is a certain nonlinear function of radius R', elastic modulus E', and pressure viscosity coefficient α. The form of this function depends on the nonlinearity of the fluid rheology, i.e., in our case on the value of constant n involved in rheological relationships (19.1) and (19.4). Below we will show that this fact is the source of the scale effect when a lubricating fluid exhibits a non-Newtonian behavior.

19.3 Application of Analytical Result to an EHL Experiment

For simplicity, we will consider the scaling effect just for one case of pure or near pure rolling in the Ertel-Grubin type of a fully flooded lubrication regime for a limiting case of high shear stresses for the Carreau-Yasuda fluid rheology (19.1) represented by the power law rheology (19.4). Solving equations (19.2) and (19.9) for the dimensional central film thickness h_c, we obtain

$$h_c = 2A_f(\tfrac{R'E'}{p_H})^{\frac{1}{3n+1}} \{[3\mu_a(u_1 + u_2)]^n G^{1-n}\alpha\}^{\frac{3}{3n+1}}. \tag{19.14}$$

Suppose we make a measurement of film thickness h_m at a scale of $R' = R_m$ in order to calculate an effective $\alpha = \alpha_e$ for the same values of E', p_H, and $\mu_0(u_1 + u_2)$ as for some system of machine elements. The parameters G and n are unknown. Therefore, for the calculation of α_e we assume that $n = 1$ which makes G to disappear from equation (19.14). Then the effective pressure viscosity coefficient α_e is determined by the formula

$$\alpha_e = \frac{h_m^{\frac{4}{3}}}{(2A_f)^{\frac{4}{3}}\mu_a(u_1+u_2)} R_m^{-\frac{1}{3}}(\tfrac{p_H}{E'})^{\frac{1}{3}}. \tag{19.15}$$

Now, let us use this effective pressure viscosity coefficient α_e to calculate the film thickness h_s of a real system of machine elements of scale $R' = R_s$ assuming that $n = 1$. The calculated film thickness is

$$h_{sc} = 2A_f \left(\frac{R_s E'}{p_H}\right)^{\frac{1}{4}} \{3\mu_a(u_1 + u_2)\alpha_e\}^{\frac{3}{4}}. \tag{19.16}$$

Substituting α_e from (19.15) into equation (19.16) gives

$$h_{sc} = h_m \left(\frac{R_s}{R_m}\right)^{\frac{1}{4}}. \tag{19.17}$$

The actual value of α is equal

$$\alpha = \frac{h_m^{\frac{3n+1}{3}} G^{n-1}}{(2A_f)^{\frac{3n+1}{3}} [\mu_a(u_1+u_2)]^n} R_m^{-\frac{1}{3}} \left(\frac{p_H}{E'}\right)^{\frac{1}{3}} \tag{19.18}$$

and the actual value of h_s is

$$h_s = h_m \left(\frac{R_s}{R_m}\right)^{\frac{1}{3n+1}}. \tag{19.19}$$

The ratio of calculated h_{sc} to actual h_s film thickness is

$$\frac{h_{sc}}{h_s} = \left(\frac{R_s}{R_m}\right)^{\frac{3(n-1)}{4(3n+1)}}. \tag{19.20}$$

TABLE 19.1
The ratio of calculated to actual film thickness for extrapolation to smaller (1/10) scale for power-law rheology (after Kudish, Kumar, Khonsari, and Bair [9]). Reprinted with permission from the ASME.

n	1	0.8	0.6	0.4	0.3
h_{sc}/h_s	1	1.11	1.28	1.60	1.89

Therefore, if the film thickness of a millimeter–scale system is calculated from an effective pressure viscosity coefficient obtained in an optical experimental measurement at centimeter scale, then the calculation will overestimate the film thickness by the ratio given in Table 19.1 for the various values of n. These ratios represent a substantial error in the calculation of film thickness. However, it must be recognized that equation (19.14) results from the assumption that the asymptotic high-shear behavior of the Carreau-Yasuda model takes place over the entire inlet zone. In reality, a portion of the inlet zone will experience nearly linear shear response and the actual error will be less than that given above.

Also, the above described scaling effect occurs in iso-viscous lubricants and lubricants with moderate pressure dependence of viscosity (pre-critical lubrication regimes) under pure or near pure rolling conditions for which the dimensionless film thickness is described by equation (19.7), which leads to

$$h_c = 2A_f \left(\frac{R'^2}{p_H^2 E'}\right)^{\frac{1}{3n+2}} \left\{[3\mu_a(u_1 + u_2)]^n G^{1-n}\alpha\right\}^{\frac{3}{3n+2}}. \qquad (19.21)$$

Now, in this regime, the exponent of R' is $2(3n+1)/(3n+2)$ times the value of the one in the previous regime and the exponent on the combined elastic modulus E' has gone from a positive value of $1/(3n+1)$ to a negative value of $-1/(3n+2)$. In most EHL rigs, glass is used for one element, resulting in $E = 1.2 \cdot 10^{11} Pa$ versus $2.1 \cdot 10^{11} Pa$ for steel on steel. Then extrapolation to a steel/steel system will also depend upon the E' effect. The lower composite modulus of the simulator will partially compensate for the scale effect if the exponent of E' is positive and vice versa.

To further quantify the error for intermediate cases between the two considered limiting regimes that results from extrapolation of optical EHL simulator measurements to smaller scales a numerical solution of the problem is obtained [9].

19.4   Numerical Solution

Now, instead of considering just the limiting cases of small and large shear stresses we will consider a full EHL solution for a compressible lubricant with the Carreau viscosity from equation (19.1) with $m = 2$ and $n = 0.4$. This numerical analysis will reveal similar scale effects as the asymptotic study of the limiting cases. Using this viscosity model, Jang et al. [12] presented extensive EHL line-contact simulations that closely agreed with published experimental results by Dyson and Wilson [6].

The solution domain in the present simulation ranges from $x = -4$ to $x = 1.5$ with a uniform grid size $\triangle x = 0.0125$. It has been verified that further mesh refinement and extension of the inlet zone cause negligible change in the results. The solution begins by assuming an initial guess for pressure distribution $\{p_i\}$ and offset film thickness. Using this, the film thickness and fluid properties are calculated and substituted in the generalized Reynolds equation

$$\frac{d}{dx}\left\{\rho\left[\frac{1}{\mu}\int\limits_{-h/2}^{h/2} zF(f + H_0 z\frac{dp}{dx})dz - h\right]\right\} = 0, \qquad (19.22)$$

in which the dimensionless variables are introduced according to equations (9.2), (19.2), $s_0 = 2\frac{u_2-u_1}{u_2+u_1}$ (see equation (9.3)) and the lubricant density ρ is scaled by its value at ambient pressure. The rest of the equations follow from (9.5)-(9.8) and equation (9.7) is rewritten in the form

$$H_0(h-1) = x^2 + \frac{2}{\pi} \int_a^c p(t) \ln \left| \frac{t}{x-t} \right| dt. \tag{19.23}$$

The lubricant density is determined according to Tait equation [2]

$$\rho = \{1 - \frac{1}{K_0'} \ln[1 + \frac{K_0'}{K_0}p]\}^{-1}, \tag{19.24}$$

where $K_0' = 11.19$ and $K_0 = 0.85\ GPa/p_H$.

The Reynolds equation (19.22) is discretized by using a mixed second order central and first order backward difference scheme. The Reynolds equation is solved along with load balance condition using the Newton-Raphson technique to obtain an improved pressure distribution and offset film thickness. The process of iterations continues until the relative error in pressure and gap distribution decreases below 10^{-5}.

For specific calculations we used the pressure dependence of viscosity in the form $\mu = \mu_a \exp(\alpha p)$, with $\mu_a = 0.08Pa \cdot s$ and $\alpha = 25GPa^{-1}$, $15GPa^{-1}$. The average contact pressure is $\frac{\pi}{4}p_H = 0.5GPa$. The rolling velocity is $0.5(u_1 + u_2) = 1m/s$ and slide-to-roll ratio $s_0 = 0$. The value of the lubricant critical stress G has been varied from $1 \cdot 10^9 Pa$ to $1 \cdot 10^5 Pa$. The intermediate values of G are selected to place the transition from the lubricant linear to non-linear response within the inlet zone. Therefore, these results are not the high shear behavior of the analysis above but rather should be representative of a real rheological response within the inlet zone. The calculated film thicknesses are given in Table 19.2 for four cases. First, the case of $R' = 0.013m$, $E' = 110GPa$, the scale and elasticity of an experimental rig at $\alpha = 25GPa^{-1}$ is considered. Then, the elastic modulus is increased to $E' = 211GPa$. This is followed by the case of $R' = 0.0013m$, $E' = 211GPa$, the scale and elasticity of, say, a small roller bearing, is investigated while the pressure viscosity coefficient is fixed at $\alpha = 25GPa^{-1}$. Finally, in order to investigate the sensitivity of pressure viscosity coefficient, the case of $R' = 0.013m$, $E' = 211GPa$, is considered with $\alpha = 15GPa^{-1}$.

Also, given in Table 19.2, are the scale, elasticity and piezo-viscous sensitivities $\triangle \ln h_c / \triangle \ln R'$, $\triangle \ln h_c / \triangle \ln E'$, and $\triangle \ln h_c / \triangle \ln \alpha$, obtained from the examples at each value of G. These sensitivities, as evident from Table 19.2, increase with decreasing values of G. This occurs because, as G is reduced, the shear-thinning response begins at the lower values of stress, which occur earlier in the inlet zone and more of the inlet experiences shear-thinning.

The trends observed in Table 19.2 are the same as concluded from the perturbation analysis above for which the scale, elasticity and piezo-viscous sensitivities $\triangle \ln h_c / \triangle \ln R'$, $\triangle \ln h_c / \triangle \ln E'$, and $\triangle \ln h_c / \triangle \ln \alpha$ are given in Table 19.3. Shear-thinning increases the sensitivity to changes in scale, and elasticity and pressure-viscosity coefficient. The classical Newtonian formulas for central film thickness from Dowson and Toyoda [13] for line contact

TABLE 19.2
The results of a full line contact numerical solution using the Carreau
form with $n = 0.4$ (after Kudish, Kumar, Khonsari, and Bair [9]).
Reprinted with permission from the ASME.

$G \ [Pa]$	$1 \cdot 10^9$	$4 \cdot 10^6$	$1 \cdot 10^6$	$3 \cdot 10^5$	$1 \cdot 10^5$
$R' = 0.013m, \ E' = 110GPa, \ \alpha = 25GPa^{-1}$ (the scale and elasticity modulus of an EHL simulator)					
$h_c \ [nm]$	577	576	561	443	257
$R' = 0.013m, \ E' = 211GPa \ \alpha = 25GPa^{-1}$					
$h_c \ [nm]$	606	605	588	467	273
$R' = 0.0013m, \ E' = 211GPa, \ \alpha = 25GPa^{-1}$ (the scale and elasticity modulus of a small roller bearing)					
$h_c \ [nm]$	267	265	235	150	79
$R' = 0.013m, \ E' = 211GPa, \ \alpha = 15GPa^{-1}$					
$h_c \ [nm]$	457	456	441	336	189
$\frac{\triangle \ln h_c}{\triangle \ln R'}$	0.35	0.36	0.40	0.49	0.54
$\frac{\triangle \ln h_c}{\triangle \ln E'}$	0.074	0.074	0.074	0.082	0.091
$\frac{\triangle \ln h_c}{\triangle \ln \alpha}$	0.55	0.55	0.56	0.65	0.72

and Hamrock and Dowson [14] for circular point contact yield the scale and
elasticity sensitivities shown in Table 19.4 for comparison. Notice again that,
here, the contact pressure rather than the normal force is used to describe the
loading.

TABLE 19.3
The sensitivities of the formulas obtained by the perturbation
analysis (after Kudish, Kumar, Khonsari, and Bair [9]).
Reprinted with permission from the ASME.

	Newtonian rheology, pre-critical regime	Newtonian rheology, over-critical regime	Power law rheology, pre-critical regime	Power law rheology, over-critical regime
$\frac{\triangle \ln h_c}{\triangle \ln R'}$	0.4	0.25	0.625 $\left(\frac{2}{3n+2}\right)$	0.455 $\left(\frac{1}{3n+1}\right)$
$\frac{\triangle \ln h_c}{\triangle \ln E'}$	-0.2	0.25	-0.313 $\left(-\frac{1}{3n+2}\right)$	0.455 $\left(\frac{1}{3n+1}\right)$
$\frac{\triangle \ln h_c}{\triangle \ln \alpha}$	0, A_f depends on α	0.75	0, A_f depends on α	1.364 $\left(\frac{3}{3n+1}\right)$

TABLE 19.4

The sensitivities of the classical Newtonian formulas
(after Kudish, Kumar, Khonsari, and Bair [9]).
Reprinted with permission from the ASME.

	$\frac{\triangle \ln h_c}{\triangle \ln R'}$	$\frac{\triangle \ln h_c}{\triangle \ln E'}$	$\frac{\triangle \ln h_c}{\triangle \ln \alpha}$
Newtonian line contact	0.31	0.070	0.56
Newtonian point contact	0.33	0.061	0.53

19.5 Application of Numerical Results to an EHL Experiment

A similar analysis of an experimental extrapolation of film thickness to that presented for the perturbation result above can be done for the numerical results of Table 19.1. Suppose we make a measurement of central film thickness h_m in an optical EHL rig simulator, for the same values of p_H and $0.5(u_1+u_2)$ as for some system of machine elements, but at an experimental scale of $R' = R_m$ in order to calculate an effective $\alpha = \alpha_e$ from some Newtonian film thickness formula. For the calculation of α_e, we assume Newtonian response and use the sensitivity to α from Table 19.3 ($1/0.56 = 1.79$) to calculate $\alpha_e = \alpha(h_m/h_N)^{1.79}$, where h_N is taken to be the film thickness for the case of $G = 1 \cdot 10^9 Pa$. The effective pressure viscosity coefficients α_e are listed in Table 19.5 for each of the lower values of G. The film thickness for the small roller bearing is now extrapolated from the film thicknesses h_m that would be obtained in the experimental rig by assuming Newtonian response

$$h_{sc} = h_m \left(\frac{R_s}{R_m}\right)^{0.35} \left(\frac{E_s}{E_m}\right)^{0.074} = h_m \left(\frac{0.0013}{0.013}\right)^{0.35} \left(\frac{211}{110}\right)^{0.074}. \qquad (19.25)$$

The extrapolated film thicknesses are listed in Table 19.5 along with the ratio of calculated to actual film thickness h_{sc}/h_s. The extrapolated film thickness may be substantially greater than the actual film thickness.

Let us sum up the results of this analysis. We have shown by a perturbation analysis and by a full EHL numerical solution that the practice of extrapolating from a laboratory scale measurement of film thickness to the film thickness of an operating contact within a real machine may substantially overestimate the film thickness in the real machine if the scale is smaller. This observation points to the need for a thorough experimental validation of the classical Newtonian film thickness calculation using accurate pressure viscosity data for a wide variety of liquid lubricants so that the limits of applicability of Newtonian calculations can be established.

TABLE 19.5
The ratio of calculated to actual film thickness
for extrapolation to smaller scale and greater
elastic modulus (after Kudish, Kumar,
Khonsari, and Bair [9]). Reprinted with
permission from the ASME.

$G \ [Pa]$	$4 \cdot 10^6$	$1 \cdot 10^6$	$3 \cdot 10^5$	$1 \cdot 10^5$
$\alpha \ [GPa^{-1}]$	25.0	23.8	15.6	5.90
$h_{sc}[nm]$	270	263	208	121
h_{sc}/h_s	1.02	1.12	1.39	1.53

19.6 Closure

A number of classic and modern EHL problems are considered. In particular, EHL problems for lubricants with Newtonian and non-Newtonian rheology are considered. The analysis is done using regular and matched asymptotic expansions as well as numerical methods. Lubricated contacts under light and heavy loading are considered. Lightly loaded EHL contacts are considered asymptotically and numerically and their results are compared. Asymptotic analysis of lightly loaded EHL contact produced formulas for film thickness and sliding and rolling frictional forces. Asymptotic analysis of heavily loaded contact lubricated by a Newtonian fluid recognizes two different regimes of lubrication: pre- and over-critical lubrication regimes. Overall, the difference between these two regimes is in how strongly the lubricant viscosity depends on pressure and how much lubricant is available at the inlet of the contact. Two different asymptotic approaches to the analysis of the pre- and over-critical regimes are developed while both are based on matched asymptotic expansions. A detailed structure of the inlet and exit zones and the Hertzian region of a lubricated contact is analyzed. A number of asymptotically based formulas for the film thickness in heavily loaded EHL contacts are derived for the cases of Newtonian and non-Newtonian lubricant rheologies. The classic Ertel-Grubin method is analyzed and its shortcomings are revealed. The numerical methods for EHL problems in the original and asymptotic formulations are proposed and these problems are solved numerically in isothermal and non-isothermal formulations. The issues of numerical precision and stability of the solutions for heavily loaded lubricated contacts of the original and asymptotic problems are discussed. A simple and effective regularization approach to solution of the original and asymptotic problems is proposed. To explain the dimple phenomenon, the elastic tangential displacements of the lubricated contact surfaces are employed. For lubricants with non-Newtonian rheologies, conditions of pure rolling and large sliding are considered. A call for caution in extrapolating the results from experimentally obtained in optical

EHL simulators data on pressure viscosity coefficient to the scale of practical applications is made.

19.7　Exercises and Problems

1. Explain the essence of the scale effect which occurs while extrapolating from measurements made using an optical EHL simulator to practical cases.

2. Describe specifically what changes in the extrapolation procedures should be made in order to get reasonably accurate results when an extrapolation of lubrication film thickness optical measurements is done to approximated contacts of larger/smaller size and with different Young's modulus.

References

[1] Dowson, D. and Higginson, G.R. 1966. *Elastohydrodynamic Lubrication*. London: Pergamon Press.

[2] Bair, S. 2007. *High − Pressure Rheology for Quantitative Elasto − hydrodynamics*. Amsterdam: Elsevier Science.

[3] Hamrock, B.J. and Dowson, D. 1977. Isothermal Elastohydrodynamic Lubrication of Point Contacts. Part III-Fully Flooded Results. *J. Lubr. Techn.* 99, No. 2:264-276.

[4] Jones, W.R. 1995. Properties of Perfluoropolyethers for Space Applications. *STLE Tribology Trans.* 38, No. 3:557-564.

[5] Bair, S., Vergne, P., and Marchetti, M. 2002. The Effect of Shear-Thinning on Film Thickness for Space Lubricants. *STLE Tribology Trans.* 45, No. 3:330-333.

[6] Dyson, A. and Wilson, A. R. 1965-6. Film Thicknesses in Elastohydrodynamic Lubrication by Silicone Fluids. *Proc. Instn. Mech. Engrs* 180:97-112.

[7] Chapkov, A. D., Bair, S., Cann, P., and Lubrecht, A. A. 2007. Film Thickness in Point Contacts under Generalized Newtonian EHL Conditions: Numerical and Experimental Analysis. *Tribology Intern.* 40:1474-1478.

[8] Liu, Y., Wang, Q. J., Bair, S., and Vergne, P. 2007. A Quantitative Solution for the Full Shear-Thinning EHL Point Contact Problem Including Traction. *Tribology Letters* 28, No. 2:171-181.

[9] Kudish, I.I., Kumar, P., Khonsari, M.M., and Bair, S. 2008. Scale Effects in Generalized Newtonian Elastohydrodynamic Films. *ASME J. Tribology*, 130, No. 4:041504-1 - 041504-8.

[10] Ertel, M.A. 1945. Hydrodynamic Calculation of Lubricated Contact for Curvilinear Surfaces. *Proc. CNIITMASh* :1-64.

[11] Grubin, A.N. 1949. The Basics of the Hydrodynamic Lubrication Theory for Heavily Loaded Curvilinear Surfaces. *Proc. CNIITMASh* 30:126-184.

[12] Jang, J.Y., Khonsari, M.M., and Bair, S. 2007. On the Elastohydrodynamic Analysis of Shear-thinning Fluids. *Proc. Royal Soc.* 463:3271-3290.

[13] Dowson, D. and Toyoda, S. 1979. A Central Film Thickness Formula for Elastohydrodynamic Line Contacts. In Proc. of Leeds-Lyon Symp. on Tribology, 5, September 19-22, 1978, Eds. D. Dowson et al., 60-65.

[14] Hamrock, B.J. and Dowson, D. 1981. *Ball Bearing Lubrication — The Elastohydrodynamics of Elliptical Contacts*. New York: Wiley-Interscience.

20

Lubrication of Line Contacts by Greases

20.1 Introduction

Grease is a semi-solid, high viscosity type of lubricant that is well-suited for use in bearings, couplings, and open gears where shock loads, high temperatures, and/or good adhesion to bearing surfaces are important performance features. Consider the problem of journal bearing lubrication in Fig. 20.1. The outer cylinder (journal) is fixed and is separated from the center rotating shaft by a layer of grease. When the shaft is at rest or turning at low speeds and/or high loads, metal-to-metal surface contact can occur. Wear protection under these conditions can be provided by lubricant decomposition products or surface-active additives which form thin, soft tribo-films which retard metal-to-metal adhesion and reduce friction. As the shaft begins to rotate at higher speeds, it climbs the journal surface in a direction opposite to the direction of rotation. Layers of grease cling to the journal and rotating shaft surfaces, the former remaining stationary and the latter moving in concert with the shaft. Additional grease is carried into the contact zone, and the system enters the hydrodynamic lubrication regime.

The choice between oil and grease lubrication depends upon the ratio of journal speed to viscosity. In general, contacts moving at relatively low relative velocity have higher viscosity requirements than surfaces moving at high velocity. Bearings designated for low speed operation usually are designed with relatively large clearance between the shaft and journal housing to facilitate introduction of high viscosity grease lubricants. High-speed bearings with small clearance are lubricated with lower viscosity oils.

Other bearings commonly lubricated by grease include rolling-element bearings such as ball bearings, cylindrical roller bearings, tapered roller bearings, spherical barrel-shaped roller bearings, and needle bearings. In these types of bearings, extremely high local pressures are formed between the relatively small rotating rolling elements and their raceways (support housings). We learned in Chapter 1 that lubricant viscosity increases rapidly at high pressures that enable the lubricating film to withstand high contact stresses while preventing contact between the rolling surfaces.

Greases are complex formulations consisting of base fluids, thickeners, structural components, and additives designed to meet specific application

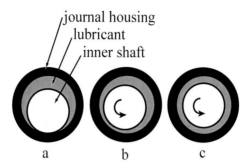

FIGURE 20.1

Development of a hydrodynamic film with increased shaft speed in a plain journal bearing: (a) at rest, (b) low rotational velocity, and (c) high rotational velocity. (after Kudish and Covitch [1]). Reprinted with permission from CRC Press.

requirements. Typical base fluids include mineral and synthetic oils. Thickeners and structural components impart shear-thinning non-Newtonian rheological properties to the grease (see Chapter 1). At rest, the lubricant is a semi-solid with extremely high apparent viscosity. As shear rate increases, the apparent viscosity falls. Thus, within the gap between a stationary journal and rotating bearing shaft, the lubricating film consists of layers of high viscosity semi-solid at the journal surface and progressively lower viscosity layers closer to the shaft surface.

Thickening agents are prepared in the presence of a base fluid by reacting a fatty acid or ester (from either animal or vegetable sources) with alkali or alkaline earth metal oxides or hydroxides. This so-called saponification process is conducted under controlled conditions of heat, pressure, and agitation. Greases are often referred to by the type of alkali or alkaline earth metal in its thickening system: sodium, calcium, lithium, aluminum, etc. The saponification process produces a three-dimensional microscopic soap fiber structure suspended in the base fluid that serves as an internal scaffolding and defines the rheology and consistency of the grease. Rheology is affected by a number of thickener-related parameters such as soap type and amount, fatty acid/ester chain length, degree of hydrocarbon branching and unsaturation (presence of double bonds) of the fatty acid/ester, presence of other polar substances (such as water), and particle size.

Alkaline Complex Greases are formulated to withstand higher operating temperatures. They are prepared by adding a lower molecular weight organic acid, also called a complexing agent, to the thickening agent mixture. For example, co-reacting 12-hydroxystearic acid (fatty acid) and azelaic acid (complexing agent) with lithium hydroxide (alkaline hydroxide) produces a more intricate lattice structure than that of a simple lithium soap.

TABLE 20.1

Applications of simple and complex greases. Dropping point - the temperature at which grease becomes soft enough to form a drop and fall, ASTM D566 and D2265; EP - Extreme Pressure properties (wear resistance under boundary lubrication); Water resistance - the ability of a grease to resist the adverse effects of water, ASTM D1264, D4049 or D1743. (after Kudish and Covitch [1]). Reprinted with permission from CRC Press.

Thickener Type	Grease Characteristics	Applications
Aluminum	Smooth, gel-like appearance Low dropping point Excellent water resistance Softening/hardening tendencies Highly shear thinning	Low-speed bearings Wet applications
Sodium	Rough, fibrous appearance Moderately high dropping point Poor water resistance Good adhesive (cohesive) properties	Older industrial equipment with frequent re-lubrication Rolling-element bearings
Calcium	Smooth, buttery appearance Low dropping point Good water resistance	Bearings in wet applications Railroad rail lubricants
Lithium	Smooth, buttery to slightly stringy appearance High dropping point Resistant to softening and leakage Moderate water resistance	Automotive chassis and wheel bearings General industrial grease Thread lubricants for oil drilling
Calcium Complex	Smooth, buttery appearance Dropping points above $500°F$ Good water resistance Inherent EP/load-carrying capability	High-temperature industrial and automotive bearings
Aluminum Complex	Smooth, slight gel-like appearance Dropping points below $500°F$ Good water resistance Resistant to softening Shorter life at high temperature	Steel mill roll neck, rolling, and plain bearings
Lithium Complex	Smooth, buttery appearance Dropping points above $500°F$ Resistant to softening and leakage Moderate water resistance	Automotive wheel bearings High-temperature industrial service incl. various rolling-element applications

Other chemical additives are formulated into grease to prevent oxidation, combat corrosion, and reduce friction and wear. Another set of chemically inert ingredients are added to modify viscosity, consistency, or tolerance to water. Examples include viscosity modifiers, pour point depressants, antifoam agents, emulsifiers, and demulsifiers.

Each grease application has its own set of requirements, and each type of grease has its own set of physical and performance properties. Selecting the right grease for a specific end-use is a matter of art, experience, and testing. A summary of industrial and automotive application areas for several families of grease may be found in Table 20.1.

In this chapter we will formulate and study isothermal and non-isothermal problems for EHL contacts in which grease was used as a lubricating medium. We will provide a general qualitative analysis of these problems and determine the conditions under which the problems for grease lubricated contacts can be reduced to the corresponding problems for contact lubricated by fluid lubricants. We will consider an analytical solution of a problem for moving rigid solids separated by grease under certain conditions as well as the asymptotic approach to isothermal and non-isothermal heavily loaded EHL contacts with greases.

20.2   Formulation of the Elastohydrodynamic Lubrication Problem for Greases (Generalized Bingham-Shvedov Viscoplastic Lubricants)

Let us formulate the basic equations pertaining to grease lubricated contacts under steady conditions. We will consider a plane non-isothermal problem about a slow motion of two infinite parallel cylinders separated by a continuous layer of incompressible grease. As in the preceding chapter, we will assume that the cylinders of radii R_1 and R_2 are made of elastic materials with Young's moduli E_i and Poisson's ratios ν_i, $i = 1, 2$. The cylinders' surfaces move with linear velocities u_1 and u_2. A normal force P presses one cylinder into another. Additionally, we will assume that the film thickness is much smaller than the size of the contact region, which, in turn, is much smaller than the cylinders' radii R_1 and R_2. The inertial forces are much smaller than the viscous ones, the heat is generated because of grease viscosity and applied shear stresses while the heat flux is mostly directed across the film thickness. Then in the coordinate system in which the x-axis is directed along the lubrication layer perpendicular to cylinders' axes and the z-axis is directed through cylinders' centers the simplified Navier-Stokes and energy equations of grease motion

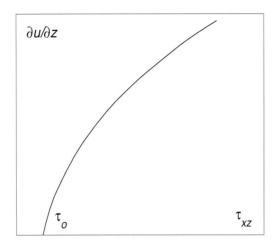

FIGURE 20.2
The general view of a rheological relationship for grease.

take the form (see equations (4.10) and (4.12))

$$\frac{\partial \tau_{xz}}{\partial z} = \frac{\partial p}{\partial x}, \ \frac{\partial p}{\partial z} = 0, \ \frac{\partial u}{\partial x} + \frac{\partial w}{\partial z} = 0, \tag{20.1}$$

$$\frac{\partial}{\partial z}\left(\lambda \frac{\partial T}{\partial z}\right) = -\tau_{xz}\frac{\partial u}{\partial z}, \tag{20.2}$$

where $p(x)$ is pressure, $\tau_{xz}(x, z)$ is the component of the shear stress in grease, $u(x, z)$ and $w(x, z)$ are the components of grease velocity along the lubrication film and across it, $T(x, z)$ is the grease temperature, $\lambda(p, T)$ is the grease heat conductivity coefficient.

Taking into account the above assumptions, the rheological relationships for greases can be represented as follows

$$\tau_{xz} = \tau_0 sign(\tau_{xz}) + \Phi\left(\mu\frac{\partial u}{\partial z}\right) \ if \ \mid \tau_{xz} \mid \geq \tau_0, \tag{20.3}$$

$$\frac{\partial u}{\partial z} = 0 \ if \ \mid \tau_{xz} \mid = 0, \tag{20.4}$$

where τ_0 is the threshold shear (yield) stress beyond which grease flows like a fluid lubricant, $\tau_0 = \tau_0(p, T) \geq 0$, μ is the grease viscosity, $\mu = \mu(p, T)$, Φ is a monotonically increasing sufficiently smooth rheological function, $\Phi(0) = 0$. The general view of such a rheological relationship is presented in Fig. 20.2.

Condition (20.4) describes the so–called core in the grease flow; i.e., it describes the region in the flow in which relative displacement of grease layers is absent. This region of the flow behaves like a rigid solid. Obviously, for

$\tau_0 = 0$, there are no cores in grease flow and the rheological model represented by equations (20.3) and (20.4) gets reduced to the rheological model of generalized non-Newtonian fluid described by equations (9.1).

As an example we can consider the Bingham-Shvedov model of grease [2, 3]

$$\tau_{xz} = \tau_0 sign(\tau_{xz}) + \mu\frac{\partial u}{\partial z} \ if \ \mid \tau_{xz} \mid \geq \tau_0; \ \frac{\partial u}{\partial z} = 0 \ if \ \mid \tau_{xz} \mid = 0. \qquad (20.5)$$

Let us derive the equations governing pressure $p(x)$ and temperature $T(x, z)$ [4, 5]. Integrating the first equation in (20.1), we find

$$\tau_{xz} = f + z\frac{dp}{dx}, \qquad (20.6)$$

where $f = f(x)$ is the sliding frictional stress. Moreover, integrating the continuity equation (last equation in (20.1)) and taking into account the no-slip conditions on solid surfaces $z = \pm\frac{h}{2}$ (see (5.3))

$$u(x, -\tfrac{h}{2}) = u_1, \ u(x, \tfrac{h}{2}) = u_2,$$

$$w(x, (-1)^j \tfrac{h}{2}) = (-1)^j \tfrac{u_i}{2}\tfrac{dh}{dx}, \ j = 1, 2, \qquad (20.7)$$

we obtain the generalized Reynolds equation in the form

$$\frac{d}{dx} \int\limits_{-h/2}^{h/2} u(x, z)dz = 0. \qquad (20.8)$$

To complete the problem formulation, we need to add to equations (20.3), (20.4), (20.6), and (20.8) the conditions of continuity of grease velocity at core boundaries in the flow. From monotonicity of function Φ and equations (20.3) and (20.6) follows that for any fixed x the derivative $\frac{\partial u}{\partial z}$ is equal to zero at no more than two values of z within the lubrication layer $\mid z \mid \leq \frac{h}{2}$. Therefore, at any x in the flow may be encountered at most one core region and all possible types of grease flow in any cross section of the flow are presented in Fig. 20.3 (cores are marked by 1 while flow with shear is marked by 2). We will call these types of flow configurations as flow of types I, IIa, IIb, and III, respectively. The no-slip conditions at the solid surfaces and the continuity conditions at the core boundaries need to be imposed on the grease velocity $u(x, z)$ for each of these flow types. For flow of type I (the core is located in the central region of the flow), we have

$$u(x, -\tfrac{h}{2}) = u_1, \ u(x, \tfrac{h}{2}) = u_2, \ u(x, z_1) = u(x, z_2),$$

$$-\tfrac{h}{2} < z_1 < z_2 < \tfrac{h}{2}, \qquad (20.9)$$

for flow of type IIa (the core is adjacent to the upper solid), we have

$$u(x, -\tfrac{h}{2}) = u_1, \ u(x, z_1) = u_2, \ -\tfrac{h}{2} < z_1 < \tfrac{h}{2} \leq z_2, \qquad (20.10)$$

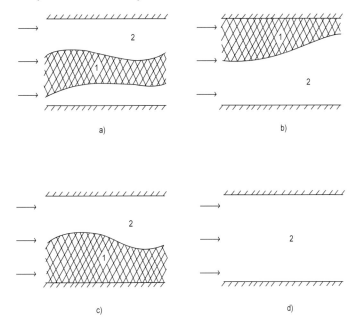

FIGURE 20.3
All possible grease flow configurations in any cross section of the flow: (a) type I, (b) type IIa, (c) type IIb, and (d) type III (1 marks grease core, 2 marks shear flow) (after Kudish [4]). Reprinted with permission from Allerton Press.

for flow of type IIb (the core is adjacent to the lower solid), we have

$$u(x, z_2) = u_1, \ u(x, \tfrac{h}{2}) = u_2, \ z_1 \leq -\tfrac{h}{2} < z_2 < \tfrac{h}{2}, \qquad (20.11)$$

for flow of type III (there is no core in the flow), we obtain

$$u(x, -\tfrac{h}{2}) = u_1, \ u(x, \tfrac{h}{2}) = u_2,$$

$$z_1 \leq z_2 \leq -\tfrac{h}{2} \ or \ -\tfrac{h}{2} < z_1 = z_2 < \tfrac{h}{2} \ or \ \tfrac{h}{2} \leq z_1 \leq z_2, \qquad (20.12)$$

where $z_1 = z_1(x)$ and $z_2 = z_2(x)$ are the lower and upper core boundaries, which are determined by equations (see (20.4))

$$\tfrac{\partial u(x, z_i)}{\partial z} = 0, \ i = 1, 2. \qquad (20.13)$$

Now, let us determine the expressions for functions $u(x, z)$ and $f(x)$. The grease flow between the solid surfaces we will treat as just a fragment of the flow in the entire xz-plane. For certainty we will assume that $u_2 - u_1 \geq 0$.

Let us consider all possible types of grease flow in an arbitrary lubrication film cross–section x and the conditions under which they are realized. First, we will consider the case when in a chosen cross–section x type I flow is realized. It can be shown that on different sides of the core in this cross section the values of $\frac{\partial u}{\partial z}$ has opposite signs. Let us analyze the case when $\frac{\partial u}{\partial z} > 0$ for $z < z_1$ and $\frac{\partial u}{\partial z} < 0$ for $z > z_2$. Taking into account that $\tau_{xz} \frac{\partial u}{\partial z} \geq 0$ from equations (20.3) and (20.6), we obtain

$$\frac{\partial u}{\partial z} = \frac{1}{\mu} F(f - \tau_0 + z\frac{dp}{dx}), \quad z \leq z_1,$$

$$\frac{\partial u}{\partial z} = \frac{1}{\mu} F(f + \tau_0 + z\frac{dp}{dx}), \quad z \geq z_2,$$

(20.14)

where F is the rheological function of the liquified grease which is a mono-tonically increasing function being the inverse to function Φ, $F(0) = 0$.

The core boundaries z_1 and z_2 can be found from equations (20.13) and (20.14)

$$z_1 = \frac{\tau_0 - f}{\frac{\partial p}{\partial x}}, \quad z_2 = -\frac{\tau_0 + f}{\frac{\partial p}{\partial x}}. \tag{20.15}$$

Moreover, using boundary conditions (20.9) and equations (20.14), we obtain the expressions for the grease velocity $u(x, z)$:

$$u(x, z) = u_1 + \int_{-h/2}^{z} \frac{1}{\mu} F(f - \tau_0 + \zeta\frac{dp}{dx}) d\zeta, \quad z \leq z_1;$$

$$u(x, z) = u(x, z_1), \quad z_1 \leq z \leq z_2, \tag{20.16}$$

$$u(x, z) = u_2 - \int_{z}^{h/2} \frac{1}{\mu} F(f + \tau_0 + \zeta\frac{dp}{dx}) d\zeta, \quad z \geq z_2.$$

From the third condition in (20.9) and equations (20.15) and (20.16), we derive the equation for $f(x)$:

$$\int_{-h/2}^{z_1} \frac{1}{\mu} F(f - \tau_0 + \zeta\frac{dp}{dx}) d\zeta + \int_{z_2}^{h/2} \frac{1}{\mu} F(f + \tau_0 + \zeta\frac{dp}{dx}) d\zeta = u_2 - u_1. \tag{20.17}$$

Employing the expressions for z_1 and z_2 from (20.15), we obtain $\frac{dp}{dx} < 0$. Finally, it follows from (20.9) and (20.15) that the flow of type I can be realized if the following inequality is satisfied

$$\tau_0 + \frac{h}{2}\frac{dp}{dx} < f < -\tau_0 - \frac{h}{2}\frac{dp}{dx}. \tag{20.18}$$

Now, let us derive the Reynolds and the heat transfer equations. Integrating equations (20.16) for $u(x, z)$ with respect to z over the corresponding intervals and using the generalized Reynolds equation (20.8), we obtain

$$\frac{d}{dx}\Big\{ \int_{-h/2}^{z_1} dz \int_{-h/2}^{z} \frac{1}{\mu} F(f - \tau_0 + \zeta\frac{dp}{dx}) d\zeta$$

$$- \int\limits_{z_2}^{h/2} dz \int\limits_{z}^{h/2} \frac{1}{\mu} F(f + \tau_0 + \zeta \frac{dp}{dx}) d\zeta$$

(20.19)

$$+ u(x, z_1)(z_2 - z_1) + u_1 z_1 - u_2 z_2 + \frac{u_1 + u_2}{2} h \bigg\} = 0.$$

By changing the order of integration in integrals involved in (20.19) and using equation (20.17), we reduce equation (20.19) to the final form

$$\frac{d}{dx} \bigg\{ \int\limits_{-h/2}^{z_1} \frac{z}{\mu} F(f - \tau_0 + z \frac{dp}{dx}) dz$$

(20.20)

$$+ \int\limits_{z_2}^{h/2} \frac{z}{\mu} F(f + \tau_0 + z \frac{dp}{dx}) dz - \frac{u_1 + u_2}{2} h \bigg\} = 0.$$

It is clear that within the core $\frac{\partial u}{\partial z} = 0$ and from equation (20.2) follows that within the core heat is not generated and the heat flux is constant. Therefore, from equations (20.2), (20.5), (20.6), and (20.16) for the energy equation, we obtain

$$\frac{\partial}{\partial z}(\lambda \frac{\partial T}{\partial z}) = -\frac{1}{\mu} F[f + \tau_0 sign(z - z_1) + z \frac{dp}{dx}](f + z \frac{dp}{dx})$$

(20.21)

$$\times \theta[(z - z_1)(z - z_2)],$$

where $\theta(x)$ is the Heavyside function such that $\theta(x) = 1$ for $x \geq 0$ and $\theta(x) = 0$ for $x \leq 0$.

In a similar fashion we can analyze the case when $\frac{\partial u}{\partial z} < 0$ for $z < z_1$ and $\frac{\partial u}{\partial z} > 0$ for $z > z_2$. Under these conditions the equations for z_1, z_2, f, and $u(x, z)$ as well as the generalized Reynolds equation coincide with equations (20.14), (20.17), (20.16), and (20.20) if τ_0 is replaced by $-\tau_0$. It can be shown that in this case $\frac{dp}{dx} > 0$. Moreover, the condition for realization of the above case coincide with (20.18) in which $\frac{dp}{dx}$ is replaced by $-\frac{dp}{dx}$.

Now, let us consider the case when in a chosen cross–section x type IIa flow is realized. It can be shown that $\frac{\partial u}{\partial z} > 0$ for $z < z_1$. Then for z_1 and z_2 we obtain formulas (20.15) while for grease velocity $u(x, z)$ we get

$$u(x, z) = u_1 + \int\limits_{-h/2}^{z} \frac{1}{\mu} F(f - \tau_0 + \zeta \frac{dp}{dx}) d\zeta, \quad -\frac{h}{2} \leq z \leq z_1,$$

(20.22)

$$u(x, z) = u_1 2, \quad z_1 \leq z \leq \frac{h}{2}.$$

Then, from equations (20.10), (20.14), and (20.22), we obtain the equation for $f(x)$:

$$\int\limits_{-h/2}^{z_1} \frac{1}{\mu} F(f - \tau_0 + \zeta \frac{dp}{dx}) d\zeta = u_2 - u_1.$$

(20.23)

From the inequality $z_1 < z_2$ and equations (20.15), we obtain that $\frac{dp}{dx} < 0$. Equations (20.15) together with inequalities in (20.10) lead to the requirement for the realization of type IIa grease flow in the form

$$\tau_0 + \frac{h}{2}\frac{dp}{dx}) < f < \tau_0 - \frac{h}{2}\frac{dp}{dx}, \quad -\tau_0 - \frac{h}{2}\frac{dp}{dx} \leq f. \tag{20.24}$$

Using the generalized Reynolds equation (20.8) and integrating the expressions for the grease velocity $u(x,z)$ from (20.22) with respect to z from $-\frac{h}{2}$ to $\frac{h}{2}$, we derive the equation

$$\frac{d}{dx}\bigg\{ \int\limits_{-h/2}^{z_1} dz \int\limits_{-h/2}^{z} \frac{1}{\mu}F(f - \tau_0 + \zeta\frac{dp}{dx})d\zeta + (u_1 - u_2)z_1$$

$$+ \frac{u_1+u_2}{2}h \bigg\} = 0. \tag{20.25}$$

By changing the order of integration in integrals involved in (20.25) and using equation (20.23), we reduce equation (20.25) to the final form

$$\frac{d}{dx}\bigg\{ \int\limits_{-h/2}^{z_1} \frac{z}{\mu}F(f - \tau_0 + z\frac{dp}{dx})dz - \frac{u_1+u_2}{2}h \bigg\} = 0. \tag{20.26}$$

Grease flow of type IIb can be analyzed in a similar fashion. In this case $\frac{\partial u}{\partial z} > 0$ for $z > z_2$. Formulas (20.15) are still valid for z_1 and z_2 if τ_0 in them is replaced by $-\tau_0$. It can be shown that in this case $\frac{dp}{dx} > 0$. For the grease velocity $u(x,z)$, we find

$$u(x,z) = u_1, \quad -\frac{h}{2} \leq z \leq z_2,$$

$$u(x,z) = u_1 + \int\limits_{z}^{z_2} \frac{1}{\mu}F(f - \tau_0 + \zeta\frac{dp}{dx})d\zeta, \quad z_2 \leq z \leq \frac{h}{2}. \tag{20.27}$$

Using equations (20.11) and (20.22), we obtain the equation for $f(x)$:

$$\int\limits_{z_2}^{h/2} \frac{1}{\mu}F(f - \tau_0 + \zeta\frac{dp}{dx})d\zeta = u_2 - u_1. \tag{20.28}$$

Besides that, the conditions for the realization of type IIb grease flow follow from conditions (20.24) if $\frac{dp}{dx}$ in them is replaced by $-\frac{dp}{dx}$.

In the fashion described above, we obtain the generalized Reynolds equation

$$\frac{d}{dx}\bigg\{ \int\limits_{z_2}^{h/2} dz \int\limits_{z_2}^{z} \frac{1}{\mu}F(f - \tau_0 + \zeta\frac{dp}{dx})d\zeta + u_1 h \bigg\} = 0, \tag{20.29}$$

which by changing the order of integration in the integral can be rewritten in the final form

$$\frac{d}{dx}\bigg\{ \int\limits_{z_2}^{h/2} \frac{z}{\mu}F(f - \tau_0 + z\frac{dp}{dx})dz - \frac{u_1+u_2}{2}h \bigg\} = 0. \tag{20.30}$$

For types IIa and IIb grease flows, the energy equation is reduced to

$$\frac{\partial}{\partial z}(\lambda \frac{\partial T}{\partial z}) = -\frac{1}{\mu}F[f - \tau_0 sign(z - z_2) + z\frac{dp}{dx}](f + z\frac{dp}{dx})\theta(z - z_2). \quad (20.31)$$

Finally, let us consider the case of type III grease flow, i.e., the flow without a core. Then for $u_2 - u_1 \geq 0$ we have $\frac{\partial u}{\partial z} \geq 0$ for all $\mid z \mid \leq \frac{h}{2}$. The case of $-\frac{h}{2} < z_1 = z_2 < \frac{h}{2}$ can be realized when either $\tau_0 = 0$ or $\tau_0 > 0$ and $\frac{dp}{dx} = 0$. This follows from the fact that z_1 and z_2 satisfy (20.14) for $\frac{dp}{dx} > 0$ and the same formulas in which τ_0 is replaced by $-\tau_0$ for $\frac{dp}{dx} < 0$. Therefore, for the first and third cases in (20.12) for $u(x, z)$ and $f(x)$, we find

$$u(x, z) = u_1 + \int\limits_{-h/2}^{z} \frac{1}{\mu}F(f - \tau_0 + \zeta\frac{dp}{dx})d\zeta, \mid z \mid \leq \frac{h}{2}, \quad (20.32)$$

$$\int\limits_{-h/2}^{h/2} \frac{1}{\mu}F(f - \tau_0 + \zeta\frac{dp}{dx})d\zeta = u_2 - u_1. \quad (20.33)$$

The condition for the realization of this type of flow is

$$\mid \frac{dp}{dx} \mid \leq \frac{2}{h}(f - \tau_0). \quad (20.34)$$

The generalized Reynolds equation will assume the form

$$\frac{d}{dx}\left\{ \int\limits_{-h/2}^{h/2} dz \int\limits_{-h/2}^{z} \frac{1}{\mu}F(f - \tau_0 + \zeta\frac{dp}{dx})d\zeta + u_1 h \right\} = 0, \quad (20.35)$$

which after changing the order of integration in the integral can be rewritten in the final form

$$\frac{d}{dx}\left\{ \int\limits_{-h/2}^{h/2} \frac{z}{\mu}F(f - \tau_0 + z\frac{dp}{dx})dz - \frac{u_1 + u_2}{2}h \right\} = 0. \quad (20.36)$$

The energy equation for grease flow of type III has the form

$$\frac{\partial}{\partial z}(\lambda \frac{\partial T}{\partial z}) = -\frac{1}{\mu}F[f - \tau_0 + z\frac{dp}{dx}](f + z\frac{dp}{dx}). \quad (20.37)$$

As it was done before (see the preceding chapter) to determine $h(x)$, we need to add to the derived system of equations and inequalities equation (5.8)

$$h(x) = h_e + \frac{x^2 - x_e^2}{2R'} + \frac{2}{\pi E'}\int\limits_{x_i}^{x_e} p(t)\ln\frac{x_e - t}{|x - t|}dt, \quad (20.38)$$

and the balance condition (5.10):

$$\int\limits_{x_i}^{x_e} p(t)dt = P, \quad (20.39)$$

where h_e is the lubrication film thickness at the exit point of the contact x_e, i.e., $h_e = h(x_e)$, x_i is the inlet point of the contact, R' and E' are the effective radius and elasticity modulus, $1/R' = 1/R_1 \pm 1/R_2$ (signs $+$ and $-$ are chosen in accordance with cylinders' curvatures) and $1/E' = 1/E'_1 + 1/E'_2$, $E'_j = E_j/(1 - \nu_j^2)$, $(j = 1, 2)$, P is the normal force applied to the cylinders.

We also need to add some boundary conditions. Within the lubricated contact, pressure is much higher than the ambient pressure. Moreover, to prevent the lubricant from cavitation at the exit of the contact, we need to require the pressure gradient to be zero (see the preceding chapter). Therefore, we impose the following boundary conditions on pressure (see conditions (5.9)):

$$p(x_i) = p(x_e) = \tfrac{dp(x_e)}{dx} = 0, \tag{20.40}$$

where the inlet coordinate x_i is considered to be known. The conditions for the temperature at the solid surfaces have the form (see (5.47))

$$T(x, -\tfrac{h}{2}) = T_{w1}(x), \ T(x, \tfrac{h}{2}) = T_{w2}(x), \tag{20.41}$$

where $T_{w1}(x)$ and $T_{w2}(x)$ are the temperatures of the lower and upper solid surfaces, respectively.

However, to solve the EHL problem for grease lubricated contact, some additional conditions are required. In particular, to complete the problem formulation at every point x_c of flow type change and at the core boundaries $z = z_1$ and $z = z_2$, it is necessary to impose some continuity conditions. These conditions follow from the requirement that the functions of pressure, gap, temperature, shear stress, and flow velocity are continuous in the entire contact region. Therefore, at the points with coordinates x_c at which a change of the flow type occurs, we have these conditions in the form

$$[p(x_c)] = [\tfrac{dp(x_c)}{dx}] = 0, \ [T(x_c, z)] = 0, \ | \ z \ | \le \tfrac{h}{2}. \tag{20.42}$$

The remaining continuity conditions on core boundaries (i.e., on the core/flow interfaces) are obtained from the requirement of continuity of temperature and heat flux and they have the form

$$[T(x, z_j)] = [\tfrac{\partial T(x, z_j)}{\partial z}] = 0, \ j = 1, 2. \tag{20.43}$$

In equations (20.42) and (20.43), the notation $[f(x_0)]$ means the jump of function $f(x)$ at $x = x_0$, i.e., $[f(x_0)] = f(x_0 + 0) - f(x_0 - 0)$.

Therefore, for the known constants x_i, u_1, u_2, R', E', P, and functions F, μ, λ, τ_0, T_{w1}, and T_{w2}, the formulated problem for pressure $p(x)$, gap $h(x)$, frictional sliding stress $f(x)$, grease temperature $T(x, z)$, and two constants: the exit film thickness h_e and exit coordinate x_e is completely formulated. In addition to that, in case of changes of grease flow type, the coordinates of points x_c at which these changes occur are also determined.

The formulated problem is a complex nonlinear problem with a number of free (unknown) boundaries. In different zones of the contact region, different flow types are realized that are described by different equations. Under certain conditions the solution of the problem in this formulation does not exist and requires certain adjustments (see the next section).

After the problem solution is found the friction force F_T, which depends on the sliding F_S and rolling F_R friction forces, can be found from formulas (for comparison see formulas (5.12) valid for $\tau_0 = 0$)

$$F_T = F_S \pm F_R, \quad F_S = \int\limits_{x_i}^{x_e} f(x)dx, \quad F_R = \frac{1}{2}\int\limits_{x_i}^{x_e} h(x)\frac{dp}{dx}dx, \qquad (20.44)$$

Finally, let us consider the formulation of the isothermal problem for

$$\frac{\partial \mu}{\partial T} = \frac{\partial \tau_0}{\partial T} = 0, \quad \lambda = \infty, \quad T_{wj} = T_0 = const \ (j = 1, 2). \qquad (20.45)$$

In the case of the simplest viscoplastic Binham-Shvedov grease model (20.5), we obtain the following equations for the isothermal EHL problem

$$\frac{d}{dx}\{[-\frac{2}{3\mu}(\tau_0 - \frac{h}{2} \mid \frac{dp}{dx} \mid)^2(2\tau_0 + \frac{h}{2} \mid \frac{dp}{dx} \mid)(\frac{dp}{dx})^{-2}$$

$$+\frac{\mu(u_2-u_1)^2\tau_0}{2}(\tau_0 - \frac{h}{2} \mid \frac{dp}{dx} \mid)^{-2}]s_p - \frac{u_1+u_2}{2}h\} = 0, \quad s_p = sign(\frac{dp}{dx}),$$

$$\tag{20.46}$$

$$f = \frac{\mu(u_2-u_1)}{h}\frac{1}{1-\frac{2\tau_0}{h}(\mid \frac{dp}{dx} \mid)^{-1}},$$

$$\mid \frac{dp}{dx} \mid> \frac{2\tau_0}{h} + \frac{\mu(u_2-u_1)}{h^2} + \frac{1}{h}\sqrt{\frac{\mu(u_2-u_1)}{h}[\frac{\mu(u_2-u_1)}{h} + 4\tau_0]},$$

$$\frac{d}{dx}\{\frac{u_2-u_1}{3}\sqrt{2\mu(u_2 - u_1)}(\mid \frac{dp}{dx} \mid)^{-1}s_p + (\frac{u_1+u_2}{2}s_p + \frac{u_1-u_2}{2})h\} = 0,$$

$$f = \tau_0 - \frac{h}{2} \mid \frac{dp}{dx} \mid +\sqrt{2\mu(u_2 - u_1) \mid \frac{dp}{dx} \mid},$$

$$\tag{20.47}$$

$$\frac{2\mu(u_2-u_1)}{h^2} <\mid \frac{dp}{dx} \mid\leq \frac{2\tau_0}{h} + \frac{\mu(u_2-u_1)}{h^2}$$

$$+\frac{1}{h}\sqrt{\frac{\mu(u_2-u_1)}{h}[\frac{\mu(u_2-u_1)}{h} + 4\tau_0]},$$

$$\frac{d}{dx}\{\frac{h^3}{12\mu}\frac{dp}{dx} - \frac{u_1+u_2}{2}h\} = 0, \quad f = \tau_0 + \frac{\mu(u_2-u_1)}{h},$$

$$\tag{20.48}$$

$$\mid \frac{dp}{dx} \mid\leq \frac{2\mu(u_2-u_1)}{h^2}, \quad z_1 = -\frac{\tau_0 s_p+f}{\frac{dp}{dx}}, \quad z_2 = \frac{\tau_0 s_p-f}{\frac{dp}{dx}},$$

$$p(x_i) = p(x_e) = \frac{dp(x_e)}{dx} = 0,$$

$$\tag{20.49}$$

$$\int\limits_{x_i}^{x_e} p(t)dt = P,$$

$$[p(x_c)] = [\frac{dp(x_c)}{dx}] = 0, \qquad (20.50)$$

$$h(x) = h_e + \frac{x^2-x_e^2}{2R'} + \frac{2}{\pi E'}\int\limits_{x_i}^{x_e} p(t)\ln\frac{x_e-t}{|x-t|}dt. \qquad (20.51)$$

20.3 Some Properties of the EHL Problem Solution for Greases. Qualitative Behavior of the Solution

For $\tau_0 = 0$ the formulated iso- and non-isothermal problems for viscoplastic lubricants are reduced to the EHL problems of the preceding chapter for fluid lubricants. For $\tau_0 \neq 0$ when conditions (20.34) are satisfied equations and solutions of the isothermal EHL problems for viscoplastic and the corresponding fluid lubricants with the same rheological function F differ just by an additive term $\tau_0(p)$ in the expression for the sliding frictional stress $f(x)$ (see equations (20.33) and (20.36)). In the case of thermal EHL problems such a relationship does not hold. This is due to the fact that for conditions satisfying (20.34) function f is involved in equations (20.33), (20.36), and (20.37) as f and $f - \tau_0$.

Moreover, for $\tau_0 \neq 0$, $u_1 = u_2$ in any contact zone where condition (20.34) is not satisfied grease exhibits its viscoplastic properties, which manifest themselves in the presence of a core(s). Cores in a grease flow behave like rigid solids. In certain cases the presence of a core(s) affects the properties of a lubricated contact greatly. Moreover, under certain conditions the formulated problem for grease does not have a solution. In particular, it happens when $u_1 = u_2$ and $\tau_0 > 0$. To alleviate such situations certain adjustments to the problem formulation should be made. They may include taking into account the compressibility of grease and the variations of the surface velocities due to the elastic tangential displacements of the solid surfaces involved in a contact. The latter is usually caused by elevated tangential stresses applied to these surfaces. Such an adjustment can be made in accordance with Section 18.2. A detailed analysis of the flow of viscoplastic medium through pipes and channels with different cross sections is presented in [6, 7].

The analysis of the formulated problem for the Bingham-Shvedov model for $u_1 = u_2 = 0$, $h(x) = h_e = const$, and $\tau_0 > 0$ shows that the nonzero flow of grease through the cross–section $[-\frac{h_e}{2}, \frac{h_e}{2}]$ of the lubrication layer exists only if $\mid \frac{dp}{dx} \mid > \frac{2\tau_0}{h}$ (see (20.18)). This condition coincides with a similar one in [7]. Also, the condition which guarantees a nonzero flow of grease through the lubrication film cross section for $u_1 = u_2 > 0$ and $\tau_0 > 0$ is identical to the one in [7]. The same condition is valid for the grease model described by relationships (20.3) and (20.4).

Let us consider the question about possible configurations of a grease flow. Taking into account the continuity of core boundaries it can be shown that the case when type III grease flow is followed by type I grease flow is impossible to realize. In fact, the only possibility of realization of such a case is the flow configuration represented in Fig. 20.4. In such a case at the tip of the core we have $z_1 = z_2$. Based on the equations (20.15) for z_1 and z_2 it is clear that it can happen only when $\tau_0 = 0$. That contradicts the fact of core existence.

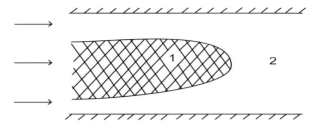

FIGURE 20.4

A grease flow configuration with a center core ending in the middle of the flow (1 marks grease core, 2 marks shear flow) (after Kudish [4]). Reprinted with permission from Allerton Press.

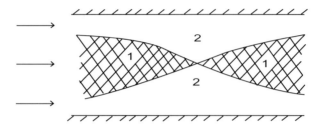

FIGURE 20.5

A grease flow configuration with a center core degenerating into a point in the middle of the flow (1 marks grease core, 2 marks shear flow) (after Kudish [4]). Reprinted with permission from Allerton Press.

This consideration shows that any configuration in which at least at one point a core of type I grease flow degenerates into a point is also impossible (see, for example, Fig. 20.5).

Also, let us show that the grease flow configuration represented in Fig. 20.6 is impossible. These type of configurations are characterized by the existence of two cross sections $x = x_0$ and $x = x_1$ such that for $x_0 < x < x_1$ the flow is of type I while to the left of x_0 and to the right of x_1 the flow is of type IIa and IIb or vice versa. Let us consider three cross sections of the flow. In the aa cross section (see Fig. 20.6), we have $\frac{dp}{dx} < 0$ while in the cc cross section we have $\frac{dp}{dx} > 0$. Assuming that $p(x)$ is a continuous function (which is in agreement with (20.42)), we obtain that between points $x = x_0$ and $x = x_1$ there is a cross section bb in which $\frac{dp}{dx} = 0$ and, therefore, the grease flow does not have a core. However, it contradicts the assumption that for all x from the interval $x_0 < x < x_1$ the flow is of type I, i.e., a grease flow with a central core detached from the solid surfaces.

Obviously, none of the configurations, which include the aforementioned two can be realized.

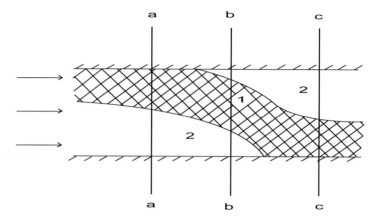

FIGURE 20.6

A grease flow configuration with a continuous core adjacent to the upper and lower contact surfaces (1 marks grease core, 2 marks shear flow) (after Kudish [4]). Reprinted with permission from Allerton Press.

The scaling of the dimensional variables for the regimes of lightly and heavily loaded grease lubricated contacts will be done in accordance with the formulas (5.13), (5.48), (5.49) (where ϑ is a certain parameter that depends on fluid/grease rheology, see Section 5.2) and (6.3), (6.4), (5.47), (7.5), respectively. For both lightly and heavily loaded regimes, the dimensionless variables for τ_0 and z_i are introduced as follows:

$$\tau_0' = \frac{\pi R'}{P}\tau_0, \quad z_i' = \frac{z_i}{h_e}, \quad i = 1, 2. \tag{20.52}$$

In further considerations primes at the dimensionless variables are omitted.

20.4   Greases in a Contact of Rigid Solids

In this section we will analyze analytically the simplest isothermal problem for grease lubricated rigid cylinders. The main goal of this study is to get an understanding of the behavior of grease in a contact and to find out the possibility of core existence and location.

20.4.1   Problem Formulation

Let us consider an isothermal problem for grease lubricated contact of two rigid cylinders. We will assume that the grease can be modeled by the Binham-

Shvedov rheological relationships (20.5). In equation (20.51) for the gap function $h(x)$ for rigid cylinders we assume that $E' = \infty$. Then in the dimensionless variables used for lightly loaded contacts (see (20.52) and Section 5.2)) from equations (20.46)-(20.51) we obtain [8]

$$\frac{d}{dx}\{[-\frac{8\gamma^2}{\mu}(\frac{\tau_0}{\gamma} - \frac{h}{2} \mid \frac{dp}{dx} \mid)^2(\frac{2\tau_0}{\gamma} + \frac{h}{2} \mid \frac{dp}{dx} \mid)(\frac{dp}{dx})^{-2}$$

$$+\frac{s_0^2}{24\gamma^2}\mu\tau_0(\frac{\tau_0}{\gamma} - \frac{h}{2} \mid \frac{dp}{dx} \mid)^{-2}]s_p - h\} = 0, \ s_p = sign(\frac{dp}{dx}), \tag{20.53}$$

$$f = \frac{s_0^2}{12\gamma^2}\frac{\mu}{h}\{1 - \frac{2\tau_0}{\gamma h}(\mid \frac{dp}{dx} \mid)^{-1}\}^{-1}, \tag{20.54}$$

$$z_1 = -(\frac{\tau_0}{\gamma}s_p + f)(\frac{dp}{dx})^{-1}, \ z_2 = (\frac{\tau_0}{\gamma}s_p - f)(\frac{dp}{dx})^{-1}, \tag{20.55}$$

$$\mid \frac{dp}{dx} \mid > \frac{1}{\gamma h}\{2\tau_0 + \frac{s_0}{12\gamma}\frac{\mu}{h} + \sqrt{\frac{s_0}{6\gamma}\frac{\mu}{h}[\frac{s_0}{24\gamma}\frac{\mu}{h} + 2\tau_0]}\}, \tag{20.56}$$

$$\frac{d}{dx}\{-\frac{s_0}{9}\sqrt{\frac{3s_0}{2}}\mu(\mid \frac{dp}{dx} \mid)^{-1}s_p + \gamma(\frac{s_0}{2}s_p - 1)h\} = 0, \tag{20.57}$$

$$f = \frac{\tau_0}{\gamma} - \frac{h}{2} \mid \frac{dp}{dx} \mid + \frac{1}{2\gamma}\sqrt{\frac{2s_0}{3}\mu \mid \frac{dp}{dx} \mid}, \tag{20.58}$$

$$z_1 = (\frac{\tau_0}{\gamma} - f)(\frac{dp}{dx})^{-1}, \ z_2 \geq \frac{h}{2} \ if \ \frac{dp}{dx} < 0,$$

$$z_1 \leq -\frac{h}{2}, \ z_2 = (\frac{\tau_0}{\gamma} - f)(\frac{dp}{dx})^{-1} \ if \ \frac{dp}{dx} > 0, \tag{20.59}$$

$$\frac{s_0}{6\gamma^2}\frac{\mu}{h^2} < \mid \frac{dp}{dx} \mid \leq \frac{1}{\gamma h}\{2\tau_0 + \frac{s_0}{12\gamma}\frac{\mu}{h} + \sqrt{\frac{s_0}{6\gamma}\frac{\mu}{h}[\frac{s_0}{24\gamma}\frac{\mu}{h} + 2\tau_0]}\}, \tag{20.60}$$

$$\frac{d}{dx}\{\frac{\gamma^2 h^3}{\mu}\frac{dp}{dx} - h\} = 0, \tag{20.61}$$

$$f = \frac{\tau_0}{\gamma} + \frac{s_0}{12\gamma^2}\frac{\mu}{h}, \tag{20.62}$$

$$\mid z_j \mid \geq \frac{h}{2}, \ j = 1, 2, \tag{20.63}$$

$$\mid \frac{dp}{dx} \mid \leq \frac{s_0}{6\gamma^2}\frac{\mu}{h^2}, \tag{20.64}$$

$$p(a) = p(c) = \frac{dp(c)}{dx} = 0, \tag{20.65}$$

$$\int_a^c p(t)dt = \frac{\pi}{2}, \tag{20.66}$$

$$[p(x_{ci})] = [\frac{dp(x_{ci})}{dx}] = 0, \ x_{ci} \in X_c, \tag{20.67}$$

$$h = 1 + \frac{x^2 - c^2}{\gamma}, \tag{20.68}$$

where X_c is the set of unknown coordinates x_{ci} of points of grease flow type changes, s_0 is the slide-to-roll ratio, a and c are the dimensionless coordinates of the inlet and exit points, and γ is the dimensionless exit film thickness (see formulas (5.19)).

Therefore, for given functions $\mu(p)$, $\tau_0(p)$, and constants a, s_0, we need to find functions $p(x)$, $h(x)$, $f(x)$, $z_j(x)$ $(j = 1, 2)$, and constants c, γ, and x_{ci}. After the solution is found the dimensionless sliding F_S and rolling F_R friction forces are calculated as follows:

$$F_S = \tfrac{2\gamma}{\pi} \int\limits_a^c f(x)dx, \ \ F_R = -\tfrac{\gamma}{\pi} \int\limits_a^c p(x)\tfrac{dh}{dx}dx. \tag{20.69}$$

20.4.2   Analysis of Possible Flow Configurations

In spite of the simplification of the problem caused by not taking into account the elastic displacements of the cylinders' surfaces (described by an integral term), the problem is still nonlinear and complex. Let us consider the full range of possible flow configurations in a grease lubricated contact. First, integrating equation (20.61) and taking into account equation (20.68) and boundary conditions (20.65) we obtain

$$\tfrac{dp}{dx} = \tfrac{h-1}{\gamma^3 h^3}\mu. \tag{20.70}$$

From (20.68) and (20.70) it follows that $\tfrac{dp}{dx} = 0$ at two points $x = \pm c$, where at $x = -c$ pressure $p(x)$ reaches its maximum value.

Let us analyze the condition of core absence in the flow given by (20.64). This condition together with equation (20.70) lead to the inequality

$$\mid h - 1 \mid \leq \tfrac{s_0}{6} h. \tag{20.71}$$

From (20.68), (20.70), and (20.71) follows that for $s_0 > 0$ the points with coordinates $x = \pm c$ at which $\tfrac{dp}{dx} = 0$ and $h(x) = 1$ belong to the considered zone of the grease flow. From (20.68) we obtain that

$$h(x) \geq 1 \ for \ x \in [a, -c], \ h(x) \leq 1 \ for \ x \in [-c, c]. \tag{20.72}$$

Now, we consider the number and location of zones in the grease lubricated contact in which relationships (20.57)-(20.60) are valid. To answer this question we need to analyze the solutions of the equation

$$\mid h(x_c) - 1 \mid = \tfrac{s_0}{6} h(x_c), \tag{20.73}$$

i.e., we need to determine the number and the coordinates of the points of grease flow type changes x_c. Using (20.72) and (20.73) we get the expressions for x_{ci}:

$$x_{c1} = -\sqrt{\Delta_1}, \ \Delta_1 = c^2 + \tfrac{\gamma s_0}{6 - s_0} \geq 0, \tag{20.74}$$

$$x_{cj} = (-1)^{j-1}\sqrt{\Delta_2}, \ \Delta_2 = c^2 - \tfrac{\gamma s_0}{6 + s_0} \geq 0, \ j = 2, 3. \tag{20.75}$$

Obviously, under the natural assumptions that $h(0) = 1 - \tfrac{c^2}{\gamma} > 0$ and $\gamma > 0$ inequality $\Delta_1 > 0$ is always valid if $0 < s_0 \leq 6$. Therefore, from formula

(20.74) follows that in the inlet zone of the contact a core is absent as long as $x_{c1} \leq a$ and it is present in the inlet zone if $x_{c1} > a$. Similarly, we obtain that in the central part of the contact a core is absent if $\Delta_2 \leq 0$ and a core is present in this part of the contact if $\Delta_2 > 0$.

Considering all possible locations of cores in the lubrication layer for which relationships (20.57)-(20.60) are valid we can classify these lubrication regimes as follows: flow configuration I (FCI) for $x_{c1} \leq a$ and $\Delta_2 \leq 0$ a core is absent in the flow, flow configuration II (FCII) for $x_{c1} > a$ and $\Delta_2 \leq 0$ there is one adjacent to a solid surface core in the inlet zone of the flow, flow configuration III (FCIII) for $x_{c1} > a$ and $\Delta_2 > 0$ there are two adjacent to solid surfaces cores in the flow, one core is located in the inlet zone of the contact while the second one is in the central part of the contact, and flow configuration IV (FCIV) for $x_{c1} \leq a$ and $\Delta_2 > 0$ there is one adjacent to a solid surface core in the central zone of the flow. In order for this classification to be valid, it is necessary that the inequality (see (20.60) and (20.64)) be satisfied

$$\mid \frac{dp}{dx} \mid \leq \frac{1}{\gamma h}\{2\tau_0 + \frac{s_0}{12\gamma}\frac{\mu}{h} + \sqrt{\frac{s_0}{6\gamma}\frac{\mu}{h}[\frac{s_0}{24\gamma}\frac{\mu}{h} + 2\tau_0]}\}. \qquad (20.76)$$

Integrating (20.57) leads to the equation

$$\frac{s_0}{9}\sqrt{\frac{3s_0}{2}\mu \mid \frac{dp}{dx} \mid^{-1}}s_p = -c_3 + \gamma(\frac{s_0}{2}s_p - 1)h. \qquad (20.77)$$

Constant c_3 coincides with the flux of grease through the gap between the surfaces. Taking into account that for $x = \pm c$ the grease flux is γ and, also, the fact that the lubricant flux through any cross section of the layer is constant from equation (20.77), we obtain

$$c_3 = \gamma. \qquad (20.78)$$

The value of constant c_3 also follows from the absence of a jump in $p(x)$ and $\frac{dp}{dx}$ at points $x = x_{cj}$ ($j = 1, 2, 3$), i.e., from conditions (20.67). Furthermore, from (20.77) and (20.78), we obtain

$$\frac{dp}{dx} = \frac{s_p s_0^3}{54\gamma^2}\mu[(\frac{s_p s_0}{2} - 1)h + 1]^{-2}. \qquad (20.79)$$

By substituting $\frac{dp}{dx}$ from equation (20.79) into inequality (20.76) after some routine calculations, we will arrive at the inequalities

$$\frac{s_0^3}{54\gamma}\mu h \leq (2\tau_0 + \frac{s_0}{12\gamma}\frac{\mu}{h})[(\frac{s_p s_0}{2} - 1)h + 1]^2,$$

$$\frac{s_0^2}{18\gamma}\mu \mid (\frac{s_p s_0}{2} - 1)h + 1 \mid \geq \mid \frac{s_0^3}{54\gamma}\mu h - 2\tau_0[(\frac{s_p s_0}{2} - 1)h + 1]^2 \mid, \qquad (20.80)$$

which are equivalent to inequality (20.76).

The flow configurations classified earlier (FCI-FCIV) are represented in Fig. 20.7 and marked by (a)-(d), respectively. Therefore, the necessary and

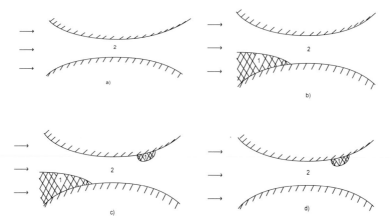

FIGURE 20.7
Grease flow configurations without a core and with cores adjacent to contact surfaces: (a) FCI, (b) FCII, (c) FCIII, and (d) FCIV (1 marks grease core, 2 marks shear flow) (after Kudish [8]). Reprinted with permission from Allerton Press.

sufficient condition of realization of the above flow configurations is equivalent to validity of one of the following inequalities:

$$s_0 \geq 6 \max\left(\frac{a^2 - c^2}{\gamma + a^2 - c^2}, \frac{c^2}{\gamma - c^2}\right), \tag{20.81}$$

$$\frac{6c^2}{\gamma - c^2} \leq s_0 < 6\frac{a^2 - c^2}{\gamma + a^2 - c^2}, \tag{20.82}$$

$$0 < s_0 < 6 \min\left(\frac{a^2 - c^2}{\gamma + a^2 - c^2}, \frac{c^2}{\gamma - c^2}\right), \tag{20.83}$$

$$6\frac{a^2 - c^2}{\gamma + a^2 - c^2} \leq s_0 < \frac{6c^2}{\gamma - c^2}. \tag{20.84}$$

Inequalities (20.81)-(20.84) follow from the inequalities for x_{c1} and Δ_2, which characterize each of the considered flow configurations.

The above analysis and continuity of $\frac{dp}{dx}$ indicate that when a core (adjacent to solid surfaces or detached from them within the lubrication layer $-\frac{h}{2} \leq z \leq \frac{h}{2}$) is present in a grease flow the core can be located only within the intervals $[a, x_{c1}]$ and $[x_{c2}, x_{c3}]$.

Now, let us establish the conditions under which grease flows described by equations (20.53)-(20.56) can be realized. From expressions (20.56), (20.76), and (20.80), it follows that the points of flow type changes for flows with cores adjacent to solid surfaces or detached from them can be found from solution of the system

$$p(x_c - 0) = p(x_c + 0), \tag{20.85}$$

$$\frac{s_0^2}{18\gamma}\mu \mid (\frac{s_p s_0}{2} - 1)h(x_c) + 1 \mid$$

$$\geq \mid \frac{s_0^3}{54\gamma}\mu h(x_c) - 2\tau_0[(\frac{s_p s_0}{2} - 1)h(x_c) + 1]^2 \mid . \tag{20.86}$$

In equation (20.85) in the right- and left-hand sides are functions $p(x)$, which satisfy equations (20.53) and (20.57).

It is important to mention that because h is determined by equation (20.68) for $\mu(p) = 1$ and $\tau_0 = const$ it is sufficient to obtain the solution of just equation (20.70).

Without getting into more detail concerning (20.85) and (20.86) let us consider flow configurations I-IV. In these flow configurations there are no points where inequality (20.56) is satisfied. Let us integrate equations (20.70), (20.79), and (20.68). For simplicity we will assume that $\mu(p) = 1$. In agreement with Section 5.2 in the region where inequality (20.64) is satisfied the solution of equation (20.70) has the form

$$p(x) = p_2(x) + c_2 = c_2 + \frac{1}{8b^3}\{\frac{\gamma - 4c^2}{b^2}[\arctan\frac{x}{b} + \frac{bx}{b^2 + x^2}] - \frac{2\gamma bx}{(b^2 + x^2)^2}\}, \tag{20.87}$$

where $b = \sqrt{\gamma - c^2}$. Similarly, by integrating equation (20.79) for $s_0 > 0$, we obtain the following expressions for pressure

$$p(x) = p_4(x) + c_4 = c_4 + \frac{4x}{27\gamma^2} \; for \; s_p = 1, \; s_0 = 2, \tag{20.88}$$

$$p(x) = p_5(x) + c_5 = c_5 + \frac{d}{2d_1^2}\{\frac{x}{x^2 + d_1^2} + \frac{1}{d_1}\arctan\frac{x}{d_1}\} \; for \; d_1^2 > 0, \tag{20.89}$$

$$p(x) = p_6(x) + c_6 = c_6 - \frac{d}{3x^3} \; for \; d_1 = 0, \tag{20.90}$$

$$p(x) = p_7(x) + c_7 = c_7 + \frac{d}{2d_2^2}\{\frac{x}{d_2^2 - x^2} + \frac{1}{2d_2}\ln\mid\frac{x + d_2}{x - d_2}\mid\} \; for \; d_2^2 > 0, \tag{20.91}$$

$$d = \frac{s_p s_0^3}{54(1 - \frac{s_p s_0}{2})^2}, \; d_1^2 = \frac{s_p s_0 \gamma}{s_p s_0 - 2} - c^2, \; d_2^2 = c^2 - \frac{s_p s_0 \gamma}{s_p s_0 - 2}. \tag{20.92}$$

Let us assume that $s_p = 1$. As it is clear from the expressions in (20.92)

$$d_1^2 > 0 \; for \; s_0 > 2; \; d_2^2 > 0 \; for \; 0 < s_0 < 2. \tag{20.93}$$

Obviously, for $s_0 > 2$ pressure $p(x)$ in (20.89) is regular in the entire contact region. For $0 < s_0 < 2$ the solution in (20.91) has a singularity at $x = - \mid d_2 \mid$ (the singularity at $x = \mid d_2 \mid$ is irrelevant as it is outside of the contact region). Therefore, to satisfy inequalities (20.80) it is necessary to require that functions $p(x)$ and $\frac{dp}{dx}$ from (20.91) are bounded. To satisfy these conditions, it is sufficient to satisfy one of the inequalities: $- \mid d_2 \mid < a$ or $- \mid d_2 \mid > x_{c1}$. Therefore, for $0 < s_0 < 2$ we obtain the inequality

$$\frac{2(a^2 - c^2)}{\gamma + a^2 - c^2} < s_0 < 2 \; for \; s_p = 1, \; d_2^2 > 0. \tag{20.94}$$

Now, let $s_p = -1$. Then we will have $d_1^2 > 0$ for $s_0 > \frac{2c^2}{\gamma - c^2}$, $d_1 = 0$ for $s_0 = \frac{2c^2}{\gamma - c^2}$, and $d_2^2 > 0$ for $0 < s_0 < \frac{2c^2}{\gamma - c^2}$. In the considered case, a core

can be realized only in the contact zone, which includes point $x = 0$. There-
fore, for $d_1 = 0$ and $s_0 = \frac{2c^2}{\gamma - c^2}$ at $x = 0 \in [x_{c2}, x_{c3}]$ functions $p(x)$ and $\frac{dp}{dx}$
from (20.90) have a singularity. Similarly, to satisfy inequality (20.80) in the
considered contact zone for $0 < s_0 < \frac{2c^2}{\gamma - c^2}$, it is necessary to require continuity of functions $p(x)$ and $\frac{dp}{dx}$ from (20.91). That is equivalent to inequality
$\mid d_2 \mid > x_{c3} = \mid x_{c2} \mid$. However, the analysis of equations (20.75) and (20.92)
shows that it is impossible to satisfy the latter inequality. Therefore, in this
case the only possibility is

$$s_0 > \frac{2c^2}{\gamma - c^2} \ for \ s_p = -1, \ d_1^2 > 0. \tag{20.95}$$

The condition under which the flow configurations I-IV are realized are inequality (20.80) and (see (20.94) and (20.95))

$$s_0 > 2\max\{\tfrac{a^2 - c^2}{\gamma + a^2 - c^2}, \tfrac{c^2}{\gamma - c^2}\}. \tag{20.96}$$

Violation of inequality (20.96) causes the existence of such zones in the contact
in which the grease flow is described by relationships (20.53)-(20.56). That
causes existence of a core in the flow detached from the solid surfaces.

In the further analysis, we will take advantage of the integrals of functions
$p(x)$ from (20.87)-(20.91):

$$I_4(x_1, x_2, c_4) = c_4(x_2 - x_1) + \tfrac{2}{27}\tfrac{x_2^2 - x_1^2}{\gamma^2}, \ s_p = 1, \ s_0 = 2, \tag{20.97}$$

$$I_5(x_1, x_2, c_5) = c_5(x_2 - x_1) + \tfrac{d}{2d_1^3}\{x_2 \arctan \tfrac{x_2}{d_1} - x_1 \arctan \tfrac{x_1}{d_1}\} \tag{20.98}$$

$$for \ d_1^2 > 0,$$

$$I_7(x_1, x_2, c_7) = c_7(x_2 - x_1) + \tfrac{d}{4d_2^3}\{x_2 \ln \mid \tfrac{x_2 + d_2}{x_2 - d_2} \mid -x_1 \ln \mid \tfrac{x_1 + d_2}{x_1 - d_2} \mid\} \tag{20.99}$$

$$for \ d_2^2 > 0,$$

$$I_2(x_1, x_2, c_2) = c_2(x_2 - x_1) + \tfrac{1}{8b^3}\{\tfrac{\gamma - 4c^2}{b^2}[x_2 \arctan \tfrac{x_2}{b}$$

$$-x_1 \arctan \tfrac{x_1}{b}] + \gamma b[\tfrac{1}{b^2 + x_2^2} - \tfrac{1}{b^2 + x_1^2}]\}. \tag{20.100}$$

To complete the determination of pressure in the contact region, it is necessary to find the values of constants c, γ, and some of the constants c_2, c_4, c_5,
and c_7. To do that we need to satisfy equations (20.65), (20.66), and the
first equation in (20.67) while using the relationships (20.74), (20.75), (20.87)-
(20.89), (20.91), (20.97)-(20.100). After the solution of this system of transcen-
dental equations is obtained it is necessary to check the validity of inequalities
(20.80) and (20.96).

Clearly, a similar analysis of the problem can be done in the case of $\mu = \mu(p)$.
Moreover, for $\tau_0 \ll 1$ using regular asymptotic expansions the solution of the

whole problem (20.53)-(20.68) can be obtained for different combinations of parameters a, τ_0, and s_0. In this case the main term of the asymptotic solution is the solution of the hydrodynamic problem for rigid solids lubricated with Newtonian fluid. This situation allows for realization of cores adjacent to and detached from the solid surfaces. The shape, size, and location of these cores as well as the next terms in the pressure expansion are the solutions of linear algebraic and differential equations. This analysis is based on the presented study of the grease flow with cores adjacent to solid surfaces. Besides that, based on the above analysis and on the method of regular asymptotic expansions [9] the solution of the elastohydrodynamic problem for elastic cylinders in case of lightly loaded contact (when $V \ll 1$) with cores adjacent to solid surfaces can be obtained.

20.4.3 Numerical Results

Let us consider the grease flow configurations defined by inequalities (20.81)-(20.84).

Suppose inequality (20.81) is satisfied. Then in the entire contact region pressure $p(x)$ is determined by equation (20.87) (see also Section 5.2). By satisfying conditions (20.65) and (20.66), we obtain the full problem solution in the absence of cores. For $\mu(p) = 1$ and $s_0 = 6$ the numerical solutions for this case are presented in Table 20.2. It is clear from this data that only the sliding frictional stress $f(x)$ and the sliding friction force F_S depend on the grease threshold stress τ_0. Also, the problem solution is independent from the slide-to-roll ration s_0 as long as inequality (20.81) is satisfied.

TABLE 20.2
Values of γ, c, and F_S versus the inlet coordinate a and threshold stress τ_0 for grease lubricated contact of rigid cylinders obtained for $\mu = 1$ and $s_0 = 6$ (after Kudish [8]). Reprinted with permission from Allerton Press.

a	τ_0	γ	c	F_S
-0.554	0	0.092	0.123	0.133
-0.554	5	0.092	0.123	0.256
-0.954	0	0.123	0.148	0.017
-0.954	5	0.123	0.148	0.038
-5	0	0.157	0.170	0.066
-5	5	0.157	0.170	0.168

Now, let us assume that inequalities (20.82) and (20.96) are satisfied. Then using (20.87) and one of the functions from (20.88), (20.89), and (20.91) and satisfying the first two conditions in (20.65) and the first condition in (20.67) for $x_{ci} = x_{c1}$ from (20.74) with the help of (20.97)-(20.100) satisfying equation (20.66) we obtain three systems of two nonlinear algebraic equations

$$p_5(x_{c1}) - p_5(a) = p_2(x_{c1}) - p_2(c),$$

$$I_5(a, x_{c1}, -p_5(a)) + I_2(x_{c1}, c, -p_2(c)) = \tfrac{\pi}{2}, \ s_0 > 2,$$

(20.101)

$$p_4(x_{c1}) - p_4(a) = p_2(x_{c1}) - p_2(c),$$

$$I_4(a, x_{c1}, -p_4(a)) + I_2(x_{c1}, c, -p_2(c)) = \tfrac{\pi}{2}, \ s_0 = 2,$$

(20.102)

$$p_7(x_{c1}) - p_7(a) = p_2(x_{c1}) - p_2(c),$$

$$I_7(a, x_{c1}, -p_7(a)) + I_2(x_{c1}, c, -p_2(c)) = \tfrac{\pi}{2}, \ s_0 < 2.$$

(20.103)

Suppose inequalities (20.83) and (20.96) are satisfied. Then using (20.87) and one of the functions from (20.88), (20.89), and (20.91) and satisfying the first two conditions in (20.65) and the first condition in (20.67) for $x_{ci} = x_{c1}, \ x_{c2}, \ x_{c3}$ from (20.74) and (20.75) with the help of (20.97)-(20.100) satisfying equation (20.66) we obtain another three systems of two nonlinear algebraic equations

$$p_5(x_{c1}) - p_5(a) = c_{21} + p_2(x_{c1}),$$

$$I_5(a, x_{c1}, -p_5(a)) + g(c_{21}, c_{51}, -p_2(c), \gamma, c) = 0, \ s_0 > 2,$$

(20.104)

$$p_4(x_{c1}) - p_4(a) = c_{21} + p_2(x_{c1}),$$

$$I_4(a, x_{c1}, -p_4(a)) + g(c_{21}, c_{51}, -p_2(c), \gamma, c) = 0, \ s_0 = 2,$$

(20.105)

$$p_7(x_{c1}) - p_7(a) = c_{21} + p_2(x_{c1}),$$

$$I_7(a, x_{c1}, -p_7(a)) + g(c_{21}, c_{51}, -p_2(c), \gamma, c) = 0, \ s_0 < 2,$$

(20.106)

where

$$g(c_{21}, c_{51}, -p_2(c), \gamma, c) = I_2(x_{c1}, x_{c2}, c_{21}) + I_5(x_{c2}, x_{c3}, c_{51})$$

$$+ I_2(x_{c3}, c, -p_2(c)) - \tfrac{\pi}{2}.$$

(20.107)

To complete each of the systems of equations (20.104)-(20.107), we need to add to them the following equations

$$c_{21} = c_{51} + p_5(x_{c2}) - p_2(x_{c2}), \ c_{51} = p_2(x_{c3}) - p_2(c) - p_5(x_{c3}). \quad (20.108)$$

Finally, let us assume that inequalities (20.84) and (20.96) are satisfied. Then using (20.87) and (20.89) and satisfying the first two conditions in

(20.65) and the first condition in (20.67) for $x_{ci} = x_{c2}$, x_{c3} from (20.75) with the help of (20.98) and (20.100) satisfying equation (20.66) we obtain another system of three nonlinear algebraic equations

$$p_2(x_{c2}) - p_2(a) = p_5(x_{c2}) + c_5, \ c_5 = p_2(x_{c3}) - p_2(c) - p_5(x_{c3}),$$

$$\text{(20.109)}$$

$$I_2(a, x_{c2}, -p_2(a)) + I_5(x_{c2}, x_{c3}, c_5) + I_2(x_{c3}, c, -p_2(c)) = \tfrac{\pi}{2}.$$

TABLE 20.3
Grease film thickness γ and the exit coordinate c versus the slide-to-roll ratio s_0 obtained for $a = -5$ and $\tau_0 = 1$ for a lubricated contact of rigid solids in the case of presence of a core flow adjacent to a solid surface (after Kudish [10]). Reprinted with permission from Allerton Press.

s_0	γ	c
4.808	0.1572	0.1702
4.424	0.1593	0.1715
3.847	0.1645	0.1748
3.482	0.1724	0.1800
3.077	0.1890	0.1906

The results of calculations done for $\mu(p) = 1$, $a = -5$, and various values of s_0 and τ_0 are given in Fig. 20.8-20.11. From Fig. 20.8 we can make a conclusion that for $\tau_0 > 0$ as the slide-to-roll ratio s_0 decreases the dimensionless grease film thickness γ and the coordinate of the exit point c increase significantly in comparison with the values for the corresponding Newtonian fluid while remaining finite. Some data for the inlet coordinate $a = -5$ and threshold shear stress $\tau_0 = 1$ is given in Table 20.3. For $\tau_0 > 0$ the sliding friction force F_S reaches its minimum at approximately $s_0 = 3$ while for a Newtonian fluid ($\tau_0 = 0$) the sliding friction force F_S is a strictly monotonically increasing function of s_0. It is important to notice that the rolling friction force F_R is by about two orders of magnitude smaller than the sliding friction force F_S.

For $\tau_0 = 0.5$ a detached core in the grease flow is present for $s_0 \leq 2.1$ while for $\tau_0 = 5$ a detached core in the grease flow is present for $s_0 \leq 1.9$. For decreasing $\mid a \mid$ (when the inlet coordinate approaches the center of the

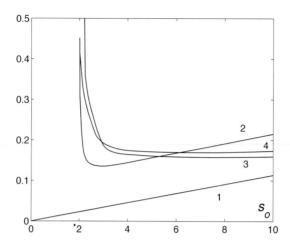

FIGURE 20.8

The dependence of the friction force F_S, film thickness γ, and exit coordinate c on the slide-to-roll ratio s_0 obtained for $\mu = 1$, $a = -5$: friction force F_S obtained for $\tau_0 = 0$ - curve 1 and for $\tau_0 = 10$ - curve 2, film thickness γ obtained for $\tau_0 = 10$ - curve 3, exit coordinate c obtained for $\tau_0 = 10$ - curve 4 (after Kudish [8]). Reprinted with permission from Allerton Press.

TABLE 20.4

Dependence of the critical value of the slide-to-roll ratio s_* on the inlet coordinate a. (after Kudish [8]). Reprinted with permission from Allerton Press.

a	-0.031	-0.164	-0.554	-0.954	-5	-10
s_*	0.96	2.64	4.56	5.28	5.96	5.99

contact region), the critical value of s_0 for which a detached core appears in the grease flow also decreases. The behavior of the critical slide-to-roll ratio value $s_* = 6\frac{a^2-c^2}{\gamma+a^2-c^2}$ beyond which cores are absent in the grease flow is similar (see Table 20.4). The core boundaries x_{c1} (marked with 1) and x_{c2} (marked with 2) as functions of s_0 are presented graphically in Fig. 20.9. These graphs show that as s_0 decreases the first core appearance occurs at the slower–moving surface. The graphs of pressure $p(x)$ distribution (see Fig. 20.10) obtained for $\tau_0 = 5$ and $s_0 = 6$ (curve 1), $s_0 = 3$ (curve 2), $s_0 = 2.5$ (curve 3), $s_0 = 2$ (curve 4), show that the pressure in the inlet and exit zones of the contact increases while in the central part of the contact it decreases significantly. The curves of the sliding frictional stress $f(x)$ presented in Fig.

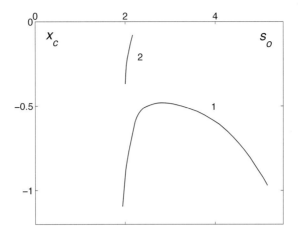

FIGURE 20.9
The dependence of the coordinates of the points of grease flow type changes x_{ci} on the slide-to-roll ratio s_0 obtained for $\mu = 1$, $a = -5$, and $\tau_0 = 10$: $x_{c1} = x_{c1}(s_0)$ - curve 1, $x_{c2} = x_{c2}(s_0)$ - curve 2 (after Kudish [8]). Reprinted with permission from Allerton Press.

20.11 are obtained for a number of different values of τ_0 and s_0 (curve 1 for $\tau_0 = 0$, $s_0 = 3$, curve 2 for $\tau_0 = 0.5$, $s_0 = 3$, curve 3 for $\tau_0 = 0.5$, $s_0 = 2.5$, curve 4 for $\tau_0 = 2.5$, $s_0 = 2$, and curve 5 for $\tau_0 = 5$, $s_0 = 2$). It follows from these graphs of $f(x)$ that its behavior changes qualitatively and quantitatively as τ_0 increases. In particular, as τ_0 increases the sliding frictional stress $f(x)$ gets distributed more evenly over the contact region and it maximum value decreases.

Based on Fig. 20.9, Tables 20.2 and 20.3 we can point out that the film thickness is greater in a grease lubricated contact compared to the one lubricated with a Newtonian fluid given all other conditions are identical. This theoretical conclusion is confirmed experimentally [11, 12, 13]. Also, in these studies it is found that under isothermal conditions the dependence of the exit film thickness h_e on the average surface velocity $0.5(u_1 + u_2)$ is similar for the grease and the base oil used for grease preparation. The same conclusion follows from the above theoretical analysis done for lightly loaded conditions because from formulas (5.19) and (5.33) for greases described by the Bingham-Shvedov model for the film thickness h_e we get the formula

$$h_e = 6\pi\gamma\frac{\mu_0(u_1+u_2)R'}{P}, \qquad (20.110)$$

which coincides with the corresponding formula for h_e for Newtonian lubricant. What is different in these two formulas is the value of the parameter γ,

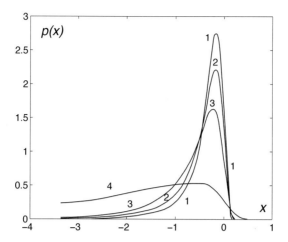

FIGURE 20.10

Pressure distribution $p(x)$ obtained for $\mu = 1$, $a = -5$, and $\tau_0 = 10$: curve 1 for $s_0 = 6$ (Newtonian fluid), curve 2 for $s_0 = 3$, curve 3 for $s_0 = 2.5$, and curve 4 for $s_0 = 2$ (after Kudish [8]). Reprinted with permission from Allerton Press.

which is greater for grease than for the corresponding fluid lubricant.

20.5 Regimes of Grease Lubrication without Cores

In the preceding section we considered some regimes of grease lubrication of rigid solids with cores adjacent to solid surfaces. In this section we will consider heavily loaded regimes of grease lubrication without cores.

For greases the rheology of which satisfies equations (20.3) and (20.4) in the dimensionless variables designed for heavily loaded regimes (see formulas (9.2), (9.3), and (10.6) in Chapter 9), we obtain the following system of integro-differential equations

$$\frac{d}{dx}\Big\{ \int_{-h/2}^{z_1} \frac{z}{\mu} F(f + \tau_0 s_p + H_0 z \frac{dp}{dx}) dz$$

$$+ \int_{z_2}^{h/2} \frac{z}{\mu} F(f - \tau_0 s_p + H_0 z \frac{dp}{dx}) dz - h \Big\} = 0, \quad s_p = sign \frac{dp}{dx}, \tag{20.111}$$

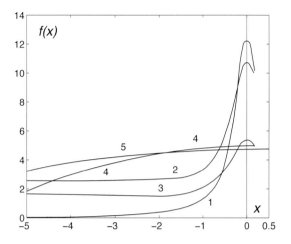

FIGURE 20.11
Sliding frictional stress distribution $f(x)$ obtained for $\mu = 1$, $a = -5$: curve 1 for $s_0 = 6$ and $\tau_0 = 0$ (Newtonian fluid), curve 2 for $s_0 = 3$ and $\tau_0 = 1$, curve 3 for $s_0 = 2.5$ and $\tau_0 = 1$, curve 4 for $s_0 = 2$ and $\tau_0 = 5$, and curve 5 for $s_0 = 2$ and $\tau_0 = 10$ (after Kudish [8]). Reprinted with permission from Allerton Press.

$$\int\limits_{-h/2}^{z_1} \frac{1}{\mu} F(f + \tau_0 s_p + H_0 z \frac{dp}{dx}) dz$$

$$+ \int\limits_{z_2}^{h/2} \frac{1}{\mu} F(f - \tau_0 s_p + H_0 z \frac{dp}{dx}) dz = s_0, \tag{20.112}$$

$$\frac{\partial}{\partial z}(\lambda \frac{\partial T}{\partial z}) = -\frac{\kappa H_0}{\mu} F[f - \tau_0 s_p sign(z - z_1) + H_0 z \frac{dp}{dx}] \tag{20.113}$$

$$\times (f + H_0 z \frac{dp}{dx})\theta[(z - z_1)(z - z_2)],$$

$$z_1 = -\frac{\tau_0 s_p + f}{H_0 \frac{\partial p}{\partial x}}, \quad z_2 = \frac{\tau_0 s_p - f}{H_0 \frac{\partial p}{\partial x}}, \tag{20.114}$$

$$\tau_0 - \frac{H_0 h}{2} \mid \frac{dp}{dx} \mid < f < -\tau_0 + \frac{H_0 h}{2} \mid \frac{dp}{dx} \mid, \tag{20.115}$$

$$\frac{d}{dx}\Big\{ \int\limits_{-h/2}^{z_1} \frac{z}{\mu} F(f - \tau_0 + H_0 z \frac{dp}{dx}) dz - h \Big\} = 0, \tag{20.116}$$

$$\int\limits_{-h/2}^{z_1} \frac{1}{\mu} F(f - \tau_0 + H_0 z \frac{dp}{dx}) dz = s_0, \tag{20.117}$$

$$\frac{\partial}{\partial z}(\lambda \frac{\partial T}{\partial z}) = -\frac{\kappa H_0}{\mu} F[f + \tau_0 sign(z - z_1) + H_0 z \frac{dp}{dx}]$$

$$\times (f + H_0 z \frac{dp}{dx})\theta(z_1 - z), \tag{20.118}$$

$$z_1 = \frac{\tau_0 - f}{H_0 \frac{\partial p}{\partial x}}, \ z_2 = -\frac{\tau_0 + f}{H_0 \frac{\partial p}{\partial x}}, \tag{20.119}$$

$$\tau_0 + \frac{H_0 h}{2} \frac{dp}{dx}) < f < \tau_0 - \frac{H_0 h}{2} \frac{dp}{dx}, \ -\tau_0 - \frac{H_0 h}{2} \frac{dp}{dx}) \leq f, \tag{20.120}$$

$$\frac{d}{dx}\left\{ \int_{z_2}^{h/2} \frac{z}{\mu} F(f - \tau_0 + H_0 z \frac{dp}{dx}) dz - h \right\} = 0, \tag{20.121}$$

$$\int_{z_2}^{h/2} \frac{1}{\mu} F(f - \tau_0 + H_0 z \frac{dp}{dx}) dz = s_0, \tag{20.122}$$

$$\frac{\partial}{\partial z}(\lambda \frac{\partial T}{\partial z}) = -\frac{\kappa H_0}{\mu} F[f - \tau_0 sign(z - z_2) + H_0 z \frac{dp}{dx}]$$

$$\times (f + H_0 z \frac{dp}{dx})\theta(z - z_2), \tag{20.123}$$

$$z_1 = -\frac{\tau_0 + f}{H_0 \frac{\partial p}{\partial x}}, \ z_2 = \frac{\tau_0 - f}{H_0 \frac{\partial p}{\partial x}}, \tag{20.124}$$

$$\tau_0 - \frac{H_0 h}{2} \frac{dp}{dx}) < f < \tau_0 + \frac{H_0 h}{2} \frac{dp}{dx}, \ -\tau_0 + \frac{H_0 h}{2} \frac{dp}{dx}) \leq f, \tag{20.125}$$

$$\frac{d}{dx}\left\{ \int_{-h/2}^{h/2} \frac{z}{\mu} F(f - \tau_0 + H_0 z \frac{dp}{dx}) dz - h \right\} = 0, \tag{20.126}$$

$$\int_{-h/2}^{h/2} \frac{1}{\mu} F(f - \tau_0 + H_0 z \frac{dp}{dx}) dz = s_0, \tag{20.127}$$

$$\frac{\partial}{\partial z}(\lambda \frac{\partial T}{\partial z}) = -\frac{\kappa H_0}{\mu} F[f - \tau_0 + H_0 z \frac{dp}{dx}](f + H_0 z \frac{dp}{dx}), \tag{20.128}$$

$$\mid \frac{dp}{dx} \mid \leq \frac{2(f - \tau_0)}{H_0 h}, \tag{20.129}$$

$$p(a) = p(c) = \frac{dp(c)}{dx} = 0, \tag{20.130}$$

$$T(x, -\frac{h}{2}) = T_{w1}(x), \ T(x, \frac{h}{2}) = T_{w2}(x), \tag{20.131}$$

$$\int_a^c p(t) dt = \frac{\pi}{2}, \tag{20.132}$$

$$H_0(h - 1) = x^2 - c^2 + \frac{2}{\pi} \int_a^c p(t) \ln \frac{c-t}{|x-t|} dt, \tag{20.133}$$

$$[p(x_{ci})] = [\frac{dp(x_{ci})}{dx}] = 0, \ x_{ci} \in X_c, \tag{20.134}$$

$$[T(x, z_i)] = [\frac{\partial T(x, z_i)}{\partial z}] = 0, \ \mid z_i \mid \leq \frac{h}{2}, \tag{20.135}$$

where parameter κ is determined in (10.6).

We will consider regimes of lubrication in which cores in the grease flow do not occur. For such regimes in the case of non-isothermal lubrication, we need to satisfy equations (20.126)-(20.134) while in the case of isothermal lubrication we need to satisfy equations (20.45), (20.126), (20.127), (20.129), (20.130), (20.132)-(20.134). As always, in EHL equations describing heavily loaded contacts, there is a small parameter ω which is usually associated with parameter $V \ll 1$ (see (9.3)) or dimensionless pressure viscosity coefficient $Q \gg 1$. The latter coefficient usually occurs in the exponential viscosity-pressure relationship, which, in dimensional variables, has the form $\mu = \mu_a \exp(\alpha p)$, where μ_a is the ambient viscosity, α is the dimensional pressure viscosity coefficient and $Q = \alpha p_H$ (p_H is the maximum Hertzian pressure in a dry contact of elastic solids). On detailed discussion of various definitions of the small parameter ω see the preceding chapter.

Let us introduce the analogue of function $\nu(x)$ from (9.94) according to the formula [14]

$$\nu(x) = \frac{H_0 h}{2(f - \tau_0)} \frac{dp}{dx}. \tag{20.136}$$

Using the definition of $\nu(x)$ from (20.136) inequality (20.129) can be rewritten as $\mid \nu(x) \mid \leq 1$. Therefore, when $\nu(x) \ll 1$ or $\nu(x) = O(1)$, $\mid \nu(x) \mid \leq 1$ for $\omega \ll 1$, in the inlet zone of the contact the asymptotic approaches developed in Chapter 6 are applicable to the EHL problem for grease. It is clear that the solution of the isothermal EHL problem for grease coincides with the solution of the EHL problem for the corresponding non-Newtonian fluid lubricant ($\tau_0 = 0$) described in detail in Chapter 6. In particular, the formulas for the film thickness H_0 in pre- and over-critical regimes in heavily loaded contacts lubricated by a grease and a non-Newtonian fluid lubricant with rheology identical to the one for the grease with $\tau_0 = 0$ are identical (see Chapter 6). The only difference between these two solutions manifests itself as additive term equal to τ_0 and term proportional to $\int_a^c \tau_0(p) dx$ (see formula (20.69)) in the sliding frictional stress $f(x)$ and force F_S, respectively.

Let us consider in more detail the case of thermal grease lubrication assuming that $\frac{\partial \tau_0}{\partial T} = 0$. The expression for the sliding frictional stress $f(x)$ and lubricant temperature $T(x, z)$ we will search in the form (for comparison see expression (9.95) and (10.15), (10.16)) [14]

$$f(x) = \tau_0(p(x)) + f_0(x) + f_1(x)\nu(x) + O(\nu^2(x)), \ \omega \ll 1, \tag{20.137}$$

$$T(x, z) = T_0(x, z) + T_1(x, z)\nu(x) + O(\nu^2(x)), \ \omega \ll 1, \tag{20.138}$$

where functions $f_0(x)$, $f_1(x)$, $T_0(x, z)$, and $T_1(x, z)$ are unknown and of the order of unity. Let us assume that the lubricant viscosity μ is determined by the equation (see (10.7))

$$\mu(p, T) = \mu^0(p) \exp(-\delta_T T), \tag{20.139}$$

where $\mu^0(p)$ is the lubricant viscosity at $T = 0$ which is dependent only on pressure p, δ_T is a positive constant or function of p, δ_T is independent from

T. Then expanding the terms of equations (20.127) and (20.128) for $\nu \ll 1$ and following the procedure of Chapter 10 we obtain

$$f_0 = \Phi\left\{\mu^0 s_0 \left(\int_{-h/2}^{h/2} e^{\delta_T T_0} dz\right)^{-1}\right\}, \tag{20.140}$$

$$\lambda\frac{\partial^2 T_0}{\partial z^2} = -\frac{\kappa H_0}{\mu^0} F(f_0)(f_0 + \tau_0)e^{\delta_T T_0}, \tag{20.141}$$

$$f_1 = -\left\{\frac{F(f_0)}{F'(f_0)}\delta \int_{-h/2}^{h/2} T_1 e^{\delta_T T_0} dz + \frac{2f_0}{h}\int_{-h/2}^{h/2} ze^{\delta_T T_0} dz\right\}$$

$$\times \left(\int_{-h/2}^{h/2} e^{\delta_T T_0} dz\right)^{-1}. \tag{20.142}$$

$$\lambda\frac{\partial^2 T_1}{\partial z^2} = -\frac{\kappa H_0}{\mu^0}F(f_0)(f_0 + \tau_0)e^{\delta_T T_0}\{\delta_T T_1$$

$$+(\frac{f_1}{f_0} + \frac{2z}{h})[\frac{f_0}{f_0+\tau_0} + \frac{f_0)F'(f_0)}{F(f_0)}]\}. \tag{20.143}$$

Let us consider a special case of equal surface temperatures $T_w(x)$ in (20.131)

$$T(x, -\tfrac{h}{2}) = T_w(x), \; T(x, \tfrac{h}{2}) = T_w(x), \tag{20.144}$$

in which solution of equations (20.141) and (20.143) is significantly simplified and it yields (the general case is considered in Chapter 10)

$$e^{\delta_T T_0} = A_0[1 - \tanh^2 Bz], \; A_0 = e^{\delta_T T_w}\cosh^2\frac{Bh}{2}, \tag{20.145}$$

$$\Phi(\frac{\mu^0 s_0 e^{-\delta_T T_w} B}{\sinh Bh}) = \frac{4\lambda}{s_0\kappa\delta_T H_0}B\tanh\frac{Bh}{2} - \tau_0, \tag{20.146}$$

$$T_1 = \frac{1}{\delta_T}[\frac{f_0}{f_0+\tau_0} + \frac{f_0 F'(f_0)}{F(f_0)}]\{\frac{\tanh Bz}{\tanh\frac{Bh}{2}} - \frac{2z}{h}\}. \tag{20.147}$$

Under the accepted conditions (20.144) we have $f_1 = 0$ and the expression for the first term in the generalized Reynolds equation (20.126) takes the form

$$\int_{-h/2}^{h/2} \frac{z}{\mu^0}F(f - \tau_0 + H_0 z\frac{dp}{dx})dz = \frac{H_0 h^3}{12\mu^0}\frac{dp}{dx}[\Phi'(\frac{\mu^0 s_0}{h})]^{-1} + O(\nu^2), \tag{20.148}$$

where the value of B is determined by equation (20.146).

As an example let us consider the case of Bingham-Shvedov model of grease (20.5). Then in dimensionless variables $\Phi(x) = \frac{V}{12H_0}x$ and equation (20.146) is reduced to

$$\frac{4\lambda}{s_0\kappa\delta_T H_0}B\tanh\frac{Bh}{2} = \frac{V}{12H_0}\frac{\mu^0 s_0 e^{-\delta_T T_w}B}{\sinh Bh} + \tau_0, \tag{20.149}$$

where V is determined in (9.3) and the expression for the integral in (20.148) involved in the generalized Reynolds equation (20.126) is reduced to (see (7.27) and (7.28))

$$\int\limits_{-h/2}^{h/2} \frac{z}{\mu^0} F(f - \tau_0 + H_0 z \frac{dp}{dx}) dz$$

(20.150)

$$= \frac{H_0^2}{V} \frac{3h^3 e^{\delta_T T_w}(1+\beta)}{\mu^0 \ln^2(\sqrt{\beta}+\sqrt{\beta+1})} \left\{ 1 + \ln(1+\beta) - \sqrt{\frac{1+\beta}{\beta}} \ln(\sqrt{\beta} + \sqrt{\beta+1}) \right.$$

$$\left. -2 \frac{\int\limits_{0}^{\ln(\sqrt{\beta}+\sqrt{\beta+1})} \ln(\cosh(t))dt}{\ln(\sqrt{\beta}+\sqrt{\beta+1})} \right\} \frac{dp}{dx} + O(\nu^2), \quad \beta = \frac{\kappa s_0^2}{8} \frac{\delta_T \mu^0}{\lambda} e^{-\delta_T T_w},$$

$$\omega \ll 1,$$

where

$$\int\limits_{0}^{z} \ln(\cosh(t))dt = \frac{z^2}{2} - z \ln 2 + \frac{\pi^2}{24} - \frac{1}{2} \sum_{k=1}^{\infty} \frac{(-1)^{k+1}}{k^2} e^{-2kz},$$

(20.151)

$$\int\limits_{0}^{z} \ln(\cosh(t))dt = \frac{z^3}{6} + O(z^4), \quad z \ll 1.$$

Assuming that $B \ll 1$ from equation (20.149), we find

$$B = \left\{ \frac{s_0 \kappa \delta_T H_0}{2\lambda h} \left(\frac{V}{12H_0} \frac{\mu^0 s_0 e^{-\delta_T T_w}}{h} + \tau_0 \right) \right\}^{1/2} \ll 1, \quad \omega \ll 1. \qquad (20.152)$$

Notice that it is easy to establish the validity conditions for formula (20.152).

Under these conditions for pre- and over-critical lubrication regimes when in the inlet (ϵ_q- or ϵ_0-) zone $\nu(x) \ll 1$, we get the same formulas for the film thickness H_0 and can conduct the asymptotic analysis of the problem equations similar to the one presented in the preceding chapter.

It can be shown that for the case of $\nu(x) \ll 1$ and $B = O(1)$, $\omega \ll 1$, in the inlet (ϵ_q-inlet) zone formulas for the film thickness H_0 still maintain the same structural form as in the preceding chapter. However, the expression for the integral involved in the generalized Reynolds equation (20.126) and, therefore, the asymptotically valid in the inlet and exit zones equations cannot be effectively simplified.

Assuming that $B \gg 1$ from equation (20.149), we find

$$B = \frac{s_0 \kappa \delta_T H_0}{4\lambda} \tau_0 \gg 1, \quad \omega \ll 1. \qquad (20.153)$$

The further analysis of the problem is similar to the one described for the case of $B \ll 1$.

Notice that the same structural formulas for the film thickness H_0 remain in force even in the case when in the inlet (ϵ_q- and ϵ_0-) zone $|\nu(x)| \leq 1$

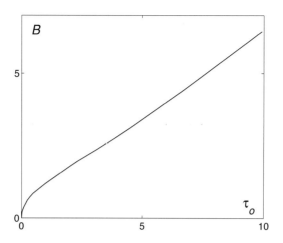

FIGURE 20.12
Dependence of parameter B on τ_0 obtained for $h = \mu^0 = \lambda = 1$, $T_w = 0$, $s_0 \kappa \delta_T H_0 = 2.62$, and $\frac{V_{s0}}{H_0} = 0.18$. (after Kudish and Covitch [1]). Reprinted with permission from CRC Press.

and $\nu(x) = O(1)$. Only the coefficient of proportionality of the order of unity changes.

In the preceding chapter it was shown that for fluid lubricants as B increases the film thickness H_0 is monotonically decreasing. For the case of grease the film thickness H_0 behaves the same way. It can be shown that as the threshold shear stress τ_0 increases the value of B decreases and, therefore, the grease film thickness H_0 decreases too. The graph of parameter B as a function of τ_0 obtained for $h = \mu^0 = \lambda = 1$, $T_w = 0$, $s_0 \kappa \delta_T H_0 = 2.62$, and $\frac{V_{s0}}{H_0} = 0.18$ is given in Fig. 20.12. The value of $B(0)$ corresponds to the value of the parameter B for a fluid lubricant with Newtonian rheology.

Therefore, under thermal contact conditions in a grease flow without cores the grease film thickness may be lower than in a similar fluid lubricant (base oil). In the presence of cores in the grease flow the relationship between the film thicknesses for grease and fluid lubricant can be different.

20.6   Closure

The formulation the problem for a grease lubricated contact is provided. It is shown that the problem is described by a series of different equations of grease flow depending on the presence/absence of cores in the grease flow and their location. Some basic properties of grease flows are established. A detailed solution for the grease lubricated contact of two rigid cylinders is provided for the cases of cores adjacent to cylinders surfaces. It is established that under certain conditions in the provided formulation the solution of the problem may not exist. The changes to the problem formulation, which allow for the problem solution, are suggested. A detailed consideration of isothermal and thermal EHL problems for grease lubricated contacts without cores in the grease flow is provided.

20.7   Exercises and Problems

1. Elaborate on what may happen if the continuity conditions (20.42) and/or (20.43) are not imposed on the problem solution.

2. Provide detailed analysis of the solution properties established in Section 20.3.

3. For rigid cylinders provide the condition that guarantees grease flow without cores. For rigid cylinders compare pressure distributions $p(x)$ for fluid lubricant and grease with cores adjacent to the solid surfaces. What is the main difference between these pressure distributions?

4. Obtain the two-term asymptotic solution for the isothermal grease lubricated contact of rigid cylinders in the case of small stress threshold $\tau_0 \ll 1$.

5. Obtain the two-term asymptotic solution for the isothermal grease lubricated lightly loaded (i.e., $V \gg 1$) contact of elastic cylinders in the case of the possible presence of cores in the grease flow adjacent to solid surfaces (cores detached from the solid surfaces are absent).

6. (a) Show that the value of B from equation (20.149) increases as the stress threshold τ_0 increases. (b) Show that when the value of B increases the film thickness decreases.

References

[1] Kudish, I.I. and Covitch, M.J. 2010. *Modeling and Analytical Methods in Tribology*. Boca Raton: CRC Press.

[2] Wilkinson, W.L. 1960 *Non − Newtonian Fluids : Fluid Mechanics, Mixing and Heat Transfer*. New York: Pergamon Press.

[3] Reiner, M. 1960. *Lectures on Theoretical Rheology*. New York: Interscience Publishers.

[4] Kudish, I.I. 1982. Plane Contact Problems with Viscoplastic Lubrication. *Soviet J. Fric. and Wear* 3, No. 6:1036-1047.

[5] Kudish, I.I. and Semin, V.N. 1983. On Formulation and Analysis of Plane Elastohydrodynamic Problem for Plastic Lubricant. *J. Solid Mech.* 6:107-113.

[6] Volarovich, M.P. and Gutkin, A.M. 1946. Flow of a Viscoplastic Medium Between Two Parallel Plates and in the Gap between Two Coaxial Cylinders. *J. Tech. Phys.* 16, No. 3:321-328.

[7] Mosolov, P.P. and Myasnikov, V.P. 1981. *Mechanics of Rigid − Plastic Media*. Moscow: Nauka.

[8] Kudish, I.I. 1984. Flow of Viscoplastic Medium in a Narrow Gap between Curvilinear Surfaces. *Soviet J. Fric. and Wear* 5, No. 5:841-852.

[9] Van Dyke, M. 1964. *Perturbation Methods in Fluid Mechanics*. New York-London: Academic Press.

[10] Kudish, I.I. 1988. About Analysis of Plane Contact Problems in the Presence of Viscoplastic Lubricant. *Soviet J. Fric. and Wear* 4, No. 3:449-457.

[11] Aihara, S. and Dowson, D. 1980. An Experimental Study of Grease Film Thickness under Elastohydrodynamic Conditions. Part 1. General Results. *J. Jpn. Soc. Lubr. Eng.* 25, No. 4:254-260.

[12] Aihara, S. and Dowson, D. 1980. An Experimental Study of Grease Film Thickness under Elastohydrodynamic Conditions. Part 2. Mechanism of Grease Film Formation. *J. Jpn. Soc. Lubr. Eng.* 25, No. 6:379-386.

[13] Jonkisz, W. and Krzeminski-Freda, H. 1982. Wlasnosti Elastohydrodynamic Znego Filmu Smaru Plastycznego. *Archiwum Budowy Maszyn* 39, No. 1:11-25.

[14] Kudish, I.I. 1983. On Analysis of Plane Contact Problems with Viscoplastic Lubricant. *Soviet J. Fric. and Wear* 3, No. 6:449-457.

21

Non-Steady EHL Problems

21.1 Introduction

In this chapter we consider several non-steady and mixed lubrication models for infinite journal bearings and non-conformal contacts. Proper modified formulations and some solutions of these problems are presented. The comparison of the solutions of the traditional and modified non-steady problems is provided. A formulation and analysis of a mixed friction problem with zones of dry and lubricant friction are considered. The necessary boundary conditions at the internal contact boundaries separating dry and lubricated zones are considered.

21.2 Properly Formulated Non-Steady Plane Lubrication Problems for Contacts of Rigid Solids

This section is dedicated to analysis of two non-steady problems: one for a conformal and the other for non-conformal lubricated contacts. Solutions for lubricated weightless joints involved in a non-steady motion have been studied by Safa and Gohar [1], Ai and Cheng [2], Chang et al. [3], Venner and Lubrecht [4, 5], and Osborn and Sadeghi [6], Cha and Bogy [7], Hashimoto and Mongkolwongrojn [8], Peiran and Shizhu [9, 10]. Non-steady lubrication problems for heavy joints were studied by San Andres and Vance [11], Larsson and Hoglund [12] and Kudish and Panovko [13].

The main purpose of the analysis of this section is to propose the proper formulations of non-steady problems for conformal and non-conformal contacts and to show the difference between the "classic" solutions and the solutions obtained based on the proper problem formulations. In both cases the approaches used and the conclusions made are very similar and follow the material in [14, 15].

21.2.1   A Proper Formulated Non-steady Plane Lubrication Problem for a Non-Conformal Contact

We will consider a non-steady line contact and show that the "classic" formulation of the problem has some significant defects that need to be fixed to adequately describe a non-steady regime of lubrication. We will see that a proper use of Newton's second law, which takes into account the inertia of the lubricated contact, removes all mentioned deficiencies.

Let us consider two smooth elastic infinite cylinders with parallel axes and radii R_1 and R_2 slowly rolling one over another and choose a motionless Cartesian coordinate system with the origin at the motionless axis of cylinder 2. In this coordinate system cylinder 1 is acted upon by force P along the z-axis (directed across the lubrication film). The linear velocities of the surface points of the cylinders are u_1 and u_2, respectively.

The contact (see Fig. 21.1) is lubricated by a Newtonian incompressible fluid with viscosity μ. The lubrication process is isothermal. The cylinders are fully separated by the lubrication film. Because of slow relative motion of the contact surfaces the inertia forces in the lubrication layer are small compared to the viscous ones. The lubrication thickness is small relative to the characteristic size of the contact and radii of the cylinders. The lubrication conditions are non-steady. That means that linear velocities u_k ($k = 1, 2$) and the applied force P as well as the coordinate of the inlet point x_i may vary with time t. The characteristic time of these variations is considered to be small compared to the period of sound waves in elastic solids. The latter requires a non-steady problem formulation only for fluid flow parameters.

Under these assumptions and in the dimensionless variables

$$(x', a, c) = \tfrac{T}{2R'}(x, x_i, x_e), \quad t' = \tfrac{(u_{10}+u_{20})T}{4R'}t, \quad (p', p'_0) = \tfrac{\pi R'}{TP_0}(p, p_0),$$

$$(h', h'_0, h'^0) = \tfrac{T^2}{2R'}(h, h_0, h^0), \quad \mu' = \tfrac{\mu}{\mu_a}, \quad u' = \tfrac{u_1+u_2}{u_{10}+u_{20}}, \quad P' = \tfrac{P}{P_0}, \qquad (21.1)$$

$$T^2 = \tfrac{P_0}{3\pi\mu_a(u_{10}+u_{20})}, \quad V = \tfrac{3\pi^2\mu_a(u_{10}+u_{20})R'E'}{P_0^2}$$

the problem can be reduced to the "classic" system of integro-differential equations (for convenience primes are omitted) [16]

$$\tfrac{\partial}{\partial x}\{\tfrac{h^3}{\mu}\tfrac{\partial p}{\partial x}\} = u(x)\tfrac{\partial h}{\partial x} + \tfrac{\partial h}{\partial t}, \qquad (21.2)$$

$$h_0(0) = h^0, \quad c(0) = c_0, \quad p(x, 0) = p_0(x), \qquad (21.3)$$

$$p(a, t) = p(c, t) = \tfrac{\partial p(c, t)}{\partial x} = 0, \qquad (21.4)$$

$$h(x, t) = h_0 + x^2 + \tfrac{2}{\pi V}\int\limits_a^c p(y, t)\ln\mid\tfrac{y}{x-y}\mid dy, \qquad (21.5)$$

$$\int\limits_a^c p(y, t)dy = \tfrac{\pi}{2}P(t). \qquad (21.6)$$

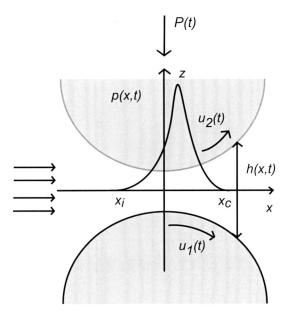

FIGURE 21.1
The general view of the lubricated contact.

In equation (21.1) x_i and x_e are the coordinates of the inlet and exit points, respectively, R' and E' are the effective radius and elastic modulus, respectively, P_0, u_{10}, and u_{20} are the characteristic force applied to the cylinders and their surface velocities, μ_a is the lubricant viscosity at ambient conditions.

In case of a purely squeezed lubrication layer $u(t) = 0$, $c = -a = \infty$ and the boundary conditions in (21.4) must be replaced by the conditions

$$p(\pm\infty, t) = \frac{\partial p(\pm\infty, t)}{\partial x} = 0. \qquad (21.7)$$

The "classic" problem formulation is completely described by equations (21.2)-(21.6) or (21.2), (21.3), (21.5)-(21.7). In addition to these equations, the relations for functions $\mu(p)$, $P(t)$, $u(t)$, $a(t)$ and h^0, c_0, and $p_0(x)$ must be given. The problem solution comprises pressure $p(x, t)$ and gap $h(x, t)$ distributions, exit boundary $c(t)$, and the central film thickness $h_0(t)$.

21.2.1.1 Lubricated Contact of Rigid Solids

Let us consider the case of cylinders made of a rigid material, i.e., $V = \infty$. For simplicity we will assume $\mu(p) = 1$. In this case the problem equations (21.2)-(21.6) can be easily reduced to

$$h_0'(t)\{h_0[\tfrac{1}{h^2(c,t)} - \tfrac{1}{h^2(a,t)}] + c[G(c,t) - G(a,t)]\}$$

$$= u(t)\{2[F(c,t) - F(a,t)] - h(c,t)[G(c,t) - G(a,t)]\},$$

$$h_0(0) = h^0,$$

$$u(t)\{\tfrac{1}{2}\ln\tfrac{h(c,t)}{h(a,t)} + H(a,c,t) - (c-a)F(a,t) \tag{21.8}$$

$$-\tfrac{h(c,t)}{2}[I(a,c,t) - (c-a)G(a,t)]\} - \tfrac{h_0'}{4}\{F(c,t) - F(a,t)$$

$$-\tfrac{2(c-a)h_0}{h^2(a,t)} + 2c[I(a,c,t) - (c-a)G(a,t)]\} = \pi P(t)h_0,$$

where

$$F(x,t) = \tfrac{x}{h(x,t)} + \tfrac{1}{\sqrt{h_0}}\arctan\tfrac{x}{\sqrt{h_0}},$$

$$G(x,t) = \tfrac{x}{h^2(x,t)} + \tfrac{3x}{2h_0 h(x,t)} + \tfrac{3}{2h_0^{3/2}}\arctan\tfrac{x}{\sqrt{h_0}},$$

$$H(a,c,t) = \tfrac{1}{\sqrt{h_0}}\{x\arctan\tfrac{x}{\sqrt{h_0}} = \tfrac{\sqrt{h_0}}{2}\ln h(x,t)\}\,|_a^c, \tag{21.9}$$

$$H(a,c,t) = \tfrac{1}{\sqrt{h_0}}\{x\arctan\tfrac{x}{\sqrt{h_0}} = \tfrac{\sqrt{h_0}}{2}\ln h(x,t)\}\,|_a^c,$$

$$I(a,c,t) = \{-\tfrac{1}{2h(x,t)} + \tfrac{3}{4h_0}\ln h(x,t)\}\,|_a^c + \tfrac{3}{2h_0}H(a,c,t).$$

After this system is solved the pressure distribution $p(x,t)$ is determined according to the formula

$$p(x,t) = \tfrac{u(t)}{2h_0}\{F(x,t) - F(a,t) - \tfrac{h(c,t)}{2}[G(x,t) - G(a,t)]\}$$

$$-\tfrac{h_0'}{4h_0}\{h_0[\tfrac{1}{h^2(x,t)} - \tfrac{1}{h^2(a,t)}] + c[G(x,t) - G(a,t)]\}. \tag{21.10}$$

The structure of system (21.8) and (21.9) allows to make a conclusion that if the external force $P(t)$ and/or the sum of linear velocities $u(t)$ experience an abrupt change (finite discontinuity) at some time moment t then $h_0(t)$, $c(t)$, and $h_0'(t)$ change abruptly as well. Therefore, for $V = \infty$ equations (21.5) and (21.10) make pressure $p(x,t)$ and gap $h(x,t)$ distributions discontinuous functions of time t.

To make this obvious let us consider the case of purely squeezed lubrication film ($u(t) = M(t) = 0$) between two rigid cylinders ($V = \infty$). In this case instead of boundary conditions (21.4), we should use conditions (21.7). The solution of the problem is represented by the formulas

$$p(x,t) = \tfrac{1}{4h^2(x,t)}\tfrac{dh_0(t)}{dt}, \tag{21.11}$$

$$h_0(t) = \frac{h^0}{[1+2\sqrt{h^0}\int\limits_0^t P(v)dv]^2}. \tag{21.12}$$

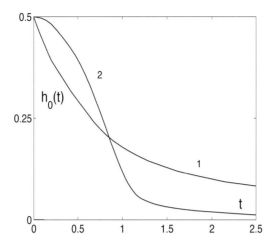

FIGURE 21.2
Graphs of the "classic" (curve marked with 1) and "modified" (curve marked with 2) solutions for $h_0(t)$ for the step load $P(t)$ (after Kudish [14]). Reprinted with permission from the STLE.

A simple analysis of the derivative of $h_0(t)$ from (21.12) shows that it is a discontinuous function of t if $P(t)$ is discontinuous. Based on (21.11) we conclude that in this case pressure $p(x,t)$ is a discontinuous function of time t as well.

The next defect of the "classic" problem formulation reveals itself for any external conditions. Because the highest time derivative in the equations of the problem is of the first order for the given initial external conditions, these equations prescribe a certain initial normal speed $h_0'(0)$ of cylinders approach to each other. This prescribed initial normal speed of approach may not correspond the actual initial speed of approach, which generally can be chosen independently of the other parameters of the problem.

The last defect of this classic problem formulation is the fact that unless the external parameters of the problem are described by oscillatory functions this problem does not provide for the possibility to describe such an oscillatory motion. It is also due to the fact that problem is described by equations with the first–order time derivative. In reality, such a lubricated system being slightly perturbed and then left alone (under constant external conditions) may be involved in a damped oscillatory motion [13].

The steady EHL problems are free of all these defects. These defects are typical for all non-steady EHL problems described based on the presented "classic" problem formulation.

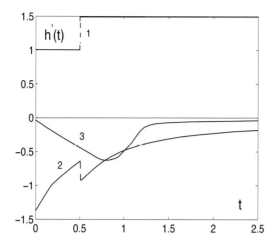

FIGURE 21.3

Graphs of force $P(t)$ (curve marked with 1) and speed $\frac{dh(t)}{dt}$ based on the "classic" (curve marked with 2) and "modified" (curve marked with 3) solutions (after Kudish [14]). Reprinted with permission from the STLE.

21.2.1.2   Lightly Loaded Lubricated Contact of Elastic Solids

Let us consider a lightly loaded regime of lubrication for which the elastic displacements of the contact surfaces are much smaller than the film thickness. Such regimes occur when $V \gg 1$. Assuming that $V \gg 1$ the solution of the problem can be found in the form of regular asymptotic expansions

$$p(x,t) = p_0(x,t) + \tfrac{1}{V}p_1(x,t) + \ldots,$$

$$h(x,t) = h^0(x,t) + \tfrac{1}{V}h^1(x,t) + \ldots, \qquad (21.13)$$

$$h_0(t) = h_0^0(t) + \tfrac{1}{V}h_0^1(t) + \ldots, \quad c(t) = c_0(t) + \tfrac{1}{V}c_1(t) + \ldots,$$

where $h_0^0(t)$, $c_0(t)$, and $p_0(x,t)$ satisfy equations (21.8) and (21.9) while $h^0(x,t)$ is determined by equation (21.5) for $V = \infty$. The next terms of the expansions such as $h_0^1(t)$, $c_1(t)$, $p_1(x,t)$, and $h^1(x,t)$ are solutions of certain linear differential equations the coefficients of which depend on $h_0^0(t)$, $c_0(t)$, $p_0(x,t)$, and $h^0(x,t)$. Moreover, these coefficients are discontinuous functions of time t if $P(t)$ and/or $u(t)$ are discontinuous. Therefore, for $V \gg 1$ and discontinuous $P(t)$ and/or $u(t)$ the solution of the "classic" problem is discontinuous.

The other way to show discontinuity of $p(x,t)$ and, therefore (see equation (21.5)), discontinuity of $h(x,t)$ is to consider equation (21.6). For any

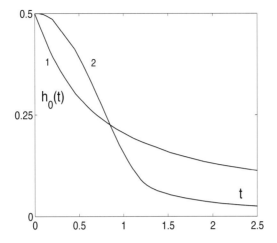

FIGURE 21.4

Graphs of the "classic" (curve marked by 1) and "modified" (curve marked by 2) solutions for $h_0(t)$ for constant load $P(t)$ (after Kudish [14]). Reprinted with permission from the STLE.

$V > 0$ if $P(t)$ is a discontinuous function of time t then, obviously, $p(x, t)$ is discontinuous.

Therefore, it is natural to expect that for any $V > 0$ in the "classic" formulation the solution of EHL problem is discontinuous in time t if $P(t)$ and/or $u(t)$ are discontinuous. This situation needs to be fixed because in reality the parameters of a lubricated contact are always continuous functions of time.

21.2.1.3 Modified Problem Formulation for Non-Steady EHL Problem

The absence of inertia in the "classic" lubrication system creates the discontinuity effects discussed above. In order to avoid these discontinuities and to "push" them up to the second–order time derivatives (i.e., acceleration) we will "soften" the balance condition in equation (21.6) and replace it by the equation naturally following from Newton's second law [14]

$$m\frac{d^2 h_0}{dt^2} = \int\limits_{x_i}^{x_e} p(x, t)dx - P(t), \qquad (21.14)$$

where m is the mass of cylinder 1 per unit length. In equation (21.14) it is assumed that cylinder 2 is either motionless or moves with a constant speed. These conditions for cylinder 2 may be guaranteed if cylinder 2 is of infinite mass.

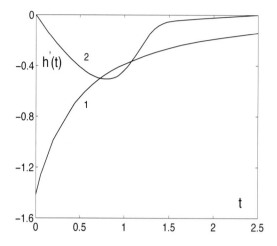

FIGURE 21.5
Graphs of speed $\frac{dh(t)}{dt}$ based on the "classic" (curve marked by 1) and "modified" (curve marked by 2) solutions for constant load $P(t)$ (after Kudish [14]). Reprinted with permission from the STLE.

To complete this problem formulation, we need to impose a different set of initial conditions on h_0 and c

$$h_0(0) = h^0, \ h_0'(0) = h^1, \ c(0) = c_0. \tag{21.15}$$

Here the value of h^1 is independent of the system internal parameters and other operating conditions.

This problem formulation, which includes equations (21.14) and (21.15), is the adequate description of the above considered non-steady EHL problem for any loading and operating conditions.

The introduction of the second–order time derivative allows for oscillatory motion, which cannot be observed based on the "classic" problem formulation. It is important to understand that when the abrupt variations in $P(t)$ and $u(t)$ are long gone and all the external parameters of the system such as $P(t)$, $u(t)$, etc., approach constant values, solutions of both the "classic" and modified EHL problem approach the same steady state. The difference is in the manner it occurs, i.e., it is in the solutions transient behavior. After values of $P(t)$, $u(t)$, and $a(t)$ have stabilized the "classic" solution approaches the steady state monotonically. Depending on the lubrication system parameters, the modified solution may approach the steady state in a monotonic or oscillatory manner.

To illustrate the difference in solutions of the problem based on the "classic" and modified formulations, let us consider the case of purely squeezed lubrication film ($u(t) = 0$) between two rigid ($V = \infty$) cylinders. The "classic" problem solution is described by formulas (21.11) and (21.12). With the help of equations (21.14) and (21.15) and the dimensionless variables from (21.1), the modified problem can be reduced to the following initial-value problem

$$\gamma \frac{d^2 h_0(t)}{dt^2} + \frac{1}{4h_0^{3/2}(t)} \frac{dh_0(t)}{dt} = -P(t), \; h_0(0) = h^0, \; h_0'(0) = h^1, \qquad (21.16)$$

where $\gamma = \frac{m(u_{10}+u_{20})^2}{8P_0 R'}$ is the dimensionless mass of cylinder 1. In this case the pressure distribution is still described by formula (21.11).

An asymptotic solution of equations (21.16) for $h^1 = 0$ and small t is represented by

$$h_0(t) = h^0 - B \int\limits_0^t P(v)[1 - e^{A(v-t)}]dv + O(t^2),$$
$$\qquad (21.17)$$

$$A = \frac{1}{\gamma B}, \; B = 4(h^0)^{3/2}.$$

The comparison of the asymptotic expansions for small t of the "classic" solution from (21.12)

$$h_0(t) = h^0 - B \int\limits_0^t P(v)dv + O(t^2), \; B = 4(h^0)^{3/2}, \qquad (21.18)$$

and the modified one (21.17) shows that for small t the modified solution is greater than the "classic" one obtained for the same values of h^0 and h^1.

The "classic" and modified solutions for $h_0(t)$ are presented in Fig. 21.2. The data for $h_0(t)$ in Fig. 21.2 is obtained with the use of Runge-Kutta method for the case of $\gamma = 1$, $h^0 = 0.5$, $h^1 = 0$, and $P(t) = 1$ for $0 \leq t < 0.5$ and $P(t) = 1.5$ for $t \geq 0.5$. The graphs of the normal velocity $h_0'(t)$ for the "classic" and modified solutions as well as for the applied normal load $P(t)$ are given in Fig. 21.3 for the same initial conditions and load $P(t)$. As it follows from Fig. 21.3, the discontinuity of the normal velocity $h_0'(t)$ at $t = 0.5$ based on the "classic" solution reflects the jump in the load $P(t)$ at the same time moment. For $\gamma = 1$, $h^0 = 0.5$, $h^1 = 0$, and $P(t) = 1$, the graphs of the central film thickness $h_0(t)$ and the normal velocity $h_0'(t)$ based on the "classic" and modified solutions are represented in Figs. 21.4 and 21.5. It follows from Figs. 21.2 and 21.4 that due to the inertia for large time t the film thickness $h_0(t)$ based on the modified solution is smaller than the one based on the "classic" solution.

In a general case of elastic solids ($0 < V < \infty$) and lubricant viscosity dependent on pressure ($\mu(p) \neq 1$), the properties of the "classic" and modified solutions are similar to the ones described above. The modified solution is represented by differentiable functions even if load $P(t)$ and/or the sum of surface speeds $u(t)$ are discontinuous functions of time t.

As a result of this analysis we can conclude that any EHL problem involving non-steady motion such as in cases of non-steady external conditions (applied

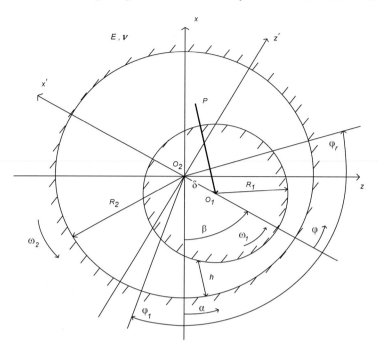

FIGURE 21.6

The general view of the lubricated conjunction (after Kudish [15]). Reprinted with permission from the ASME.

load, surface speeds, supply of lubricant, etc.), rough surfaces, surfaces with dents and/or bumps, surface and/or subsurface cracks should be considered based on a modified problem formulation, which takes into account the system inertia. A similar situation takes places in spatial contacts involved in non-steady motion.

21.2.2   Properly Formulated Non-Steady Plane Lubrication Problems for Journal Bearings

Let us consider a non-steady problem for a conformal lubricated contact of two infinite parallel cylindrical solids. We will show that the "classic" formulation of the non-steady EHL problem leads to discontinuous solutions in the cases of abrupt changes in applied load and surface linear velocities. Moreover, we will show the inability of the "classic" solution to accommodate an arbitrary value of the shaft normal initial velocity. We will perform an analytical analysis of the dynamic bearing response to abrupt changes in external load.

Our main goal is to propose a modified formulation of the EHL problem free from the aforementioned defects. The modified problem will be reduced to a system of nonlinear integro-differential Reynolds and integral equations and equations following from Newton's second law.

21.2.2.1 General Assumptions and "Classic" Problem Formulation

Let us consider a shaft with radius R_1 slowly rolling over a busing with radius R_2 (see Fig. 21.6). The shaft and bushing with smooth surfaces are made of the same material with elastic modulus E and Poison's ratio ν. Their axes are parallel and radii are approximately equal, i.e., $\Delta = R_2 - R_1 \ll R_1$. A Cartesian coordinate system (x, y) with the x-axis across the lubrication film and the z-axis perpendicular to it (and to the cylinders axes) is used. The shaft is acted upon by force $P_0 = (X_0, Z_0)$. The angular velocities of the shaft and bushing are ω_1 and ω_2. The joint is lubricated by a Newtonian incompressible fluid with viscosity μ. The shaft and bushing are fully separated by an isothermal lubrication layer. Inertia forces in the lubrication layer can be neglected compared to viscous forces. The lubrication film thickness is small relative to the size of the contact region and radii of the shaft and bushing. EHL conditions are non-steady due to variations in time t of shaft and bushing angular velocities ω_i, applied force P, and/or the inlet boundary α_1. The characteristic time of these variations is small compared to the period of sound waves in elastic material and, therefore, a non-steady problem is required only for the fluid related parameters.

Let us consider the problem in a moving Cartesian coordinate system (x', z') such that the x'-axis passes through the shaft and bushing centers and the z'-axis is tangent to the bushing surface. The angle between the x- and x'-axes is measured in the counterclockwise direction and is equal $\beta(t) = \arctan \frac{\delta_z(t)}{\delta_x(t)}$, where $\delta_x(t)$ and $\delta_z(t)$ are the x- and z-components of the shaft eccentricity vector. The frictional stresses $\tau_i(\varphi, t)$ and forces Y^i_{fr}, $(i = 1, 2)$ are equal to

$$\tau_1(\varphi, t) = \frac{\mu R(\omega_2 - \omega_1)}{h} - \frac{h}{2R} \frac{\partial p}{\partial \varphi}, \quad \tau_2(\varphi, t) = -\frac{\mu R(\omega_2 - \omega_1)}{h} - \frac{h}{2R} \frac{\partial p}{\partial \varphi},$$

$$(21.19)$$

$$Y^i_{fr}(t) = R \int\limits_{\varphi_l}^{\varphi_r} \tau_i(\theta, t) \cos \theta d\theta,$$

where φ_l and φ_r are the angles that correspond to the inlet and exit points of the contact in the moving coordinate system.

Under the stated assumptions and in any stationary Cartesian coordinate system with the x-axis across the film, the Reynolds equation can be derived in a manner described in the previous chapters (also see [16]). Taking into account that $z = R\alpha$ and $R = 0.5(R_1 + R_2) \approx R_1 \approx R_2$ one can get ($\omega(t) = \omega_1(t) + \omega_2(t)$)

$$\frac{\partial}{\partial \alpha} \left\{ \frac{h^3}{12\mu} \frac{\partial p}{\partial \alpha} \right\} = R^2 \frac{\omega}{2} \frac{\partial h}{\partial \alpha} + R^2 \frac{\partial h}{\partial t}. \quad (21.20)$$

If $\omega(t)$ remains positive, then at the contact boundaries $\alpha = \alpha_l(t)$ and $\alpha = \alpha_r(t)$ it is necessary to impose the conditions

$$p(\alpha_l, t) = p(\alpha_r, t) = 0 \quad (21.21)$$

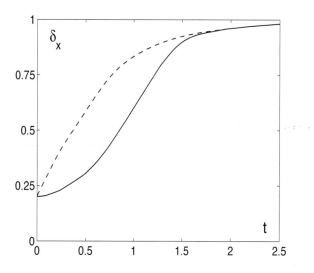

FIGURE 21.7

Graphs of the "classic" (dashed curve) and "modified" (solid curve) solutions for the shaft displacement $\delta_x(t)$ (after Kudish [15]). Reprinted with permission from the ASME.

as well as the cavitation boundary condition

$$\frac{\partial p(\alpha_r,t)}{\partial \alpha} = 0 \qquad (21.22)$$

and the initial conditions

$$\delta_x(0) = \delta_{x0}, \ \delta_z(0) = \delta_{z0}, \ \alpha_r(0) = \alpha_{r0}. \qquad (21.23)$$

If $\omega(t) = 0$, then contact boundaries $\alpha_r(t) = -\alpha_l(t) = \pi$, and it is necessary to impose the conditions

$$p(\pm\pi, t) = \frac{\partial p(\pm\pi,t)}{\partial \alpha} = 0. \qquad (21.24)$$

Now, let us derive the equation for gap $h(\varphi, t)$ in a moving coordinate system. Using the Hertzian assumptions for the radial elastic displacement of the shaft w_{1r}, we get (see [17])

$$w_{1r} + \frac{\partial^2 w_{1r}}{\partial \varphi^2} = \frac{R(1-\kappa)}{4G} p(\varphi, t) + \frac{R(1+\kappa)}{8\pi G} \int\limits_{\varphi_l}^{\varphi_r} \cot \frac{\varphi - \theta}{2} \frac{\partial p(\theta, t)}{\partial \theta} d\theta$$

$$- \frac{1}{2\pi G}(X_0 \cos \varphi + Z_0 \sin \varphi), \qquad (21.25)$$

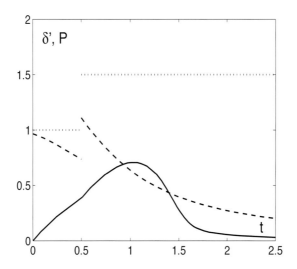

FIGURE 21.8
Graphs of force $P_x(t)$ (dotted curve) and shaft normal speed $d\delta_x(t)dt$ based on the "classic" (dashed curve) and "modified" (solid curve) solutions (after Kudish [15]). Reprinted with permission from the ASME.

where $G = \frac{E}{2(1+\nu)}$, $\kappa = 3 - 4\nu$ for plane deformation and $\kappa = \frac{3-\nu}{1+\nu}$ for generalized plane stress state. Similarly, for the radial elastic displacement of the bushing w_{2r} we have [17]

$$w_{2r} + \frac{\partial^2 w_{2r}}{\partial \varphi^2} = \frac{R(1-\kappa)}{4G} p(\varphi, t) - \frac{R(1+\kappa)}{8\pi G} \int_{\varphi_l}^{\varphi_r} \cot \frac{\varphi - \theta}{2} \frac{\partial p(\theta, t)}{\partial \theta} d\theta$$

$$\tag{21.26}$$

$$+ \frac{R(1+\kappa)}{8\pi G} \int_{\varphi_l}^{\varphi_r} p(\theta, t) d\theta + \frac{\kappa}{2\pi G} (X_0 \cos \varphi + Z_0 \sin \varphi).$$

The film thickness equation can be derived based on the fact that the curvature radii

$$\rho_1(\varphi, t) = R_1 + w_{1r}(\varphi, t), \quad \rho_2(\varphi, t) = R_2 + w_{2r}(\varphi, t) - h(\varphi, t) \tag{21.27}$$

of the two contact surfaces are equal. Thus, the equation for gap h between the shaft and bushing follows from

$$\Delta = w_{1r} + \frac{\partial^2 w_{1r}}{\partial \varphi^2} - w_{2r} - \frac{\partial^2 w_{2r}}{\partial \varphi^2} + h + \frac{\partial^2 h}{\partial \varphi^2}. \tag{21.28}$$

Assuming that functions w_{1r} and w_{2r} are known the latter equation can be considered as a differential equation for h. To solve it for h two initial conditions must be used

$$h(\pi, t) = \Delta + \sqrt{\delta_x^2 + \delta_z^2}, \quad \frac{\partial h(\pi, t)}{\partial \varphi} = 0. \tag{21.29}$$

Integrating equations (21.25)-(21.29) we arrive at the equation

$$h(\varphi, t) = \Delta - \sqrt{\delta_x^2 + \delta_z^2} \cos \varphi$$

$$+ 4R\lambda \int_{\varphi_l}^{\varphi_r} p(\theta, t) \cos(\theta - \varphi) \ln \left| \frac{\cos \frac{\theta}{2}}{\sin \frac{\theta - \varphi}{2}} \right| d\theta$$

$$+ R\lambda(1 + \cos \varphi) \int_{\varphi_l}^{\varphi_r} p(\theta, t) d\theta + 2R\lambda \sin \varphi \int_{\varphi_l}^{\varphi_r} p(\theta, t) \sin \theta d\theta \tag{21.30}$$

$$- 4R\lambda \cos \frac{\varphi}{2} \int_{\varphi_l}^{\varphi_r} p(\theta, t) \frac{\cos \frac{\theta - \varphi}{2}}{\cos \frac{\theta}{2}} d\theta, \quad \lambda = \frac{1 + \kappa}{8\pi G}.$$

To complete the problem formulation, the balance conditions

$$\int_{\varphi_l}^{\varphi_r} p(\theta, t) \cos \theta d\theta = \frac{X_0 \cos \beta + Z_0 \sin \beta}{R},$$

$$\int_{\varphi_l}^{\varphi_r} p(\theta, t) \sin \theta d\theta = \frac{-X_0 \sin \beta + Z_0 \cos \beta}{R} \tag{21.31}$$

and the initial conditions for the components $\delta_x(0)$ and $\delta_z(0)$ of the eccentricity vector and the angle $\varphi_r(0)$

$$\delta_x(0) = \delta_{x0}, \quad \delta_z(0) = \delta_{z0}, \quad \varphi_r(0) = \varphi_{r0} \tag{21.32}$$

must be added to the latter equation. Based on (21.30) and (21.32) and using the given value of $\varphi_l(0)$ the initial condition for h can be calculated.

Equations (21.20)-(21.24) can be rewritten in the moving coordinate system (x', y') taking into account the fact that the line through the shaft and bushing centers is at an angle of $\beta(t)$ (see Fig. 21.6) and

$$\alpha = \varphi + \beta(t), \quad \beta(t) = \arctan \frac{\delta_z(t)}{\delta_x(t)}. \tag{21.33}$$

Using the following dimensionless variables

$$p' = \frac{p}{p_0}, \quad (h', \delta_x', \delta_z') = \frac{1}{\Delta}(h, \delta_x, \delta_z), \quad \mu' = \frac{\mu}{\mu_a}, \quad \omega' = \frac{\omega}{\omega_0},$$

$$(P_x, P_z) = \frac{1}{X_{00}}(X_0, Z_0), \quad s = \frac{\omega_2 - \omega_1}{\omega_0}, \quad t' = \frac{t}{t_0}, \quad p_0 = \frac{X_{00}}{\pi R},$$

$$\omega_0 = \omega_{10} + \omega_{20}, \quad t_0 = \frac{2}{\omega_0}, \quad V = \frac{6\pi \mu_a \omega_0 R^3}{\Delta^2 X_{00}}, \quad \lambda_0 = \frac{(1 + \kappa) X_{00}}{8\pi \Delta G}, \tag{21.34}$$

$$\epsilon = \frac{\Delta}{2R}$$

and omitting primes we scale the problem equations

$$\frac{\partial}{\partial\varphi}\left\{\frac{h^3}{12\mu}\frac{\partial p}{\partial\varphi}\right\} = V\left(\omega - \frac{d\beta}{dt}\right)\frac{\partial h}{\partial\varphi} + V\frac{\partial h}{\partial t}, \quad \beta = \arctan\frac{\delta_z}{\delta_x}, \tag{21.35}$$

$$p(\varphi_l, t) = p(\varphi_r, t) = \frac{\partial p(\varphi_r, t)}{\partial\varphi} = 0, \tag{21.36}$$

$$\delta_x(0) = \delta_{x0}, \ \delta_z(0) = \delta_{z0}, \ \varphi_r(0) = \varphi_{r0}, \tag{21.37}$$

$$h(\varphi, t) = 1 - \sqrt{\delta_x^2 + \delta_z^2}\cos\varphi$$

$$+\frac{4\lambda_0}{\pi}\left\{\int_{\varphi_l}^{\varphi_r} p(\theta, t)\cos(\theta - \varphi)\ln\left|\frac{\cos\frac{\theta}{2}}{\sin\frac{\theta-\varphi}{2}}\right|d\theta\right.$$

$$+\frac{1+\cos\varphi}{4}\int_{\varphi_l}^{\varphi_r} p(\theta, t)d\theta + \frac{\sin\varphi}{2}\int_{\varphi_l}^{\varphi_r} p(\theta, t)\sin\theta d\theta \tag{21.38}$$

$$\left.-\cos\frac{\varphi}{2}\int_{\varphi_l}^{\varphi_r} p(\theta, t)\frac{\cos\frac{\theta-\varphi}{2}}{\cos\frac{\theta}{2}}d\theta\right\},$$

$$\int_{\varphi_l}^{\varphi_r} p(\theta, t)\cos\theta d\theta = \pi[P_x(t)\cos\beta(t) + P_z(t)\sin\beta(t)],$$

$$\tag{21.39}$$

$$\int_{\varphi_l}^{\varphi_r} p(\theta, t)\sin\theta d\theta = \pi[-P_x(t)\sin\beta(t) + P_z(t)\cos\beta(t)].$$

For a rolling/sliding contact the problem is completely described by equations (21.35)-(21.39). For a purely squeezed lubrication layer, $\omega(t) = \beta(t) = 0$ and boundary conditions (21.36) must be replaced by

$$p(\pm\pi, t) = \frac{\partial p(\pm\pi, t)}{\partial\varphi} = 0. \tag{21.40}$$

In equations (21.35)-(21.40) functions $\mu(p)$, $\omega(t)$, $P_x(t)$, $P_z(t)$, $\varphi_l(t)$ and constants V, λ_0, δ_{x0}, δ_{z0}, and φ_{r0} are given. The problem solution consists of functions $p(\varphi, t)$, $h(\varphi, t)$, $\delta_x(t)$, $\delta_z(t)$, and $\varphi_r(t)$. The further analysis except for formulas (21.48) is presented in dimensionless variables for $\varphi_l(t) = \varphi_{l0}$.

21.2.2.2 Purely Squeezed Lubrication Layer between Rigid Solids

Suppose the shaft and bushing are made of a rigid material, $\lambda_0 = 0$. For a purely squeezed lubrication layer $P_z(t) = \omega(t) = \beta(t) = \delta_z(t) = 0$. For simplicity let us assume that $\mu(p) = 1$. In this case instead of conditions (21.36), we should use formulas (21.40). Hence, the problem is described by the following equations

$$\frac{d\delta_x(t)}{dt} = \frac{1}{V}P_x(t)[1 - \delta_x^2]^{3/2}, \ \delta_x(0) = \delta_{x0}. \tag{21.41}$$

The solution of this problem is given by formulas [16]

$$p(\varphi, t) = \frac{V}{2\delta_x(t)} \frac{d\delta_x(t)}{dt} [(1 - \delta_x \cos \varphi)^{-2} - (1 + \delta_x)^{-2}], \tag{21.42}$$

$$\delta_x(t) = \frac{Z(t)}{\sqrt{1+Z^2(t)}}, \quad \delta_x'(t) = \frac{P_x(t)}{V[1+Z^2(t)]^{3/2}},$$

$$\tag{21.43}$$

$$Z(t) = \frac{\delta_{x0}}{\sqrt{1-\delta_{x0}^2}} + \frac{1}{V} \int_0^t P_x(u) du.$$

Analysis of $\delta_x'(t)$ from (21.43) shows that it is a discontinuous function of t if $P_x(t)$ is discontinuous. Based on equations (21.42), (21.43), and (21.38) (for $\lambda_0 = 0$), we can conclude that $p(\varphi, t)$ and $\frac{\partial h(\varphi,t)}{\partial t}$ are discontinuous functions in time t.

21.2.2.3   Lubricated Contact of Rolling Rigid Solids

Again, suppose the shaft and bushing are made of a rigid material, $\lambda_0 = 0$. As before we assume that $P_z(t) = \beta(t) = \delta_z(t) = 0$ and $\mu(p) = 1$. In this case instead of conditions (21.36), we should use formulas (21.40). Hence, the problem is described by the following equations (21.35)-(21.39) can be easily reduced to the following system of differential and algebraic equations

$$\delta_x'(t) = -\omega(t) \frac{r(\delta_x, \varphi_l, \varphi_r)}{q(\delta_x, \varphi_l, \varphi_r)}, \quad \delta_x(0) = \delta_{x0},$$

$$\tag{21.44}$$

$$f(\delta_x, \varphi_l, \varphi_r) - g(\delta_x, \varphi_l, \varphi_r) \frac{r(\delta_x, \varphi_l, \varphi_r)}{q(\delta_x, \varphi_l, \varphi_r)} = \frac{\pi P_x(t)}{V \omega(t)},$$

$$p(\varphi, t) = V\omega(t) U(\delta_x, \varphi_l, \varphi_r) |_\varphi^{\varphi_l} + V\delta_x'(t) W(\delta_x, \varphi_l, \varphi_r) |_\varphi^{\varphi_l}, \tag{21.45}$$

where f, g, r, q, U, and W are certain continuous functions of δ_x, φ, φ, and φ_l. The structure of system (21.44) and (21.45) allows for a very important conclusion to be drawn: if the external force $P_x(t)$ and/or the sum of angular velocities $\omega(t)$ experience an abrupt change (i.e., a finite discontinuity) at some moment in time t then $\delta_x(t)$, $\varphi_r(t)$, and $\delta_x'(t)$ change abruptly too. Therefore, formula (21.45) makes pressure $p(\varphi, t)$ a discontinuous function of time t.

21.2.2.4   Lightly Loaded Lubricated Contact of Rolling Elastic Solids

For a lightly loaded lubricated contact of elastic shaft and bushing ($\lambda_0 \ll 1$), we assume that $P_z(t) = \omega(t) = \beta(t) = \delta_z(t) = 0$ and $\mu(p) = 1$. Assuming that $\lambda_0 \ll 1$ the solution of the problem can be found in the form of a regular perturbation series in powers of λ_0

$$p = p_0(\varphi, t) + \lambda_0 p_1(\varphi, t) + \dots, \quad h = h_0(\varphi, t) + \lambda_0 h_1(\varphi, t) + \dots,$$

$$\tag{21.46}$$

$$\delta_x = \delta_x^0(t) + \lambda_0 \delta_x^1(t) + \dots, \quad \varphi_r = \varphi_r^0(t) + \lambda_0 \varphi_r^1(t) + \dots,$$

where $\delta_x^0(t)$, $\varphi_r^0(t)$, and $p_0(\varphi, t)$ satisfy equations (21.44) and (21.45) while $h^0(\varphi, t)$ is determined by equation (21.38) for the case of rigid solids $\lambda_0 = 0$.

The next terms of these series such as $\delta_x^1(t)$, $\varphi_r^1(t)$, $p_1(\varphi, t)$ and $h^1(\varphi, t)$ are solutions of certain linear problems described by equations which coefficients depend on $\delta_x^0(t)$, $\varphi_r^0(t)$, $p_0(\varphi, t)$ and $h^0(\varphi, t)$. Taking into account that the latter functions are discontinuous if $P_x(t)$ and/or $\omega(t)$ are discontinuous we can conclude that $p_1(\varphi, t)$ and $h^1(\varphi, t)$ are discontinuous functions of time t as well. Therefore, under such external conditions $h(\varphi, t)$ (see equation (21.38)) is discontinuous in time t. Therefore, in this formulation the problem solution for $\lambda_0 \ll 1$ is discontinuous in time t.

21.2.2.5 Arbitrarily Loaded EHL Contact of Rolling Solids

There is another way to show that the solution of the problem in the "classic" formulation is discontinuous for discontinuous $P_x(t)$. If $P_x(t)$ is discontinuous at some moment t, then it follows from equations (21.39) that the function of pressure $p(\varphi, t)$ is also discontinuous. As a result of that for any $V > 0$ and $\lambda_0 > 0$, the function of gap $h(\varphi, t)$ determined by equation (21.38) is discontinuous in time t as well.

It follows from the Reynolds equation (21.35) that when $\omega(t) - \frac{d\beta(t)}{dt}$ is a discontinuous function of time t functions $p(\varphi, t)$ and $h(\varphi, t)$ are discontinuous in time as well.

21.2.2.6 Other Defects of the "Classic" EHL Problem Formulation

The second defect of the "classic" problem formulation besides just established discontinuity of its solution in some cases can be noticed even for continuously varying input parameters, including $P_x(t)$ and $\omega(t) - \frac{d\beta(t)}{dt}$. The "classic" problem formulation is based on the initial-value problem for the first order differential equation. Therefore, only one initial condition can be imposed on the solution of this equation, namely, the condition of the initial position of the shaft. In other words, the "classic" solution prescribes a certain initial shaft velocity for its particular initial position. It follows from the Reynolds equation (21.35). Moreover, the solution of the problem does not exist if the shaft initial velocity is different from the one prescribed by the Reynolds equation. For example, for the case of a purely squeezed lubrication layer, the prescribed initial shaft velocity is (see equation (21.43))

$$\delta_x'(0) = \frac{1}{V} P_x(0)(1 - \delta_{x0}^2)^{3/2}, \tag{21.47}$$

which, apparently, depends on V, the shaft initial position δ_{x0}, and force $P_x(0)$ applied to the shaft at the initial time moment $t = 0$. However, in reality, the shaft initial velocity $\delta_x'(0)$ in most cases is independent of the above parameters and, therefore, is different from the one prescribed by the "classic" solution (21.43). Thus, the solution of the "classic" problem does not exist for an arbitrary initial velocity of the shaft $\delta_x'(0)$. Clearly, this defect of the "classic" problem formulation impairs the very possibility of analysis of such a problem for various realistic initial conditions.

The third defect of the "classic" problem formulation is that the "classic" problem formulation does not allow for oscillatory motion without a presence of an oscillating driving force. For instance, depending on the relationship between the problem parameters a small perturbation in values of external parameters may cause a damped oscillatory behavior of the lubricated system (see the next section and [13]).

Therefore, the "classic" EHL problem solution may be discontinuous if $P_x(t)$ and/or $\omega(t) - \frac{d\beta(t)}{dt}$ change abruptly, it cannot accommodate arbitrary normal initial velocity of the shaft $\delta'_x(0)$, and it does not allow for oscillatory motion. Obviously, such a situation cannot be considered acceptable. Therefore, the "classic" EHL problem is unsatisfactory and it must be modified in such a way that : (a) discontinuity in $p(\varphi, t)$ and $h(\varphi, t)$ would be eliminated, (b) the solution of the problem would exist for any combination of the initial normal shaft position $\delta_x(0)$ and its velocity $\delta'_x(0)$, and (c) the solution would allow for oscillatory motion.

21.2.2.7   Modified Problem Formulation for Non-Steady EHL Problem

It is necessary to mention that none of the above–mentioned defects is present in solutions of any steady EHL problem because the acceleration of the system is zero. However, these defects are typical for non-steady EHL problems such as in the cases of non-steady external conditions (applied load, surface velocities, etc.), rough surfaces, surface dents, and/or bumps, etc.

In the considered "classic" problem formulation, it is assumed that the bushing is stationary and the weightless shaft is involved in a non-steady motion. The absence of inertia of a lubricated system causes the described defects. In order to avoid them, we will "soften" the balance conditions (21.39) and replace them by the equations following from Newton's second law

$$m\frac{d^2\delta_x}{dt^2} = X_0(t) - R \int_{\varphi_l}^{\varphi_r} p(\theta, t)\cos(\theta + \beta)d\theta,$$

$$m\frac{d^2\delta_y}{dt^2} = Z_0(t) - R \int_{\varphi_l}^{\varphi_r} p(\theta, t)\sin(\theta + \beta)d\theta,$$

(21.48)

where m is the mass of the shaft per unit length while the bushing is assumed to be of infinite mass.

The introduction of the second-order time derivatives allows to prevent solution discontinuities, to accommodate an arbitrary normal initial velocity of the shaft, and to consider oscillatory motion.

To illustrate the difference in the problem solutions based on the "classic" and modified formulations, let us consider a purely squeezed lubrication layer $(P_z(t) = \omega(t) = \beta(t) = \delta_z(t) = \lambda_0 = 0)$ between two solids made of a rigid material. The "classic" problem is described by equations (21.42) and

(21.43). With the help of the first equation from (21.48) and the dimensionless variables from (21.34), the modified problem can be reduced to

$$\gamma\frac{d^2\delta_x}{dt^2} + \frac{V\delta_x'(t)}{[1-\delta_x^2(t)]^{3/2}} = P_x(t), \ \delta_x(0) = \delta_{x0}, \ \delta_x'(0) = \delta_{x1}, \tag{21.49}$$

where $\gamma = \frac{\pi m\Delta}{2X_{00}}$ is the dimensionless mass of the shaft and δ_{x1} is the initial shaft normal velocity. In both cases of "classic" and modified problems the pressure distribution is given by formula (21.42).

Now, let us consider an example for $V = 1$, $\gamma = 1$, $\delta_{x0} = 0.2$, and $P_x(t) = 1$ for $t < 0.5$ and $P_x(t) = 1.5$ for $t \geq 0.5$. To obtain the modified problem solution, a Runge-Kutta method is used. The shaft initial velocity $\delta_x'(0)$ dictated by the "classic" solution is 0.9406. Assuming the same value for the shaft initial velocity in the modified solution we obtain that the maximum difference between the "classic" and the modified values of $\delta_x(t)$ is 11%. The shaft velocity based on the modified problem formulation is a smooth function of time t and the one based on the "classic" formulation is discontinuous at $t = 0.5$ when the applied force $P_x(t)$ experiences a discontinuity. At $t = 0.5$ the difference between the shaft velocities in these two cases is 38% and the maximum difference is approximately 166%. As it was shown earlier the "classic" solution does not exist for the shaft initial velocity different from the one given by equation (21.47). If the shaft initial velocity is different from the latter, we may see even greater difference between the modified and the "classic" solutions. For example, for $\delta_{x1} = 0$ and the same as earlier other initial and external parameters the graphs of the "classic" and modified shaft displacement $\delta_x(t)$ are given in Fig. 21.7. The graphs of the shaft normal velocity $\delta_x'(t)$ for the "classic" and modified solutions are presented in Fig. 21.8 for the same values of the parameters V, γ, γ $P_x(t)$, and the initial conditions. It is obvious from Fig. 21.8 that the discontinuity of the shaft speed $\delta_x'(t)$ at $t = 0.5$ is caused by the jump in the load $P_x(t)$.

Numerical calculations show that for a heavier shaft the modified solution differs from the "classic" one more than for the case of a lighter shaft. Heavy shafts (large γ) accelerate slowly. Therefore, for heavy shafts the modified solution is significantly different from the "classic" one. For example, for the solutions presented in Fig. 21.7 the maximum difference between $\delta_x(t)$ based on the modified and "classic" solutions is 0.308. For heavier shaft with $\gamma = 10$ and all other parameters, the same as before the same value is 0.580. Depending on the initial conditions, the changes in the shaft velocity $\delta_x'(t)$ due to the shaft inertia may be even more significant (see Fig. 21.8).

Therefore, any EHL problem involving non-steady motion such as in the cases of non-steady external conditions (applied load, surface velocities, rough surfaces, surfaces with dents and/or bumps, surface and/or subsurface cracks, etc.) should be analyzed based on the problem formulation similar to the modified one, i.e., taking into account the inertia of the shaft/bushing joint and Newton's second law. The same situation takes place in spatial EHL problems for conformal and non-conformal lubricated contacts involved in a non-steady motion.

21.3   Non-Steady Lubrication Problems for a Journal Bearing with Dents and Bumps

Numerous papers have been dedicated to solution of steady EHL problems for journal bearings (Hamrock [16], Allair and Flack [18], Ghosh et al. [19], Lin and Rylander [20]). Obviously, solutions of non-steady EHL problems are as important as solutions of steady ones. In many practical cases bearing durability and premature failure are to a great extent determined by regimes of starting, stopping, acceleration, deceleration, and riding on a bumpy road. Therefore, it is important to be able to analyze these truly non-steady cases. There have been published very few papers on non-steady motion of journal bearings (see references in the preceding section). Even fewer papers treat the non-steadiness in a proper manner, taking into account the shaft and bushing masses involved in a non-steady motion. For non-conformal lubricated contacts such problems are considered by Wijnant et al. [21].

The approach to the problem is based on the "modified" problem formulation proposed in the preceding section and used in [22]. The problem solution is free of certain defects such as discontinuity and independence from the initial shaft speed. The problem is reduced to a system of the nonlinear Reynolds equation and integral equations describing lubrication and contact interaction between elastic solids. The additional conditions include initial and boundary conditions and Newton's second law applied to the shaft motion. The main emphasis of the paper is threefold: the analysis of the transient dynamics of the system under constant external conditions, the analysis of the transient dynamics of the system due to abrupt changes in applied load, and the system behavior in the case of a bump/dent presence on the shaft surface. The numerical analysis shows that usually the solution exhibits oscillatory behavior while approaching a steady state. The specific features of the solution behavior depend on the relationship of such parameters as shaft mass, speed, applied load, lubricant viscosity, etc. Under constant external conditions it is observed that in the process of transient motion the radial displacement of the shaft center may vary by no more than 2.5% while the maximum pressure varies by at least 15% and by as much as 350%. Moreover, the variations of pressure are greater for stiffer materials and heavier shafts.

21.3.1   General Assumptions and Problem Formulation

Let us consider a lubricated contact of a smooth shaft with radius R_1 and mass m per unit length slowly moving with angular velocity ω over a motionless bushing with radius R_2. The axes of the shaft and bushing are parallel and the radii are approximately equal, i.e., $\Delta = R_2 - R_1 \ll R_1$. The shaft is acted upon by force $\vec{F}$ and by moment M_0 about the shaft axis.

To simplify the further analysis, let us assume that both the shaft and bushing are made from the same elastic material, for example, steel ($G = 0.5E/(1 + \nu)$, $\kappa = 3 - 4\nu$ for plane deformation and $\kappa = (3 - \nu)/(1 + \nu)$ for generalized plane stress state). It is necessary to point out that in practice the materials of the shaft and bushing are generally different. In case of different materials the problem equations and the analysis are slightly more complicated, however, it does not change significantly the predictions derived from the case of same materials. The joint (see Fig. 21.6) is lubricated by a Newtonian incompressible fluid with viscosity μ. The lubrication process is isothermal. The shaft and bushing are fully separated by the lubrication layer. Because of slow relative motion of the contact surfaces the inertial forces in the lubrication layer are small compared to the viscous forces and can be neglected. The lubrication film thickness is considered to be small relative to the size of the contact region and radii of the shaft and bushing. Under non-steady lubrication conditions, the shaft angular velocity ω, applied force $\vec{F}$, and moment M_0 may vary in time t. The characteristic time of these variations is small compared to the period of sound waves in elastic solids. The latter requires a non-steady problem formulation only for fluid flow parameters.

It is convenient to consider this problem in a moving coordinate system (x', y') shown in Fig. 21.6 with the origin at the bushing center, the x'-axis along the line connecting the centers of the shaft and bushing (across the film thickness), and the y'-axis tangent to the bushing surface. Suppose the center line is moving and is currently at an angle of $\beta = \beta(t)$ with the fixed vertical line shown in Fig. 21.6. The angle φ is measured from the x-axis in the counterclockwise direction. Let us suppose that the total force applied to the shaft $\vec{F}$ is characterized by two components: X_0 - along the x-axis and Y_0 - along the y-axis, i.e., $tan\beta(t) = Y_0/X_0$. Then, assuming that $\omega(t)$ and $\beta(t)$ are given moment $M_0(t)$ applied to the shaft can be determined after the solution is obtained by calculating the moment of friction force applied to the shaft (integral of $p(\varphi, t)$ from (21.56)) and the first derivative of $\omega(t)$ (see Kudish et al. [15]). Let us introduce the dimensionless variables as follows

$$(p', p'_*, \tau'_1, \tau'_2) = \tfrac{1}{p_0}(p, p_*, \tau_1, \tau_2), \ (h', \delta', \delta'_*) = \tfrac{1}{\Delta}(h, \delta, \delta_*),$$

$$(21.50)$$

$$\mu' = \tfrac{\mu}{\mu_a}, \ \omega' = \tfrac{\omega}{\omega_0}, \ X = \tfrac{\sqrt{X_0^2 + Y_0^2}}{X_{00}}, \ Y'_f r^i = \tfrac{Y_f r^i}{X_{00}}, \ t' = \tfrac{t}{t_0}$$

$$\delta'_{1*} = \tfrac{t_0}{\Delta}\delta_{1*}, \ p_0 = \tfrac{X_{00}}{\pi R}, \ \omega_0 = \omega(0), \ t_0 = \tfrac{2}{\omega_0},$$

$$V = \tfrac{6\pi\mu_a\omega_0 R^3}{\Delta^2 X_{00}}, \ \lambda_0 = \tfrac{(1+\kappa)X_{00}}{8\pi\Delta G}, \ \varepsilon = \tfrac{\Delta}{2R}, \ \gamma = \varepsilon\tfrac{m\omega_0^2 R}{2X_{00}},$$

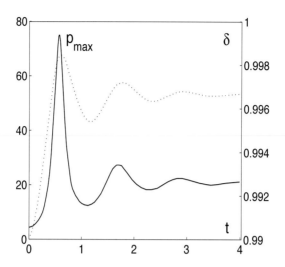

FIGURE 21.9

Distribution of the maximum pressure p_{max} (solid curve) and shaft displacement δ (dotted curve) versus time t for rigid materials, $X = 1$, $\delta_* = 0.99$, $\delta_{1*} = 0.005$, and $\gamma = 10$ (after Kudish [22]). Reprinted with permission from the ASME.

$(R = 0.5(R_1 + R_2) \approx R_1 \approx R_2)$, and write down the problem governing equations Kudish et al. [15] in the moving coordinate system in the dimensionless form (primes at dimensionless variables are omitted)

$$\frac{\partial}{\partial \varphi}\{\frac{h^3}{\mu}\frac{\partial p}{\partial \varphi}\} = V(\omega - \frac{d\beta}{dt})\frac{\partial h}{\partial \varphi} + V\frac{\partial h}{\partial t}, \qquad (21.51)$$

$$p(\varphi, 0) = p_*(\varphi), \ \varphi_l(0) = \varphi_{l*}, \ \varphi_r(0) = \varphi_{r*}, \qquad (21.52)$$

$$p(\varphi_l, t) = p(\varphi_r, t) = 0, \ \frac{\partial p(\varphi_r, t)}{\partial \varphi} = 0, \qquad (21.53)$$

$$h(\varphi, t) = 1 - \delta \cos\varphi + \frac{2}{\pi}\lambda_0 \int_{\varphi_l}^{\varphi_r} p(\theta, t)L_1(\theta, \varphi)d\theta,$$

$$L_1(\theta, \varphi) = 2cos(\theta - \varphi)\ln|\ \frac{\cos\frac{\theta}{2}}{\sin\frac{\theta-\varphi}{2}}\ | -2\cos\frac{\varphi}{2}\frac{\cos\frac{\theta-\varphi}{2}}{\cos\frac{\theta}{2}} \qquad (21.54)$$

$$+\frac{1+\cos\varphi}{2} + \sin\varphi\sin\theta,$$

$$\gamma\frac{d^2\delta}{dt^2} = X - P, \ P = \frac{1}{\pi}\int_{\varphi_l}^{\varphi_r} p(\theta, t)\cos\theta d\theta, \ \delta(0) = \delta_*, \ \frac{d\delta(0)}{dt} = \delta_{1*}, \qquad (21.55)$$

where $\varphi_l(t)$ and $\varphi_r(t)$ are the inlet and exit angular coordinates of the contact region in the moving coordinate system, $\delta(t) = \delta_x(t)$ and $\delta_y(t) = 0$. In

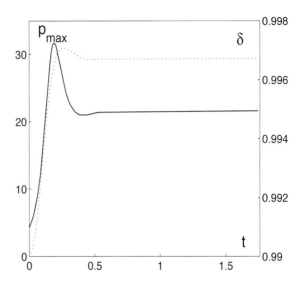

FIGURE 21.10
Distribution of the maximum pressure p_{max} (solid curve) and shaft displacement δ (dotted curve) versus time t for rigid materials, $X = 1$, $\delta_* = 0.99$, $\delta_{1*} = 0.005$, and $\gamma = 1$ (after Kudish [22]). Reprinted with permission from the ASME.

equations (21.51)-(21.55) functions $\mu(p)$, $\omega(t)$, $\beta(t)$, $X(t), \varphi_l(t)$, $p_*(\varphi)$ and constants V, λ_0, δ_*, δ_{1*}, φ_{l*}, φ_{r*} are given. The solution for this problem consists of functions $p(\varphi,t)$, $h(\varphi,t)$, $\delta(t)$, and $\varphi_r(t)$. In the further analysis, the inlet angle coordinate $\varphi_l(t)$ is considered to be equal to a constant φ_{l*}.

After the problem is solved, the frictional stresses $\tau_i(\varphi,t)$ and friction forces Y_{fr}^i $(i=1,2)$ are determined from equations

$$\tau_{1,2}(\varphi,t) = -\frac{\varepsilon V}{3}\frac{\mu\omega}{h} \mp \varepsilon h \frac{\partial p}{\partial \varphi},$$

$$\{Y_{fr}^1, Y_{fr}^2\} = \frac{1}{\pi} \int_{\varphi_l}^{\varphi_r} \{\tau_1(\theta,t), \tau_1(\theta,t)\} \cos\theta d\theta. \tag{21.56}$$

21.3.2 Case of Rigid Materials

First, let us consider rigid materials, i.e., $\lambda_0 = 0$. In the case of constant viscosity $\mu = 1$, angular velocity $\omega(t) = 1$, and angle $\beta(t) = 0$, the solution of the problem (21.51)-(21.55) for $p(\varphi,t)$ can be obtained in an analytical form

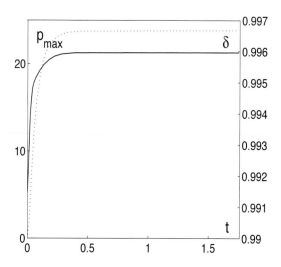

FIGURE 21.11

Distribution of the maximum pressure p_{max} (solid curve) and shaft displacement δ (dotted curve) versus time t for rigid materials, $X = 1$, $\delta_* = 0.99$, $\delta_{1*} = 0.005$, and $\gamma = 0.1$ (after Kudish [22]). Reprinted with permission from the ASME.

$$p(\varphi, t) = \frac{V\delta}{2(1-\delta^2)}[\cos\varphi_r H_1(\delta, \varphi_l, \varphi) - H_2(\delta, \varphi_l, \varphi)]$$

$$+ \frac{V\delta'}{2(1-\delta^2)}[\sin\varphi_r H_1(\delta, \varphi_l, \varphi) + \frac{1-\delta^2}{\delta}(\frac{1}{h^2} - \frac{1}{h_l^2})],$$

$$H_1(\delta, \varphi_l, \varphi) = \delta[\frac{\sin\varphi}{h^2} - \frac{\sin\varphi_l}{h_l^2} + \frac{3}{1-\delta^2}(\frac{\sin\varphi}{h} - \frac{\sin\varphi_l}{h_l})]$$

$$+ \frac{2+\delta^2}{1-\delta^2}H_3(\delta, \varphi_l, \varphi),$$

$$H_2(\delta, \varphi_l, \varphi) = \frac{\sin\varphi}{h^2} - \frac{\sin\varphi_l}{h_l^2} + \frac{1+2\delta^2}{1-\delta^2}(\frac{\sin\varphi}{h} - \frac{\sin\varphi_l}{h_l})$$

$$+ \frac{3}{1-\delta^2}H_3(\delta, \varphi_l, \varphi), \quad H_3(\delta, \varphi_l, \varphi) = \frac{2}{1-\delta^2}\arctan(\sqrt{\frac{1+\delta}{1-\delta}}\tan\frac{\varphi}{2})\,|^\varphi_{\varphi_l},$$

(21.57)

where $h_l = h(\varphi_l, t)$, $h_r = h(\varphi_r, t)$, and $\delta(t)$ and $\varphi_r(t)$ satisfy the following system of equations (see equations (21.53) and (21.55))

$$\delta'[\sin\varphi_r H_1(\delta, \varphi_l, \varphi_r) + \frac{1-\delta^2}{\delta}(\frac{1}{h_r^2} - \frac{1}{h_l^2})] \quad\quad (21.58)$$

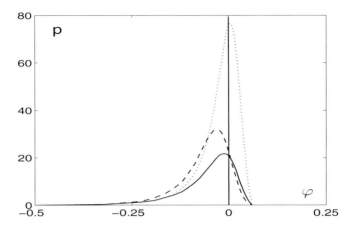

FIGURE 21.12

Pressure distribution $p(\varphi, t)$ for $\gamma = 10$ (dotted curve), (b) $\gamma = 1$ (dashed curve), and (c) $\gamma = 0.1$ (solid curve) for rigid materials, $X = 1$, $\delta_* = 0.99$, $\delta_{1*} = 0.005$ at the time moments t at which the maximum pressure reaches its maximum value (after Kudish [22]). Reprinted with permission from the ASME.

$$+\delta[\cos\varphi_r H_1(\delta, \varphi_l, \varphi_r) - H_2(\delta, \varphi_l, \varphi_r)] = 0,$$

$$\gamma\delta'' = X - P, \ \delta(0) = \delta_*, \ \delta'(0) = \delta_{1*},$$

$$P = \tfrac{V\delta}{2\pi(1-\delta^2)}[\cos\varphi_r H_{1i}(\delta, \varphi_l, \varphi_r) - H_{2i}(\delta, \varphi_l, \varphi_r)]$$

$$+\tfrac{V\delta'}{2\pi(1-\delta^2)}\{\sin\varphi_r H_{1i}(\delta, \varphi_l, \varphi_r) + \tfrac{1-\delta^2}{\delta}[H_i(\delta, \varphi_l, \varphi_r) - \tfrac{\sin\varphi_r - \sin\varphi_l}{h_l^2}]\},$$

$$H_{1i}(\delta, \varphi_l, \varphi_r) = \delta[I_2(\delta, \varphi_l, \varphi_r) + \tfrac{3}{1-\delta^2}I_1(\delta, \varphi_l, \varphi_r)$$

$$-\tfrac{\sin\varphi_l}{h_l^2}(1 + \tfrac{3h_1}{1-\delta^2})(\sin\varphi_r - \sin\varphi_l)] + \tfrac{2+\delta^2}{1-\delta^2}I_3(\delta, \varphi_l, \varphi_r),$$

$$H_{2i}(\delta, \varphi_l, \varphi_r) = I_2(\delta, \varphi_l, \varphi_r) + \tfrac{1+2\delta^2}{1-\delta^2}I_1(\delta, \varphi_l, \varphi_r) \tag{21.59}$$

$$-\tfrac{\sin\varphi_l}{h_l^2}(1 + \tfrac{1+2\delta^2}{1-\delta^2}h_1)(\sin\varphi_r - \sin\varphi_l) + \tfrac{3\delta}{1-\delta^2}I_3(\delta, \varphi_l, \varphi_r),$$

$$H_i(\delta, \varphi_l, \varphi_r) = \tfrac{1}{1-\delta^2}[\tfrac{\sin\varphi_r}{h_r} - \tfrac{\sin\varphi_l}{h_l}]$$

$$+\tfrac{2\delta}{(1-\delta^2)^{3/2}}\arctan[\sqrt{\tfrac{1+\delta}{1-\delta}}\tan\tfrac{\theta}{2}]\mid_{\varphi_l}^{\varphi_r},$$

$$I_1(\delta, \varphi_l, \varphi_r) = \tfrac{1}{\delta^2}[-h_r + h_l + \ln\tfrac{h_r}{h_l}],$$

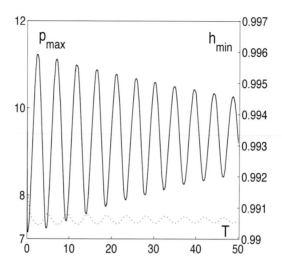

FIGURE 21.13

Distribution of the maximum pressure p_{max} (solid curve) and the minimum film thickness h_{min} (dotted curve) versus time T for $\gamma = 10$, $\delta_{1*} = 0$, and $X = 1$ (after Kudish [22]). Reprinted with permission from the ASME.

$$I_2(\delta, \varphi_l, \varphi_r) = \tfrac{1}{\delta^2}\left[-\tfrac{1}{h_r} + \tfrac{1}{h_l} - \ln\tfrac{h_r}{h_l}\right],$$

$$I_3(\delta, \varphi_l, \varphi_r) = \sin\varphi_r H_3(\delta, \varphi_l, \varphi_r) - \tfrac{1}{\delta}\ln\tfrac{h_r}{h_l}.$$

The numerical solution of equations (21.58) and (21.59) can be performed by reducing these equations to a system of first–order differential equations that are discretized using the trapezoidal rule. Let us discuss the numerical results obtained for the case of a constant load $X = 1$, $V = 0.025$, $\varphi_l = -1.309$, $\delta_* = 0.99$, and $\delta_{1*} = 0.005$. Figures 21.9-21.11 illustrate the transient behavior of the maximum pressure p_{max} with time t for the system inertia $\gamma = 10$, 1, 0.1. It is clear from Fig. 21.9 that fluctuations of pressure are greater for greater system inertia γ. Moreover, small variations of the shaft displacement δ of about 0.7% of its steady value ($\delta = 0.99665$) cause variations in the maximum pressure p_{max} of about 350% for $\gamma = 10$ (see Fig. 21.9) and 150% for $\gamma = 1$ (see Fig. 21.10) of its steady value ($p_{max} = 21.26056$), respectively. For $\gamma = 10$, 1, 0.1, the pressure distributions $p(\varphi, t)$ versus φ at the time moments t at which the maximum pressure p_{max} reaches its maximum value are given in Fig. 21.12. In Fig. 21.12, for $\gamma = 0.1$ the pressure distribution $p(\varphi, t)$ versus φ is the steady pressure distribution.

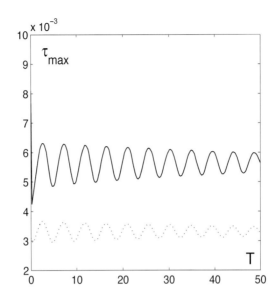

FIGURE 21.14
Distribution of the maximum frictional stresses τ_1 and τ_2 applied to both contact surfaces versus time T for $\gamma = 10$, $\delta_{1*} = 0$, and $X = 1$ (solid curve - lower surface, dotted curve - upper surface) (after Kudish [22]). Reprinted with permission from the ASME.

21.3.3 Contact Region Transformation

Equations (21.51)-(21.55) are written for a contact region with one free boundary - the exit coordinate $\varphi_r(t)$. The existence of the free boundary makes the numerical solution of the problem for elastic solids (i.e., for $\lambda_0 > 0$) complicated. To avoid that we introduce a simple transform

$$T = t, \quad \varphi = \frac{\varphi_r + \varphi_l}{2} + \frac{\varphi_r - \varphi_l}{\pi} \psi. \tag{21.60}$$

For $\varphi_l(t) = \varphi_{l*}$ by using transform (21.60) equations (21.51)-(21.55) can be rewritten in the form

$$\frac{\partial}{\partial \psi} \{ \frac{h^3}{\mu} \frac{\partial p}{\partial \psi} \} = V \frac{\varphi_r - \varphi_l}{\pi} [\omega - \beta' - \frac{\varphi_r' + \varphi_l'}{2} - \frac{\varphi_r' - \varphi_l'}{\pi} \psi] \frac{\partial h}{\partial \psi}$$
$$+ V (\frac{\varphi_r - \varphi_l}{2})^2 \frac{\partial h}{\partial T}, \tag{21.61}$$

$$p(\psi, 0) = p_*(\psi), \quad \varphi_r(0) = \varphi_{r*}, \tag{21.62}$$

$$p(\pm \frac{\pi}{2}, t) = 0, \quad \frac{\partial p(\frac{\pi}{2}, t)}{\partial \psi} = 0, \tag{21.63}$$

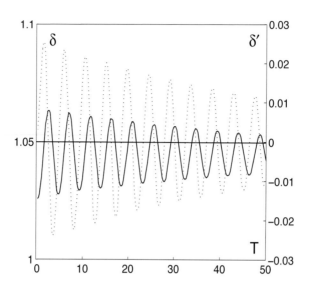

FIGURE 21.15
Distribution of shaft radial displacement δ (solid curve) and its radial speed
δ' (dotted curve) versus time T for $\gamma = 10$, $\delta_{1*} = 0$, and $X = 1$ (after Kudish
[22]). Reprinted with permission from the ASME.

$$h = 1 - \delta(T) \cos \Lambda(\psi)$$

$$+2\lambda_0 \frac{\varphi_r - \varphi_l}{\pi^2} \int_{-\pi/2}^{\pi/2} p(\Theta, T) L_1(\Theta, \psi, \varphi_l, \varphi_r) d\Theta,$$

$$L_1(\Theta, \psi, \varphi_l, \varphi_r) = 2\cos\left(\frac{\varphi_r - \varphi_l}{\pi}(\Theta - \psi)\right) \ln \mid \frac{\cos\frac{\Lambda(\Theta)}{2}}{\sin[\frac{\varphi_r - \varphi_l}{2\pi}(\Theta - \psi)]} \mid \qquad (21.64)$$

$$-2\cos[\frac{\varphi_r - \varphi_l}{2\pi}(\Theta - \psi)]\frac{\cos\frac{\Lambda(\psi)}{2}}{\cos\frac{\Lambda(\Theta)}{2}} + \frac{1}{2}\cos\Lambda(\psi) + \frac{1}{2}$$

$$+\sin\Lambda(\psi)\sin\Lambda(\Theta),\ \Lambda(\psi) = \frac{\varphi_r + \varphi_l}{2} + \frac{\varphi_r - \varphi_l}{\pi}\psi,$$

$$\gamma\frac{d^2\delta}{dT^2} = X - P,\ \delta(0) = \delta_*,\ \frac{d\delta(0)}{dT} = \delta_{1*},$$

$$P = \frac{\varphi_r - \varphi_l}{\pi^2} \int\limits_{-\pi/2}^{\pi/2} p(\Theta, T) \cos\Lambda(\Theta)d\Theta, \qquad (21.65)$$

where $\beta'(T) = d\beta(T)/dT$, etc.

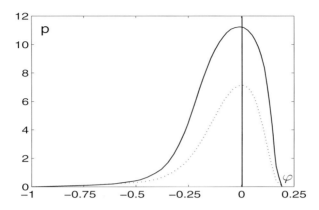

FIGURE 21.16
The maximum pressure distribution at the moments when the reaction force P applied to the surfaces reaches its minimum ($T = 0.2896$, $p_{max} = 7.0965$ - dotted curve) and maximum ($T = 2.4432$, $p_{max} = 11.2482$ - solid curve), $\gamma = 10$, $\delta_{1*} = 0$, and $X = 1$ (after Kudish [22]). Reprinted with permission from the ASME.

21.3.4 Quadrature Formula and Discretization

Let us introduce two sets of nodes: $\{\psi_k\}$, $k = 1, \ldots, N$, $\psi_1 = -\pi/2$, $\psi_N = \pi/2$, and $\{\psi_{j+1/2}\}$, $\psi_{j+1/2} = 0.5(\psi_j + \psi_{j+1})$, $j = 1, \ldots, N-1$. Further, the low index indicates the node number. We make use of the second–order accuracy trapezoidal quadrature formula for computation of the integrals involved in equations (21.64) and (21.65). Based on that we obtain

$$\int_{-\pi/2}^{\pi/2} L(\Theta, \psi_{j+1/2})p(\Theta, T)d\Theta = \frac{1}{2}\sum_{k=1}^{N-1}(\psi_{k+1} - \psi_k)$$

$$\times [L(\psi_k, \psi_{j+1/2})p_k(T) + L(\psi_{k+1}, \psi_{j+1/2})p_{k+1}(T)],$$

(21.66)

where $p_k(T) = p(\psi_k, T)$ and $L(\Theta, \psi)$ is a given kernel.

Let us introduce a set of time moments $\{T_i\}$, $T_0 = 0$, $T_{i+1} = T_i + \triangle T_i$, $i = 0, 1, \ldots$. To provide for better numerical stability, an implicit numerical scheme of the first order of accuracy with respect to time and the second order of accuracy with respect to angle is used. Therefore, when integrating equation (21.61) with respect to time T over the interval $[T_{i-1}, T_i]$, we use the following approximations:

$$\int_{T_{i-1}}^{T_i} f(\psi, T)dT \approx \triangle T_{i-1}f(\psi, T_i),$$

$$\int_{T_{i-1}}^{T_i} \beta'(T)f(\psi, T)dT \approx [\beta(T_i) - \beta(T_{i-1})]f(\psi, T_i).$$

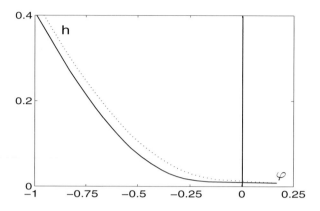

FIGURE 21.17
Gap distributions at the moments when the reaction force P applied to the surfaces reaches its minimum ($T = 0.2896$, $p_{max} = 7.0965$ - dotted curve) and maximum ($T = 2.4432$, $p_{max} = 11.2482$ - solid curve), $\gamma = 10$, $\delta_{1*} = 0$, and $X = 1$ (after Kudish [22]). Reprinted with permission from the ASME.

In a similar fashion, an implicit numerical scheme of the second order of accuracy with respect to both time and angle can be obtained by using the trapezoidal rule of integration not only with respect to angle but also with respect to time. However, that scheme will practically double the amount of necessary calculations without much improvement in solution stability.

The second–order differential equation (21.65) for δ is reduced to a system of two first–order differential equations and then discretized using the trapezoidal rule

$$\delta_i = \delta_{i-1} + \eta_{i-1}\triangle T_{i-1}$$

$$+[X(T_i) + X(T_{i-1}) - P(T_i) - P(T_{i-1})]\tfrac{\triangle T_{i-1}^2}{4\gamma},$$

$$\eta_i = \eta_{i-1} + [X(T_i) + X(T_{i-1}) - P(T_i) - P(T_{i-1})]\tfrac{\triangle T_{i-1}}{2\gamma},$$

$$\quad\quad\quad\quad\quad\quad\quad (21.67)$$

$$i = 1, 2, \ldots,$$

where $\delta_0 = \delta_*$ and $\eta_0 = \delta_{1*}$ are given in (21.65) and $\triangle T_i = T_i - T_{i-1}$.

Now, let us integrate equation (21.61) with respect to T over the interval $[T_{i-1}, T_i]$ ($i = 1, 2, \ldots$) and, then, integrate the obtained result with respect to ψ over interval $[\psi_{j-1/2}, \psi_{j+1/2}]$ ($j = 2, \ldots, N-1$). Discretizing in time and space the obtained equation and equations (21.62)-(21.64) give the following formulas:

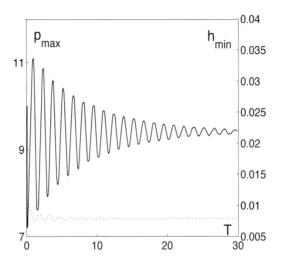

FIGURE 21.18

Distributions of the maximum pressure p_{max} (solid curve) and the minimum film thickness h_{min} (dotted curve) versus time T for $\gamma = 1$, $\delta_{1*} = 0$, and $X = 1$ (after Kudish [22]). Reprinted with permission from the ASME.

$$\frac{\triangle T_i}{V}\left\{\frac{h_{j+1/2}^3}{\mu_{j+1/2}}\frac{p_{j+1}-p_j}{\psi_{j+1}-\psi_j} - \frac{h_{j-1/2}^3}{\mu_{j-1/2}}\frac{p_j-p_{j-1}-}{\psi_j-\psi_{j-1}}\right\}$$

$$= \frac{\varphi_r-\varphi_l}{\pi}\left\{[\Omega - \frac{\varphi_r+\varphi_l-\Phi}{2}](h_{j+1/2}-h_{j-1/2}) - \frac{\varphi_r-\varphi_l-\Psi}{\pi}\left[\psi_{j+1/2}h_{j+1/2}\right.\right.$$

$$\left.-\psi_{j-1/2}h_{j-1/2} - \int\limits_{\psi_{j-1/2}}^{\psi_{j+1/2}} hd\psi\right]\right\} + (\frac{\varphi_r-\varphi_l}{2})^2[\int\limits_{\psi_{j-1/2}}^{\psi_{j+1/2}} hd\psi - F_j],$$

$$\Omega = (\omega - \beta')\triangle T_{i-1}, \ \ \Phi = \varphi_r + \varphi_l, \ \ \Psi = \varphi_r - \varphi_l, \tag{21.68}$$

$$F_j = \int\limits_{\psi_{j-1/2}}^{\psi_{j+1/2}} hd\psi, \ \ \int\limits_{\psi_{j-1/2}}^{\psi_{j+1/2}} hd\psi = [\psi - \frac{\pi\delta}{\varphi_r-\varphi_l}\sin\Lambda(\psi)] \mid_{\psi_{j-1/2}}^{\psi_{j+1/2}} \tag{21.69}$$

$$+\lambda_0\frac{2}{\pi}\int\limits_{-\pi/2}^{\pi/2_{j+1/2}} p(\Theta,T)L_2(\Theta,\psi_{j-1/2},\psi_{j+1/2},\varphi_l,\varphi_r)d\Theta,$$

$$L_2(\Theta,\psi_{j-1/2},\psi_{j+1/2},\varphi_l,\varphi_r) = \left\{\sin\psi\ln\frac{\cos^2\frac{\Lambda(\Theta)}{2}}{|\sin\frac{\psi}{2}|} + \frac{\psi+\sin\psi}{2}\right.$$

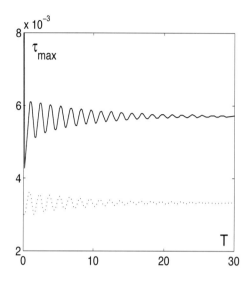

FIGURE 21.19

Distribution of the maximum frictional stresses τ_1 and τ_2 applied to both contact surfaces versus time T for $\gamma = 1$, $\delta_{1*} = 0$, and $X = 1$ (solid curve - lower surface, dotted curve - upper surface) (after Kudish [22]). Reprinted with permission from the ASME.

$$-\frac{\sin[\psi+\frac{\Lambda(\Theta)}{2}]}{\cos\frac{\Lambda(\Theta)}{2}}\Bigg\} \Big|_{a_{j-1/2}}^{a_{j+1/2}}$$

$$+[\tfrac{1}{2}\sin\Lambda(\psi) - \tfrac{\varphi_r-\varphi_l}{2\pi}\psi - \cos\Lambda(\psi)\sin\Lambda(\Theta)]\Big|_{\psi_{j-1/2}}^{\psi_{j+1/2}},$$

$$a_{j\pm1/2} = \tfrac{\varphi_r-\varphi_l}{\pi}(\psi_{j\pm1/2}-\Theta),\ j=2,\ldots,N-1,$$

$$p_j(0) = p_{*j},\ j=2,\ldots,N-1;\ \delta(0)=\delta_*, \tag{21.70}$$

$$p_1 = p_N = 0,\ p_{N-2}-4p_{N-1}=0, \tag{21.71}$$

$$P(T_i) = \tfrac{\varphi_r-\varphi_l}{\pi^2}\int\limits_{-\pi/2}^{\pi/2} p(\Theta,T_i)L_3(\Theta,\varphi_l,\varphi_r)d\Theta, \tag{21.72}$$

$$L_3(\Theta,\varphi_l,\varphi_r) = \cos\Lambda(\Theta),$$

$$h_{j+1/2} = 1 - \delta\cos\Lambda(\psi_{j+1/2})$$

$$+2\lambda_0\tfrac{\varphi_r-\varphi_l}{\pi^2}\int_{-\pi/2}^{\pi/2}p(\Theta,T)L_1(\Theta,\psi_{j+1/2},\varphi_l,\varphi_r)d\Theta, \tag{21.73}$$

where index i, indicating the time step, is omitted and $\varphi_r = \varphi_r(T_i)$, $\Omega = \Omega(T_i)$, $\Phi = \Phi(T_{i-1})$, $\Psi = \Psi(T_{i-1})$, and $X_i = X(T_i)$. In equations (21.69) the

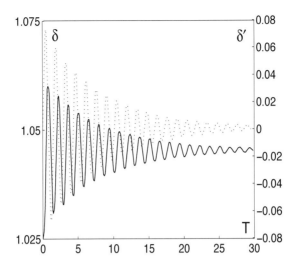

FIGURE 21.20
Distribution of shaft radial displacement δ (solid curve) and its radial speed δ' (dotted curve) versus time T for $\gamma = 1$, $\delta_{1*} = 0$, and $X = 1$ (after Kudish [22]). Reprinted with permission from the ASME.

value of F_j is calculated based on the solution at the previous time moment T_{i-1}, and the integrals in equations (21.68), (21.69), (21.72), and (21.73) are calculated according to formula (21.66). At any time moment T_i, equations (21.68)-(21.73) represent a system of $N + 2$ nonlinear algebraic equations for $N + 2$ unknowns: $\{p_j\}$ ($j = 1, \ldots, N$), δ, and φ_r.

21.3.5   Iterative Numerical Scheme

To solve the system of nonlinear equations (21.68)-(21.73) Newton's method is used. It allows to determine the new iterates of $\{p_j\}$ ($j = 1, \ldots, N$), δ, and φ_r simultaneously. The system of equations is linearized at each consequent iteration and then solved for the new unknowns:

$$q_j^{l+1} = p_j^{l+1} - p_j^l \ (j = 1, \ldots, N), \ \triangle\delta = \delta^{l+1} - \delta^l,$$

$$\triangle\varphi_r = \varphi_r^{l+1} - \varphi_r^l, \tag{21.74}$$

where l indicates the current solution iteration. After the quantities of $\{q_j^{l+1}\}$ ($j = 1, \ldots, N$), $\triangle\delta$, and $\triangle\varphi_r$ for the time moment $T = T_i$ are obtained using formulas (21.74), the new iterates p_j^{l+1}, δ^{l+1}, and φ_r^{l+1} can be calculated. The iterations must be continued until a sufficient accuracy is reached.

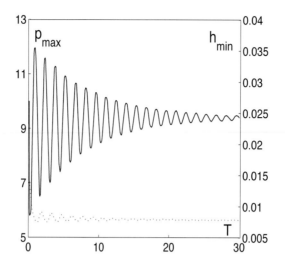

FIGURE 21.21
Distributions of the maximum pressure p_{max} (solid curve) and the minimum
film thickness h_{min} (dotted curve) versus time T for $\gamma = 1$, $\delta_{1*} = -0.06$, and
$X = 1$ (after Kudish [22]). Reprinted with permission from the ASME.

After that the solution for the next time moment $T = T_{i+1}$ can be calculated
in the same manner.

An adaptive numerical procedure is employed for choosing the size of the
time step $\triangle T_i$. It is done in order to overcome a poor convergence and stability
of the numerical solution. In particular, the method allows to decrease or
increase the time step $\triangle T_i$, depending on the behavior of the solution iterates
at the previous and current time steps.

21.3.6   Analysis of Numerical Results

In this section we consider a number of different non-steady problems for a
journal bearing such as a transient oscillatory motion under constant external
conditions, motion with abruptly changing load, and a shaft motion affected
by a bump/dent on its surface. For $\lambda_0 > 0$ all calculations are done with
double precision. The number of nodes used is $N = 50$ and the absolute
precision used in calculations is 10^{-4}. The CPU time required for one run of
the program on a PC is high due to relatively poor convergence and stability
of an EHL problem with such a conformal geometry. The CPU time depends
on the problem parameters. The longest computation time is required for the
oscillatory process for a shaft with large mass ($\gamma = 10$) and certain initial

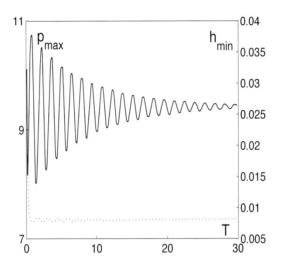

FIGURE 21.22
Distributions of the maximum pressure p_{max} (solid curve) and the minimum film thickness h_{min} (dotted curve) versus time T for $\gamma = 1$, $\delta_{1*} = 0.06$, and $X = 1$ (after Kudish [22]). Reprinted with permission from the ASME.

values of δ and δ' to reach the steady state.

Below, all examples of numerical solutions are given for constant viscosity $\mu = 1$, angular velocity $\omega(T) = 1$ and the following problem parameters: $V = 0.025$, $\lambda_0 = 0.005$, $\varepsilon = 0.003$, $\varphi_l(T) = -1.309$, $\beta(T) = 0$.

21.3.6.1 Transient Oscillatory Motion under Constant External Conditions

Let us consider a transient motion of the shaft subjected to constant external conditions $X(T) = 1$ with the given initial values for parameters $\delta_* = \delta_H + 0.01V^{3/5}/(1 + \lambda_0 V)$, $\varphi_{r*} = 1.16\varphi_H$, and p_{*j} is taken equal to a slightly smoothed out Hertzian pressure distribution $p_H(\varphi(\psi_j)), j = 1, \ldots, N$ (where $\varphi(\psi)$ is given by equation (21.57)). The Hertzian values for a dry frictionless elastic contact of a shaft and bushing made from the same material, namely, the shaft radial displacement δ_H, the contact half-width φ_H, and the dimensionless pressure distribution $p_H(\varphi)$ are as follows (see Teplyi [17], pp. 20-23, 47)

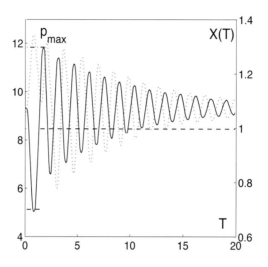

FIGURE 21.23
Distributions of the maximum pressure p_{max} and the external force X versus time T for $\gamma = 1$, $\delta_{1*} = 0$: maximum pressure for case (a) solid curve, external force for case (a) dashed curve, maximum pressure for case (b) dotted curve, external force for case (b) dash-dotted curve (after Kudish [22]). Reprinted with permission from the ASME.

$$p_H(\varphi) = \tfrac{1}{\sqrt{a^2+1}}\{X[\tfrac{\sqrt{a^2+1}-1}{\sqrt{a^2+1}}\chi + \tfrac{2\alpha(\varphi)}{1+\xi^2}] - \tfrac{1}{2}(W + \tfrac{1}{2\lambda_0})\chi\},$$

$$\chi = \ln\frac{\sqrt{a^2+1}-\alpha(\varphi)}{\sqrt{a^2+1}+\alpha(\varphi)}, \tag{21.75}$$

$$W = \frac{2X\{a^2+(\sqrt{a^2+1}-1)[2-\ln(a^2+1)]\}}{\sqrt{a^2+1}[2-\ln(a^2+1)]} + \frac{2(\sqrt{a^2+1}-1)+\ln(a^2+1)}{2\lambda_0[2-\ln(a^2+1)]},$$

$$\frac{2a^4+\ln(a^2+1)-2}{a^4+a^2} = -\frac{1}{\lambda_0 X}, \quad \alpha(\varphi) = \sqrt{a^2-\xi^2}, \quad \xi = \tan\frac{\varphi}{2},$$

$$a = \tan\frac{\varphi_H}{2}, \tag{21.76}$$

$$\delta_H = 1 - \lambda_0\frac{2}{\pi}\int\limits_{-\varphi_H}^{\varphi_H} p_H(\Theta)\{1 + 2\cos(\Theta)\ln(\tan\frac{|\Theta|}{2})\}d\Theta.$$

To consider the oscillatory motion of the shaft in detail, we analyze solutions for different values of parameters γ and δ_{1*}. First, let us discuss the solutions obtained for $\delta_{1*} = 0$. For $\gamma = 10$ the maximum pressure and minimum film thickness, maximum frictional stresses applied to both contact

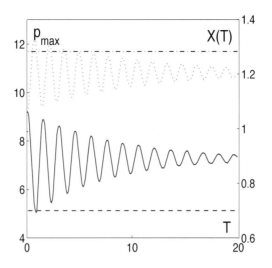

FIGURE 21.24
Distributions of the maximum pressure p_{max} and the external force X versus time T for $\gamma = 1$, $\delta_{1*} = 0$: maximum pressure for case (c) solid curve, external force for case (c) dashed curve, maximum pressure for case (d) dotted curve, external force for case (d) dash-dotted curve (after Kudish [22]). Reprinted with permission from the ASME.

surfaces, and shaft displacement and its normal speed versus time, respectively, are presented in Figs. 21.13-21.15. For this case, the pressure and gap distributions at the moments when the reaction force P reaches its minimum values ($T = 0.2896$, $p_{max} = 7.0965$, $\delta = 1.025937$) and maximum ($T = 2.4432$, $p_{max} = 11.2482$, $\delta = 1.062896$) are given in Figs. 21.16 and 21.17, respectively. In the steady state $p_{max} = 9.3955$ and $\delta = 1.04494$ as it follows from the numerical results for large times (see also Fig. 21.18). From Figs. 21.11 and 21.6 it is clear that the maximum contact pressure p_{max} varies within the range from 75.5% to 119.7% while the shaft displacement δ varies within a much smaller range from 98.2% to 101.7% of their steady values, respectively, and p_{max} and δ slowly approach the steady state values. Therefore, small perturbations in the shaft displacement δ cause large variations in pressure p. It could have been anticipated based on the fact that solution for pressure p of the equation for gap h is an ill-conditioned problem, i.e., a problem the solution of which is very sensitive to small variations in the values of the input data.

Similarly, for $\gamma = 1$ the maximum pressure and minimum film thickness, maximum frictional stresses applied to both contact surfaces, and shaft

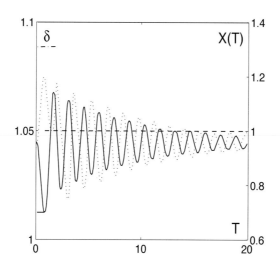

FIGURE 21.25

Distributions of the shaft displacement δ and the external force X versus time T for $\gamma = 1$, $\delta_{1*} = 0$: shaft displacement for case (a) solid curve, external force for case (a) dashed curve, shaft displacement for case (b) dotted curve, external force for case (b) dash-dotted curve (after Kudish [22]). Reprinted with permission from the ASME.

displacement and its normal speed versus time, respectively, are presented in Figs. 21.18 - 21.20. In these cases after a certain time, there is a phase shift of $\pi/2$ between δ and δ' and a phase shift of π between the maximum pressure and minimum film thickness; the maximum frictional stresses on the surfaces are in phase. In particular, it means that pressure reaches its maximum and minimum values when the shaft normal speed is zero and the shaft displacement is minimal and maximal, respectively.

It is obvious from Figs. 21.13 - 21.15 and 21.18 - 21.20 that the frequency of oscillations and, especially, damping decrease rapidly as the system inertia increases. After a relatively short period of time, the lubricated system behaves like a free harmonic (linear) oscillator with small amplitude and damping. Therefore, a simplified representation of the system can be done by the well-known equation

$$\delta'' + 2\nu_1\delta' + \nu_0^2\delta = 0, \qquad (21.77)$$

where ν_0 is the natural frequency of undamped oscillations, ν_1 is the coefficient of proportionality in the damping term. The natural frequencies and damping coefficient of the lubricated system can be determined by considering small perturbation of a steady solution and by trying to find the characteristics

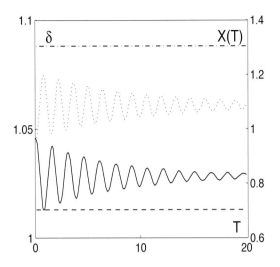

FIGURE 21.26
Distributions of the shaft displacement δ and the external force X versus time T for $\gamma = 1$, $\delta_{1*} = 0$: shaft displacement for case (c) solid curve, external force for case (c) dashed curve, shaft displacement for case (d) dotted curve, external force for case (d) dash-dotted curve (after Kudish [22]). Reprinted with permission from the ASME.

of the oscillatory motion. If $p_0(\varphi)$, $h_0(\varphi)$, δ_0, and φ_{r0} represent the steady solution, then the non-steady solution can be found in the form

$$p(\varphi, T) = p_0(\varphi) + p_1(\varphi)e^{i\nu T} + \dots,$$

$$h(\varphi, T) = h_0(\varphi) + h_1(\varphi)e^{i\nu T} + \dots, \quad \delta(T) = \delta_0 + \delta_1 e^{i\nu T} + \dots, \qquad (21.78)$$

$$\varphi_r(T) = \varphi_{r0} + \varphi_{r1}e^{i\nu T} + \dots, \quad \nu = \nu_0 + i\nu_1,$$

where $p_1(\varphi)$, $h_1(\varphi)$, δ_1, and φ_{r1} are small perturbation values, i is the imaginary unit. Then equations (21.61)-(21.65) can be linearized for $p_1(\varphi)$, $h_1(\varphi)$, δ_1, and φ_{r1} in the vicinity of $p_0(\varphi)$, $h_0(\varphi)$, δ_0, and φ_{r0} and after that reduced to an algebraic equation for ν with complex coefficients. It is important to keep in mind that in this case the initial conditions should be dropped.

The actual derivation of this equation for ν is difficult and not necessary if we have the numerical results presented earlier. It follows from these results that both ν_0 and ν_1 are nonnegative. If $\nu_0 \neq 0$ we have an underdamped motion and if $\nu_0 = 0$ an overdamped one. In particular, using the formula for the frequency of damped oscillations f_{osc} (following from (21.77))

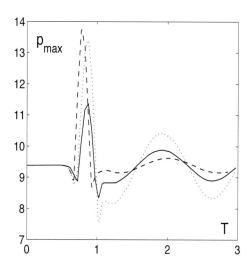

FIGURE 21.27
Distributions of the maximum pressure p_{max} versus time T for bumps 1-3 (bump 1: $a = 0.004$, $\omega_b = 10$ - solid curve; bump 2: $a = 0.008$, $\omega_b = 10$ - dashed curve; bump 3: $a = 0.004$, $\omega_b = 25$ - dotted curve) and $\gamma = 1$, $\delta_{1*} = 0$ (after Kudish [22]). Reprinted with permission from the ASME.

$$f_{osc} = \sqrt{\nu_0^2 - \nu_1^2}$$

and numerical data (see Figs. 21.15 and 21.18) we obtain the values of ν_0 and ν_1 presented in Table 21.1. It follows from Table 21.1 that ν_0 is proportional to $\gamma^{-1/2}$ and ν_1 increases rapidly as γ decreases. Therefore, as the mass of the lubricated system decreases its frequency and damping increase and vice versa. It is important to mention that for a heavy shaft (large γ) damping is a very slow process (small ν_1) and oscillatory motion is sustained for a long period of time. On contrary, for a light shaft (small γ) damping is fast (large ν_1) and oscillations vanish quickly. This trend is in agreement with the physical intuition and can be clearly seen in Figs. 21.13 - 21.15 and 21.18 - 21.20.

The maximum pressure and minimum film thickness for $\gamma = 1$, $\delta_{1*} = -0.06$ and $\delta_{1*} = 0.06$ and the same values of other parameters are presented in Figs. 21.21 and 21.22, respectively. For $\delta_{1*} = -0.06$ the pressure maximum p_{max} varies between 5.6101 and 12.0822 and for $\delta_{1*} = 0.06$ it varies between 7.9395 and 10.8289. Therefore, all EHL parameters, including pressure and film thickness, vary significantly in time for different initial values of δ_0 and δ_{1*}.

This analysis indicates that small perturbations of steady conditions caused

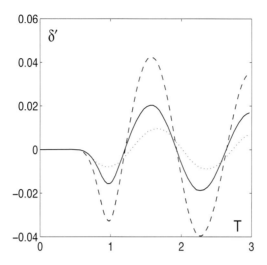

FIGURE 21.28
Distributions of the shaft radial velocity δ' versus time T for bumps 1-3 (bump 1: $a = 0.004$, $\omega_b = 10$ - solid curve; bump 2: $a = 0.008$, $\omega_b = 10$ - dotted curve; bump 3: $a = 0.004$, $\omega_b = 25$ - dashed curve) and $\gamma = 1$, $\delta_{1*} = 0$ (after Kudish [22]). Reprinted with permission from the ASME.

by instabilities in external load, surface velocities, or surface roughness lead to a relatively long damped oscillatory motion. The analogy presented by equation (21.77) allows to make a conclusion that the potentially most harmful small perturbation (causing large amplitudes of pressure oscillations) is the periodic perturbation with frequency of $(\nu_0^2 - 2\nu_1^2)^{1/2}$. As it was mentioned such a situation can be created not only by external conditions but also by surface waviness and/or roughness developed in the process of work. The data from Table 21.1 indicates that the lighter the shaft the greater effect such small perturbations may cause. The comparison of the solutions for rigid and elastic materials shows that oscillations are damped much faster in the case of rigid materials than the elastic ones (see Figs. 21.11, 21.13, 21.15, 21.18, 21.20 - 21.22). This can be easily understood by taking into account two facts: (1) some of the energy is stored in the elastic materials in the form of the potential energy of elastic deformation and (2) only the kinetic portion of the system's total energy is dissipated. The second important conclusion, which follows from the above comparison, is that in spite of very small variations in the shaft displacement δ the maximum pressure p_{max} may vary within very large margins. Therefore, in practice, the fact that the measured shaft displacement varies insignificantly while the shaft is involved in a non-steady motion does

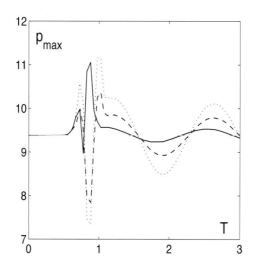

FIGURE 21.29
Distributions of the maximum pressure p_{max} versus time T for dents 1-3 (dent 1: $a = 0.004$, $\omega_b = 10$ - solid curve; dent 2: $a = 0.008$, $\omega_b = 10$ - dashed curve; dent 3: $a = 0.004$, $\omega_b = 25$ - dotted curve) and $\gamma = 1$, $\delta_{1*} = 0$ (after Kudish [22]). Reprinted with permission from the ASME.

not mean that the variations of pressure are small as well. In other words, in the case of a non-steady motion, it is necessary to take into account Newton's second law (see equation (21.55)) instead of just the balance condition on the external force X, i.e.,

$$X = \frac{1}{\pi} \int\limits_{\varphi_l}^{\varphi_r} p(\theta, T) \cos\theta d\theta.$$

Finally, the account for elasticity of materials is very important as it not only changes the dynamics of the system described above but also changes significantly the steady solutions. In fact, for rigid materials the steady parameters of the lubricated contact are $p_{max} = 21.2606$ and $\delta = 0.99665$ while for the contact of elastic materials they are $p_{max} = 9.3955$ and $\delta = 1.04494$.

21.3.6.2 Transient Oscillatory Motion under Action of a Step Load

Let us consider a transient motion of the shaft subjected to a step load. The initial parameters of the problem correspond to the solution of the problem for the steady state conditions ($\gamma = 1$, $\delta_{1*} = 0$) described in the preceding section. The applied load $X(T)$ for four cases is given by the formulas: case (a) $X(T) = 1$ if $T < 0.1$ and $T > 1.1471794$, and $X(T) = 0.7$ if $0.1 \leq T \leq$

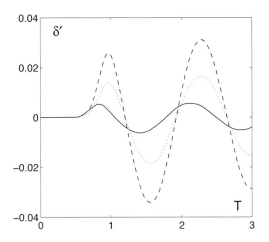

FIGURE 21.30
Distributions of the shaft radial velocity δ' versus time T for dents 1-3 (dent 1: $a = 0.004$, $\omega_b = 10$ - solid curve; dent 2: $a = 0.008$, $\omega_b = 10$ - dashed curve; dent 3: $a = 0.004$, $\omega_b = 25$ - dotted curve) and $\gamma = 1$, $\delta_{1*} = 0$ (after Kudish [22]). Reprinted with permission from the ASME.

TABLE 21.1
Dependence of the oscillation frequency and damping on the system inertia (after Kudish [22]). Reprinted with permission from the ASME.

γ	0.1	1	10
ν_0	13.96	4.33	1.37
ν_1	0.41	0.124	0.017

1.1471794; case (b) $X(T) = 1$ if $T < 0.1$ and $T > 2.1743589$, and $X(T) = 1.3$ if $0.1 \leq T \leq 2.1743589$; case (c) $X(T) = 1$ if $T < 0.1$ and $X(T) = 0.7$ if $T \geq 0.1$; and case (d) $X(T) = 1$ if $T < 0.1$ and $X(T) = 1.3$ if $T \geq 0.1$. The numerical solutions for the maximum pressure and shaft displacement obtained for cases (a)-(d) are presented in Figs. 21.23 - 21.26, respectively. The data presented in Figs. 21.23 - 21.26 indicate that after the external load experienced a jump the shaft is involved in oscillatory motion. The maximum pressure varies within the range from 55% to 135% of the steady value. The relative variations of the shaft displacement are not as great, however, the solution of the problem is very sensitive even to small changes in the shaft displacement. As before, the rate of oscillation damping depends primarily on the system inertia, i.e., the greater is inertia, the slower is damping.

21.3.6.3   Oscillatory Motion under Constant External Conditions Due to a Bump/Dent Presence

Suppose there is a bump/dent on the surface of the shaft outside of the contact, which, at a certain time moment, enters the contact region and disturbs the steady process of lubrication. In reality, in the presence of a bump/dent, generally, the initial conditions in the contact region are not steady. In spite of that, the described situation gives a clear understanding as of how in practice a lubricated system would respond to the presence of a bump/dent in the contact. Let us consider a motion of a shaft subjected to a constant force $X(T) = 1$. As the initial contact parameters, let us take those used in the preceding two sections and $\gamma = 1$, $\delta_{1*} = 0$. Let the shape of a bump/dent is described by function $A(\psi)$, where A is a nonnegative function. As a bump/dent passes through the contact its relative position changes. Therefore, the dimensionless gap equation can be presented in the form

$$h(\psi, T) = h_{ideal}(\psi, T) \pm A(\psi - 2T), \qquad (21.79)$$

where h_{ideal} is the gap without a bump/dent present (see equation (21.64)) and the upper/lower sign is used in the case of a dent/bump, respectively. In particular calculations function A was defined as follows:

$$A(\psi) = b\{1 + \sin[\omega_b(\psi - \psi_0)]\} \; if \; -\tfrac{\pi}{2} \le \omega_b(\psi - \psi_0) \le \tfrac{3\pi}{2},$$

$$A(\psi) = 0 \; otherwise, \qquad (21.80)$$

where $2b$ and ψ_0 are the bump/dent amplitude and initial phase shift ($\psi_0 = \psi_1 - 1.5\pi/\omega_b - 0.1$), respectively, ω_b is the parameter controlling the bump/dent width, which is equal to $2\pi/\omega_b$. At the initial time moment $T = 0$ the bump/dent is completely outside of the contact region $-\pi/2 \le \varphi \le \pi/2$.

For three different bumps/dents (bump/dent 1: $b = 0.004$, $\omega_b = 10$; bump/dent 2: $b = 0.008$, $\omega_b = 10$; bump/dent 3: $b = 0.004$, $\omega_b = 25$), the graphs of the maximum pressure and shaft normal velocity are given in Figs. 21.27 - 21.30, respectively. At some point in time the bump/dent enters the contact region and perturbs the steady state of a lubricated contact causing shaft displacement and pressure oscillations. These oscillations seem to be greater when the bump/dent is higher/deeper and narrower. For bumps 1-3, the maximum pressure varies within the range from 94.0% to 121.3%, 92.0% to 147.0%, and from 79.8% to 143.2% of the steady maximum pressure $p_{max} = 9.3955$, respectively. Similarly, for dents 1-3 the maximum pressure varies within the range from 83.5% to 110.7%, from 77.9% to 119.2%, and from 94.8% to 118.2% of the steady maximum pressure $p_{max} = 9.3955$, respectively. Obviously, the influence of a bump presence in the contact is stronger than that of a dent with the same shape. The major changes in the maximum pressure take place when the bump/dent center approaches the point of minimum gap. After the bump/dent leaves the contact, the parameter perturbations slowly subside due to viscous damping until the bump/dent enters the contact region the next time. In case when the bump/dent enters the contact multiple times the situation gets more complicated. The behavior of the lubrication parameters depends not only on the bump/dent and contact parameters but also on the value of the angular velocity ω.

Let us summarize the results of the section. A non-steady problem for a conformal EHL contact of two infinite cylindrical surfaces with parallel axes is considered. It takes into account the elasticity of cylinders, lubricant viscosity, contact surface velocities, and applied load. The problem is solved based on the "modified" formulation, which includes Newton's second law of motion. The main emphasis in the section is put on the analysis of the transient dynamics of the system under constant and abruptly changing external load as well as on the system behavior in the case of a bump/dent presence on the shaft surface. A strong dependence of non-steady EHL solutions on the system inertia (mass) and elasticity of materials is established. Generally, the numerical solutions exhibit damped oscillatory behavior. The transient values of pressure are extremely sensitive to small variations in the shaft displacement. This leads to the conclusion that any non-steady motion of a lubricated system must be considered based on Newton's second law.

21.4   Closure

The traditional ("classic") non-steady EHL problems for line conformal and concentrated EHL contacts are formulated and analyzed. Based on the analysis of rigid contact solids and lightly loaded contacts it is shown that the "classic" EHL problem formulations possess a number of serious defects. To mitigate these defects "modified" formulations of non-steady EHL problems are proposed. The main difference between the "classic" and "modified" problem formulations is that the "modified" EHL problem formulations take into account the inertia of the system. The advantages of the "modified" problem formulation are shown. The "modified" problem formulation allows for lubricated system oscillatory motion which is impossible in a system considered based on the "classic" non-steady EHL problem formulation. A non-steady EHL problem for an infinite journal bearing in "modified" formulation is solved numerically for numerous combinations of the problem input data including cases of abruptly varying load, bumps, and dents on the contact surface, etc.

21.5   Exercises and Problems

1. (a) Based on the modified problem formulation obtain a two-term asymptotic solution for a lightly loaded non-steady non-conformal lubricated contact (i.e., for the case of $V \gg 1$) for a lubricant with constant viscosity. Show that all components of the solution such as pressure $p(x,t)$, gap $h(x,t)$, etc., are continuous functions of time t not only for continuous external load $P(t)$ and average surface velocities $u(t)$ but also for finite discontinuous ones.

(b) Repeat the same analysis for a lightly loaded non-steady conformal lubricated contact, i.e., for $\lambda_0 \ll 1$.

2. Solve equation (21.27) and (21.28) to obtain the expression for $h(\varphi, t)$ from (21.29).

3. Describe what influence system inertia has on the behavior of a lubricated contact. What can be expected for relatively small and large system mass?

References

[1] Safa, M.M.A. and Gohar, R. 1086. Pressure Distribution Under a Ball Impacting a Thin Lubricant Layer. *ASME J. Tribology* 108, No. 3:372-376.

[2] Ai, X. and Cheng, H.S. 1994. A Transient EHL Analysis for Line Contacts with Measured Surface Roughness Using Multigrid Technique. *ASME J. Tribology* 116, No. 3:549-558.

[3] Chang, L., Webster, M.N., and Jackson, A. 1994. A Line-Contact Micro-EHL Model with Three Dimensional Surface Topography. *ASME J. Tribology* 116, No. 1:21-28.

[4] Venner, C.H. and Lubrecht, A.A. 1994. Transient Analysis of Surface Features in an EHL Line Contact in Case of Sliding. *ASME J. Tribology* 116, No. 2:186-193.

[5] Venner, C.H. and Lubrecht, A.A. 1994. Numerical Simulation of Transverse Ridge in a Circular EHL Contact Under Rolling/Sliding. *ASME J. Tribology* 116, No. 4:751-761.

[6] Osborn, K.F. and Sadeghi, F. 1992. Time Dependent Line EHL Lubrication Using the Multigrid/Multilevel Technique. *ASME J. Tribology* 114, No. 1:68-74.

[7] Cha, E. and Bogy, D.B. 1995. A Numerical Scheme for Static and Dynamic Simulation of Subambient Pressure Shaped Rail Sliders. *ASME J. Tribology* 117, No. 1:36-46.

[8] Hashimoto, H. and Mongkolwongrojn, M. 1994. Adiabatic Approximate Solution for Static and Dynamic Characteristics of Turbulent Partial Journal Bearing with Surface Roughness. *ASME J. Tribology* 116, No. 4:672-680.

[9] Peiran, Y. and Shizhu, W. 1991. Pure Squeeze Action in an Isothermal Elastohydrodynamically Lubricated Spherical Conjunction. Part 1. Theory and Dynamic Load Results. *Wear* 142, No. 1:1-16.

[10] Peiran, Y. and Shizhu, W. 1991. Pure Squeeze Action in an Isothermal Elastohydrodynamically Lubricated Spherical Conjunction. Part 2. Constant Speed and Constant Load Results. *Wear* 142, No. 1:17-30.

[11] San Andres, L.A. and Vance, J.M. 1987. Effect of Fluid Inertia on Squeeze Film Damper Forces for Small Amplitude Motions about an Off-Center Equilibrium Position. *ASLE Tribology Trans.* 30, No. 1:63-68.

[12] Larsson, R. and Hoglund, E. 1995. Numerical Simulation of a Ball Impacting and Rebounding a Lubricated Surface. *ASME J. Tribology* 117, No. 1:94-102.

[13] Kudish, I.I. and Panovko, M.Ya. 1992. Oscillations of a Deformable Lubricated Cylinder Rolling Along a Rigid Half-Space. *Soviet J. Fric. and Wear* 13, No. 5:1-11.

[14] Kudish, I.I. 1999. On Formulation of a Non-Steady Lubrication Problem for a Non-Conformal Contact. *STLE Tribology Trans.* 42, No. 1:53-57.

[15] Kudish, I.I., Kelley, F., and Mikrut, D. 1999. Defects of the Classic Formulation of a Non-steady EHL Problem for a Journal Bearing. New Problem Formulation. *ASME J. Tribology* 121, No. 4:995-1000.

[16] Hamrock, B.J. 1991. *Fundamentals of Fluid Film Lubrication.* Cleveland: NASA, Reference Publication 1255.

[17] Teplyi, M.I. 1983. *Contact Problems for Regions with Circular Boundaries.* Lvov: Naukova Dumka.

[18] Allaire, P.E. and Flack, R.D. 1980. Journal Bearing Design for High Speed Turbomachinery. In *Bearing Design–Historical Aspects, Present Technology and Future Problems, Proc. Intern. Conf. "Century 2 - Emerging Technology,"* Ed. W.J. Anderson, ASME, New York. 111-160.

[19] Ghosh, M.K., Hamrock, B.J., and Brewe, D. 1985. Hydrodynamic Lubrication of Rigid Non-Conformal Contacts in Combined Rolling and Normal Motion. *ASME J. Tribology* 107:97-103.

[20] Lin, C.R. and Rylander, H.G., Jr. 1991. Performance Characteristics of Compliant Journal Bearings. *ASME J. Tribology* 113, No. 3:639-644.

[21] Wijnant, Y.H., Venner, C.H., Larsson, R., and Erickson, P. 1999. Effects of Structural Vibrations on the Film Thickness in an EHL Circular Contact. *ASME J. Tribology* 121, No. 2:259-264.

[22] Kudish, I.I. 2002. A Conformal Lubricated Contact of Cylindrical Surfaces Involved in a Non-steady Motion. *ASME J. Tribology* 124, No. 1:62-71.

22

Lubricant Starvation and Mixed Friction Problems for Point Contacts

22.1 Introduction

A numerical analysis of a steady point EHL problem is conducted. The main focus of the analysis is the influence of lubricant starvation of contact parameters. As a continuation of this analysis a problem for mixed lubrication is formulated and an analytical study of the mixed lubrication problem is undertaken. A specific structure of the contact region is revealed which consists of alternating lubricated and dry contact stripes. Boundary friction is considered. Additional analytical analysis is done for the case when the contact is elongated in the direction perpendicular to the lubricant flow. It is shown that in a dry elongated contact it is possible to propose an optimal shape of the solids in contact.

22.2 Starved Lubrication of a Lightly Loaded Spatial EHL Contact. Effect of the Shape of Lubricant Meniscus on Problem Solution

In practice, because of different reasons (starvation, instability of lubrication regime, proximity to critical velocities, influence of lubricant surface tension, etc.) the input oil meniscus is not always far from the center of the contact. Moreover, very often the meniscus is relatively close to it and changes its configuration in time. Usually, this is reflected on graphs of frictional stresses versus time under conditions seemed to be stationary. The oscillations of the friction stress in such a situation may be dramatically large.

Interference methods allow to observe this behavior of EHL film thickness in relation to the configuration (location) of the input oil meniscus. It was registered that in such cases the profile of EHL film thickness (gap between contacting solids) is far from that which may be observed for truly stationary or fully flooded conditions and it depends on the shape of the inlet oil

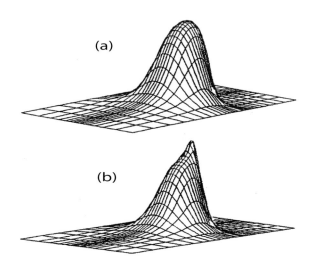

FIGURE 22.1

Pressure distribution for fully flooded lubrication regime: (a) $Q_0 = 0$, $p_{max} = 3.82$, (b) $Q_0 = 0.6$, $p_{max} = 5.53$ (after Kudish and Panovko [1]). Reprinted with permission from the ASME.

meniscus. In particular, due to the inlet meniscus shape the gap between lubricated contacting solids may approach or reach zero (direct dry contact of elastic solids) in some zones of the contact area.

In this section we provide a conservative numerical method and some numerical solutions for this problem taking into account a complex shape of the input oil meniscus. The particular physical phenomena (such as instability of lubrication regime, proximity to critical velocities, influence of lubricant surface tension, etc.) which caused the oil meniscus to approach the boundary of the dry contact are not considered here.

The material presented in this section is different from the studies published earlier by Oh [2], Evans and Hughes [3], Hamrock and Dowson [4], Lubrecht and Ten Napel [5] in two ways: (1) the finite-difference equations of the EHL problem are obtained based on the fluid flux conservation considerations. (2) the finite-difference equations of the EHL problem take into account the complex shape of the input oil meniscus. The second provision allows to model some lubrication regimes that are far from fully flooded conditions. The further analysis is based on [1].

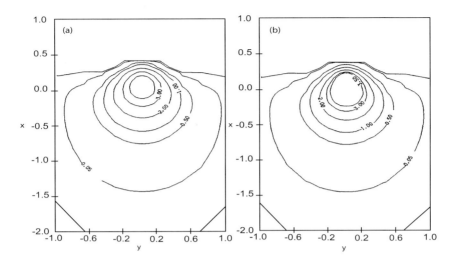

FIGURE 22.2
Isobars for fully flooded lubrication regime: (a) $Q_0 = 0$, (b) $Q_0 = 0.6$ (after
Kudish and Panovko [1]). Reprinted with permission from the ASME.

Some numerical results for pressure, gap, frictional stress, and subsurface
stress distributions in the EHL contact are presented.

22.2.1 Problem Formulation

Let us assume that two moving (rolling) solids with smooth surfaces made
of the same elastic material are lubricated by an incompressible Newtonian
viscous fluid under isothermal conditions. The thickness of the lubrication
layer is considered to be small compared to characteristic size of the contact
area and size of the solids [6]. The solids are in a concentrated (non-conformal)
contact and they experience an external compressive force P.

Let us introduce a moving rectangular coordinate system: the z-axis passes
through the solids' centers of curvature, and the xy-plane is equidistant from
the solids.

It is assumed that the solids' stationary motion is slow and occurs with
linear velocities $\overrightarrow{u}_1 = (u_{x1}, v_{y1})$ and $\overrightarrow{u}_2 = (u_{x2}, v_{y2})$. The inertia forces in
the lubrication layer are considered to be small compared with the viscous
ones.

In the following dimensionless variables [7]

$$(x', y') = \frac{\vartheta}{2R_x}(x, y), \ (p', \tau') = \frac{8\pi R_x^2}{3P\vartheta^2}(p, \tau), \ h' = \frac{h}{h_0}, \ \mu' = \frac{\mu}{\mu_a},$$

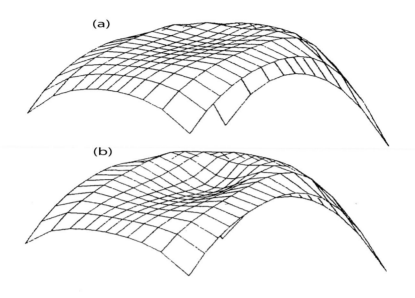

FIGURE 22.3

Gap distribution for fully flooded lubrication regime: (a) $Q_0 = 0$, $h_{min} = 0.726$, (b) $Q_0 = 0.6$, $h_{min} = 0.694$ (after Kudish and Panovko [1]). Reprinted with permission from the ASME.

$$(\vec{u'}_1, \vec{u'}_2) = \frac{(\vec{u}_1, \vec{u}_2)}{|\ \vec{u}_1 + \vec{u}_2\ |}, \ \gamma = \frac{\vartheta^2}{2R_x}h_0, \ \vartheta = \frac{P}{8\pi\mu_a\ |\ \vec{u}_1 + \vec{u}_2\ |\ R_x},$$

which are characteristic for lightly loaded EHL contacts the EHL problem equations are as follows (further, primes are omitted at the dimensionless variables)

$$L(p) = \{\nabla \cdot [\gamma^2 \tfrac{h^3}{\mu}\nabla p - \vec{u}h]\} = 0, \ \nabla = (\tfrac{\partial}{\partial x}, \tfrac{\partial}{\partial y}), \ \vec{u} = (u_x, v_y), \qquad (22.1)$$

$$h = 1 + \tfrac{x^2 + \rho y^2}{\gamma} + \tfrac{2}{\pi V\gamma}\int\int_\Omega \left[\frac{1}{\sqrt{(\xi - x)^2 + (\eta - y)^2}} - \frac{1}{\sqrt{\xi^2 + \eta^2}}\right]p(\xi, \eta)d\xi d\eta,$$
$$(22.2)$$

$$\int\int_\Omega p(\xi, \eta)d\xi d\eta = \tfrac{2\pi}{3},$$

$$p\ |_\Gamma = 0, \qquad\qquad\qquad (22.3)$$

$$\gamma = \tfrac{\theta_0^2}{2R_x}h_0, \ V = \tfrac{16\pi R_x^2 E'}{3P\theta_0^3}, \ \rho = \tfrac{R_x}{R_y}, \ \vec{u} = \tfrac{\vec{u}_1 + \vec{u}_2}{2}, \qquad (22.4)$$

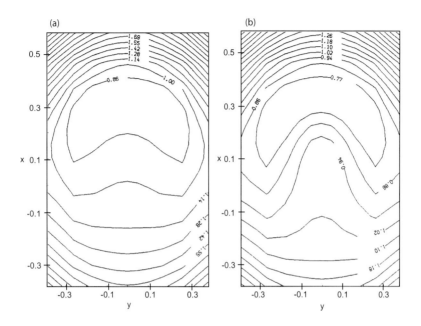

FIGURE 22.4
Level curves of gap distribution for fully flooded lubrication regime: (a) $Q_0 = 0$, (b) $Q_0 = 0.6$ (after Kudish and Panovko [1]). Reprinted with permission from the ASME.

where $p(x, y)$ and $h(x, y)$ are the pressure and gap in the contact, Ω is the contact region, γ is the dimensionless central lubrication film thickness, Γ is the boundary of the contact, V and ρ are the given dimensionless parameters, $\vec{u}$ is the average velocity of the solid surfaces in contact. The dimensionless variables are introduced in such a way that for a similar hydrodynamic problem (i.e., for a lubrication problem for rigid solids) $p = O(1)$ and $x^2 + y^2 = O(1)$.

Also, the problem solution must satisfy the boundary conditions for pressure p, which are different at different portions of the boundary of the contact region: the inlet and exit boundaries. Let us assume that Γ_i and Γ_e are the inlet and exit portions of the contact boundary, which are given and unknown in advance, respectively. The whole boundary of the contact Γ is $\Gamma_i \bigcup \Gamma_e$. On the contact inlet boundary, we have the condition:

$$\Gamma_i \ is \ given \ if \ \vec{Q}_f \cdot \vec{n_\Gamma} \ |_{\Gamma_i} < 0, \tag{22.5}$$

representing the fact that lubricant enters the contact through this bound-

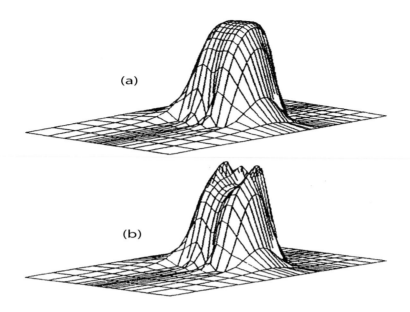

FIGURE 22.5
Pressure distribution for notched input oil meniscus: (a) $Q_0 = 0$, $p_{max} = 3.92$, $p_{min} = 3.915$, (b) $Q_0 = 0.6$, $p_{max} = 4.64$, $p_{min} = 3.28$ (after Kudish and Panovko [1]). Reprinted with permission from the ASME.

ary as well as the boundary condition (22.3) must be satisfied. At the same time, the exit boundary is also defined by condition (22.3) and the cavitation condition

$$\nabla p \cdot \vec{n_\Gamma} \mid_{\Gamma_e} = 0 \ if \ \vec{Q_f} \cdot \vec{n_\Gamma} \mid_{\Gamma_e} > 0, \tag{22.6}$$

where $\vec{Q_f}$ is the vector of fluid flux given by the expression

$$\vec{Q_f} = h\vec{u} - \gamma^2 \frac{h^3}{\mu} \nabla p, \tag{22.7}$$

and $\vec{n_\Gamma}$ is the external normal vector to the contact boundary Γ.

The location of the contact exit boundary can also be found using complementarity considerations

$$L(p) = 0 \ for \ p > 0, \ i.e., \ within \ the \ contact \ region,$$

$$\tag{22.8}$$

$$L(p) < 0 \ for \ p = 0 \ outside \ the \ contact \ if \ \vec{Q_f} \cdot \vec{n_\Gamma} \mid_{\Gamma_e} \geq 0,$$

which are described in detail by Kostreva [8] and Oh [2]. Oh et al. [9] showed that the complementarity approach leads to the solution of the problem satisfying the cavitation condition on the exit boundary (see equation (22.6)).

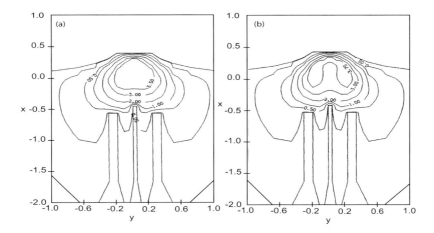

FIGURE 22.6
Isobars for notched input oil meniscus: (a) $Q_0 = 0$, (b) $Q_0 = 0.6$ (after Kudish and Panovko [1]). Reprinted with permission from the ASME.

In the system of equations and inequalities (22.1)-(22.8), the geometric shape of the input oil meniscus, parameters V (for lightly loaded EHL contact $V \gg 1$), ρ, u_x, v_y, and the function of viscosity $\mu(p)$ are considered to be known. In the presented below numerical results, the exponential relationship for $\mu(p)$ is used

$$\mu(p) = \exp(Qp), \tag{22.9}$$

where Q is the dimensionless pressure viscosity coefficient.

The solution of system (22.1)-(22.9) is represented by pressure $p(x,y)$ and gap $h(x,y)$ distributions, dimensionless EHL film thickness γ, and the location of the exit contact boundary Γ_e.

It is necessary to mention that boundary conditions (22.5), (22.7), and (22.8) are of local nature. Thus, some pieces of the contact boundary Γ are known and some are not.

After the solution of the EHL problem is found, the friction stress distributions on the contact surfaces 1 and 2 can be found according to the formulas:

$$\vec{\tau}_{1,2} = \frac{\mu \vec{s}}{12\gamma h} \mp \frac{h}{2} \gamma \nabla p, \tag{22.10}$$

where $\vec{s}$ is the dimensionless sliding velocity, $\vec{s} = 2(\vec{u}_2 - \vec{u}_1)/ \mid \vec{u}_2 + \vec{u}_1 \mid$. The first term in (22.10) is the friction stress due to sliding $\vec{\tau}_s$, the second term is the frictional stress due to rolling $\vec{\tau}_r$. The stress components $\sigma_{ij}(x,y,z)$ in the subsurface layers of the contact solids are determined by well–known

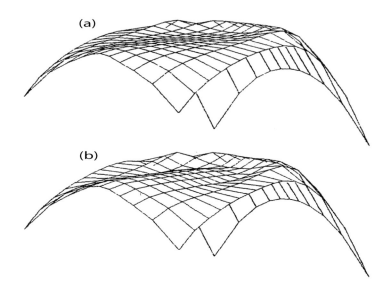

FIGURE 22.7

Gap distribution for notched input oil meniscus: (a) $Q_0 = 0$, $h_{min} = 0.797$, (b) $Q_0 = 0.6$, $h_{min} = 0.896$ (after Kudish and Panovko [1]). Reprinted with permission from the ASME.

Boussinesq relations [10]

$$\sigma_{ij}(x, y, z) = \int\int_{\Omega} G_{ij}(x - \xi, y - \eta, z)p(\xi, \eta)d\xi d\eta, \qquad (22.11)$$

where $G_{ij}(x, y, z)$ is the Green's tensor obtained from the Boussinesq problem [10]. To characterize the subsurface stress distributions, we will use the octahedral normal and tangential stresses. The latter are determined by the formulas

$$\sigma^{oct} = \frac{\sigma_{11} + \sigma_{22} + \sigma_{33}}{3},$$

$$\tau^{oct} = \frac{1}{3}[(\sigma_{11} - \sigma_{22})^2 + (\sigma_{22} - \sigma_{33})^2 + (\sigma_{11} - \sigma_{33})^2 \qquad (22.12)$$

$$+\tau_{12}^2 + \tau_{23}^2 + \tau_{13}^2]^{1/2}.$$

22.2.2   Numerical Method

Let us consider solution of the problem equations in a rectangular region $\Omega = \{(x, y) \mid a \leq x \leq b, \ c \leq y \leq d\}$, where a, b, c, and d are certain

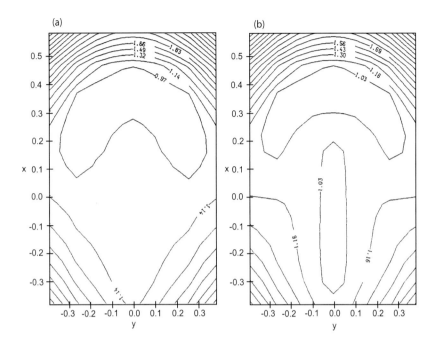

FIGURE 22.8

Level curves of gap distribution for notched input oil meniscus: (a) $Q_0 = 0$, (b) $Q_0 = 0.6$ (after Kudish and Panovko [1]). Reprinted with permission from the ASME.

constants. Using a transform based on a hyperbolic sine, we can introduce a nonuniform grid with the nodes in the xy-plane with the following coordinates:

$$x_i \ (i = 0, \ldots, i_{max} + 1), \ x_{i-1/2} = (x_i + x_{i-1})/2 \ (i = 1, \ldots, i_{max} + 1),$$

$$y_j \ (j = 0, \ldots, j_{max} + 1), \ y_{j-1/2} = (y_j + y_{j-1})/2 \ (j = 1, \ldots, j_{max} + 1),$$

$$(x_{m0-1/2}, y_{n0-1/2}) = (0, 0).$$

The pressure distribution will be determined at nodes (x_i, y_j) and the gap distribution will be determined at nodes $(x_{i-1/2}, y_{j-1/2})$.

Integrating equation (22.1) over the cell around node (x_i, y_j) gives its integral analogue in the form

$$L_1(p) = \oint_{l_{ij}} \{\gamma^2 \frac{h^3}{\mu} \nabla p \cdot \vec{n} - h\vec{v} \cdot \vec{n}\} dl = 0, \tag{22.13}$$

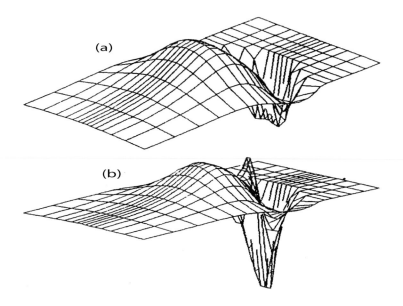

FIGURE 22.9
Rolling frictional stress distribution τ_{rx} for fully flooded lubrication regime:
(a) $Q_0 = 0$, $\tau_{rx,max} = 0.672$, $\tau_{rx,min} = -0.932$, (b) $Q_0 = 0.6$, $\tau_{rx,max} = 0.958$,
$\tau_{rx,min} = -2.61$ (after Kudish and Panovko [1]). Reprinted with permission
from the ASME.

where l_{ij} is the rectangular boundary of the cell surrounding the node (x_i, y_j)
and $\vec{n}$ is the external normal vector to the boundary of the cell.

In the complementarity conditions (22.8), operator $L(p)$ should be replaced
by operator $L_1(p)$ from equation (22.13).

The iterative process for the EHL problem is based on Newton's method.
Equations (22.13), (22.3), and (22.5) linearized near the k-th iterates of the
solution $(p_k(x, y), h_k(x, y), \gamma_k, \Gamma_{e,k})$ have the form

$$\oint_{l_{ij}} \{ [2\gamma_k \frac{h_k^3}{\mu_k} \frac{\partial p_k}{\partial \vec{n}} - \vec{v} \cdot \vec{n} \frac{\partial h}{\partial \gamma} |_k + \gamma_k^2 \frac{3h_k^2}{\mu_k} \frac{\partial p_k}{\partial \vec{n} \frac{\partial h}{\partial \gamma}} |_k] \triangle\gamma_{k+1}$$

$$- \gamma_k^2 \frac{h_k^3}{\mu_k^2} \frac{\partial mu}{\partial p} |_k \frac{\partial p_k}{\partial \vec{n}} \triangle p_{k+1} + \gamma_k^2 \frac{3h_k^2}{\mu_k} \frac{\partial h}{\partial p} |_k \triangle p_{k+1} \partial p_k \partial\vec{n} \qquad (22.14)$$

$$+ \gamma_k^2 \frac{h_k^3}{\mu_k} \frac{\partial \triangle p_{k+1}}{\partial \vec{n}} - \vec{u} \cdot \vec{n} \frac{\partial h}{\partial p} |_k \triangle p_{k+1} \} = - \oint_{l_{ij}} [\gamma_k^2 \frac{h_k^3}{\mu_k} \frac{\partial p_k}{\partial \vec{n}} - \vec{u} \cdot \vec{n} h_k] dl,$$

$$\int\int_\Omega \triangle p_{k+1}(\xi, \eta) d\xi d\eta = \frac{2\pi}{3} - \int\int_\Omega p_k(\xi, \eta) d\xi d\eta, \qquad (22.15)$$

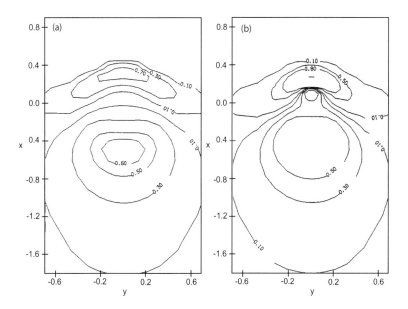

FIGURE 22.10
Level curves of rolling frictional stress distribution τ_{rx} for fully flooded lubrication regime: (a) $Q_0 = 0$, (b) $Q_0 = 0.6$ (after Kudish and Panovko [1]). Reprinted with permission from the ASME.

$$\triangle p_{k+1} \mid_{\Gamma_k} = 0, \qquad (22.16)$$

where k is the iteration number, $\triangle p_{k+1} = p_{k+1} - p_k$, $\triangle \gamma_{k+1} = \gamma_{k+1} - \gamma_k$, $\frac{\partial h}{\partial \gamma} \mid_k$ is the partial derivative of h_k with respect to γ_k calculated using equation (22.2), $\frac{\partial \mu}{\partial p} \mid_k$ is the partial derivative of μ_k with respect to p_k calculated using equation (22.9); $\frac{\partial h}{\partial p} \mid_k \triangle p_{k+1}$ is the linear operator representing the derivative of h_k with respect to p_k applied to $\triangle p_{k+1}$

$$\frac{\partial h}{\partial p} \mid_k \triangle p_{k+1} = \frac{2}{\pi V \gamma} \int \int_{\Omega} \left[\frac{1}{\sqrt{(\xi-x)^2 + (\eta-y)^2}} \right.$$

$$\left. - \frac{1}{\sqrt{\xi^2 + \eta^2}} \right] \triangle p_{k+1}(\xi, \eta) d\xi d\eta. \qquad (22.17)$$

Integration over the contour l_{ij} is performed along the intervals $[x_{i-1/2}, x_{i+1/2}]$ and $[y_{j-1/2}, y_{j+1/2}]$ taking into account the direction of integration.

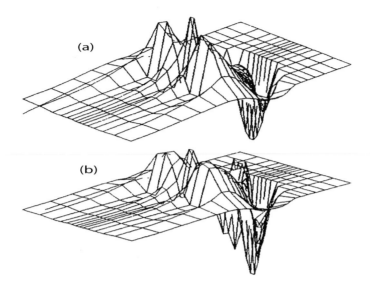

FIGURE 22.11
Rolling frictional stress distribution τ_{rx} for notched input oil meniscus: (a) $Q_0 = 0$, $\tau_{rx,max} = 1.056$, $\tau_{rx,min} = -1.34$, (b) $Q_0 = 0.6$, $\tau_{rx,max} = 1.034$, $\tau_{rx,min} = -2.12$ (after Kudish and Panovko [1]). Reprinted with permission from the ASME.

The following approximations are used for derivatives

$$\frac{\partial p(x_{i\pm1/2}, y_j)}{\partial x} \approx \frac{\pm p(x_{i\pm1}, y_j) \mp p(x_i, y_j)}{\pm x_{i\pm1} \mp x_i},$$

$$\frac{\partial p(x_i, y_{j\pm1/2})}{\partial y} \approx \frac{\pm p(x_i, y_{j\pm1}) \mp p(x_i, y_j)}{\pm y_{j\pm1} \mp y_j}.$$

The coefficients of $\triangle p_{k+1}$ and $\triangle \gamma_{k+1}$ are calculated at the points with coordinates $(x_{i\pm1/2}, y_j)$ and $(x_i, y_{j\pm1/2})$. The singular integral in equation (22.13) is calculated according to the following cubature formula (Belotcerkovsky and

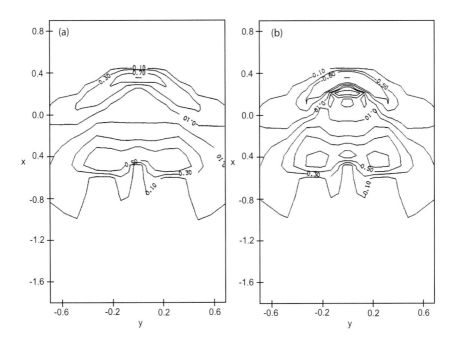

FIGURE 22.12
Level curves of rolling frictional stress distribution τ_{rx} for notched input oil meniscus: (a) $Q_0 = 0$, (b) $Q_0 = 0.6$ (after Kudish and Panovko [1]). Reprinted with permission from the ASME.

Lifanov [11]):

$$\int \int_{\Omega} \left[\frac{1}{\sqrt{(\xi - x_{m-1/2})^2 + (\eta - y_{n-1/2})^2}} - \frac{1}{\sqrt{\xi^2 + \eta^2}} \right] p(\xi, \eta) d\xi d\eta$$

$$= \sum_{j=1}^{j_{max}} \sum_{i_1(j)}^{i_2(j)} \left[\frac{1}{\sqrt{(x_i - x_{m-1/2})^2 + (y_j - y_{n-1/2})^2}} \right.$$

$$\left. - \frac{1}{\sqrt{(x_i - x_{m0-1/2})^2 + (y_j - y_{n0-1/2})^2}} \right] p(x_i, y_j)$$

$$\times (x_{i+1/2} - x_{i-1/2})(y_{j+1/2} - y_{j-1/2}),$$

where $i_1(j)$ and $i_2(j)$ are the integer arrays describing the location of the inlet and exit boundaries, respectively, $m = 1, \ldots, i_{max} + 1$, $n = 1, \ldots, j_{max} + 1$, node $(x_{m0-1/2}, y_{n0-1/2})$ is the origin of the coordinate system. The array

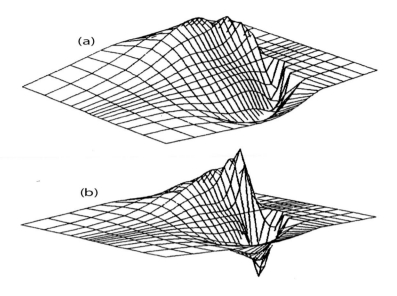

FIGURE 22.13

Rolling frictional stress distribution τ_{ry} for fully flooded lubrication regime:
(a) $Q_0 = 0$, $\tau_{ry,max} = 0.687$, $\tau_{ry,min} = -0.687$, (b) $Q_0 = 0.6$, $\tau_{ry,max} = 1.904$,
$\tau_{ry,min} = -1.904$ (after Kudish and Panovko [1]). Reprinted with permission
from the ASME.

of indexes $i_1(j)$ is known as the inlet boundary Γ_i is given and the array of
indexes $i_2(j)$ is found after each iteration from the complementarity conditions
(22.8). The values of h at points $(x_{i\pm 1/2}, y_j)$ and $(x_i, y_{j\pm 1/2})$ are obtained by
interpolation. A similar cubature formula is used for calculation of $\frac{\partial h}{\partial p}\mid_k$
$\triangle p_{k+1}$ (see equation (22.17)).

As an initial approximation for the iterative process, the solution of a similar lubrication problem for rigid solids can be used. The latter problem is
solved for the given geometry of the input meniscus and γ_0 by the method
of upper relaxation. The aforementioned value of dimensionless film thickness
γ_0 is approximately determined from the solution of an equation for the total
load applied to the contact $P(\gamma) = 2\pi/3$. The left-hand side of this equation
is obtained from the solutions of the lubrication problem for rigid solids for
several values of γ. As an initial approximation for the iterative process, the
existing solution of a similar lubrication problem for different regime parameters can be used as well.

The first step of the iteration process is represented by the process of solving

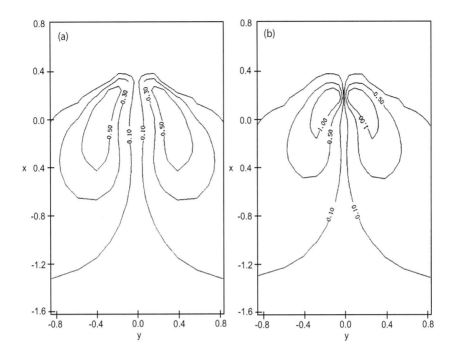

FIGURE 22.14
Level curves of rolling frictional stress distribution τ_{ry} for fully flooded lubrication regime: (a) $Q_0 = 0$, (b) $Q_0 = 0.6$ (after Kudish and Panovko [1]). Reprinted with permission from the ASME.

of the finite-difference analog of equations (22.14)-(22.18) for $\triangle p_{k+1}(x_i, y_j)$ and $\triangle \gamma_{k+1}$ by the Gaussian elimination with partial pivoting. After that, we reconstruct the values of $p_{k+1}(x_i, y_j) = p_k(x_i, y_j) + \triangle p_{k+1}(x_i, y_j)$ and $\gamma_{k+1} = \gamma_k + \triangle \gamma_{k+1}$ and the new location of the exit boundary $\Gamma_{e,k+1}$ is determined according to the complementarity conditions. At this stage the gap distribution $h(x_{i-1/2}, y_{j-1/2})$ is calculated using formula (22.2) and (22.18). The described process of calculations is repeated until the desired relative accuracy of the solution is reached: $| p_{k+1}(x_i, y_j)/p_k(x_i, y_j) - 1 | < \delta$, etc.

22.2.3 Numerical Results

The described numerical process is used for calculations of a lightly loaded point EHL contact of two spheres with equal radii ($\rho = 1$) rolling along the x-axis ($u_x = 1$, $v_y = 0$) with relative sliding $s_x = 0.01$ and $s_y = 0$. The solutions are obtained for varied geometry of the input meniscus and values

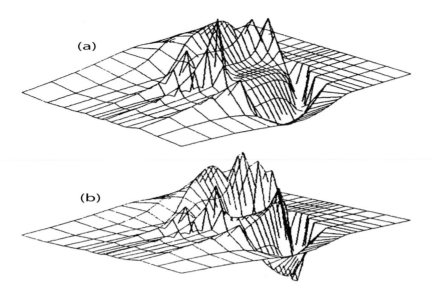

FIGURE 22.15

Rolling frictional stress distribution τ_{ry} for notched input oil meniscus: (a) $Q_0 = 0$, $\tau_{ry,max} = 1.16$, $\tau_{ry,min} = -0.688$, (b) $Q_0 = 0.6$, $\tau_{ry,max} = 1.41$, $\tau_{ry,min} = -1.41$ (after Kudish and Panovko [1]). Reprinted with permission from the ASME.

of parameters V and Q. The calculations are conducted with the relative accuracy $\delta = 10^{-5}$ with more detailed grid at the center of the contact. The entire grid is represented by 26×16 nodes with $\triangle x_{min} = 0.04305$, $\triangle x_{max} = 0.3778$, $\triangle y_{min} = 0.04815$, and $\triangle y_{max} = 0.2647$. All calculations were performed with double precision. The oil meniscus shapes used for calculations had one and three notches. The presented numerical solutions are obtained for $V = 15$, $Q = 0$, and $Q = 0.6$.

For testing purposes series of calculations using two grids 26×16 and 30×30 ($\triangle x_{min} = 0.03158$, $\triangle x_{max} = 0.3698$, $\triangle y_{min} = 0.02419$, and $\triangle y_{max} = 0.1606$) were performed. The obtained results for $V = 15$ and different values of Q are as follows: $\gamma = 0.12$, $h_{min} = 0.726$, $p_{max} = 3.82$, $Q = 0$ and $\gamma = 0.203$, $h_{min} = 0.694$, $p_{max} = 5.53$, $Q = 0.6$ for grid 26×16 and $\gamma = 0.118$, $h_{min} = 0.701$, $p_{max} = 3.805$, $Q = 0$ and $\gamma = 0.200$, $h_{min} = 0.700$, $p_{max} = 5.473$, $Q = 0.6$ for grid 30×30. These results indicate that the relative precision of the solutions obtained for grids 26×16 and 30×30 is in the range of $2\% - 3\%$ for γ, h_{min} and in the range of $0.5\% - 1\%$ for p_{max}. For more detailed grids $N \times M$, the time required for solution of the problem

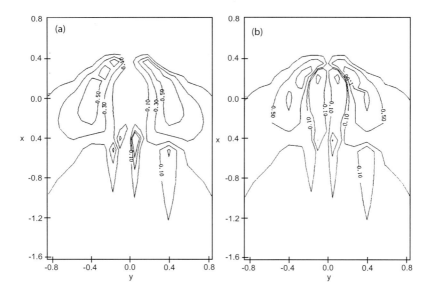

FIGURE 22.16
Level curves of rolling frictional stress distribution τ_{ry} for notched input oil meniscus: (a) $Q_0 = 0$, (b) $Q_0 = 0.6$ (after Kudish and Panovko [1]). Reprinted with permission from the ASME.

is proportional to $(NM)^3$.

The numerical experiments showed stability of the described numerical method in the following range of the input parameters of the EHL problem for lightly loaded contacts: $15 \leq V < \infty$ and $0 \leq Q \leq 0.6$.

22.2.3.1   Pressure, Gap, and Frictional Stress Distributions

The distributions of pressure $p(x, y)$ and gap $h(x, y)$ for the case of fully flooded lubrication (the input oil meniscus is far enough from the center of the contact and coincides with the coordinate lines bounding the contact region) are shown in Figs. 22.1 - 22.4. For convenience the distributions of gap $h(x, y)$ are presented only for the portion of the contact region with noticeable pressure $p(x, y)$. The distribution of pressure $p(x, y)$ for $Q = 0.6$ features a sharp peak near the exit boundary, which is absent for the case of $Q = 0$ (see Figs. 22.1 and 22.2). The difference in the pressure maxima for these cases is substantial (see Fig. 22.1). The distribution of gap $h(x, y)$ is more "concave" for the case of $Q = 0.6$ than for $Q = 0$, which is related to the differences in the pressure distributions (see Figs. 22.3 and 22.4). The location of isobars

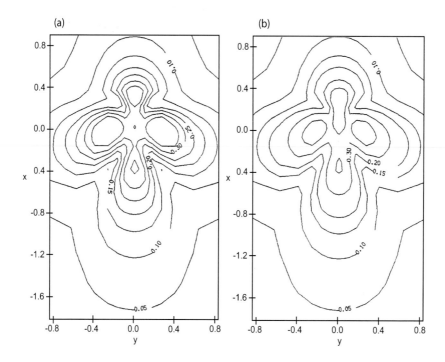

FIGURE 22.17
Level curves of tangential octahedral stress τ^{oct} in plane $z = 0.0027$ for fully flooded lubrication regime: (a) $Q_0 = 0$, (b) $Q_0 = 0.6$ (after Kudish and Panovko [1]). Reprinted with permission from the ASME.

varies with the pressure viscosity coefficient Q. However, the exit boundary remains almost unchanged (see Fig. 22.2). A noticeable change in the EHL film thickness γ is related to the change in Q: $\gamma = 0.12$ for $Q = 0$ and $\gamma = 0.203$ for $Q = 0.6$.

The numerical results for the case of an input oil meniscus with complex geometry are represented in Figs. 22.5 - 22.8. The changes in the geometry of the input oil meniscus employed in calculations (see Fig. 22.6) are accompanied by a decrease in the fluid flux through the contact that leads to a decrease in the EHL film thickness: $\gamma = 0.0834$ for $Q = 0$ and $\gamma = 0.107$ for $Q = 0.6$. The distribution of pressure $p(x, y)$ varies moderately for $Q = 0$ and much more significantly for $Q = 0.6$ exhibiting several sharp peaks compared to that for fully flooded lubrication regime (compare Figs. 22.1, 22.2 and 22.5, 22.6). The pressure maximum p_{max} for $Q = 0.6$ decreases substantially compared to the case of fully flooded regime: $p_{max} = 4.64$ and $p_{max} = 5.53$, respectively. Similar changes occur with the distributions of gap $h(x, y)$ (compare Figs.

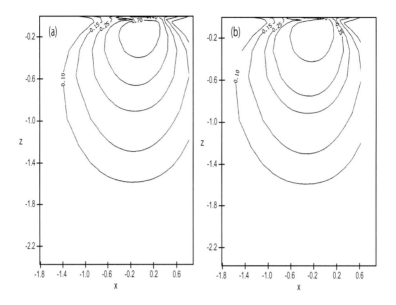

FIGURE 22.18
Level curves of tangential octahedral stress τ^{oct} in plane $y = 0$ for fully flooded lubrication regime: (a) $Q_0 = 0$, (b) $Q_0 = 0.6$ (after Kudish and Panovko [1]). Reprinted with permission from the ASME.

22.3, 22.4 and 22.7, 22.8). These results resemble the behavior of the pressure distribution in a long and narrow EHL contact with a notched input meniscus considered in the next section.

The behavior of the x-component of the sliding frictional stress τ_{sx} is somewhat similar to the behavior of the pressure distribution in the corresponding cases. The extremum values of τ_{sx} increase as Q increases while condition $s_y = 0$ leads to $\tau_{sy} = 0$.

Figures 22.9 - 22.16 represent the behavior of the components of the rolling frictional stress τ_{rx} and τ_{ry} for the described above cases. The formulas for the components of the rolling frictional stress are as follows:

$$\tau_{rx} = \frac{\gamma h}{2}\frac{\partial p}{\partial x}, \quad \tau_{ry} = \frac{\gamma h}{2}\frac{\partial p}{\partial y}.$$

The behavior of the frictional stress due to rolling $\vec{\tau}_r(x, y)$ is more sophisticated than that for the sliding frictional stress as it is proportional to ∇p. The graphs of $\tau_{rx}(x, y)$ and $\tau_{ry}(x, y)$ are presented in Figs. 22.9 - 22.16. In some portions of the contact, these functions are positive and in others negative. The behavior of $\tau_{rx}(x, y)$ and $\tau_{ry}(x, y)$ is more irregular for the case of

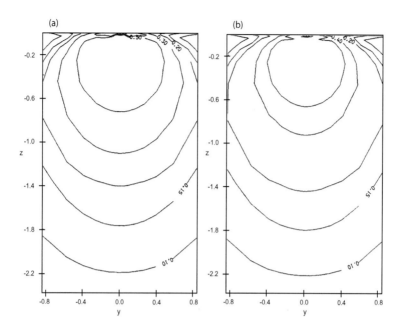

FIGURE 22.19

Level curves of tangential octahedral stress τ^{oct} in plane $x = 0.0601$ for fully flooded lubrication regime: (a) $Q_0 = 0$, (b) $Q_0 = 0.6$ (after Kudish and Panovko [1]). Reprinted with permission from the ASME.

$Q = 0.6$ than for $Q = 0$. The shape of these distributions depends on the geometry of the input oil meniscus and the value Q. It may be pointed out that, in general, ranges for $\tau_{rx}(x, y)$ and $\tau_{ry}(x, y)$ increase as Q increases. Besides, the rolling frictional stress distributions are affected by the geometry of the input oil meniscus stronger for $Q = 0$ than for $Q = 0.6$. In particular, the maximum values of the rolling frictional stress may change by 100% or even more (see Figs. 22.9, 22.11 and 22.13, 22.15).

In general, in a lightly loaded EHL contact the rolling frictional stresses prevail over the sliding friction stresses.

22.2.3.2　Octahedral Stresses in the Subsurface Layer

The octahedral subsurface stresses are calculated according to formulas (22.12) using the values for stresses $\sigma_{ij}(x, y, z)$ obtained from formula (22.11). The employed cubature formula is similar to formula (22.18). The calculations

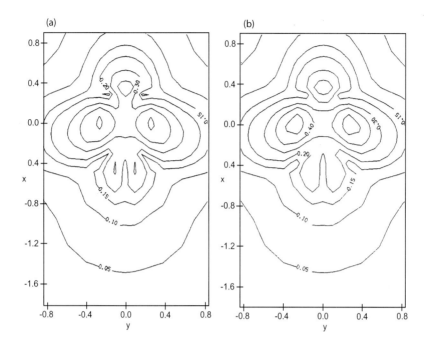

FIGURE 22.20
Level curves of tangential octahedral stress τ^{oct} in plane $z = 0.0027$ for notched input oil meniscus: (a) $Q_0 = 0$, (b) $Q_0 = 0.6$ (after Kudish and Panovko [1]). Reprinted with permission from the ASME.

are conducted for the typical value of Poisson's ratio $\nu = 0.3$. The calculated results are presented in Figs. 22.17 - 22.22 in the form of level curves in the planes $z = const$, $y = 0$, and $x = const$. The plane $z = const$ ($z = 0.0027$) passes through the node on the z-axis closest to the surface. The location of the plane $x = const$ ($x = 0.0601$) is related to the location of the pressure maximum in the contact.

Analysis of the level curves of the normal octahedral stress σ^{oct} leads to a conclusion that in the subsurface layer σ^{oct} reaches negative extremums. In case of the fully flooded lubrication regime $\mid \min(\sigma^{oct}) \mid$ increases as Q increases. For example, $\min(\sigma^{oct}) \approx -2.85$ for $Q = 0$ (reached at $x = 0.0165$, $y = 0$, $z = -0.0447$) and $\min(\sigma^{oct}) \approx -3.54$ for $Q = 0.6$ (reached at $x = 0.106$, $y = 0$, $z = -0.0447$). In case of the input oil meniscus with several deep notches that effect is almost negligible: $\min(\sigma^{oct}) \approx -3.01$ for $Q = 0$ (reached at $x = 0.0165$, $y = 0$, $z = -0.0447$) and $\min(\sigma^{oct}) \approx -2.96$ for $Q = 0.6$ (reached at $x = 0.06$, $y = 0.172$, $z = -0.0447$). However, in the

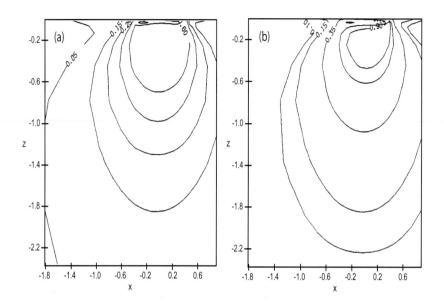

FIGURE 22.21
Level curves of tangential octahedral stress τ^{oct} in plane $y = 0$ for notched input oil meniscus: (a) $Q_0 = 0$, (b) $Q_0 = 0.6$ (after Kudish and Panovko [1]). Reprinted with permission from the ASME.

latter case the level curves of σ^{oct} are deformed in the vicinity of a notch. The quantitative changes in these distributions are insignificant.

The behavior of the tangential octahedral stress τ^{oct} is of more sophisticated nature (see Figs. 22.17 - 22.22). It can be clearly seen in the z-plane where several positive local extrema of τ^{oct} exist. Increase in the number of notches in the input oil meniscus causes an increase in the number of the aforementioned extrema of τ^{oct} and further deformation of the level curves near notches. Similar to σ^{oct}, in case of the fully flooded lubrication regime, $\max(\tau^{oct})$ increases as Q increases. For example, $\max(\tau^{oct}) \approx 1.1$ for $Q = 0$ (reached at $x = 0.106$, $y = 0$, $z = -0.148$) and $\max(\tau^{oct}) \approx 1.47$ for $Q = 0.6$ (reached at $x = 0.157$, $y = 0$, $z = -0.0825$). The values of the absolute minimum also rise: $\min(\tau^{oct}) \approx 0.141$ for $Q = 0$ (reached at $x = -0.23$, $y = 0.172$, $z = -0.0447$) and $\min(\tau^{oct}) \approx 0.18$ for $Q = 0.6$ (reached at $x = -0.3$, $y = 0.172$, $z = -0.0447$).

For the input oil meniscus with several deep notches that effect is al-

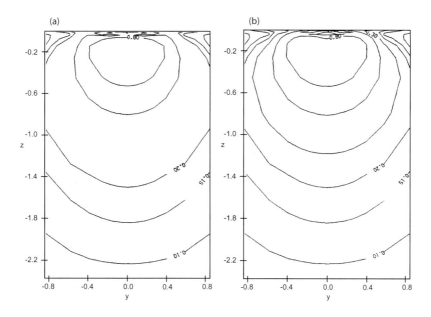

FIGURE 22.22
Level curves of tangential octahedral stress τ^{oct} in plane $x = 0.0601$ for notched input oil meniscus: (a) $Q_0 = 0$, (b) $Q_0 = 0.6$ (after Kudish and Panovko [1]). Reprinted with permission from the ASME.

most negligible: $\max(\tau^{oct}) \approx 1.19$ for $Q = 0$ (reached at $x = 0.06$, $y = 0$, $z = -0.26$) and $\max(\tau^{oct}) \approx 1.25$ for $Q = 0.6$ (reached at $x = 0.157$, $y = -0.172$, $z = -0.148$). However, the absolute minimum of τ^{oct} continues to grow as Q increases: $\min(\tau^{oct}) \approx 0.417$ for $Q = 0$ (reached at $x = -0.0263$, $y = 0$, $z = -0.0228$) and $\min(\tau^{oct}) \approx 0.509$ for $Q = 0.6$ (reached at $x = 0.157$, $y = 0.103$, $z = -0.0447$).

It should be noted that the locations of the local extrema of functions p, h, τ_{sx}, τ_{sy}, τ_{rx}, τ_{ry}, σ^{oct}, and τ^{oct} substantially depend on the value of the pressure viscosity coefficient Q.

The material presented in this section is only the beginning in understanding of the major mechanisms governing starved lubrication regimes and mixed friction. The numerical results show strong dependence on the geometry of the input oil meniscus and value of the pressure viscosity coefficient Q. For instance, the maximum values of pressure may change by 20% while the rolling frictional stress may change by 100% or even more. The input oil meniscus geometry irregularities are reflected in irregularities and complicated structure of pressure, gap, frictional stress, subsurface stresses, and their extrema and may cause significant qualitative and quantitative changes in the distributions of the lubricated contact parameters. Increase in the number of notches in the input oil meniscus causes an increase in the number of extrema of τ^{oct}. In practice, this may cause developing of multiple spots of damage (fatigue, wear) on the contact surface and in subsurface layers.

22.3   Formulation and Analysis of a Mixed Lubrication Problem

In practice due to different reasons (starvation, instability of lubrication regime, approaching critical velocities, influence of lubricant surface tension, etc.), the input oil meniscus is not always sufficiently far away from the boundary of purely elastic dry contact. Moreover, very often the meniscus is close to it and changes its configuration in time. Usually, this is reflected on graphs of frictional stresses versus time under conditions seemed to be stationary. The oscillations of frictional stress in such a situation may be dramatically large.

Optical interference methods allow to observe this behavior of lubrication film thickness in relation with the configuration (location) of the input oil meniscus. It was registered that in such cases the profile of the lubrication film thickness is far from the one which may be observed for truly stationary or fully flooded conditions and it depends on the shape of the inlet oil meniscus. In particular, due to the inlet meniscus shape the gap between lubricated contacting solids may approach or reach zero in some regions of contact area.

This phenomenon can be easily understood from the consideration of solutions of a plane EHL problem. The general behavior of lubrication film thickness in a plane lubricated contact is well known. If oil meniscus is outside of the Hertzian dry elastic contact, the lubrication film thickness is greater than zero and it vanishes as the meniscus approaches the Hertzian elastic contact. Therefore, if the contact is long and narrow in the directions perpendicular and parallel to the direction of lubricant flow, respectively, in different sections of the contact area, the situation is similar to the described one for a plane EHL problem. It depends on relative location of the input oil meniscus and boundary of the Hertzian contact. In particular, it can result in the existence of alternating lubricated and non-lubricated stripes in the contact area.

In this section we provide a mathematical formulation of this problem taking into account the existence of the zones with different types of (dry and fluid) friction. An EHL problem for partially lubricated solids is formulated and analyzed. The consideration of such a problem is unavoidable under mixed friction conditions when the elastic solids are separated by a lubricant film in one part of the contact area while in the other part they are in direct elastic contact without presence of a lubricant. A dry frictional contact occurs with the presence of slippage and adhesion zones. Such severe conditions appear in most practical cases (bearings, gears, etc.). The problem is reduced to a system of alternating nonlinear Reynolds' equation and integral equations and inequalities valid in the contact area. The inlet contact boundary is considered to be known and located close to the boundary of the purely elastic contact subjected to the same conditions. The location of the exit contact boundary must be determined from the problem solution as well as a number of internal contact boundaries separating the zones of dry and lubricated contact. The conditions of continuity of (dry and fluid) friction stresses on these internal boundaries are formulated. The special cases of this problem formulation are the classic contact problem of elasticity with dry friction taking into account slippage and adhesion zones and the EHL problem for completely lubricated contact.

The theoretical analysis of a partially lubricated contact is given for the case of a narrow contact significantly elongated in the direction orthogonal to the direction of lubricant motion. Under these conditions, in the main part of the contact region the problem analysis can be reduced to the analysis of plane elastic and EHL problems. These problems can be successfully analyzed by asymptotic methods. The specific features of the problem solution, i.e., the location and magnitude of dry and lubricated zones, friction stresses, and pressure are mainly dependent on the configuration of the inlet contact boundary.

The particular physical phenomena (such as instability of lubrication regime, approaching critical velocities, influence of lubricant surface tension, etc.) are not considered here.

22.3.1   Problem Formulation

Let us assume that two moving (rolling) solids with smooth surfaces made of the same elastic material are lubricated by a Rivlin-type incompressible non-Newtonian viscous fluid under isothermal conditions. Consideration of non-Newtonian fluid is essential because it is necessary to take into account transition from fluid to dry friction mechanisms. The lubrication layer is considered to be small compared with the characteristic sizes of the contact region and solids' sizes. The solids form a concentrated (non-conformal) contact and they experience an external compressive force (see Fig. 22.23). Under these assumptions the solids can be replaced by two contacting elastic half-spaces. The further analysis is based on [12].

Let us introduce a moving coordinate system: the z-axis passes through the centers of curvature of the solids and the xy-plane is equidistant from the solid surfaces.

It is assumed that the solids are involved in steady slow motion with linear velocities $\vec{u}_1 = (u_1, v_1)$ and $\vec{u}_2 = (u_2, v_2)$. The slippage velocity $\vec{s}$ is assumed

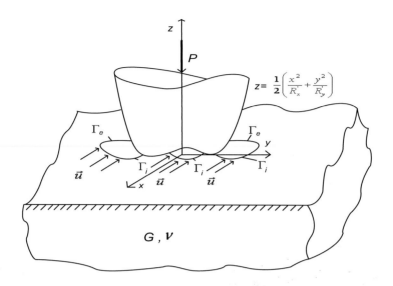

FIGURE 22.23

The general view of the solids involved in a contact with mixed lubrication. (after Kudish and Covitch [13]). Reprinted with permission from CRC Press.

to be small compared with the rolling velocity $0.5(\vec{u}_1 + \vec{u}_2)$. The inertial forces in lubrication layer are considered to be small compared with viscous ones.

Under these assumptions the tangential stress vector $\vec{\tau}(x, y)$ in the lubrication layer is proportional to the gradient of the lubricant linear velocity $\frac{\partial \vec{u}(x,y)}{\partial z}$

$$\mu \frac{\partial u(x,y)}{\partial z} = F[\tau_{xz}(x, y)], \ \ \mu \frac{\partial v(x,y)}{\partial z} = F[\tau_{yz}(x, y)] \ or$$

$$\tau_{xz}(x, y) = \Phi[\mu \frac{\partial u(x,y)}{\partial z}], \ \ \tau_{yz}(x, y) = \Phi[\mu \frac{\partial v(x,y)}{\partial z}], \tag{22.19}$$

where $\mu = \mu(p)$ is the lubricant viscosity at pressure p, F and Φ are given inverse to each other monotonic smooth enough functions describing the lubricant rheology, $F(0) = \Phi(0) = 0$. The regions of a dry contact are represented by the adhesion and slippage zones in which the vector of the relative solid slippage $\vec{s}(x, y)$ is zero and different from zero, respectively. Moreover, the friction stress and slippage in these zones satisfy Coulomb's law

$$\vec{\tau} = \lambda p \frac{\vec{s}}{|\vec{s}|} \ for \ |\vec{s}| > 0; \ |\vec{\tau}| \le \lambda p \ for \ |\vec{s}| = 0, \tag{22.20}$$

where p is the contact pressure and $\lambda = \lambda(p, |\vec{s}|)$ is the coefficient of dry friction.

22.3.1.1 Fluid Film Lubrication

First, let us derive the equations governing the considered process in the lubricated regions. Under conditions of a thin lubricant layer and slow motion the equations of fluid motion are [6]

$$\frac{\partial \tau_{xz}}{\partial z} = \frac{\partial p}{\partial x}, \ \frac{\partial \tau_{yz}}{\partial z} = \frac{\partial p}{\partial y}, \ \frac{\partial p}{\partial z} = 0, \ \nabla \cdot \vec{w} = 0, \tag{22.21}$$

where $\vec{w} = (u, v, w)$ is the vector of fluid velocity. The latter leads to the relationship $p = p(x, y)$.

Taking into account the no slippage assumption at the contact surfaces and that the gap $h(x, y)$ gradient and $\mid \vec{s} \mid / \mid \vec{u}_1 + \vec{u}_2 \mid$ are small in comparison with 1 the components of the fluid particle velocity $\vec{w} = (u, v, w)$ at the contact surfaces are

$$\vec{u} = \vec{u}_1, \ w = -\frac{1}{2}\vec{u}_1 \cdot \nabla h \ for \ z = -\frac{h}{2},$$

$$\vec{u} = \vec{u}_2, \ w = \frac{1}{2}\vec{u}_2 \cdot \nabla h \ for \ z = \frac{h}{2}. \tag{22.22}$$

Integrating the continuity equation $\nabla \cdot \vec{w} = 0$ with respect to z from $-h/2$ to $h/2$ and using boundary conditions (22.22), we obtain (Q_x and Q_y are the x- and y-components of the fluid flux)

$$\frac{\partial Q_x}{\partial x} + \frac{\partial Q_y}{\partial y} = 0, \ \vec{Q} = (Q_x, Q_y) = \int\limits_{-h/2}^{h/2} \vec{u}dz. \tag{22.23}$$

Integrating the first two equations in (22.19) with boundary conditions from (22.22), we find

$$u = u_1 + \frac{1}{\mu} \int\limits_{-h/2}^{z} F(f_x + t\frac{\partial p}{\partial x})dt, \ v = v_1 + \frac{1}{\mu} \int\limits_{-h/2}^{z} F(f_y + t\frac{\partial p}{\partial y})dt, \tag{22.24}$$

where $\vec{f} = (f_x, f_y)$ is the vector of the sliding frictional stress the components of which satisfy the equations

$$\frac{1}{\mu} \int\limits_{-h/2}^{h/2} F(f_x + t\frac{\partial p}{\partial x})dt = s_x, \ \frac{1}{\mu} \int\limits_{-h/2}^{h/2} F(f_y + t\frac{\partial p}{\partial y})dt = s_y. \tag{22.25}$$

Finally, using equations (22.23) and (22.24), we derive the generalized Reynolds equation

$$\frac{\partial}{\partial x}\Big\{\frac{1}{\mu} \int\limits_{-h/2}^{h/2} zF(f_x + z\frac{\partial p}{\partial x})dz\Big\} + \frac{\partial}{\partial y}\Big\{\frac{1}{\mu} \int\limits_{-h/2}^{h/2} zF(f_y + z\frac{\partial p}{\partial y})dz\Big\} \tag{22.26}$$

$$= \frac{1}{2}(\vec{u}_1 + \vec{u}_2) \cdot \nabla h \ if \ h(x, y) > 0,$$

where the fluid flux components can be calculated from

$$Q_x = h\frac{u_1+u_2}{2} - \frac{1}{\mu} \int\limits_{-h/2}^{h/2} zF(f_x + z\tfrac{\partial p}{\partial x})dz,$$

$$Q_y = h\frac{v_1+v_2}{2} - \frac{1}{\mu} \int\limits_{-h/2}^{h/2} zF(f_y + z\tfrac{\partial p}{\partial y})dz. \qquad (22.27)$$

Obviously, equation (22.25)-(22.27) hold only in the lubricated zones of the contact region.

Now, let us consider the difference in elastic displacements $\triangle\vec{W} = (\triangle U, \triangle V, \triangle W)$ of the surface points of the two (upper and lower) contact solids. The well–known formulas give [14]

$$\triangle U = \frac{1}{2\pi G} \int\int\limits_{\Omega_\tau} \{\tfrac{1-\nu\sin^2\theta}{R}(\tau_{xz}^+ - \tau_{xz}^-) + \tfrac{\nu\sin\theta\cos\theta}{R}(\tau_{yz}^+ - \tau_{yz}^-)\}d\xi d\eta,$$

$$\triangle V = \frac{1}{2\pi G} \int\int\limits_{\Omega_\tau} \{\tfrac{\nu\sin^\theta\cos\theta}{R}(\tau_{xz}^+ - \tau_{xz}^-) + \tfrac{1-\nu\cos^2\theta}{R}(\tau_{yz}^+ - \tau_{yz}^-)\}d\xi d\eta,$$

$$\triangle W = \frac{1-2\nu}{4\pi G} \int\int\limits_{\Omega_\tau} \{\tfrac{\cos\theta}{R}(\tau_{xz}^+ + \tau_{xz}^-) + \tfrac{\sin\theta}{R}(\tau_{yz}^+ + \tau_{yz}^-)\}d\xi d\eta \qquad (22.28)$$

$$+\frac{1-\nu}{\pi G} \int\int\limits_{\Omega_p} \tfrac{p}{R}d\xi d\eta, \quad R = \sqrt{(x-\xi)^2 + (y-\eta)^2},$$

$$\sin\theta = \tfrac{y-\eta}{R}, \quad \cos\theta = \tfrac{x-\xi}{R},$$

where Ω_p and Ω_τ are the normal and tangential contact area at the boundaries of which pressure p and the frictional stress $\vec{\tau}$ vanish (regions Ω_p and Ω_τ not necessarily coincide), G and ν are the shear elastic modulus and Poisson's ratio of the solid material, $G = \frac{E}{2(1+\nu)}$ (E is Young's modulus), superscripts $+$ and $-$ are related to the upper and lower solids, respectively. Moreover, $\tau^+ = (\tau_{xz}^+, \tau_{yz}^+)$ and $\tau^- = (\tau_{xz}^-, \tau_{yz}^-)$ are the tangential stresses acting on the surfaces of the upper and lower solids, which for dry contact zones satisfy conditions (22.20) and for lubricated regions are determined by the equations (see equations (22.19), (22.23), and (22.26))

$$\vec{\tau}^\pm = \mp\vec{f} - \tfrac{h}{2}\nabla p. \qquad (22.29)$$

Using well–known considerations applied to contact problems of elasticity and the expression for difference in vertical displacements $\triangle W$, we get the relation for gap h between the contacting solids

$$h = h_0 + \frac{x^2}{2R_x} + \frac{y^2}{2R_y} + \frac{1-\nu}{\pi G} \underset{\Omega_p}{\int \int} \frac{p d\xi d\eta}{R}$$

$$- \frac{1-2\nu}{4\pi G} \underset{\Omega_\tau}{\int \int} \{\cos\theta \frac{\partial p}{\partial x} + \sin\theta \frac{\partial p}{\partial y}\} \frac{h d\xi d\eta}{R},$$

(22.30)

where h_0 is an unknown constant, R_x and R_y are the effective radii of the surface curvature for the contact solids in the directions of the x- and y-axes, respectively. Equation (22.30) can be simplified for the case of an incompressible elastic material for which $\nu = 1/2$ and

$$h = h_0 + \frac{x^2}{2R_x} + \frac{y^2}{2R_y} + \frac{1-\nu}{\pi G} \underset{\Omega_p}{\int \int} \frac{p d\xi d\eta}{R}.$$

(22.31)

22.3.2   Fluid Friction in Lightly and Heavily Loaded Lubricated Contacts

Let us consider the case of a lightly loaded contact of elastic solids lubricated by an incompressible fluid with non-Newtonian rheology. Scaling the problem similar to Sections 5.2 and 22.2 and applying regular perturbation methods would allow to obtain the following structural formula for the dimensionless film thickness $\gamma = \frac{\vartheta^2}{2R_x} h_c$ (compare to formulas (5.27) and (5.32)):

$$\gamma = \gamma_0 + \frac{1}{V}\gamma_1 + \ldots,$$

(22.32)

where $V = \frac{8\pi R_x^2 E'}{3P\vartheta^3}$ and $\vartheta = \vartheta(\mu_a, u_1, u_2, R', P, \ldots)$ is a certain dimensionless parameter that depends on the fluid specific rheology and is independent from the effective elastic modulus E' ($1/E' = 1/E_1' + 1/E_2'$, $E_j' = E_j/(1 - \nu_j^2)$, $j = 1, 2$) while parameters γ_0 and γ_1 are independent of V and may depend on the viscosity pressure coefficient α (assuming that the lubricant viscosity $\mu = \mu_a e^{\alpha p}$, μ_a is the ambient lubricant viscosity), the ratios of velocities $(v_1 + v_2)/(u_1 + u_2)$ and surface radii R_x/R_y, the slide-to-roll ratio $s_0 = 2 \mid \vec{u}_2 - \vec{u}_1 \mid / \mid \vec{u}_2 + \vec{u}_1 \mid$, the lubricant rheology, and the shape of the inlet oil meniscus Γ_i. In case of Newtonian rheology $\vartheta = \frac{P}{8\pi\mu_a|\vec{u}_1+\vec{u}_2|R_x}$.

Let us consider the case of a heavily loaded contact of elastic solids lubricated by an incompressible fluid with Newtonian rheology. By scaling the problem formulated above using the Hertzian semi-axes a_H, b_H, and the maximum pressure p_H we will see that the problem solution depends on the dimensionless parameters V (see (22.50))

$$V = \frac{24\mu_a|\vec{u}_1+\vec{u}_2|R_x^2}{p_H a_H^3}, \quad Q = \alpha p_H,$$

where it is assumed that the lubricant viscosity $\mu = \mu_a e^{\alpha p}$ (μ_a is the ambient lubricant viscosity and α is the viscosity pressure coefficient).

As before, for heavily loaded lubrication regimes we have $V \ll 1$ and/or $Q \gg 1$. Based on the analysis conducted in Chapter 12 for pre-critical starved lubrication regimes (see definitions (6.75), (6.57), and (6.59)) for the dimensionless central film thickness $H_{00} = \frac{2R_x h_c}{a_H^2}$ (h_c is the dimensional central film thickness) we get the formula (compare with formula (6.55))

$$H_{00} = A(V\epsilon_q^2)^{1/3}, \ A = O(1), \ \epsilon_q \ll V^{2/5}, \tag{22.33}$$

while for pre-critical fully flooded lubrication regimes (see definitions (6.58) and (6.60)) the formula for the central film thickness H_0 will get the form (compare with formula (6.63))

$$H_{00} = AV^{3/5}, \ A = O(1), \ \epsilon_q = V^{2/5}. \tag{22.34}$$

In formulas (22.33) and (22.34), ϵ_q is the characteristic distance between the inlet oil meniscus and the boundary of the Hertzian region, the values of constants A are independent of V but may depend on dimensionless viscosity pressure coefficient Q, the ratios of velocities $(v_1 + v_2)/(u_1 + u_2)$ and surface radii R_x/R_y, and the shape of the inlet meniscus Γ_i of region Ω_p.

For over-critical lubrication regimes (see definition (6.76)), in cases of starved lubrication (see definition (6.57)) the central film thickness is determined by the formula (compare with formula (6.141))

$$H_{00} = A(VQ\epsilon_q^{5/2})^{1/3}, \ A = O(1), \ \epsilon_q \ll (VQ)^{1/2}, \tag{22.35}$$

while for fully flooded lubrication regimes (see definition (6.58)) the formula for the central film thickness H_{00} will get the form (compare with formula (6.149))

$$H_{00} = A(VQ)^{3/4}, \ A = O(1), \ \epsilon_q = (VQ)^{1/2}. \tag{22.36}$$

In formulas (22.35) and (22.36), ϵ_q is the characteristic distance between the inlet oil meniscus and the boundary of the Hertzian region, the values of constants A are independent of V and Q but may depend on the ratios of velocities $(v_1 + v_2)/(u_1 + u_2)$ and surface radii R_x/R_y, and the shape of the inlet meniscus Γ_i of region Ω_p.

In formulas (22.33)-(22.36), the particular values of coefficients A (different for different formulas) can be determined numerically or experimentally.

Similar results were obtained earlier for the cases of lubricants with non-Newtonian rheology.

22.3.3　Boundary Friction

Let us derive the equations governing the process in the dry zones. First, we will consider the normal problem. In the dry zones the gap between the contact solids is zero, i.e.,

$$h_0 + \frac{x^2}{2R_x} + \frac{y^2}{2R_y} + \frac{1-\nu}{\pi G} \int\int\limits_{\Omega_p} \frac{pd\xi d\eta}{R}$$

$$-\frac{1-2\nu}{4\pi G} \int\int\limits_{\Omega_\tau} \{\cos\theta \frac{\partial p}{\partial x} + \sin\theta \frac{\partial p}{\partial y}\} \frac{hd\xi d\eta}{R} = 0 \ for \ h(x,y) = 0. \tag{22.37}$$

Under the conditions of a slow stationary motion and small slippage, the equation for the slippage velocity $\vec{s}$ can be obtained by applying the differential operator $0.5(\vec{u}_1 + \vec{u}_2) \cdot \nabla$ to both sides of the first two equations in (22.28). That operation means differentiation with respect to time. Therefore, the equation for $\vec{s}$ can be expressed in the form

$$\vec{s} = -B(\vec{\tau}) + \vec{u}_2 - \vec{u}_1,$$

$$B(\vec{\tau}) = \frac{u_1 + u_2}{2} \int\int_{\Omega_\tau} \mathbf{B}_x \vec{\tau}(\xi, \eta) d\xi d\eta + \frac{v_1 + v_2}{2} \int\int_{\Omega_\tau} \mathbf{B}_y \vec{\tau}(\xi, \eta) d\xi d\eta,$$

$$\mathbf{B}_x = \mathbf{D}_x(x - \xi, y - \eta), \quad \mathbf{B}_y = \mathbf{D}_y(x - \xi, y - \eta),$$

$$B_{x11} = -\frac{\cos\theta(3\nu\sin^2\theta - 1)}{\pi G R^2}, \quad B_{x12} = B_{x21} = -\frac{\nu\sin\theta(1 - 3\cos^2\theta)}{\pi G R^2},$$

$$B_{x22} = -\frac{\cos\theta(\nu - 1 - 3\nu\sin^2\theta)}{\pi G R^2}, \quad B_{y11} = -\frac{\nu\sin\theta(\nu - 1 - 3\cos^2\theta)}{\pi G R^2},$$

$$B_{y12} = B_{y21} = -\frac{\nu\cos\theta(1 - 3\sin^2\theta)}{\pi G R^2}, \quad B_{y22} = -\frac{\sin\theta(3\nu\cos^2\theta - 1)}{\pi G R^2},$$

$$(22.38)$$

Here $\vec{\tau}$ is the sliding frictional stress that coincides with the frictional stress in the dry zones (determined by equations (22.20)) and is equal to the sliding frictional stress $\vec{f}$ in the lubricated zones (determined by equations (22.25)).

22.3.3.1   Partial Lubrication

When in a contact region in some zones contact takes place directly between smooth surfaces or asperities while in other zones of the contact solid surfaces are separated by fluid film partial lubrication (sometimes referred to as "mixed lubrication") occurs. The behavior of a contact in partial lubrication regime is governed by a combination of boundary friction and fluid film effects.

Everywhere in the contact region contact pressure $p(x, y)$ must be nonnegative

$$p(x, y) \geq 0. \tag{22.39}$$

Several additional conditions must be imposed on the parameters characterizing a partially lubricated contact. One of them is the static condition

$$\int\int_{\Omega_p} p(x, y) dx dy = P, \tag{22.40}$$

where P is the external force applied to the contact solids. Everywhere at the contact boundary Γ_p of the region Ω_p

$$p \mid_{\Gamma_p} = 0, \tag{22.41}$$

parts of Γ_p where $h \mid_{\Gamma_p} = 0$ are unknown.

Note that boundary condition (22.41) is the local one. Therefore, some pieces of the contact boundary Γ_p are known and others are unknown.

On the other hand, on the boundary Γ_τ of the region Ω_τ the tangential stress should vanish, i.e.,

$$\vec{\tau}\,|_{\Gamma_\tau} = \vec{0}. \tag{22.42}$$

Also, the problem solution must satisfy the boundary conditions for pressure p different at different zones of the contact region boundary belonging whether to lubricated or dry zones, inlet or exit zones of the contact. Suppose that Γ_i and Γ_e are the inlet and exit parts of the contact region Ω_p boundary, which are given and unknown in advance, respectively, $\Gamma_p = \Gamma_i \bigcup \Gamma_e$. The boundary condition for the dry contact zone boundary is given in (22.41). Let us consider the lubricated contact zones. In this case different boundary conditions are required. Here are the conditions that we need to impose on the problem solution:

$$\Gamma_i \text{ is given if } h\,|_{\Gamma_i} > 0 \text{ and } \vec{Q} \cdot \vec{n}\,|_{\Gamma_i} < 0,$$

$$\frac{dp}{d\vec{n}}\,|_{\Gamma_e} = 0 \text{ if } h\,|_{\Gamma_e} > 0 \text{ and } \vec{Q} \cdot \vec{n}\,|_{\Gamma_e} \geq 0, \tag{22.43}$$

where $\vec{n}$ is the vector of the external normal to boundary Γ_p.

The presented boundary conditions (22.41) and (22.43) must be combined with condition (22.39).

Now, the formulation of the boundary conditions imposed on the external boundaries of the contact is complete. In the case of mixed friction conditions, the contact area may contain some internal boundaries between dry and lubricated zones, which, apparently, coincide with flow lines of the lubricant. Therefore, it is necessary to formulate the boundary conditions that must be satisfied on the internal boundaries. As we know the frictional stresses for lubricated and dry conditions are described by equations (22.19) and (22.20) (see also (22.29)). These additional boundary conditions must be imposed on the frictional stress and represent the requirement of the frictional stress continuity on the internal boundaries. Suppose l_i and $\vec{n}_i = (n_{xi}, n_{yi})$ are the i-th internal boundary and a unit normal vector to it, respectively. Therefore, the continuity condition can be expressed in the form

$$\lim_{\epsilon \to 0} \{\vec{\tau}(x - \epsilon n_{xi}, y - \epsilon n_{yi}) - \vec{\tau}(x + \epsilon n_{xi}, y + \epsilon n_{yi})\} = \vec{0} \tag{22.44}$$

$$for\ (x, y) \in l_i.$$

Let us consider the physical nature of the transition from fluid to dry friction. A fundamental experimental study [15] shows that a continuous transition of the frictional stress from fluid friction to dry friction occurs in a small number of fluid molecular layers on a solid surface (see Fig. 22.24).

The specific features of this transition have not been studied extensively, and they depend on the adsorption properties of the lubricant-solid surface interface. We will assume that the equations of the continuous fluid mechanics

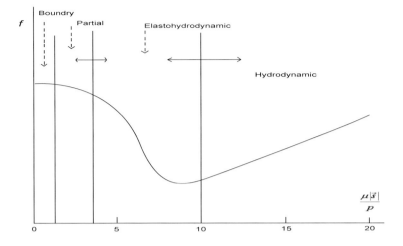

FIGURE 22.24
Variation of friction coefficient with film parameter $\Lambda = \frac{\mu|\vec{s}|}{p}$. (after Kudish and Covitch [13]). Reprinted with permission from CRC Press.

such as (22.29) can be extended on thin molecular layer of fluid. Let us consider the internal boundary between lubricated and dry regions on which slippage occurs, i.e., $|\vec{s}| > 0$. Then using equation (22.44) together with (22.20) and (22.29), we obtain

$$\lim_{h \to 0} f(p, h, \mu, |\vec{s}|) = \lambda p \frac{\vec{s}}{|\vec{s}|} \ for \ \ |\vec{s}| > 0. \tag{22.45}$$

It means that due to adsorption effects, the lubricant boundary layers show the properties of structural anisotropic fluids, and in such thin (almost monomolecular) layers lubricant viscosity $\mu = \mu(p, h, |\vec{s}|)$, i.e., besides pressure p it also depends on film thickness h and sliding speed $|\vec{s}|$. For a particular case of Newtonian fluid with $F(x) = x$, equation (22.45) becomes more transparent

$$\lim_{h \to 0} \frac{\mu}{h} = \frac{\lambda p}{|\vec{s}|} \ for \ \ |\vec{s}| > 0. \tag{22.46}$$

Therefore, we have to assume that the fluid viscosity μ in thin lubrication films is proportional to the film thickness h. That may be related to the property of lubricity of lubricants in thin layers.

Finally, the problem for partially lubricated contact is completely described by equations (22.25), (22.26), (22.29), (22.30) for the lubricated zones, and by equations (22.20), (22.37), (22.38) for the dry zones together with additional conditions (22.39)-(22.44). In addition to these equations and inequalities, the relationships for functions F, Φ, μ, λ, and the inlet boundary Γ_i must be given. It is important that the functions of lubricant viscosity $\mu = \mu(p, h, |\vec{s}|)$

and the coefficient of dry friction $\lambda = \lambda(p, |\vec{s}|)$ satisfy the continuity relationships on the boundaries l_i of dry and lubricated zones. The problem solution consists of functions $p(x, y)$, $h(x, y)$, $\vec{\tau}(x, y)$, $\vec{s}(x, y)$, the exit boundary Γ_e and some parts of the inlet boundary Γ_i of the region Ω_p, which represent the boundaries of the dry zones, the internal boundaries l_i between the lubricated and dry contact zones, the boundary of region Ω_τ, and the central film thickness h_0.

Obviously, if everywhere in the contact region $h(x, y) > 0$, the formulated problem is reduced to the EHL problem, which, for incompressible materials ($\nu = 1/2$), coincides with the EHL problems considered in the preceding chapters. If $h(x, y) = 0$, then the problem is reduced to an analog of a Hertz problem. In this case for incompressible materials ($\nu = 1/2$), the normal problem precisely coincides with the Hertzian one while the tangential problem coincides with the one considered for a dry contact with stick and slip in [16].

In the case of small influence of elastic deformations on slippage, the normal and tangential problems can be decoupled and solved separately. In this case it means that $\vec{s} = \vec{u}_2 - \vec{u}_1 + o(\vec{u}_2 - \vec{u}_1)$, and the normal problem becomes independent of the tangential problem. Thus, first, must be solved the normal problem for $p(x, y)$ and $h(x, y)$, and after that must be solved the tangential problem for $\vec{\tau}(x, y)$ and $\vec{s}(x, y)$.

After the problem is solved, the frictional force $\vec{F}_T^\pm$ applied to the upper/lower contact surfaces can be calculated according to the formulas

$$\vec{F}_T^\pm = \int\!\!\int_{\Omega_\tau} \vec{\tau}^\pm d\xi d\eta = \vec{F}_f^\pm + \vec{F}_d^\pm, \quad \vec{F}_f^\pm = \mp \vec{F}_{fs} - \vec{F}_{fr},$$

$$\vec{F}_{fs} = \int\!\!\int_{\Omega_{\tau f}} \vec{f} d\xi d\eta, \quad \vec{F}_{fr} = \tfrac{1}{2} \int\!\!\int_{\Omega_{\tau f}} h\nabla p d\xi d\eta, \qquad (22.47)$$

$$\vec{F}_d^\pm = \int\!\!\int_{\Omega_{\tau d}} \vec{\tau}^\pm d\xi d\eta,$$

where $\Omega_{\tau f}$ and $\Omega_{\tau d}$ are the zones of fluid ($h(x, y) > 0$) and dry ($h(x, y) = 0$) friction in the contact region Ω_τ. Here subscripts f and d indicate fluid and dry friction conditions while subscripts s and r indicate sliding and rolling frictional conditions.

As it was mentioned above the shape of the inlet meniscus depends on many different factors which in most cases cannot be registered. Therefore, it is appropriate to treat the shape of the inlet meniscus Γ_i as a random function depending on several parameters, for example, $\Gamma_i(a_1, a_2, \ldots, a_n)$. In majority of practical cases it is important to know the average values of such characteristics of the contact as pressure, frictional stress, slippage, and frictional force. Assuming that the probability density function $f(a_1, a_2, \ldots, a_n)$ is known it is easy to do averaging by integrating the function/constant to be averaged over the domain A of the set of parameters $a_1, a_2, \ldots, a_n$.

22.3.4 Partial Lubrication of a Narrow Contact

Let us consider the case of an incompressible elastic material ($\nu = 0.5$) and a long in the direction of the y-axis and narrow in the direction of the x-axis partially lubricated contact. This condition is equivalent to the inequality $R_x/R_y \ll 1$. It means that the eccentricity of the Hertzian (dry) contact ellipse $e = \sqrt{1 - \delta^2} \approx 1$ and it satisfies the equation

$$\frac{\delta^2 D(e)}{K(e) - D(e)} = \frac{R_x}{R_y}, \quad D(e) = \frac{K(e) - E(e)}{e^2}, \tag{22.48}$$

where δ is the relative width of the contact, $\delta = a_H/b_H$ (a_H and b_H are the smaller and larger semi-axes of the Hertzian contact, $a_H = b_H\sqrt{1 - e^2}$, $b_H = \sqrt[3]{\frac{3E(e)}{1 - e^2} \frac{PR_x}{\pi E'(1 + \rho)}}$, $p_H a_H b_H = \frac{3P}{2\pi}$, p_H is the maximum Hertzian pressure), $K(e)$ and $E(e)$ are the complete elliptic integrals of the first and second kind.

It can be shown that $\delta \ll 1$ for $R_x/R_y \ll 1$. For $\delta \ll 1$ we have $e - 1 \ll 1$. We will use the asymptotic expansions [17, 18] for the integrals over an elongated in the direction of the y-axis narrow contact region $\Omega = \{(x, y) \mid a(y) \leq x \leq c(y), -\delta^{-1} \leq y \leq \delta^{-1}\}$

$$\iint_\Omega \frac{p(\xi, \eta) d\xi d\eta}{\sqrt{(x - \xi)^2 + (y - \eta)^2}} = 2 \int_a^c p(\xi, y) \ln \frac{1}{|\xi - x|} d\xi$$

$$+ F_p(y) \ln[4(\delta^{-2} - y^2)] + \int_{-\delta^{-1}}^{\delta^{-1}} \frac{F_p(\eta) - F_p(y)}{|\eta - y|} d\eta + O(\delta^2 \ln \tfrac{1}{\delta}),$$

$$\tag{22.49}$$

$$\iint_\Omega \frac{(x - \xi) p(\xi, \eta) d\xi d\eta}{\sqrt{(x - \xi)^2 + (y - \eta)^2}} = \pi \int_a^c p(\xi, y) \mathrm{sign}(x - \xi) d\xi + O(\delta),$$

$$\iint_\Omega \frac{(y - \eta) p(\xi, \eta) d\xi d\eta}{\sqrt{(x - \xi)^2 + (y - \eta)^2}} = \int_{-\delta^{-1}}^{\delta^{-1}} \frac{F_p(\eta) d\eta}{y - \eta} + O(\delta), \quad F_p(y) = \int_a^c p(\xi, y) d\xi.$$

Then, for $\delta \ll 1$ let us introduce the following dimensionless variables (further primes are omitted at the dimensionless variables)

$$x' = \frac{x}{a_H}, \; y' = \frac{y}{b_H}, \; z' = \frac{z}{h_e}, \; F' = \frac{2h_e}{\mu_a |\bar{u}_1 + \bar{u}_2|} F, \; \mu' = \frac{\mu}{\mu_a},$$

$$\tag{22.50}$$

$$p' = \frac{p}{p_H}, \; (\tau', f') = \frac{a_H^2}{\mu_a |\bar{u}_1 + \bar{u}_2| R_x}(\tau, f), \; \vec{s}' = \frac{\vec{s}}{|\bar{u}_1 + \bar{u}_2|},$$

$$H_0 = \frac{2R_x h_e}{a_H^2}, \; V = \frac{24\mu_a |\bar{u}_1 + \bar{u}_2| R_x^2}{p_H a_H^3}, \; \theta = \frac{p_H a_H^2}{\mu_a |\bar{u}_1 + \bar{u}_2| R_x}, \; \eta = \frac{2}{\pi \delta^2 D} \frac{a_H}{R_x},$$

where μ_a is the fluid viscosity at ambient pressure, and h_e is the film thickness at the exit point, $h_e = h_e(y)$.

Using these dimensionless variables and the asymptotic expansions (22.49) for the integrals involved in the problem equations, the normal problem for the main asymptotic terms for $H_0 > 0$ and $H_0 = 0$ can be found in the form

$$\frac{\partial}{\partial x}\left\{\frac{1}{\mu}\int\limits_{-h/2}^{h/2} zF(f_x + zH_0\frac{\partial p}{\partial x})dz - h\right\} = 0, \tag{22.51}$$

$$\frac{1}{\mu}\int\limits_{-h/2}^{h/2} F(f_x + zH_0\frac{\partial p}{\partial x})dz = s_x \ for \ H_0 > 0, \tag{22.52}$$

$$\frac{2}{\pi\delta^2 D}\int\limits_{a_p}^{c_p} p(t,y)\ln\mid \frac{c_p-t}{x-t}\mid dt = c_p^2 - x^2 \ for \ H_0 = 0, \tag{22.53}$$

$$\int\limits_{a_p}^{c_p} p(t,y)dt = \frac{\pi}{2}P_0(y), \tag{22.54}$$

$$p(a_p,y) = p(c_p,y) = 0 \ for \ H_0 \geq 0, \tag{22.55}$$

$$\frac{\partial p(c_p,y)}{\partial x} = 0 \ for \ H_0 > 0, \tag{22.56}$$

$$H_0(h-1) = x^2 - c_p^2 + \frac{2}{\pi\delta^2 D}\int\limits_{a_p}^{c_p} p(t,y)\ln\mid \frac{c_p-t}{x-t}\mid dt$$

$$+\frac{H_0\epsilon_x}{\delta^2 D}\int\limits_{x}^{c_\tau} h(t,y)\frac{\partial p(t,y)}{\partial t}dt \ for \ H_0 > 0, \tag{22.57}$$

where $\epsilon_x = \frac{1-2\nu}{1-\nu}\frac{a_H}{8R_x}$. In these equations $[a_p, c_p]$ is the interval $p(x,y) \geq 0$ for the particular y, $a_p = a_p(y)$ and $c_p = c_p(y)$ are the inlet and exit points of this interval/contact region, and $P_0(y)$ is a positive function equal to the force applied to the contact region $[a_p, c_p]$. Also, $[a_\tau, c_\tau]$ is the interval where $\vec{\tau}(x,y) \neq \vec{0}$ for the particular y, $a_\tau = a_\tau(y)$ and $c_\tau = c_\tau(y)$ are the inlet and exit points of this interval (see below). It is important to notice that in equations (22.51)-(22.57) constant h_0 is replaced by function $h_e = h_e(y)$ in such a way that $h = h_e$ for $(x,y) \in \Gamma_e$, and by the exit film thickness $H_0 = H_0(y)$. Besides that, for $\delta \ll 1$ the conditions $h(x,y) > 0$ and $h(x,y) = 0$ on the interval $[a_p(y), c_p(y)]$ are equivalent to the conditions $H_0 > 0$ and $H_0 = 0$, respectively.

In a similar fashion, for $\delta \ll 1$ taking into account that s_y, $v_y = O(\delta)$ the tangential problem for the main terms of the asymptotic expansions can be obtained in the form

$$\tau_x = f_x \ for \ H_0 > 0, \tag{22.58}$$

$$\tau_x = \theta\lambda p\frac{s_x}{|s_x|}$$

$$if \mid s_x \mid > 0 \ and \mid \tau_x \mid \leq \theta\lambda p \ if \mid s_x \mid = 0 \ for \ H_0 = 0, \tag{22.59}$$

$$s_x = -\eta\int\limits_{a_\tau}^{c_\tau} \frac{\tau_x(t,y)dt}{t-x} + v_x, \ \tau_x(a_\tau,y) = \tau_x(c_\tau,y) = 0 \ for \ H_0 \geq 0. \tag{22.60}$$

Therefore, the space problem for mixed friction is reduced to a family of plane problems for lubricated and dry conditions. Here it is important to mention again that if $H_0(y_*) > 0$ for some y_* then the whole segment $[a_p(y_*), c_p(y_*)]$ is lubricated while if $H_0(y_*) = 0$ for some y_* then the whole segment $[a_p(y_*), c_p(y_*)]$ is dry. This follows from the fact that the lubricant volume flow flux is constant in a plane case. The general view of such a partially lubricated contact with one dry strip and two lubricated ones is given in Fig. 22.23.

For $\delta \ll 1$ the systems of equations (22.51)-(22.57) and (22.58)-(22.60) hold outside of the small vicinities, of order of δ, of the points $(x, y) = (x, -d_l)$ and $(x, y) = (x, d_u)$ (which represent the lower and upper tips of the contact boundary Γ) and the points where the radius of inlet boundary Γ_i is of order of δ. In the considered case the contact area is represented by alternating contact strips (bounded by straight lines $y = const$) in which fluid or dry friction occurs.

In general, systems of equations (22.51)-(22.57) and (22.58)-(22.60) must be solved simultaneously. For incompressible materials ($\nu = 1/2$), the normal and tangential problems get decoupled. In such cases, first the normal problem has to be solved and after that the tangential one. For dry contacts the latter problem was considered in detail in a number of studies such as [10] while for lubricated contact it was analyzed numerically in Chapter 6 lubricated soft elastic materials. Here we will concentrate on the normal problem.

Let us assume that the cross section $y = y_*$ is lubricated, i.e., $H_0(y_*) > 0$. By introducing the following transformation of variables:

$$(x, a_p, c_p) = \sigma(x_1, a_{1p}, c_{1p}), \ \ p = p_0 p_1, \ \ H_0 = H_* H_1, \ \ V = V_0 V_1,$$

$$p_0 = \delta \sqrt{P_0 D}, \ \ \sigma = \frac{1}{\delta} \sqrt{\frac{P_0}{D}}, \ \ H_* = \frac{P_0}{\delta^2 D}, \ \ P_0 = \frac{V_0}{H_*}, \ \ V_0 = \frac{P_0^2}{\delta^2 D}, \qquad (22.61)$$

$$\epsilon_x = \frac{\epsilon_{x1}}{\sigma},$$

equations (22.51)-(22.57) can be reduced to the equations of a plane EHL problem in the form used before. For simplicity, let us consider the case of Newtonian fluid, i.e., the case with $F(x) = x$. Then in the introduced variables equations (22.51), (22.52), (22.54)-(22.57) are reduced to the following ones

$$\frac{\partial}{\partial x_1}\{\frac{H_1^2}{V_1} \frac{h^3}{\mu} \frac{\partial p_1}{\partial x_1} - h\} = 0, \qquad (22.62)$$

$$p_1(a_{1p}, y) = p(c_{1p}, y) = 0, \ \ \frac{p_1(c_{1p}, y)}{\partial x_1} = 0, \qquad (22.63)$$

$$H_1(h - 1) = x_1^2 - c_{1p}^2 + \frac{2}{\pi} \int\limits_{a_{1p}}^{c_{1p}} p_1(t, y) \ln \mid \frac{c_{1p} - t}{x_1 - t} \mid dt$$

$$\qquad (22.64)$$

$$+ H_1 \epsilon_{x1} \int\limits_{x_1}^{c_{1\tau}} h(t, y) \frac{\partial p_1(t, y)}{\partial t} dt,$$

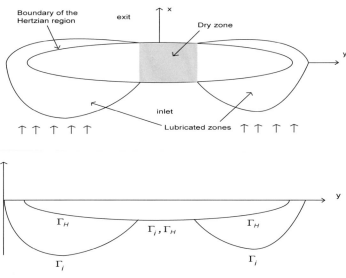

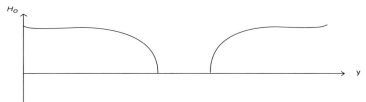

FIGURE 22.25
The view of a partially lubricated contact region with indicated boundary of
the Hertzian (dry) contact Γ_H and the boundary of the partially lubricated
contact region Γ_i (upper and middle sketches) and the film thickness H_0
distribution along the y-axis (bottom sketch). (after Kudish and Covitch [13]).
Reprinted with permission from CRC Press.

$$\int_{a_{1p}}^{c_{1p}} p_1(t, y)dt = \tfrac{\pi}{2}. \tag{22.65}$$

Equations (22.62)-(22.65) describe a familiar problem for a line EHL contact. The only difference of these equations from the ones studied in Chapter
6 is the presence of the last term in equation (22.64) proportional to ϵ_{x1},
which represents the influence of the fluid friction on the other parameters of
a lubricated contact. For incompressible solid materials ($\nu = 1/2$), we have
$\epsilon_{x1} = 0$ and the lubrication problem described by equations (22.62)-(22.65)
is identical to the EHL problem for a Newtonian fluid studied in Chapter 6.
Therefore, all conclusions and solutions obtained for this problem in Chapter
6 can be extended to the problem at hand. For compressible solid materials
($\nu < 1/2$), we have $\epsilon_{x1} \neq 0$. In this case the problem still can be analyzed by

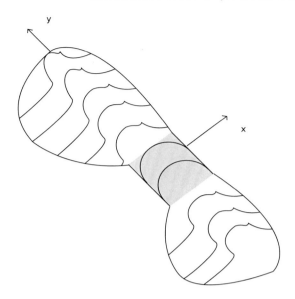

FIGURE 22.26
The schematic pressure distribution in the lubricated and dry zones of the contact region. (after Kudish and Covitch [13]). Reprinted with permission from CRC Press.

the same asymptotic methods developed in Chapter 6. Moreover, for starved and fully flooded pre-critical lubrication regimes ($\epsilon_q = O(V_1^{2/5})$, see (6.61) and (6.62)) the analysis (and the results) for the problem described by equations (22.62)-(22.65) is identical to the one used (and the results received) in Chapter 6. In particular, based on formulas (6.84) for $\mu = 1$, we can conclude that for starved lubrication conditions we would have

$$H_0(d) = H_{0*} \mid \tfrac{d}{d_0} \mid^{2/3} \theta(d), \ c_p(d) = \sigma(1 + B \mid \tfrac{d}{d_0} \mid \theta(d)), \qquad (22.66)$$

where H_{0*} is the film thickness when the oil meniscus is located outside of the Hertzian region at a distance of d_0 from the inlet side of the Hertzian contact boundary, d is the distance of the oil meniscus located outside of the Hertzian region from the inlet side of the Hertzian contact boundary ($d > 0$) when the oil meniscus is outside of the Hertzian region, B is a positive constant, $\theta(x)$ is a step function, $\theta(x) = 0$, $x \leq 0$ and $\theta(x) = 1$, $x > 0$. This behavior of the lubrication film thickness H_0 as a function of the distance d of the oil meniscus from the Hertzian contact was clearly observed in optical experimental studies of starved lubrication [19].

Based on (22.66) it is obvious that when the oil meniscus touches the boundary of the Hertzian region the film thickness H_0 becomes equal to zero and the exit coordinate of the contact coincides with the Hertzian one. Qualitatively the behavior of the film thickness H_0 and the exit boundary c_p remains the same for cases of the lubricant viscosity μ varying with pressure p. The described behavior of the inlet boundary and the film thickness H_0 is illustrated in Fig. 22.25. The schematic pressure distribution in the lubricated and dry zones of the contact region is presented in Fig. 22.26.

Usually, under equal conditions dry frictional stress is greater than fluid one. Therefore, for mixed friction the friction force is higher than for purely fluid regime. The increase in friction force depends on a relative portion of the contact region occupied by the dry friction zones that negatively affects contact fatigue life.

To conclude the problem solution function $P_0(y)$ should be determined. We should keep in mind that function $P_0(y)$ for the case of a lubricated contact differs from the one for a dry contact by a small value of $O(\epsilon_q^{3/2}, H_0\epsilon_q^{1/2}\frac{\epsilon_x}{\delta^2 D})$ for pre-critical lubrication regimes and by $O(\epsilon_0^{1/2}\epsilon_q, H_0\epsilon_q^{1/2}\frac{\epsilon_x}{\delta^2 D})$ for over-critical lubrication regimes. Therefore, in most cases for determining function $P_0(y)$ it is sufficient to consider the corresponding dry contacts. That is done in the next section.

The study presented in this section is only the beginning in the understanding of major mechanisms governing the phenomenon of mixed friction.

22.4 Dry Narrow Contact of Elastic Solids

Let us consider a narrow (along the x-axis) and significantly stretched along the y-axis contact of a rigid indenter with an elastic half-space. The shape of the indenter surface can be approximated by a paraboloid. In the direction of the x- and y-axes the radii of curvature of the indenter surface are R_x and R_y, respectively. The indenter is impressed in the half-space by a normal force P. In this case $R_x/R_y \ll 1$ and, therefore, $\delta = a_H/b_H \ll 1$ and the eccentricity e of the contact ellipse is close to 1, i.e., $e - 1 \ll 1$ (see equation (22.48)). Using the asymptotic expansions [17, 18] for the integrals over an elongated in the direction of the y-axis narrow contact region in dimensionless variables (22.50) for pressure $p(x,y)$ in any cross section of the contact $y = conts$ at a distance greater than 1 away from the tips of the contact $(0, \pm\delta^{-1})$ we obtain equations (see equations (22.53)-(22.55))

$$\frac{2}{\pi\delta^2 D} \int\limits_{a_p}^{c_p} p(t,y) \ln\left|\frac{c_p-t}{x-t}\right| dt = c_p^2 - x^2, \quad \int\limits_{a_p}^{c_p} p(t,y)dt = \frac{\pi}{2}P_0(y), \qquad (22.67)$$

where $a_p = a_p(y)$ and $c_p = c_p(y)$ are the boundaries of the contact, constants $\delta \ll 1$ and D satisfy equation (22.48), $P_0(y)$ is an unknown function equal to the force applied to this cross section of the contact (see (22.49)).

In this formulation we can consider the cases of the indenter with smooth and sharp edges. The difference between these cases is that the contact boundaries a_p and c_p are unknown and given, respectively.

The solution of this problem can be represented in the form [14]

$$p(x, y) = \delta^2 D q(x, y) + \frac{P_0(y)}{2\sqrt{(x-a_p)(c_p-x)}}, \quad \int_{a_p}^{c_p} q(x, y) dx = 0,$$

$$(22.68)$$

$$q(x, y) = \sqrt{(x - a_p)(c_p - x)} + \frac{2a_p c_p + (c_p - a_p)^2/4 - (c_p + a_p)x}{2\sqrt{(x-a_p)(c_p-x)}}.$$

Outside of the contact in its cross section $y = const$ at a distance of $x = O(\delta^{-1}) \gg 1$, the details of the pressure $p(x, y)$ distribution along the x-axis can be neglected and the contact region can be replaced by a medium line loaded with pressure $p(x, y) = \frac{\pi}{2} P_0(y) \delta(x)$, where $\delta(x)$ is the Dirac's delta-function. Then, at these distances from the contact region the vertical displacement of the indenter w is

$$w(x, y) = -\frac{1}{2\delta^2 D} \int_{-d}^{d} \frac{P_0(\eta) d\eta}{\sqrt{x^2 + (y-\eta)^2}} + w_0, \quad (22.69)$$

where $2d = O(\delta^{-1})$ is the diameter of the contact region, i.e., the distance between two most distanced points of the contact, w_0 is the indenter displacement far away from the contact.

Let us determine the asymptotic of the integral in (22.69) for $x = O(1)$. We will have

$$\int_{-d}^{d} \frac{P_0(\eta) d\eta}{\sqrt{x^2 + (y-\eta)^2}} = \int_{-d}^{d} \frac{[P_0(\eta) - P_0(y)] d\eta}{\sqrt{x^2 + (y-\eta)^2}} + P_0(y) \int_{-d}^{d} \frac{d\eta}{\sqrt{x^2 + (y-\eta)^2}}$$

$$(22.70)$$

$$= \int_{-d}^{d} \frac{[P_0(\eta) - P_0(y)] d\eta}{|y-\eta|} + 2P_0(y)[\ln 2 + \ln \sqrt{d^2 - y^2} - \ln |x|] + O(1).$$

Therefore, from equations (22.69) and (22.70) for $w(x, y)$, we obtain

$$w(x, y) = -\frac{1}{2\delta^2 D} \int_{-d}^{d} \frac{[P_0(\eta) - P_0(y)] d\eta}{|y-\eta|} - \frac{P_0(y)}{\delta^2 D} \ln \frac{2\sqrt{d^2 - y^2}}{|x|} + w_0 + \dots \quad (22.71)$$

From the first equation of (22.67), it is clear that in the region $x = O(1)$ in this cross–section $y = const$ the elastic displacement of the half-space surface $w(x, y)$ has the form

$$w(x, y) = h_w - \frac{2}{\pi \delta^2 D} \int_{a_p}^{c_p} p(t, y) \ln \left| \frac{1}{x-t} \right| dt + \dots \quad (22.72)$$

For $x \gg 1$ from (22.72) and the second equation in (22.67), we obtain the asymptotic representation for $w(x, y)$

$$w(x, y) = h_w - \frac{P_0(y)}{\delta^2 D} \ln \frac{1}{|x|} + \dots \quad (22.73)$$

Applying the principle of matched asymptotic expansions [21] to the expressions for $w(x, y)$ from (22.71) and (22.73) in the intermediate region $1 \ll |x| \ll \delta^{-1}$, we obtain the equation for function $P_0(y)$

$$\frac{1}{2} \int_{-d}^{d} \frac{[P_0(\eta) - P_0(y)]d\eta}{|y - \eta|} + P_0(y) \ln[2\sqrt{d^2 - y^2}]$$

$$= \delta^2 D(w_0 - c_p^2 - \rho y^2) - \frac{2}{\pi} \int_{a_p}^{c_p} p(t, y) \ln \left| \frac{1}{c_p - t} \right| dt. \tag{22.74}$$

This equation involves the unknown constant w_0, which is determined by the force balance equation for the entire contact (see the dimensional analog (22.40))

$$\iint_{\Omega_p} p(\xi, \eta)d\xi d\eta = \frac{2\pi}{3} \delta^{-1}. \tag{22.75}$$

Using the second equation in (22.67) equation (22.75) is easily transformed into

$$\int_{-d}^{d} P_0(y)dy = \frac{4}{3\delta}. \tag{22.76}$$

By substituting (22.68) into (22.74), we obtain

$$\frac{1}{2} \int_{-d}^{d} \frac{[P_0(\eta) - P_0(y)]d\eta}{|y - \eta|} + P_0(y) \ln \frac{8\sqrt{d^2 - y^2}}{c_p - a_p}$$

$$= \delta^2 D[w_0 - c_p^2 - \rho y^2 - \frac{2}{\pi} \int_{a_p}^{c_p} q(t, y) \ln \frac{1}{c_p - t} dt], \tag{22.77}$$

where in the case of $q(x, y)$ from (22.68) we have

$$\frac{8}{\pi} \int_{a_p}^{c_p} q(t, y) \ln \frac{1}{c_p - t} dt = -3c_p^2 + 2c_p a_p + a_p^2. \tag{22.78}$$

22.4.1   Examples of Dry Narrow Contacts

First, let us consider the case of an indenter with an ellipsoidal shape such that $c_p = -a_p = \sqrt{1 - \frac{y^2}{d^2}}$ and $d = \delta^{-1} \gg 1$. Then the problem is reduced to equations (see equations (22.76)-(22.78))

$$\frac{1}{2} \int_{-d}^{d} \frac{[P_0(\eta) - P_0(y)]d\eta}{|y - \eta|} + P_0(y) \ln(4d) = \delta^2 D(w_0 - \rho y^2), \tag{22.79}$$

$$\int_{-d}^{d} P_0(y)dy = \frac{4}{3\delta},$$

where constant δ satisfies equations (22.48).

For $d \gg 1$ away from points $y = \pm d$ from the first of equations (22.79), we obtain

$$P_0(y) = \frac{\delta^2 D(w_0 - \rho y^2)}{\ln(4d)}. \tag{22.80}$$

After that using the second equation in (22.79) and equation (22.80), we find

$$P_0(y) = \frac{2}{3} + \rho \frac{\delta^2 D(\frac{1}{3}d^2 - y^2)}{\ln(4d)}, \quad w_0 = \frac{\rho d^2}{3} + \frac{2}{3} \frac{\ln(4d)}{\delta^2 D}. \tag{22.81}$$

Now, let us consider the case of an indenter with an ellipsoidal shape or the case of two elliptic elastic solids. In such cases $c_p = -a_p = l(y)$ and $d = \delta^{-1} \gg 1$. Moreover, function $l(y)$ satisfies the equations

$$p(\pm l, y) = 0. \tag{22.82}$$

Using equations (22.68) and (22.82), we find that

$$l(y) = \sqrt{\frac{P_0(y)}{\delta^2 D}}, \quad p(x, y) = \delta^2 D \sqrt{l^2(y) - x^2}. \tag{22.83}$$

Substituting the expressions for $l(y)$ and $p(x, y)$ from (22.83) into equations (22.77) and (22.78) for $P_0(y)$, we obtain the equations

$$\frac{1}{2} \int_{-d}^{d} \frac{[P_0(\eta) - P_0(y)]d\eta}{|y - \eta|} + P_0(y) \ln \frac{4\delta D^{1/2}\sqrt{d^2 - y^2}}{P_0^{1/2}(y)} = \delta^2 D(w_0 - \rho y^2),$$

$$\tag{22.84}$$

$$\int_{-d}^{d} P_0(y)dy = \frac{4}{3\delta}.$$

The approximate solution of equations (22.84) we will try to find in the form

$$P_0(y) = A^2(d^2 - y^2), \tag{22.85}$$

where A is an unknown constant. Substituting (22.85) into (22.84) and equating the coefficients at y^0 we find $A = d^{-1}$ and

$$P_0(y) = \frac{d^2 - y^2}{d}, \quad w_0 = \frac{1}{\delta^2 D}[\ln(4D^{1/2}) - \frac{1}{2}]. \tag{22.86}$$

It can be shown that the indenter displacement within the contact region is described by a paraboloid of the second degree. Moreover, the error created by this approximation decreases [17] from 15% for $\delta = 0.5$ to 3% for $\delta = 0.2$.

It is worth mentioning that the approximate solution (22.86) for $p(x, y)$ coincides with the exact solution of this problem [20]. As it was mentioned in the preceding section for lubricated contacts function $P_0(y)$ is very close to the one obtained for corresponding dry contacts (see (22.86)).

22.4.2 Optimal Shape of Solids in a Normal Narrow Contact

Now, let us consider the problem of optimal design of contact surfaces in bearings and gears. Solution of this problem is important for increasing fatigue life of ball and roller bearings as well as gears.

Let us consider this problem on an example of a roller bearing working under normal load without skewness. For simplicity we will assume that the surface of a relatively long roller in the direction of the y-axis is described by the equation $z = f_0(y)$. This roller made of an elastic material is normally indented in an elastic half-space. In each cross section $y = const$ the roller radius $R(y)$ is given by $R(y) = R_0 - f_0(y)$, where R_0 is the roller radius in its central cross–section $y = 0$. Then the gap between the non-deformed roller and half-space is described by function

$$f(x, y) = f_0(y) + \frac{x^2}{2[R_0 - f_0(y)]}. \tag{22.87}$$

Let us introduce the dimensionless variables

$$(f', f_0') = \frac{2R_0}{a_H^2}(f, f_0), \quad \epsilon_* = \frac{1}{2}\left(\frac{a_H}{R_0}\right)^2 \tag{22.88}$$

in addition to the ones introduced in (22.50). In (22.88) a_H is the smaller semi-axis of the elliptic contact region of an indenter with radii R_0 and $[f_0''(0)]^{-1}$ along the x- and y-axes, respectively, $e = \sqrt{1 - \delta^2}$ is the eccentricity of the ellipse, which satisfies equation (22.48).

In these dimensionless variables (for simplicity primes are omitted) in cross sections of the contact $y = const$ away from points $(0, \pm d)$, we get equations

$$\frac{x^2 - c_p^2}{1 - \epsilon_* f_0(y)} + \frac{2}{\pi \delta^2 D} \int_{a_p}^{c_p} p(t, y) \ln \left| \frac{c_p - t}{x - t} \right| dt = 0,$$

$$\int_{a_p}^{c_p} p(t, y)dt = \frac{\pi}{2} P_0(y), \tag{22.89}$$

where d is determined by the equations $l(\pm d) = 0$.

The solution of of equations (22.89) has the form

$$p(x, y) = \frac{\delta^2 D}{1 - \epsilon_* f_0(y)} q(x, y) + \frac{P_0(y)}{2\sqrt{(x - a_p)(c_p - x)}}, \quad \int_{a_p}^{c_p} q(x, y)dx = 0, \tag{22.90}$$

$$q(x, y) = \sqrt{(x - a_p)(c_p - x)} + \frac{2a_p c_p + (c_p - a_p)^2/4 - (c_p + a_p)x}{2\sqrt{(x - a_p)(c_p - x)}}.$$

By taking $c_p = -a_p = l(y)$ and satisfying equations (22.82), we find

$$l(y) = \sqrt{[1 - \epsilon_* f_0(y)]\frac{P_0(y)}{\delta^2 D}}, \quad p(x, y) = \frac{\delta^2 D}{1 - \epsilon_* f_0(y)}\sqrt{l^2(y) - x^2}. \tag{22.91}$$

Let us consider the pressure distribution along the roller medium line $x = 0$ at which the maximum of pressure is reached. From (22.91) we have

$$p(0, y) = \sqrt{\frac{\delta^2 D P_0(y)}{1 - \epsilon_* f_0(y)}}. \tag{22.92}$$

Because contact fatigue life is inverse proportional to a certain positive power of the maximum contact pressure (see [13]), the optimal pressure distribution is the one that is constant along the medium roller line. Therefore, the optimal roller configuration we will obtain from the condition $p(0, y)$ is equal to a constant (see (22.92))

$$\frac{P_0(y)}{1 - \epsilon_* f_0(y)} = P_0(0), \tag{22.93}$$

where we took into account that $f_0(0) = 0$. From equation (22.93) we find

$$f_0(y) = \frac{1}{\epsilon_*}\{1 - \frac{P_0(y)}{P_0(0)}\}. \tag{22.94}$$

In the above dimensionless variables using the relationships (22.87) and (22.91), we derive the equation for $P_0(y)$ in the form

$$\frac{\epsilon_*}{2} \int_{-d}^{d} \frac{[P_0(\eta) - P_0(y)]d\eta}{|y - \eta|} + P_0(y)\{\epsilon_* \ln[\frac{4\delta}{P_0(y)}\sqrt{DP_0(0)(d^2 - y^2)}] \tag{22.95}$$

$$- \frac{\delta^2 D}{P_0(0)}\} = \delta^2 D(\epsilon_* w_0 - 1), \int_{-d}^{d} P_0(y)dy = \frac{4}{3\delta}.$$

For $\epsilon_* \ll 1$ for solution of latter equations let us use the regular perturbation method [21]. The solution of equations (22.95) we will try to find in the form

$$P_0(y) = P_{00}(y) + \epsilon_* P_{01}(y) + O(\epsilon_*^2), \quad w_0 = w_{00} + O(\epsilon_*). \tag{22.96}$$

For $\epsilon_* \ll 1$ away from points $(0, \pm d)$, we have $f_0(y) = O(1)$ and, therefore, from equation (22.93) we obtain $P_{00}(y) = P_{00}(0)$. Using that from the integral condition in (22.95), we get

$$P_{00}(y) = \frac{2}{3\delta d}. \tag{22.97}$$

By equating the coefficients at ϵ_* in the expansion of equation (22.95) in series of powers of ϵ_*, we derive the equation for $P_{01}(y)$:

$$P_{00} \ln\{4\delta\sqrt{\frac{D(d^2 - y^2)}{P_{00}}}\} + \frac{\delta^2 D}{P_{00}}[P_{01}(0) - P_{01}(y)] = \delta^2 D w_{00}, \tag{22.98}$$

$$\int_{-d}^{d} P_{01}(y)dy = 0.$$

The solution of equations (22.98) is

$$P_{01} = \frac{4}{9d^2\delta^4 D}\{1 - \ln 2 + \tfrac{1}{2}\ln(1 - \tfrac{y^2}{d^2})\},$$

$$(22.99)$$

$$w_{00} = \frac{2}{3d\delta^3 D}\ln\{4d\delta\sqrt{\tfrac{3d\delta D}{2}}\}.$$

Therefore, using equations (22.94), (22.96), (22.97), and (22.99), we obtain

$$f_0(y) = -\frac{\ln(1 - \tfrac{y^2}{d^2})}{3d\delta^3 D + 2\epsilon_*(1 - \ln 2)} + \dots \qquad (22.100)$$

A simple analysis of problem (22.95) and its solution show that it is a singularly perturbed problem. That is the reason why constant d cannot be determined from the solution of equation $l(d) = 0$ if the solution expansion (22.96), (22.97), and (22.99) is used. Nevertheless, using the terms of this expansion we can estimate the size of the boundary layers adjacent to points $(0, \pm d)$ which is of the order of $\exp(-\frac{3d\delta^3 D}{\epsilon_*})$.

Therefore, if the value of d is known, then the optimal roller shape for which $p(0, y)$ remains constant is given by formula (22.100). In particular, if the contact is realized along the whole length $2L$ of the roller then $d = L$ and the roller shape is completely determined. In spite of the fact that inaccuracy of the obtained approximate optimal shape of the roller increases as we approach the roller end this approximation is acceptable for practical applications.

22.5   Closure

Solution of a spatial EHL problem for starved lightly loaded contact with a complex shape of the inlet meniscus is considered. It is shown that the solution varies considerably depending on the shape of the inlet meniscus. In connection with the latter problem the formulation and analysis of a mixed lubrication problem are considered. The problem formulation takes into account such essential conditions as frictional stress continuity across the boundaries between dry and lubricated zones of the contact. A detailed analysis of the problem is proposed for the case of a contact extended in the direction perpendicular to the motion. The problem is considered asymptotically. Using the fatigue considerations an optimal shape of a roller in a rolling bearing is proposed.

22.6   Exercises and Problems

1. Describe the influence of the shape and distance of the inlet oil meniscus on contact pressure $p(x, y)$, frictional $\tau_{sx}(x, y)$, $\tau_{rx}(x, y)$, and $\tau_{ry}(x, y)$, and octahedral $\sigma^{oct}(x, y, z)$ stresses.

2. Elaborate on why the rheology of fluids in thin films differs from the one of fluids in thick films. To substantiate your analysis use the continuity relationships (22.45) and (22.46).

3. Graph and analyze the optimal shape of a long roller from formula (22.100).

References

[1] Kudish, I.I. and Panovko, M.Ya. 1997. Influence of an Inlet Oil Meniscus Geometry on Parameters of a Point Elastohydrodynamic Contact. *ASME J. Tribology* 119, No. 1:112-125.

[2] Oh, K.P. 1984. The Numerical Solution of Dynamically Loaded Elastohydrodynamic Contacts as a Nonlinear Complementarity Problem. *ASME J. Tribology* 106, No. 1:88-94.

[3] Evans, H.P. and Hughes, T.G. 2000. Evaluation of Deflection in Semi-infinite Bodies by a Differential Method. *Proc. Instn. Mech. Engrs.* 214, Part C: 563-584.

[4] Hamrock, B.J. and Dowson, D. 1981. *Ball Bearing Lubrication – The Elastohydrodynamics of Elliptical Contacts.* New York: Wiley-Interscience.

[5] Lubrecht, A.A., Ten Napel, W.E., and Bosma, R. 1987. Multigrid Alternative Method of Solution for Two-Dimensional Elastohydrodynamically Lubricated Point Contact Calculations. *ASME J. Tribology* 109, No. 3:437-443.

[6] Hamrock, B.J. 1991. *Fundamentals of Fluid Film Lubrication.* Cleveland: NASA, Reference Publication 1255.

[7] Kudish, I.I. 1981. Some Problems of Elastohydrodynamic Theory of Lubrication for a Lightly Loaded Contact. *J. Mech. of Solids* 16, No. 3:75-88.

[8] Kostreva, M.M. 1984. Elastohydrodynamic Lubrication: A Nonlinear Complementarity Problem. *Intern. J. Numerical Methods in Fluids* 4:377-397.

[9] Oh, K.P., Li, C.H., and Goenka, P.K. 1987. Elastohydrodynamic Lubrication of Piston Skirts. *ASME J. Tribology* 109, No. 2:356-365.

[10] Johnson, K. 1985. *Contact Mechanics.* Cambridge: Cambridge University Press.

[11] Belotcerkovsky S.M. and Lifanov I.K. 1992. *Method of Discrete Vortices.* Boca Raton: CRC Press.

[12] Kudish, I.I. 1983. On Formulation and Analysis of Spatial Contact Problem for Elastic Solids Under Conditions of Mixed Friction. *J. Appl. Math. and Mech.* 47, No. 6:1006-1014.

[13] Kudish I.I. and Covitch M.J. 2010. *Modeling and Analytical Methods in Tribology.* Boca Raton: CRC Press.

[14] Galin, L.A. 1980. *Contact Problems of Elasticity and Viscoelasticity*. Moscow: Nauka.

[15] Akhmatov, A.S. 1963. *Molecular Physics of Boundary Friction*. Moscow: Fizmatgiz.

[16] Goldstein, R.V., Zazovsky, A.F., Spector, A.A., and Fedorenko, R.P. 1979. *Solution of Spatial Rolling Contact Problems with Stick and Slip by a Variational Method*. Reprint of the Inst. for Problems in Mech., USSR Academy of Sciences, Moscow, No. 134.

[17] Kalker, J.J. 1972. On Elastic Line Contact. *ASME Trans., Ser. E, J. Appl. Mech.* 33, No. 4:1125-1132.

[18] Kalker, J.J. 1977. The Surface Displacement of an Elastic Half-Space Loaded in a Slender, Bounded, Curved Surface Region with Application to the Calculation of the Contact Pressure under a Roller. *J. Inst. Maths. Applics.* 19:127-144.

[19] Bakashvili, D.L., Berdenikov, A.I., Imerlishvili, T.V., Manucharov, Yu.S., Mikhailov, I.G., and Shvatsman, V.Sh. 1985. Study of Lubricant Film of Liquids with High Bulk Viscosity in EHL Contact. *Soviet J. Fric. and Wear* 6, No. 2:54-59.

[20] Lurye, A.I. 1955. *Spatial Problems in Elasticity*. Moscow: Gostekhizdat.

[21] Van-Dyke, M. 1984. *Perturbation Methods in Fluid Mechanics*. New York-London: Academic Press.

23

Final Remarks

23.1 Introduction

The aim of this chapter is to provide a reader with some additional ideas and reading sources which would augment the material presented in the preceding chapters. A proper analysis of lubricated contacts is extremely important for successful application and operation of machinery employing such contacts such as bearings, gears, etc. From a machinery designer point of view the most important factors derived from lubricated contacts are lubrication film thickness, friction forces, lubricant temperature, and stress fields created in lubricated materials. These factors are used to determine the load capacity of joints and their longevity. Therefore, the logical question would be: How to use the developed EHL framework in more complex situations and how to use the factors following from the EHL analysis to predict joint longevity? In what follows we will try to provide some guide lines which may help to answer some of these questions.

23.2 EHL Contacts for Rough Surfaces

EHL problems for elastic solids with rough surfaces are non-steady problems and their solutions depend on time which determines the specific geometry of contact surfaces at any specific moment. These are even more complex problems than the steady ones. To get an idea of the lubricated contact parameters behavior in this case the problem can be simplified by considering a contact of moving smooth surface and stationary rough surface. Then for line isothermal or thermal contacts lubricated by Newtonian or non-Newtonian fluids the same asymptotic methods as described in Chapters 4-10 can be used as long as the surfaces are completely separated by lubricant. In the latter case the role of the pressure in the Hertzian region will be played by pressure obtained from the solution of a contact problem of elasticity for a dry contact of the same solids with a fixed boundary. This pressure can be obtained analytically by expanding surface roughness in Chebyshev orthogonal polynomials [1, 2].

669

The rest of the asymptotic and numerical analysis of EHL problems for rough surfaces can be done along the lines of the analysis for smooth surfaces.

For point contacts EHL problems for rough surfaces also can be considered in a manner similar to the case of smooth surfaces (as it was done in Chapters 12, 15-17) by using the same simplification as indicated above for line contacts if the contact surfaces are completely separated by lubricant. More specifically, it can be done in cases when the the surface roughness is longitudinal or transverse with respect to the direction of lubricant entrainment. In these cases as before the EHL problem for point contacts can be reduced to solution of the corresponding line EHL problems in the central parts of the lubricated contacts. The latter can be first analyzed asymptotically and then numerically.

23.3 Lubricant Degradation

Lubricants involved in moving joints are subject to degradation which adversely affects the performance of joints. Lubricant degradation is caused by a number of factors among which are stresses acting in lubricated contacts, oxidation processes caused by elevated temperatures and aggressive environment, lack or additive replenishment, contamination, etc. There are some empirical approaches to measure the effect of lubricant degradation as well as its effect on joint performance. The theory of stress-induced lubricant degradation has been recently developed and presented in [3]. It considers degradation of two types of polymer additives (viscosity improvers) - polymers with linear structure and star polymer additives. Depending on the polymer structure the problem is reduced to solution of a kinetic equation for linear polymers and a system of kinetic equations for the densities of polymer additive distribution(s). Degradation of lubricant affects lubricant viscosity which results in the corresponding changes of all lubrication parameters including lubrication film thickness and friction stresses. A conceptually similar approach can be used for modeling of lubricant oxidation, kinetics of additives reacting with contact surfaces, etc.

23.4 Contact Fatigue

One of the very important limitations of joint performance is its longevity. Joint longevity can be limited by many processes including lubricant degradation. Aside of lubricant degradation joint longevity is limited by contact fatigue of contact surfaces, their wear, scuffing, etc. Each of the mechanisms affecting joint longevity depends on the stresses developed in lubricated solids. These stresses can be calculated based on the formulas presented in [4, 5] as long as the contact pressure and frictional stresses acting on contact surfaces and calculated from solution of EHL problems as well as residual stresses in the solids are known. All mechanisms of joint longevity are of statistical nature including the processes of lubricant degradation.

One of the most important characteristics of joint longevity is contact fatigue. Two- and three-dimensional statistical models of contact (as well as structural) fatigue have been developed in [3]. They take into account the initial statistical defectiveness of materials of contact solids, material fatigue and elastic properties, and loading conditions. In addition to that lubricant composition may have a significant effect on contact fatigue. These models provide simple formulas for fatigue life as a function of survival probability which also depends on material initial defectiveness, material fatigue and elastic properties, and loading. It is important to realize that besides external normal loading the frictional stresses play an extremely important part in fatigue crack growth as they create zones in material subjected to resulting tensile stresses. These tensile stresses represent the driving force of crack growth and, ultimately, fatigue failure. It is also important to remember that in lubricated contacts these frictional forces result from solution of the corresponding EHL problems.

References

[1] *Handbook of Mathematical Functions with Formulas, Graphs and Mathematical Tables*, Eds. M. Abramowitz and I.A. Stegun, National Bureau of Standards, 55, 1964.

[2] Kudish, I.I., Cohen, D.K., and Vyletel, B., 2012. Perfect Mechanical Sealing in Rough Elastic Contacts: Is It Possible?, $posted July 07, 2012. doi : 10.1115/1.4007085$.

[3] Kudish, I.I. and Covitch, M.J. 2010. *Modeling and Analytical Methods in Tribology*. Boca Raton: Chapman & Hall/CRC.

[4] Galin, L.A. 1980. *Contact Problems of Elasticity and Viscoelasticity*. Moscow: Nauka.

[5] Lurye, A.I. 1955. *Spatial Problems in Elasticity*. Moscow: Gostekhizdat.

Index